ANNUAL REVIEW OF NUCLEAR SCIENCE

EMILIO SEGRÈ, *Editor*
University of California, Berkeley

J. ROBB GROVER, *Associate Editor*
Brookhaven National Laboratory

H. PIERRE NOYES, *Associate Editor*
Stanford University

VOLUME 19

1969

ANNUAL REVIEWS, INC.
4139 EL CAMINO WAY
PALO ALTO, CALIFORNIA, U.S.A.

ANNUAL REVIEWS, INC.
PALO ALTO, CALIFORNIA, U.S.A.

© 1969 BY ANNUAL REVIEWS, INC.
ALL RIGHTS RESERVED

Standard Book Number 8243-1519-7
Library of Congress Catalogue Number: 53-995

FOREIGN AGENCY

Maruzen Company, Limited
6 Tori-Nichome, Nihonbashi
Tokyo

PRINTED AND BOUND IN THE UNITED STATES OF AMERICA BY
GEORGE BANTA COMPANY, INC.

RAPID CHEMICAL SEPARATIONS

BY GÜNTER HERRMANN AND HANS OTTO DENSCHLAG

Institut für Anorganische Chemie und Kernchemie der Universität Mainz
Mainz, Germany

CONTENTS

1. INTRODUCTION

First and foremost, in a review of rapid radiochemical separations a tribute should be paid to the pioneers of radioactivity who developed and applied such procedures more than 60 years ago when they unraveled the natural decay series (1), beginning with the isolation of 55-sec ^{220}Rn by Rutherford (2) in 1900 and culminating in the direct observation of 1.8-msec ^{215}Po by Moseley & Fajans (3) in 1911. Later, similar and a few novel techniques were applied to artificially produced short-lived nuclides, as may be seen in the compilation published by Kusaka & Meinke (4) in 1961.

Most of the rapid separations reported until that time, however, concern favorable cases. The short-lived nuclide was either separated from a long-lived ancestor or measured indirectly via a long-lived descendant. In both

1

cases, only a few elements were to be separated rapidly since either the ancestor or the descendant can be purified by lengthy procedures. Direct observation of short-lived nuclides occurring in complex mixtures of reaction products succeeded rather rarely. An early example is Hahn & Strassmann's (5) discovery of 57-sec ^{91}Rb among the fission products, a more recent one the observation of 1.6-sec ^{90}Br by Perlow & Stehney (6), also in fission.

Efforts are now being made to proceed to nuclides far off the stability line. The number of still unknown nuclei is believed to equal the number of nuclei studied so far (7). The reasons for current interest in such nuclei were discussed by Bergström (7) and, in more detail, during the Lysekil meeting (8): How extrapolate nuclear properties into this region far off stability, and are new phenomena to be expected? One particularly exciting aspect, the access to islands of superheavy nuclei, is covered in recent reviews (9, 10). Since the unknown nuclides should have half-lives ranging from some minutes down to a few milliseconds (7), and since conceivable production processes lead to very complex mixtures, new experimental techniques have to be developed.

One approach, rapid chemical separations (11, 12), is treated here. Special attention is given to procedures that can be performed within a few seconds, but separations requiring up to ~1 min are included since they may be the basis for faster versions. Techniques often considered "physical" will be referred to if they rely on chemical properties. For example, separations using the deposition of recoil atoms thermalized in gases are considered chemical, those using differences in the recoil range in matter are considered physical. Typical applications will also be reported. The literature is surveyed until December 1968; work described in an earlier compilation (4) is generally not quoted.

Most of the rapid separations were worked out by groups which are mainly interested in the application of such techniques to the study of short-lived nuclides and their formation processes. Therefore, the field is rather unsystematically covered, and more systematic work is strongly desired.

Additional information on the use of such techniques may be obtained from reviews on fission (13) and on heavy elements (9, 10, 14). Online isotope separation is discussed in detail by Klapisch (15). Since many rapid separations are based on standard radiochemical procedures or make use of certain nuclear phenomena like recoil effects, the consultation of review articles on radiochemical separations (16–18), solvent extraction (19), ion exchange (20), distillation (21), recoil (22), and recoil reactions (23, 24) is recommended.

2. DEMANDS ON RAPID RADIOCHEMICAL SEPARATIONS

The speed and selectivity required in a rapid separation depend on the half-lives of the nuclides involved, on the complexity of the sample, and on the aim of the work, as may be illustrated by common production processes for nuclides far off stability (25).

Among such processes, heavy-ion-induced reactions yield moderately complex mixtures of neutron-deficient nuclides, typically containing five to six elements, each with a few isotopes (26). Low-energy fission results in ~500 neutron-rich nuclides distributed over 40 elements; half of the products are expected to have half-lives of <10 sec (13). In high-energy spallation reactions, the number of isotopes per element increases to about two dozen, extending from the neutron-deficient to the neutron-rich region (25, 27).

Even for such complex mixtures, rather unselective separations can be used when a short-lived component is detected indirectly, i.e., when its half-life, elemental and mass assignment, or formation cross section is deduced by counting a long-lived descendant. Then only those few elements that contribute to the decay chain involved have to be separated quickly; other elements can be removed later by purifying the descendant.

For direct observation of short-lived activities, separation methods of moderate selectivity and speed suffice if combined with specific detection techniques permitting the detection of minor components even after considerable decay; an example is the high-resolution spectrometry of alpha particles or gamma rays employing solid-state detectors. Such measurements yield information not only on properties just listed but also on the principal radiations emitted from the nuclide of interest. Coincidence techniques may improve selectivity; especially, coincidences with characteristic X rays of the desired element may help to suppress contributions from contaminants in gamma-ray spectra. But even simple counting devices may be appropriate for selective detection when uncommon decay modes such as neutron emission or spontaneous fission are involved.

The most severe requirements in purity and speed have to be met in detailed decay-scheme studies including weak transitions. However, even a perfect chemical separation may not be sufficient because of serious interference from accompanying isotopes. In this case, isotopic separation becomes essential, preceded by a chemical step which need not be very selective or rapid since the mass discrimination further contributes to the separation effect.

Another important aspect is continuous or discontinuous operation of the whole sequence: production, separation, and measurement of a short-lived nuclide, including transport from the production place to the detector over a distance depending on shielding requirements. When a relatively small number of experiments gives the desired information, discontinuous procedures are convenient. When such an experiment would have to be repeated too often, a continuous process delivering a steady source of short-lived activity is preferred. Alternatively, a quasicontinuous, cyclic operation may be applied.

In the latter cases, the target is permanently exposed to the beam, and the element of interest is intermittently or continuously extracted and transported to the detector. These requirements practically restrict one to solid targets and volatilization methods. In discontinuous operation, on the other hand, the irradiated target can be projected to a separation apparatus placed close to the detector. This gives more flexibility in the physical state of the target and, hence, in the choice of the separation method. Separations from aqueous solutions can be applied as well as multistep procedures.

Some other general aspects may best be illustrated by specific examples which may also show the practical application of such procedures. To indicate the efficiency of a particular technique, the shortest-lived nuclide observed so far will sometimes be given.

3. EXEMPLARY APPROACHES

3.1 *Indirect detection.*—Indirect techniques will be illustrated by the detection of 4.1-sec ^{92}Rb via its descendant 3.5-h ^{92}Y to obtain half-life and yield from thermal-neutron fission of ^{235}U, as reported by Wahl, Norris & Ferguson (28). The corresponding decay chain is

$$1.9\text{-sec }^{92}\text{Kr} \xrightarrow{\beta^-} 4.1\text{-sec }^{92}\text{Rb} \xrightarrow{\beta^-} 2.7\text{-h }^{92}\text{Sr} \xrightarrow{\beta^-} 3.5\text{-h }^{92}\text{Y}$$

In a series of short irradiations, the time interval between irradiation and separation of the growing-in daughter product ^{92}Sr was varied. The ^{92}Sr was separated from ^{92}Rb by passing the irradiated solution through a filter bed of strontium carbonate. Later, ^{92}Y was isolated from both the filter bed and the filtrate, purified, and counted.

With increasing time interval, the fraction of ^{92}Y found in the filtrate decreases because ^{92}Rb decays before separation. By plotting this fraction versus the time interval, the half-life of ^{92}Rb is obtained, as shown in Figure 1. The intercept indicates what fraction of the chain is originally produced as either ^{92}Kr or ^{92}Rb. From this value and from the known yield of ^{92}Kr, the fractional independent yield of ^{92}Rb is deduced.

This example shows the characteristic features of indirect techniques: The separation problem is reduced to the separation of two components. The nuclide of interest can be detected after it has decayed appreciably, and its decay can be followed over several half-lives. Accompanying isotopes do not interfere since they normally pass into decay products of quite different properties; in the present case, 6-sec ^{93}Rb passes into 10-h ^{93}Y, and 2.6-sec ^{94}Rb into 20-min ^{94}Y. On the other hand, each experiment contributes often only a single point to the decay curve.

3.2 *Discontinuous procedure: separation from aqueous solution.*—Separation from aqueous solution is still the most frequently used radiochemical technique. Normally, such procedures proceed rather slowly since considerable time is lost until, for example, a precipitate has formed and is filtered or, in the case of solvent extraction, until two liquid phases have separated.

It is, however, possible to adapt conventional techniques to rapid work. Precipitation procedures are superseded by rapid reactions with preformed

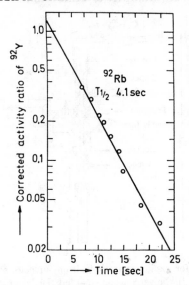

FIG. 1. Indirect determination of the half-life of ^{92}Rb and its fractional independent yield in thermal-neutron fission of ^{235}U via the descendant 3.5-h ^{92}Y. After Wahl, Norris & Ferguson (28).

precipitates: the solution is quickly sucked through a thin layer of a reactive precipitate, prepared before irradiation, which takes up the desired radionuclide by adsorption (29) or heterogeneous exchange (30). For solvent extraction, quasisolid agents are used, i.e., the organic phase is adsorbed on a fine-grained carrier, and the aqueous solution is passed through a layer of this material (31).

Schüssler et al. (32, 34) have demonstrated that such techniques permit the direct observation of short-lived fission products within a few seconds after irradiation if the whole sequence is operated automatically (35). Irradiations are carried out in a pneumatic tube system installed in a beam hole of a reactor. The solutions are sealed in polystyrene capsules. After irradiation the capsule is projected within 0.1 to 0.2 sec into the separation apparatus, Figure 2, placed ∼4 m away outside the reactor shield. There, the capsule gets smashed when hitting the walls of a polycarbonate vessel. The solution is quickly filtered through the reactive layer, and washing solution is injected from a syringe. This sequence can be completed within 1.5 sec. By using selective detection techniques, the activity retained in the reactive layer can be observed immediately with detectors placed nearby.

When higher purity and lower background are required, washing is repeated and the reactive layer is pneumatically transported to a shielded detector, as shown in the right part of Figure 2. These steps require another second. Both versions were applied to the study of delayed-neutron precur-

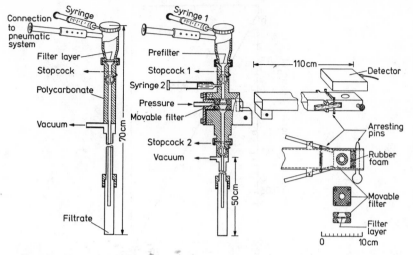

Fig. 2. Apparatus for rapid separations from aqueous solutions by filtration through reactive layers. In the version shown at left the detector is placed close to the filter layer; in the one at right the layer is pneumatically transported after separation to a shielded detector. From Schüssler et al. (35).

sors in low-energy fission, e.g. of 0.8-sec ^{140}I separated by exchange with silver iodide (32, 34).

3.3 *Cyclic operation: deposition of recoil atoms.*—An example for cyclic operation is the helium-jet technique developed by MacFarlane & Griffioen (36) for the study of alpha-particle emitters produced by heavy-ion bombardment. Figure 3 shows a recent version (26). During a beam burst, the whole mixture of reaction products is promptly ejected out of a thin target by nuclear recoil. The recoil atoms are thermalized in helium of atmospheric pressure. The thermalized atoms are swept by the helium through a small orifice into another chamber evacuated by a high-speed diffusion pump. The helium jet is intercepted by a metal collector drum and the recoil atoms stick to the surface. Then, the drum is rotated to bring the activity opposite a detector placed nearby. Alpha-particle spectra and half-lives are measured until the next burst provides a fresh sample.

With the apparatus shown in Figure 3, more than 90 per cent of the recoil atoms are collected within 400 msec and a good fraction within 100 msec, if the cylinder in front of the orifice is extended into the region of thermalization. The drum movement requires 70 msec. Thus, activities with half-lives down to 50 msec can be observed, e.g. 70-msec ^{155}Lu produced by ^{144}Sm$(^{20}$Ne, $p8n)$.

In most applications, the helium jet serves as a rapid technique for transportation and preparation of thin counting samples; elemental assignments of the observed activities are deduced from the production processes. Crude

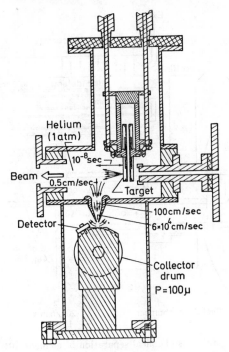

FIG. 3. System for collection and detection of nuclear recoil atoms by the helium-jet technique. After MacFarlane (26).

separations may, however, be achieved since volatile elements will not be collected at room temperature (26).

3.4 *Continuous chemical separation: formation of volatile species.*—Recoil out of solid targets can also be used in cyclic or continuous procedures requiring more selective separations if, in a subsequent chemical step, the element of interest is selectively converted into a volatile species and transported in this form to the detector. This principle has been utilized by Zvara et al. (37) to prove chemically the assignment to element 104 of a 0.3-sec spontaneous fissioning activity produced in bombardments of ^{242}Pu with ^{22}Ne ions.

The experiment, illustrated in Figure 4, is based on the actinide concept: the behavior of element 104, the first transactinide element, should be analogous to that of hafnium; thus element 104 should form a volatile chloride when brought into contact with certain metal chloride vapors whereas actinide and lanthanide elements remain nonvolatile. Details on the separation method (38), apparatus (39), and results (40) have been recently reported.

Thermalized recoil atoms are swept out of the target chamber with nitrogen gas. Downstream, niobium pentachloride and zirconium tetrachloride

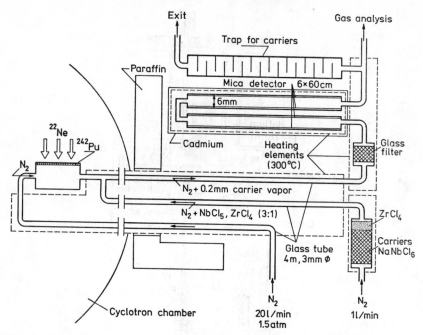

Fig. 4. System for continuous chemical separation of element 104 by conversion of its thermalized recoil atoms into a volatile chloride. After Zvara et al. (40).

vapor in a 3:1 ratio are added to a partial pressure of 0.2 torr to act as a chlorinating agent and as a carrier for the microcomponent. The mixture is pushed through a long, narrow glass tube and a glass filter where nonvolatile components deposit. Then it passes through two pair of mica detectors for 0.7 sec where the fission fragments emitted in the decay of the 0.3-sec activity are detected. The gas tract including the detector chamber is kept at 300° C.

The experiment was performed with extremely low activities: under the best conditions, ten fission events were detected after ~6 days of operation. By comparing this number with the production rate, a transfer time of ~1 sec was estimated for the 0.3-sec activity compared to values of 0.15 sec calculated for the nitrogen carrier gas and 0.2 sec measured for hafnium as microcomponent. Zvara et al. (40) have pointed out that this delay agrees with the adsorption behavior expected for the tetrachloride of element 104.

3.5 *Continuous isotopic separation: evaporation out of solids.*—For online operation of electromagnetic isotope separators, the element of interest has to be introduced continuously into an ion source operated at pressures below 0.01 torr. This is difficult to accomplish with the separation and transportation techniques outlined in the preceding sections. One approach is to use a

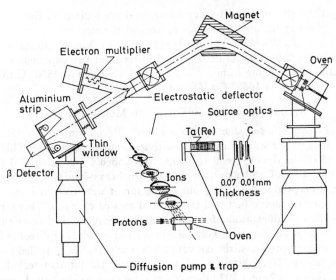

FIG. 5. Mass spectrometer, ion source, and target for continuous isotopic separation of alkali metals using rapid diffusion out of a uranium-graphite target and surface ionization at tantalum and rhenium foils. After Amarel et al. (45).

thick target, from which the element of interest escapes by selective diffusion plus evaporation, and to maintain target and transport line at low pressure; however, when the target must be kept several meters from the ion source, transportation of even noble gases will require several seconds (41), apart from the time for diffusion out of the target.

Faster operation can be achieved if the bombarding beam can be focused on a target placed inside the ion source, as Klapisch & Bernas (42, 43) have demonstrated for alkali metals produced in spallation reactions with 150-MeV protons. The unique diffusion behavior of alkalies in hot graphite and their selective surface ionization were used as separation steps. The Orsay group was able to follow directly the decay of 0.18-sec ^{9}Li (42) and 0.13-sec ^{97}Rb (44) produced in carbon and uranium targets, respectively.

The instrument (45), a 22-cm Nier-type mass spectrometer, is shown schematically in Figure 5. An electrostatic deflector is placed in the path of the separated ions to bend the beam towards the electron multiplier for mass determination and ion counting, or to permit straight travel to an aluminum catcher for decay studies. When necessary, a moving tape is used as collector to eliminate accumulating daughter-product activities.

The target, shown as an insert in Figure 5, consists of a stack of uranium layers, 1 to 8 mg/cm^2 thick, or of other metals, deposited on 70-μ-thick graphite slabs. These slabs are heated to $\sim$1600° C in a tantalum or rhenium oven. The reaction products recoil into the graphite. Alkali elements diffuse rapidly through the graphite, evaporate, and are ionized at the surface of the tantalum or rhenium foil wrapped around the target.

Diffusion studies (41–45) showed that graphite is superior to other media. The diffusion behavior can be described in terms of two exponential modes with half-times ranging from 5 and 28 msec for sodium at 1550° C to 3 and 65 sec for cesium at 1650° C (45).

4. Separations from Solid and Molten Targets

4.1 *Electrostatic collection of recoil atoms.*—Recoil atoms thermalized in a gas atmosphere normally retain a positive charge for some time. This effect was used in the double-recoil technique developed by Ghiorso (46) for indirect detection of alpha-particle emitters in heavy-ion-induced reactions. The reaction products recoiling out of the target were collected on a negatively charged moving belt and transported to a catcher foil. There, half of the daughter-product atoms formed by alpha decay recoiled from the belt to the foil kept at a negative voltage relative to the belt. For direct observation of primary products, the conveyor belt was periodically pulled to solid-state detectors (47). Ghiorso et al. have applied the indirect technique to identify the new element 102 (48) and the direct one to identify element 103 (47).

With increasing beam intensities, however, it became difficult to extract recoil atoms electrostatically from the plasma generated in the stopping gas by the heavy-ion beam. MacFarlane (49) could collect at best only 30 per cent of rare-earth recoils; surprisingly, a positive collecting voltage was more effective. Even lower efficiencies were reported by Mikheev (50) in a systematic study of this technique.

Donets, Shchegolev & Ermakov (51) have therefore modified the double-recoil method by using diffusion to deposit the primary products on the walls of a rotating disk. Yields of ~70 per cent were achieved with argon as the stopping gas. In the subsequent alpha decay, the recoil atoms were again electrostatically collected on a catcher foil. The method was applied first to the 3-sec ^{256}No produced by the reaction ^{238}U$(^{22}$Ne,$4n)^{256}$No and detected via the daughter product 23-h ^{252}Fm.

Recently, Donets & Shchegolev (52) have extended the double-recoil technique to the case in which the daughter product collected on the catcher foil is also a short-lived alpha emitter. Then half of its decay product, the granddaughter of the primary product, is projected into the foil and incorporated. This fraction does not dissolve when the surface of the foil is etched chemically. The technique was checked with 0.12-sec ^{157}Hf which decays by two successive alpha emissions into 4.1-h ^{149}Tb. But attempts to discover alpha decay of 0.3-sec ^{260}104 via the incorporation of its granddaughter 23-h ^{252}Fm were unsuccessful (52). Incorporation in the support was also observed for ^{208}Tl originating by alpha decay from a ^{212}Bi source electroplated on platinum (53).

Experience with fission products shows that chemical properties may play a role in electrostatic collection. Iodine and tellurium were separated as negatively charged ions formed in nitrogen and argon (54). Noble gases could not be collected; obviously, they rapidly become neutralized (55). Unexpected behavior was noted when fission products were stopped close to the collection electrode: the yields were independent of the polarity of the applied voltage (56). Transportation of thermalized fission-product ions through a long drift-tube system was also studied (57, 58).

Electrostatic deposition of beta-decay products is used in a technique for indirect half-life measurement of noble-gas isotopes (59). The gases are swept through a long tube by a carrier gas of known flow rate, and the decay products are deposited on negatively charged steel grids. Patzelt & Herrmann (60) applied this method to fission products with half-lives down to 1.0 sec, viz. ^{143}Xe detected via 33-h ^{143}Ce. When fission halogens were passed through the tube in the form of volatile species, the growth of noble-gas isotopes from halogen precursors was observed along the tube. For example, the existence of a decay chain 0.4-sec ^{141}I→1.7-sec ^{141}Xe was deduced from the 32-d ^{141}Ce activities found on the grids (61). Aerosol filters have also been used to deposit beta-decay products in a similar technique (62). Electrostatic collection was applied to remove the cesium daughter products continuously in direct half-life measurements of xenon isotopes (63).

Beta recoil from solid supports has not yet been utilized in separations of short-lived nuclides, presumably because of the low efficiency reported for this process in the older literature. However, Mundschenk (64) has found that ∼40 per cent of the ^{212}Bi atoms formed in beta decay of ^{212}Pb can be collected on a catcher foil if the source is kept at 100° C in a moderate vacuum. In this system, the average beta-recoil energy amounts only to 0.4 eV. For recoil atoms of ∼10 eV average energy a collection efficiency of ∼10 per cent was measured at room temperature (56). The yields were not affected by applying a collecting voltage (56, 64).

4.2 *Gas-jet technique.*—Some features of the gas-jet technique were already implicit in a method reported by Friedman & Mohr (65) who deposited thermalized reaction recoils by pumping the stopping gas through an aerosol filter.

The technique described in Section 3.3 became widely used in heavy-ion reactions after Griffioen & MacFarlane (66) demonstrated its advantages in a study of francium isotopes including 2.2-sec ^{204}Fr. Mikheev (50) investigated the influence of several parameters on the collection of rare-earth and fermium recoil atoms. In his device, a yield of 60 per cent was achieved under optimum conditions, irrespective of the intensity of the heavy-ion beam. Helium was superior to argon as the stopping medium but argon was employed occasionally (67). An admixture of 1 per cent air to the helium reduced the collection efficiency of rare earths by a factor of two (50) but increased that of astatine by an order of magnitude (68). Obviously, chemical effects are involved in the collection process.

Different versions of the method have been reported, employing, for example, sequential counting of the activity with several solid-state detectors (69), or operation inside a cyclotron chamber (70, 71). Construction details may be seen in these and other publications (72–74). The technique was also applied to delayed-proton emitters (26, 75, 76). A combination with a range measurement of the recoiling daughter atom was used to assign a 19-sec delayed-proton precursor to ^{111}Te (76).

Recently, faster versions were reported by Flerov et al. (77), Valli & Hyde (78), and MacFarlane (79). The volume of the stopping gas is kept as small as the range of the recoil atoms permits (78, 79), and the mechanical transportation of the collected activity is circumvented by conducting the nozzle through an annular solid-state detector under which the collector is placed (77, 78). Measurable amounts of recoil atoms can be detected within 0.5 msec, and maximum yields can be reached within 2 msec after a beam burst (79). Such techniques were used in a search for element 105 (77) and in studies of alpha-particle emitters in the 126 neutron-shell region (78–81) with half-lives down to 50 μsec, viz. ^{217}Th (78, 79). The observation that the collection of volatile species depends on the temperature of the collector may extend the applicability of the gas-jet technique (82).

4.3 *Emanation of noble gases.*—It has been known for a long time that noble gases escape efficiently at room temperature from metal soaps such as barium stearate. This behavior is described in terms of the emanating power, defined as the fraction of radioactive gas atoms that escape from a solid in which they are formed by a nuclear transformation. Two processes contribute: a prompt, temperature-independent process arising from the recoil, and a diffusive mode of escape (83).

Emanation techniques became of interest for rapid separations when Wahl et al. (84–88) found that 4.0-sec ^{219}Rn (85, 87, 88) and 1.9-sec ^{92}Kr (84, 86, 87) escape quantitatively from thin layers of the stearates of barium, rare earths, and uranium kept in a moderate vacuum or swept by an air stream. In such systems, high emanating power will extend to thick samples, as shown for uranyl palmitate (89). The rapidity of this process was demonstrated by Grover, Lebowitz & Baker (90) who found that at least 30 per cent of 35-msec ^{218}Rn escapes within <10 msec from uranyl and lanthanum stearate.

Such salts of organic acids, however, have a drawback: they undergo radiation damage during intense exposures to fission fragments (63, 87), protons (91), and heavy ions (92); this leads to the formation of gaseous decomposition products (87, 91) and to a decrease of emanating power (63, 92). Inorganic compounds which are more resistant to radiation will normally show low emanating power at room temperature. However, a few exceptions are known. Fiedler et al. (93) and Patzelt (94, 95) have found that the hydrated dioxides of zirconium (93–95), titanium, cerium(IV), and thorium (94, 95) are good emanators at high exposures. Alkali diuranates may also be suitable (96).

The emanating power of metal soaps was utilized by Wahl (84) in the first yield measurements of short-lived fission products: When fission products recoiling from a thin target are stopped in a metal soap, noble-gas isotopes escape through the stopping layer into the evacuated irradiation container, and their descendants deposit on the walls of the container. By comparing this activity with the activity of the same nuclide remaining in the stopping medium, the fractional cumulative yield of the noble-gas isotope is deduced. Yields of isotopes with half-lives down to $\sim$1 sec (84, 88) were measured in spontaneous (87) and thermal-neutron fission (84, 86–88). Thick targets of uranyl stearate (97) or multilayer arrangements (98) were used in 14-MeV neutron fission.

In studies of decay properties, ^{235}U was irradiated in the form of uranyl stearate (60) or of uranium oxide coated with barium stearate (99, 100) when low thermal-neutron fluxes were applied, and zirconium dioxide loaded with uranyl ions was used in high fluxes (63).

For direct observation of noble-gas isotopes formed in fission, a mutual separation of krypton and xenon is needed. Pure krypton fractions can rapidly be obtained by removing the xenon in a liquid air trap (99, 100) or a short charcoal column (101). The larger range of krypton fragments in matter may also be used (102). Rapid chromatographic separations of the two elements were reported by Ockenden & Tomlinson (103) and Lepold (104), using charcoal columns at room temperature and helium (103) or nitrogen (104) as carrier gases. Under optimum conditions, the krypton peak appeared after 3 sec, that of xenon after 25 sec. Slower by a factor of about four are separations in a long empty glass tube cooled with liquid nitrogen (63). Such techniques were applied in direct observations of 1.9-sec ^{92}Kr (99), 8-sec ^{91}Kr (105), and 14-sec ^{140}Xe (63, 103).

Various emanating sources were utilized in recent online isotope separations of noble gases. For the study of fission products, uranyl stearate was irradiated in low thermal-neutron fluxes (106) or fission fragments recoiling from a ^{252}Cf source were stopped in barium stearate (107, 108). A uranium dioxide target was placed inside a reactor (109, 110) where it was self-heated to $\sim$600° C (111). For the study of krypton, xenon, and radon isotopes produced by spallation, the oxide hydrates of zirconium, cerium(IV), and thorium, respectively, were bombarded with high-energy protons (27, 95). The transfer from the target to the separator through long, evacuated tubes needed about 10 to 20 sec (95, 106, 109). Nonetheless, products with half-lives of a few seconds were studied, e.g. 1.2-sec ^{142}Xe (110) and 3-sec ^{201}Rn (27). Molecular flow of short-lived nuclides through narrow tubes was treated theoretically by Grover (112).

Among the lighter noble gases, continuous separations of 37-sec ^{23}Ne (113, 114) and 0.8-sec ^6He (115, 116) produced by neutron irradiation of sodium carbonate (113), sodium aluminum silicate (114), or beryllium oxide (115, 116) were reported. The noble gases were swept by either sulfur hexafluoride (113, 115) or water vapor (114, 116) as carrier, to be frozen out before purification of the desired activity. Rapid emanation of helium out of cotton wool and plastic foils was observed by Poskanzer, Esterlund & McPherson (117) and used in a study of 0.12-sec ^{8}He produced by spallation.

4.4 *Diffusion and evaporation.*—Rapid escape of less volatile elements from solid targets implies that both the diffusion to the surface and the subsequent evaporation therefrom are rapid processes. In spite of a large body of diffusion data, predictions are difficult because systems interesting for nuclear studies were seldom investigated. Even if such data are available, they may not be valid at tracer levels, and radiation damage may cause further complications. Hence, conditions for rapid escape have to be sought experimentally. The boiling point of the desired element may give some idea about the temperature needed, and the formulas derived by Rudstam (118) may serve to estimate escape rates.

Cowan & Orth have studied the escape of fission products (119), lanthanides, and actinides (120) from hot graphite. Temperatures around 2000° C are required to evaporate rapidly substantial fractions of the alkalies, alkaline earths, halogens, arsenic, silver, cadmium, tin, antimony, and tellurium. Obviously, the process is not very selective. Interesting fractionations were found among the lanthanide and actinide elements. The rapid diffusion of alkali elements (41–45) has already been mentioned in Section 3.5; additional information is given in (121–123).

Refractory materials such as magnesium oxide (119) or uranium oxides (124) offer, as far as data exist, no advantages for rapid release of fission products.

Rapid diffusion of rare-earth elements out of thin foils of tantalum and neighboring metals heated to ∼2300° C was found by Andersen, Nielsen & Scharff (125, 126). Tarantin et al. (127, 128) have investigated the reemission from heated recoil catchers of recoil atoms produced by heavy-ion bombardments. Dysprosium was reemitted from tantalum kept at 1750° C with a half-time of 0.1 sec; hafnium escaped more slowly (127). Oxide catchers were much inferior for dysprosium but gave good results for polonium (128). This technique will be utilized inside the ion source of an isotope separator (127).

The application of rapid evaporation for online isotope separation of alkali metals is described in Section 3.5. Overall yields of ∼1 per cent were achieved (43). The selectivity of the surface ionization at hot tantalum and rhenium foils was demonstrated by Philippe et al. (121). Among the applications, the detection of new alkali isotopes (44, 122, 123, 129) and yield measurements in spallation reactions (122, 123) may be cited.

Valli, Nurmia & Hyde (92) have found that radon escapes continuously from platinum and gold foils bombarded with intense heavy-ion beams. They used this effect to study radon isotopes including 3-sec ^{201}Rn.

Molten metals were applied as targets (91, 130) by the Isolde group (27) for online isotope separation of cadmium and mercury isotopes produced by spallation of tin and lead, respectively. In spite of a transfer time of ∼35 sec (95), 9-sec ^{183}Hg could be observed. A circulating-liquid system was proposed

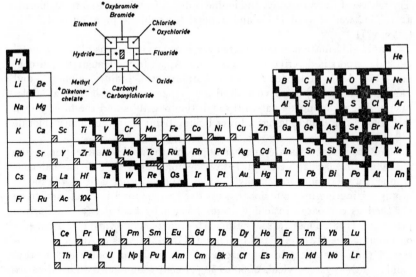

FIG. 6. Volatile compounds with boiling or sublimation points below 350° C. Solid areas refer to compounds indicated without asterisks in the insert, upper left. If these compounds are not volatile, shaded areas are used to refer to compounds with an asterisk in the insert. From Herrmann (12).

to achieve fast evaporation by breaking molten targets into fine droplets (131).

4.5 *Formation of volatile compounds.*—For most elements, species exist that are volatile at moderate temperatures, as is illustrated in Figure 6 for certain classes of compounds. Such species may be formed either within homogeneous targets or in heterogeneous systems, i.e., after recoil of the nuclide from the target into a suitable medium. When the species is produced in a solid, diffusion and evaporation will govern the speed of the separation process. These limitations will not exist for reactions with gaseous partners but such reactions often have the drawback of low yields. The chemical step may be based on a conventional reaction or on a nuclear effect. In the latter case, the kinetic energy of recoiling atoms, or radicals produced during slowing-down of recoils and in radiation fields may initiate at room temperatures reactions normally requiring much higher temperatures.

Only a few examples are known for rapid release of volatile species from homogeneous solids. Escape of 20-min ^{11}C in the form of methane and acetylene was observed from polyethylene (132). Escape of fission halogens from metal soaps, being relatively small in low neutron fluxes (89, 100), becomes comparable to the escape of noble gases in high neutron fluxes (61); an application (61) is mentioned in Section 4.1. Efficient emanation of 32-msec ^{217}At from lanthanum stearate was observed, probably as hydroastatic acid (90).

The release of tin, antimony, and iodine isotopes from tellurium tetrachloride at 150° C was studied (133) and applied for online separation of 1.5-min ^{109m}Sn (27).

Rapid chlorination of thermalized recoil atoms in the gas phase by metal chloride vapors is described in Section 3.4. Apart from element 104, zirconium (134, 135), molybdenum (135), hafnium, tin, indium, niobium, and vanadium (38, 136) were volatilized and separated from less volatile elements. The radionuclide passes through the long tube and the filter shown in Figure 4 if its chloride has a lower boiling point than the carrier chloride (38, 135, 136).

Hot-atom reactions are another approach to rapid volatilization in the gas phase. Denschlag et al. (137–139) demonstrated that fission halogens stopped in methane form methyl halides. About 10 per cent of the iodine atoms originating in the fission process itself were found to react, and to be enriched as compared to iodine atoms formed by beta decay of tellurium precursors (138). This effect was utilized for independent-yield measurements of fission iodine (137, 140) and bromine (141, 142). Paiss & Amiel (143) observed the same effect in the reaction between fission iodine and methyl iodide; in this system, yields of 30 to 50 per cent were reported for primarily formed iodine atoms (100). Additional information about such reactions was obtained by Silbert & Tomlinson (141, 142) who studied the formation of methyl bromide in methane.

The same authors (141, 142) used this reaction to separate 4-sec ^{89}Br from fission products. For this purpose, methyl iodide was removed from the gas stream by reaction with solid silver nitrate and methyl bromide by adsorption in a molecular sieve (141). Similar techniques were applied by K. L. Kratz (34) to observe directly fission halogens with half-lives down to 0.4 sec. Rapid gas chromatography of alkyl halides was used to measure 16-sec ^{88}Br directly (139). Amiel et al. (100) were able to separate methyl bromide and methyl iodide within 8 sec on a tricresyl phosphate column. Cram & Brownlee (144) developed a rapid gas-chromatographic technique for the study of organic compounds labeled with 11-sec ^{20}F.

Attempts to extend such recoil reactions to other elements have been unsuccessful. No volatilization of lead (145), selenium, antimony, and tellurium (146) was observed in methane, nor of antimony in hydrogen (147). Obviously, compounds like antimony hydride SbH_3, which cannot originate from a single collision, are not formed in gases.

This may change in condensed phases since the recoil atom will come to rest in an environment of high radical density. For example, the reaction yield of fission iodine increases to a value of 20 to 30 per cent in liquid (137, 138, 148) and to 60 per cent in solid organic media (138), and a 40 per cent yield is found in solid ammonium acetate (149). Reactions of other elements were indeed reported, namely for tellurium with n-pentane (138); for tin, antimony, and tellurium with neohexane (150); and for germanium, selenium, and arsenic with $tert$-butyl alcohol (148).

An alternative technique for rapid formation of volatile species in solids, the replacement of metal atoms in metal complexes, was introduced by Baumgärtner & Reichold (151). When fission recoils are stopped in chromium hexacarbonyl, the chromium is selectively replaced by molybdenum with a yield of 60 per cent (151). The so-formed molybdenum hexacarbonyl can rapidly be sublimed together with the chromium compound, as was applied for indirect detection of 42-sec [105]Mo (152). Replacement of iron in ferrocene by fission ruthenium (153, 154) and of metal atoms in acetylacetonates by fission products (155) has also been studied. There is evidence that such reactions are exchange processes in the thermal region (154, 155). Replacement of the lead atom in gaseous tetramethyl lead by recoil atoms of lead and bismuth was observed with yields of a few per cent (156).

A different mechanism may be involved in the formation of diphenyl telluride (157, 158) and triphenyl antimony (158) from fission fragments and solid tetraphenyl tin. Rapid release of various fission products out of an irradiated mixture of uranium monocarbide and ruthenium trichloride was found by Blachot & Cavallini (159).

To separate technetium, zirconium, niobium, molybdenum, antimony, and tellurium released at 550° C from these samples, differential adsorption in a glass tube, kept in a temperature gradient, was used (159). The behavior of metal chlorides in such a tube was studied at reduced (160) and normal pressures (38). The technique was also applied to the separation of phenyl compounds formed as mentioned above (158).

Systematic work on the volatilization of nuclear reaction products out of nonvolatile host compounds would be of great interest to develop continuous procedures of this kind.

In a rapid dry volatilization of fission iodine, uranium peroxide was irradiated in solid periodic acid mixed with activated carbon, and elemental iodine was liberated by heating this mixture. Yields of 70 per cent were achieved within 4 sec (161).

Differences in the valence state of fission products formed directly in fission compared to those originating in beta decay were found for iodine (162) and tin (163). The primarily formed atoms prefer lower valence states.

5. Separations from Aqueous Solutions

5.1 *Handling of irradiated samples.*—The first step in rapid separations from aqueous solutions is the removal of the solution from the irradiation capsule. One procedure is described in Section 3.2. An alternative method was developed by Greendale & Love (164). The rabbit is impaled on two hypodermic needles which penetrate through a diaphragm. Emptying and rinsing is performed within 2 sec. Similar devices were described by other workers (165, 166).

5.2 *Volatilization and distillation.*—Sweeping of noble gases from aqueous solutions was used for detection of the fission product 1.3-sec [93]Kr (167). The release is delayed for several seconds (98) and is therefore incomplete for very short-lived nuclides (85).

Considerable work has been carried out on the volatilization of hydrides since Greendale & Love (168) showed that arsenic and antimony are rapidly reduced to the hydrides when their solutions in sulfuric acid are dropped onto granular zinc. Yields of 70 to 80 per cent were achieved within 10 sec. To separate both elements, antimony hydride was decomposed to the metal at 600° C, arsenic hydride at $\sim$1000° C. The method was used for fission-yield measurements of antimony isotopes via iodine descendants (169, 170). With modified versions, del Marmol & Nève de Mévergnies (171, 172) and Tomlinson & Hurdus (173) have directly observed the fission products 2.0-sec ^{85,86}As and 1.7-sec ^{135}Sb. Other groups have decomposed antimony hydride in liquid absorbents (174–176) or have produced it by reduction with sodium borohydride (176).

A few per cent of the germanium present was also found in the metal mirrors (171, 173). Covolatilized selenium can be recovered in solid calcium sulfate (177) or in water (178) with a yield of $\sim$2 (177) or 20 per cent (178) in 5 sec. This method was used for indirect detection of 6-sec ^{87}Se (177).

Yields of 80 to 90 per cent of the hydrides of arsenic, selenium, antimony, and tellurium can be achieved within a second in strong hydrochloric acid with zinc powder, as Folger, Kratz & Herrmann (179) have found. Selenium and tellurium were selectively absorbed in sodium hydroxide solution and separated by solvent extraction. By this method, the fission products 2.2-sec ^{88}Se and 3.5-sec ^{137}Te were directly detected (34).

Rapid volatilization of tin hydride was described by Greendale & Love (180) using sodium borohydride as reducing agent and sodium hydroxide-impregnated asbestos for removal of antimony hydride. The method was applied in fission-yield measurements (169). Some modifications were given by Treytl (181).

Electrolytic generation of hydrides has also been investigated with mercury (182) and graphite electrodes (183, 184) and utilized in fission studies, but the yields decreased substantially in strong radiation fields (185, 186).

Rapid volatilization of halogens was used by Perlow & Stehney (6) in the first direct observation of very short-lived fission products, viz. 1.6-sec ^{90}Br and 2.7-sec ^{139}I. A burst of air loaded with iodine or bromine was blown through the uranium solution and then through carbon tetrachloride serving as an absorbent for the halogens. By choice of the medium, the undesired halogen was kept in a nonvolatile form. Modifications of this method, e.g. the use of other absorbents, were reported by Shpakov et al. (187). An adaptation to online isotope separation was discussed by Stehney (188): A solution of uranium in concentrated sulfuric acid, suspended in the holes of thin platinum grids, is proposed as a target which can be maintained in vacuum for sufficiently long periods.

Rapid distillation of ruthenium tetroxide was studied by Riccato & Strassmann (189). At a temperature of 160°C, 40 per cent was distilled within 5 sec from strong perchloric acid. The fission product 35-sec ^{109}Ru, previously indirectly detected by a distillation method (190), was directly observed.

A rapid sublimation procedure was worked out for indium by Weiss &

Ballou (191). Indium acetylacetonate was precipitated and filtered on a steel wire mesh which was then placed on a molten alloy maintained at 425° C. About 25 per cent of the indium compound sublimed to a cold finger within 10 sec. The method was applied in indirect yield measurements of the fission product 3.1-min ^{121}In via 27-h ^{121}Sn (191).

5.3 *Precipitation procedures.*—Nearly all recent studies on short-lived nuclides with precipitation procedures concern indirect determinations of half-lives and fission yields. For this purpose, either the nuclide of interest or its daughter product may be precipitated. For example, fission alkalies like 2.6-sec ^{94}Rb and 1.6-sec ^{143}Cs were indirectly detected by Fritze, Kennett & Prestwich (192–194) after a rapid precipitation of alkali perchlorates, whereas Wahl et al. (87) precipitated the nitrates of the alkaline-earth daughters for detection of 57-sec ^{91}Rb and 24-sec ^{141}Cs.

In similar studies, long-lived daughter products were precipitated in the form of the following species: hydroxides of yttrium (195), zirconium, niobium (196), and lanthanum (87, 197); iodates of cerium(IV) (87) and zirconium (198); tin sulfides (199), elemental tellurium (169), and oxalates of indium and tin (200). A few examples are the detection of 10-sec ^{99}Nb (196), 12-sec ^{144}Ba (197), and 13-sec ^{121}Cd (200).

The short-lived nuclide, on the other hand, was precipitated in the following cases: 37-sec ^{115}Pd was precipitated by dimethylglyoxime (201), 10-sec ^{106}Mo by oxine (202), and fission technetium was coprecipitated with tetraphenyl-arsonium perrhenate (202).

Devices for rapid performance of such procedures were described by Sam & Love (203) who applied high-pressure filtration in a rapid precipitation of indium by pyridine, and by Vallis & Perkin (204) who continuously removed freshly formed solids by means of a special centrifuge cup.

5.4 *Deposition in preformed precipitates.*—Rapid uptake of ions from solutions into preformed precipitates, already discussed in Section 3.2, was demonstrated for trace amounts, e.g. of barium ions in barium sulfate, by Rai, Nethaway & Wahl (29) and Eckhardt, Schüssler & Herrmann (30). In certain systems, the whole solid phase or most of it takes part in such reactions, for example in the exchange between silver ions and silver iodide (30) or silver sulfide (205).

Ion exchange reactions can also be used if the exchange leads to a solid phase less soluble than the original precipitate. Hence, both bromide and iodide ions are taken up by silver chloride, but bromide passes through silver iodide layers (30). When pure solids react too slowly, fast exchange may be achieved by coating cellulose powder with the solid, as was shown for the exchange of fission products with various sulfides (205).

In addition to these systems, quantitative uptake was reported for rubidium and cesium into salts of heteropoly acids (30, 206, 207), for barium (30) and strontium (29) into their carbonates, for palladium into palladium cyanide (207), and for antimony(III) into antimony trioxide (207).

Rapid separation procedures based upon such reactions were worked out for the halogens (30) and for cesium (206) from fission-product mixtures.

Some applications in fission studies are told in Sections 3.1 and 3.2. In addition, the direct observation of 2.6-sec ^{94}Rb after exchange with ammonium phosphomolybdate may be cited (32, 33), as well as the use of reactions between selenium (208, 209) and tellurium ions (210) with their elemental forms.

The reverse reaction has also been observed, namely rapid release into solution of a decay product formed in a solid in which its parent substance is incorporated. Ruddy & Pate (211) washed 0.3-sec ^{136m}Ba continuously out of thallium phosphototungstate labeled with 13-d ^{136}Cs. Stenström & Jung (212) washed 7-sec ^{161m}Ho from erbium phthalocyanine containing its parent 3.1-h ^{161}Er. Repeated milking of silver and cadmium descendants out of palladium dimethylglyoxime was used to detect 37-sec ^{115}Pd (213).

5.5 *Solvent extraction.*—Conventional solvent extraction of individual fission products can be performed within 20 to 30 sec, as Hübscher (214) has shown in decay studies of 24-sec ^{107}Tc which was extracted into a solution of tetraphenyl arsonium chloride after the halogens had been removed by exchange with silver chloride. Other examples are the simultaneous extraction of bromide and iodide using triphenyltin hydroxide (215) and the indirect detection of short-lived ^{79}Ge by extraction into carbon tetrachloride (216).

Various techniques have been developed to speed up solvent extraction by improving the phase separation. The centrifuge cup mentioned in Section 5.3 was recommended for this purpose (204). Centrifugal action is also utilized in the separation stirrers developed by Münzel et al. (217, 218). Examples of their application are the continuous milking of 5-sec ^{118}In from 50-min ^{118}Cd (218), and a separation procedure for fission cadmium by dithizone extraction (218, 219).

Viscous solvents can be used in a quasisolid form, as mentioned in Section 3.2. This was demonstrated by Denig, Trautmann & Herrmann (31) for the extraction of pentavalent antimony and protactinium into diisobutylcarbinol and di-(2-ethylhexyl)-orthophosphoric acid adsorbed on plastic grains. The latter solvent was used for direct detection of the fission product 2-sec ^{98,99}Y (34). The 31-sec ^{98}Zr was separated from fission products with tri-*n*-butyl phosphate as the solvent, and its decay product 2.9-sec ^{98}Nb was continuously milked from the quasisolid phase with strong nitric acid (220). The same solvent was applied for the separation of selenium from tellurium mentioned in Section 5.2 (34, 179).

Such techniques may also be useful for isolating somewhat longer-lived nuclides if multistep separations are required. For example, four extraction and two precipitation steps are necessary to obtain clean samples of 2.3-min ^{238}Pa produced by fast-neutron bombardment of uranium in a side reaction to fission. With techniques outlined here, this procedure can be completed within 2 min (221).

The common extraction chromatography of rare-earth elements was tested at large flow rates but a few minutes were still required to separate individual elements (222).

5.6 *Ion exchange.*—Cation exchange of rubidium ions with the ion exchange resin Dowex 50 was found to occur within a few seconds (223). Wish (224) has utilized such a process for detection of 17-sec ^{77}Ga via germanium and arsenic descendants. Gallium was separated by passing a fission-product solution in hydrochloric acid through a layer of Dowex 50 within a second; 80 to 90 per cent of the gallium was retained.

Progress has also been made in the fast separation of rare-earth elements by elution from cation exchange resins with complex-forming agents. About 3 min were required to separate one component from a mixture of three adjacent rare earths (225), and about 8 min to isolate promethium and neodymium from fission products (226). The Szilard-Chalmers effect may be used to reduce the amount of inactive carrier when rare-earth elements are the target material (227).

Continuous milking of short-lived decay products from parent substances adsorbed on ion exchange columns was reported for 2.9-sec ^{98}Nb from 31-sec ^{98}Zr (228) and 5-sec ^{102}Tc from 11-min ^{102}Mo (229).

5.7 *Electrochemical techniques.*—Rapid electrochemical deposition of various elements on copper powder was observed by Weiss & Reichert (230) who passed nitric acid solutions through layers of this material within less than a second. Silver, palladium, mercury, and gold were retained with at least 70 per cent yield. In applying this technique to fission products, complex-forming agents were added to suppress the deposition of silver when palladium was of interest (231, 232), and vice versa (233). In this way, 5-sec ^{117}Pd (232) and 5-sec ^{118}Ag (233) were indirectly detected.

Okashita (234, 235) has observed a rapid deposition of noble metals, selenium, and tellurium into metallic mercury under ultrasonic agitation; for mercury ions, half-exchange times of 3 sec were achieved. Similar results were reported by Monnier et al. (236–238). Extraction into sodium amalgam was used to separate 1.2-min ^{142}Eu from samarium targets (239, 240).

Conditions for rapid electrolysis were studied and a minimal half-time of 14 sec was measured for the deposition of bismuth on steel grids (241).

Electrophoretic ion focusing was found to be too slow for rapid separations of fission rare earths; about 30 min were required to isolate individual elements (242).

6. SEPARATIONS FROM GASEOUS TARGETS

Continuous separations of 17-sec ^{19}Ne and 1.8-sec ^{35}Ar produced via the reactions ^{19}F$(p,n)^{19}$Ne and ^{32}S$(\alpha,n)^{35}$Ar by bombarding sulfur hexafluoride were carried out (243) as described at the end of Section 4.3. Iodine isotopes produced by (n,p) reactions on gaseous xenon deposited on the walls of the container and were washed out after bombardment (244). The 54-sec ^{86}Br formed by bombardment of solid ^{86}Kr was also found there after the krypton had been quickly evaporated (245).

7. APPLICATIONS OF RAPID CHEMICAL SEPARATIONS IN NUCLEAR CHEMISTRY

Tables I and II summarize the application of rapid separations in studies

TABLE I

SUMMARY OF DIRECT STUDIES OF SHORT-LIVED NUCLIDES
BY RAPID SEPARATIONS

Element[a]	Production process[b]	Separation technique and references[c]	Shortest half-life[d]
He	C+p	4.3 (117)	0.12-s ^{8}He
Li	C+p	3.5, 4.4 (42, 43, 45)	0.18-s ^{9}Li
Na	Si+p	3.5, 4.4 (43, 45)	0.4-s ^{20}Na
	Vs+p	3.5, 4.4 (122)	1.0-s ^{26}Na
As	U(n,f)	5.2 (171, 172, 173, 185, 186)	2.0-s ^{85}As[e]
Se	U(n,f)	5.2 (34)	2.2-s ^{88}Se
Br	U(n,f)	5.2 (6, 187, 246, 247)	1.6-s ^{90}Br
		4.5 (34, 139, 141, 142)	0.5-s ^{91}Br[f]
		3.2, 5.4 (32, 34)	1.6-s ^{90}Br
Kr	Zr+p	4.3 (27)	13-s ^{81m}Kr
	U(n,f)	4.3 (99, 100, 101, 103–106, 109, 110)	1.3-s ^{93}Kr
		5.2 (167)	1.3-s ^{93}Kr
Kr (Rb)	U(n,f)	5.2 (167)	6-s ^{98}Rb
		4.3 (99, 100, 109, 110)	2.6-s ^{94}Rb
Rb	U(n,f)	5.4 (32, 34)	2.6-s ^{94}Rb
	U(p,f)	3.5, 4.4 (43, 44, 45, 129)	0.13-s ^{97}Rb
	Vs+p	3.5, 4.4 (123)	0.23-s ^{96}Rb
Rb (Sr)	U(p,f)	3.5, 4.4 (129)	4-s ^{96}Sr
Sr	U(n,f)	5.3 (193)	26-s ^{95}Sr
Y	U(n,f)	5.5 (34)	2-s ^{99}Y[g]
Zr	U(n,f)	5.5 (220)	31-s ^{98}Zr
Nb	Zr(β)	5.5 (220), 5.6 (228)	2.9-s ^{98}Nb
Tc	U(n,f)	5.5 (214, 248)	27-s ^{107}Tc
Mo (Tc)	U(n,f)	4.5 (249)	50-s ^{108}Tc
Ru	U(n,f)	5.2 (189)	35-s ^{109}Ru
Ag	U(d,f)	5.4 (250)	42-s ^{117}Ag
Sb	U(n,f)	5.2 (173, 185, 186)	1.7-s ^{135}Sb
Te	U(n,f)	5.2 (34)	3.5-s ^{137}Te
	Ru(O,xn)	3.3, 4.2 (26, 251)	2.2-s ^{107}Te
	Pd(C,xn)	4.2 (75, 76, 252)	4.2-s ^{109}Te
I	U(n, f)	5.2 (6, 187, 246)	2.7-s ^{139}I
		5.4 (32, 34)	0.8-s ^{140}I
		4.5 (34, 140)	0.4-s ^{141}I
Xe	U(n,f)	4.3 (63, 101, 103, 106, 109, 110, 253)	1.2-s ^{142}Xe
	Cf(sp,f)	4.3 (108)	1.2-s ^{142}Xe
	Ce+p	4.3 (27)	19-s ^{115}Xe
Xe (Cs)	U(n,f)	4.3 (63, 109, 110)	1.6-s ^{143}Cs
	Cf(sp,f)	4.3 (108)	1.6-s ^{143}Cs
Cs	U(n,f)	5.4 (206)	24-s ^{141}Cs
	U(p,f)	3.5, 4.4 (44, 45, 129)	1.1-s ^{144}Cs
	Vs+p	3.5, 4.4 (123)	1.6-s ^{143}Cs

TABLE I (*Continued*)

Element[a]	Production process[b]	Separation technique and references[c]	Shortest half-life[d]
Cs (Ba)	U(p,f)	3.5, 4.4 (129)	12-s ^{144}Ba
La	U(p,f)	3.5, 4.4 (129)	41-s ^{144}La
Ho	Pr(O,xn)	3.3, 4.2 (254)	20-s ^{150}Ho
Er	Nd(O,xn)	3.3, 4.2 (255)	11-s ^{152}Er
Yb	$Vs+HI$	3.3, 4.2 (256)	0.4-s ^{154}Yb
Lu	Sm(F,xn)	3.3, 4.2 (257)	70-ms ^{155}Lu
Hf	Sm(Ne,xn)	3.3, 4.2 (257)	0.12-s ^{157}Hf
Ir	$Vs+HI$	4.2 (258)	1.0-s ^{171}Ir
Pt	$Vs+HI$	4.2 (*259*, 260)	0.7-s ^{174}Pt
Hg	Pb+p	4.4 (27)	9-s ^{183}Hg
	$Vs+HI$	4.2 (260)	3-s ^{179}Hg
Au	$Vs+HI$	4.2 (260, *261*)	1.4-s ^{177}Au
Po	$Vs+HI$	4.2 (262)	0.6-s ^{194}Po
	U(HI,f)	4.2 (67)	25-s ^{211m}Po
At	Re(Ne,xn)	4.2 (68)	0.3-s ^{196}At
At (Bi)	Re(Ne,xn)	4.2 (68)	14-s ^{190}Bi[h]
Rn	Th+p	4.3 (27)	12-s ^{202}Rn
	$Vs+HI$	4.4 (95)	3-s ^{201}Rn
Fr	$Vs+HI$	4.2 (66, *78*, *80*, 263)	4-ms ^{214}Fr
Ra	$Vs+HI$	4.2 (*78*, *79*, 264)	1.6-ms ^{215}Ra
Ac	$Vs+HI$	4.2 (74, *78*, *79*)	0.4-ms ^{216}Ac
Th	Pb(O,xn)	4.2 (*78*, *79*, 81)	50-μs ^{217}Th
Fm	$Vs+HI$	4.2 (70, 265)	1.4-s ^{246}Fm
No	$Vs+HI$	4.2 (47, *69*, 71, 72, 73, 266, 267, *268*, 269, *270*)	0.8-s ^{251}No
Lw	$Vs+HI$	4.2 (*47*, 71, 271)	8-s ^{257}Lw
104	Pu(Ne,xn)	3.4, 4.5 (37, 40)	0.3-s ^{260}104[i]

[a] The separated element is given; when a decay product was studied, its symbol is given in parentheses.

[b] Abbreviations: *Vs* various target elements, *HI* various heavy ions, *p* high-energy protons, *f* fission, *sp* spontaneous, *α*, *β* alpha or beta decay.

[c] The separation technique is indicated by the corresponding Section number of this article. Literature up to (245) is covered in the text but the aspect under which the reference appears in the Table is not necessarily discussed.

[d] For illustration, the shortest-lived nuclide yet detected with a particular technique is given with its adopted half-life rather than the value reported in the cited work. The corresponding reference is italicized in the preceding column unless the nuclide indicated was studied in all references cited there.

[e] Mass number 85 or 86.

[f] Probably ^{91}Br plus ^{92}Br.

[g] Mass number 98 or 99.

[h] Mass number 190 or 194.

[i] Mass number uncertain (9, 10, 52).

HERRMANN & DENSCHLAG

TABLE II

SUMMARY OF INDIRECT STUDIES OF SHORT-LIVED NUCLIDES
BY RAPID SEPARATIONS

Element[a]	Production process[b]	Separation technique and references[c]	Shortest half-life[d]
Ga	$U(n,f)$	5.6 (224)	17-s ^{77}Ga
Ge	$U(n,f)$	5.5 (216)	10-s ^{79}Ge[e]
Se	$U(n,f)$	5.2 (177)	6-s ^{87}Se
		5.4 (208)	16-s ^{86}Se
Se (As)	$U(n,f)$	5.2 (34)	1.0-s ^{87}As
Br	$U(n,f)$	4.1 (61)	0.6-s ^{91}Br
Kr	$U(n,f)$	4.3 (60, 61, 84, 87, 88, 97, _100_)	0.4-s ^{94}Kr
Rb	$U(n,f)$	5.3 (192, _193_), 5.4 (28)	2.7-s ^{94}Rb
Y (Sr)	$U(n,f)$	5.3 (195)	26-s ^{95}Sr
Zr (Y)	$U(n,f)$	5.3 (198)	6-s ^{97}Y
Nb	$U(n,f)$	5.3 (196)	10-s ^{99}Nb
Mo	$U(n,f)$	4.5 (152), 5.3 (_202_)	10-s ^{106}Mo
Tc	$U(n,f)$	5.5 (_248, 272_), 5.3 (202)	27-s ^{107}Tc
Tc (Mo)	$U(n,f)$	5.5 (272)	42-s ^{105}Mo
Ru	$U(n,f)$	5.2 (190)	35-s ^{109}Ru
Pd	$U(n,f)$	5.3 (201, 213), 5.4 (_273_), 5.7 (231, _232_)	5-s ^{117}Pd
Ag	$U(n,f)$	5.4 (216, _233_, 273)	5-s ^{118}Ag
In	$U(n,f)$	5.2 (191), 5.5 (_216_)	18-s ^{125}In
In (Cd)	$U(n,f)$	5.3 (200)	13-s ^{121}Cd
Sn	$U(n,f)$	5.2, 5.3 (169)	55-s ^{133}Sn
Sn (In)	$U(n,f)$	5.3 (199), 5.4 (_274_)	2.5-s ^{125}In
Sb	$U(n,f)$	5.2 (169, 170, 174, _175_, 176)	1.7-s ^{135}Sb
Te	$Pd(C,xn)$	4.2 (76)	19-s ^{111}Te
I	$U(n,f)$	4.1 (61)	0.4-s ^{141}I
I (Te)	$U(n,f)$	4.5 (140), 5.4 (_275_)	18-s ^{135}Te
Xe	$U(n,f)$	4.3 (60, _61_, 62, _84_, 86, _88_, 97, 98)	1.2-s ^{144}Xe
	$Cf(sp,f)$	4.3 (87)	1.2-s ^{144}Xe
Cs	$U(n,f)$	5.3 (194)	1.6-s ^{143}Cs
Ba (Cs)	$U(n,f)$	5.3 (87)	24-s ^{141}Cs
La (Ba)	$U(n,f)$	5.3 (87, _197_)	12-s ^{144}Ba
Tb	$Er(\alpha)$	4.1 (255)	36-s ^{153}Er
Hf	$Sm(Ne,xn)$	4.1 (52)	0.12-s ^{157}Hf
Fm	$U(O,xn)$	4.1 (276)	38-s ^{248}Fm
No	$Vs+HI$	4.1 (_48_, _51_, 276, 277)	3-s ^{252}No
Lw	$Am(O,xn)$	4.1 (278)	35-s ^{256}Lw

[a, b, c, d] See Table I.
[e] Qualitatively detected, half-life estimated.

of nuclides with <1 min half-life produced by complex nuclear reactions. Table I refers to direct detection of the short-lived nuclide, Table II to its indirect detection via long-lived descendants.

ACKNOWLEDGMENTS

It is a pleasure to thank Miss Rita Weis for her invaluable help in literature search. We should also like to thank Dr. S. M. Qaim for reading the manuscript. Finally, and not least, we are indebted to the many who have communicated results before publication.

LITERATURE CITED

1. Meyer, S., Schweidler, E., *Radioaktivität*, 2nd ed. (Teubner, Leipzig, 721 pp., 1927)
2. Rutherford, E., *Phil. Mag.*, **49**, 1–14 (1900)
3. Moseley, H. G. J., Fajans, K., *Phil. Mag.*, **22**, 629–38 (1911)
4. Kusaka, Y., Meinke, W. W., *Rapid Radiochemical Separations* (Natl. Acad. Sci.–Natl. Res. Council, Nucl. Sci. Ser. NAS-NS 3104, Washington, D. C., 125 pp., 1961)
5. Hahn, O., Strassmann, F., *Naturwissenschaften*, **28**, 54–61 (1940)
6. Perlow, G. J., Stehney, A. F., *Phys. Rev.*, **113**, 1269–76 (1959)
7. Bergström, I., *Nucl. Instr. Methods*, **43**, 116–28 (1966)
8. *Nuclides Far Off the Stability Line*, Proc. Intern. Symp., Lysekil, Sweden, August 1966 (Forsling, W., Herrlander, C. J., Ryde, H., Eds., Almqvist & Wiksell, Stockholm, 686 pp., 1967) [=*Arkiv Fysik*, **36** (1967)]
9. Seaborg, G. T., *Ann. Rev. Nucl. Sci.*, **18**, 53–152 (1968)
10. Herrmann, G., Seyb, K. E., *Naturwissenschaften* (In press, 1969)
11. Amiel, S., in *Nuclear Chemistry*, **2**, 251–94 (Yaffe, L., Ed., Academic Press, New York, 409 pp., 1968)
12. Herrmann, G. (See Ref. 8), 111–26
13. Herrmann, G., *Radiochim. Acta*, **3**, 169–85 (1964), **4**, 173–88 (1965) [Engl. transl. *At. Energy Res. Establ. Harwell Rept. AERE-Trans 1036* (37 pp., 1965), *AERE-Trans 1057* (44 pp., 1966)]
14. Flerov, G. N., *At. Energ. (USSR)*, **24**, 5–17 (1968); *Ann. Phys.*, **2**, 311–23 (1967)
15. Klapisch, R., *Ann. Rev. Nucl. Sci.*, **19**, 33–60 (1969)
16. Stevenson, P. C., Hicks, H. G., *Ann. Rev. Nucl. Sci.*, **3**, 221–34 (1953)
17. Finston, H. L., Miskel, J., *Ann. Rev. Nucl. Sci.*, **5**, 269–96 (1955)
18. *The Radiochemistry of the Elements* (Natl. Acad. Sci.–Natl. Res. Council, Nucl. Sci. Ser. NAS-NS 3001–3057, ed. by Subcomm. on Radiochem, Washington, D. C., 1960–1965)
19. Freiser, H., Morrison, G. H., *Ann. Rev. Nucl. Sci.*, **9**, 221–44 (1959)
20. Kraus, K. A., Nelson, F., *Ann. Rev. Nucl. Sci.*, **7**, 31–46 (1957)
21. DeVoe, J. R., *Application of Distillation Techniques to Radiochemical Separations* (Natl. Acad. Sci.–Natl. Res. Council, NAS-NS 3108, Washington, D. C., 29 pp., 1962)
22. Harvey, B. G., *Ann. Rev. Nucl. Sci.*, **10**, 235–58 (1960)
23. Wolfgang, R., *Progr. Reaction Kinetics*, **3**, 97–169 (1965)
24. Harbottle, G., *Ann. Rev. Nucl. Sci.*, **15**, 89–124 (1965)
25. Rudstam, G. (See Ref. 8), 9–17
26. MacFarlane, R. D. (See Ref. 8), 431–43
27. Hansen, P. G., Hornshøj, P., Nielsen, H. L., Wilsky, K., Kugler, H., Astner, G., Hagebø, E., Hudis, J., Kjelberg, A., Münnich, F., Patzelt, P., Alpsten, M., Andersson, G., Appelqvist, Aa., Bengtsson, B., Naumann, R. A., Nielsen, O. B., Beck, E., Foucher, R., Husson, J. P., Jastrzebski, J., Johnson, A., Alstad, J., Jahnsen, T., Pappas, A. C., Tunaal, T., Henck, R., Siffert, P., Rudstam, G., *Phys. Letters*, **28B**, 415–19 (1969)
28. Wahl, A. C., Norris, A. E., Ferguson, R. L., *Phys. Rev.*, **146**, 931–34 (1966)
29. Rai, R. S., Nethaway, D. R., Wahl, A. C., *Radiochim. Acta*, **5**, 30–34 (1966)
30. Eckhardt, W., Herrmann, G., Schüss-

ler, H. D., *Z. Anal. Chem.*, **226**, 71–88 (1967)

31. Denig, R., Trautmann, N., Herrmann, G., *Z. Anal. Chem.*, **216**, 41–50 (1966)

32. Herrmann, G. (See Ref. 33), 147–66

33. *Delayed Fission Neutrons*, Proc. of Panel, Vienna, April 1967 (IAEA, Vienna, 249 pp., 1968)

34. Schüssler, H. D., Ahrens, H., Folger, H., Franz, H., Grimm, W., Herrmann, G., Kratz, J. V., Kratz, K. L., *Delayed Neutron Precursors in Fission of* ^{235}U *by Thermal Neutrons* (To be presented at 2nd Symp. Phys. and Chem. of Fission, Vienna, July 1969)

35. Schüssler, H. D., Grimm, W., Weber, M., Tharun, U., Denschlag, H. O., Herrmann, G., *Nucl. Instr. Methods* (In press, 1969)

36. MacFarlane, R. D., Griffioen, R. D., *Nucl. Instr. Methods*, **24**, 461–64 (1963)

37. Zvara, I., Chuburkov, Yu.T., Caletka, R., Zvarova, T. S., Shalaevskii, M. R., Shilov, B. V., *At. Energ. (USSR)*, **21**, 83–84 (1966) [= *Soviet J. At. Energy*, **21**, 709–10 (1966)]

38. Zvara, I., Chuburkov, Yu.T., Zvarova, T. S., Caletka, R., *Joint Inst. Nucl. Res. Dubna Rept. D6-3281* (19 pp., 1967)

39. Chuburkov, Yu.T., Zvara, I., Shilov, B. V., *Joint Inst. Nucl. Res. Dubna Rept. P7-4021* (30 pp., 1968)

40. Zvara, I., Chuburkov, Yu.T., Caletka, R., Shalaevskii, M. R., *Joint Inst. Inst. Nucl. Res. Dubna Rept. P7-3783* (26 pp., 1968)

41. Talbert, W. L., McConnell, J. R. (See Ref. 8), 99–105

42. Klapisch, R., Bernas, R., *Nucl. Instr. Methods*, **38**, 291–95 (1965)

43. Klapisch, R., Chaumont, J., Philippe, C., Amarel, I., Fergeau, R., Salome, M., Bernas, R., *Nucl. Instr. Methods*, **53**, 216–28 (1967)

44. Amarel, I., Gauvin, H., Johnson, A., *J. Inorg. Nucl. Chem.*, **31**, 577–84 (1969)

45. Amarel, I., Bernas, R., Chaumont, J., Foucher, R., Jastrzebski, J., Johnson, A., Klapisch, R., Teillac, J. (See Ref. 8), 77–89

46. Ghiorso, A., *At. Energ. (USSR)*, **7**, 338–50 (1959) [= *Soviet J. At. Energy*, **7**, 819–29 (1959)]

47. Ghiorso, A., Sikkeland, T., Larsh,

A. E., Latimer, R. M., *Phys. Rev. Letters*, **6**, 473–75 (1961)

48. Ghiorso, A., Sikkeland, T., Walton, J. R., Seaborg, G. T., *Phys. Rev. Letters*, **1**, 18–21 (1958)

49. MacFarlane, R. D., *Phys. Rev.*, **126**, 274–76 (1962)

50. Mikheev, V. L., *Pribory Tekhn. Eksperim.*, **1966**, *No. 4*, 22–26 [= *Instr. Exp. Tech.*, **1966**, 785–89]

51. Donets, E. D., Shchegolev, V. A., Ermakov, V. A., *At. Energ. (USSR)* **16**, 195–207 (1964) [= *Soviet J. At. Energy*, **16**, 233–45 (1964)]

52. Donets, E. D., Shchegolev, V. A., *Joint Inst. Nucl. Res. Dubna Rept. P7-3835* (14 pp., 1968)

53. Weber, M., *Über das elektrochemische Verhalten des Urans als Kathode und die elektrolytische Ablösung von Spaltprodukten* (Dissertation, Univ. Mainz, 1965)

54. Fröhner, F. H., *Z. Physik*, **170**, 62–75 (1962)

55. Boos, A., *Versuche zur Trennung von Spaltprodukten in einem mit Wasserstoff gefüllten Plattenkondensator* (Dissertation, Univ. Mainz, 1967)

56. Jung, D., *Untersuchungen über die Trennung von Spaltprodukten durch β-Rückstoss* (Dissertation, Univ. Mainz, 1969)

57. Stevenson, P. C., Cambey, L. A., Smith, T. (See Ref. 8), 279–85

58. Cambey, L. A., Smith, T. C., *Univ. Calif. Rept. UCRL-13260* (48 pp., 1966)

59. Rodenbusch, H., Herrmann, G., *Z. Naturforsch.*, **16a**, 577–82 (1961)

60. Patzelt, P., Herrmann, G., in *Physics and Chemistry of Fission*, Symp. Salzburg, 1965, **2**, 243–52 (IAEA, Vienna, 469 pp., 1965)

61. Patzelt, P., Ahrens, H., Herrmann, G., *Proc. Radioanal. Conf.*, Stary Smokovec, April 1968 (In press, 1968)

62. Cordes, O. L., Cline, J. E., Reich, C. W., *Nucl. Instr. Methods*, **48**, 125–30 (1967)

63. Archer, N. P., Keech, G. L., *Can. J. Phys.*, **44**, 1823–45 (1966)

64. Mundschenk, H., *Eine Untersuchung über die Abtrennung kurzlebiger Nuklide durch β-Rückstoss* (Dissertation, Univ. Mainz, 1962)

65. Friedman, A. M., Mohr, W. C., *Nucl. Instr. Methods*, **17**, 78–80 (1962)

66. Griffioen, R. D., MacFarlane, R. D., *Phys. Rev.*, **133**, B1373–80 (1964)

67. Kuznetsov, I. V., Mal'tseva, N. S.,

Oganesyan, Yu.Ts., Sukhov, A. M., Shchegolev, V. A., *Yad. Fiz.*, **8**, 448–53 (1968)

68. Treytl, W., Valli, K., *Nucl. Phys.*, **A97**, 405–16 (1967)

69. Ghiorso, A., Sikkeland, T., Nurmia, M. J., *Phys. Rev. Letters*, **18**, 401–4 (1967)

70. Akap'ev, G. N., Demin, A. G., Druin, V. A., Imaev, E. G., Kolesov, I. V., Lobanov, Yu.V., Pashchenko, L. P., *At. Energ.* (*USSR*), **21**, 243–46 (1966) [=*Soviet J. At. Energy*, **21**, 908–11 (1966)]

71. Flerov, G. N., Akap'ev, G. N., Demin, A. G., Druin, V. A., Lobanov, Yu.V., Fefilov, B. V., *Yad. Fiz.*, **7**, 977–83 (1968) [=*Soviet J. Nucl. Phys.*, **7**, 588–91 (1968)]

72. Zager, B. A., Miller, M. B., Mikheev, V. L., Polikanov, S. M., Sukhov, A. M., Flerov, G. N., Chelnokov, L. P., *At. Energ.* (*USSR*), **20**, 230–32 (1966) [=*Soviet J. At. Energy*, **20**, 264–66 (1966)]

73. Mikheev, V. L., Ilyushchenko, V. I., Miller, M. B., Polikanov, S. M., Flerov, G. N., Kharitonov, Yu.P., *At. Energ.* (*USSR*), **22**, 90–97 (1967) [=*Soviet J. At. Energy*, **22**, 93–100 (1967)]

74. Valli, K., Treytl, W. J., Hyde, E. K., *Phys. Rev.*, **167**, 1094–1104 (1968)

75. Bogdanov, D. D., Darotsi, Sh., Karnaukhov, V. A., Petrov, L. A., Ter-Akopyan, G. M., *Yad. Fiz.*, **6**, 893–900 (1967) [=*Soviet J. Nucl. Phys.*, **6**, 650–55 (1968)]

76. Bogdanov, D. D., Bacho, I., Karnaukhov, V. A., Petrov, L. A., *Yad. Fiz.*, **6**, 1113–16 (1967) [=*Soviet J. Nucl. Phys.*, **6**, 807–9 (1968)]

77. Flerov, G. N., Druin, V. A., Demin, A. G., Lobanov, Yu. V., Skobelev, N. K., Akap'ev, G. N., Fefilov, B. V., Kolesov, I. V., Gavrilov, K. A., Kharitonov, Yu.P., Chelnokov, L. P., *Joint Inst. Nucl. Res. Dubna Rept. P7-3808* (16 pp., 1968)

78. Valli, K., Hyde, E. K., *Phys. Rev.*, **176**, 1377–89 (1968)

79. MacFarlane, R. D., *Texas A & M Univ. Rept. ORO-3820-1* (17 pp., 1969)

80. Torgerson, D. F., Gough, R. A., MacFarlane, R. D., *Phys. Rev.*, **174**, 1494–99 (1968)

81. Torgerson, D. F., MacFarlane, R. D.,

Bull. Am. Phys. Soc., **13**, 1370, AF13 (1968)

82. MacFarlane, R. D. (Personal communication, 1969)

83. Wahl, A. C., in *Radioactivity Applied to Chemistry*, 284–310 (Wahl, A. C., Bonner, N. A., Eds., Wiley, New York, 604 pp., 1951)

84. Wahl, A. C., *J. Inorg. Nucl. Chem.*, **6**, 263–71 (1958)

85. Wahl, A. C., Daniels, W. R., *J. Inorg. Nucl. Chem.*, **6**, 278–87 (1958)

86. Wolfsberg, K., Nethaway, D. R., Malan, H. P., Wahl, A. C., *J. Inorg. Nucl. Chem.*, **12**, 201–5 (1960)

87. Wahl, A. C., Ferguson, R. L., Nethaway, D. R., Troutner, D. E., Wolfsberg, K., *Phys. Rev.*, **126**, 1112–27 (1962)

88. Wolfsberg, K., *Phys. Rev.*, **137**, B929–35 (1965)

89. Henzel, N., Herrmann, G., Patzelt, P. (Unpublished work cited in Ref. 12)

90. Grover, J. R., Lebowitz, E., Baker, E., *J. Inorg. Nucl. Chem.* (In press, 1969)

91. Hagebø, E., Sundell, S., *Target for the CERN Isotope Separator On-Line, Isolde*, Preprint Europ. Organ. Nucl. Res. Geneva (35 pp., 1968)

92. Valli, K., Nurmia, M. J., Hyde, E. K., *Phys. Rev.*, **159**, 1013–21 (1967)

93. Fiedler, H. J., Archer, N. P., Kennett, T. J., Keech, G. L. (Unpublished work cited in Ref. 63)

94. Patzelt, P., *The Hydroxides of Zr, Ce(IV), and Th for the On-Line Production of Rare Gases* (Presented at Seminar on Isolde Chem. Problems, Geneva, Nov. 1967)

95. Hagebø, E., Kjelberg, A., Patzelt, P., Sundell, S., *Performance of the Isolde Target System*, Preprint Europ. Organ. Nucl. Res. Geneva (20 pp., 1968)

96. Erdelen, G., *Die Temperaturabhängigkeit des Emaniervermögens einiger Uranverbindungen* (Dissertation, Univ. Mainz, 1960)

97. Grimm, W., Herrmann, G. (To be published)

98. Apollonova, A. N., Krisyuk, I. T., Ushatskii, V. N., *Radiokhimiya*, **4**, 587–91 (1962)

99. Amiel, S., Gilat, J., Notea, A., Yellin, E. (See Ref. 8), 169–76

100. Amiel, S., Gilat, J., Notea, A., Yellin, E. (See Ref. 33), 115–45

101. Wahlgreen, M. A., Meinke, W. W., *J. Inorg. Nucl. Chem.*, **24**, 1527–38 (1962)
102. O'Kelley, G. D., Lazar, N. H., Eichler, E., *Phys. Rev.*, **102**, 223–27 (1956)
103. Ockenden, D. W., Tomlinson, R. H., *Can. J. Chem.*, **40**, 1594–1604 (1962)
104. Lepold, M. F., *Kernforschungszentrum Karlsruhe Rept. KFK-365* (65 pp., 1965)
105. Goodman, R. H., Kitching, J. E., Johns, M. W., *Nucl. Phys.*, **54**, 1–16 (1964)
106. Borg, S., Fägerquist, U., Holm, G., Kropff, F., *Nucl. Instr. Methods*, **38**, 296–98 (1965)
107. Sidenius, G., Gammon, R. M., Naumann, R. A., Thomas, T. D., *Nucl. Instr. Methods*, **38**, 299–302 (1965)
108. Alväger, T., Naumann, R. A., Petry, R. F., Sidenius, G., Thomas, T. D., *Phys. Rev.*, **167**, 1105–16 (1968)
109. Talbert, W. L., Tucker, A. B., Day, G. M. *Phys. Rev.*, **177**, 1805–16 (1969)
110. Carlson, G. C., Schick, W. C., Talbert, W. L., Wohn, F. K., *Nucl. Phys.*, **A125**, 267–75 (1969)
111. Day, G. M., Tucker, A. B., Talbert, W. L. (See Ref. 33), 103–13
112. Grover, J. R., *J. Inorg. Nucl. Chem.* (In press, 1969)
113. Burman, R. L., Herrmannsfeldt, W. B., Allen, J. S., Braid, T. H., *Phys. Rev. Letters*, **2**, 9–11 (1959)
114. Carlson, T. A., *Phys. Rev.*, **130**, 2361–65 (1963)
115. Herrmannsfeldt, W. B., Burman, R. L., Stähelin, P., Allen, J. S., Braid, T. H., *Phys. Rev. Letters*, **1**, 61–63 (1958)
116. Bienlein, J. K., Pleasonton, F., *Nucl. Phys.*, **37**, 529–34 (1962)
117. Poskanzer, A. M., Esterlund, R. A., McPherson, R., *Phys. Rev. Letters*, **15**, 1030–33 (1965)
118. Rudstam, G., *Nucl. Instr. Methods*, **38**, 282–90 (1965)
119. Cowan, G. A., Orth, C., *Proc. 2nd Intern. Conf. Peaceful Uses At. Energy*, **7**, 328–34 (UN, Geneva, 844 pp., 1958)
120. Orth, C. J., *Nucl. Sci. Eng.*, **9**, 417–20 (1961)
121. Philippe, C., Amarel, I., Bernas, R., Chaumont, J., Klapisch, R., *Experience with Surface Ionization Ion Sources for On-Line Studies at Orsay* (Presented at Seminar on Isolde Chem. Problems, Geneva, Nov. 1967)
122. Klapisch, R., Philippe, C., Suchorzewska, J., Detraz, C., Bernas, R., *Phys. Rev. Letters*, **20**, 740–42 (1968)
123. Klapisch, R., Chaumont, J., Jastrzebski, J., Bernas, R., Simonoff, G. N., Lagarde, M., *Phys. Rev. Letters*, **20**, 743–45 (1968)
124. Röder, E., Herrmann, G., *Z. Anal. Chem.*, **219**, 93–102 (1966)
125. Andersen, M. L., Nielsen, O. B., Scharff, B., *Nucl. Instr. Methods*, **38**, 303–5 (1965)
126. Andersen, M. L., Scharff, B. (See Ref. 8), 151–53
127. Tarantin, N. I., Gordeev, V. V., Dem'yanov, A. V., *At. Energ.* (*USSR*), **22**, 280–85 (1967) [= *Soviet J. At. Energy*, **22**, 352–57 (1967)]
128. Amov, B., Permyakov, V. P., Tarantin, N. I., *Joint Inst. Nucl. Res. Dubna Rept. P13-3019* (14 pp., 1966)
129. Amarel, I., Bernas, R., Foucher, R., Jastrzbski, J., Johnson, A., Teillac, J., Gauvin, H., *Phys. Letters*, **24B**, 402–4 (1967)
130. Hagebø, E., Kjelberg, A., Sundell, S. (See Ref. 8), 127–32
131. Chackett, K. F. (See Ref. 8), 133–50
132. Cumming, J. B., Poskanzer, A. M., Hudis, J., *Phys. Rev. Letters*, **6**, 484–85 (1961)
133. Jahnsen, T., Pappas, A. C., Tunaal, T., *Target Material (TeCl4) and Chemical Separation Method for Studying Tin and Antimony Isotopes* (Presented at Seminar on Isolde Chem. Problems, Geneva, Nov. 1967)
134. Zvara, I., Tarasov, L. K., Krzhivanek, M., Su, Hung-kuei, Zvarova, T. S., *Dokl. Akad. Nauk SSSR*, **148**, 63 (1963) [= *Soviet Phys.-Doklady*, **8**, 63–64 (1963)]
135. Zvara, I., Zvarova, T. S., Krzhivanek, M., Chuburkov, Yu. T., *Radiokhimiya*, **8**, 77–84 (1966) [= *Soviet Radiochem.*, **8**, 72–78 (1966)]
136. Zvara, I., Zvarova, T. S., Caletka, R., Chuburkov, Yu. T., Shalaevskii, M. R., *Radiokhimiya*, **9**, 231–39 (1967) [= *Soviet Radiochem.*, **9**, 226–33 (1967)]
137. Denschlag, H. O., Henzel, N., Herrmann, G., *Radiochim. Acta*, **1**, 172–73 (1963)
138. Denschlag, H. O., zur Heide, F., Hen-

zel, N., Herrmann, G., Hübscher, D., Kratz, K. L. (To be published, cited in Ref. 12)

139. Denschlag, H. O., Gordus, A. A., Z. *Anal. Chem.*, 226, 62–71 (1967)
140. Wunderlich, F., *Radiochim. Acta*, 7, 105–14 (1967)
141. Silbert, M. D., Tomlinson, R. H., *Radiochim. Acta*, 5, 217–23 (1966)
142. Silbert, M. D., Tomlinson, R. H., *Radiochim. Acta*, 5, 223–27 (1966)
143. Paiss, Y., Amiel, S., *J. Am. Chem. Soc.*, 86, 2332 (1964)
144. Cram, S. P., Brownlee, J. L., *J. Gas Chromatog.*, 5, 353–58 (1967)
145. Schwarz, E., Denschlag, H. O., Herrmann, G., *Radiochim. Acta*, 5, 53–54 (1966)
146. Kratz, K. L. (Unpublished)
147. Ahrens, H., Folger, H., Herrmann G. (Unpublished work cited in Ref. 12)
148. Tsoukatos, M. P., *Thermal Neutron Irradiation of Uranium-235, Bromine, Iodine in Presence of Organic Liquid* (Ph. D. thesis, Univ. Michigan, Ann Arbor, 1967)
149. Chen, Teng-Yueh, Yeh, Si-Jung, *Nuclear Science*, 4, 65–67 (1964) (Ed. Inst. Nucl. Sci., Natl. Tsing Hua Univ., Taiwan)
150. Prussin, S. G., *Radiochemical Separations Through Recoil Reactions and Partial Characterization of* ^{133}Te (Ph. D. thesis, Univ. Michigan, Ann Arbor, 1964)
151. Baumgärtner, F., Reichold, P., Z. *Naturforsch.*, 16a, 945–48 (1961)
152. Kienle, P., Weckermann, B., Baumgärtner, F., Zahn, U., *Naturwissenschaften*, 49, 295–96 (1962)
153. Baumgärtner, F., Reichold, P., Z. *Naturforsch.*, 16a, 374–79 (1961)
154. Baumgärtner, F., Schön, A., *Radiochim. Acta*, 3, 141–45 (1964)
155. Meinhold, H., Reichold, P., *Naturwissenschaften*, 55, 344–45 (1968)
156. Denschlag, H. O., Willis, J. H. (Unpublished)
157. Blachot, J., Vargas, J. I., *Centre Études Nucl. Grenoble Rept. INT/ CHN/67-01* (12 pp., 1967)
158. Blachot, J., Carraz, L. C., *Radiochim. Acta*, 11, 45–49 (1969)
159. Blachot, J., Cavallini, P. (Personal communication, 1969)
160. Westgaard, L., Rudstam, G., Jonsson, O. C., *J. Inorg. Nucl. Chem.* (In press, 1969)
161. Greendale, A. E., Love, D. L., Deluc-

chi, A. A., *Anal. Chim. Acta*, 34, 32–40 (1966)
162. Hall, D., Walton, G. N., *J. Inorg. Nucl. Chem.*, 19, 16–26 (1961)
163. Brown, L. C., Wahl, A. C., *J. Inorg. Nucl. Chem.*, 29, 2133–45 (1967)
164. Greendale, A. E., Love, D. L., *Nucl. Instr. Methods*, 23, 209–12 (1963)
165. Johnson, N. R., Eichler, E., O'Kelley, G. D., in *Technique of Inorganic Chemistry*, 2, 70 (Jonassen, H. B., Weissberger, A., Eds., Interscience, New York, 202 pp., 1963)
166. Bemis, C. E., Irvine, J. W., *Nucl. Instr. Methods*, 34, 57–60 (1965)
167. Stehney, A. F., Perlow, G. J., *Bull. Am. Phys. Soc.*, 6, 62, TA 11 (1961)
168. Greendale, A. E., Love, D. L., *Anal. Chem.*, 35, 632–35 (1963)
169. Strom, P. O., Love, D. L., Greendale, A. E., Delucchi, A. A., Sam, D., Ballou, N. E., *Phys. Rev.*, 144, 984–93 (1966)
170. Delucchi, A. A., Greendale, A. E., Strom, P. O., *Phys. Rev.*, 173, 1159–65 (1968)
171. del Marmol, P., Nève de Mévergnies, M., *J. Inorg. Nucl. Chem.*, 29, 273–79 (1967)
172. del Marmol, P., *J. Inorg. Nucl. Chem.*, 30, 2873–80 (1968)
173. Tomlinson, L., Hurdus, M. H., *J. Inorg. Nucl. Chem.*, 30, 1649–61 (1968)
174. Troutner, D. E., Wahl, A. C., Ferguson, R. L., *Phys. Rev.*, 134, B1027–29 (1964)
175. Bemis, C. E., Gordon, G. E., Coryell, C. D., *J. Inorg. Nucl. Chem.*, 26, 213–18 (1964)
176. Parsa, B., Gordon, G. E., Wenzel, A., *J. Inorg. Nucl. Chem.*, 31, 585–90 (1969)
177. Tomlinson, L., Hurdus, M. H., *J. Inorg. Nucl. Chem.*, 30, 1995–2002 (1968)
178. del Marmol, P., Van Tigchelt, H., *Radiochim. Acta* (In press, 1969)
179. Folger, H., Kratz, J. V., Herrmann, G., *Radiochem. Radioanal. Letters* (In press, 1969)
180. Greendale, A. E., Love, D. L., *Anal. Chem.*, 35, 1712–15 (1963)
181. Treytl, W. J., *U. S. Naval Radiol. Defense Lab. San Francisco Rept. NRDL-TR-118* (22 pp., 1968)
182. Ahrens, H., Hollstein, M., Herrmann, G. (Unpublished)
183. Tomlinson, L., *Anal. Chim. Acta*, 31, 545–51 (1964)

184. Tomlinson, L., *Anal. Chim. Acta*, **32**, 157–64 (1965)
185. Tomlinson, L., *J. Inorg. Nucl. Chem.*, **28**, 287–301 (1966)
186. Tomlinson, L., Hurdus, M. H., *J. Inorg. Nucl. Chem.*, **30**, 1125–38 (1968)
187. Shpakov, V. I., Kostochkin, O. I., Petrzhak, K. A., Aron, P. M., *Radiokhimiya*, **7**, 96–103 (1965) [= *Soviet Radiochem.*, **7**, 94–100 (1965)]
188. Stehney, A., *Wet Chemistry for On-Line Bombardments* (Presented at Seminar on Isolde Chem. Problems, Geneva, Nov. 1967)
189. Riccato, M. T., Strassmann, F. (Personal communication, 1969)
190. Griffiths, K., Fritze, K., *Z. Anal. Chem.*, **226**, 122–24 (1967)
191. Weiss, H. V., Ballou, N. E., *J. Inorg. Nucl. Chem.*, **27**, 1917–23 (1965)
192. Fritze, K., Kennett, T. J., *Can. J. Phys.*, **38**, 1614–22 (1960)
193. Fritze, K., Kennett, T. J., Prestwich, W. V., *Can. J. Chem.*, **39**, 675–80 (1961)
194. Fritze, K., *Can. J. Chem.*, **40**, 1344–49 (1962)
195. Norris, A. E., Wahl, A. C., *Phys. Rev.*, **146**, 926–31 (1966)
196. Troutner, D. E., Ferguson, R. L., O'Kelley, G. D., *Phys. Rev.*, **130**, 1466–70 (1963)
197. Runnalls, N. G., Troutner, D. E., Ferguson, R. L., *Phys. Rev.*, **179**, 1188–93 (1969)
198. Niece, L. H., *Oak Ridge Natl. Lab. Rept. ORNL-TM-1333* (111 pp., 1965)
199. Wahl, A. C., Nethaway, D. R., *Phys. Rev.*, **131**, 830–31 (1963)
200. Weiss, H. V., *Phys. Rev.*, **139**, B304–6 (1965)
201. Roche, M. F., *The Independent Yield of* ¹¹²*Ag and the Cumulative Yield of* ¹¹⁵*Pd in Thermal-Neutron Induced* ²³³*U and* ²³⁵*U Fission* (Dissertation, Univ. Missouri, Columbia, Mo., 1968)
202. Hastings, J. D., Troutner, D. E., *Radiochim. Acta*, **11**, 51–56 (1969)
203. Sam, D., Love, D. L., *Anal. Chim. Acta*, **35**, 154–61 (1966)
204. Vallis, D. G., Perkin, J. L., *J. Inorg. Nucl. Chem.*, **22**, 1–5 (1961)
205. Schüssler, H.-D., Herrmann, G., *Radiochim. Acta* (In press, 1969)
206. Fiedler, H. J., Archer, N. P., *Z. Anal. Chem.*, **226**, 114–22 (1967)
207. Eckhardt, W., Grimm, W., Schüssler, H.-D., Weinreich, R., Herrmann, G., *Proc. Radioanal. Conf.*, Stary Smokovec, April 1968 (In press, 1968)
208. Sattizahn, J. E., Knight, J. D., Kahn, M., *J. Inorg. Nucl. Chem.*, **12**, 206–22 (1960)
209. Rengan, K., *Rapid Radiochemical Separations and Studies of Some Nuclear Properties* (Ph. D. thesis, Univ. Michigan, Ann Arbor, 1966)
210. Johnson, N. R., Eichler, E., O'Kelley, G. D., Chase, J. W., Wasson, J. T., *Phys. Rev.*, **122**, 1546–58 (1961)
211. Ruddy, F., Pate, B. D., *Nucl. Phys.*, **69**, 471–76 (1965)
212. Stenström, T., Jung, B., *Radiochim. Acta*, **4**, 3–6 (1965)
213. Kjelberg, A., Pappas, A. C., Tunaal, T., *J. Inorg. Nucl. Chem.*, **30**, 337–44 (1968)
214. Hübscher, D., *Zerfallsstudien an kurzlebigen neutronenreichen Technetiumisotopen* (Dissertation, Univ. Mainz, 1969)
215. Herrmann, G., Fiedler, H. J., Benedict, G., Eckhardt, W., Luthardt, G., Patzelt, P., Schüssler, H. D., in *Physics and Chemistry of Fission*, Symp. Salzburg, 1965, **2**, 197–213 (IAEA, Vienna, 469 pp., 1965)
216. Fritze, K., Griffiths, K., *Radiochim. Acta*, **7**, 59–60 (1967)
217. Koch, L., Münzel, H., Thoma, H., *Radiochim. Acta*, **2**, 33–36 (1964)
218. Schwarzbach, E., Münzel, H., *Radiochim. Acta*, **10**, 20–26 (1968)
219. Münzel, H., Koch, L., *Radiochim. Acta*, **4**, 188–91 (1965)
220. Herzog, W., Trautmann, N., Denig, R., Herrmann, G. (To be published)
221. Trautmann, N., Denig, R., Rabsch, U. (To be published)
222. Riccato, M. T., Herrmann, G. (To be published)
223. Schwarzbach, E., Münzel, H., *Radiochim. Acta*, **9**, 31–36 (1968)
224. Wish, L., *Phys. Rev.*, **172**, 1262–66 (1968)
225. Bleyl, H. J., Münzel, H., *Radiochim. Acta*, **9**, 149–53 (1968)
226. Seyb, K. E. (Private communication, 1968)
227. Beyer, G. J., Khalkin, V. A., Grosse-Ruyken, H., Pfrepper, G., *Joint Inst. Nucl. Res. Dubna Rept. P12-3887* (12 pp., 1968, submitted to *J. Inorg. Nucl. Chem.*)

228. Hübenthal, K., *Compt. Rend. Ser. A et B*, **264**, 1468–71 (1967)
229. Baeckmann, A. von, *Radiochim. Acta*, **7**, 1–3 (1967)
230. Weiss, H. V., Reichert, W. L., *Anal. Chim. Acta*, **34**, 119–22 (1966)
231. Weiss, H. V., Reichert, W. L., J. *Inorg. Nucl. Chem.*, **28**, 2067–70 (1966)
232. Weiss, H. V., Elzie, J. L., Fresco, J. M., *Phys. Rev.*, **172**, 1269–71 (1968)
233. Weiss, H. V., Fresco, J. M., Reichert, W. L., *Phys. Rev.*, **172**, 1266–69 (1967)
234. Okashita, H., *Radiochim. Acta*, **7**, 81–85 (1967)
235. Okashita, H., *Radiochim. Acta*, **7**, 85–89 (1967)
236. Loepfe, E., Monnier, D., *Anal. Chim. Acta*, **41**, 467–74 (1968)
237. Monnier, D., Loepfe, E., *Anal. Chim. Acta*, **41**, 475–81 (1968)
238. Monnier, D., Gross, P., Haerdi, W., *Z. Anal. Chem.*, **236**, 519–24 (1968)
239. Malan, H. P., Münzel, H., *Radiochim. Acta*, **5**, 20–23 (1966)
240. Malan, H. P., Münzel, H., Pfennig, G., *Radiochim. Acta*, **5**, 24–28 (1966)
241. Fahland, J., Herrmann, G., *Z. Anorg. Allg. Chem.*, **316**, 141–53 (1962)
242. Günther, E. W., Starke, K., *Chem. Zvesti*, **21**, 561–70 (1967)
243. Allen, J. S., Burman, R. L., Herrmannsfeldt, W. B., Stähelin, P., Braid, T. H., *Phys. Rev.*, **116**, 134–43 (1959)
244. Denig, R., Wahl, A. C. (Private communication, 1969)
245. Stehney, A. F., Steinberg, E. P., *Phys. Rev.*, **127**, 563–69 (1962)
246. Aron, P. M., Kostochkin, O. I., Petrzhak, K. A., Shpakov, V. I., *At. Energ. (USSR)*, **16**, 368–70 (1964) [=*Soviet J. At. Energy*, **16**, 447–49 (1964)]
247. Cojocaru, V., Nachman, M., *Inst. At. Phys. Bucharest Rept. IFA-NR-24* (11 pp., 1966)
248. Baeckmann, A. von, Feuerstein, H., *Radiochim. Acta*, **5**, 234–35 (1966)
249. Kienle, P., Wien, K., Zahn, U., Weckermann, B., *Z. Physik*, **176**, 226–36 (1963)
250. Bahn, E. L., *Decay Studies of Silver Isotopes of Masses 115, 116, and 117* (Ph. D. thesis, Washington Univ., St. Louis, Mo., 1962)

251. MacFarlane, R. D., Siivola, A., *Phys. Rev. Letters*, **14**, 114–15 (1965)
252. Karnaukhov, V. A., Ter-Akopyan, G. M. (See Ref. 8), 419–29
253. Talbert, W. L., Cook, J. W., Schick, W. C., *Bull. Am. Phys. Soc.*, **13**, 71, EG13 (1968)
254. MacFarlane, R. D., Griffioen, R. D., *Phys. Rev.*, **130**, 1491–98 (1963)
255. MacFarlane, R. D., Griffioen, R. D., *Phys. Rev.*, **131**, 2176–81 (1963)
256. MacFarlane, R. D., *Phys. Rev.*, **136**, B941–47 (1964)
257. MacFarlane, R. D., *Phys. Rev.*, **137**, B1448–52 (1965)
258. Siivola, A., *Nucl. Phys.*, **A92**, 475–80 (1967)
259. Siivola, A., *Nucl. Phys.*, **84**, 385–97 (1966)
260. Demin, A. G., Fényes, T., Mahunka, I., Subbotin, V. G., Trón, L., *Nucl. Phys.*, **A106**, 337–49 (1968)
261. Siivola, A., *Nucl. Phys.*, **A109**, 231–35 (1968)
262. Siivola, A., *Nucl. Phys.*, **A101**, 129–37 (1967)
263. Valli, K., Hyde, E. K., Treytl, W., *J. Inorg. Nucl. Chem.*, **29**, 2503–14 (1967)
264. Lobanov, Yu. V., Druin, V. A., *Joint Inst. Nucl. Res. Dubna Rept. P7-3613* (6 pp., 1968, submitted to *Yad. Fiz.*)
265. Flerov, G. N., Polikanov, S. M., Mikheev, V. L., Ilyushchenko, V. I., Miller, M. B., Shchegolev, V. A., *At. Energ. (USSR)*, **22**, 342–46 (1967) [=*Soviet J. At. Energy*, **22**, 434–37 (1967)]
266. Flerov, G. N., Polikanov, S. M., Mikheev, V. L., Ilyushchenko, V. I., Kushniruk, V. F., Miller, M. B., Sukhov, A. M., Shchegolev, V. A., *Yad. Fiz.*, **5**, 1186–91 (1967) [=*Soviet J. Nucl. Phys.*, **5**, 848–51 (1967)]
267. Druin, V. A., Akap'ev, G. N., Demin, A. G., Lobanov, Yu. V., Fefilov, B. V., Flerov, G. N., Chelnokov, L. P., *At. Energ. (USSR)*, **22**, 127–28 (1967) [=*Soviet J. At. Energy*, **22**, 135–36 (1967)]
268. Akap'ev, G. N., Demin, A. G., Druin, V. A., Flerov, G. N., Korotkin, Yu. S., Lobanov, Yu. V., *Joint Inst. Nucl. Res. Dubna Rept. E7-3261* (8 pp., 1967)
269. Flerov, G. N., Demin, A. G., Druin, V. A., Lobanov, Yu. V., Mikheev, V. L., Polikanov, S. M., Shchegolev,

V. A., *Yad. Fiz.*, **7**, 239–41 (1968) [=*Soviet J. Nucl. Phys.*, **7**, 168–69 (1968)]

270. Sikkeland, T., Ghiorso, A., Nurmia, M. J., *Phys. Rev.*, **172**, 1232–38 (1968)
271. Flerov, G. N., Korotkin, Yu. S., Mikheev, V. L., Miller, M. B., Polikanov, S. M., Shchegolev, V. A., *Nucl. Phys.*, **A106**, 476–80 (1968)
272. Baeckmann, A. von, Feuerstein, H., *Radiochim. Acta*, **4**, 111–12 (1965)
273. Kiefer, W., Weinreich, R., Herrmann, G. (Unpublished)
274. Erdal, B. R., *Nuclear Charge and Mass Distribution in Fission: Yields and Genetic Histories of* ^{121}Sn, ^{123}Sn, and ^{125}Sn *from Thermal-Neutron Fission of* ^{235}U (Ph.D. thesis, Washington Univ., St. Louis, Mo., 1966)
275. Denschlag, H. O., *J. Inorg. Nucl. Chem.* (In press, 1969)
276. Donets, E. D., Shchegolev, V. A., Ermakov, V. A., *Yad. Fiz.*, **2**, 1015–23 (1965) [=*Soviet J. Nucl. Phys.*, **2**, 723–29 (1966)]
277. Donets, E. D., Shchegolev, V. A., Ermakov, V. A., *At. Energ. (USSR)*, **20**, 223–30 (1966) [=*Soviet J. At. Energy*, **20**, 257–63 (1966)]
278. Donets, E. D., Shchegolev, V. A., Ermakov, V. A., *At. Energ. (USSR)*, **19**, 109–11 (1965) [=*Soviet J. At. Energy*, **19**, 995–99 (1965)]

MASS SEPARATION FOR NUCLEAR REACTION STUDIES

By R. Klapisch

Institut de Physique Nucléaire et Centre de Spectométrie Nucléaire et de Spectrométrie de Masse, Orsay, France

CONTENTS

INTRODUCTION

Complex nuclear reactions (fission, spallation, and other high-energy reactions) result in the production of a variety of nuclides. While some are radioactive and very familiar to radiochemists, others are stable or have very short half-lives. Among the latter many are as yet unidentified.

Cross sections as a function of target, projectile energy etc. . . are fundamental data in trying to ascribe a particular mechanism to a given reaction. Also, cross-section data are needed to test current models on the synthesis of elements in stars or the origin of heavy cosmic rays.

Various techniques have been used to study these reactions: radiochemistry, nuclear emulsion and other visual techniques, counters (notably solid-

33

state), and mass-spectrometric arrangements. This review treats of the various possible ways by which mass separation has, to date, been helpful in the study of complex nuclear reactions.

Apart from mass spectrometry proper which is in a sense self-sufficient, isotope separators have been used to add a mass-selective step to otherwise conventional radiochemical procedures. Both aspects will be reviewed. However, isotope separators have been used extensively for the preparation of sources for nuclear spectroscopy and this broad field will remain outside the scope of my review.

An extensive review was given by Hintenberger (1) in 1962 on the techniques and applications of high-sensitivity mass spectrometry. More recently Bernas (2) reviewed the use of mass spectrometry to study high-energy nuclear reactions. Isotope separator techniques are the subject of regular symposia (3a to d), the last of which was in 1965 in Aarhus (3d).

Based on the measurements of differential energy loss (dE/dx) and total kinetic energy E, electronic techniques have recently become capable of identifying both Z and M of light nuclei ($Z \lesssim 8$) produced in nuclear reactions. Although this method gives excellent results which are in some cases closely related to the topics here discussed, a detailed discussion would have been rather remote from the main subject and has not, therefore, been included in this review.

METHODS
MASS SPECTROMETRY (OFFLINE)

Classical method used to study fission products.—Low-energy fission products being predominantly neutron-excess, measurements of the yields of stable endproducts after β decay will give the total yield at a particular mass number. Following the early work of Thode & Graham (4) in 1947 on Kr and Xe, the Canadian group of Thode, Tomlinson and their co-workers (5–7) has determined with the mass spectrometer practically all the major cumulative yields and products produced in slow-neutron fission. See Hyde (8) for a discussion of the results. Before we discuss more recent work in subsequent sections, it is appropriate to recall for the sake of comparison some of the features of this classical work.

After chemical separation from the target, the fission products are deposited on a filament for mass-spectrometric analysis. The chemical selectivity of surface ionization is sufficient in most cases while the noble gases Kr and Xe are purified before being assayed in an electron bombardment ion source. Absolute yields are obtained by isotope dilution or by cross-calibration between adjacent isobars. With quantities $\sim 10^{-8}$ g (corresponding to an irradiation in a flux of 10^{17} neutrons) emphasis is on high precision and yields have been determined to within 1 per cent.

Difficulties appear, however, if one wants to extend these measurements to very low yields of shielded isotopes (9) or to charged-particle nuclear reactions. Owing to available fluxes at accelerators and the lower cross sec-

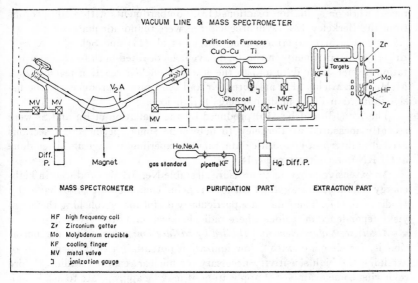

FIG. 1. Experimental setup used by the Brookhaven group (11) for the measurement of rare-gas cross sections at GeV energies.

tions (often in the region of millibarns), one deals here with quantities of stable products of less than 10^{-9} or even 10^{-12} g. Problems that arise are connected both with the sensitivity of the instrument and with special chemical methods to avoid contamination of the small samples. As a result, the method ceases to be generally applicable and we now turn to a review of the work done in three very distinct cases.

High-sensitivity mass spectrometry of rare gases.—Because of obvious similarities the technique owes much to the development of measurements of trace quantities of rare gases in meteorites (10). Figure 1 shows an example of a typical experimental setup (11). The target is melted to extract gases and volatile products which are then passed through a series of filters and cold traps to remove the active portions before introduction into the mass spectrometer. A simple sector machine with a conventional electron impact ion source is used, but special attention is given to vacuum problems. The apparatus is completely bakable and pressures of $<10^{-9}$ torr are routinely achieved.

The sensitivity is very high if the apparatus is run in the static mode (all pumps being closed after the introduction of the sample) and the ions are detected at the collector with an electron multiplier. While analyses have been performed with samples as small as 5.10^{5} atoms of Xe (12), it is generally safer to work with somewhat higher quantities because of "memory effects" in the apparatus. However, the method is extremely sensitive and in recent work on spallation reactions (13), bombardments with 3.10^{13} protons from the Berkeley 184" synchrocyclotron were found adequate.

Since the first experiments of Mayne et al. (14) and Schaeffer & Zähringer (10), experiments in this field have been devoted both to the interpretation of meteoritic data and to the study of nuclear reaction mechanisms. McHugh & Michel (15) measured Kr and Xe isotopes produced in the fission induced in Th by medium-energy helium ions.

The Xe (13) and Kr (16) produced by the spallation of Ba and Sr were recently measured at Berkeley to determine some specific cross sections needed to interpret meteoritic data. Similar experiments have also been done at CERN in recent years (17, 18).

An extensive program to measure all stable Ne, Ar, Kr produced in high-energy reactions in targets of Cu to U is being completed at Brookhaven by Hudis et al. (11). These data are particularly useful for establishing the mass-yield dependence in regions where radiochemical data are very scarce.

Alkali (and other elements) studied by surface ionization.—Surface ionization is, for elements with a low ionization potential, a convenient way of attaining the high sensitivity necessary for nuclear reaction studies. Before reviewing some examples of application, it may be appropriate to ask if high sensitivity is really the main characteristic which is required from a mass spectrometer used for these purposes. It would seem that not only high sensitivity but also high dispersion is indispensable if one wants to measure a very low yield when a neighboring mass isotope has a very high one. Such is the case in the determination of the low independent yield of a shielded isotope like ^{86}Rb as compared to the high cumulative yield of ^{87}Rb in thermal-neutron fission. McHugh (9) has used a very dispersive three-stage mass spectrometer capable of measuring isotopic ratios of 10^9:1 on adjacent masses to measure ^{86}Rb/^{87}Rb; ^{132}Cs/^{137}Cs; ^{134}Cs/^{137}Cs and ^{136}Cs/^{137}Cs.

This is, however, an exceptional case. One can even argue that the ratio of independent ^{86}Rb/^{87}Rb could be measured (e.g. online) with an abundance sensitivity ($\backsim 1/10$) compatible with an apparatus of moderate dispersion. Thus, the most general problem is that of a high sensitivity and the most severe limitation comes from the possibility of a contamination by stable natural isotopes. Considering the small quantities formed during the nuclear reaction, care must be taken not only of impurities in the target but also of impurities introduced at the later stage of extracting and purifying the reaction products.

Friedlander, Friedman, Gordon & Yaffe (19) made a very detailed study of the stable or long-lived isotopes of Rb and Cs produced in the fission of U by high-energy protons. Following a method developed earlier (20) the target was dissolved and the U separated from the fission products by organic solvent extraction.

McHugh & Michel (15) used ion exchange resins to separate the fission products from the target in their study of Rb, Cs, and rare earths produced in fission induced by medium-energy charged particles. Cation-exchange resin was used in a subsequent step to separate the rare earths.

To study the production of stable Li and Na in high-energy reactions the Orsay group has developed several techniques (21) in which the reaction products are separated from the target by vacuum distillation. The chemical selectivity for alkalies is also provided here by surface ionization.

In the "oven technique" the reaction products are transferred to a cooled mass-spectrometer filament placed above a heated tantalum oven containing the irradiated target. Thus, Nguyen-Long-Den has measured the production cross sections of ^{22}Na, ^{23}Na, ^{24}Na in the spallation of Al by 150 MeV protons (22) and studied extensively the spallation in La and Pr leading to Cs isotopes through ^{133}La($p,3pxn$) and ^{141}Pr($p,5pxn$) (23a).

Bernas et al. (24) have measured the production of ^{6}Li and ^{7}Li from ^{12}C in a broad range of energies. The importance of these and other (25) measurements in connection with astrophysical problems is discussed in the next chapter, especially on p. 56 et seq.

While experiments on V, Nb, and Ta targets with the oven technique have also been reported (26), another method is preferred for these target elements and consists of direct irradiation of mass-spectrometer filaments in the internal beam of a synchrotron. Although the filaments are thin, the production is high enough because of the well-known effect of multiple traversals. The successive steps of the experiment (27) are explained in Figure 2, the advantage being an efficient control of any contamination. Recently, Nguyen-Long-Den et al. (23d) have caught on Re filaments the Eu recoils from (^{12}C,xn) reactions.

Li, Be, and B studied by the sputtering mass spectrometer.—To measure Li, Be, and B production in the spallation of light targets, Yiou et al. (25) have used an entirely different technique of ionization based on sputtering phenomena. This is more broadly applicable than surface ionization but still has a high enough intrinsic sensitivity (28). Samples of 10^{-13} to 10^{-12} g of Be and B produced by the spallation of C and O in proton bombardments of a few microampere-hours can be measured, an improvement of several orders of magnitude over more conventional techniques (25).

The apparatus is a simplified version of the ion microprobe built by Castaing & Slodzian (29) and is depicted in Figure 3. Under the impact of a focused primary ion beam, low-energy secondary Li^+, Be^+, and B^+ are emitted from the sample plate. These ions are then accelerated, focused and magnetically analyzed. Primary ion beams of Cs^+ (2 keV at 10^{-9} Å/mm^2, 2 mm^2 cross section) have been used. Detection at the collector is achieved by counting individual ions with an electron multiplier (27, 30) and statistical enhancement techniques (27, 31) are used to minimize fluctuations, as is now current practice.

The purity problem was solved by the preparation by fractional crystallization of a target of very pure water containing less than 10^{-12} g/g of either Li or Be and less than 10^{-11} g/g of B (32). Care was taken that during the irradiation and subsequent analysis, contamination by natural B or Li remained several times less than the quantities produced in the reaction.

Although this ionizing technique is not as chemically selective as surface

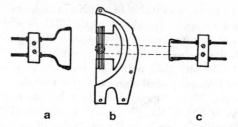

a—Filament is spot-welded on holder for cleaning and checking purity before irradiation.
b—Same, mounted for internal irradiation in a proton synchrotron.
c—After irradiation the filament is mounted for analysis.

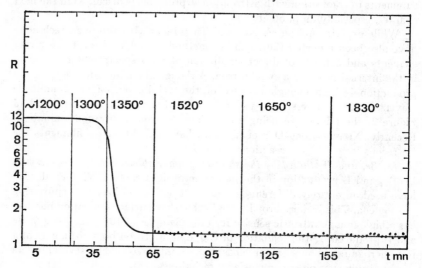

d—With progressive raising of the temperature of the filament the superficial contamination is first evaporated and the Li isotopic ratio is $R = 12$ as in natural
Li. As the temperature and time increase, the "fragmentation" Li diffuses out
and the ratio falls to $R = 1.3$.

FIG. 2. Filament technique used at Orsay to measure $^7Li/^6Li$ in Pt and other
targets irradiated at high energies (21).

ionization, its efficiency still varies significantly from one element to another.
Typically the efficiency for Li, Be, B is like 80:1:0.1. However, at one given
mass position, one cannot in general ascribe a unique element, as is apparent
from Figure 4.

The contribution at mass 8 is due to formation of a hydride $^7BeH^+$. Use
can be made of this property—which is peculiar to Be—to separate the Li
and Be contribution at mass 7. The situation at masses 10 and 11 is more com-

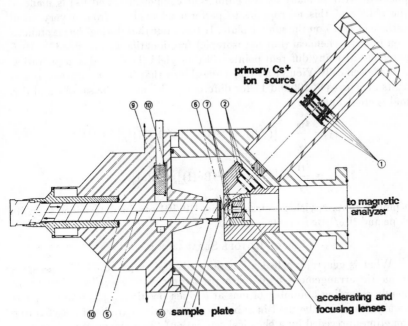

FIG. 3. Ion sputtering source used by Yiou et al. (25) to measure cross sections for production of stable Li, Be, and B in the spallation of light elements.

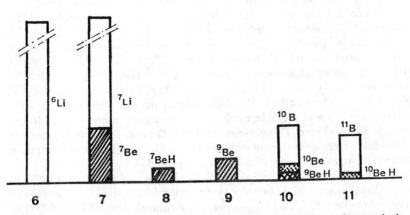

FIG. 4. Schematic mass spectrum showing contributions to the various peaks in irradiated water (25). Notice especially the consequences of BeH formation at masses 8, 10, 11 (see text).

plicated, with as many as three different components, and use is made of the ability of this microprobe to perform an analysis from a very small surface (1 mm²) on the sample plate. It turns out that during the crystallization process, chemical (but not isotopic) fractionation can occur (33). Measurement at many different points will thus yield the desired isotopic ratios. More specifically Yiou et al. (25) have shown that if one measures ion currents at masses 9, 10, and 11 for different positions on the sample and if r_1 and r_2 are defined as:

$$r_1 = \frac{i_{10}}{i_9} = \left[(^{10}\text{B}) + (^{10}\text{Be}) + (^9\text{BeH}) \right] / \left[(^9\text{Be}) \right]$$

$$r_2 = \frac{i_{11}}{i_9} = \left[(^{11}\text{B}) + (^{10}\text{BeH}) \right] / \left[(^9\text{Be}) \right]$$

a plot of r_1 vs. r_2 yields a straight line. $^{10}\text{B}/^{11}\text{B}$ can be determined from its slope and $^9\text{Be}/^{10}\text{Be}$ from its intercept.

USE OF ISOTOPE SEPARATOR (OFFLINE)

What is generally called a laboratory isotope separator is a mass-spectrometric arrangement used for a preparatory purpose. Thus instead of merely recording a number of ions at a given position, a physical sample is obtained and can be used in subsequent operations (radioactive counting sometimes preceded by a chemical separation). As a means of studying nuclear reactions it thus has all the advantages (and some of the limitations) of the more conventional radiochemical methods.

From their former purpose—to separate weighable quantities of stable isotopes—the present generation of machines still retains the distinctive features of high ion currents through the use of plasma ion sources and the ability to collect simultaneously ions of different masses.

Plasma ion sources yield routinely efficiencies of 0.1 to several per cent (and sometimes more) (34) for practically all the chemical elements. While their inherently high current (microamperes to milliamperes) would hardly seem necessary to separate the trace amounts of radioactive isotopes produced in nuclear reactions, this high current density is indispensable for their operation and a stable carrier gas (e.g. argon) is normally added to sustain the discharge. This of course prohibits the use of an isotope separator to measure trace quantities of stable isotopes. The use of a surface ionization source for an isotope separation has been reported (23b,c).

The simultaneous collection of all the isotopes of interest necessitates high-dispersion machines in order that the physical spacing of masses in the focal plane be sufficient. In high-energy fission work done at CERN (35) or at Brookhaven (38) it was found necessary, because of the large number of isotopes produced, to collect simultaneously a span of masses with a 20 per cent relative mass difference.

While the most numerous studies have been directed towards the production of pure sources for nuclear spectroscopy, isotope separation has also been applied to nuclear reaction studies at Uppsala (36), Orsay (37), and more recently at CERN and Brookhaven (38). A discussion of this method was given in 1963 by Andersson & Rudstam (35) and in 1967 by Ewald (39).

Some more specific requirements will now be examined:

(a) Ion source problems. Strictly speaking, an isotope separation must be supplemented by some kind of chemistry if one is to collect pure nuclidic samples. While ionization by a plasma source is not very selective, a great simplification is introduced when isobars can be resolved after the separation, by means of their radioactive characteristics.

Separation of the reaction products from the target can often be done by simple heating (40). When the products of interest are not volatile enough, a volatile compound (generally a halide) is used (41). To study rare-earth products of high-energy fission (38) of U, anion exchange chromatography is first used to remove the bulk U. The mixed rare-earth fraction is then introduced in the form of oxides into the ion source either immediately or after a further cation exchange separation.

It is convenient to add a stable "mass marker" either elemental or molecular, to provide for the necessary tuning of the beam. Care must be taken that this does not introduce a mass discrimination by intense sputtering at certain mass positions.

To study relatively short-lived activities (10–20 min) two-oven ion sources have been developed (42, 43) where the radioactive material can be added to the ion source already running with stable carrier. Separation time is thus decreased to a few minutes and other steps (transfer from the accelerator, chemistry) therefore become the most time consuming. The practical limit for offline machines is ∽30 min.

(b) Separation and collection of the samples. The degree of isotopic purity that can be achieved depends on the dispersion and on the beam profile. The sharpness of the lines depends on the running conditions but there will always be a more or less pronounced tail due to scattering processes that will cause some contamination (44, 45). The use of narrow slits is to be avoided when yields of various isotopes are to be compared since these will cause a loss which can hardly be made constant over the whole mass spectrum. Normally the contamination from neighboring masses can be kept below a few per cent and one can easily correct it, while the losses to more distant masses are usually assumed to be negligible.

Figure 5 shows the collector arrangement used at CERN, with a series of collector foils standing perpendicular to their beam. A detailed knowledge of the beam geometry in the collector region is necessary in order to be able to work with collectors in predetermined positions.

(c) Applications of the technique and conclusion. Apart from earlier work, the most extensive use of this technique has been made by the CERN group to study high-energy fission and spallation. Rudstam & Sørensen (46) studied yields of the iodine isotopes, Brandt (47) reports cross-section

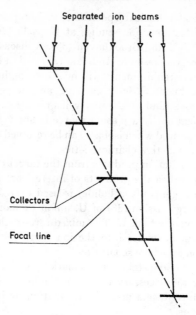

FIG. 5. Schematic representation of the collector arrangement used at the isotope separator at CERN (35).

and range measurement of Br isotopes, Hagebø (48) investigated Sb produced by U fission and Chackett (49) K isotopes produced by spallation reactions.

The Brookhaven group is presently measuring rare-earth products from high-energy U fission, extending previous charge dispersion measurements to a region where radiochemical measurements alone would be inadequate (38).

As we said earlier, this method is always associated with some kind of radiochemical measurement. Indeed, with the advent of high-resolution Ge(Li) γ-ray spectrometers there are many cases where a mass separation (with a loss of 2–3 orders of magnitude of activity) is found to be less advantageous than straightforward chemistry. The usefulness of mass separation remains, however, when chemical separation is difficult (e.g. rare-earth) or when a low-yield product has to be separated from a very abundant one that would otherwise contaminate it.

The limiting factor for the accuracy of cross-section determinations is found to be knowledge about the relevant decay schemes. In most favorable cases the accuracy can be as good as 1 or 2 per cent but considerable systematic errors can be introduced if the decay schemes are not sufficiently well known. With low-level counting techniques the sensitivity can be good and, typically, cross sections as low as 100 μb have been measured. See Figure 11 for a comparison with other methods.

The Online Mass Spectrometer

High-energy nuclear reactions (50) result in products that can extend very far on both sides—neutron-excess and neutron-deficient—of the stability valley. There is in this case a distinct advantage in using an online mass spectrometer capable of recording isotopes of practically any half-life from milliseconds to stability.

In the method which has been developed at Orsay (51), the ion source of the mass spectrometer is directly bombarded by a beam of particles (protons, neutrons etc. . .). One relies on high-temperature diffusion to extract the reaction products from the target rapidly and on surface ionization to perform an efficient (and at the same time chemically selective) ionization of alkali elements.

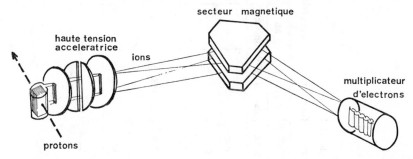

FIG. 6. Principle of the online mass spectrometer used by the Orsay group to measure cross sections for the production of Li, Na, K, Rb, and Cs isotopes in nuclear reactions (51).

Figure 6 illustrates the principle of the method: a succession of thin foils of the target element and of thin graphite slabs are enclosed in a cylindrical Re or Ta foil heated by joule effect to some 1800° C. These are bombarded by (e.g.) high-energy protons, and the nuclear reaction products, recoiling out of the target, will be stopped in the graphite catcher. Alkali elements—and possibly others—diffuse very fast out of heated graphite; however, only alkalies are efficiently ionized by contact with the hot metallic surface. The ions are then accelerated by a 3 kV potential difference and mass-analyzed by a sector magnetic field. At a given mass setting, the ions traversing a slit in the focal plane will be counted by an electron multiplier.

A brief discussion of the possibilities and limitations of this method will now be given.

Sensitivity.—The number of ions detected at the collector depends on the reaction cross section, the thickness of the target, the flux of incident particles, and the efficiency of the spectrometer.

The efficiency of surface ionization is given by the Saha-Langmuir relation $(n^+/n_0 = A \exp (W-1)/KT$ and a few examples of interest are given in Table I. One sees readily that the use of Re is advantageous for Li and Na.

TABLE I

Efficiency of and Chemical Separation by Surface Ionization on Heated Metallic Surfaces

| | Li^+ | $Li^+/(Li+Li^+)$ | Na^+ | $Na^+/(Na+Na^+)$ | K^+ | $K^+/(K+K^+)$ | Rb^+ | $Rb^+/(Rb+Rb^+)$ | Cs^+ | $Cs^+/(Cs+Cs^+)$ |
	$Li+Li^+$	$Be^+/(Be+Be^+)$	$Na+Na^+$	$Al^+/(Al+Al^+)$	$K+K^+$	$Ca^+/(Ca+Ca^+)$	$Rb+Rb^+$	$Sr^+/(Sr+Sr^+)$	$Cs+Cs^+$	$Ba^+/(Ba+Ba^+)$
Ta 1500°	2.10^{-4}	10^{10}	10^{-3}	350	0.15	10^4	0.38	6000	0.78	1000
Ta 1800°	6.10^{-4}	3.10^8	$2.6.10^{-3}$	350	0.17	5000	0.37	2000	0.73	360
Re 1500°	$6.6.10^{-2}$	3.10^{10}	0.29	140	0.99	1000	0.99	76	1	4.5
Re 1800°	$8.6.10^{-2}$	10^9	0.30	65	0.98	500	0.99	45	1	4

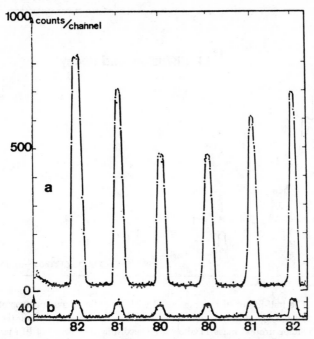

FIG. 7. A portion of the mass spectrum of Rb isotopes from a Th target bombarded by 10 GeV protons (59). Spectrum **a** taken just after the proton burst for some 60 to 200 msec represents the fast-diffusing part of the reaction products. Spectrum **b** is taken 8 sec later and represents the residual due to slow diffusion. The symmetry reflects the triangular modulation of the accelerating potential.

On the contrary, in the case of Rb and Cs, Ta is preferable if one wants to retain high efficiency and remain selective towards the alkaline earths Sr and Ba.

Note that the useful target thickness depends on the range of the reaction products. One of the advantages of this method is the use of relatively thick targets. This should be kept in mind if one wants to adapt this technique to the study of heavy-ion reactions (63).

The lower limit of sensitivity depends on the signal/noise ratio rather than on the absolute number of recorded events. The sources of background include residual contamination of natural isotope at some mass numbers, a continuum between peaks due to stray ions or to neutrons and gammas activating or otherwise reacting with the dynodes of the electron multiplier, etc . . . To minimize the fluctuations of this background one makes use of statistical enhancement techniques (27). See Figure 7.

Finally, the present experimental limit of sensitivity is around a few μb and could probably still be improved. See Figure 11 for a comparison with the other techniques.

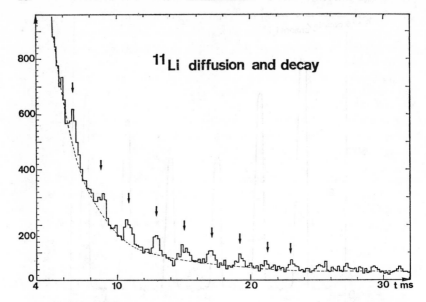

FIG. 8. The half-life of ^{11}Li as measured with an online mass spectrometer (53). Following the CERN proton-synchrotron burst, a fast modulation of the ion-accelerating voltage around mass 11 allows the recording of a series of ^{11}Li peaks at 2 msec time intervals. Arrows indicate time of theoretical occurrence of peaks.

The decrease with time of the peak intensity is due to the combined effect of diffusion from the target and radioactive decay. From a comparison with stable ^{6}Li one deduces the half-life of ^{11}Li to be 8.5 ± 1 msec. The dotted line represents the radiation background following the accelerator pulse. The half-lives of the Na isotopes of masses 27 to 31 have been measured by the same method.

Short half-lives and diffusion.—Diffusion as a means of quick separation of nuclear reaction products from a target is discussed extensively by Herrmann & Denschlag (52). The diffusion of alkali elements is readily studied by using the online mass-spectrometer setup in a pulsed accelerator, and typical experimental results are shown in Figure 8.

The diffusion curve can be fitted by least squares techniques to an expression of the form:

$$i(T) = \sum_i A_i \exp\left\{-\lambda_d{}^i T\right\}$$

The notations λ and T are analogous to the usual radioactive ones and are self-explanatory. The subscripts "d" and "β" refer to diffusion and β decay respectively.

This result is understood in terms of the mathematical theory of diffusion (27, 51) and is used to calculate the loss by radioactive decay during diffusion. While a correction is usually applied for isotopes of half-lives $T_\beta \le 2$ sec, there is no "delay" before diffusion and owing to the fast initial diffusion, a sizable proportion of an isotope of 10 msec half-life can be recorded (52). This can be applied to the search for new isotopes as no β radioactive isotopes are shorter lived.

To the extent that the diffusion of progenitors is fast enough compared to their half-lives, the recorded yields will be independent rather than cumulative (51, 53).

Experiments done with this technique.—The instruments are of moderate size and can be easily moved to different accelerators. Experiments have so far been performed at Orsay (150 MeV protons), Saclay (3 GeV protons), and CERN (10 and 24 GeV protons), and with the EL3 reactor at Saclay. The results can be summarized as follows: (a) Light fragment emission. The cross sections for all the Li isotopes from ^{12}C with 150 MeV protons were first reported (27, 54). Following was a study of Na isotopes from heavier targets at 3 GeV (Saclay), 10 GeV (55a), and 24 GeV (55b) (CERN). (b) Fission and spallation. The isotopic yields of all Rb and Cs isotopes have been measured in U^{238} (56–58) and Th, at 150 MeV, 10 GeV (59), and 24 GeV (58). Comparison between spallation and fission has been made at 10 GeV (59).

Rb and Cs are complementary fission fragments in thermal-neutron fission of U^{235}. Specific information on fission can thus be obtained by a detailed knowledge of isotopic cross sections for these elements (58).

ONLINE ISOTOPE SEPARATION

The idea of connecting an isotope separator to an accelerator beam is not new and as early as 1950, Kofoed-Hansen & Nielssen (60) studied fission product Kr isotopes at the Copenhagen cyclotron. But only in recent years have projects been made to connect laboratory isotope separators to accelerators (61–63) or reactors (64, 65). A review of plans and realizations as of 1966 is given by G. Andersson (66). The last two years have seen the effective completion of the facility at CERN (ISOLDE).

This effort is primarily intended to yield well-separated sources for nuclear spectroscopy of short-lived isotopes. So far, cross-section measurements have been a marginal preoccupation, but preliminary measurements of Xe isotopes from the spontaneous fission of Cf have demonstrated (67) the feasibility of this type of experiments. See also (64).

A brief inspection of the ISOLDE project reveals the following steps in the process: (a) separation of reaction products from target, (b) transport to ion source of isotope separator, (c) ionization, (d) mass analysis, (e) collection and study of isotopes.

Apart from some features (shielding, direct study at the collector) which are not relevant to this discussion, one will see that steps (a) and (b) are the ones that really distinguish this method from the offline separator reviewed on pp. 40–42. Methods used so far for (a) include emanating substances (67, 68) and molten targets (68). These and other separation methods are discussed in detail by Herrmann & Denschlag (52).

Low-pressure gas flow using a rare-gas carrier has been used so far for step (*b*). Some progress has been reported in the use of high-pressure (1 atm) transport (64).

Steps (*a*) and (*b*) take time. The present holdup time at ISOLDE is ~10–20 sec (69). In some cases this will certainly be improved in the future, but the prerequisite of precise cross-section measurements would be to study the time of transport to be able to correct for decay losses.

Attention should also be paid to mass discriminating effects due to molecular flow during transport in the pipe as has indeed been discussed by Andersson & Rudstam for the offline case (35). For nuclear spectroscopy purposes, emphasis is on the simultaneous collection of beams at different mass numbers. When this is obtained by ion beam handling, steering magnets, etc. (65) care must be taken that the transmission and focus conditions are really the same in different mass channels.

Finally if one is interested in determining independent yields of isotopes, a fast and selective chemical step is indispensable and is now only available in a limited number of cases (rare gases, Hg . . .). Also a good knowledge of the decay schemes and branching ratios is indispensable if one is to distinguish parent from daughter activities at the collector.

Online isotope separators have recently been built in conjunction with heavy-ion accelerators at Dubna (63) and Berkeley (87). The latter uses a quadrupole mass filter of the Paul type (88). Although this device has many interesting features, it is not, in its present form, well suited for cross-section studies because of strong mass-discriminating effects [see Figure 24 of (87)].

To conclude, one may say that isotope separators online are at present especially suited to the spectroscopy of short-lived nuclides (in itself a subject of considerable interest). One can anticipate that with the development of "fast chemistry" and other techniques, and with the increasing knowledge of decay schemes of isotopes far from stability, these instruments will become useful to study cross sections for the production of isotopes far from stability.

For the time being, the online mass spectrometer, although limited to alkalies, is much better adapted to precision studies and very short-lived nuclides.

DIRECT SEPARATION OF UNSLOWED FISSION FRAGMENTS

In the preceding sections we have reviewed methods of separating reaction products that have previously been brought to rest. We now turn to methods of separating fragments with their full kinetic energy.

Fission fragments of a given mass not only come out with a spectrum of different kinetic energies (50 to 110 MeV) but also with a spectrum of effective ionic charge states ($e = 15$ to 25). The mass separation of these fragments is therefore exceedingly difficult. A considerable effort has been made in recent years by two groups in Jülich (70) and Garching (71) and the problem has been approached along three different directions.

The gas-filled isotope separator.—This method was proposed by Cohen & Fulmer (72) in 1958 but has been most extensively used for experiments by Armbruster and co-workers (70, 73, 74) and recently at Dubna by Karnaukhov et al. (109).

The most probable charge of a heavy ion of atomic number Z with an equilibrium ionic charge distribution can be shown to be:

$$\bar{e}(v, Z) = \frac{v}{v_0} Z^{1-\gamma} \qquad\qquad 1.$$

where v is the velocity of the ion, v_0 is the velocity of an electron in a Bohr orbit, and γ is an empirical parameter independent of v.

As a consequence, the deflection in a magnetic field will become velocity independent:

$$B\rho = \frac{Mv}{\bar{e}} = v_0 M Z^{\gamma-1}$$

where ρ is the radius of curvature; and as Z is proportional to M within 1 per cent for fission fragments,

$$B\rho \propto M^\gamma$$

In fact Equation 1 is only an approximation. A more elaborate treatment (75) leads, however, to the same qualitative conclusion that the deflection in a magnetic field depends neither on the velocity nor on the ionic charge state of the fission product.

Thus, if fission fragments would have a narrow distribution around their mean ionic charge it would be possible to separate them in a magnetic field despite their velocity spectrum. Unfortunately, this is not the case because the charge spectrum is so broad. Cohen & Fulmer showed that if the space along the magnetic deflection is filled with low-pressure gas and the fission products undergo charge-changing collisions with the gas molecules, the fluctuation of the average radius of curvature around its mean value would be decreased. This is due to the averaging, over all different charge states, of the magnetic deflection along the path of the particle. One finds however that beyond a certain optimum an increase in the number of collisions brings on a broadening of the beam by multiple scattering.

As a result the maximum resolution attainable is about $\Delta M/M = 1/25$ with a corresponding mass purity of about 1 per cent. The luminosity is high because all velocities and all charge states are accepted and because the mass resolution is low. The admixtures from adjacent masses can be overcome if a particular isotope can be identified by a characteristic nuclear radiation. Following work on ^{84}Sr (74), new data on identification of several new short lived-fission products were recently reported (76). By the same technique, the mass and time dependence of the delayed-neutron emission for thermal ^{235}U fission were investigated (77). The β chain length of fission products was measured in order to determine the primary charge distribution in fission (78, 79).

A new gas-filled mass separator is presently under construction at Jülich (80). Owing to the use of a higher dispersion magnet system, the mass resolution should be around $\Delta M/M = 1/75$. The beam intensity should be improved two or three orders of magnitude by the use of a "charged wire fission products guide" (92).

The high-resolution Mattauch-Herzog mass spectrometer.—As is well known, in a combination of magnetic field B and electric field E charged particles can only be sorted according to their velocities v and e/M values. The deflection of a charged particle in an electric field E and magnetic field B respectively can be written:

$$B\rho_m = \frac{Mv}{e} \quad \text{and} \quad eE = \frac{Mv^2}{\rho_e}$$

from which we deduce

$$v = \frac{E}{B} \frac{\rho_e}{\rho_m} \quad \text{and} \quad \frac{e}{M} = \frac{E}{B^2} \frac{\rho_e}{\rho_m{}^2}$$

As e (the ionic charge) and M only take discrete values, about 2000 distinct e/M values are taken by the 160 fission products. To attain a high purity, a high resolution is necessary however ($\Delta M/M = 1/800$ to obtain 99.5 per cent purity). Moreover this high resolution is to be obtained for all velocities.

For the heavy fission fragment a good mass purity can be achieved with a resolution sufficient to separate adjacent masses ($M/\Delta M = 150$). This is due to a fortunate dependence of the kinetic energy of fission fragments on mass in this region as has been shown by Konecny (81 a, b).

A Mattauch-Herzog-type mass spectrograph (toroidal condenser and homogeneous magnetic field) has been working in Garching (near Munich) according to this principle since 1964 (82a). Figure 9 shows a spectrum of fission products. Note that this apparatus only accepts narrow energy intervals of about 1 per cent. Correspondingly the intensity is low (about 1 count per sec on the most abundant lines) and up to now this has restricted the studies to heavy fragments near the maximum of the distribution (82b).

If the fission fragments are caught in a nuclear emulsion and allowed to decay, one can count under the microscope the number of β tracks and thus estimate the length of chain before stability is reached (83a, b). Characteristics of the charge distribution in fission have thus been studied and fine structure effects in the mass distribution correlated with fragment kinetic energies (84 a, b).

The focusing parabola spectrograph.—An improvement over the Mattauch-Herzog spectrograph is the focusing parabola spectrograph capable of higher dispersion and higher luminosity. The construction of such

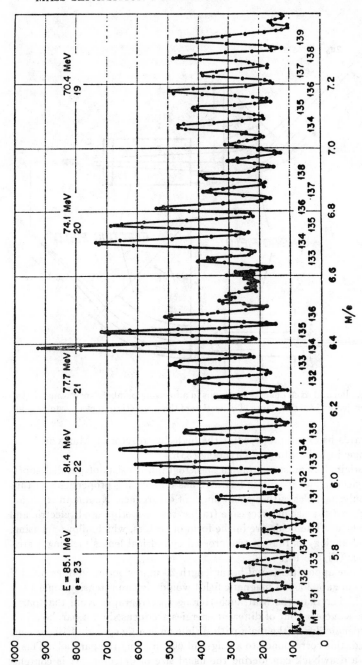

FIG. 9. Fission-particle spectrum obtained with a Mattauch-Herzog mass spectrometer (81a). Note the spread of a particular mass number over different values of the charge state e and the energy.

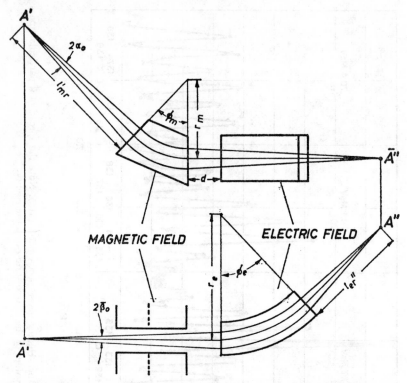

FIG. 10. Radial and axial deflections of ions in a focusing parabola spectrometer (trajectories projected into the planes of deflection of the magnetic and electrostatic field).

an apparatus has been proposed at the high-flux reactor of Max von Laue-Paul Langevin Laboratory in Grenoble (85).

The original Thomson parabola spectrograph consists of parallel superimposed electric and magnetic fields. It has no focusing properties and thus accepts only particles with a small angle of divergence. Neumann & Ewald (86) suggested the use of successive (rather than superimposed) electric and magnetic fields. The fields are in the form of sectors, which allows focusing. An optical analogy is that of two crossed cylindrical lenses yielding a stigmatic focus (Figure 10).

This huge apparatus (total beam length 20 m, deflecting voltage 500 kV over a 25 cm gap, 3000 G magnetic field) would give an intensity some 50 to 70 times higher than the Mattauch-Herzog spectrograph. Note that particles of the same e/m but of different energies are focused on a parabola in a plane perpendicular to the main path. While an energy spread of 10 per cent is accepted, this corresponds to a length of 60 cm along the parabola. This is certainly a drawback considering the usual size of detectors. It is contem-

TABLE II

COMPARISON OF DIFFERENT FISSION PRODUCTS SEPARATORS FOR $U^{235}+10^{14}$ n/cm^2 sec

Type	Target thickness mg/cm²	Luminosity[a] 10^{-5} cm²	Mass resolution	$N(e)^b$ ΣN(e)	$N(E)^b$ ΣN(E)	Purity % light heavy		Intensity for a 6% yield isotope (counts/sec)
Gas-filled	1.6	20	25	1	1	25		9.10⁵
Mattauch-Herzog	0.4	10	150	0.15	0.09	50	95	2.4.10³
		0.5	800				99	1.7.10²
Focusing parabola	0.4	10	150			50	95	2.10⁴
		0.5	800	0.15	0.76		99	1.10³

a Luminosity is a product of target surface by transmission.
b $N(e)/ΣN(e)$ and $N(E)/ΣN(E)$ refer to the portion of the charge spectrum and energy spectrum that are accepted.

plated to refocus the ions of different velocities from a given e/m parabola by using a second magnetic field.

It is difficult to assess the merits of an apparatus before it has actually been built. The extrapolations from actual experience are given in Table II and will form the basis of the following comparison with other techniques.

(a) The most specific information that can be obtained with this technique is the kinetic energy of mass-separated fragments. There is, however, no separation in Z and the situation in this respect is similar to that with conventional isotope separators.

(b) The method is universal rather than restricted to certain chemical elements, thus enabling continuous surveys.

(c) Even with a high-flux reactor, the activities obtained are small compared to the gas-filled separator or to devices which work with stopped fragments, this being due to the use of thin targets (0.5 mg/cm²) inherent to the method, and because one particular mass is spread over some ten charge states.

(d) For investigating very short-lived nuclides, a transit time of microseconds is not a distinct advantage over the milliseconds that are necessary in the case of stopped alkalies. As short-lived isotopes are usually produced with low cross sections, the loss of intensity is a distinct disadvantage.

CONCLUSION

Since the last review by Hintenberger in 1962 (1) the introduction of new techniques has brought rapid progress in this field.

Spark mass spectrometry, which seemed very promising for the analysis of solids (1), proved not to be sensitive enough at the level of <1 ppb that is necessary for nuclear reaction studies (89). On the other hand the sputtering ion source was shown to be a very valuable tool.

In 1962, Hintenberger compared the sensitivity of mass spectrometry and radioactivity measurements to measure trace quantities. He pointed out that for long (or even moderately long) half-lives, mass spectrometry was the most sensitive. While this conclusion remains of course valid, it seems interesting that the introduction of online techniques has extended to very short half-lives the applicability of the mass spectrometer to nuclear reaction cross-section measurements. Figure 11 shows a comparison of different methods in this respect.

The task remains to extend to other elements a technique which is, so far, restricted to alkalies. A foreseeable line of development would be to adapt to the online mass separator or mass spectrometer the rapid methods of transport of recoil products in He gas that have been developed by Ghiorso (90) and McFarlane (91). The difficulty is to get rid of the excess He gas pressure so as to permit normal ion source operation without at the same time losing

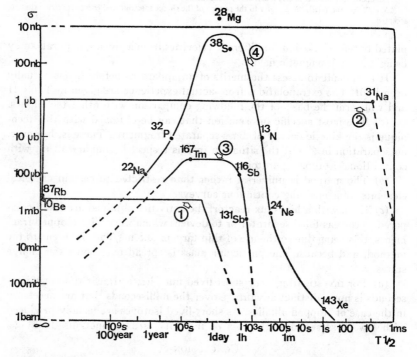

FIG. 11. Comparison of the sensitivity of different methods discussed in this review for cross-section measurements of nuclides of various half-lives. For each method I have tried to plot the lowest cross sections that have been reported for a nuclide of a given half-life.
1—offline mass spectrometry (rare gases, alkalies, $Z \leq 5$ by sputtering)
2—online mass spectrometry (restricted to alkali elements)
3—offline isotope separator (applicable to most elements)
4—radiochemistry (for comparison)
Attention is drawn to the purely empirical character of these curves.

activity. Ionization could conceivably be performed with a sputtering ion source similar to the one used in (25).

Finally, the properties of multipole mass filters (87) should be further investigated.

APPLICATIONS TO PROBLEMS OF CURRENT INTEREST

DISTRIBUTION OF HIGH-ENERGY REACTION PRODUCTS

The mass distribution of products from low-energy fission has been studied extensively and is now fairly well known in good part through measurements on stable isotopes done by mass spectrometry (5, 7, 8). The situation is far from being the same in high-energy nuclear reactions, because of complications, both experimental and theoretical. Although new reaction features undoubtedly appear at high energies (50) they have not yet been proven to be due to an entirely new reaction mechanism. Substantial progress has been made in the past 5 years. While the techniques first used were radiochemistry and conventional mass spectrometry, new data were produced by the extensive use of the isotope separator (CERN and Brookhaven) and more recently by online mass spectrometry and counter experiments.

For the heavy fragments (fissionlike), the situation was reviewed in 1965 in Salzburg by Friedlander (93) and Rudstam (94): At high energy the charge distribution of fission products is much broader, extending well into the neutron-deficient side of stability. Indications are that the curves show a structure with the neutron-excess being connected to moderate-energy fission and the neutron-deficient being, at least in part, reminiscent of a very extensive spallation.

The detailed results obtained for Rb and Cs with online mass spectrometry essentially confirm that picture (58, 59) while recent Brookhaven results (38) in the rare-earth region show that the charge dispersion curve for slightly heavier masses ($A = 146$) has an even more pronounced effect (Figure 12).

For the lighter fragments a sizable amount of data also came in the last years. Measurement of stable rare gases helped to establish the mass-yield curve in some cases and pointed to an increase in cross section for fragments of mass ≤ 40, a feature very different from ordinary spallation that has been ascribed to some kind of direct fragmentation or evaporation of those heavy entities (50).

A detailed and precise isotopic yield measurement was brought about for the first time by online mass spectrometry of all sodium isotopes (55a, b). The relative measurement was precise enough to reveal an unexpected odd-even effect: the emission of even-neutron number isotopes is favored in this process. A similar effect with proton number has also been observed (95).

NEW ISOTOPES

The subject of new isotopes far from stability and of the limit of nuclear stability has recently regained interest with both new mass predictions (96–98) and experimental findings (see 99).

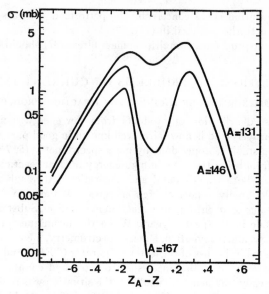

FIG. 12. Charge distribution of the products of the fission of U^{238} induced by 30-GeV protons for different mass numbers (38). Note the very pronounced structure for $A = 146$. The peak for neutron-rich isotopes is reminiscent of moderate-excitation-energy fission.

To the extent to which nuclei far from stability (either on neutron-excess or neutron-deficient side) are usually formed in complex nuclear reactions, very selective methods are necessary to sort them. Mass separation sometimes supplemented by the measurement of another characteristic is of obvious importance in this respect.

Isotopes of Na, Rb, and Cs have been found by online mass spectrometry. Sometimes it has also been possible by radiation measurement at the collector to determine decay characteristics (β half-life; or delayed neutron emission) (100).

Similarly new isotopes of Ar, I, Xe, Au, Hg, Rn, and Fr were reported at CERN by the ISOLDE collaboration (101).

For at least three elements (Hg, Cs, and Rb) it seems within the possibilities of online methods to establish experimentally the limit of nuclear stability on the neutron-deficient side. This would set very clear-cut constraints on the mass formula.

Experimental difficulties are the very sharp decrease in spallation cross section and also the short half-life (in the case of the isotope separator).

APPLICATION TO ASTROPHYSICAL PROBLEMS

The Li, Be, B problem.—These three elements through their extreme scarcity in the universe, compared to their neighbors, are an interesting probe for the occurrence of spallation reactions in astrophysical settings. In the last few years cross-section measurements have led to progress on two main topics.

(a) The origin of Li, Be, B in the solar system. It is generally accepted that Li, Be, and B (whose abundances in the solar system are many orders of magnitude less than those of the neighboring elements) were formed by the spallation of the more abundant heavier elements by medium- to high-energy protons in the early stages of the history of the solar system.

A detailed description of such a process depends rather critically on the value of the spallation ratios ^{7}Li/^{6}Li, ^{11}B/^{10}B, and Li/Be.

Former models based on estimated cross-section values, in particular that proposed by Fowler, Greenstein & Hoyle (102) in 1962, have had to be drastically revised on the basis of the experimental data obtained by the Orsay group (see pp. 8–10). From these data, Gradsztajn (103), and Bernas et al. (104) proposed a different model in which the spallation formation ratio ^{11}B/^{10}B did not have to be altered by subsequent nuclear processes in order to yield the corresponding values observed on earth and in the meteorites. In particular, assumptions made in former models of a strong thermal-neutron irradiation are not necessary. The ^{7}Li/^{6}Li ratio ($\sim$2.5 from spallation) has to be brought to the natural value of 12.5 but thermonuclear (p,α) reactions on Li at the base of the convective zone of the early sun can, according to these authors, account for this transformation, through selective burning of ^{6}Li, without interfering with the Be or B present.

(b) The chemical composition of the galactic cosmic rays at the source. In cosmic rays the abundance of Li, Be, and B relative to C, N, O, and Ne is 10^5 times greater than in the solar system. Their presence is interpreted as a product of the spallation of high-energy C, N, O, Ne, Si, and Fe on the interstellar H at rest. Because of their high abundance, C, N, and O are however the most effective progenitors of Li, Be, and B.

Beck & Yiou (105) and Shapiro & Silberberg (106), using the measured cosmic-ray composition near the earth and the new cross-section data, have calculated the quantity of matter traversed by cosmic rays to be 5 ± 1 g/cm^2. One can also draw conclusions about the composition of cosmic rays at the source (in particular the quasi-absence of nitrogen) and about the stability of ^{7}Be in galactic cosmic rays (105).

Finally the low values of ^{10}Be/^{9}Be cross-section ratios (25) will make it more difficult than expected to use the relative abundance of ^{10}Be ($T_{1/2} = 2.10^6$ years) in the cosmic radiation to determine the "age" of this radiation (i.e. its propagation time).

Study of spallation reactions necessary to understand meteoritic data.—One would think that spallation reactions of medium and heavy elements are now reasonably well understood. While this is true in general and the predictions of the Rudstam formula (107) (see also 108) give a good overall fit of the cross sections, it is always better to have an experimental measurement when a particular value is needed.

The yields of Xe and Kr isotopes from the spallation of Sr and Ba by 730 MeV protons (11, 13) have been measured recently. There were striking differences with theoretical predictions, in particular at masses 130 and 132, which point out the importance of such measurements. The object of this experiment was to compare the isotopic spectra of rare gases found in meteorites that have been bombarded by high-energy cosmic rays with the pure spallation spectra of elements which are considered to be the main cosmic-ray targets in the meteorites. This was also partially the motive of other rare-gas cross-section measurements at Brookhaven (11) and at CERN (17, 18).

There has also been some extensive use of the spallation-produced K to deduce radiation ages of meteorites (1). Thanks to the low natural abundance of ^{40}K, contamination problems can be circumvented. It would be strongly desirable anyway to have spallation cross sections for the K isotopes. Although the measurement of stable K isotopes is not available at present, Chackett has extrapolated values from his measurements of ^{38}K, ^{42}K, ^{43}K done with the CERN isotope separator (49).

This review was written while I was a visitor at Princeton University. I am grateful to T.D. Thomas and the Chemistry Department at Princeton for hospitality during this period.

I thank R. Bernas and E. Roeckl for a critical reading of the manuscript and many valuable remarks.

LITERATURE CITED

1. Hintenberger, H., *Ann. Rev. Nucl. Sci.*, **12**, 435 (1962)
2. Bernas, R., *Advan. Mass Spec.*, **4** (In press)
3a. Smith, L. M., *Electromagnetically Enriched Isotopes and Mass Spectrometry* (Butterworth, London, 1956)
3b. Koch, J., Dawton, R. V. M., Smith, M. L., Walcher, W., *Electromagnetic Isotopes and Applications of Enriched Isotopes* (North-Holland, Amsterdam, 1958)
3c. Higatsberger, M. J., Viehbeock, F. R., *Electromagnetic Separation of Radioactive Isotopes* (Springer-Verlag, Wien, 1961)
3d. Koch, J., Nielsen, K. O., *Nucl. Instr. Methods*, **38** (1965)
4. Thode, H. G., Graham, R. L., *Can. J. Res.*, A25 1 (1947)
5. Farrar, H., Tomlinson, R. H., *Nucl. Phys.*, **34**, 367 (1962)
6. Thode, H. G., McMullen, C. C., Fritze, K., *Advan. Inorg. Chem. Radiochem.*, **2** (1960)
7. Fickel, H. R., Tomlinson, R. H., *Can. J. Phys.*, **37**, 916 (1959)
8. Hyde, E. K., *The Nuclear Properties of Heavy Elements*, **III** (Prentice-Hall, 1964)
9. McHugh, J. A., *J. Inorg. Nucl. Chem.*, **20**, 1787 (1966)
10. Schaeffer, O. A., Zähringer, J., *Z. Naturforsch.*, **13a**, 346 (1958)
11. Hudis, J., Kirsten, T., Schaeffer, O. A., Stoenner, R. W., *Symp. High Energy Nucl. Reactions, 153rd ACS Meeting, Miami Beach, 1967, Abstr. 63*
12. Reynolds, J. H., *Rev. Sci. Instr.*, **22**, 928 (1956)
13. Funk, H., Podosek, F., Rowe, M. W., *Earth Planet. Sci. Letters*, **3**, 193 (1967)
14. Martin, G. R., Mayne, K. I., Thomson, S. J., Wardle, G., *Phil. Mag.*, **45**, 410 (1954)
15a. McHugh, J. A., Michel, M. C., *Phys. Rev.*, **172**, 1160 (1968)
15b. McHugh, J. A., *Univ. Calif. Rad. Lab. Rept. UCRL 10 673* (1963) (Unpublished)
16. Funk, H., Rowe, M. W., *Earth Planet. Sci. Letters*, **2**, 215 (1967)
17. Bieri, R. H., Rutsch, W., *Helv. Phys. Acta*, **35**, 553 (1962)
18. Goebel, K., Schulz, H., Zähringer, J., *Rept. CERN 64-12* (1964) (Unpublished)
19. Friedlander, G., Friedman, L., Gordon, B. M., Yaffe, L., *Phys. Rev.*, **129**, 1809 (1963)
20. Gordon, B. M., Friedman, L., *Phys. Rev.*, **108**, 1053 (1957)
21. Bernas, R., Gradsztajn, E., Klapisch,

R., Yiou, F., Nguyen-Long-Den, *Nucl. Instr. Methods*, 37, 141 (1965)

22. Nguyen-Long-Den Borot, M., *Phys. Letters*, 5, 92 (1963)

23a. N'guyen-Long-Den, (Thesis, Inst. Phys. Nucl., Orsay, 1966); *Advan. Mass Spec.*, 3, 615 (1966)

23b. Krupa, J. C., Nguyen-Long-Den, *J. Phys. Radium*, 29, 205 (1968)

23c. Krupa, J. C., Nguyen-Long-Den (To be published 1969, *J. Inorg. Nucl. Chem.*)

23d. Nguyen-Long-Den, de Saint-Simon, M., Bouissiere, G. (To be published); (Private communication)

24. Bernas, R., Epherre, M., Gradsztajn, E., Klapisch, R., Yiou, F., *Phys. Letters*, 15, 147 (1965)

25. Yiou, F., Baril, R., Dufaure de Citres, J., Fontes, P., Gradsztajn, E., Bernas, R., *Phys. Rev.*, 166, 968 (1968)

26. Klapisch, R., Gradsztajn, E., Yiou, F., Epherre, M., Bernas, R., *Advan. Mass Spec.*, 3, 547 (1966)

27. Klapisch, R. (Thesis, Inst. Phys. Nucl., Orsay, 1966)

28. Collins, T. L., McHugh, J. A., *Advan. Mass Spec.*, 3, 169 (1966)

29. Castaing, R., Slodzian, G., *J. Microscop.*, 1, 395 (1962)

30. White, F. A., Collins, T. L., *Appl. Spec.*, 8, 17 (1954)

31. Barton, G. W., Jr., Gibson, L. E., Tolman, L. F., *Anal. Chem.*, 32, 1599 (1960)

32. Dufaure de Citres, J., *Bull. Soc. Chim. France* (1969) (In press)

33. Dufaure de Citres, J. (Thèse 3e Cycle, Inst. Phys. Nucl., Orsay, 1967)

34. Chavet, I., Bernas, R., *Nucl. Instr. Methods*, 51, 77 (1967)

35. Andersson, G., Rudstam, G., *Nucl. Instr. Methods*, 29, 93 (1964)

36. Andersson, G., *Phil. Mag.*, 45, 621 (1954); *Arkiv Fysik*, 12, 331 (1957)

37. Poffe, N., Albouy, G., Gusakow, M., *J. Phys. Radium*, 23, 213A (1962)

38. Friedlander, G. L., Chu, Y. Y. (Private communication, 1969)

39. Ewald, H., *Advan. Mass Spec.*, 4 (In press)

40. Albouy, G., Gusakow, M., Poffe, N., *J. Phys. Radium*, 21, 751 (1960)

41. Sidenius, G., Skilbreid, O., in Higatsberger & Viehbock, Ref. 3c

42. Ühler, J., Alvager, T., *Arkiv Fysik*, 14, 473 (1958)

43. Sarrouy, J. L., Camplan, J., Dionisio, J. S., Fournet-Fayas, J., Levy, G., Obert, J., *Nucl. Instr. Methods*, 38, 29 (1965)

44. Menat, M., *Can. J. Phys.*, 42, 164 (1964)

45. Freeman, J. H., *Nucl. Instr. Methods*, 38, 49 (1965)

46. Rudstam, G., Sørensen, G., *J. Inorg. Nucl. Chem.*, 28, 771 (1966)

47. Brandt, R., *Proc. 1965 Salzgurg IAEA Symp. Fission*, II, 329

48. Hagebø, E., *J. Inorg. Nucl. Chem.*, 29, 2515 (1967)

49. Chackett, K. F., *J. Inorg. Nucl. Chem.*, 27, 2493 (1965)

50. Miller, J. M., Hudis, J., *Ann. Rev. Nucl. Sci.*, 9, 159 (1959)

51. Klapisch, R., Chaumont, J., Philippe. C., Amarel, I., Fergeau, R., Salome, M., Bernas, R., *Nucl. Instr. Methods*, 53, 216 (1967)

52. Herrmann, G., Denschlag, H. O., *Ann. Rev. Nucl. Sci.*, 19, 1 (1969)

53. Klapisch, R., Philippe, C., Detraz, C., Chaumont, J., Bernas, R., Beck, E. (To be published 1969)

54. Klapisch, R., Bernas, R., *Nucl. Instr. Methods*, 38, 291 (1966)

55a. Klapisch, R., Philippe, C., Suchorzewska, J., Detraz, C., Bernas, R., *Phys. Rev. Letters*, 20, 740 (1968)

55b. Philippe, C., Klapisch, R., Chaumont, J., Bernas, R. (To be published)

56. Amarel, I. (Thesis, Inst. Phys. Nucl., Orsay, 1967)

57. Amarel, I., Bernas, R., Chaumont, J., Foucher, R., Jastrzebski, J., Johnson, A., Klapisch, R., Teillac, J., *Arkiv Fysik*, 77, 36 (1967)

58. Chaumont, J., Roeckl, E., Nir-el, Y., Klapisch, R. (Unpublished—Submitted for communication to the Vienna IAEA Symp., 1969)

59. Klapisch, R., Chaumont, J., Jastrzebski, J., Bernas, R., Simonoff, G. N., Lagarde, M., *Phys. Rev. Letters*, 20, 743 (1968)

60. Kofoed-Hansen, O., Nielsen, K. O., *Kgl. Danske Mat. Fys. Medd.*, 26, No. 7 (1951)

61. Rudstam, G., *Nucl. Instr. Methods*, 38, 282 (1969)

62. Borg, S., Fägerquist, U., Holm, G., Kropff, F., *Nucl. Instr. Methods*, 38, 296 (1965)

63. Tarantin, V. I., Demianov, A. V., Ivanov, N. S., Kabachenko, A. P. (Unpublished—Preprint P13 (4061) JINR, Dubna, 1968)

64. Amiel, S., *Arkiv Fysik*, 36, 73 (1968)

65. Talbert, W. L., Jr., McConnell, J. R., *Arkiv Fysik*, 36, 99 (1967)

66. Andersson, G., *Arkiv Fysik*, 36, 61 (1967)

67. Sidenius, G., Gammon, R. M.,

Naumann, R. A., Thomas, T. D., *Nucl. Instr. Methods*, **38**, 299 (1965)
68. Hagebø, E., Kjelberg, A., Sundell, S., *Arkiv Fysik*, **36**, 127 (1967)
69. ISOLDE Group (To be published); (Private communication of A. Kjelberg)
70. Armbruster, P., Eidens, J., Roeckl, E., *Arik Fysik*, **36**, 293 (1967)
71. Ewald, H., *Arkiv Fysik*, **36**, 311 (1967)
72. Cohen, B. L., Fulmer, C. B., *Nucl. Phys.*, **6**, 547 (1958)
73. Armbruster, P., *Nukleonik*, **3**, 188 (1961)
74. Hovestadt, D., Armbruster, P., Eidens, J., Z. *Physik*, **178**, 226 (1964)
75. Betz, H. D., Hortig, G., Leischner, E., Schmelzer, C., Stadler, B., Weihrauch, J., *Phys. Letters*, **22**, 643 (1966)
76. Eidens, J., Roeckl, E., Armbruster, P. (Submitted to *Nucl. Phys.*, 1969)
77. Roeckl, E., Eidens, J., Armbruster, P., Z. *Physik*, **220**, 101 (1969)
78. Armbruster, P., Neister, H., Z. *Physik*, **170**, 274 (1962)
79. Sistemich, K., Eidens, J., Roeckl, E., Armbruster, P. (Submitted for communication to the Vienna IAEA Symp., 1969)
80. Armbruster, P., et al. (Private communication by E. Roeckl)
81a. Konecny, E., Opower, H., Ewald, H., Z. *Naturforsch.*, **19a**, 200 (1964)
81b. Konecny, E., Siegert, G., Z. *Naturforsch.*, **21a**, 192 (1966)
82a. Ewald, H., Konecny, E., Opower, H., Rösler, H., Z. *Naturforsch.*, **19a**, 194 (1964)
82b. Ewald, H., Konecny, E., Opower, H., *Proc. IAEA Symp Phys. and Chem. Fission, Salzburg*, **1**, 505 (1965)
83a. Konecny, E., Gunther, H., Siegert, G., *Arkiv Fysik*, **36**, 319 (1967)
83b. Konecny, E., Opower, H., Gunther, H., Gobel, H., *Proc. IAEA Symp. Phys. and Chem. Fission, Salzburg*, **1**, 401 (1965)
84a. Konecny, E., Gunther, H., Siegert, G., Winter, L., *Nucl. Phys.*, **A100**, 465 (1967)
84b. Gunther, H., Siegert, G., Gebert, R., Kerr, D., Konecny, E., Z. *Naturforsch.*, **22a**, 1808 (1967)
85. Armbruster, P., Ewald, H., Fiebig, G., Konecny, E., Lawin, H.,

Wollnik, H., *Arkiv Fysik*, **36**, 305 (1967)
86. Neumann, S., Ewald, H., Z. *Physik*, **169**, 224 (1962)
87. Nitshke, J. M., *Univ. Calif. Rad. Lab. Rept. UCRL 18 463* (Unpublished)
88. Paul, W., Raether, M., Z. *Physik*, **140**, 262 (1955)
89. Irsa, A. P., Friedman, L., *13th Ann. ASTM Meeting (E-14)*, **36** (1965)
90. Ghiorso, A. (Unpublished); cf. *Phys. Rev.*, **167**, 1094 (1968)
91. MacFarlane, R. D., Griffioen, R. D., *Nucl. Instr. Methods*, **24**, 461 (1963)
92. Oakey, N. S., MacFarlane, R. D., *Nucl. Instr. Methods*, **49**, 220 (1967)
93. Friedlander, G., *Proc. IAEA Symp. Fission, Salzburg*, **II** (1965)
94. Rudstam, G., *Proc. IAEA Symp. Fission, Salzburg*, **II** (1965)
95. Thomas, T. D., Raisbeck, G. M., Boerstling, P., Garvey, G. T., Lynch, R. P., *Phys. Letters*, **27B**, 504 (1968)
96. Baz, A. I., Goldanskii, V. I., Zeldovich, Y. B., *Soviet Phys.*, **8**, 177 (1965); *Soviet Phys. Usp.*, **8**, 177 (1965)
97. Garvey, G. T., Kelson, I., *Phys. Rev. Letters*, **16**, 197 (1966)
98. Garvey, G. T., *Ann. Rev. Nucl. Sci.*, **19**, 433 (1969)
99. Cerny, J., *Ann. Rev. Nucl. Sci.*, **18**, 27 (1968)
100. Amarel, I., Bernas, R., Foucher, R., Jastrzebski, J., Johnson, A., Teillac, J., Gauvin, H., *Phys. Letters*, **24B**, 402 (1967)
101. The ISOLDE Collaboration—CERN, *Phys. Letters*, **28B**, 415 (1969)
102. Fowler, W. A., Greenstein, J., Hoyle, F., *Geophys. J.*, **6**, 148 (1962)
103. Gradsztajn, E., *Ann. Phys.*, **10**, 791 (1965)
104. Bernas, R., Gradsztajn, E., Reeves, H., Schatzman, E., *Ann. Phys.*, **44**, 426 (1967)
105. Beck, F., Yiou, F., *Ap. Letters*, **1**, 75 (1968)
106. Shapiro, M., Silberberg, R., *Can. J. Phys.*, **46**, S561 (1968)
107. Rudstam, G., Z. *Naturforsch.*, **21a**, 1027 (1966)
108. Audouze, J., Epherre, M., Reeves, H., *Nucl. Phys.*, **A97**, 144 (1967)
109. Karnaukhov, et al. (Unpublished); (Private communication of G. N. Flerov)

THE THREE-NUCLEON PROBLEM

By R. D. Amado[1]

Department of Physics, University of Pennsylvania
Philadelphia, Pennsylvania

CONTENTS

1. Introduction

The classic purpose of three-nucleon studies is to shed light on how nucleons interact. In particular one hopes to learn about those aspects of the nucleon interaction that cannot be studied in nucleon-nucleon scattering, such as off-shell behavior, i.e. form of the two-body potential; and one hopes to learn about those aspects which are not additive, such as three-body forces, meson exchange effects. All these aspects of nucleon interactions manifest themselves in other systems as well; for example, off-shell effects are important in nuclear matter and in nucleon-nucleon bremsstrahlung, three-body forces occur in nuclear matter, and meson exchange effects are present in any complex nucleus. The three-nucleon system is singled out because it is hoped that the problem will prove sufficiently tractable to allow one to investigate these questions. Until recently this hope has been largely

[1] Supported in part by the National Science Foundation.

61

unfulfilled, but in the last few years advances in theoretical techniques and computing capabilities have made it possible to manage the three-body aspects of the problem and have placed us on the threshold of understanding some of these other questions. Unfortunately, there is not the space here to review these theoretical developments. (See Section 5.)

The purpose of this article is to present a critical review of the present state of our knowledge of the three-nucleon system. The general picture that emerges is one of broad qualitative understanding. There are no outstanding puzzles or surprises. The major components of the two-body force combined with good three-body theory seem capable of accounting for the major features of the three-body system. To be sure, there are effects that require meson currents and there is room for three-body forces and potential dependence, but the effect of these is certainly not large. Perhaps the most interesting thing to emerge is the extent to which these features are intertwined. Traditional approaches have been largely fragmentary, and in the treatment of meson current effects and three-body forces quite primitive. In view of the sophisticated efforts and excellent results in the three-body problem with two-body potentials, the time would seem ripe for a more unified and sophisticated approach to the remaining features of the problem.

Part of the purpose of the review is to distinguish which aspects of the problem seem to be fairly model independent and which give promise of yielding more detailed information. Within the domain of purely nuclear systems, the binding energies of H^3 and He^3 are the most likely candidates for providing detailed information. The calculations are already at a point where selection among the predictions of various potential forms awaits better analysis of three-body forces and relativistic effects. Nucleon-deuteron elastic and inelastic scattering, on the other hand, seems explicable without recourse to detailed assumptions except perhaps at very low energy. Further information comes from probing the three-nucleon system with electromagnetic or weak interactions. Here the detailed questions of the classical three-nucleon problem, like wavefunction components, seem always to have their answers in those situations where meson exchange effects also play an important role. Although our understanding of these is primitive, there seem to be no outstanding anomalies. However, without better theories of the nonadditive effects, it is difficult to make the data yield detailed three-body information.

2. BOUND STATES

There are, unfortunately, only two bound states of the three-nucleon system, the ground states of He^3 and H^3. These bound states have spin $\frac{1}{2}$ positive parity and binding energies of 8.482 MeV for H^3 and 7.718 for He^3 (1). They form an isotopic spin doublet and thus for most purposes may be considered to be one state. No excited states of He^3 or H^3 have been established. The evidence against $T=\frac{3}{2}$ states (n^3) also seems very strong.

Hence the theorist of the three-nucleon bound states is forced to search in a relatively featureless landscape.

In this section we shall consider the theoretical attempts to obtain the binding energies of H^3 and He^3 and discuss briefly the evidence against other states. We shall also discuss the wavefunction of the bound states, although many of the data on the form of these are involved in the question of weak and electromagnetic probes, treated in Section 5.

Classification of states.—In an L-S scheme, the simplest state of the three-nucleon system commensurate with a $J=\frac{1}{2}^+$ state is $^2S_{1/2}$. Of the possible symmetries for such a state, totally symmetric in space is most likely. The totally symmetric S state would be the only component if nuclear forces were purely central and spin independent—pure Wigner force— and since the Wigner force is certainly the major part of the nuclear force, the symmetric S state is the major component of the three-nucleon bound state. Much of the effort in three-body theory has gone to obtaining the remaining components of the wavefunction. Of these, by far the most important are the S state of mixed permutation symmetry (known as the S' state) and the D states (cf. 2, 3). The S' state arises from the spin dependence of the central force. In an $^2S_{1/2}$ state some nucleon pairs are in $S=0$ states and some in $S=1$ states. Since the force between nucleons is somewhat different in these two states, the wavefunction cannot be completely symmetric with respect to them. The precise fraction of S' state in the three-nucleon bound state is a point of some dispute, as we shall see, but most estimates put it between 1 and 4 per cent with 2 per cent emerging as probably preferred. The D states are mixed in by the tensor force, just as the D state is in the deuteron. However, the fact that the three-nucleon bound state is more spin symmetric than the deuteron means that the contribution of the tensor force to the binding energy is relatively smaller for three nucleons than for two. Hence a two-body force which binds the deuteron by virtue of a large tensor force will have a harder job giving a correct three-nucleon binding than will one with little or no tensor force. In fact purely central forces fitted to the deuteron usually overbind in the three-nucleon case. The percentage of all D states in the three-nucleon bound state is between 5 and 9 per cent. $^2P_{1/2}$ and $^4P_{1/2}$ states are possible, but they are mixed in only by the two-body L-S force and the tensor force is higher order and hence they are relatively unimportant.

An important theoretical activity in the three-nucleon problem has been an enumeration of *all* possible states with $J=\frac{1}{2}^+$. The first to attempt this systematically were Gerjuoy & Schwinger in 1943 (4), and many others have carried the program further, finding new states, and removing linearly dependent ones (2, 3). In recent times the Australian group of Blatt et al. have devoted much attention to the problem (5–7). There are in all ten linearly independent states. They can be classified by giving the values of L,S,T, the symmetries of each of these parts, and any additional quantum numbers required. Much, if not all of this, can be done without explicit reference to a coordinate system. The permutation group on three objects has three irreducible representations: one is totally symmetric, one is totally antisymmetric, and one is a two-dimensional representation of mixed symmetry. The total three-body wavefunction must be totally antisymmetric, but each of the LST parts can, in principle, belong to any of the three permutation representations. In fact, since the spin wavefunction and the isospin wavefunction are compounded from three objects, each with only two degrees of freedom, only the symmetric and mixed-symmetry states are possible. These are S (or T) $=\frac{3}{2}$ and $\frac{1}{2}$, respectively. For $J=\frac{1}{2}^+$ there are three states of $L=0$, one with each type of permutation symmetry. The totally symmetric and the mixed-symmetry ones we discussed above. The totally antisymmetric one is physically not very likely in the bound three-nucleon system. The $^2P_{1/2}$ states have $L=1$ states of all three permutation varieties, but only the mixed-symmetry $L=1$ state occurs for $^4P_{1/2}$. There are three $L=2$ states, all of mixed symmetry. They are conveniently distinguished by choosing a coordinate system and forming explicit states (5). However, theory is not yet sufficiently precise to distinguish them in the bound state, hence they are usually taken together in describing the wavefunction. Although states of all ten types of symmetry can be and have been constructed, in the present state of our knowledge of the three-nucleon system it is enough to form a wavefunction from the symmetric S state, the S' state, and the D states taken together.

Various authors have addressed themselves to the problem of whether the Coulomb force and other T noninvariant terms including perhaps even an isovector two-body force can admix sizable $T=\frac{3}{2}$ into the bound state. So far no very convincing argument for a $T=\frac{3}{2}$ component much larger than .05 per cent has been given (8–10), but possible difficulties with the Coulomb energy difference may suggest problems in this area.

Binding energy.—The oldest and most extensively studied technique for obtaining three-body binding energies is the variational method. In fact, one of the first such calculations produced a new prediction about the two-body force. In 1935 Thomas used this method to rule out a zero-range force (then compatible with the two-body data) by showing that it would lead to infinite binding for the three-body system (11).

The variational method is deceptively simple in principle. One chooses a parametrized trial function, calculates the expected value for some predetermined Hamiltonian, and the minimum is an upper bound on the true energy. Carrying this program out for anything but the simplest potentials and simplest trial wavefunctions is very complicated, and there is a great tradition of heroic calculation in this field (cf. 12). For example, in the 1962 calculation of Blatt, Derrick & Lyness (13), obtaining an energy bound for a given two-body potential choice is reported to have taken 100 hr of IBM 7090 time and to have involved the variation of 55 nonlinear parameters. The complexity of these calculations comes about in part from the complex nature of the modern two-body potentials. These include large tensor forces, short-range repulsion and $L \cdot S$ and quadratic $L \cdot S$ forces. Further in the presence of such potentials all ten symmetry types of the three-body wavefunction in principle can contribute. The most persistent and serious attempts to combine all these aspects into one calculation have been made by the Australian group of Blatt and co-workers (13–17). They have concentrated their efforts in recent years on the Yale and Hamada-Johnston two-body potentials, and after some discouraging first attempts they have recently found rather good and stable answers (17). The improvements have resulted largely from adding scattering-like parts to the trial function (15) and from stressing a systematic improvement scheme based on *linear* variational parameters (18). In this way a triton binding energy for the Hamada-Johnston potential of 6.7 ± 1.0 MeV has recently been obtained (17). (The error comes from numerical uncertainties.) This value is very stable against adding linear parameters and seems to have converged. It therefore presumably represents the true Hamada-Johnston binding. This binding energy is well within reach of the experimental value considering the uncertainties in three-body forces and relativistic corrections.

Efforts to estimate the effects of three-body forces are still in a relatively schematic state. Even the sign of the effect is doubtful, although some recent calculations using standard meson theory methods have given an increase in the binding energy $\sim.5$ to 1 MeV (19, 20). Arguments based on the partially conserved axial-vector current, however, indicate that the longest-range component of the three-body force is zero (21). In nuclear matter it has been shown that the short-range component of the three-body force and the part due to higher resonances make an appreciable contribution to the binding (22, 23). The three-nucleon system is probably partway between nuclear matter and very long range and a careful estimate of the three-body force contribution to the binding energy using modern methods remains to be done. Relativistic corrections to the binding energy have been estimated by Gupta, Bhakar & Mitra to be of the order of $-.5$ MeV, but no systematic investigation of the model dependence of these results has been undertaken (24). As calculations of the binding energy approach—with great effort—within 1 MeV of the correct answer, it would be useful if a corresponding effort were made to evaluate these two important corrections, particularly as the calculations now make good three-body wavefunctions available. If the corrections make it necessary to replace the Hamada-Johnston potential, one would probably look for one with less tensor force. Such a potential, assuming it could be made to fit the two-body data, would give more triton binding, as we discussed above. Noyes & Fiedeldey have investigated in detail the binding-energy dependence on two-body parameters and have suggested that it should be possible to vary the triton binding energy by 2–4 MeV while retaining any given fit to the two-body data (25). If so, the triton binding energy will be a fruitful place to distinguish potentials but this range of variation seems optimistic. Meanwhile the recent results on the

Hamada-Johnston potential are reassuring and sensible but by no means end the story. The remaining conclusions coming from the variational calculation are: (a) The rms radius will come out right if the binding energy does (15, 16). (b) The combined D states make up about 9 per cent of the wavefunction and the mixed-symmetry S state (S') about 2 per cent. Except for the symmetric S state all remaining components are negligible (17). Further one finds that the various components of the two-body force by no means contribute equally to the binding energy and some, in particular the triplet odd forces, are essentially negligible.

An alternative approach to the binding energy is via separable potentials. An n-component separable potential between pairs reduces the three-body problem to the solution of n-coupled one-dimensional integral equations. These equations are easily solved "exactly" on a computer, particularly in the bound-state case since the integral equation kernels are then real and free of singularities. Whether for a two-body force of a given complexity the numerical errors of the potential calculations can be made smaller or controlled better than those of a variational calculation with a local potential is a detailed question for numerical analysts which in the present state of our knowledge is largely academic. Separable-potential calculations of the binding are easier, require less sophistication, and take far less computing machine time. For largely these reasons, they have been attempted by many groups but not by any one group on the heroic scale of the variational calculations.

Before presenting the results of separable-potential calculations of the binding, let us review the physical assumptions involved in their use, since they are unfamiliar. There are many reasons for at first sight preferring local to nonlocal potentials, and certainly to separable ones. It is now well established that the longest-range part of the nucleon force, the one-pion-exchange region, can be characterized by a local Yukawa-type potential. Furthermore, our intuition and training bias our understanding in terms of local potentials. On the other hand, in the three-body bound state, the long-range component of the force is relatively unimportant. Calculations with three identical bosons interacting via Yukawa potentials of the nucleon range and strength show that one gets nearly all the binding from the relative S waves of the pairs (26). There is no evidence experimental or theoretical that the S-wave nucleon-nucleon potential is local (in fact there is some evidence to the contrary) (cf. 27). A dominant feature of the S wave is a short-range repulsion, but an infinite local repulsion at .4 F is certainly not a *fundamental* feature of the "true" interaction. So long as there is no fundamental theory of the interaction for the low partial waves, we are left to proceed empirically. The question therefore is how to characterize those aspects of the nucleon-nucleon force that are relevant to the three-body problem, or better to decide what aspects *are* relevant. Presumably the one universal is that all potentials must fit the two-body data, that is, give the correct two-body t-matrix on-shell (cf. 28–30). This can certainly be arranged for a separable potential. How far off-shell is important for the three-nucleon bound state remains unknown. For the scattering problem it is clear physically as well as from the structure of the Faddeev equations (31) (which are more appropriate to this problem than Schrödinger equations) that the on-shell t-matrix element is more relevant than the potential form. For the bound-state problem, there is no clear way known to choose, and in particular there seems no reason to prefer a given potential form. The advantage of the separable potential is the ease with which calculations can be done and therefore parameters varied, and potential forms tested. Until a deeper theoretical understanding of the two-body system is available, it is only by comparing three-body calculations with different potential forms that we can shed light on these off-shell questions.

The first three-nucleon calculations with separable potentials used only central spin-dependent S-wave forms fitted to the low-energy data and hence gave too much binding, as one expects (32–35). However they also showed that with the introduction of one parameter, not only could one fit the binding energy, but the value of that parameter was reasonable in terms of the effects expected from tensor forms and short-range repulsion (34, 35). The effects of these components have also been studied, usually in model situations, again with reasonable results, as has the effect of various functional forms for the separable potential (36–41). Recently Schrenk & Mitra have reported a calculation including the effects of the tensor force and a short-range repulsion in the ^1S channel (42). Because of an unnecessarily narrow numerical mesh they were not able to include other important components of the force, like the attractive central D wave and repulsion in the ^3S channel. Nevertheless for a reasonable range of tensor force parameters they obtain a binding energy of about 9 MeV for the triton. A similar calculation including tensor forces and short-range repulsion in the ^1S state has recently been reported by Dabrowski & Dworzecka and they find about 8.8 MeV for the triton binding (43). Wavefunctions from these calculations are not available, but previous calculations with just hard cores, or just tensor forces, indicated that one will get about 2 per cent S' and at least 5 per cent D wave.

We see that separable potentials can account for the three-nucleon binding energy within the uncertainties of the potential used and of the three-body forces and relativistic corrections. Unfortunately no one has yet attempted a full calculation using all the important components of a force from a separable potential made to fit the two-body data. Such a calculation is by no means technically impossible, as even more difficult three-body calculations have been done for other systems (44). In spite of the uncertainties in the binding-energy corrections, such a calculation seems at least as worthwhile as the full local-potential calculation.

The fact that the best variational calculations underestimate the binding whereas the best separable-potential calculations give a little too much binding may be significant, but until a better separable-potential calculation is done with a potential comparable in sophistication to the Hamada-Johnston potential, firm conclusions are premature.

Coulomb energy.—The Coulomb energy of the three-nucleon system is the difference in binding of He^3 and H^3. Experimentally this is .76 MeV (1). Many efforts have been made to calculate it, usually in perturbation theory, by taking the expectation value in the best wavefunction of the proton-proton Coulomb interaction. The simplest form involves using point protons. Many such calculations, including some of the best variational ones, have not been able to get a big enough Coulomb energy difference (15, 45–48). This difficulty has been stressed particularly by Okamoto & Lucas, who claim that .63 MeV is the best that they can do, including the finite proton size, and that the remaining difference should be ascribed to an isovector component of the strong interaction (49–51). However, Gupta & Mitra, using a wavefunction from a separable-potential calculation with a spreadout proton charge distribution, get .8 MeV (52). This result does not include hard-core effects and presumably the three-body wavefunction is therefore too "small," but as Mitra has recently argued, it probably still indicates the importance of the correlations in the exact wavefunction (53). Similarly Pappademos obtains a reasonable Coulomb energy using a wavefunction made to include hard-core effects and to have the correct asymptotic form (54). Phillips finds a similar result (55). Thus the question of the sensitivity of the Coulomb energy to the form of the three-body wavefunction requires further investigation. In particular, it should be stressed that the wavefunction from a variational calculation of the energy is not guaranteed to be good for other things. There is also the question of the assumed form of the Coulomb energy. Adya has recently studied a soluble model to determine the validity of using first-order perturbation theory (56). Although his results are too schematic to bear direct application to He^3, he finds that the higher orders could easily increase the Coulomb difference by .1 MeV. Folk has studied the problem of the correct operator to use (57). He points out that there are polarization effects on the proton charge cloud, and meson charge current effects that could also give as much as .1 MeV.

There is certainly an isovector component of the nucleon force due to Coulomb corrections, mass differences, etc., but in view of the uncertainties in the calculation of the Coulomb energy of He^3, it seems premature to blame as much as .1 MeV on it. Presumably attempts to learn about this component of the forces from He^3 will require better wavefunctions, and a better understanding of the meson current effects.

Excited states.—Although there have been reports of excited states of He^3 (58) and particle stable states of three neutrons (n^3) (59), all subsequent efforts to confirm these reports have failed and it now seems fairly well established experimentally that these states do not exist (60–69).

Theoretically, no excited states of He^3 have been predicted. For n^3 the situation is confused. The best candidate for a stable n^3 state would be a P state. The most important pair-force is then ^3P, which is attractive. Variational attempts to bind n^3 have failed, although the forces used have not always been ones that fit the ^3P data well (70, 71). On the other hand, Mitra & Bhasin claim n^3 could be bound by a ^3P separable potential that is compatible with the two-body data (72), and recently Jacob & Gupta have made an even stronger claim (73). They find n^3 binding with a separable potential only one half as strong as the one needed to fit the ^3P two-body data. In view of the variational estimates, the general systematics of neutron systems, and the lack of experimental evidence for n^3, this seems very implausible and may well be an artifact of their unusual potential form and method of solution.

Other methods.—A number of other approaches to the three-nucleon bound state have been attempted. One of recurrent interest is the independent-pair approximation, in which one tries to find the best wavefunction made up of products of pair wavefunctions (74–79). Another method exploits the relatively compact nature of the bound state by expanding in six-dimensional spherical polar coordinates. In this way one gets *ordinary* differential equations in the six-dimensional radial coordinates, but coupled together by the interaction terms which are not diagonal in the six-dimensional "angular momentum." The compact nature of the bound state leads one to hope that this coupling will decrease rapidly so that few states are significantly coupled (80). The mathematics for carrying such a scheme out has been developed, including the classifying of states by their symmetries (81), and preliminary calculations are promising (82, 83). In general, however, no scheme has yet been carried to the level of sophistication or of results comparable to the variational and separable-potential approaches.

3. SCATTERING AND BREAKUP

For the three-nucleon system the only two-body scattering channels are n-d and p-d. Apart from Coulomb effects these two are presumably the same and since the deuteron has isospin zero, lead to scattering in the $T = \frac{1}{2}$ three-body channel only. There are two spin channels for n-d and p-d scattering, doublet ($S = \frac{1}{2}$) and quartet ($S = \frac{1}{2}$). The presence of noncentral two-body forces (and in principle three-body forces) couples the spin and orbital angular momentum so that the scattering amplitude is not strictly diagonal in the L-S representation. However, at low energies the mixing, as reflected in polarization, for example, seems to be fairly small. At essentially all energies the dominant qualitative features of n-d and p-d scattering arise from the large size and rather fragile nature of the deuteron and the identity of the incident particle and one of the deuteron constituents. The first of these leads to the rapid onset of high partial waves in the scattering, relatively little contribution from virtual collisions of high momenta, and considerable inelasticity. The second makes the dominant scattering mechanism nucleon exchange or pickup. As we shall see, a theory incorporating these features yields the major features of n-d and p-d scattering up to moderate energies.

Scattering lengths.—For n-d scattering at zero kinetic energy only S waves are important and the spin channels are uncoupled. Hence there are two n-d scattering lengths $a_2(S=\frac{1}{2})$ and $a_4(S=\frac{3}{2})$. In the doublet channel the S-wave force is attractive (the triton is in this channel) and the scattering length may be expected to be a sensitive reflection of the three-body dynamics. In the quartet channel the spins are all parallel, and the Pauli principle keeps the particles apart. Hence the effective S-wave n-d force is repulsive in the quartet channel and the scattering length is relatively insensitive to the details of the interactions.

Experimentally the scattering-length situation is now both much clearer and somewhat more confused than it has been in the recent past. For some time two sets of scattering lengths were consistent with experiment (84).

Set A $a_2 = 0.7 \pm 0.3$ F $a_4 = 6.38 \pm 0.6$ F

or

Set B $a_2 = 8.26 \pm .12$ F $a_4 = 2.6 \pm .2$ F

This ambiguity results from combining the Fermi-Marshall value for the cross section of "epithermal" neutrons on deuterium $(\sigma = 4\pi(\frac{2}{3}a_4^2 + \frac{1}{3}a_2^2)$ $= 3.44 \pm 0.06 \times 10^{-24}$ cm²) (85) with a measurement by Hurst & Alcock (84) of cold neutrons on deuterium which, from the coherent molecular scattering, gives a value for the ratio of the scattering lengths. In a recent heroic experiment, Alfimenkov et al. (86) have resolved the ambiguity by measuring the cross section for polarized slow neutrons on polarized deuterons. Their result unambiguously rules out set B. However, more recently still, Van Oers & Seagrave (87, 88) have examined new data on coherent and incoherent n-d scattering. They obtain a new scattering-length set.

Set A' $a_2 = 0.12 \pm 0.07$ F $a_4 = 6.11 \pm 0.06$ F

These imply a "free" n-d total cross section of $3.13 \pm 0.06 \times 10^{-24}$ cm², which is in better agreement with zero-energy extrapolations of low-energy n-d data, but well outside the Fermi-Marshall value (85). The present situation, therefore, is that it is definitely established that $a_2 \sim 0$ F and $a_4 \sim 6$ F, and presumably set A' is preferable to the older set A, but the experiments by no means overdetermine the numbers and we can probably expect to see more small fluctuations, in a_2 particularly, in the future.

The small positive-doublet scattering length has led to some confusion of interpretation in the past. The scattering length is minus the slope of the phase shift with wavenumber at zero energy. For a weak attraction this slope is positive (negative a) and grows to infinity as the force becomes strong enough to produce one bound state. For a force just barely producing one bound state the phase shift has a large negative slope (positive a) and as the force grows the slope grows, finally becoming large and positive again just before the onset of the second bound state. If the slope of the phase shift is near zero (scattering length near zero), the force is roughly halfway between that required to bind one and two states and this is the situation for the

n-d doublet channel. This small scattering length has two important consequences: doublet scattering is anomalously small at low energies; $k \cot \delta$ is very large as $k \to 0$. This large value of $k \cot \delta$ is associated with a pole of $k \cot \delta$ very near to $k = 0$, as was first noted many years ago by J. Gammel (unpublished). This pole blocks the usual linear extrapolation of the low-energy data to get a_2 as Christian & Gammel emphasized (89). Plots of $k \cot \delta$ showing the effect of the pole can be found in Van Oers & Seagrave (87). They also show that the best values for the low-energy p-d parameters with the Coulomb effects removed are consistent with set A'.

The calculation of the scattering lengths parallels very closely the bound-state calculations. There exists a variational principle for bounds on the scattering lengths by a method similar to the Rayleigh-Ritz bound for the binding energy (90). Hence any triton binding variational calculation can be converted to a scattering-length calculation (cf. 91). Any separable-potential calculation of the binding is also easily adopted to give the scattering length, particularly as the separable-potential calculations even for the bound state are usually done in the framework of scattering integral equations (cf. 34).

Nearly all modern calculations, variational (92–96) or separable potential (cf. 33–35, 97, 98), give a quartet scattering length in good agreement with set A or A', even those carried out with purely central forces. As we noted before, in this state the symmetry keeps the particles far apart and hence the parameters are insensitive to the details of the force. Since by a numerical accident quartet scattering dominates the low-energy n-d system, this insensitivity is a warning against expecting much detailed two-body information from the n-d continuum. Calculations of the doublet scattering length, on the other hand, are very sensitive to the two-body force, particularly (because of the accidentally small value of the scattering length) if one looks at percentage changes. This is also no surprise since the triton binding energy is also very sensitive and these two numbers go hand in hand. In fact, calculations of the doublet scattering length with potentials that do not fit or do not attempt to fit the triton binding energy are not very instructive. Unfortunately the simple connection between bound-state pole and scattering length that can be made for the deuteron is more complicated here because the bound-state pole is farther from the elastic threshold and because of the importance of the nearby inelastic threshold. Some interesting attempts have been made to correlate these features in a dispersion theoretic approach and the results are generally correct, but much more model dependent in detail and much less informative than in the two-body case (99–101).

The best variational bound on the doublet scattering length is 1.2 ± 1 F corresponding to a triton binding energy bound of 6.7 ± 1 MeV. This comes from the recent calculation of Delves, Blatt, Pask & Davies with the Hamada-Johnston potential (17). However, it is not clear that the scattering-length bound has converged as well as the binding energy. Earlier variational bounds have varied from 1.6 to 4.8 F (15, 16, 102).

The simplest separable-potential calculations (*S*-wave attractive spin-dependent forces only) give too small (algebraically) a doublet scattering length, just as they overbind the triton (33–35, 97, 98). However when one parameter is introduced to simulate the leftout components of the force, and this parameter is adjusted to give the correct triton binding, the doublet scattering length comes out to be 0.7 F (34, 35) which is the correct value for set A, but not the more modern set A'.

Using more sophisticated separable potentials does not substantially change things. Schrenk & Mitra find with tensor forces and short-range repulsion that a binding energy of 9 MeV goes with a scattering length of about −.15 F (42). In fact Phillips points out that a plot of all the separable-potential doublet scattering lengths results against triton binding energy gives a straight line passing through .7 F at the experimental binding (103). The set A' scattering length of .1 F corresponds to a binding energy value of 9.5 MeV on that straight line. The significance of this remains obscure, particularly in the absence of a calculation with a separable potential of sophistication comparable to the Hamada-Johnston potential. Note, however, that the most recent variational calculation of the scattering length and binding energy (17) also seems to give values lying on Phillips' straight line. Should the discrepancy remain, there are many possible explanations. One is that the set A' is not correct. Another is that three-body force and relativistic correction effects are more important in the bound state than for scattering (as they very likely are) and that they are repulsive so that the binding in their absence is around 9 MeV. Finally, there is the optimistic hope that the scattering length and binding energy together are sufficiently restrictive to distinguish potential forms and that local potentials give (algebraically) smaller scattering lengths for the same binding. Delves & Phillips (104) point out that a hint of this may be found in comparing the local central-potential calculations of Humberston (95) with the separable-potential calculation of Sitenko et al. (39) for the same scattering parameters. Much detailed work remains to be done before these questions can be answered.

Elastic scattering.—Although considerable data, much of it very high quality, on elastic *n-d* and *p-d* scattering has existed for some time, reasonable theoretical understanding has only come with the development of separable interactions which provide a method for obtaining numbers based on the physics of the problem and of the Faddeev equations which provide a sound formal framework (31, 105, 106). Previous theoretical work, mostly using variational methods, or the resonating group method yielded little quantitative success or qualitative understanding (cf. 107). An exception to this is the pioneering and largely neglected 1953 paper of Christian & Gammel (89). They point out that the first Born approximation for *n-d* scattering is nucleon exchange, represented in Figure 1. The contribution of this term depends only on the deuteron binding energy and the low-energy *n-p* parameters, and not on the details of the nuclear force. Furthermore, this is the

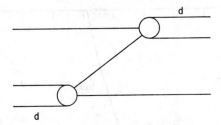

FIG. 1. The nucleon exchange Born diagram for
nucleon-deuteron scattering.

longest-range component of the n-d interaction. Christian & Gammel exploit this fact to make a partial-wave analysis of n-d scattering. The S waves are left free to be fit, but all other waves are fit with the nucleon exchange Born term. Their fit is good, and the S-wave phase shifts extracted are comparable with modern S-wave fits. Thus they showed that nucleon exchange was the dominant mechanism for elastic n-d scattering and that this in turn depends only on the low-energy n-p parameters. It would seem that this idea of using the longest-range part in Born approximations for the high partial waves, an idea later exploited with the one-pion-exchange term in nucleon-nucleon scattering, appeared first here. Recently this semiphenomenological approach to the phase-shift analysis of n-d and p-d scattering has been revived by Van Oers & Brockman (108). They have used better data and made allowances for inelasticities, and their phase shifts are in substantial agreement with the separable-potential calculations that we now turn to.

The importance of the nucleon exhange term to n-d scattering was rediscovered in the separable-potential approach (105). This term is the dominant mechanism or driving term ("potential") for n-d scattering in all waves, but not in Born approximation. For example, the S-wave contribution of this term violates the unitarity bound by a considerable factor. We wish then to unitarize this term, preserving the three-body dynamics. In other words, we wish to solve the Schrödinger equation for a problem in which the nucleon exchange term is the first approximation. The simplest such theory is one which has the $d \rightleftarrows n + p$ vertex, or, what is the same thing, an n-p interaction through the deuteron. This is just a separable potential. Including an n-p interaction in the ^1S state to reflect the other major low-energy nucleon-nucleon interaction, we have a fairly complete description of the low-energy nucleon-nucleon interaction in a three-body theory of n-d scattering that includes a correct description of the asymptotic states, a correct description of the primary interaction mechanism, three-body dynamics, and is soluble numerically since it reduces the equations to coupled one-dimensional integral equations after partial-wave decomposition. Such a simple theory overbinds the triton somewhat and gives a correspondingly wrong doublet scattering length, since it includes only central attractive S-

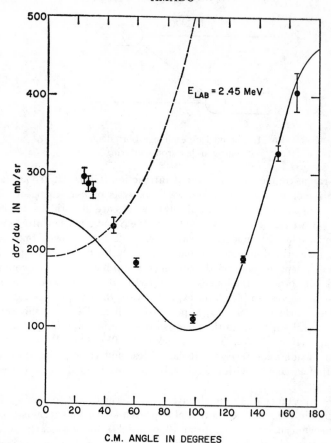

FIG. 2. Experimental (109) and theoretical (34) angular distribution for *n-d* scattering at 2.45 MeV. The dashed line is the Born approximation cross section corresponding to Fig. 1.

wave components. This can be repaired by introducing a parameter as we discussed above, but for positive energies the cross-section predictions are nearly independent of this parameter. This is in part a consequence of the dominance of the quartet channel.

The fits to *n-d* scattering obtained in this way by Aaron, Amado & Yam (34) for 2.45, 3.27, 5.5, 9.0, and 14 MeV are shown in Figures 2–6. Similar fits have been obtained by Phillips (35), but they are not quite as good since he does not include the high partial waves.

We see that in general the fit is quite good. At the smallest momentum transfers the separable-potential fit is low. In the forward direction the most important mechanism is the impulse approximation which in turns depends in this case on the forward-scattering part of the nucleon-nucleon *t* matrix.

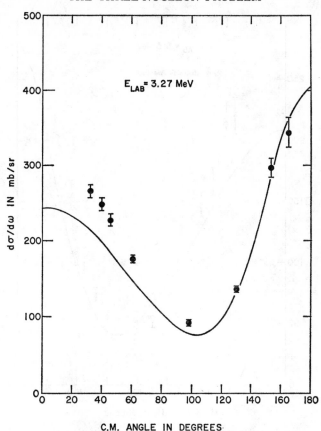

FIG. 3. Experimental (109) and theoretical (34) angular distribution for *n-d* scattering at 3.27 MeV.

The S-wave separable t matrix is not a good theory of this since the higher partial waves, or high-impact-parameter components are important in the forward direction. This defect could presumably be easily corrected by including a unitarized impulse-approximation correction. The small angle discrepancy appears to go to smaller angles or even to go away at higher energies because the cross section is plotted as a function of scattering angle rather than momentum transfer, since the same range of momentum transfer shrinks in angle with increasing energy. Also shown in Figure 2 is the Born term only. It has the characteristic backward peak of an exchange diagram, but is clearly not the whole story. The fact that a Born term is the dominant mechanism for a process does not make it a good approximation by itself. The Born terms for the quartet and doublet channels differ only by a multiplicative factor, but the effect of solving the full equation is markedly different in these two channels, as is seen in Figure 7 where the quartet and

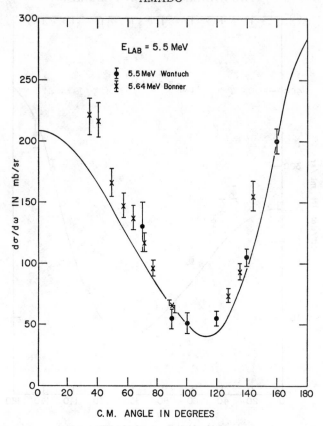

FIG. 4. Experimental (110, 111) and theoretical (34) angular distribution
for *n-d* scattering at 5.5 MeV.

doublet contributions are plotted separately for the 14.1 MeV case. The
quartet is clearly dominant. It has the backward peak of the exchange term.
Unitarity reduces that peak and adds a forward peak. The doublet channel
has lost all evidence of its backward Born-term origins and is diffractive,
presumably reflecting the importance of the virtual breakup process to
elastic scattering in this channel.

   The total *n-d* elastic and inelastic cross sections calculated with the
spin-dependent S-wave separable interactions are shown in Figure 8 compared with the data. The one-parameter adjustment discussed above has
been made so as to give the correct zero-energy cross section. The elastic
part is slightly low owing to the neglect of the large-impact-parameter parts
of the two-body force, which in turn leads to an underestimate of the forward cross section, but the general order of magnitude and trends are ex-

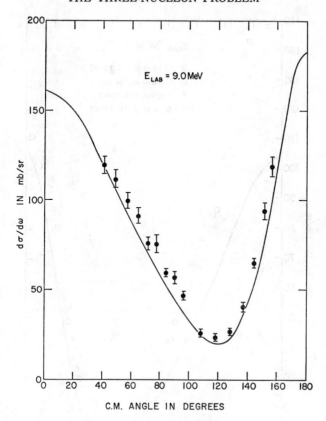

FIG. 5. Experimental (111) and theoretical (34) angular distribution
for *n-d* scattering at 9.0 MeV.

cellent, again indicating that the major features of the scattering are accounted for.

The surprisingly good account of low-energy *n-d* scattering given by the
simple *S*-wave spin-dependent separable-potential calculation is a mixed
victory. It comes about largely, as Christian & Gammel foresaw, from the
spreadout nature of the deuteron, and from the dominance of the relatively
uninteresting quartet channel. These features, coupled with the fragile
nature of the deuteron which makes virtual breakup important, mean that
a theory with simple dynamics, but correct treatment of the three-body
features, like the separable interaction theory, can and does account for
most of the physics. What is disappointing is that so little is then left for the
more sophisticated questions one might like to ask. In other words, low- and
medium-energy *n-d* elastic scattering does not appear to be a fruitful ground

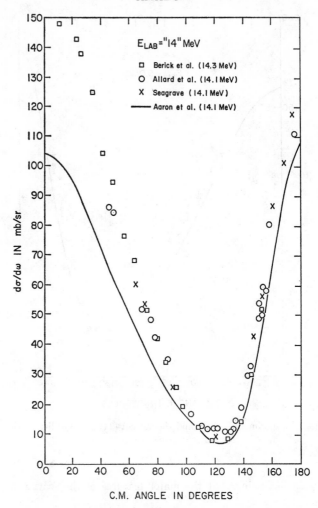

$E_{LAB} = "14" MeV$

☐ Berick et al. (14.3 MeV)
○ Allard et al. (14.1 MeV)
✕ Seagrave (14.1 MeV)
— Aaron et al. (14.1 MeV)

$d\sigma/d\omega$ IN mb/sr

C.M. ANGLE IN DEGREES

FIG. 6. Experimental (112–114) and theoretical (34) angular distribution for *n-d* scattering at "14" MeV.

for learning about the detailed features of three-body systems, or about the off-shell nature of two-body forces.

Breakup.—Deuteron breakup by protons or neutrons is more difficult than elastic scattering experimentally and theoretically, and has been less thoroughly explored. Experimentally there are many problems, not the least of which is the fact that the cross section is a function of many variables. Theoretically, there are serious problems of principle as well as practice, only some of which are solved by the Faddeev approach. So far the only

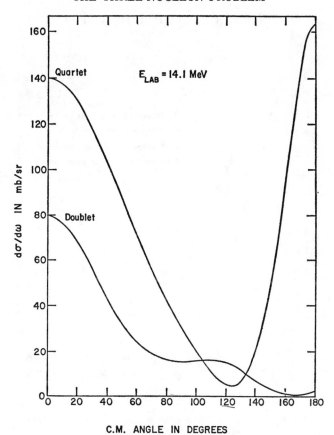

FIG. 7. Theoretical doublet and quartet contributions to the cross section of Fig. 6.

serious calculations have been for the reaction $n+d \rightarrow n+n+p$ at 14.4 MeV using S-wave spin-dependent separable interactions (117, 118). Even here there are technical problems connected with singularities of the kernel. Aaron & Amado (117), by adopting the contour deformation techniques first used for elastic scattering by Hetherington & Schick (119), were able to circumvent these difficulties and their breakup calculation must be considered more reliable than that of Phillips (118), who did not use this method.

The major feature of the deuteron breakup reaction by neutrons is the strong final-state interaction between the neutrons in the ^1S state. This interaction greatly enhances the cross section for the ejection of fast forward protons, and it has been hoped that the size and shape of this enhancement could provide information on the n-n scattering length. Previous attempts to

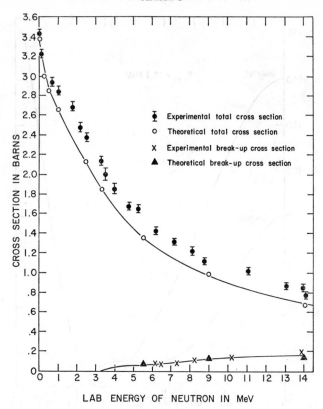

FIG. 8. Total and breakup cross sections. Theory from (34) and data from (115) and (116).

extract that scattering length from the breakup reaction have used some form of impulse approximation or Watson theory (cf. 120). One of the motivations of the exact three-body calculations of the breakup was to attempt to shed light on the scattering-length question, either by fitting the data for a particular neutron-neutron scattering length, or at least by testing in the context of an exact theory, the validity of the approximate Watson analysis.

The results of the 14.4 MeV calculation (117) are compared with the experimental results of Cerineo et al. in Figures 9 and 10 (120, 121). The effect of the strong *n-n* final-state interaction is clearly seen, particularly in the proton spectrum at the forward angle. At that angle the theory is shown for two popular values of the *n-n* scattering length and the fit is certainly not good enough to distinguish between them. How well this could be done with a better theory is unclear since the relatively poor experimental resolu-

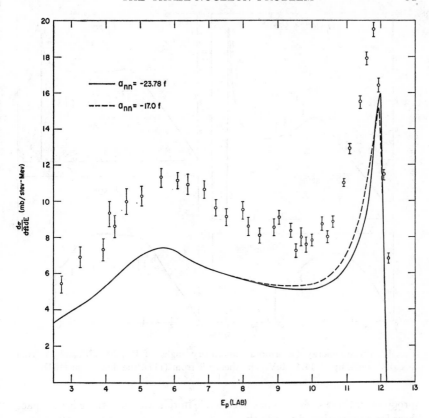

FIG. 9. Proton energy spectrum at 4.8° in the laboratory from deuteron breakup by neutrons at 14.4 MeV. The theoretical curves corresponding to two n-n scattering-length possibilities are from (117), the data from (120).

tion has not been folded into the theoretical curves. The general features of the breakup are given by the theory at all angles, but the detailed fit is not as good for the elastic data. This probably reflects the greater importance of the more sensitive doublet channel in the breakup. The low cross section at forward angles is also related to the low forward elastic cross section. Since the theory gives the correct total breakup cross section, but is low at forward proton angles, it must be high at backward angles and correct in between. This presumably explains the agreement in magnitude at 10°.

Since the theory is not sufficiently sophisticated to fit the data but is nonetheless an exact three-body calculation, it can be a testing ground for the usual phenomenological schemes to extract two-body parameters from multibody final states, and to shed light on the question of how two-body information is distributed over such a state (117, 122). Considerable

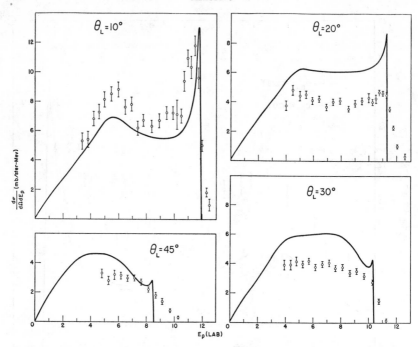

FIG. 10. Proton energy spectrum at laboratory angles of 10°, 20°, 30°, and 45° from
n-d break up at 14.4 MeV. The theory is from (117), the data from (121).

progress on these questions has been made in this way, but there is not space
here for a discussion of this work.

   Polarization.—After an uncertain start, nucleon polarization data in
n-d and p-d scattering is now fairly well known over a wide range of energies.
At low energy the polarization is quite small, as we mentioned above (123–
125), but it attains reasonable values and structure at medium energies
(126–134).[2] Even deuteron tensor polarizations are now known (135). Until
recently very little serious theoretical effort has been given to the polarizations. It might be hoped that careful theoretical study of the polarizations
may yield detailed information about two- and three-particle interactions,
but this possibility appears remote in view of the recent work of Purrington
& Gammel (136). They repeat the phase-shift analysis of Christian &
Gammel (89) using the nucleon exchange Born term of Figure 1. This time,
however, they include a D-wave component in the deuteron due to the
tensor force. They fit all but the lowest waves with the Born term, leaving
the lowest free. They get a very good fit to the polarization in p-d scattering

   [2] There are still some discrepancies in the small-angle polarization around 12 MeV
(131, 133, 134).

at 9 MeV in this way. Since the Born term is not sensitive to the details of the two-body force, but only to the rather well-known deuteron parameters, the success of Purrington & Gammel would indicate that the polarizations are not a sensitive source of two-body information, but like the elastic scattering, are a reflection of the gross properties of the deuteron and of the well-known low-energy nucleon-nucleon parameters.

Medium and high-energy scattering.—There has been considerable work on p-d and some work on n-d scattering at energies higher than we have been discussing (cf. 132). The medium-energy range (40–100 MeV) is the classic range for the impulse approximation. In fact, it was just for this problem that it was invented (137). Modern calculations using the impulse approximation do a fairly good job of accounting for the elastic scattering and polarization data. At the lower energies (40 MeV) one must take account of the multiple scattering corrections and off-mass shell effects, but at the higher energies the straight impulse approximation does quite well, particularly if the effects of the deuteron D state are included (138–141). There is some problem throughout this region with fitting the backward-scattering results, since there the exchange term of Figure 1. is dominant rather than the impulse term. Including this term in Born approximation makes a considerable qualitative improvement, but some disagreement remains to indicate that one should do better (140, 141).

Recent work in p-d scattering at much higher energies (1–2 BeV) (142, 143) has been analyzed using the Glauber approximation (144), a semiclassical method intended for strongly forward-peaked cross sections such as one finds in this case. Both the deuteron D state and double-scattering corrections seem to be important in the analysis, but when these are taken into account, the data is well represented (cf. 145, 146). In the backward direction there is also a peak in p-d scattering at these energies and it is tempting to assign it to the familiar exchange graph (Figure 1). However, with the usual deuteron wavefunctions this graph, even in Born approximation, gives too little backward cross section. Kerman & Kisslinger (147) have pointed out that a very small component of $N^*(1688)$ in the deuteron can account for about one half of the backward p-d scattering at this energy and hence solve the problem. In this way, backward p-d scattering may be a very sensitive measure of these esoteric components of the deuteron.

The work in these energy regions, although of great interest, seems to shed little light on the problems traditionally associated with the three-nucleon problem, and we will therefore not go into it further here.

4. ELECTROMAGNETIC AND WEAK PROBES

Since the electromagnetic and weak interactions can be treated in Born approximation, we can hope to get strong-interaction information about the three-body system by probing it with these interactions. To do this we must, of course, understand the form of the coupling between the probe and the three-body system. In electromagnetism it is convenient to dis-

tinguish electric and magnetic interactions. The forms of the electric couplings are well known, but their matrix elements turn out to contain little detailed three-body information. The magnetic terms always seem to be on the
verge of making important contributions on three-body questions when
their detailed form enters and meson exchange currents or interaction currents become important. Not nearly enough work has yet been done on this
potentially fruitful subject to allow detailed conclusions in most cases. Similar difficulties plague detailed understanding of weak interactions in the
three-nucleon system, although some recent work promises to shed light on
nonadditive contributions in β decay and μ capture.

Magnetic moments.—Since the three-nucleon bound states have spin $\frac{1}{2}$,
the only static moments are magnetic dipole. These are (in nuclear magnetons) for He^3 -2.12755 and for H^3 2.97885 (148). Further, since there are
no excited states of either nucleus there are no transition moments. The
simplest theory of these systems is one for which the magnetic-moment
operator is just the usual sum of single-nucleon operators

$$\mathfrak{y} = \sum_i \frac{1}{2}(1 - \tau_{3i})\mu_n \,\delta_i + \sum_i \frac{1}{2}(1 + \tau_{3i})(\mu_p \,\delta_i + l_i) \qquad 1.$$

where $\frac{1}{2}(1 \pm \tau_{3i})$ is the proton/neutron isotopic spin projection operator and
μ_p and μ_n are the proton/neutron moments. The major component of the
bound-state wavefunction is symmetric S state. For this case the orbital
part of μ is zero, and the like nucleons have their spins coupled to zero.
Hence this part of the wavefunction predicts $\mu_{He^3} = \mu_n$ and $\mu_{H^3} = \mu_p$. This is
obviously a good first orientation. In fact $\mu_{He^3}/\mu_n = 1.112$ and $\mu_{H^3}/\mu_p = 1.067$.
The simplest way to hope to account for the remaining discrepancy is with
other states, the most promising of which is the D state, since it adds an
orbital component, although the S' state also changes the magnetic moment
from the pure symmetric S prediction (cf. 2, 3, 149). Presumably the P
states are too small to matter. In fact adding in the D state *decreases* the
magnitude of the magnetic moments from the single-nucleon values, whereas
in both cases the moments are large. The most reasonable explanation of
this is that the magnetic-moment operator given in Equation 1 is too simple
and we must add in meson exchange corrections. Some insight into their
form comes from studying the sums of the moments. Because the nuclei form
an isodoublet, the sum picks out the isoscalar part of the moments. One gets

$$\mu_{H^3} + \mu_{He^3} = (\mu_p + \mu_n) \sum \delta_k + \sum l_k \qquad 2.$$

For the S state only (independent of the symmetry) $\mu_H + \mu_{He^3} = \mu_p + \mu_n$. In
fact, one has $(\mu_H + \mu_{He^3})/(\mu_p + \mu_n) = .968$. In this case the D state goes in the
correct direction. A 4 per cent D state (with no P state) gives the correct sum
of moments. Four per cent is somewhat small for the D state and therefore
there is presumably some small exchange-current-moment correction to the
sum, but no reliable estimate of the isoscalar exchange moment seems to

exist. The major correction to the individual moments is isovector. This is not surprising since the longest-range part of the meson current is carried by the isovector π mesons. The fact that the leading moment correction is isovector also accounts for the absence of a substantial interaction current-moment correction in the deuteron, which is $T=0$.

There have been a number of attempts to construct general phenomenological forms for the interaction moments (cf. 150) but there seem to be few serious attempts to derive the form and magnitude from an underlying meson theory. In 1947, Villars, using the meson theory and parameters known at the time, obtained in isovector interaction moment of the correct sign and order of magnitude (151). More recently, Brennan, Frank & Padgett (152, 153) have verified this result and extended it to include the effects of two-meson currents and currents connecting the S and D states. Arenhovel & Danos (154) have recently attempted to explain the moment anomaly in terms of a small admixture of $N^*(1470)$ in the three-body wavefunction. This is similar in spirit to the work of Kerman & Kisslinger (147) discussed in the previous section. So far these results for the moments are at the level of phenomenological plausibility, but clearly this is an interesting and novel approach that deserves further study. In general the results on meson effects are satisfactory and indicate that the moments can be accounted for in this general way, but no serious attempt to explore the model dependence of the results or to determine the effects of these calculations back on our knowledge of the three-body wavefunctions has been attempted. Hence a great deal remains to be done on this problem and hopefully something remains to be learned.

Form factors.—Elastic electron scattering from H^3 and He^3 can be interpreted in terms of electric and magnetic form factors just as in the nucleon case since in both cases we have spin $\frac{1}{2}$ systems. These form factors are essentially the Fourier transforms of the charge and magnetization density. As such, they should give detailed information on the structure of the bound states. However, since even the free proton and neutron have themselves charge and magnetic form factors and since, when they combine to form nuclei, the interactions between them and the meson currents can further change matters, the connection between the charge and magnetization density of the nucleus, and the nucleon position density, as described by a usual nonrelativistic wave function, is not straightforward.

Experiments to determine the form factors are not simple, since He^3 and tritium targets are difficult to come by and difficult to handle, but they have been carried out by the Stanford group (155, 156) The results are qualitatively easy to understand (155, 157). The charge and magnetic form factors of H^3 are quite similar, and are in turn rather similar to the magnetic form factor of He^3 whereas the charge form factor of He^3 is consistently smaller than the magnetic one. This is borne out in the rms radii shown in Table I. In H^3 the charge and magnetization are carried predominantly by the odd proton. In He^3 the magnetization is carried predominantly by the odd neu-

TABLE I

Root Mean Square Radii in Fermis[a]

	Charge	Magnetic
He[3]	$1.87 \pm .05$	$1.74 \pm .10$
H[3]	$1.70 \pm .05$	$1.70 \pm .05$

[a] Reference 156.

tron but the charge by the protons. If the wavefunctions of the bound state were completely symmetric S state, all the four form factors would be alike. That they are not is evidence against a completely symmetric state. So long as the wavefunctions are mirror identical (pure $T=\frac{1}{2}$), the spatial distribution of the odd or even nucleons in each will be the same, but since the force between like nucleons is somewhat weaker than between unlike nucleons, one expects the spatial distribution of like nucleons to be somewhat more extended. This makes the magnetic radius of He[3] larger than its charge radius and makes the magnetic form factor smaller than the charge form factor, in agreement with experiment. A corresponding analysis in terms of the difference of binding of the like or odd nucleon gives the same qualitative features of the radii and form factors (158). The average form factor out to moderate momentum transfers ($\sim4F^{-2}$) and the average rms radius seem to be fit by any reasonable model wavefunction or variational wavefunction (15, 16, 55, 159–161).[3] This is particularly so if short-range repulsion effects are included, since they tend to push the wavefunction out slightly and reduce the central nucleon density—hence increasing the form factor at larger momentum transfers (158, 162). At the very largest values of the momentum transfers (6–8 F^{-2}), relativistic effects (recoil) begin to come in and the relation of the form factor to the three-dimensional Fourier transform of the Schrödinger density becomes obscure.

More detailed wavefunction information can only come from a better analysis of the form factors. The simplest such analysis assumes that only the nucleons contribute and that they do so without mutual interference of distortion. This analysis is particularly appropriate to the charge form factors since it neglects exchange currents. These clearly do not contribute to the static charge form factors (zero momentum transfer) and it can be hoped that extrapolation of Siegert's theorem (163) to off-mass shell photons suppresses electric exchange effects for finite momentum transfers. Sarker (164–166) has raised doubts about the absence of exchange-current effects in the charge form factors at the larger momentum transfers, but most analyses

[3] The unusually large S' probability reported in the separable-potential calculation of form factors and wavefunctions by Amado (150) is due to a normalization error of 4. With this correction the S' probability becomes 2 per cent, in keeping with other estimates.

proceed without them (155–157, 167, 168). In part this is a necessity since if exchange-current contributions are included without a theory of their form, detailed wavefunction information cannot be extracted. If one adds to the impulse or additivity assumption the assumption that the wavefunctions of He³ and H³ are exactly mirror symmetric so the odd and like nucleons have the same Fourier transform of position density in each, F_0 and F_L respectively, the charge form factors can be written

$$F_{ch}(He^3) = F_{ch}(p)F_L + \tfrac{1}{2}F_{ch}(n)F_0$$

$$F_{ch}(H^3) = F_{ch}(p)F_0 + 2F_{ch}(n)F_L$$

3.

where $F_{ch}(p/n)$ are free proton/neutron charge form factors. Equation 3 can be used to construct F_0 and F_L from the data. However, since $F_{ch}(n)$ is not precisely known, there remain ambiguities in doing the analysis. To first approximation $F_{ch}(n) \sim 0$ and $F_L/F_0 = F_{ch}(He^3)/F_{ch}(H^3)$. For the reason we gave above this ratio is not one, and the difference between F_0 and F_L is a measure of the nonsymmetric components of the wavefunction (always subject to the assumption about exchange currents). The most appealing component is S' since it has spin $\tfrac{1}{2}$ and gives a nonvanishing cross term in the charge form factor with the predominant S state. Using reasonable-model wavefunction, Schiff found (157, 169) 4 per cent S' was needed to fit F_0 and F_L. (This was particularly disturbing since at that time variational calculations were giving only about 1 per cent; more recently they find around 2 per cent as do most separable-potential calculations.) More recently Gibson (168) has shown that the D states, even though they have no cross term with the S state in the charge form factors because of spin orthogonality, make an important contribution at small momentum transfers and that by using about 2 per cent S' and some small $T = \tfrac{3}{2}$ ($\sim$.25 per cent) component he can fit F_0 and F_L. This may be a rather large amount of $T = \tfrac{3}{2}$, although estimates of this are very difficult (8–10, 170), but Gibson shows that it is really the combination of S' and $T = \tfrac{3}{2}$ that matters and not each individually. Further, his analysis is subject to uncertainties in $F_{ch}(n)$, in the exchange currents, and in the understanding of the wavefunction model dependence. All one can say at present then is that reasonable wavefunctions of the type discussed in Section 2, are consistent with F_0 and F_L, but that better understanding of $F_{ch}(L)$ and the exchange currents is needed before the charge form factors can be made to yield detailed constraints on the three-body wavefunction.

The magnetic form factors are more difficult to analyze. In this case nucleon orbital motion and exchange currents make contributions to the static nucleon moments, and hence must be included in the form factor analysis. The simplest assumption is that the nucleon position density can be obtained from the charge form factor analysis and the magnetic form factor used to extract exchange-current information. Levinger & Srivastava (167) do this and show that the isovector and isoscalar exchange form factors needed to fit the data are reasonable and in particular the isovector one fits the general form calculated by Brennan, Frank & Padgett (153). However, there remains much more to be done in turning this information into detailed understanding of the three-body system.

n-d *Capture.*—The first electromagnetic reaction in the three-nucleon system we consider is thermal n-d capture. This reaction has a very small cross section. For example for 2.2 km/sec neutrons, the capture cross section is $\sim$.5 mb whereas the n-p capture cross section at this velocity is 330 mb (171). This small rate for n-d capture is a consequence of the dominant symmetric S nature of the triton as was first realized by Schiff (172). Thermal capture proceeds from the S state of the n-d system and is M-1 in character. The M-1 operator is related to the static magnetic dipole operator, and if we assume nucleon additivity has the form given in Equation 1. For capture to the S-wave part of the triton the orbital term does not contribute and the operator acts only in spin space. If the nuclear Hamiltonian were spin dependent, the spin and orbit parts would factor exactly. Then since the orbit parts of the bound state and the continuum doublet would belong to the same Hamiltonian, but correspond to different energies, capture from the doublet would be zero, owing to the orthogonality of the spatial wavefunctions. The spin part of the quartet continuum state is totally symmetric, and the isospin part for the n-d system is of mixed symmetry, therefore the orbital part of the quartet has mixed symmetry and is orthogonal to the completely symmetric S state of the triton. The presence of spin dependence in the Hamiltonian mixes S' into the triton, making the quartet capture state different from zero. It also spoils the factorization of the doublet states (since there is more than one linearity-independent spin state of $S=\frac{1}{2}$), and hence gives capture from the doublet continuum state. In both cases, since the capture proceeds by virtue of the spin dependence of the nuclear Hamiltonian, the rate can be expressed in terms of the S' probability in the triton. Hence in the absence of exchange currents, the capture rate is a measure of this probability.

Simple-model calculation of the capture rate without exchange currents and neglecting or making very simple assumptions about the effects of interactions in the n-d state allows very little S' (1–.01 per cent), with considerable variation depending on the assumptions of the model (cf. 173–175). However, n-d interactions are important, and have been estimated by a elastic-unitarity-dispersion-theory approach to suppress the no interaction rate by factors $\sim$.03 (176, 177). Furthermore, exchange-current effects are also important (178, 179). In a simple model of the capture rate, they are of the same order of, and opposite in sign to, the normal contribution from the S' state (175). However, this model neglects the detailed effects of distortions in the n-d state. In an exact separable-potential calculation, Phillips (180) finds that interactions in the initial state reduce the S' contribution so much as to actually reverse the sign and make it interfere constructively with the exchange contribution. This result may itself be very model dependent, but what does seem clear is that interaction effects in the n-d system reduce the S' contribution to n-d thermal capture so much that the rate cannot be used to learn about this state. It may be a fertile source of information through the exchange current, but few serious efforts have been made to extract that information.

Photobreakup.—Photons on the three-nucleon bound state can lead to two-body (nucleon plus deuteron) or three-body final states. The usual reasons combined with the special dynamical suppression of M-1 transition discussed in the previous section make the dominant mechanism E-1 and as the energy goes up E-2. Siegert's theorem (163, 181) then guarantees that the interaction is "$E \cdot r$" and that there are no exchange terms, etc. The fact that "$E \cdot r$" rather than "$p \cdot A$" should be used is particularly relevant to the separable-potential calculation since the nonlocal potentials will give extra terms if the usual $p \rightarrow p\text{-}eA$ prescription is used.

Since "r" is such a simple operator, one will expect the photobreakup reactions to be fairly model independent. Mostly total cross sections have been measured (a comprehensive review of results and theoretical attempts to explain them is in Fetisov et al.) (182, 183) and it is found that two- and three-body breakup cross sections are of the same order of magnitude with roughly the same energy dependence. Early theoretical attempts to fit the data for the two-body breakup do give the general shape and the absolute magnitude to within 20 or 30 per cent (184–187). These calculations usually neglect the final nucleon-deuteron interaction since it is a p-wave state and also neglect deuteron distortion. Unfortunately, many calculations also attempt to introduce more sophisticated bound-state wavefunctions to see if agreement can be improved. Although it is certainly the case that the "r" operator makes the matrix element sensitive to wavefunctions of the correct asymptotic form, the effects of final-state interaction even in the two-body channel must be considered at the same time as more sophisticated bound-state wavefunctions before it is possible to draw deep conclusions about them. In fact, we would hope that a correct treatment of the final state and of the ground state would explain the data without the need of any sophisticated components in the bound states. Barbour & Phillips (188) have done just such a calculation using a final-state wavefunction generated by a spin-dependent S-wave separable-potential calculation and the symmetric S part of the bound state, and find excellent agreement with the two-body breakup cross section in energy dependence and magnitude.

For three-body photobreakup the neglect of final-state interactions is much more disastrous. Plane-wave calculations overestimate the cross section by factors of 2 or 3 (182, 184). Nucleon-nucleon correlation data on the final state makes it clear that ^1S interactions are important, as we would expect in view of the ^1S scattering length. Introducing them in an impulse approximation improves agreement with the shape of the correlation, but leaves the magnitude of the total cross section overestimated by a factor of 2 (189, 190). Some improvement can be achieved by careful treatment of the asymptotic form of the bound state (187, 191, 192), but this is an illusory gain as has recently been pointed out by Gerasimov and Barton (193, 194). They apply the well-known bremsstrahlung weighted sum rule to the total E-1 cross section. This gives

$$J_T = \int_0^\infty \sigma_{E1}(E_\gamma) \frac{dE_\gamma}{E_\gamma} = \frac{4}{3} \pi^2 \alpha \left(\frac{2}{3} r_{pn} \right)^2 \qquad 4.$$

where α is the fine-structure constant and r_{pn} the rms distance between the center of the neutron and proton distributions in the bound state. This sum rule depends only on the applicability of Siegert's theorem to the E-1 operator. For a spatially symmetric state $\frac{2}{3} r_{pn}$ is the rms charge radius for point protons. To evaluate J_T from the data, one must remove the E2 cross section. Fetisov et al. (182) evaluate the sum to an upper limit of 170 MeV, which should be enough, and find $J_T = 2.53 \pm .19$ mb. This gives, if we assume a symmetric wavefunction, a charge radius of 1.62 ± 0.06 F or, including the finite proton size, 1.81 ± 0.06 F, in general agreement with the electron scattering results. It can also be written

$$J_T = J_2 + J_3 \qquad 5.$$

where J_2 is the contribution for two-body breakup and J_3 from three-body breakup. Since all quantities are positive, $J_T > J_3$. Further, since the plane waves form a complete set by themselves, calculating the sum rule with plane-wave final-state cross sections gives J_T not J_3. Thus unless one uses a final-state wavefunction that "knows" about the two-body channel, one will always overestimate the three-body cross section. In fact it turns out that $J_3 \simeq J_2$, so that $J_T \simeq 2J_3$, which accounts for the large error of the calculations that neglect the two-body channel. Gerasimov & Barton also analyze the sum rules by isotopic spin. The two-body channel is pure $T = \frac{1}{2}$, the three-body channel can be $T = \frac{1}{2}$ or $\frac{3}{2}$. They write

$$J_T = J_{1/2} + J_{3/2} \qquad 6.$$

The equality of J_3 and J_2 and the dominant symmetry of the ground state lead to the conclusion that $J_{1/2} = J_2$ and $J_3 = J_{3/2}$, so that the $T = \frac{1}{2}$ component of the three-body breakup is zero. This surprising result has recently been explained by Prats, Lehman & O'Connell (195) in terms of the symmetries of the continuum wavefunction. Of course, if one does a full three-body calculation, that correctly accounts for the two-body and three-body final state, all these problems disappear. Barbour & Phillips (188) using their S-wave spin-dependent separable-potential wavefunctions calculate three-body breakup, and get substantially correct results. Thus we return to our original prejudice that the photodisintegration reactions are relatively model independent and do not yield detailed information on the three-body system, so long as the calculation is done reasonably.

Electrodisintegration.—Inelastic electron scattering from the three-nucleon bound state leading to nucleon plus deuteron or three free nucleons is analogous to photodisintegration. The important difference is that since the virtual photon need not have zero mass, one gets added information about the transition form factors in the three-body system. The experiments are very difficult and few have been carried out (196, 197). As usual these have been analyzed in terms of various bound-state models, but with neglect of final-state interactions (198–200). This is probably more justified in the two-body than the three-body disintegration. The conclusions are that the dominant features can be fit by a reasonable wavefunction, and are probably independent of detailed, if reasonable, assumptions. Other features of the data may well depend on the details of the three-body system, but neither the analysis nor the experiment is yet in a state to extract them.

Beta decay.—The neutron-proton mass difference just overcomes the Coulomb repulsion in three-body nuclei so that the triton beta decays to He^3 with a very small energy release [18.6 keV (1)]. The ft value for the decay is 1137 ± 20 sec (201), with the major uncertainty coming from electron endpoint energy determination. Since the decay proceeds between mirror states of spin and isotopic spin $\frac{1}{2}$, it should closely resemble neutron decay. The best value for the neutron ft is 1131 ± 18 sec (202), with the error reflecting the difficulty of the lifetime measurement itself. Since the triton bound state is not pure symmetric S state, this agreement of neutron and triton ft values is actually too good. Using standard weak-interaction theory (cf. 203) the ratio of the ft values can be written

$$\frac{(ft)_n}{(ft)_{H^3}} = \frac{|M_v|_{H^3}^2 G_v^2 + |M_A|_{H^3}^2 G_A^2}{|M_v|_n^2 G_v^2 + |M_A|_n^2 G_A^2} \qquad 7.$$

where $|M_v|^2$ and $|M_A|^2$ are the appropriate vector and axial-vector matrix elements, and G_v^2 and G_A^2 the coupling constants. For the neutron $|M_v|^2 = 1$ and $|M_A|^2 = 3$. The assumption of conserved vector current (*CVC*) makes $|M_v|_{H^3}^2 = 1$ (up to very small Coulomb corrections). In particular there are no meson exchange-current corrections to $|M_v|^2$. Since the ratio of ft's is nearly one, one finds $|M_A|_{H^3}^2 = 3 \pm .12$. In the simplest form of weak-interaction theory, the decay is due to the decay of "free" nucleons (impulse approximation) and the axial-vector operator for allowed decays is σ. In this case $|M_A|^2 = 3$ only for pure symmetric S state and is < 3 for any reasonable mixture of states in which the symmetric S state (S) dominates. For a mixture of S, S', and D one finds (204)

$$\left| M_A \right|^2_{H^3\ \text{impulse}} = 3 \left(1 - \frac{4}{3} P_{S'} - \frac{2}{3} P_D \right)^2 \qquad 8.$$

where $P_{S'}$ and P_D are the S' and D probabilities. With $P_{S'} = 2$ per cent and $P_D = 7$ per cent, reasonable values as we have seen in Section 2, we get $|M_A|_{H^3\ \text{impulse}}^2 = 2.57$. To bring the axial-vector matrix element back to 3, one must call on meson exchange contributions (205). Early estimates of these seemed to indicate they may be too small. This difficulty was particularly acute in phenomenological analysis based on earlier neutron ft values (206–208). Recently the meson exchange effects have been re-estimated with the inclusion of known baryon and meson resonances by Cheng (209). He finds that the meson exchange effects are of the right order of magnitude and sign and therefore there is no discrepancy. A more ambitious program to evaluate these exchange effects has been undertaken by Kim & Primakoff (210, 211) using elementary-particle methods. They use partial conservation of the axial current ($PCAC$) and the Goldberger-Treiman relations to calculate the axial current "renormalization." This they relate to π-nucleus coupling which they in turn express in terms of the isovector anomalous magnetic moments. For H^3 this gives $|M_A|_{H^{32}} = 3.06$ including the effects of the mixed-symmetry components in the wavefunction. The systematic application of these techniques to complex nuclei is a very interesting new avenue which brings in many new and largely unexplored topics, π-nucleus scattering for example, but a discussion of it would take us too far afield. A similarly promising approach for investigating the exchange effects currently under investigation by Cheng is the use of soft-pion techniques. The eventual unraveling of all these questions not only will shed light on the corrections to nucleon additivity, but will give further information on the structure and components of the three-body bound state.

μ *Capture.*—The capture of μ^- mesons on He^3 can lead to three final states

$$\mu^- + He^3 \rightarrow H^3 + \nu$$
$$\rightarrow n + d + \nu$$
$$\rightarrow n + n + p$$

The first reaction is the easiest to study, both experimentally and theoretically. Two recent measurements of its rate are 1465 ± 67 sec^{-1} (212) and 1505 ± 46 sec^{-1} (213); the total rate $\mu^- + He^3 \rightarrow$ anything has also been determined as $2170 \pm^{170}_{430}$ sec^{-1} (213).

Assuming μ-e universality, μ capture is closely related to tritium beta decay. The main difference is the large momentum transfer involved. This has two effects. Firstly, form factors appear, and secondly, new invariant terms, of which the most important is the induced pseudoscalar coupling, also appear. The capture can be analyzed by parametrizing these in terms of a phenomenological coupling as in the standard theory of beta decay (cf. 214), and one can get reasonable answers in this way by making reasonable wavefunction assumptions (215–218). Even the two- and three-body breakup channels can be treated, although as usual plane-wave final-state calculations are probably suspect (219). However, the model dependence of many of these calculations is often not carefully investigated.

An interesting approach to the capture which stresses the model independence has been given by Kim & Primakoff using their "elementary-particle" approach to weak interactions (211, 220). For capture leading to H^3 they use CVC to relate the form factors to the electromagnetic form factors, μ-e universality to relate the coupling strength to the triton beta decay, and $PCAC$ and the Goldberger-Treiman relation to get at the pseudoscalar term. The results are excellent. They find 1510 ± 4 sec^{-1} for the capture rate to H^3, and are presently using the formalism to test detailed forms of the assumption (221). Perhaps more interesting in our context than the good results is the correlation they achieve between the various three-body aspects. In one sense, relating the μ capture to the beta decay and electromagnetic form factors gives no three-body information, but in another sense, it points out what there is to explain. Unfortunately many three-body investigations do not attempt to relate the various aspects of the problem in this way. The elementary-particle theory, combined with current algebra, can also be made to give the total μ-capture rate with good agreement with experiment (222), but a detailed discussion of this would take us too far afield.

5. OTHER REVIEWS

The important theoretical developments which have, in large measure, been responsible for the results presented here have recently been reviewed in a number of articles. A general review of modern theoretical developments in the nonrelativistic three-body problem is contained in my 1967 Brandeis Summer School Lectures (223) and in the recent monograph by Watson & Nuttall (224). A review of the "state of the art" can be found in the proceedings of the three-body conference of 1968 (225). Duck (226) has also reviewed three-body theory with particular emphasis on separable-potential methods. Two recent reviews on the three-nucleon problem have also been prepared. Mitra (227) has reviewed in considerable detail the accomplishments of his separable-potential approach for the three-body bound state. Delves & Phillips (104), in a much more comprehensive review, give an excellent survey of the entire problem.

ACKNOWLEDGMENTS

I am very grateful to Dr. L. M. Delves for sending me the results of their latest variational calculation prior to publication and to Dr. Delves and Dr. A. C. Phillips for a prepublication copy of their fine review of the three-nucleon problem. Many conversations with Professor H. Primakoff, particularly on the role of the weak interactions in the three-body system, have been most enlightening.

LITERATURE CITED

1. Mattauch, J. H. E., Thiele, W., Wapstra, A. H., *Nucl. Phys.*, **67**, 1 (1965)
2. Sachs, R. G., *Nuclear Theory* (Addison-Wesley, Cambridge, Mass., 1953)
3. Verde, M., The Three-Body Problem in Nuclear Physics, in *Enclycopedia of Physics* (Flügge, S., Ed., Springer-Verlag, Berlin, 1957)
4. Gerjuoy, E., Schwinger, J., *Phys. Rev.*, **61**, 138 (1942)
5. Derrick, G. H., Blatt, J. M., *Nucl. Phys.*, **8**, 310 (1958)
6. Derrick, G. H., *Nucl. Phys.*, **18**, 303 (1960)
7. Derrick, G. H., *Nucl. Phys.*, **16**, 405 (1960)
8. Werntz, C., Valk, H. S., *Phys. Rev. Letters*, **14**, 910 (1965)
9. Ohmura, H., *Progr. Theoret. Phys.*, **38**, 626 (1967)
10. Ohmura, H., *Contrib. Intern. Conf. Nucl. Struct. Tokyo, 1967*, 24
11. Thomas, L. H., *Phys. Rev.*, **47**, 903 (1935)
12. Pease, R. L., Feshbach, H., *Phys. Rev.*, **81**, 1422 (1951); **88**, 945 (1952)
13. Blatt, J. M., Derrick, G. H., Lyness, J. N., *Phys. Rev. Letters*, **8**, 323 (1962)
14. Blatt, J. M., Delves, L. M., *Phys. Rev. Letters*, **12**, 544 (1964)
15. Delves, L. M., Blatt, J. M., *Nucl. Phys.*, **A98**, 503 (1967)
16. Davis, B., *Nucl. Phys.*, **A103**, 165 (1967)
17. Delves, L. M., Blatt, J. M., Pask, C., Davies, B., *Phys. Letters*, **26B**, 472 (1969)
18. Delves, L. M., in *Three-Particle Scattering in Quantum Mechanics* (See Ref. 225)
19. Loiseau, B. A., Nogami, Y., *Nucl. Phys.*, **B2**, 470 (1967)
20. Pask, C., *Phys. Letters*, **25B**, 78 (1967)
21. Brown, G. E., Green, A. M., Gerace, W. J., *Nucl. Phys.*, **A115**, 435 (1968)
22. McKellar, B. H. J., Rajaraman, R., *Phys. Rev. Letters*, **21**, 450 (1968)
23. Choudhury, S. R., *Phys. Rev. Letters*, **22**, 234 (1969)
24. Gupta, V. K., Bhakar, B. S., Mitra, A. N., *Phys. Rev. Letters*, **15**, 974 (1965)
25. Noyes, H. P., Fiedeldey, H., Calculations of Three-Nucleon Low Energy Parameters, in *Three-Particle Scattering in Quantum Mechanics* (See Ref. 225)
26. Humberston, J. W., Hall, R. L., Osborn, T. A., *Phys. Letters*, **27B**, 195 (1968)
27. Noyes, H. P., in *Proc. 2nd Intern. Symp. Polarization Phenomena of Nucleons, Karlsruhe, 1965*, 238 (Hubert, P., Schopper, H., Eds., Birkhauser Verlag, Basle und Stuttgart, 1966)
28. Tabakin, F., *Ann. Phys. (N. Y.)*, **30**, 51 (1964)
29. Strobel, G. L., *Nucl. Phys.*, **A116**, 465 (1968)
30. Mongan, T. R., *Phys. Rev.*, **175**, 1260 (1968)
31. Faddeev, L. D., *Zh. Eksperim. Teor. Fiz.*, **39**, 1459 (1960) [Engl. transl., *Soviet Phys. JETP*, **12**, 1014 (1961)]; Faddeev, L. D., *Mathematical Aspects of the Three-Body Problem in the Quantum Scattering Theory* (Israel Program for Sci. Transl., Jerusalem, 1965)
32. Mitra, A. N., *Nucl. Phys.*, **32**, 529 (1962)
33. Sitenko, O. G., Kharchenko, V. F., *Nucl. Phys.*, **49**, 15 (1963)
34. Aaron, R., Amado, R. D., Yam, Y. Y., *Phys. Rev*, **140**, B1291 (1965)
35. Phillips, A. C., *Phys. Rev.*, **142**, 984 (1966)
36. Tabakin, F., *Phys. Rev.*, **137**, B75 (1965)
37. Bhakar, B. S., Mitra, A. N., *Phys. Rev. Letters*, **14**, 143 (1965)
38. Borysowicz, J., Dabrowski, J., *Phys. Letters*, **24B**, 125 (1967)
39. Sitenko, O. G., Kharchenko, V. F., Petrov, N. M., *Phys. Letters*, **21**, 54 (1966)
40. Petrov, N. M., Storozhenko, S. A., Kharchenko, V. K., *J. Nucl. Phys. (USSR)*, **6**, 466 (1967) [Transl., *Soviet J. Nucl. Phys.*, **6**, 340 (1968)]
41. Kharchenko, V. F., Petrov, N. M., Storozhenko, S. A., *Nucl. Phys.*, **A106**, 464 (1968)
42. Schrenk, G. L., Mitra, A. N., *Phys. Rev. Letters*, **19**, 530 (1967)
43. Dabrowski, J., Dworzecka, M., *Phys. Letters*, **26B**, 4 (1968)
44. Shanley, P. E., *Phys. Rev. Letters*, **21**, 627 (1968)
45. Ohmura, H., Ohmura, T., *Phys. Rev.*, **128**, 729 (1962)
46. Tang, Y. C., Herndon, R. C., *Phys. Letters*, **18**, 43 (1965)

47. Khanna, F. C., *Nucl. Phys.*, **A97, 417** (1967)
48. Mathur, V. S., Lagu, A. V., *Nucl. Phys.*, **A118**, 369 (1968)
49. Okamoto, K., Lucas, C., *Nucl. Phys.*, **B2**, 347 (1967)
50. Okamoto, K., Lucas, C., *Phys. Letters*, **26B**, 188 (1968)
51. Stevens, M. St. J., *Phys. Letters*, **19**, 499 (1965)
52. Gupta, V. K., Mitra, A. N., *Phys. Letters*, **24B**, 27 (1967)
53. Mitra, A. N., *Nuovo Cimento*, **58B**, 344 (1968)
54. Pappademos, J. N., *Nucl. Phys.*, **42**, 122 (1963); **56**, 351 (1964)
55. Phillips, A. C., *Phys. Rev.*, **145**, 733 (1966)
56. Adya, S., *Phys. Rev.*, **166**, 991 (1968)
57. Folk, R. T., *Phys. Letters*, **28B**, 159 (1968)
58. Kim, C. C., Bunch, S. M., Devins, D. W., Forster, H. H., *Phys. Letters*, **22**, 314 (1961)
59. Ajdacic, V., Cerineo, M., Lalovic, B., Paic, G., Slaus, I., Tomas, P., *Phys. Rev. Letters*, **14**, 444 (1965)
60. Anderson, J. D., Wong, C., McClure, J. W., Pohl, B. A., *Phys. Rev. Letters*, **15**, 66 (1965)
61. Cookson, J. A., *Phys. Letters*, **22**, 612 (1966)
62. Thornton, S. T., Bair, J. K., Jones, C. M., Willard, H. B., *Phys. Rev. Letters*, **17**, 701 (1966)
63. Fuschini, E., Maroni, C., Uguzzoni, A., Verondini, E., Vitale, A., *Nuovo Cimento*, **48B**, 190 (1967)
64. Debertin, K., Rossel, E., *Nucl. Phys.*, **A107**, 693 (1968)
65. Fujikawa, K., Morinaga, H., *Nucl. Phys.*, **A115**, 1 (1968)
66. Harbison, S. A., Kingston, F. G., Johnston, A. R., McClatchie, E. A., *Nucl. Phys.*, **A103**, 478 (1968)
67. Ohlsen, G. G., Stokes, R. H., Young, P. G., *Phys. Rev.*, **176**, 1163 (1968)
68. Olsen, D. K., Brown, R. E., *Phys. Rev.*, **176**, 1192 (1968)
69. Tombrello, T. A., Slobodrian, R. J., *Nucl. Phys.*, **A111**, 236 (1968)
70. Okamoto, K., Davies, B., *Phys. Letters*, **24B**, 18 (1967)
71. Barbi, M., *Nucl. Phys.*, **A99**, 522 (1967)
72. Mitra, A. N., Bhasin, V. S., *Phys. Rev. Letters*, **16**, 523 (1966)
73. Jacob, H., Gupta, V. K., *Phys. Rev.*, **174**, 1213 (1968)
74. Delves, L. M., Derrick, G. H., *Ann. Phys. (N.Y.)*, **23**, 133 (1963)
75. Bodmer, A. R., Ali, S., *Nucl. Phys.*, **56**, 657 (1964)
76. Murphy, J. M., Rosati, S., *Nucl. Phys.*, **63**, 625 (1965)
77. Rosati, S., Barbi, M., *Phys. Rev.*, **147**, 730 (1966)
78. Folk, R., *Nucl. Phys.*, **85**, 449 (1966)
79. Homan, D. H., Kok, L. P., Van Wageningen, R., *Nucl. Phys.*, **A117**, 231 (1968)
80. Smith, F. T., *Phys. Rev.*, **120**, 1058 (1960)
81. Dragt, A. J., *J. Math. Phys.*, **6**, 533 (1965)
82. Simonov, Yu. A., Badalyan, A. M., *J. Nucl. Phys. (USSR)*, **5**, 88 (1967) [Engl. transl., *Soviet J. Nucl. Phys.*, **5**, 60 (1967)]
83. Amado, R. D., Schneider, T. (To be published)
84. Hurst, D. G., Alcock, J., *Can. J. Phys.*, **29**, 36 (1951)
85. Fermi, E., Marshall, L., *Phys. Rev.*, **75**, 578 (1949)
86. Alfimenkov, V. P., Lushchikov, V. I., Nikolenko, V. G., Taran, Yu. V., Shapiro, F. L., *Phys. Letters*, **24B**, 151 (1967)
87. Van Oers, W. T. H., Seagrave, J. D., *Phys. Letters*, **24B**, 562 (1967)
88. Seagrave, J. D., Van Oers, W. T. H., On the Neutron-Deuteron Scattering Lengths and S-wave Phase Shifts, in *Few Body Problems, Light Nuclei and Nuclear Reactions* (Paic, G., Slaus, T., Eds., Gordon & Breach, New York, 1968)
89. Christian, R. S., Gammel, J. L., *Phys. Rev.*, **91**, 100 (1953)
90. Spruch, L., Rosenberg, L., *Phys. Rev.*, **116**, 1034 (1959); Rosenberg, L., Spruch, L., O'Malley, T. F., *Phys. Rev.*, **118**, 184 (1960); Rosenberg, L., Spruch, L., *Phys. Rev.*, **121**, 1720 (1961); Han, Y., O'Malley, T. F., Spruch, L., *Phys. Rev.*, **130**, 381 (1963)
91. Delves, L. M., Lyness, J. N., *Nucl. Phys.*, **45**, 296 (1963)
92. Sartori, L., Rubinow, S. T., *Phys. Rev.*, **112**, 214 (1958)
93. Efimov, Y. N., *J. Exptl. Theoret. Phys. (USSR)*, **35**, 137 (1958) [Engl. transl. *Soviet Phys. JETP*, **8**, 98 (1959)]
94. Burke, P. G., Haas, F. A., *Proc. Roy. Soc. A*, **252**, 177 (1959)
95. Humberston, J. W., *Phys. Letters*, **10**, 207 (1964)
96. Pett, T. G., *Phys. Letters*, **24B**, 25 (1967)

97. Mitra, A. N., Bhasin, V. S., *Phys. Rev.*, **131**, 1265 (1963)
98. Bhasin, V. S., Schrenk, G. L., Mitra, A. N., *Phys. Rev.*, **137**, B398 (1965)
99. Segre, G., *Nuovo Cimento*, **38**, 422 (1965)
100. Phillips, A. C., Barton, G., *Phys. Letters*, **26B**, 378 (1969)
101. Reiner, A. S. (To be published)
102. Delves, L. M., Lyness, J. N., Blatt, J. M., *Phys. Rev. Letters*, **12**, 542 (1964)
103. Phillips, A. C., *Nucl. Phys.*, **A107**, 209 (1968)
104. Delves, L. M., Phillips, A. C., The Present Status of the Nuclear Three-Body Problem, *Rev. Mod. Phys.* (To be published)
105. Amado, R. D., *Phys. Rev.*, **132**, 485 (1963)
106. Lovelace, C., in *Strong Interactions and High Energy Physics* (Moorhouse, R. G., Ed., Plenum Press, New York, 1964); *Phys. Rev.*, **135**, B1225 (1964)
107. de Brode, A. H., Massey, H. S. W., *Proc. Phys. Soc. (London)*, *A*, **68**, 769 (1955)
108. Van Oers, W. T. H., Brocklman, K. W., Jr., *Nucl. Phys.*, **A92**, 561 (1967)
109. Seagrave, J., Cranberg, L., *Phys. Rev.*, **105**, 1816 (1957)
110. Wantuch, E., *Phys. Rev.*, **84**, 169 (1951)
111. Bonner, B. E. (Thesis, Rice Univ., 1965, unpublished)
112. Allard, J. C., Armstrong, A. H., Rosen, L., *Phys. Rev.*, **91**, 90 (1953)
113. Seagrave, J. D., *Phys. Rev.*, **97**, 757 (1954)
114. Berick, A. C., Riddle, R. A. J., York, C. M., *Phys. Rev.*, **174**, 1105 (1968)
115. Howerton, R. J., *Tabulated Neutron Cross Sections, Univ. Calif. Radiation Lab. Rept. UCRL 5573* (1961, unpublished)
116. Catron, H. C., Goldberg, M. D., Hill, R. W., LeBlanc, J. M., Stoering, J. P., Taylor, C. J., Williamson, M. A., *Phys. Rev.*, **123**, 218 (1961)
117. Aaron, R., Amado, R. D., *Phys. Rev.*, **150**, 857 (1966)
118. Phillips, A. C., *Phys. Letters*, **20**, 50 (1966)
119. Hetherington, J. H., Schick, L. H., *Phys. Rev.*, **137**, B935 (1965)
120. Cerineo, M., Ilakovac, K., Slaus, I., Tomas, P., Valkovic, V., *Phys. Rev.*, **133**, B948 (1964)
121. Ilakovac, K., Kuo, L. G., Petravic, M., Slaus, I., Tomas, P., *Nucl. Phys.*, **43**, 254 (1963)
122. Amado, R. D., *Phys. Rev.*, **158**, 1414 (1967)
123. Cranberg, L., *Phys. Rev.*, **114**, 174 (1959)
124. Brullmann, M., Gerber, H. J., Meierand, D., Scherrer, P., *Helv. Phys. Acta*, **32**, 511 (1959)
125. Chalmers, R. A., Cox, R. S., Seth, K. K., Strait, E. N., *Nucl. Phys.*, **62**, 497 (1965)
126. Walber, R. L., Kelsey, C. A., *Nucl. Phys.*, **46**, 66 (1963)
127. Conzett, H. E., Goldberg, H. S., Shield, E., Slobodrian, R. J., Yamabe, S., *Phys. Letters*, **11**, 68 (1964)
128. Conzett, H. E., Igo, G., Knox, W. J., *Phys. Rev. Letters*, **12**, 222 (1969)
129. Malanify, J. J., Simmons, J. E., Perkins, R. B., Walter, R. L., *Phys. Rev.*, **146**, 632 (1966)
130. Gruebler, W., Haeberli, W., Extermann, P., *Nucl. Phys.*, **77**, 394 (1966)
131. McKee, J. S. C., Luccio, A. U., Slobodrian, R. J., Tivol, W. F., *Nucl. Phys.*, **A108**, 177 (1968)
132. Bunker, S. N., Cameron, J. M., Carlson, R. F., Richardson, J. R., Tomas, P., Van Oers, W. T. H., Verba, J. W., *Nucl. Phys.*, **A113**, 401 (1968)
133. Clegg, T. B., Plattner, G. R., Haeberli, W., *Nucl. Phys.*, **A119**, 238 (1968)
134. Faivre, J. C., Garreta, D., Jungerman, J., Papineau, A., Sura, J., Tarrats, A., *Nucl. Phys.* (To be published)
135. Young, P. G., Ivanovitch, M., *Phys. Letters*, **23**, 361 (1968)
136. Purrington, R. D., Gammel, J. L., *Phys. Rev.*, **168**, 1174 (1968)
137. Chew, G. F., *Phys. Rev.*, **80**, 196 (1950); Chew, G. F., Wick, G. C., *Phys. Rev.*, **85**, 636 (1952); Chew, G. F., Goldberger, M. L., *Phys. Rev.*, **87**, 778 (1952)
138. Kowalski, K. L., Feldman, D., *Phys. Rev.*, **130**, 276 (1963)
139. Queen, N. M., *Nucl. Phys.*, **55**, 177 (1964)
140. Benoist-Gueutal, P., Gomez-Gimeno, F., *Phys. Letters*, **13**, 68 (1964); *J. Phys. Radium*, **26**, 403 (1965)
141. Kottler, H., Kowalski, K. L., *Phys. Rev.*, **138**, B619 (1965)
142. Coleman, E., Heinz, R. M., Overseth, O. E., Pellett, D. E., *Phys. Rev. Letters*, **16**, 761 (1966); *Phys. Rev.*, **164**, 1655 (1967)

143. Bennett, G. W., Friedes, J. L., Palev-sky, H., Sutter, R. J., Igo, G. J., Simpson, W. D., Phillips, G. C., Stearns, R. L., Corley, D. M., *Phys. Rev. Letters*, **19**, 387 (1967)

144. Glauber, R. J., *Lectures in Theoretical Physics*, **1**, 315 (1959)

145. Franco, V., *Phys. Rev. Letters*, **21**, 1360 (1968).

146. Harrington, D. R., *Phys. Rev. Letters*, **21**, 1496 (1968)

147. Kerman, A., Kisslinger, L., *Phys. Rev.* (In press)

148. Lindegren, I., Table of Nuclear Spins and Moments, Appendix 4 of *Alpha-Beta-and Gamma Ray Spectroscopy* (Siegbahn, K., Ed., North-Holland, Amsterdam, 1965)

149. Sachs, R. G., *Phys. Rev.*, **72**, 312 (1947)

150. Foldy, L., Osborne, R., *Phys. Rev.*, **79**, 795 (1950)

151. Villars, F., *Helv. Phys. Acta*, **20**, 476 (1947)

152. Padgett, D. W., Frank, W. M., Brennan, J. G., *Nucl. Phys.*, **73**, 424 (1965)

153. Brennan, J. G., Frank, W. M., Padgett, D. W., *Nucl. Phys.*, **73**, 445 (1965)

154. Arenhovel, H., Danos, M., *Phys. Letters*, **28B**, 299 (1968)

155. Schiff, L. I., Collard, H., Hofstadter, R., Johansson, A., Yearian, M. R., *Phys. Rev. Letters*, **11**, 387 (1963)

156. Collard, H., Hofstadter, R., Hughes, E. B., Johansson, A., Yearian, M. R., Day, R. B., Wagner, R. T., *Phys. Rev.*, **138**, B57 (1965)

157. Schiff, L. I., *Phys. Rev.*, **133**, B802 (1964)

158. Dalitz, R. H., Thacker, T. W., *Phys. Rev. Letters*, **15**, 204 (1965)

159. Srivastava, B. K., *Phys. Rev.*, **133**, B545 (1964)

160. Amado, R. D., *Phys. Rev.*, **141**, 902 (1966)

161. Gupta, V. K., Bhakar, B. S., Mitra, A. N., *Phys. Rev.*, **153**, 1114 (1967)

162. Tang, Y. C., Herndon, R. C., *Phys. Letters*, **18**, 42 (1965)

163. Siegert, A. J. F., *Phys. Rev.*, **52**, 787 (1937)

164. Sarker, A. Q., *Phys. Rev. Letters*, **13**, 375 (1964)

165. Sarker, A. Q., *Nuovo Cimento*, **36**, 392 (1965)

166. Sarker, A. Q., *Nuovo Cimento*, **36**, 410 (1965)

167. Levinger, J. S., Srivastava, B. K., *Phys. Rev.*, **137**, B426 (1965)

168. Gibson, B. F., *Phys. Rev.*, **139**, B1153 (1965)

169. Gibson, B. F., Schiff, L. I., *Phys. Rev.*, **138**, B26 (1965)

170. Griffy, T. A., *Phys. Letters*, **11**, 155 (1964)

171. *Neutron Cross Sections*, 2nd ed. (Hughes, D. J., Harvey, J A., Eds., USAEC, Washington, D. C., 1955)

172. Schiff, L. I., *Phys. Rev.*, **52**, 242 (1937)

173. Verde, M., *Helv. Phys. Acta*, **23**, 453 (1950)

174. Meister, N. T., Radha, T. K., Schiff, L. I., *Phys. Rev. Letters*, 12, 509 (1964)

175. Radha, T. K., Meister, N. T., *Phys. Rev.*, **136**, B388 (1964) Errata *Phys. Rev.*, **138**, AB7 (1965)

176. Barucchi, G., Bosco, B., Nata, P., *Phys. Letters*, **15**, 253 (1965)

177. Erdas, F., Milani, C., Pompei, A., Seatzu, S., *Nuovo Cimento*, **45B**, 72 (1966)

178. Austern, N., Sachs, R. G., *Phys. Rev.*, **81**, 710 (1951)

179. Austern, N., *Phys. Rev.*, **85**, 147 (1952)

180. Phillips, A. C., in *Few Body Problems, Light Nuclei and Nuclear Reactions* (Pais, G., Slaus, I., Eds., Gordon & Breach, New York, 1968)

181. Sachs, R. G., Austern, N., *Phys. Rev.*, **81**, 705 (1951)

182. Fetisov, V. N., Gorbonov, A. N., Varfolomeev, A. T., *Nucl. Phys.*, **71**, 305 (1965)

183. Gerstenberg, H. M., O'Connell, J. S., *Phys. Rev.*, **144**, 834 (1966)

184. Eichmann, U., *Z. Physik*, **175**, 115 (1963)

185. Boesch, R., Lang, J., Müller, R., Woelfli, W., *Phys. Letters*, **15**, 243 (1965)

186. Gibson, B. F., *Nucl. Phys.*, **B2**, 501 (1967)

187. Knight, J. M., O'Connell, J. S., Prats, F., *Phys. Rev.*, **164**, 1354 (1967)

188. Barbour, I. M., Phillips, A. C., *Phys. Rev. Letters*, **19**, 1388 (1968)

189. Fetisov, V. N., *J. Nucl. Phys. (USSR)*, **4**, 720 (1966) [Engl. transl., *Soviet J. Nucl. Phys.*, **4**, 513 (1967)]

190. O'Connell, J. S., Prats, F., *Phys. Letters*, **26B**, 197 (1968)

191. Fetisov, V. N., *Phys. Letters*, **21**, 52 (1966)

192. Knight, J. M., O'Connell, J. S., Prats, F., *Phys. Letters*, **22**, 322 (1966) Erratum *Phys. Letters*, **23**, 491 (1966)

193. Gerasimov, S. B., *Zh. ETP Pis'ma*, **5**, 412 (1967) [Engl. transl., *Soviet Phys. JETP Letters*, **5**, 337 (1967)]

194. Barton, G., *Nucl. Phys.*, **A104**, 289 (1967)

195. Prats, F., Lehman, D. R., O'Connell, J. S., *Bull. Am. Phys. Soc.*, **13**, 1401 (1968)
196. Johansson, A., *Phys. Rev.*, **136**, B1030 (1964)
197. Hughes, E. B., Yearian, M. R., Hofstadter, R., *Phys. Rev.*, **151**, 841 (1966)
198. Griffy, T. A., Oakes, R. J., *Phys. Rev.*, **135**, B1161 (1964)
199. Griffy, T. A., Oakes, R. J., *Rev. Mod. Phys.*, **37**, 402 (1965)
200. Gibson, B. F., West, G. B., *Nucl. Phys.*, **B1**, 349 (1967)
201. Porter, F. T., *Phys. Rev.*, **115**, 450 (1959)
202. Christensen, C. J., Nielsen, A., Bahnsen, A., Brown, W. K., Rustad, B. M., *Phys. Letters*, **26 B**, 11 (1967)
203. Blin-Stoyle, R. J., Nair, S. C. K., *Advan. Phys.*, **15**, 493 (1966)
204. Blatt, J. M., *Phys. Rev.*, **89**, 86 (1952)
205. Blin-Stoyle, R. J., Gupta, V., Primakoff, H., *Nucl. Phys.*, **11**, 444 (1959)
206. Blin-Stoyle, R. J., *Phys. Rev. Letters*, **13**, 55 (1964)
207. Blin-Stoyle, R. J., Papageorgiou, S., *Nucl. Phys.*, **64**, 1 (1964)
208. Blin-Stoyle, R. J., Tint, M., *Phys. Rev.*, **160**, 803 (1967)
209. Cheng, W. K. (Thesis, Univ. Pennsylvania, 1966, unpublished)
210. Kim, C. W., Primakoff, H., *Phys. Rev.*, **139**, B1447 (1965)
211. Primakoff, H., Nuclei as Elementary Particles in *High Energy Physics and Nuclear Structure* (Alexander, G., Ed., North-Holland, Amsterdam, 1967)
212. Clay, D. R., Keuffel, J. W., Wagner,

R. L., Jr., Edelstein, R. M., *Phys. Rev.*, **140**, B586 (1965)
213. Auerbach, L. B., Esterling, R. J., Hill, R. E., Jenkins, D. A., Lach, J. T., Lipman, N. H., *Phys. Rev.*, **138**, B127 (1965)
214. Primakoff, H., *Rev. Mod. Phys.*, **31**, 802 (1959)
215. Yano, A. F., *Phys. Rev. Letters*, **12**, 110 (1964)
216. Pascual, P., Pascual, R., *Nuovo Cimento*, **48**, A963 (1967)
217. Lim, T. K., *Nucl. Phys.*, **A109**, 641 (1968)
218. Petersen, E. A., *Phys. Rev.*, **167**, 971 (1968)
219. Wong, I. T., *Phys. Rev.*, **139**, B1544 (1965)
220. Kim, C. W., Primakoff, H., *Phys. Rev.*, **140**, B566 (1965)
221. Frazier, J. F., Kim, C. W. (To be published)
222. Kim, C. W., Ram, M., *Phys. Rev. Letters*, **18**, 327 (1967)
223. Amado, R. D., in *Lectures in Theoretical Physics* (Brandeis Summer School 1967, to be published)
224. Watson, K. M., Nuttall, J., *Topics in Several Particle Dynamics* (Holden-Day, San Francisco, Calif., 1967)
225. Gillespie, J., Nuttall, J., Eds., *Three-Particle Scattering in Quantum Mechanics* (Benjamin, New York, 1968)
226. Duck, I., Three Particle Scattering, *Advan. Nucl. Phys.*, **1** (Plenum Press, New York, 1968)
227. Mitra, A. N., The Nuclear Three-Body Problem, *Advan. Nucl. Phys.*, **3** (To be published)

COUPLED-CHANNEL APPROACH TO NUCLEAR REACTIONS[1,2]

By Taro Tamura

Center for Nuclear Studies, University of Texas
Austin, Texas

CONTENTS

I. INTRODUCTION

After large, high-speed computers became available to the nuclear physicists around 1960, extensive numerical calculations concerning several aspects of nuclear reaction and structure problems were performed. Perhaps the most extensive use of the computer was made with the distorted-wave Born approximation ($DWBA$) and the coupled-channel calculations (CC): $DWBA$ (1–3) has mainly been used to analyze transmutation reaction data, but it has also been used for inelastic scattering data; CC has mainly been used to analyze inelastic scattering data, but it has also been used recently for more complicated processes.

At least as far as the inelastic scattering process is concerned, CC has several advantages over $DWBA$. Except for truncating the number of channels to be considered explicitly, CC treats the interaction exactly, while $DWBA$ treats it only to its first power. Therefore, if the interaction is very strong, $DWBA$ may become a poor approximation, while CC does not. Because $DWBA$ is the first-order Born approximation, it cannot describe the multiple-excitation processes, while CC can. Note that once a program is

---

[1] Supported in part by the United States Atomic Energy Commission.

[2] Abbreviations used: CC = coupled-channel calculation; $DWBA$ = distorted-wave Born approximation; $CCBA$ = coupled-channel Born approximation.

coded for a CC calculation of one-step processes, its extension to multiple-excitation processes is rather easy. For these reasons, CC has been used most extensively to calculate the cross section of inelastic scattering from various collective nuclei. There the interaction is strong, and the multiple-excitation cross sections are appreciable.

The purpose of the present article is to review the use of CC in explaining various reaction data. It thus summarizes many works reported on various occasions, including a few previous review articles (4–7). CC is formulated in Section 2, very briefly, since quite detailed accounts of the formulation have been given elsewhere (4, 7, 8). Application of CC to inelastic scattering data is presented in Section 3. Since collective states are excited, the nuclear model underlying these calculations is that of Bohr & Mottelson (9) or its extensions (7, 8). Section 4 discusses the application of CC to problems other than straightforward inelastic scattering.

After Sections 3 and 4 have shown that CC has successfully explained various experimental data, Section 5 discusses the scope of future work: How can the CC itself be improved and what structure calculations are needed for full use of what has been achieved by CC so far, in understanding nuclear structure?

2. FORMULATION OF THE COUPLED-CHANNEL CALCULATIONS

We shall formulate the coupled-channel calculations in a general way, though the exchange of the nucleons in the projectile with those in the target is neglected. The wavefunction ψ of the total system, the projectile plus the target, may then be written as

$$
\begin{aligned}
\Psi &= r^{-1} \sum_{J n l_n j_n} R_{J n l_n j_n}(r)(y_{l_n j_n} \otimes \Phi_{I_n})_{JM} \\
&= r^{-1} \sum_{J n l_n j_n m_j M_n} R_{J n l_n j_n}(r)(j_n I_n m_j M_n \mid JM) y_{l_n j_n m_j} \Phi_{I_n M_n} \\
&= r^{-1} \sum_{J n l_n j_n m_j M_n m_l m_s} R_{J n l_n j_n}(r)(j_n I_n m_j M_n \mid JM) \\
&\quad \times (l_n s m_l m_s \mid j_n m_j) i^l Y_{l_n m_l}(\theta, \phi) \chi_{s m_s}(\sigma) \Phi_{J_n M_n}(\xi)
\end{aligned}
\qquad 2.1
$$

Here $\Phi_{I_n}(\xi)$ stands for the normalized eigenfunction of the nth excited state of the target ($n=1$ corresponds to the ground state) with the eigenenergy ϵ_n and the spin I_n, while $y_{l_n j_n}$ describes the angular momentum part of the projectile wavefunction. Its exact definition can be seen from the last expression of 2.1; note that $\chi_{s m_s}(\sigma)$ is the spin wavefunction of the projectile. The function $R_{J n l_n j_n}(r)$ is the radial part of the wavefunction of the projectile or, more precisely, that of the relative motion between the projectile and the target. Finally J is the angular momentum of the total system, as can be seen from the arguments of the Clebsch-Gordan coefficients introduced in 2.1. Note that 2.1 is nothing but an expansion of the wavefunction Ψ in terms of the complete set $(y_{l_n j_n} \otimes \Phi_{I_n})_{JM}$, with unknown coefficients $R_{J n l_n j_n}(r)$.

Corresponding to the wavefunction 2.1, the Hamiltonian H of the total system may be written as

$$H = H_N + K_p + V \qquad 2.2$$

where H_N is the Hamiltonian that describes the internal motion of the target, and thus satisfies

$$H_N \Phi_{I_n}(\xi) = \epsilon_n \Phi_{I_n}(\xi) \qquad 2.3$$

We choose the energy scale so that $\epsilon_1 = 0$. In 2.2, K_p is the kinetic energy of the projectile, while V is the interaction between the projectile and the target. In the simple optical model (10), $V = V_{op}(r)$ is a function only of the coordinate r (including the spin σ of the projectile if necessary) and not of the coordinate ξ of the target. Therefore V is diagonal with respect to the complete set $(y_{l_n j_n} \otimes \Phi_{I_n})_{JM}$ introduced in 2.1. The extension of this simple optical model to CC now consists in assuming that V is a function also of the coordinate ξ, as well as all the coordinates of the projectile (r, θ, ϕ). We shall write this extended potential as

$$V(r, \theta, \phi; \xi) = V_{\text{diag}} + V_{\text{coupl}} \qquad 2.4$$

where V_{diag} is the original optical potential and is diagonal, while the second term is nondiagonal in the complete set of 2.1.

The Schrödinger equation of the total system is

$$(H - E)\Psi = 0 \qquad 2.5$$

Insert 2.1 and 2.4 into 2.5, multiply it from the left by $(y_{l_n j_n} \otimes \Phi_{I_n})_{JM}$ and then integrate it over all the coordinates excepting r. One then gets the following set of equations:

$$\left[\frac{\hbar^2}{2m} \left(\frac{d^2}{dr^2} - \frac{l_n(l_n + 1)}{r^2} - V_{\text{diag}} + E_n \right) \right] R_{J n l_n j_n}(r)$$
$$= \sum_{n' l_n' j_n'} \langle (y_{l_n j_n} \otimes \Phi_{I_n})_{JM} | V_{\text{coupl}} | (y_{l_n' j_n'} \otimes \Phi_{I_n'})_{JM} \rangle R_{J n' I_n' j_n'}(r) \qquad 2.6$$

with $E_n = E - \epsilon_n$. Here E_1 is the energy of the incident projectile, while $E_n (n \neq 1)$ is the kinetic energy with which the projectile leaves the target, after exciting the latter to its nth state. Equation 2.6 is the coupled equation to be solved to calculate the inelastic scattering cross sections. It is clear that 2.6 reduces to the simple optical-model equation if $V_{\text{coupl}} = 0$.

To use 2.6 for any practical calculation, one has to give explicit forms to the potentials V_{diag} and V_{coupl} and the wavefunction Φ_{I_n}. Among them, V_{diag} is fixed by fitting the elastic scattering data, while a specific model has to be introduced, depending on the nature of the target, to fix V_{coupl} and Φ_{I_n}. A very detailed account of how this model is to be made has been given elsewhere (4). Most of the calculations so far made use the model of Bohr & Mottelson (9) or its extension (7, 8); thus the nuclei are classified into two major groups: the spherical (vibrational) and deformed (rotational) nuclei, the parameter $\beta_\lambda (\lambda = 2, 3, 4, \text{——})$ describing the collectivity or deformability of these nuclei. By fitting the experimental data with the theory thus fixed, important information on the structure of the target nucleus can be extracted.

A very specific example of 2.6 is the Lane equation (11), in which V_{coupl} is given as

$$V_{\text{coupl}} = - (t \cdot T) U_1(r) \qquad 2.7$$

and 2.6 is reduced into the following very simple form (separately for each set of l and j)

$$\left[\frac{\hbar^2}{2m} \left(\frac{d^2}{dr^2} - \frac{l(l+1)}{r^2} \right) - U_0 - \frac{1}{2} T_0 U_1 \right] R_p(r)$$

$$= - \sqrt{\frac{1}{2} T_0} \, U_1 R_n(r)$$

$$\left[\frac{\hbar^2}{2m} \left(\frac{d^2}{dr^2} - \frac{l(l+1)}{r^2} \right) - U_0 + \frac{1}{2} (T_0 - 1) U_1 \right] R_n(r)$$

$$= - \sqrt{\frac{1}{2} T_0} U_1 R_p(r) \qquad 2.8$$

The radial functions $R(r)$ are suffixed by p and n to show explicitly that the channels coupled are proton and neutron channels describing, respectively, the motion of a proton with respect to a core nucleus C, and that of a neutron with respect to the isobaric analog nucleus A of C; $A = T_- C$. In 2.8, U_0 is the isospin-independent part of the optical potential, while T_0 is the isospin of C. The energy E_p in 2.8 may be interpreted as E_1 of 2.6, while E_n in 2.8 is interpreted E_2 of 2.6, its value being equal to $E_p - \Delta_c$, where Δ_c is the Coulomb energy of a single proton in the target. There will be no danger of confusion by using the same notation E_n in both 2.6 and 2.8, with somewhat different meanings.

3. FITTING SCATTERING DATA BY COUPLED-CHANNEL CALCULATIONS

3.1 SPHERICAL NUCLEI

The most accurate and abundant data of the inelastic scattering processes from spherical, vibrational nuclei exist for those in the so-called cadmium region, notably for the isotopes [106,108,110]Pd and [112,114]Cd (12–15). These nuclei seem to be the most typical ones exhibiting vibrational spectra of higher multiplicity, if not the "ideal" spectrum of the Bohr-Mottelson theory (9). To investigate and understand the deviation of the experimental facts from those expected from ideal vibrational nuclei has been one of the most exciting objects of nuclear structure theory. Historically, the accumulation of experimental knowledge of the excited states of these nuclei was started with Coulomb excitation experiments; see Alder et al. (16) and deBoer & Eichler (17) for a review of the Coulomb excitation processes. Accurate data on (p,p') processes from the same nuclei, and their analyses in terms of CC, may supply information in line with, or complementary to Coulomb excitation data, as we shall see below.

Before starting a comparison of the (p,p') and Coulomb excitation processes, it will be of interest to see how rapidly progress has been made in (p,p') experiment and theory in the past few years: we shall compare some old work with a new work devoted to the same subject. Figure 1 shows the experimental data of $^{114}Cd(p,p')$ at $E_p=14$ MeV, and a CC fit made in 1964 (18). Level B is the one-phonon 2^+ state at 556 keV, while levels C, D, and E are the two-quadrupole-phonon triad states with $I^\pi=0^+$, 2^+, and 4^+ and with energies 1133, 1208, and 1282 keV, respectively. The experiment was made by using a cyclotron and nuclear emulsion. Compare these results with those given in Figure 2, again of $^{114}Cd(p,p')$, as reported in 1968 (14). The levels shown there are also two-phonon triads, and the data were taken by using a tandem Van de Graaff and solid-state detectors. The improved quality of the data is clearly seen. The improved data then required more sophistication in CC to obtain a satisfactory fit. See, e.g., the curve D of Figure 1, which seems to explain the 2_2^+ (second 2^+) cross section there fairly nicely. However, if this curve is superposed upon the new 2_2^+ cross-section data in Figure 2, the fit is distressingly poor.

The major improvement made in CC in obtaining the good fit to the new 2_2^+ data, as shown in Figure 2, was to consider (8) the admixing of the one-phonon state with an amplitude α, say, to the wavefunction of the 2_2^+ state, by introducing a new parameter $\beta_{02}''=\alpha\beta_2$. In the previous calculation (18) the parameter β_{02}', which describes the strength of the quadratic coupling term, was treated as adjustable (19), but in the new calculation (14) it was fixed as $\beta_{02}'=(\beta_2\beta_{22})^{1/2}$, β_{22} being the coupling strength between the 2_1^+ and 2_2^+ states (8). Therefore the number of parameters has not been increased in the new calculation, as compared to the old ones. Nevertheless, the much improved agreement obtained with the more accurate new data convinces us that the physical picture on which our new calculation is based is intrinsically correct.

The sign of β_{02}'', as well as its magnitude, is very important in fitting the data. Thus if β_{02}'' were -0.050, say, instead of 0.050 as in Figure 2, the cross section would have peaked at around 110°, in complete disagreement with experiment. As discussed by Tamura (20), the magnitude and the sign of β_{02}'' are intimately related to and consistent with an amplitude with which the two-phonon 2^+ state is admixed into the 2_1^+ state. This latter amplitude was required in a model by Tamura & Udagawa (21) to explain the magnitude and sign of the static quadrupole moment of ^{114}Cd, extracted from recent data of the Coulomb excitation experiment (17, 22–24). Since the model of

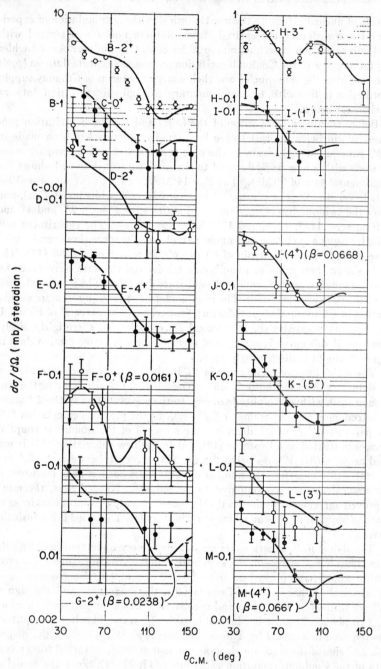

FIG. 1. Scattering of 14 MeV protons by ^{114}Cd; from (18).

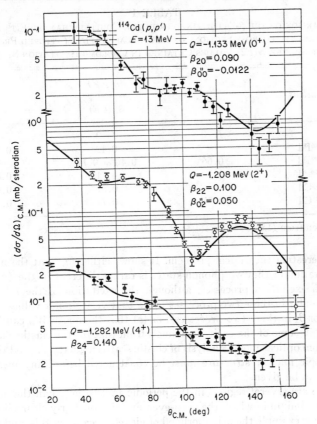

FIG. 2. Scattering of 13 MeV protons by ^{114}Cd; from (14).

(21) is rather crude and more sophisticated models would have to be used to explain the static quadrupole moment (25, 26), it is premature to draw any strong conclusion from the above argument. It is nevertheless gratifying to see that these two processes are consistent in this way.

Quite an interesting relationship between the Coulomb excitation and (p,p') results is seen when the magnitudes of the parameters $\beta_{2I}(I=0,\ 2,$ and 4) (8), extracted in both ways, are compared. Such a comparison is made in Table I (15) for five nuclei in the cadmium region, and it is seen that $\beta_{2I}(p,p')$ is systematically smaller than $\beta_{2I}(CE)$ by 20 to 50 per cent. Note that in obtaining β_{2I} (p,p') in Table I, the value of W_D, the strength of the imaginary potential, in the two-phonon channels was taken to be 1.2 times W_D in the elastic channel. It was then found that if W_D were taken to be 1.5 or 1.6 times that in the elastic channel, and at the same time $\beta_{2I}(p,p')$ were taken to be equal to the corresponding $\beta_{2I}(CE)$, almost equally good fits to the data, as shown in Figure 2, were recovered. This experience indicates a

TABLE I

COMPARISON OF THE β_{2I}'s OBTAINED FROM THE COUPLED-CHANNEL FITS TO THE
DIFFERENTIAL CROSS SECTIONS OF THE INELASTICALLY SCATTERED PROTONS
AND FROM THE $B(E2)$ VALUES OBTAINED IN COULOMB EXCITATION STUDIES[a]

Nucleus	$\dfrac{\beta_{2I}(p,p')}{\beta_{2I}(CE)}$		
	$I=0$	$I=2$	$I=4$
^{106}Pd	—	0.42	0.57
^{108}Pd	0.66	0.46 or 0.79	—
^{110}Pd	0.70	0.81	0.62
^{112}Cd	—	0.57	0.63
^{114}Cd	0.77	0.71	0.81

[a] Taken from (15).

very interesting theoretical problem. It shows either (a) that the nature of the excitation of the two-phonon states is rather different in the (p,p') and Coulomb excitation processes, although the excitation of the 2_1^+ state is rather similar in these two processes (note that very similar values were obtained for β_2 in these two processes); or (b) that the depth W_D can be quite different in some of the excited channels from that for the elastic channel as found from optical-model analysis of elastic scattering data. A good theory of microscopic nuclear structure must be constructed before any meaningful answer can be given.

There are several regions in the periodic table other than the neighborhood of cadmium in which the even-even nuclei show fairly clear vibrational spectra, for example the neighborhood of nickel. Because of this, and perhaps because of the comparative ease in preparing the target, a large amount of data for inelastic scattering from Ni isotopes has been reported. The data (27) of ^{60}Ni(d,d') shown in Figure 3 are fitted by CC, and the fit is quite good except in the large-angle regions beyond 140°. The CC calculations were made in very much the same way as those for Cd isotopes. A few theoretical works (e.g. 28), based on a microscopic description of the nuclear states, indicate that the multiple quadrupole-phonon states cannot be well developed in this mass region, since the single-particle orbits having large multiplicities, i.e. large angular momenta, are very scant. If such is indeed the case, a straightforward application of the phonon model of these nuclei may be questioned, but we shall leave this question untouched. Figure 3 shows that, among the three angular distributions from the triad states, the one from the 0^+ state shows a marked pattern and thus is quite different from those of the other two. This feature seems to be rather characteristic of any 0^+ angular distribution excited by the (d,d') processes (29), and this fact may make the (d,d') process a nice tool in locating 0^+ states.

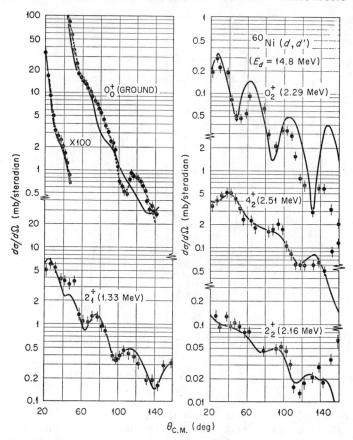

FIG. 3. Scattering of 14.8 MeV deuterons by ^{60}Ni; from (27).

Figure 4 shows similar data, but this time induced by the bombardment of α particles (30). The CC fit (8) is again rather good, except to the 2_2^+ state, whose angular distribution is rather unusual. Putting this point aside, we note that the theoretical curves all show very marked diffraction patterns, which is caused by the strong nuclear absorption of the α particles. It is interesting to observe in Figures 1–4 that the characteristic pattern of the angular distribution changes gradually in going from nucleon to deuteron, and then from deuteron to α particle.

Figure 5 shows again the results of an (α,α') process, but this time from ^{62}Ni. The data were obtained by Meriwether et al. (31), and the most interesting feature is the gradual change of the relative phase of the angular distributions with increased bombarding energy. We have mentioned above that the α particles are strongly absorbed by nuclei, and for such strongly

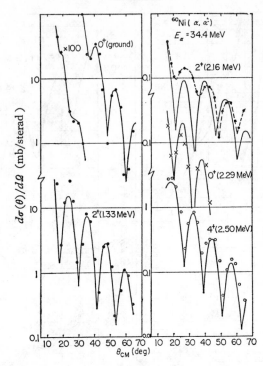

FIG. 4. Scattering of 34.4 MeV α particles by ^{60}Ni; from (8).

absorbed projectiles a very simple theory developed by Inopin (32) and
Blair (33) is known to work quite well. One prediction of this theory, however, is that the elastic and the two-phonon angular distributions have the
same phase, while the one-phonon distribution is exactly out of phase, and
this relation should hold regardless of the bombarding energy. The data in
Figure 5 then show that this prediction does not hold in nature. As is seen
in this figure, CC reproduces the angular distribution, and thus the phase relation at all the energies considered (34). This result indicates that although
nuclear absorption of the α particles is rather strong, it is not extremely
strong. Note that in obtaining the CC curve in Figure 5, V and W, the depths
of the real and imaginary parts of the optical potential were changed (monotonously) as the bombarding energy was increased, so that a good fit to the
elastic scattering cross section was obtained at each energy, but no other
ad hoc devices were invented to modify the phase relations.

Austern & Blair (35) recently modified Blair's previous theory (33)
in order to loosen some of the restrictive assumptions in the latter. The
numerical calculation involved in this new theory is only slightly more in-

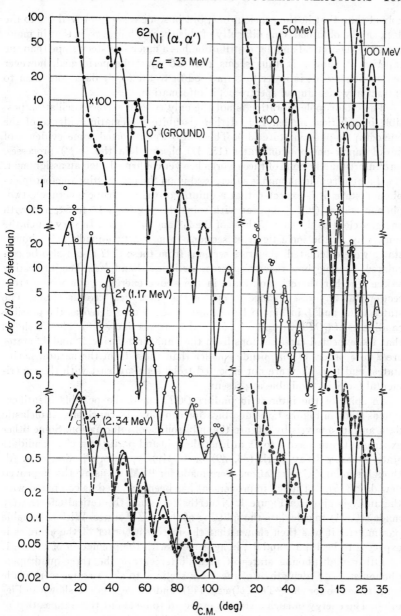

FIG. 5. Energy variation of the phase relations of the angular distributions in the scattering of α particles by ^{62}Ni; from (34).

volved than that in the previous one, but in some cases improved fits to the data can be obtained. One difficulty of this theory, however, is that in many cases the theoretical angular distribution has a much less steep slope than the experiment shows. It thus seems that CC must be performed, however tedious it may be, in order to get a good fit to the (α,α') data, and then to extract any meaningful spectroscopic information.

The quadrupole vibration is not the only type of surface oscillation possible in spherical nuclei. Indeed there is quite systematic evidence of the one-octupole-phonon state (36). There also was reported some evidence of hexadecapole-one-phonon states (15, 37) observed in the (p,p') processes, but unfortunately none of these works has yet confirmed such an assignment.

If a quadrupole phonon is superposed upon the pre-existing one-octupole-phonon state, it is expected that a quintet of the quadrupole-octupole-two-phonon states $(QOTP)$ appear with spins ranging from 1^- to 5^-, and with energies close to the sum of those of the 2_1^+ and 3_1^- states. In very much the same way as the inelastic cross sections from the two-quadrupole-phonon states were calculated, the cross sections from these $QOTP$ can also be calculated. Having these theoretical cross sections in hand, it is then interesting to see whether there is any state in the above energy region whose cross section has any resemblance to one of the theoretical cross sections. The states I, K, and L in Figure 1 are those that are considered as the possible candidates of $QOTP$ with $I^\pi = 1^-, 5^-$, and 3^-, respectively (18). The CC shows that the expected cross sections from the unnatural parity 2^- and 4^- states are 2 to 3 orders of magnitude smaller than those from the natural parity states, and thus they are not expected to be seen, at least with the experimental facilities available at present.

In spite of the interest in finding $QOTP$, and the possible identification as made in Figure 1, its identification can never be conclusive, if inelastic data and its analysis by CC are the only source of information. Some other experiments are needed to fix at least the spin and parity of such a candidate state. Some γ-decay work (38, 39) or $(p,p'\gamma)$-type work is desirable. Although similar CC calculations were made for ^{110}Pd by using the improved new data (15), not much improvement has been achieved as compared with the analysis given in Figure 1. Nevertheless, some theoretical efforts were made (40, 41) to calculate the splitting of the $QOTP$ states. With reasonable parameters, it was then shown that the increasing order of the energies is expected to be 1^-, 5^-, and 3^-, in agreement with the prediction of Figure 1.

Other multiphonon states of great interest are the three-quadrupole phonon states. Calculations of the cross sections to these states were made some time ago for the ^{126}Te(p,p') case (42), and the results are shown in Figure 6. The energy of such a state is expected to be about three times that of the 2_1^+ state, and the search for possible candidates in this energy region was made in very much the same way as for $QOTP$. As is seen in Figure 6, candidates were found for all the positive parity states excepting 6^+, although a similar analysis made in ^{110}Pd (15) located two possible candidates for the

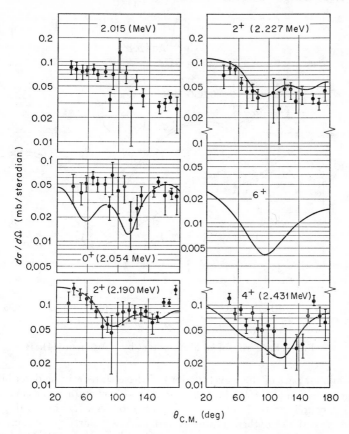

FIG. 6. Possible assignments of the three-phonon states in ^{126}Te; from (42).

6^+ state. The inconclusiveness of the assignment of the three-phonon states thus made is more or less the same as that of the $QOTP$. Nevertheless, as far as the 6^+ state is concerned, the situation could be better. For instance, (α, xn) or similar types of experiments (43) are known to be rather powerful tools in exciting the member of the vibrational states whose spin is the highest for a given phonon number. The energy of the 6^+ state can be fixed fairly conclusively this way. Then the (p,p') cross section from that level can be compared with the CC result, and some spectroscopic information may be extracted from such a comparison.

This subsection will be closed by mentioning the inelastic scattering processes from odd-A spherical nuclei. An example is ^{107}Ag(p,p') (44) and the cross sections from the 780 and 950 keV states are shown in Figure 7: since the ground state of ^{107}Ag has spin $\frac{1}{2}$ and the energies of these two states are close to the energy (1194 keV) of the 2_2^+ state of ^{106}Pd core, it is expected that

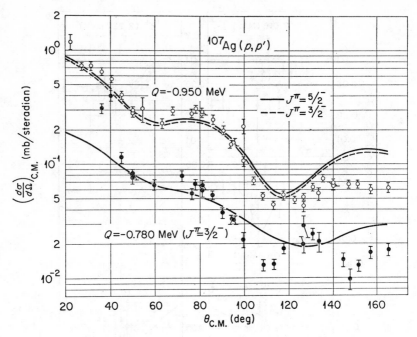

FIG. 7. Scattering of 13 MeV protons by ¹⁰⁷Ag; from (44).

the dominant part of the wavefunctions of these states will be described as the superposition of the $p_{1/2}$ proton upon this 2_2^+ core state, if the concept of the weak coupling model (45, 46) is applicable to ¹⁰⁷Ag. It is interesting that the cross sections to these states indeed have angular distributions very similar to the 2_2^+ cross section in the ¹⁰⁶Pd(p,p') process (12). This means that the excitation mechanism in Ag is essentially the same as it was in Pd, the valence proton participating as a spectator. The CC calculation can then be made very much as it is made for even-even nuclei, and it has been shown (44) that ¹⁰⁷Ag(p,p') can be fairly satisfactorily analyzed in this way. It was found, nevertheless, that the assumption of the weak coupling model is not completely satisfied, and thus some disagreement with experiment has inevitably been noticed, when the spectroscopic knowledge extracted from the CC calculation was used to calculate quantities like the magnetic moment and the $M1$ transition probabilities.

3.2 Deformed Nuclei

Historically, the first use of CC to fit actual data was for deformed nuclei [though Yoshida's work (47) for spherical nuclei with rather crude numerical calculation preceded it], with the objective of explaining the mass-number dependence of the strength function, which is essentially the total-reaction cross section of almost zero-energy neutrons. If the nucleus were "black," i.e. if it were perfectly absorptive to neutrons, the strength function would be a monotonously increasing function of the mass number A. Experiment (48) showed that such is not the case, and indeed the strength function oscillates as a function of A. Since it is known that nuclei are not so strongly absorptive to neutrons and that the elastic scattering is well described by the optical model (49), this oscillatory nature of the strength function should no longer be surprising. The embarrassing fact, nevertheless, was that the strength function calculated with the optical model has a broad peak centered at $A \approx 160$, with minima at $A \approx 100$ and 200, while the experiment shows that there are two peaks within this mass range centered at $A \approx 140$ and 180. Margolis et al. (50) and Chase et al. (51) showed that this fact can quite naturally be explained if it is noticed that most of the nuclei that lie in this mass range are deformed, and thus that CC should be used, instead of the optical model, to calculate the strength function. Later a more detailed calculation of the strength functions was made by Buck & Perey (52) over the entire mass region, and a quite good fit to the more abundant data (48, 53–57) was obtained. See Figure 8 for the fit of S-wave strength function; see also (58) for more recent analyses.

A very interesting experiment that one can attempt with deformed nuclei, but not with spherical nuclei, is the observation of the so-called deformation effect in the total cross section σ_t of the neutrons. Take an odd-A deformed nucleus of spin I, and prepare out of it an oriented target. Here, an oriented target has the probability b_M, say, of occupying a magnetic substate M (the direction z of the quantization being appropriately chosen to define a space-fixed coordinate system), which satisfies the relation $b_M = b_{-M}$, but $b_M \neq b_M'$, if $M \neq M'$. In the unoriented target $b_M = 1/(2I+1)$ for all M. The polarized target has the property that $b_M \neq b_{-M}$.

When this deformed target is in its ground state, the projection K of its spin I along the symmetry axis of deformation, which is taken as the z' axis of the body-fixed coordinate system, is equal to I; $K = I$. If the nucleus is in the magnetic substate M in the space-fixed coordinate system, the theory of space rotation shows that the spatial distribution of the direction of the nuclear symmetry axis is proportional to the rotation function $D_{MI}{}^I(\theta_i)$ (59), the argument θ_i standing for the Euler angles that describe the relative orientation of the two coordinate systems. The nature of the function $D_{MI}{}^I(\theta_i)$ is such as to make the direction of the symmetry axis get closer to the z axis, the larger $|M|$ is. Thus, take the case in which the deformation of the nucleus is cigar-shaped (prolate deformation), and the orientation is such as to make $b_{|M|} > b_{|M'|}$ for $|M| > |M'|$. In this situation, the target exposes its smaller (geometrical) cross-sectional area more often than its larger area, when it is seen along the z axis. Therefore, if this target is bombarded with neutrons impinging along the z axis, and if the actual total cross section σ_t is proportional to the geometrical cross-sectional area, σ_t will be observed to be smaller than when the observation is made with an unoriented target.

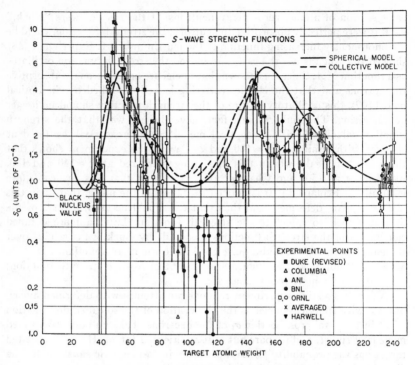

FIG. 8. *S*-wave strength function as a function
of the mass number; from (52).

The difference of σ_t with oriented and with unoriented target is usually called the deformation effect and is denoted as $\Delta\sigma_{\text{def}}$; $\Delta\sigma_{\text{def}} = (\sigma_t)_{\text{or}} - (\sigma_t)_{\text{unor}}$. In the situation described just above, we see that $\Delta\sigma_{\text{def}} < 0$. Its *sign* will be changed if the deformation is of a pancake shape (oblate deformation), or if the orientation is such as to make $b_{|M|} < b_{|M'|}$ for $|M| > |M'|$.

The assumption of the proportionality of σ_t and the geometrical cross-sectional area, made for simplicity in the above considerations, is (approximately) correct only if the target is black to neutrons. Since the target is an optical medium to neutrons, various interference effects can play important roles, and the situation may be more complicated than was described above. The magnitude and even the sign of $\Delta\sigma_{\text{def}}$ will probably oscillate with the neutron energy E_n (60), even if the target is kept fixed in a given orientation. The deformation effect is nonetheless a calculable quantity, once the deformation parameter, the orientation of the target, and a set of the optical parameters are known or assumed. By fitting the thus-calculated deformation effect to experiment, important information concerning the above parameters will be extracted.

The first observation of the deformation effect was made by Wagner et al. (61) in 1965 for neutrons with $E_n = 0.35$ MeV, with oriented ^{165}Ho as target. Afterwards the experiment was extended to $E_n = 14$ MeV by Marshak et al. (62), and to 8 and 15 MeV by Fisher et al. (63), the target in each case ^{165}Ho. The measured $\Delta\sigma_{\text{def}}$ was found to be negative at all these energies, in agreement with the prediction of the black nucleus model, although its magnitude indeed depended on E_n, and did not agree with the black nucleus value (32). This situation is somewhat unexpected because, as we stated above, it is more natural that $\Delta\sigma_{\text{def}}$ should oscillate with E_n and even change its sign for certain values of E_n. That such behavior is the case has now been proved by two experiments recently performed (64, 65). Indeed Marshak and Langsford developed an ingenious experimental device so that σ_t can be measured easily for E_n ranging from 2 to 135 MeV; their results for σ_t and for $\Delta\sigma_{\text{def}}$ are shown in Figures 9 and 10. Both σ_t and $\Delta\sigma_{\text{def}}$ are oscillatory functions of E_n as was

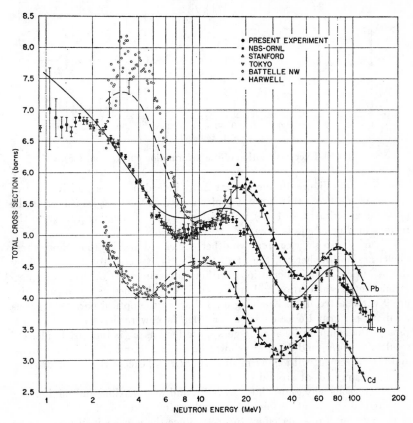

FIG. 9. Energy variation of the total neutron cross sections of Cd, Ho, and Pb; from (65).

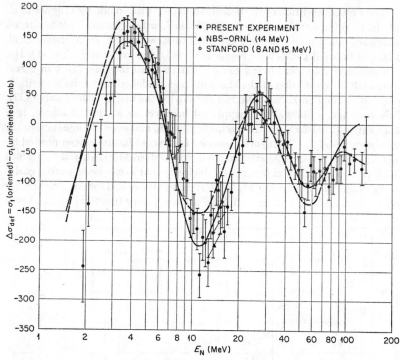

Fig. 10. Energy variation of the deformation effect; from (65).

expected, and Figure 10 shows that $\Delta\sigma_{def}$ indeed becomes positive in two regions of E_n. From this figure it is also seen that the energies chosen in the previous experiments (61–63) were unfortunate in belonging always to the regions in which $\Delta\sigma_{def}$ is negative.

Figure 9 also shows σ_t for two other nuclei, Cd and Pb (66). Since these nuclei are both spherical, the fit to these σ_t is expected to be achieved by simple optical-model calculations, without appealing to CC. Perey (67) performed such a calculation and very good fit was obtained as is shown in Figure 9 with dotted lines. Knowing a good set of the optical parameters thus fixed, to perform CC calculation for ^{165}Ho is now rather straightforward, though the computational time gets quite long as E_n is increased. The CC results (65) of σ_t and $\Delta\sigma_{def}$ are shown in Figures 9 and 10 with solid lines, and the agreement with experiments is very good, except that some discrepancy appears in σ_t at around 10 MeV.

The reasons why σ_t oscillates with E_n, and why maxima and minima are expected to appear at certain E_n are rather simply supplied by the concept of the nuclear Ramsauer effect (68, 69). Since the nucleus is dispersive, the wavenumber of the wave that passes through the nucleus is different from that of the wave that goes by the nucleus. This difference Δk times R, the nuclear radius, then gives rise to a difference of the phases of these two waves, and if this phase difference equals an odd(even)-integer multiple of π, σ_t will have a maximum (minimum). Since Δk is easily calculable once E_n, R, and V, the depth of the potential, are given, the positions of the maxima and minima are easily obtained. If the target is then oriented, the effective path-length R will be different from what it was for an unoriented target, and it will result in a shift of the positions of the maxima and minima from those in the unoriented target. Because of this it is expected that $\Delta\sigma_{def}$ will also be oscillatory. One defect of the concept of a nuclear Ramsauer effect is that it cannot predict the magnitude of σ_t and thus that of $\Delta\sigma_{def}$. However, if it is assumed that the energy dependence of σ_t (within a small energy range) is the same as that in the black nucleus model, i.e. proportional to $\pi(R+\lambda)^2$, it is possible to work out a simple formula to associate $\Delta\sigma_{def}$ with the experimental value of $(\sigma_t)_{unor}$. The dotted line in Figure 10 was calculated this way (65), and the agreement is again seen to be quite good.

Another subject of interest that can be investigated when one has a polarized odd-A target, deformed or spherical, is to extract the sign and the strength of the so-called spin-spin interaction, if such an interaction ever exists. Such an interaction is proportional to the operator $(\mathbf{\delta}\cdot\mathbf{I})$, $\mathbf{\delta}$ and $\mathbf{I}$ being the spin operators of the projectile and the target, respectively. Suppose the direction of the target polarization is at one time along the $+z$ axis and at another time along the $-z$ axis. Suppose also that the bombarding neutron is polarized in the $+z$ axis all the time and impinges from a direction perpendicular to the z axis. If the strength V_{ss}, say, of the spin-spin interaction is weak, its contribution to σ_t will be simply proportional to $\langle\mathbf{\delta}\rangle\langle\mathbf{I}\rangle$, where the bracket means the expectation value of the spin operator with respect to the spin wavefunction. Because of the assumption of the fixed direction of the polarization of the projectile, the sign and value of $\langle\mathbf{\delta}\rangle$ are fixed, while the sign of $\langle\mathbf{I}\rangle$ will be opposite in the above two experimental situations. If V_{ss} is not extremely weak, the σ_t obtained under these two situations will have a measurable difference, and from this difference the sign and the magnitude of V_{ss} may be extracted.

There is one reason to expect that V_{ss} will be rather small. Rosen et al. (70) once compared the polarization of the elastically scattered protons from an even-even nucleus and a neighboring odd-A nucleus. If V_{ss} were strong, the polarizations observed in these two experiments should show some difference but such was not the case. Nevertheless, the experiments with polarized target and projectile were performed in the way described above (61, 63, 71), and it has been reported that V_{ss} is zero within the experimental error. Keep in mind, however, that these results were obtained with [165]Ho as a target. It may be that V_{ss} becomes stronger, as the target gets lighter. See e.g. (72) for the case with [59]Co taken as target.

The CC analyses of straightforward inelastic scattering from deformed nuclei have also been made in several cases; Figure 11 gives an example.

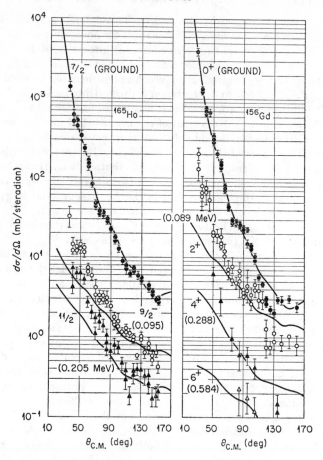

FIG. 11. Scattering of 17 MeV protons by two typical
deformed nuclei; from (76).

The targets chosen here were ^{165}Ho and ^{156}Gd, and they were bombarded
with 17 MeV protons (73). The states excited in this experiment are all
members of the ground-state band, thus their nature is supposed to be quite
simple. For performing CC, the only parameter needed in addition to the
optical-model parameters is the deformation parameter β_2, and with this
single parameter one should be able to fit the cross section to all the states.
The theoretical results shown in Figure 11 with solid lines are obtained by
using the adiabatic CC (4, 74, 75), since it is applicable to these cases, and
the calculations can be done several times faster than with the usual nonadiabatic CC. The agreement with experiment obtained (76) in Figure 11 is
rather satisfactory, except for some discrepancy in most of the inelastic cross
sections at smaller angles. It was noted (76), however, that the unfolding of

the contribution of the high, elastic scattering peak (in the spectrum of the scattered proton) from the inelastic peaks might have been made insufficiently, and thus this disagreement could be only apparent.

An extensive program of (α, α') experiments with several targets in the rare-earth region, and the corresponding CC analyses, were undertaken recently by a Berkeley group (77), and an example is shown in Figure 12. A remarkable thing to be noted here is that in order to fit all the 0^+, 2^+, 4^+, and 6^+ cross-section data, a nonvanishing value with a definite sign had to be assigned to the parameter β_4, in addition to β_2. Similar nonzero values of β_4 were extracted in the same way for other nuclei in this mass region, and the

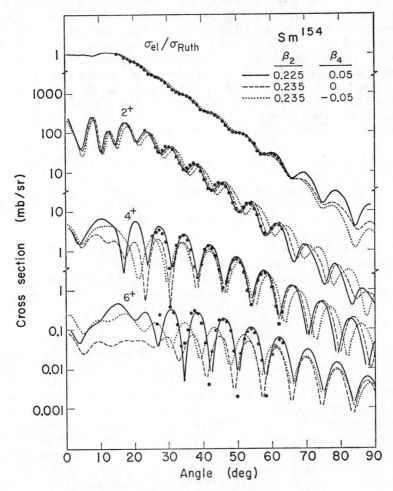

FIG. 12. CC fit to the scattering of 50 MeV α particles by ^{154}Sm, to extract the value of β_4; from (77).

values found are expressed approximately by a linear function of A, being equal to 0.08 for $A \approx 150$, and to -0.08 for $A \approx 180$. The value of β_4, as well as β_2, is calculable if the Nilsson model (78) is taken, and the effects of some of the residual interactions are considered. Such a calculation was made by Möller et al. (79), and the above experimental data were nicely reproduced.

It has been known (e.g. 80) that there exist highly deformed nuclei in the lighter mass region in the periodic table, in addition to the heavier nuclei in the rare-earth and actinide regions. They are found in part of the $2s$-$1d$ shell ($A \approx 19$ to 27, say), and in the $1p$ shell. There are also quite a large number of inelastic scattering data for these nuclei, but the optical model usually does not work very well for such light nuclei, unless the projectile energy is rather high. Nevertheless some CC analyses were made, and one example is that for $^{19}\text{F}(p,p')$ data (81). The fit was attempted to the elastic scattering from the ground-state $\frac{1}{2}^+$ state and the second excited $\frac{5}{2}^+$ state: they were assumed to belong to the ground-state band, and adiabatic CC were used. Moderately good fits to these cross sections were simultaneously obtained for $E_p = 4.26$, 5.96, and 6.87 MeV, by varying only the depth of the imaginary part of the optical potential with E_p. However, the quality of the fit was not so impressive as that achieved for heavier nuclei. A similar analysis was made also for $^{12}\text{C}(n,n')$ data (82), this time considering the excitation of some of the internally excited states, in addition to the members of the ground-state band. The quality of the agreement was more or less similar to that for $^{19}\text{F}(p,p')$.

When the excitation of the internal states is considered, one interesting analysis so far made is that of $^{24}\text{Mg}(\alpha,\alpha')$, particularly in relation to the magnitude of the cross section for excitation of the unnatural parity 3^+ state, at 5.23 MeV. Among the lowest states in ^{24}Mg, the 0^+ (ground), 2^+ (1.37 MeV), and 4^+ (4.12 MeV) states may very well be considered as members of the ground band, while the 3^+ state may be considered as the second member of the γ-vibrational band (9), which has the 2^+ state at 4.23 MeV as its bandhead. The 3^+ cross section is interesting because it can never be excited by a one-step (α,α') process, since the α particle has zero spin and thus in this process the parity and the angular momentum cannot be conserved simultaneously. However, if two-step processes are considered, in which one of the low-lying 2^+ states is excited first, and then the 3^+ state is excited, the simultaneous conservation of the parity and the angular momentum can be satisfied; thus it is no surprise to have a nonvanishing 3^+ cross section. The question is, however, whether the theory can predict the magnitude of that cross section, with reasonable parametrization of the theory. In principle the only new parameter that comes into this calculation, after the others are fixed by fitting the cross sections from the members of the ground band, is the strength of the interaction that couples the channel in the ground and the γ-vibrational band. However, this strength can be fixed if the cross section from the 2^+ state at 4.23 MeV is first fitted. The evaluation of the 3^+ cross section can then be performed as a calculation with *no* adjustable parameter.

If the fit to the data is obtained this way, it shows that the two-step process is indeed the correct explanation of the 3^+ cross section. A first calculation along this line was made by Tamura (83) by fitting the data of Kokame et al. (84). In this experiment, however, the cross sections from the two close-lying 4^+ and 2_2^+ states were not separated. Independent fit to the latter cross section was not possible, and only the sum of the two theoretical cross sections from these two states could be compared with experiment. Nevertheless it could be concluded that the theory predicts the 3^+ cross section with a correct order of magnitude. In a recent improved experiment the cross sections of the above two states were separated, and a CC calculation became possible in the way just described (85). The results are shown in Figure 13. The two sets of curves, solid and dotted lines, drawn in Figure 13 are the CC results thus calculated, with two different sets of optical parameters. Although neither set gives perfect agreement with all the cross sections simultaneously, the agreement achieved here is sufficiently good to make one conclude that the two-step process indeed explains the 3^+ cross section.

The situation is quite different if the unnatural parity state is excited by the inelastic scattering of protons, rather than that of the α particles, because the proton can flip the spin, and thus the simultaneous satisfaction of the angular momentum and parity can be achieved in a one-step process. As is expected in this way, the data of ^{24}Mg(p,p') indeed shows (86, 87) that the 3^+ cross section is comparable to the 2_2^+ cross section. No detailed CC analysis of such data has yet been made.

The final subject in this section is a comparison of the (p,p') data from spherical and deformed nuclei in the same mass region. Let us assume that the intrinsic diffuseness of the optical potential is the same for these two nuclei. If the deformed nucleus forms an unoriented target, however, the effective diffuseness seen by the projectile is much larger than its intrinsic diffuseness, since the geometrical radius is distributed between its longer and shorter axes; the radius does not take a single value as it does for a spherical nucleus. Consequently, one may expect that the pattern of the scattering cross section made by an (unoriented) deformed nucleus is much more blurred than that by a spherical nucleus. That this is indeed the case is seen from Figure 14, where the (p,p') data (88) from ^{148}Sm (spherical) and ^{152}Sm (deformed) are compared. The CC analysis (89) made with the same intrinsic diffuseness agrees with both data quite nicely. This result shows, then, that the scattering data can sometimes be used in determining whether a given nucleus is spherical or deformed.

4. APPLICATION OF THE COUPLED-CHANNEL CALCULATIONS TO PROBLEMS OTHER THAN STRAIGHTFORWARD SCATTERING PROCESSES

4.1 RESONANCE DUE TO CHANNEL COUPLING

The subject of the present subsection differs little from that of the preceding section: we still consider the cross sections of the scattering processes,

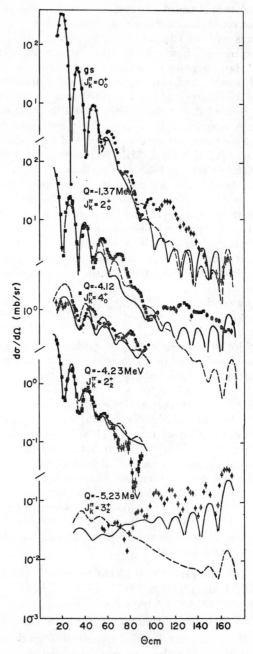

FIG. 13. Excitation of the unnatural-parity 3$^+$ state in ^{24}Mg in the inelastic scattering of 42 MeV α particles. The curves are CC fit with two sets of parameters; from (85).

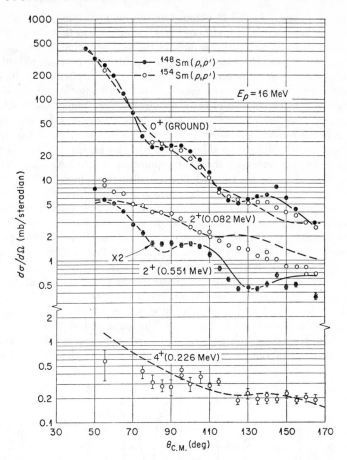

FIG. 14. Comparison of the angular distributions of the scattering of 16 MeV protons by spherical [148]Sm and deformed [154]Sm nuclei; from (5, 89).

but the emphasis is on anomalous behavior of these cross sections as a function of the incident energy of the projectile. Such anomalous behavior can appear if a resonance is involved in one way or the other, but what we want to show here is that the resonance can be predicted, under certain circumstances, by straightforwardly solving the coupled equations, already given in Section 2.

Take as an example the elastic scattering of a proton by [12]C. It is known experimentally (90–95) that the single-particle resonances of the $s_{1/2}$, $d_{5/2}$, and $d_{3/2}$ waves appear in this channel at $E_p = 0.42$, 1.61, and 6.2 MeV, respectively. Since the first excited 2^+ state of ^{12}C lies at 4.43 MeV above the ground state, the energy $E_{p'}$ in the 2^+ channel is given as $E_{p'} = E_p - 4.43$ MeV. Therefore, if $E_p \approx 5$ MeV, $E_{p'} \approx 0.4$ MeV, thus an $s_{1/2}$ single-particle resonance will occur in the 2^+ channel at this energy region. The spin and parity of the resonant state is either $\frac{3}{2}^+$ or $\frac{5}{2}^+$, which means that the amplitude of states of this spin and parity is large in the nuclear region. If the coupling between the 0^+ and 2^+ channels is then considered, the largeness of this amplitude causes the amplitude of a wave in the elastic channel of the same spin and parity to become large also. In other words, the $d_{3/2}$ or $d_{5/2}$ single-particle wave in the elastic channel acquires a large amplitude inside the nucleus and thus behaves quite anomalously in this energy region. This is nothing but a resonance observed in the elastic channel, as well as in the inelastic channel. Note that this occurred in the energy region where no single-particle resonance of the above quantum number was expected to appear. Note also that the coupled-channel calculations have been developed with the concept of the direct process primarily in mind, thus no specific device to predict a resonance is built in the formalism. It is thus quite interesting to see that a resonance comes about in the above way.

The possibility of the appearance of this type of resonance was first pointed out by Tombrello & Phillips (96) and by Okai & Tamura (97), but no detailed comparison with experiment was attempted. By having accurate experimental data, in particular the result of the phase-shift analysis of the elastic scattering cross sections (90–95, 98), a closer test of this idea became feasible recently, and such a calculation was done by Bernard (99). A part of his results is reproduced in Figure 15. The lower and upper curves in this figure show, respectively, δ_r and $\exp(-\delta_i)$, where δ_r and δ_i are the real and imaginary parts of the phase shift of the $d_{3/2}$ wave in the elastic channel.

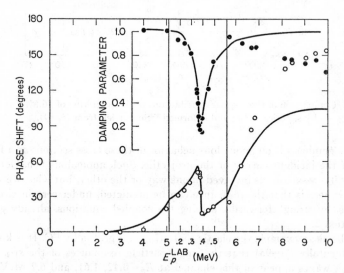

FIG. 15. Energy variation of the real and imaginary parts of the $d_{5/2}$ wave in the scattering of protons by ^{12}C; from (99).

The appearance of an anomaly at $E_p \approx 5$ MeV ($E_p{}^{lab} \approx 5.4$ MeV) is clearly seen experimentally, and it is also seen that the theory fits the data quite nicely. The agreement with experiment gets worse beyond $E_p \approx 6$ MeV, probably because Bernard used a square well for the optical potential, and also restricted to two the number of the partial waves that form the coupled equations. Similar calculations, but with a more realistic potential, and without artificial restriction of the number of the coupled partial waves, were performed by Reynolds et al. (100), and quite good fits to the data were obtained for the elastic scattering of neutrons by ^{12}C. See (101, 102) for similar works.

Another example in which this concept of the resonance is of interest is the analysis of the proton elastic scattering via the isobaric analog resonances (103–105). The coupled equation that comes into play in this case is the Lane equation given in Equation 2.8. We are interested in the solution of this equation when $E_p > 0$, but $\Delta_c - E_p > 0$, so that $E_n < 0$. Under this situation, it may happen that for some value of E_p, E_n will become an eigenenergy of a single-particle neutron state in the potential given by the part of the Lane potential. Since the Lane equations are to be solved always with the boundary condition that there exists a unit-amplitude proton-plane wave at infinity, the amplitude of the neutron wave inside the nucleus is usually small. The latter will become rather large, however, for a value of E_p for which the corresponding E_n equals the above eigenenergy. Because of the coupling between the proton and neutron channels, this situation is reflected back into the proton channel, causing an anomalous behavior to appear there. This is the analog resonance, as understood in the framework of the Lane equation. An example (103) for ^{92}Mo(p,p_0) data (106) is given in Figure 16 and the agreement is seen to be quite good. The validity of the Lane model for isobaric analog resonances can, however, be questioned, and this point will be discussed in Section 5.

4.2 Coupled-Channel Calculations Applied to Bound-State Problems

In constructing the coupled-channel equations of 2.6, there has been no assumption that all the energies E_n must be positive. Indeed in the example of the analog resonance given in the last subsection, one energy, the neutron energy, was taken to be negative. Here we consider the case in which all the E_n are negative. To solve the coupled equations under this situation is an eigenvalue problem, rather than a scattering problem.

Let us consider the following example. Take a spherical even-even nucleus in its ground state Φ_0, and add to it a proton occupying an eigenstate with spin (jm) and wavefunction ψ_j, so as to form the ground state of some odd-A nucleus. The latter's wavefunction may be written as

$$\Phi_0 \psi_{jm} \qquad\qquad 4.1$$

Let the separation energy of this proton be B, so that the radial part of ψ_j behaves for $r \to \infty$ as exp $(-\kappa_B r)$, where $\kappa_B = (2mB/\hbar^2)^{1/2}$. We next consider

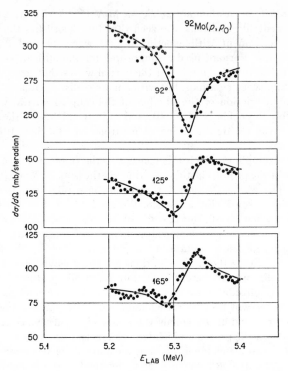

FIG. 16. CC fit to the elastic scattering of protons by ^{92}Mo via an isobaric analog resonance; from (103).

the excitation of the even-even (core) nucleus to a vibrational 2^+ state with an energy $E(2^+)$ and the wavefunction Φ_2. If the concept of the weak coupling model (45, 46) is to apply here, an improved wavefunction of the ground state of our odd-A nucleus may be written as a linear combination of 4.1 and

$$(\Phi_2 \otimes \psi_j)_{jm} \qquad\qquad 4.2$$

Rigorously speaking (107), the use of ψ_j in 4.2 is not correct, in particular in the asymptotic region, because ψ_j corresponds to the binding energy B, while the single-particle proton wavefunction $\hat{\psi}_j$, say, which should have appeared in 4.2 in place of ψ_j, should have corresponded to the binding energy $\hat{B} = B - E(2^+)$; i.e. it should behave for $r \to \infty$ as exp $(-\kappa_{\hat{B}} r)$ with $\kappa_{\hat{B}} = (2m \hat{B}/\hbar^2)^{1/2}$, and not as exp $(-\kappa_B r)$. In the usual nuclear-structure calculations, the error induced by using $\hat{\psi}_j$ for ψ_j in 4.2 is not serious. However, when $\hat{\psi}_j$ is to be used, e.g., as the form factor of the proton in a $DWBA$ calculation for a $(^3$He,$d)$ reaction, say, the incorrect asymptotic behavior of $\hat{\psi}_j$ may be serious, because the dominant contribution to the reaction amplitude comes from the part of $\hat{\psi}_j$ in the region beyond the nuclear radius.

It is, of course, possible in the framework of the above procedure to let both ψ_j and $\hat{\psi}_j$ have correct asymptotic forms simultaneously. Since the calculation used to obtain these wavefunctions is an eigenvalue problem, however, it means that one assumes different potentials in the 0^+ and 2^+ channels, a procedure which introduces too much arbitrariness into the theory.

All these difficulties will be avoided if we start with the coupled equation 2.6 and let the solutions $R_{Jnlj}(r)$ play the role of the radial parts of ψ_j and $\hat{\psi}_j$. Because of the relation $E_n = E_1 - \epsilon_n$, as given below in 2.6, it is clear that the above-requested asymptotic behavior of the radial wavefunctions is automatically satisfied if E_1 is taken equal to $-B$ in Equation 2.6. The problem is again an eigenvalue problem, thus the depth of the potential has to be adjusted to make $-B$ an eigensolution of Equation 2.6. The same depth of the potential is used, nevertheless, in all the channels in this approach.

An example in which this technique was nicely used is obtaining the spectroscopic factor $S_>$, associated with the single-neutron pickup reactions ending up with an isobaric analog state. It has been known that (108) if $S_>$ is derived from the usual $DWBA$ procedure, it is too large by a factor $1.5 \sim 2.0$ as compared to what is expected from the French-MacFarlane sum rule (109). Since the final state of the concerned reaction is an isobaric analog state, however, a little thought allows us to see that the form factor of the picked-up neutron had better be obtained as a part of the solution of the Lane equation 2.8, solved with E_p, $E_n < 0$, rather than being taken as a simple, single-particle solution as usual in $DWBA$. If the former has greater amplitude in the asymptotic region than does the latter, the use of the former makes the derived $S_>$ smaller than what is derived with the use of the latter. An example (110, 111) of such a comparison is shown in Figure 17, and as is seen the solution of the Lane equation does have a larger amplitude than the usual form factor. A similar result was obtained by Rost (112), who also solved Equation 2.6 with all $E_n < 0$ for a few deformed nuclei. A similar calculation for an odd-A spherical nucleus was also reported by Tamura (113).

Once the use of the coupled Equation 2.6 is found to give a correct way of calculating the form factors, it may be worthwhile to try to solve it approximately, rather than exactly. We shall not go into detail of this problem here, but refer the readers to (114–119). In many of these works, the coupling interaction V_{coupl} in 2.6 is taken as a two-body residual interaction, rather than the phenomenological one of the Bohr-Mottelson model (9). Thus the equation to be solved is in general a set of coupled integrodifferential equations, rather than a set of coupled differential equations.

4.3 COUPLED-CHANNEL BORN APPROXIMATION ($CCBA$)

In the usual $DWBA$ (1–3), the amplitude e.g. of a (d,p) process is written, in a very brief notation, as

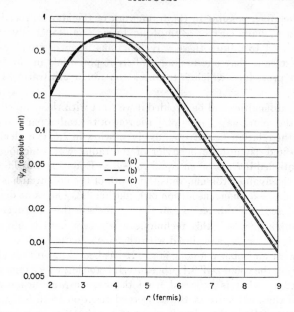

r (fermis)

FIG. 17. Form factors of a neutron to be picked up from ^{52}Cr to form an isobaric analog state in ^{51}Cr. Curve (a) is the normal shell-model wavefunction; curves (b) and (c) are results of CC with, respectively, volume and surface type form factors for the (t· T) interaction of the Lane model; from (110, 111).

$$\langle \chi_p^{(-)} \phi_n \mid V_{np} \mid \chi_d^{(+)} \rangle \qquad\qquad 4.3$$

Here $\chi_d^{(+)}$ and $\chi_p^{(-)}$ are the wavefunctions, the distorted waves, that describe the elastic scattering of the deuterons and the protons, respectively, by the target and the residual nucleus. The superscript $(+)$ attached to $\chi_d^{(+)}$ means that it has a boundary condition such that it consists of a plane wave propagating in a direction $\mathbf{k}_d$, say, and outgoing spherical waves. Correspondingly, the superscript $(-)$ on $\chi_p^{(-)}$ means that it consists of a plane wave that propagates in a direction $\mathbf{k}_p$, say, and incoming spherical waves.

In the practical calculations, these distorted waves are separated into partial waves, each specified by the quantum numbers (lj). The neutron that is captured also goes into an orbit with a definite set of these quantum numbers, and thus its wavefunction ϕ_n in 4.3 may be written more explicitly as $\phi_{n;j_n l_n}$. The amplitude 4.3 may now be written as

$$\sum_{j_p l_p j_d l_d} \langle \chi_{p;j_p l_p}^{(-)} \phi_{n;j_n l_n} \mid V_{np} \mid \chi_{d;j_d l_d}^{(+)} \rangle \qquad\qquad 4.4$$

In the form of 4.4, the angular-momentum and parity selection rules for each matrix element are clear, since the interaction V_{np} is a scalar operator. However, the roles that the spins I and the parities π of the target and residual nuclei play are still to be considered. If the wavefunctions of these nuclei are written respectively, as $\Phi_{I_i M_i}{}^{\pi i}$ and $\phi_{I_f M_f}{}^{\pi f}$, 4.4 is further rewritten as

$$\sum_{J j_p l_p j_d l_d} \langle (\chi_{p; j_p l_p}{}^{(-)} \otimes \Phi_{I_f}{}^{\pi f})_{JM} \mid V_{np} \mid (\chi_{d; j_d l_d}{}^{(+)} \otimes \Phi_{I_i}{}^{\pi i})_{JM} \rangle$$

$$= \sum_{J j_p l_p j_d l_d} \langle (\chi_{p; j_p l_p}{}^{(-)} \otimes (\phi_{n; j_n l_n} \otimes \Phi_{I_i}{}^{\pi i}) I_f)_{JM} \mid V_{np} \mid (\chi_{d; j_d l_d}{}^{(+)}$$

$$\otimes \Phi_{I_i}{}^{\pi i})_{JM} \rangle \qquad 4.5$$

and the selection rules that are to be satisfied in each matrix element are now summarized as follows:

$$I_i + j_n = I_f, \ I_i + j_d = J = I_f + j_p$$

$$\pi_f = \pi_i \times (-)^{l_n}, \ \pi_f \times (-)^{l_p} = \pi_i \times (-)^{l_d} \qquad 4.6$$

Consider, e.g., the $^{24}\text{Mg}(d,p)^{25}\text{Mg}^*(\frac{7}{2}^+)$ process. Since $I_i{}^{\pi i}=0^+$ and $I_f{}^{\pi f}=\frac{7}{2}^+$, the selection rules of 4.6 require that $(j_n l_n)=(\frac{7}{2}, 4)$, i.e. that the neutron must be captured into a $g_{7/2}$ orbit. It is well known that in light nuclei such as ^{25}Mg, the single-particle orbit $g_{7/2}$ lies so high that its amplitude, if any, is very small, at least for the low-lying states which we are interested in. Therefore the amplitude 4.5 for the above process is predicted to be very small, and contradicts experiment (120) which shows that the cross section to the $\frac{7}{2}^+$ state in ^{25}Mg is not small.

Consider, then, that the deuteron is subjected to an inelastic scattering before the stripping reaction occurs. As is seen from Equation 2.1, the ket vector in 4.5 can be rewritten as a sum of terms that belong to different (scattering) channels. We may thus rewrite 4.3 as

$$\sum_{I_i, I_f} (\text{amplitude of } 4.3) \qquad 4.7$$

Taking again the above $^{24}\text{Mg}(d,p)^{25}\text{Mg}^*(\frac{7}{2}^+)$ process, let us consider two terms in 4.5 that correspond to $I_i{}^{\pi i}=0^+$ and 2^+; i.e. we consider the excitation of the first excited 2^+ state in ^{24}Mg, in addition to the elastic scattering. The first term is very small since it is the same as in 4.5. However, if $I_i{}^{\pi i}=2^+$ is used in the selection rule 4.6, it is clear that $(j_n l_n)$ can now take the values $(\frac{3}{2}, 2)$, $(\frac{5}{2}, 2)$, ... The corresponding term can then be fairly large, since the amplitudes with which a neutron occupies the $d_{3/2}$ and $d_{5/2}$ orbits in ^{25}Mg can be rather large. Note that the amplitude of the distorted wave $\chi_{j_d l_d}{}^{(+)}$ will be smaller in the excited channel than it is in the elastic channel. Therefore the cross section obtained will not be as large as what one gets in the usual $DWBA$, with its leading term nonvanishing in spite of any selection rule. Still the correction term due to $CCBA$ can be much larger than the $DWBA$ term, if the latter is suppressed by some selection rules.

Detailed formulation of $CCBA$ was first made by Penny & Satchler (121), and some numerical calculation was made by Penny (122). In the examples considered by Penny, however, the leading $DWBA$ term was fairly large, so that the consideration of $CCBA$ led to only a small correction to it; thus the result was not particularly exciting (see also 123). A more interesting application was then attempted by Iano & Austern (124), to analyze the above-mentioned $^{24}Mg(d,p)^{25}Mg^*(\frac{7}{2})$ data (120) by taking properly into account the fact that ^{24}Mg and ^{25}Mg are highly deformed nuclei, and also taking into account the contribution of the inelastic scattering of protons, before they leave the nuclear region. In spite of these improvements, they used, instead of CC, the approximation of Austern & Blair (35), which will be a rather poor approximation when weakly absorbed particles like protons and deuterons are considered. Thus the obtained cross section was not in very good agreement with experiment, though not very much off in magnitude.

Perhaps the best calculation so far reported along this line is that of Dupont & Chabre (125) who analyzed the data of $^{12}C(d,^3He)^{11}B^*(\frac{7}{2}^-)$ for $E_d = 28$ and 50 MeV (126, 127). Since the selection rules 4.6 require that the picked-up proton must have occupied an $f_{7/2}$ orbit, and since such an orbit can hardly be occupied by nuclei as light as ^{12}C, the necessity of $CCBA$ is obvious. The above authors performed the $CCBA$ calculation in exactly the way described by 4.7, and the results are compared with experiment in Figure 18. The agreement with experiment, though not perfect, is sufficiently good to convince us that $CCBA$ is indeed effectively working. Note that the coupling strength between the 0^+ and 2^+ channels in the initial $d+^{12}C$ system has been fixed in fitting the inelastic cross section to the 2^+ state (upper portion of Figure 18), while the strength of V_{pn} in 4.7, or more precisely in 4.5, has been fixed by fitting the $(d,^3He)$ cross section to the $\frac{3}{2}^-$ ground state of ^{11}B. Therefore the $CCBA$ for the $\frac{7}{2}^+$ state is a calculation without any adjustable parameter, and the agreement to this cross section seen in the lower portion of Figure 18 is indeed significant.

The application of $CCBA$ has allowed us to perform rather interesting calculations to analyze data of various transmutation reactions, to or via the isobaric analog states. If we use once again the Lane equations to couple the proton and neutron channels, it is not very difficult to see that the $CCBA$ amplitude 4.7 takes the following very simplified form (128, 129)

$$\langle \chi_p^{(-)}\phi_n \mid V_{np} \mid \chi_d^{(+)} \rangle - \frac{1}{\sqrt{2T_0+1}} \langle \chi_n^{(-)}\phi_p \mid V_{np} \mid \chi_d^{(+)} \rangle \qquad 4.8$$

In order to explain the meaning of 4.8 a little more closely, let us take the $^{90}Zr(d,p)^{91}Zr_{gr}$ process as an example. The first term of 4.8 is just the usual $DWBA$ amplitude for this process. On the other hand, the second term describes the amplitude of a process in which a (d,n) process occurs first to form a final state $^{91}Nb^A = T_-{}^{91}Zr_{gr}$, but the neutron produced in this way is converted into a proton before it leaves the nuclear region, owing to the charge exchange interaction, the nucleus $^{91}Nb^A$ being transformed into $^{91}Zr_{gr}$ at the same time. In this way, the second term of 4.8 works as a correction term to the usual $DWBA$ amplitude.

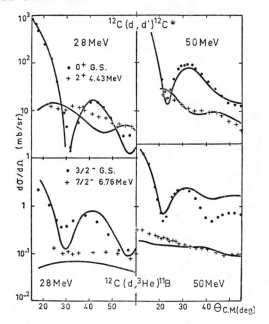

FIG. 18. Lower portion; *CCBA* fit to $^{12}C(d,^3He)^{11}B$ with 28 and 50 MeV deuterons. Upper portion; corresponding *CC* fit of the scattering. From (125).

In most cases the contribution of the second term of 4.8 is not significant. A very interesting thing happens, however, if $E_d = \Delta_c - Q$, because as can easily be seen, E_n, the energy of the neutron produced in the above process, is about zero for this E_d. In other words this E_d is the threshold of the above (d,n) process. Thus the second term of 4.8 might behave anomalously, giving rise to some anomaly in the (d,p) cross section in this energy region. That such is indeed the case was found by Moore et al. (130–141), and a typical datum is shown in Figure 19. In this figure the theoretical result (142) of using 4.8 is also shown, and the fit is quite satisfactory. Calculation was made (142) also for $^{90}Zr(d,n)^{91}Nb^4$, and good agreement with data (143) was obtained there too.

If the energy E_d were somewhat smaller than it was in the above threshold region, the corresponding E_n may become equal to an eigenenergy in its potential. If the corresponding proton channel is considered at the same time, this is just the situation in which the isobaric analog resonance is formed; see the end of Section 4.1. It is thus expected that some anomaly in the (d,p) cross section will appear in this energy region too. Such an expectation was answered affirmatively by an experiment by Hamburger (144), and the results are shown in Figure 20. In this figure theoretical calculations by Tamura (5) to fit the (p,p_0) and (d,p) data simultaneously are shown, and although the fit is not very good, it succeeded in reproducing the qualitative

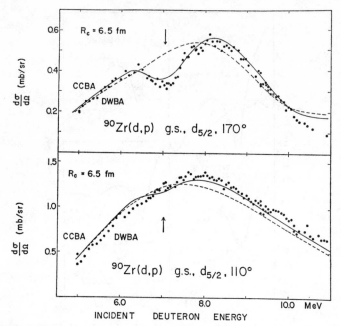

FIG. 19. Threshold anomaly in ^{90}Zr(d,p). The solid lines are CC fit; the dotted lines are normal $DWBA$ fit; from (142).

nature of the data, which suggests that the $CCBA$ approach is basically correct. A further discussion on this point will be given in Section 5.

5. SCOPE OF FUTURE WORK

We have seen in Sections 3 and 4 that CC has been carried out rather successfully in explaining many of the existing data of various nuclear reactions, and then in extracting information pertaining to nuclear structure. However, these calculations, in particular those of the straightforward inelastic scattering cross sections, were made almost exclusively by using the phenomenological model of Bohr & Mottelson, and its extensions. Since great progress has been made in the past years in the theory of the nuclear many-body problem, including the possibility of describing the collective states in a microscopic way, it would be very natural to investigate the possibility of performing the above sort of calculations in a microscopic manner.

In the past few years, indeed, a large amount of work along this line has been reported (145–152). It is my opinion, however, that comparatively little has been achieved, except perhaps for cases with very high-energy projectiles, where the use of the impulse approximation is possible (153, 154). Generally speaking, the success of the microscopic calculation would be expected only if one has a very *good* knowledge of the microscopic wavefunctions of the target states, and of the effective two-body interactions that cause the inelastic scattering. However, at present, the knowledge of none of these seems sufficiently good, although it is known that the random-phase approximation (155–158) allows one to describe the one-phonon-type excited state fairly nicely. The situation looks better with high-energy projectiles, as stated above, because in using the impulse approximation, the t matrix derived from known two-body interaction is used to play the role of a *good* effective interaction. No corresponding situation exists so far for the lower-energy projectiles and thus very phenomenological two-body interactions are used. One possible way out of this difficulty may be to extend, to the continuum, the work of Kuo & Brown (159), which has to date been quite successful in deriving, from the known two-body interaction, the effective interactions to be used in bound-state structure problems.

In spite of the above criticism, there are cases in which the microscopic approach has played an important role. An example is found in the discussion of inelastic scattering with spin-flip, such as the case of $^{24}Mg(p,p')$ to the 3^+ state (87) considered in Section 3.2. It should nevertheless be noted

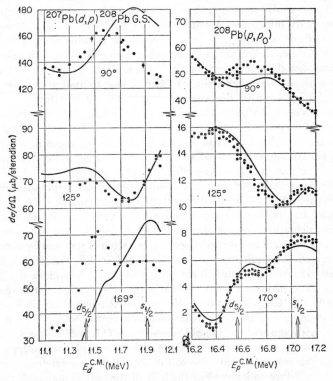

FIG. 20. *CC* explanation of the Hamburger effect; from (5).

that the spin-flip can also be described fairly nicely within the framework of the phenomenological model, if deformation of the spin-orbit potential is also considered, together with the rest of the optical potential (160). A subject which can be discussed only in the microscopic theory is the contribution of the exchange effect to the inelastic scattering amplitude, which according to (150, 152) seems to be rather significant. It does not seem clear yet, however, how seriously such a conclusion is to be taken, because the calculation has again been made with a phenomenological two-body interaction. Note that in either $DWBA$ or CC the elastic scattering is described in terms of the optical model, where no exchange effect is considered explicitly, with the understanding that it is already somehow embodied in the optical potential. However, the optical potential takes into account not only what happens in the elastic channel, but also what occurs in all the channels including the inelastic scattering channels. Therefore, in a sense, both the exchange effect and the nonexchange effect occurring in the elastic process must have been included somehow in the optical potential. It thus seems quite premature to draw any strong conclusions about the above question, until one knows how to derive the effective interaction from bare two-body interactions, at the same time deriving the optical potential from the same bare two-body interaction. We have seen in Section 3.1 that a phenomenological analysis of $^{114}Cd(p,p')$ data gave rise to an interesting problem of deriving the imaginary part of the optical potential from first principles. It is thus seen that whatever approach one may start with, one is likely to end up with the very challenging problem of how to understand the optical potential in a very much deeper sense than it is now understood.

Recently several attempts have been made to extend the coupled-channel calculations to processes other than scattering (161–163). We do not mean by this the $CCBA$ discussed in Section 4.3, but an approach in which a set of coupled equations is derived so as to couple the incoming deuteron and the outgoing proton channels together, if a (d,p) process is considered as an example. As is easily seen the coupled equation thus derived is not a differential equation, but an integrodifferential equation, making the numerical task much harder than otherwise. So far no numerical results were reported on such calculations, except those in which several drastic approximations were made so that the integrodifferential equation is replaced by a differential equation. The numerical problem involved is then the same as it is in the scattering problem, but it is rather difficult to see how much significance one can place with such a calculation. Another relevant point is that it is often claimed (162, 163) that the optical potential, particularly the imaginary part of it, to be used to fit the deuteron elastic scattering data is very much different with and without the above coupling. This is of course what is expected, and the numerical value of the difference is of some interest. Still, if this is the *only* conclusion one can derive after a tremendous numerical calculation, before understanding the bulk of the optical potential, the sense of putting forth such an effort may be doubted. Note that in (d,p) there are usually a large number of equally strong outgoing channels. Thus the truncation of the number of channels which was allowed with a fairly high confidence in the application of CC for the scattering problem will not be applicable here. In other words, to draw any meaningful conclusions out of such a calculation, a huge set of coupled integrodifferential equations would have to be solved all the time. It is my suspicion that a more authentic three-body approach (e.g. 164, 165) rather than this sort of coupled-channel calculation might be a more profitable way to improve $DWBA$.

We saw in Section 4 that the Lane equation was quite useful in explaining various reactions to or via isobaric analog states, but that it nevertheless failed sometimes to allow for very accurate discussions of the reaction mechanisms involved. For example, in explaining the Hamburger effect, the concept of isospin-selection rule plays a vital role, but the Lane equation embodies this concept only vaguely. It is thus very desirable that a theory is constructed so that various delicate features of the isobaric analog state are more accurately and more closely described. Such a work is now under way at the University of Texas. See the works (168, 169).

Similar comment may also apply to the use of CC to produce resonant scattering. While it is an interesting calculation, it may become too tedious if resonances to be considered are more complicated than those discussed in Section 4.1. In that event the use of more conventional resonance theory might be more effective. See for example a work of Buttle (166) who analyzed the resonant scattering of protons by ^{12}C by using the R-matrix theory (167).

The author is indebted to Dr. W. R. Coker for a careful reading of the manuscript.

LITERATURE CITED

1. Tobocman, W., *Theory of Direct Nuclear Reactions* (Oxford Univ. Press, 1961)
2. Austern, N., in *Selected Topics in Nuclear Theory* (Janouch, F., Ed., IAEA, Vienna, 1963)
3. Satchler, G. R., *Nucl. Phys.*, **55**, 1 (1964)
4. Tamura, T., *Rev. Mod. Phys.*, **37**, 679 (1965)
5. Tamura, T., *Proc. Intern. Conf. Nucl. Struct. Tokyo, 1967*, 288 (Sanada, J., Ed.)
6. Tamura, T., *Proc. Intern. Symp. Nucl. Struct. Dubna, 1968* (To be published)
7. Tamura, T., *Oak Ridge Natl. Lab. Rept. ORNL 1111* (1967)
8. Tamura, T., *Progr. Theoret. Phys. (Kyoto)*, *Suppl.*, **37/38**, 386 (1966)
9. Bohr, A., Mottelson, B. R., *Kgl. Danske Mat.-Fys. Medd.*, **27**, No. 16 (1953)
10. Hodgson, P. E., *Ann. Rev. Nucl. Sci.*, **17**, 1 (1967)
11. Lane, A. M., *Nucl. Phys.*, **35**, 676 (1962)
12. Robinson, R. L., Ford, J. L. C., Jr., Stelson, P. H., Satchler, G. R., *Phys. Rev.*, **146**, 816 (1966)
13. Ford, J. L. C., Jr., Wong, C. Y., Tamura, T., Robinson, R. L., Stelson, P. H., *Phys. Rev.*, **158**, 1194 (1967)
14. Stelson, P. H., Ford, J. L. C., Jr., Robinson, R. L., Wong, C. Y., Tamura, T., *Nucl. Phys.*, **A119**, 14 (1968)
15. Robinson, R. L., Ford, J. L. C., Jr., Wong, C. Y., Stelson, P. H., Tamura, T. (Unpublished data)
16. Alder, K., Bohr, A., Huus, T., Mottelson, B. R., Winther, A., *Rev. Mod. Phys.*, **28**, 432 (1955)
17. deBoer, J., Eichler, H., *Advan. Nucl. Phys.*, **1**, 1 (1968)

18. Sakai, M., Tamura, T., *Phys. Letters*, 10, 323 (1964)
19. Buck, B., *Phys. Rev.*, 130, 712 (1963)
20. Tamura, T., *Phys. Letters*, 28B, 90 (1968)
21. Tamura, T., Udagawa, T., *Phys Rev.*, 150, 783 (1966)
22. deBoer, J., Stockstad, R. G., Symons, G. D., Winther, A., *Phys. Rev. Letters*, 14, 564 (1965)
23. Stelson, P. H., Milner, W. T., Ford, J. L. C., Jr., McGowan, F. K., Robinson, R. L., *Bull. Am. Phys. Soc.*, 10, 427 (1965)
24. Simpson, J. J., Eccleshall, D., Yates, M. J. L., Freeman, N. J., *Nucl. Phys.*, A94, 177 (1967)
25. DoDang, G., Dreizler, R., Klein, A., Wu, C. S., *Phys. Rev. Letters*, 17, 709 (1966)
26. Kisslinger, L. S., Kumar, K., *Phys. Rev. Letters*, 19, 1239 (1967)
27. Jolly, R. K., Tamura, T., *Phys. Letters*, 18, 295 (1965)
28. Hsu, L. S., *Nucl. Phys.*, A96, 624 (1967)
29. Jolly, R. K., *Phys. Rev.*, 139B, 318 (1965); (Personal communication)
30. Kumabe, I., Ogata, H., Tomita, S., Inoue, M., Okuma, Y., *Phys. Letters*, 17, 45 (1965)
31. Meriwether, J. R., Bussiere de Nercy, A., Harvey, B. G., Horen, D. J., *Phys. Letters*, 11, 229 (1964)
32. Inopin, E. V., *Zh. Eksperim. Teor. Fiz.*, 31, 901 (1956) [Engl. transl., *Soviet Phys. JETP*, 4, 784 (1957)]
33. Blair, J. S., *Phys. Rev.*, 115, 928 (1958)
34. Tamura, T., *Nucl. Phys.*, 73, 81 (1965)
35. Austern, N., Blair, J. S., *Ann. Phys. (N. Y.)*, 33, 15 (1965)
36. Matsuda, K., *Nucl. Phys.*, 33, 536 (1962)
37. Koike, M., Nonaka, I., Kokame, J., Kamitsubo, H., Awaya, Y., Wada, T., Nakamura, H., *Proc. Intern. Conf. Nucl. Struct. Tokyo, 1967*, 696
38. Baba, C. V. K., Ewan, G. T., Saurez, J. F., *Nucl. Phys.*, 43, 264, 285 (1963)
39. Metzger, F. R., *Phys. Rev.*, 137B, 1415 (1967)
40. Bhatt, K. H., *Phys. Letters*, 17, 282 (1965)
41. Lipas, P. O., *Nucl. Phys.*, 82, 91 (1966)
42. Pramila, G. C., Middleton, R., Tamura, T., Satchler, G. R., *Nucl. Phys.*, 61, 448 (1965)
43. Lark, N. L., Morinaga, H., *Nucl. Phys.*, 63, 466 (1965)
44. Ford, J. L. C., Jr., Wong, C. Y.,

Tamura, T., Robinson, R. L., Stelson, P. H., *Phys. Rev.*, 158, 1194 (1967)
45. Choudhuri, D. C., *Kgl. Danske Mat.-Fys. Medd.*, 28, No. 1 (1954)
46. de-Shalit, A., *Phys. Rev.*, 122, 1530 (1960)
47. Yoshida, S., *Proc. Roy. Soc. (London)*, A, 69, 668 (1956)
48. Hughes, D. J., Zimmerman, R. L., Chrien, R. E., *Phys. Rev. Letters*, 1, 461 (1958)
49. Feshbach, H., Porter, C. E., Weisskopf, V. F., *Phys. Rev.*, 96, 448 (1954)
50. Margolis, B., Troubetzkoy, E. S., *Phys. Rev.*, 106, 105 (1957)
51. Chase, D. M., Wilets, L., Edmonds, A. R., *Phys. Rev.*, 110, 1080 (1958)
52. Buck, B., Perey, F. G., *Phys. Rev. Letters*, 11, 444 (1962)
53. Saplakoglu, A., Bollinger, L. M., Coté, R. E., *Phys. Rev.*, 109, 1258 (1958)
54. Coté, R. E., Bollinger, L. M., LeBlanc, J. M., *Phys. Rev.*, 111, 288 (1958)
55. LeBlanc, J. M., Coté, R. E., Bollinger, L. M., *Nucl. Phys.*, 14, 120 (1959/60)
56. Weston, L. W., Seth, K. K., Bilpuch, E. G., Newson, H. W., *Ann. Phys. (N.Y.)*, 10, 477 (1960)
57. Desjardins, J. S., Rosen, J., Havens, W. W., Jr., Rainwater, L. J., *Phys. Rev.*, 120, 2214 (1960)
58. Divadeenam, M. (Doctoral thesis, Duke Univ., Durham, N. C., 1967)
59. Brink D. M., Satchler, R. G., *Angular Momentum* (Oxford Univ. Press, 1962)
60. Davies, K. T. R., Satchler, G. R., Drisko, R. M., Bassel, R. H., *Nucl. Phys.*, 44, 607 (1963)
61. Wagner, R., Miller, P. D., Tamura, T., Marshak, H., *Phys. Letters*, 10, 316 (1964); *Phys. Rev.*, 139B, 29 (1965)
62. Marshak, H., Richardson, A. C. B., Tamura, T., *Phys. Rev.*, 150, 996 (1966)
63. Fisher, T. R., Safrata, R. S., Shelley, E. G., McCarthy, J., Austin, S. M., Barrett, R. C., *Phys. Rev.*, 157, 1149 (1967)
64. McCarthy, J. S., Fisher, T. R., Shelley, E. G., Safrata, R. S., Healey, D., *Phys. Rev. Letters*, 20, 502 (1968)
65. Marshak, H., Langsford, A., Wong, C. Y., Tamura, T., *Phys. Rev. Letters*, 20, 554 (1968)
66. Bowen, P. H., Scanlon, J. P., Stafford, G. H., Thresher, J. J., Hodg-

son, P. E., *Nucl. Phys.*, **22**, 640 (1961)

67. Perey, F. G. (Unpublished data)

68. Lawson, J. D., *Phil. Mag.*, **44**, 102 (1953)

69. Peterson, J. M., *Phys. Rev.*, **125**, 955 (1962)

70. Rosen. L., Brolley, J. E., Jr., Stewart, L., *Phys. Rev.*, **121**, 1423 (1961)

71. Kobayashi, S., Kamitsubo, H., Katori, K., Uchida, A., Imaizumi, A., Nagamine, K., *J. Phys. Soc. Japan*, **22**, 368 (1967)

72. Kobayashi, S., Nagamine, K., Imaizumi, M., Uchida, A., Katori, K., *Proc. Intern. Conf. Nucl. Struct. Tokyo, 1967*, 699

73. Lieber, A., Whitten, C. A., *Phys. Rev.*, **132**, 2582 (1963)

74. Drozdov, S. I., *Zh. Eksperim. Teor. Fiz.*, **28**, 735, 736 (1955) [Engl. transl., *Soviet Phys. JETP*, **1**, 588, 591 (1955)]

75. Barrett, R. C., *Nucl. Phys.*, **51**, 27 (1964)

76. Tamura, T., *Phys. Letters*, **9**, 334 (1964)

77. Hendrie, D. L., Glendenning, N. K., Harvey, B. G., Jarvis, O. N., Duhm, H. H., Sansinos, J., Mahoney, J., *Phys. Letters*, **26B**, 127 (1968)

78. Nilsson, S. G., *Kgl. Danske Mat.-Fys. Medd.*, *No. 16* (1955)

79. Möller, P., Nilsson, B., Nilsson, S. G., Sobiezewski, A., Szymanski, Z., Wycech, S., *Phys. Letters*, **26B**, 418 (1968)

80. Mottelson, B. R., Nilsson, S. G., *Danske Videnskab. Selskab Mat.-Fys. Skr.*, **1**, *No. 8* (1959)

81. Thompson, W. J., Edwards, S., Tamura, T., *Nucl. Phys.*, **A105**, 678 (1967)

82. Grin, G. A., Joseph, C., Wong, C. Y., Tamura, T., *Phys. Letters*, **25**, 387 (1967)

83. Tamura, T., *Nucl. Phys.*, **73**, 241 (1965)

84. Kokame, J., Fukunaga, K., Inoue, N., Nakamura, H., *Phys. Letters*, **8**, 342 (1964)

85. Vincent, J. S., Boschitz, E. T., Priest, J. R., *Phys. Letters*, **25B**, 81 (1967)

86. Crawley, G. M., Garvey, G. T., *Phys. Rev.*, **160**, 981 (1967)

87. Rush, A. A., Ganguly, N. K., *Nucl. Phys.*, **A117**, 101 (1968)

88. Stoler, P., Slagowitz, M., Makofske, W., Kruse, T., *Phys. Rev.*, **155**, 1334 (1967)

89. Slavik, C. J., Tamura, T. (Unpublished data)

90. Jackson, H. L., Galonsky, A. I., *Phys. Rev.*, **89**, 370 (1953)

91. Reich, C. W., Phillips, G. C., Russel, J. L. J., Jr., *Phys. Rev.*, **104**, 143 (1956)

92. Moss, S. J., Haeberli, W., *Nucl. Phys.*, **72**, 417 (1965)

93. Barnard, A. C. L., Swint, J. B., Clegg, T. B., *Nucl. Phys.*, **86**, 130 (1966)

94. Barnard, A. C. L., Duval, J. S., Jr., Swint, J. B., *Phys. Letters*, **20**, 412 (1966)

95. Duval, J. S., Jr., Barnard, A. C. L., Swint, J. B., *Nucl. Phys.*, **A93**, 177 (1967)

96. Tombrello, T. A., Phillips, G. C., *Nucl. Phys.*, **20**, 648 (1960)

97. Okai, S., Tamura, T., *Nucl. Phys.*, **31**, 183 (1962)

98. Barnard, A. C. L., Swint, J. B., Clegg, T. B., *Nucl. Phys.*, **86**, 130 (1966)

99. Barnard, A. C. L., *Phys. Rev.*, **155**, 1135 (1967)

100. Reynolds, J. T., Slavik, C. J., Lubitz, C. R., Francis, N. C., *Phys. Rev.*, **176**, 1213 (1968)

101. Pisent, G., Saruis, A. M., *Nucl. Phys.*, **A91**, 561 (1967)

102. Roeder, J. L., *Ann. Phys. (N. Y.)*, **43**, 382 (1967)

103. Tamura, T., in *Isobaric Spin in Nuclear Physics*, 447 (Robson, D., Fox, J. D., Eds., Academic Press, New York., 1966)

104. Bondorf, J. P., Jagare, S., Lutken, H., *Phys. Letters*, **21**, 185 (1966)

105. Auerbach, E. H., Dover, C. B., Kerman, A. K., Lemmer, R. H., Schwarcz, E. H., *Phys. Rev. Letters*, **17**, 1184 (1966)

106. Moore, C. F., Richard, P., Watson, C. E., Robson, D., Fox, J. D., *Phys. Rev.*, **141**, 1166 (1966)

107. Pinkston, W. T., Satchler, G. R., *Nucl. Phys.*, **72**, 641 (1965)

108. Sherr, R., Bayman, B. F., Rost, E., Rickey, M. E., Hoot, C. G., *Phys. Rev.*, **129**, B1272 (1965)

109. French, J. B., MacFarlane, M. H., *Nucl. Phys.*, **26**, 168 (1961)

110. Stock, R., Tamura, T., *Phys. Letters*, **22**, 304 (1966)

111. Stock, R., Bock, R., David, P., Duhm, H. H., Tamura, T., *Nucl. Phys.*, **A104**, 136 (1967)

112. Rost, E., *Phys. Rev.*, **154**, 994 (1967)

113. Tamura, T., *Phys. Letters*, **22**, 644 (1966)

114. Austern, N., *Phys. Rev.*, **136**, 1743 (1964)

115. Prakash, A., *Phys. Rev. Letters*, **20**, 864 (1968)
116. Kawai, M., Yazaki, K., *Progr. Theoret. Phys.* (*Kyoto*), **37**, 638 (1967)
117. Sugawara, K., *Nucl. Phys.*, **A110**, 305 (1968)
118. Philpott, R. J., Pinkston, W. T., Satchler, G. R., *Nucl. Phys.*, **A119**, 241 (1968)
119. Hamamoto, I., *Nucl. Phys.* (In press)
120. Hamburger, E. W., Blair, A. G., *Phys. Rev.*, **119**, 777 (1960)
121. Penny, S. K., Satchler, G. R., *Nucl. Phys.*, **53**, 145 (1964)
122. Penny, S. K. (Doctoral thesis, Univ. Tennessee, 1966, unpublished)
123. Kozlowsky, B., de-Shalit, A., *Nucl. Phys.*, **77**, 215 (1966)
124. Iano, P., Austern, N., *Phys. Rev.*, **151**, 853 (1966)
125. Dupont, Y., Chabre, M., *Phys. Letters*, **26B**, 362 (1968)
126. Gaillard, M., Gaillard, P. (Quoted in Ref. 125)
127. Haase, E., Brueckmann (Quoted in Ref. 125)
128. Zaidi, S. A. A., von Brentano, P., *Phys. Letters*, **23**, 466 (1966)
129. Tamura, T., Watson, C. E., *Phys. Letters*, **25**, 186 (1967)
130. Moore, C. F., Watson, C. E., Zaidi, S. A. A., Kent, J. J., Kulleck, J. G., *Phys. Rev. Letters*, **17**, 926 (1966)
131. Coker, W. R., Moore, C. F., *Phys. Letters*, **25**, 271 (1967)
132. Heffner, R., Ling, C., Cue, N., Richard, P., *Phys. Letters*, **26**, 150 (1968)
133. Moore, C. F., *Phys. Letters*, **25**, 408 (1967)
134. Alexander, E. F., Watson, C. E., Shelton, N. (In press)
135. Weil, J. L., Cosack, M., McEllistrem, M. T., Norman, J. C., Stauberg, M. M., *Bull. Am. Phys. Soc.*, **12**, 1196 (1967)
136. Martin, D. G., Zaidi. S. A. A., *Bull. Am. Phys. Soc.*, **12**, 1196 (1967)
137. Michelman, L. S., Moore, C. F., *Phys. Letters*, **26B**, 446 (1968)
138. Clarkson, R. G., Coker, W. R., *Bull. Am. Phys. Soc.*, **13**, 631 (1968) (To be published)
139. Coker, W. R., Moore, C. F., *Bull. Am. Phys. Soc.*, **13**, 631 (1968)
140. Clarkson, R. G. (Doctoral thesis, Univ. Texas, 1968, unpublished)
141. Clarkson, R. G., Coker, R. G., Griffy, T. A., Moore, C. F., *Tech. Rept. No. 4* (Center Nucl. Studies, Univ. Texas, 1968)
142. Coker, W. R., Tamura, T., *Phys. Rev.* (In press)
143. Cue, N., Richard, P., *Phys. Rev.*, **173**, 1108 (1968)
144. Hamburger, E. W., *Phys. Rev. Letters*, **19**, 36 (1967)
145. Amos, K. A., Madsen, V. A., McCarthy, I. E., *Phys. Rev.*, **135**, B330 (1964)
146. Madsen, V. A., Tobocman, W., *Phys. Rev.*, **139**, B864 (1965)
147. Glendenning, N. K., Veneroni, M., *Phys. Rev.*, **144**, 839 (1966)
148. Satchler, G. R., *Nucl. Phys.*, **77**, 481 (1966)
149. Satchler, G. R., *Nucl. Phys.*, **A95**, 1 (1967)
150. Amos, K. A., *Nucl. Phys.*, **A103**, 657 (1967)
151. Faessler, A., Glendenning, N. K., Plastino, A., *Phys. Rev.*, **159**, 846 (1967)
152. Agassi, D., Schaeffer, R., *Phys. Letters*, **26**, 703 (1968)
153. Kerman, A. K., McManus, H., Thaler, R. M., *Ann. Phys.* (*N.Y.*), **8**, 551 (1959)
154. Haybron, R. M., McManus, H., *Phys, Rev.*, **140**, B638 (1965)
155. Barranger, M., *Phys. Rev.*, **120**, 957 (1960)
156. Kobayashi, M., Marumori, T., *Progr. Theoret. Phys.*, **23**, 387 (1960)
157. Arvieu, R., Veneroni, M., *Compt. Rend.*, **250**, 992, 2155 (1960)
158. Kisslinger, L. S., Sorensen, R. A., *Rev. Mod. Phys.*, **35**, 853 (1963)
159. Kuo, T. T. S., Brown, G. E., *Nucl. Phys.*, **85**, 40 (1966)
160. Sherif, H., Blair, J. S., *Phys. Letters*, **26B**, 489 (1968)
161. Stamp, A. P., *Nucl. Phys.*, **83**, 232 (1966)
162. Rawitscher, G. H., *Phys. Rev.*, **163**, 1223 (1967)
163. Ohmura, T., Imanishi, B., Ichimura, M., Kawai, M. (In press)
164. Noble, J. V., *Phys. Rev.*, **157**, 939 (1967)
165. Reiner, A. S., Jaffe, A. I., *Phys. Rev.*, **161**, 935 (1967)
166. Buttle, P. I. A., *Phys. Rev.* **160**, 719 (1967)
167. Lane, A. M., Thomas, R. G., *Rev. Mod. Phys.*, **30**, 254 (1958)
168. Tamura, T., *Phys. Rev.* (In press)
169. Coker, W. R., Tamura, T., *Bull. Am. Phys. Soc.*, **14**, 590 (1969)

MECHANISMS OF RADIATION MUTAGENESIS IN CELLULAR AND SUBCELLULAR SYSTEMS

By B. A. Bridges

Medical Research Council Radiobiology Unit, Harwell
Didcot, Berkshire, England

CONTENTS

INTRODUCTION

Seventeen years have passed since the field of radiation genetics was reviewed in this series by Kimball (1). Since then interest in the subject has grown steadily and it is necessary to restrict the scope of the present review to a consideration of the mechanisms by which genetic mutations arise within populations exposed to various radiations: ionizing, ultraviolet, visible, and far violet ("black light"). Advances in the field of ultraviolet mutagenesis in particular have been far reaching and are leading to a much greater understanding of ionizing-radiation mutagenesis. In many instances the nature of the genetic change being assayed in a mutation system is not known. This is particularly true where the inactivation of a gene is involved and a large number of possible alterations could be responsible. In a few instances the nature of point mutations is established. They are of two types: (*a*) base substitutions, either transitions where a purine-pyrimidine base pair is replaced by the other purine-pyrimidine base pair (AT↔GC), or transversions, where a purine-pyrimidine base pair is replaced by a pyrimidine-purine base pair (AT↔TA↔GC↔CG); and (*b*) frameshift mutations where insertion or deletion of a small number of base pairs alters the reading frame and makes missense of messenger RNA. Although the term frameshift is strictly applicable only to loci specifying messenger RNA it will, for convenience, also be used for loci specifying ribosomal or transfer RNA to indicate insertion or deletion of a small number of base pairs.

139

The Mechanism—Chemical or Biological?

The question is not meant to imply that biology is not ultimately a chemical process. The intention is to distinguish between mutations which arise directly as a result of a chemical change and those which are produced from some initial chemical damage by means of biological (metabolic, enzymic) processes. It will become clear in subsequent sections that biological processes most certainly intervene between the exposure to radiation and the establishment of the mutation. Let us, however, first consider how a mutation might occur directly, without biological intervention.

The most simple case would be the direct conversion of one base into another. The radiation chemistry of nucleic acids and their constituent bases is exceedingly complex but at least one example is known, the conversion of cytosine to uracil in dilute aqueous solution by ionizing radiation(2).

The next simplest case would be the conversion of a base to some photoproduct which, while not a natural DNA base, would be expected to specify an incorrect base at the next replication. In dilute aqueous solution adenine is converted (with low efficiency) to hypoxanthine by ionizing radiation (3). If this were to occur in DNA also, it would be expected to result in a transition to GC at an AT base pair. Possibly more important is the photoproduct cytosine hydrate which is recognized as thymine by DNA polymerase and as uracil by RNA polymerase (4). Grossman suggests this as the most probable mutagenic lesion in single-stranded regions of DNA, but there does not appear to be any experimental evidence yet to suggest that it is of importance *in vivo*.

One of the problems arising here is that where only one base is changed, only one daughter chromatid is expected to emerge as mutant from the subsequent replication whereas very often both are observed to be mutant (see below). This is not an insuperable objection since the same result is obtained with mutagenesis by known base analogues such as 2-aminopurine and 5-bromouracil (5, 6).

A change which would be expected to result in an alteration in coding specificity for both strands simultaneously is the tautomeric shift of a base pair. Watson & Crick (7) first suggested this as a possible cause of spontaneous mutations and Löwdin (8) calculated that it might occur spontaneously as a consequence of two-way proton tunnelling, and lead ultimately to a transition. Ionizing radiation might lead to an anion-cation tautomerism, the anion would not code for anything and could thus lead to a deletion, the cation could pair with any of the four bases and might thus lead to a base-pair mutation at the next replication.

Subsequently Ladik (9), using the Huckel approximation, calculated that an excitation produced by ultraviolet might lead to a tautomerization from G–C to G*–C* by one-way proton tunnelling which would subsequently code for T and A at replication. The essential feature of this hypothesis seems to be that the excited state permits the tautomerization and that this is 'fixed' by the resumption of the ground state. Subsequent calculations using the more refined SCF-LCAO-MO semiempirical approximations showed, however, this was not likely in the excited state (10). Much more likely was one-way proton tunnelling to give an anion-cation tautomerism. This could also occur as a result of ionization (11). It should be realized that these calculations are highly sensitive to the assumptions one makes and their limitations have been stressed by Löwdin (12). Pullman (13) has also concluded that base-pair tautomerism could occur in excited states; he used an entirely different basis, namely a study of the stabilization of the base pairs through resonance or delocalization energies.

So far the reactions and calculations described have very little if any *in vivo* evidence to support them. One characteristic of premutational damage is, in many instances, its reparability, and it might be imagined that this would argue against any direct mechanisms of mutagenesis. While this might be valid for tautomeric changes, particularly two-way proton tunnelling should it occur, it is not necessarily so for base changes confined to a single strand. To explain repair it would be necessary first for an enzyme to be present capable of recognizing a single mismatched base pair. While such an enzyme is not known, it has been found desirable to postulate its existence to explain other rather puzzling observations regarding mutation, gene conversion, and transformation [e.g. (13a); for a discussion see Bridges & Munson (14)].

The most weighty evidence against direct chemical change being important in mutagenesis, at least in bacteria, is the observation that the *presence* of a particular repair system (specified by the gene *exr*) is necessary for nearly all of the mutagenic effect of ultraviolet and around 90 per cent of that of ionizing radiation. It is possible, of course, that mutations which are independent of *exr* in bacteria, and perhaps those in other systems, are caused by direct chemical change and work should be pursued with this in mind.

In the absence of definite evidence to the contrary, it is more realistic to consider mutations as arising indirectly from nonspecific lesions (premutational lesions) of which some are known and more probably remain to be identified. Repair processes may then get to work to remove or modify these lesions. Such repair processes may not be strictly part of the mutation process but they frequently play a great part in determining the outcome of mutation experiments. Lesions which escape repair may then be fixed or established as heritable mutations, possibly by the action of other "last-ditch" repair systems which are programmed to produce a mistake (mutation) rather than leave a lethal lesion. The process of DNA replication may possibly be involved. All these aspects will be considered below in connection with mutagenesis by visible, black, and ultraviolet light, and by ionizing radiation.

VISIBLE AND NEAR-VISIBLE LIGHT

Mutations induced in microorganisms by visible light in the absence of any dye were first reported in 1956 by Kaplan & Kaplan (15). The results were somewhat unusual in that mutations (to inability to form pigment in *Serratia marcescens*) occurred only when light was shone on bacteria which had been dried and rehydrated. Neither dried cells nor wet cells were mutable by visible light.

Subsequently Ritchie (16, 17) studied the production of *rII* mutants in phage T4 by visible light under conditions where there was no inactivation. The mutants were revertible by growth in the presence of acridine but not of 5-bromodeoxyuridine, and were therefore presumed to be frameshift mutations.

More recently, workers at the Argonne National Laboratory have reported the detection of a mutagenic action on bacteria of black light (320–400 nm) in the absence of added dye (18, 19). The system used was the production in chemostat cultures of *Escherichia coli* of mutants resistant to the bacteriophage T5. The nature of such mutations at the DNA level is not established but it is not unlikely that they involve the inactivation of a gene specifying receptors for T5 on the cell surface. Mutation rates were found to be proportional to irradiance and were up to 40 times greater than the spontaneous rate. Unlike spontaneous mutation, but like that produced by ultraviolet and ionizing radiation, mutation by black light showed no dependence of mutation rate upon growth rate.

The mutation rate by black light is about two to three times greater for cultures growing in the presence of oxygen than in its absence. This suggested to both sets of workers that a photodynamic component might be involved in mutagenesis and Webb & Malina (19) suggest riboflavin, vitamin K, and phorphyrins as possible natural chromophores. It should not be forgotten, however, that these cultures are grown for long periods of time in the presence or absence of oxygen and are doubtless in rather different physiological states. The possibility should therefore be borne in mind that oxygen might influence the induced mutation rate indirectly through altering the efficiency of a repair system or of some other physiological system involved in mutagenesis. Furthermore Webb & Malina (19) point out that any postulated chromophore might be depleted during anoxic growth. It is interesting that the absolute mutation rates for black light-induced mutations are almost identical in the work of Kubitschek (18) and Webb & Malina (19, although Kubitschek was working with an ultraviolet-sensitive strain (defective in excision repair) derived from *E. coli* B/r. It would therefore appear the black light damage leading to mutation is not the sort of DNA damage susceptible to the excision-repair system.

Webb & Malina (19) have also observed mutation by "white" light with <1 per cent emission below 380 nm. Mutation appeared to be completely dependent upon the presence of oxygen. Leff & Krinsky (20) have also reported the induction of mutants of *Euglena gracilus* lacking chloroplasts by treatment with red light (>610 nm.) The authors suggested that chlorophyll itself was the chromophore. These mutants also lacked chloroplast DNA. Thus these mutations, like those to T5 resistance in *E. coli*, appear to be due to the inactivation or deletion of a segment of DNA.

The only evidence that black or white light might induce base-change mutations comes from the work of Ashwood-Smith, Copeland & Wilcockson (21) who detected the induction of presumed base-change mutations to prototrophy in *E. coli* WP2 *try* when the cells were maintained at −60° but not at 29° C. There is a high probability that the ultraviolet content of sunlight was responsible for these mutations as the mutation frequency was greatly reduced when the quartz cover was replaced by glass. Moreover, black light (310–380 nm) is entirely nonmutagenic in this system.

That visible light could be mutagenic in the presence of certain dyes (photodynamic action) was first reported by Kaplan (22) using the dye erythrosine and following the production of mutants of *Serratia marcescens* having abnormal colonial morphology. He established a multihit (∼2-hit) dose-response curve which almost all subsequent works have confirmed. Mutation of specific loci has been reported by other workers in yeasts (23), bacteria (6, 24–26), and bacteriophages (27–31) and with a number of other dyes (e.g. acridine orange, acridine yellow, methylene blue, toluidine blue, proflavine, and thiopyronine).

Little is known about the initial damage produced by photodynamic action although it appears to be the result of an oxidative process since the induced mutation rate, at least with acridine orange, is considerably reduced (but not to zero) in the absence of oxygen (19, 25). No photoreactivation has been detected (28) so that the lesions are not cyclobutane-type pyrimidine dimers as found commonly with ultraviolet.

There is evidence that inactivation by light plus methylene blue may be caused largely by the destruction of guanine (and to a lesser extent thymine) (32–35). Brendel & Kaplan (28), however, believe that mutations are not produced from guanine lesions on three grounds: 1. the dose-response kinetics are different (single-hit versus multihit); 2. light plus methylene blue gives a different spectrum of mutation response in *Serratia* phage κ from that produced by guanine-specific mutagens ethane methanesulphonate and low pH; 3. T4 phages believed to have a GC base pair at the mutable site have the same induced mutation rate as those with an AT base pair. This latter point is contradicted by Baricelli & Del Zoppo (31).

Following the demonstration that single-strand breaks are produced in DNA treated with light plus methylene blue or acridine orange (35a, 36), the way is open for a hypothesis for the mechanism of mutation similar to that suggested below (and in 37) for ultraviolet and ionizing radiation. It would seem likely that approaches similar to those used with ultraviolet (see below) would be profitable.

We know almost nothing about the mechanism by which mutations are produced from the initial lesions or even whether direct photochemical change of one DNA base into another can occur. There is no strong evidence for repair of mutagenic photodynamic damage. Ito & Hieda (23), for example, were unable to detect any recovery from damage leading to forward mutations at the *ad* locus in yeast when the cells were held for some time in distilled water before plating. Such liquid holding recovery was marked after ultraviolet irradiation. They point out however that repair may have been so rapid as to occur on the plate without need for liquid holding.

Uretz (38) found that lethal damage to *E. coli* by acridine orange and visible light was the same in strains B/r, B, and B$_{s-1}$ which have vastly different capacities for repairing or bypassing damage inflicted by ultraviolet or ionizing radiation. He concluded that photodynamic lethal damage is irreparable. Mutations, however, appeared to be lost in broth-grown (but not minimal grown) cells when these were held in minimal medium before plating. Although premutational and lethal lesions could therefore differ, some common mode of formation is suggested by the strong dependence of rates of mutation and killing upon the molarity of the buffer in which they are suspended.

There is no information about the molecular nature of the mutations induced in bacteria by photodynamic action. With phage treated extracellularly, however, strong evidence suggests that point mutations may be induced. Firstly, Brendel (30) found that methylene blue and light induced presumed base-change back mutations at a number of both AT and GC sites at the *rII* locus in T4 and also reverted two out of three frameshift sites.

Secondly, Ritchie (16) induced forward mutations at the *rII* locus in T4 by proflavine and visible light and found about half were revertible by 5-bromodeoxyuridine (and thus presumable base-change transition mutants) and the other half by growth in the presence of acridine (and thus presumably frameshift mutants). Subsequently (17) he showed that either light alone or proflavine alone induced presumed frameshift mutations and that the additional mutations induced by light in the presence of proflavine were base-change transitions. The possibility must therefore be considered that the frameshift mutants of Brendel (and possibly also the AT base-change mutants) might not be photodynamic in origin but due to light or dye alone.

Drake & McGuire (29) found little or no evidence for frameshift mutation by thiopyronin plus light, but only transitions and presumed transversions.

Kubitschek (6) was unable to detect segregation of mutants to T7 resistance during growth of *E. coli* after exposure to light plus methylene blue or acridine orange. With the assumption of the conventional Watson-Crick model for DNA replication, this means (on the simplest interpretation) that both strands of DNA are mutated in a complementary manner, something difficult to reconcile with an initial lesion presumed to be confined to only one strand. Kubitschek preferred to postulate a new mode of DNA replication to explain his findings (39). We shall discuss this problem again in connexion with mutation induction by ionizing radiation.

The induction of mutations by black light in the presence of psoralens has received little attention until recently. The action is quite different from that of the photodynamic dyes in being oxygen independent (39). Base substitution mutations have been found to be induced in both *E. coli* (39) and phage T4 (29) but little or no frameshift mutation was found in the latter organism. Igali and his colleagues (39b) conclude that damage produced by 8-methoxypsoralen and black light is not photoreversible, but is excisable in Hcr$^+$ strains and gives rise to mutations via the exr-dependent pathway (see below).

ULTRAVIOLET LIGHT

Viruses.—It has been known for many years that ultraviolet irradiation of phage-bacteria complexes leads to the induction of phage mutations, particularly in temperate and semitemperate phages such as λ (40, 41), S13 (42), T3 (43), and T1 (44). More recently Drake (45) isolated many rII forward mutants from coliphage T4 and was able to characterize them by the specificity of their reversion by chemical mutagens. About half were presumed to be base-pair transitions, mostly from GC to AT, and half were frameshift mutations. It would therefore appear that ultraviolet induced mostly point mutations.

Ultraviolet mutagenesis of extracellular phage is a fairly recent achievement. Although Krieg (46) had reported in 1959 reversion of rII mutants by ultraviolet irradiation of extracellular T4, he was not able to obtain forward r mutations consistently [quoted by Drake (47)]. Folsom (48) and Drake (47) were successful in inducing forward mutations, the latter worker reporting that they were not expressed unless a very soft overlay agar was used. Both base-pair transitions and frameshift mutations were found by Drake (47) to be induced linearly with dose, the spectrum of mutations being very similar to that previously reported by him for intracellularly irradiated phage (45). Folsom's collection (48), however, was rather different and the spectrum of mutations more nearly resembled that occurring spontaneously.

Mutations induced extracellularly by ultraviolet have also been reported in *Serratia* phage κ (49) and coliphage C_d (50). Drake (51) records that mutations were induced less efficiently when phages were exposed to ultraviolet outside rather than inside the host cell, and ϕX174 isolated DNA was only half as mutable by ultraviolet as the intact extracellular phage (52). Mutation of isolated DNA by ultraviolet has also been reported with the *Bacillus subtilis* transformation system (53).

Pyrimidine dimers seem to be important in the induction of phage mutants, as they are in many other ultraviolet-induced effects. We are fortunate that there is a test for the involvement of pyrimidine dimers which is believed to be reasonably specific, their monomerization by a photoreactivation enzyme under the action of visible light. Although an early investigation (51) failed to detect any photoreversal of rII forward mutations in T4, subsequent work with intracellularly irradiated T4v (which has a larger photoreversible sector) showed that the induction both of base-pair transitions and of frameshift mutations was photoreversible; this indicates that pyrimidine dimers were the initial lesions in both cases (54).

In the κ phage system, Winkler (55) found that c mutants, which are induced with two-hit kinetics, were photoreversible. If one assumes that both hits are photoreverse at odnce, then the photoreversible fraction of premutational damage is 71 per cent, almost identical with that of lethal damage

(75 per cent). If, however, the two hits are assumed to be photoreversed separately, then the photoreversible fraction of damage is 45 per cent. There is no evidence to suggest which assumption is correct. Possibly the non-photoreversible lesions are not pyrimidine dimers, but their nature remains to be determined.

Pyrimidine dimers are known to be largely excised in wild-type bacteria and premutational lesions are likewise repaired by host cell systems that appear to include excision repair. Winkler (56) found that repair of lesions leading to c mutations in κ phage was reduced if caffeine or trypaflavine was present after exposure. These are believed to inhibit excision repair. Adsorption to an Hcr⁻ (presumed excision-defective) *Serratia* strain after ultraviolet had a similar effect. On the assumption that both hits of the two-hit induction process are repaired at once, the Hcr system acts with equal effectiveness on premutational and lethal damage. Caffeine did not affect the yield of rII mutants from extracellularly irradiated T4 (51). Host cell repair does not operate in T4 but the phage itself has a gene v which specifies a relatively small amount of excision repair. This repair is equally effective on premutational and lethal damage, if I interpret Drake (47) correctly.

An unusual Hcr⁻ strain of *Serratia* (strain 91) is reported by Winkler (57) which renders κ phage more sensitive to the lethal effect of ultraviolet, but at the same time completely eliminates ultraviolet mutability. Winkler interprets this as indicating that a bacterial repair system in strain 91 has a changed specificity in opposite directions for premutational and lethal damage, so that premutational lesions are completely removed whereas lethal damage is repaired less well. An alternative explanation, however, could be advanced along the lines of that proposed for *Exr*⁻ bacterial mutants (58). Strain 91 would be deemed to be defective in a repair system (or component thereof) which makes mistakes (i.e. mutations); it would thus be more sensitive but immutable. There is at present no other evidence for such a mistake-prone repair system involved in the mutagenic process with phage.

The fact that pyrimidine dimers can give rise to such different mutations as GC to AT transitions and frameshift mutations (54) suggests that the premutational lesion is relatively nonspecific and may initiate mutational changes at base pairs other than those at which the dimer is situated. Ultraviolet is known to be highly recombinogenic and because of its efficiency one may assume that pyrimidine dimers are involved. If gaps are left in newly replicated phage DNA synthesized from dimer-containing DNA [as happens with bacteria (59)], then these could well initiate recombination with a finite possibility of a mistake resulting. A multiplicity of infection of no more than one would be required. Drake (47) has already shown that the frequency of rII mutants is the same whether or not multiplicity reactivation is taking place. He has also suggested that most of the mutations in temperate and semitemperate phages could occur as a result of recombination events with the host's chromosome [cf. Stent (60)]. So far, however, there is no concrete evidence which would enable us to establish the role (if any) of recombination in ultraviolet mutagenesis of T4 phage, although the material is flexible, well studied, and obviously suitable for future work along these lines.

A rather different mechanism has been proposed for phage λ which is only mutable by ultraviolet if the host bacterium is also irradiated (40). Devoret (61) found that mutants are not produced in a preirradiated Hcr⁻ host. He suggested that mutations arise from the incorporation of degradation products released during excision of the DNA of the host cells. This is in accord with the observation that irradiation of the host alone can cause mutations in unirradiated phage (40). An alternative explanation might be sought along the lines suggested above, with lesions in either host or phage acting as initiating sites for recombination. Indeed, more recent work suggests that the presence of the λ attachment site at the *gal* locus on the host genome may be a prerequisite for mutation (61a).

Bacteria.—The induction of mutations in bacteria by ultraviolet light has received much attention during the last two decades. There is now so much information that a single comprehensive hypothesis to cover all mutation systems seems most unlikely. Discussion of earlier work will not be attempted here and the reader is referred to (62–64) for the opinions of various authors of the state of the art around 1966. Recent observations, however, are leading us towards a better understanding of a process at the molecular level and it is these upon which we shall concentrate here.

Three types of mutation system are in general use, mutation to bacteriophage resistance, to antibiotic resistance, and to independence from some nutritional requirement (auxotrophy to prototrophy). All these changes can be selected for, after mutagenic treatments which produce very little inviability (although not all workers have taken advantage of this latter property). Only in the case of some of the mutations to prototrophy has the nature of the genetic changes been characterized at the molecular level and I will describe one such system as it and others related to it are now favourite tools for those interested in mechanisms of mutagenesis.

Escherichia coli WP2 *try* is typical of a class of auxotrophic organisms (i.e., unable to make some amino acid or other essential metabolite) which are readily mutated by ultraviolet to the prototrophic state (i.e., they have regained the ability to make the metabolite). Many newly induced mutants are unable to produce colonies on unsupplemented minimal medium. Sufficient of the required metabolite is therefore included in the plating medium to enable expression. A large number (up to 5×10^8) of auxotrophs can be applied to the surface of a plate, but only prototrophic mutants grow into discrete colonies, the others being starved of their required metabolite (in this case tryptophan) after a few generations. A small correction has to be made for mutants arising spontaneously during these residual divisions of the auxotrophs. When only a hundred or so auxotrophs are plated, they grow into small discrete colonies, so the same medium can be used to assay both the viable cells of the total population and the mutant fraction.

It has been suspected for some time that many of the prototrophic mutants are not true back mutations at the locus specifying the requirement, but external suppressor mutants at some other site. Witkin (65), for example, found that two different requirements could be eliminated at a single mutational step, and Hill (66) showed that many prototrophic mutant cultures tended to change back to the auxotrophic state, a phenomenon which she interpreted as due to episome-chromosome interaction but which is, perhaps, more simply seen as a reflection of the selective disadvantage which suppressor-containing strains show under certain conditions (62).

Recently two groups of workers have shown that many of the induced prototrophs were ochre suppressor mutations (67, 68). These suppressor mutants are able to support the growth of T4 bacteriophages bearing certain ochre and amber mutations thus enabling them to be distinguished from true reversions at the ochre triplet itself. The true revertants are believed to be AT→GC transitions at the first position of the ochre triplet (69). The ochre suppressors are largely mutations of glutamine tRNA molecules. If, as suggested by Person & Osborn (70), these are mutations at the anticodon, they should be GC→AT transitions. It can be seen, then, that *E. coli* WP2, used in conjunction with T4 ochre and amber phages, provides the best-characterized cellular system for the study of mutation at the molecular level.

Using this system Bridges, Dennis & Munson (71) showed that ultraviolet light induced both true and (at higher frequency) suppressor mutations. There is thus little doubt that ultraviolet produces base-change mutations in abundance. This is in agreement with Witkin's analysis of some of Yanofsky's ultraviolet-induced missense mutations (72). All could be attributed to a single base-pair substitution (transition or transversion). One frameshift mutation that was analyzed appeared to be a single base-pair deletion.

Ultraviolet does not appear to be a very specific mutagen. Although suppressor mutations (presumed GC→AT) are induced at a higher frequency in WP2 than true reversions (presumed AT→GC), it is not clear how much of this difference is due to the special characteristics of the tRNA locus. Zampieri & Greenberg (73) found little evidence of specificity in the induction of *Lac*+ mutants in *Lac*− strains of *E. coli* S.

Unstable mutations to *Lac*+ are sometimes observed in *E. coli* K-12 and one interpretation of the results involves the postulate of an episome-like controlling agent (74, cf. 75).

Our knowledge of the initial photochemical products which lead to mutation is very limited and based largely upon the photoreversibility of premutational damage. In Hcr− strains it is clear that at low ultraviolet doses, photoreversible pyrimidine dimers are involved in the induction of mutations to streptomycin resistance (62), and to prototrophy in *E. coli* WP2, both true revertants and suppressors (62, 71). Photoreactivation action spectra support this conclusion (76). It is not excluded that some other photoproduct (e.g. cytosine hydrate?) in addition to a pyrimidine dimer is involved in the production of a mutation and this may even be considered likely in view of the dose response both for true reversions and for suppressor mutations, which approximates a two-hit curve (71).

In Hcr$^+$ strains most pyrimidine dimers are excised and there has been some controversy as to whether pyrimidine dimers are involved in mutation. Hill (77) concluded that Try^+ mutations arise from thymine dimers on the basis of the differential sensitivity to mutation and killing in Hcr$^+$ and Hcr$^-$ strains of WP2. It has been shown, however, that photoreversal of mutation in an Hcr$^+$ strain is an indirect process mediated by a dark-repair system ("mutation frequency decline") (78). That pyrimidine dimers are involved is indicated because photoreversal of mutations is more efficient in a Phr$^+$ than in a Phr$^-$ strain although the overall extent is the same (79). Action spectra for photoreversal confirm that dimer splitting is involved in a Phr$^+$ strain (80). Even in such a strain, the effect of photoreversing light can be abolished by subsequent application of a dark-repair inhibitor such as caffeine or acriflavine (81). It has been argued (62, 63) that the best interpretation of these data is that prototrophic mutations in Hcr$^+$ strains are caused by lesions which are not enzymically photoreversible but which are removed by dark repair at a rate inversely proportional to the number of pyrimidine dimers in the DNA. Photoreactivating light removes dimers and so speeds up dark repair. If bacteria are plated on media without broth supplementation where natural dark repair of mutations is maximal, there is little or no photoreversal of either true reversions or suppressor mutations (71). The suppressor mutation process in Hcr$^+$ strains may thus be thought of as a two-event process (63) with dark repair of the mutagenic lesions being progressively inhibited at increasing ultraviolet doses by the other lesions (postulated to be pyrimidine dimers). The dose-response curve for suppressor mutations does approximate a two-hit model while that for true reversions (which do not show this sort of dark repair) is linear (71).

The nature of the nonphotoreversible lesion is unknown. It would seem unlikely to be a pyrimidine dimer whose photoreversibility is impaired by its location, since pyrimidine dimers at the same sites are nearly all photoreversible in the Hcr$^-$ strain. (On the other hand it is possible that the photoreversibility is limited by the much greater number of dimers at mutagenic doses in Hcr$^+$ strains.) Obviously a lesion such as cytosine hydrate is a strong possibility but until we have more information as to the latter's stability and susceptibility to removal by dark repair, it would be wise to reserve judgement.

Results essentially similar to those of Witkin and the present author and his colleagues have been reported by Doudney (64). He envisages a two-event mechanism but favours the photoreversible dimer as the mutagenic lesion.

The experiments with WP2 described above may be relevant only to suppressor mutations although somewhat similar results have been reported for mutation to low-level streptomycin resistance (82). For mutation to high-level streptomycin resistance in B/r there is clear evidence that pyrimidine dimers are involved since all photoreversibility is lost in a Phr⁻ strain (78). Horneck-Witt & Kaplan (83) have presented evidence that the lesion in *E. coli* B Phr⁻ leading to low-level streptomycin resistance is different from that leading to a lethality (largely pyrimidine dimers). Mutations to *Lac⁻* in *E. coli* WP2 arise from lesions which are excisable by dark repair but are not photoreversible (84). This must mean either that dimers produced in the *Lac* operon are not photoreversible or that some other lesion is involved. It can be seen then that pyrimidine dimers do not tell the whole story as far as mutation is concerned.

Hainz & Kaplan (85) reported that *Serratia* were more sensitive to the lethal effect of ultraviolet in the frozen state below −25° C but induction of colour mutations was not enhanced at low temperature. Ashwood-Smith & Bridges (86), however, found that both lethality and induction of prototrophic mutants of *E. coli* WP2 were markedly enhanced when the temperature of frozen bacteria was lowered from −5 to −79° C. Marked differences between ultraviolet mutagenesis of frozen and unfrozen bacteria led to the conclusion that there is a different mutagenic lesion produced at −79° which is less readily excised (86) and shows very little "mutation frequency decline" (see below) (87). Moreover, the dose-response curve is more nearly linear for both true and suppressor mutations and there are more nearly equal proportions of the two types of mutation (71).

Biochemical studies have recently shown that the formation of cyclobutane-type pyrimidine dimers is drastically reduced in the frozen state and that thymine-containing photoproducts are formed similar or identical to those in bacterial spores at room temperature (88–91). Zamenhof & Reddy (92) have also concluded that the lethal and mutagenic injury in *Bacillus* spores is different from that in vegetative cells (pyrimidine dimers).

In the foregoing discussion on the nature of the initial mutagenic lesion, it has not been possible to hide the fact that there is a great deal of evidence that premutational lesions induced by ultraviolet can be repaired. The damage leading to mutation in Hcr⁻ strains is largely excisable as judged by the differential sensitivity of Hcr⁺ and Hcr⁻ strains which differ in their ability to excise thymine dimers from their DNA. This is true for ultraviolet damage produced both at room temperature (62, 76, 77) and at −79° (86). Whether the damage which persists to give rise to mutations in Hcr⁺ strains is of a different type, essentially nonexcisable, is not clear. Such evidence as exists perhaps suggests the opposite (62, 63).

A further type of presumed repair is "mutation frequency decline," the irreversible loss of mutations that occurs when RNA and protein synthesis is inhibited after ultraviolet. It appears to be restricted to ultraviolet damage and, in the only case so far examined (WP2), it occurred with suppressor mutations but not with true revertants (71). The most common assumption (62–64, 93) is that mutation frequency decline reflects the operation of the excision-repair system. The evidence for this is twofold: Hcr⁻ (i.e. excision-deficient) strains of *E. coli* do not show mutation frequency decline (84, 93) and mutants have been isolated which do not show the phenomenon (Mfd^-) and which have a reduced rate of excision of thymine dimers (84). Following Witkin's suggestion (84) that the excision of photoproducts at suppressor loci might be more efficient in the repressed state, Igali & Bridges have isolated a relaxed strain of *E. coli* WP2 in which the repression of RNA synthesis is much less complete than in its parent. Subsequent investigations showed that mutation frequency decline of ultraviolet-induced mutations during tryptophan starvation is significantly less than the normal rate (94). This is entirely consistent with Witkin's suggestions, but the detailed mechanism of mutation frequency decline must still be regarded as uncertain. In particular one would like to know if it is solely a characteristic of suppressor loci, and if so, what it is that distinguishes loci specifying tRNA from loci specifying mRNA.

Many workers have reported experiments with chemicals which are believed to inhibit dark repair, notably caffeine (cf. 95) and acriflavine (cf. 96). As expected, in Hcr^+ strains, not only are the bacteria more sensitive to ultraviolet but suppressor mutations are induced by lower ultraviolet doses when acriflavine or caffeine is present in the plating medium. The exact magnitude for the enhanced mutability depends upon the locus (97–99). Generally the two compounds give effects of similar magnitudes although the concentrations required to produce the same effect differ greatly, caffeine being much less efficient that acriflavine (99, 100). One group of workers (98), however, reported acriflavine to produce a somewhat greater effect than caffeine at three loci.

Both caffeine and acriflavine have been reported to inhibit the excision of thymine dimers (101, 102). Both compounds have little or no effect on ultraviolet mutagenesis in an Hcr^- strain which has little or no excision-repair capacity (99, 103). It therefore seems likely that inhibition of excision repair is responsible for a major part of the mutation-enhancing effect of these compounds. The exact mechanism is still obscure; several authors have suggested that acriflavine might act by binding to damaged DNA and caffeine by binding to the excision enzyme (96, 99, 100, 104).

An interesting difference between the two compounds, reported by Shankel & Kleinberg (100), was that the effect of caffeine, but not acriflavine, could be largely reversed by photoreactivating white light. It is not clear from their abstract whether the compounds were present during the light treatment. If they were, the result could be taken as evidence that acriflavine, unlike caffeine, binds to damaged DNA and inhibits both excision and photoreactivation.

Caffeine does not inhibit dark repair under all conditions, for example it has no effect on the induction of low-level streptomycin-resistant mutants in *E. coli* B Phr$^-$ MG_2 in rich medium (although it does under starvation conditions) (83). There is also recent evidence that mild heating (105) inhibits or inactivates excision-repair capacity for premutational lesions.

A further phenomenon which may be mentioned briefly is the low-temperature mutation loss which was first studied by Berrie (106) and Witkin (107). More recently it has been examined by Munson & Bridges (93) who have concluded that the greater part of the loss of mutations which occurs when plates are incubated at 16° after ultraviolet is due to mutation frequency decline which is not inhibited at the low temperature by the presence of broth in the plating medium. There is a further fall by a factor of 2 (it constitutes the only loss in an Hcr⁻ strain) which is apparently identical with that found after exposure to ionizing radiation. It does not appear to involve the excision-repair system; indeed there is little evidence to establish whether it reflects repair or death of potential mutants.

Whereas the dark repair of premutational lesions has received much attention, it has yielded little information about the process of mutagenesis itself. It is clear that mutations do not arise to a significant extent as mistakes in excision repair (62, 63, 84) but rather as mistakes induced by those lesions which are unrepaired. This has recently been challenged by Doudney & Nishioka (108) who find that loss of photoreversibility for mutation to streptomycin resistance in Hcr⁺ strains correlates with pyrimidine dimer excision rather than with DNA replication. As the reviewer has seen only an abstract of this work it would seem to be wise to reserve judgement. One should, however, mention the possibility that mutations are induced during the slow replication of DNA very near the replication point and that as photoproducts are progressively removed, DNA synthesis speeds up, and the DNA being replicated contains progressively fewer photoproducts and generates fewer mutations upon replication. One could thus conceive of a situation where mutations are induced largely before DNA synthesis resumes at its normal rate, and loss of photoreversibility would appear to correlate better with dimer excision than with DNA synthesis.

Advance towards an understanding of how unexcised lesions give rise to mutations has come from two approaches. That of Bridges & Munson has been to study the mutagenic process in a strain which is unable to excise pyrimidine dimers (*E. coli* WP2 Hcr⁻). Their philosophy is based on the belief that almost every experiment carried out with Hcr⁺ strains is in effect observing the response of the excision-repair system, which thus masks any response of the mutagenic process. There is a further advantage in using an Hcr⁻ strain because almost all the mutations arise from photoreversible pyrimidine dimers (62, 71). This means that exposure to visible light at any time after exposure to ultraviolet gives a measure of the rate at which dimers give rise to mutations. Using this approach they have been able to show that pyrimidine dimers may persist for several generation times after ultraviolet and that they may give rise to mutations with a low probability (∼1–5 per cent) per replication cycle (69, 109). This fact in itself implies that when a dimer passes through the replication point, the information needed to specify the purines on the complementary daughter strand must come either from the parental strand opposite the dimer, or from the daughter strand which is made complementary to the parental strand opposite the dimer. In the latter case some recombination-like process would have to be involved.

Mutations were not established (did not lose their photoreversibility) when DNA replication was halted at the end of a cycle (69), which is in agreement with earlier data with Hcr⁺ strains suggesting that DNA replication is involved in the establishing of mutations. The work of Rupp & Howard-Flanders (59) indicated a way in which mutations might occur. They found that when pyrimidine dimers were allowed to pass through the replication point in an Hcr⁻ strain of K-12, the newly synthesized DNA had a lower than normal molecular weight. Their calculations showed there was approximately one break in the daughter DNA for every dimer on the parental DNA which passed through the replication point, which strongly suggests that a gap was left in the daughter strand opposite the dimer. These gaps disappeared during subsequent incubation. It was suggested (71, 110) that mutations might occur as mistakes during the filling of these gaps. Calculations by Bridges & Munson (69) suggest strongly (but not conclusively) that once a mutation is established (opposite a dimer?), both daughter duplexes are mutants at the next replication. This is in agreement with the idea discussed above that in replicating past a dimer, information is normally obtained from the other parental strand or from the daughter strand formed complementary to the other parental strand by some recombination-like process.

The other recent step towards an understanding of such a process came from the discovery by Witkin (110) that *E. coli* strains bearing the *exr⁻* allele from B$_{s-1}$ or B$_{s-2}$ were almost completely unmutable by ultraviolet although they mutated spontaneously at a normal rate. Strains bearing the *exr⁻* allele are up to three times more sensitive to ultraviolet than wild-type strains. Witkin (110) proposed a model with two mechanisms for dealing with unexcised dimers at the replication point, one error-free, and one (*exr*-dependent) error-prone. More recently she has modified this model in terms of a recombinational mechanism for the repair of dimer-induced replication gaps (72). The *exr* gene function is seen as increasing the efficiency of gap-filling while at the same time increasing the probability of error. According to Witkin (72) there is a good correlation between the recombination ability of strains and their ability to cope with unexcised dimers. Moreover Exr⁻ strains appear to be rather poor recipients in conjugation experiments.

The suggestion that there are error-prone and error-free components in postreplication repair enables an explanation to be formulated for various observations hitherto unexplained. For example, the mutation loss which occurs when plates are incubated at low temperatures after ultraviolet could simply reflect a temperature-dependent change in the balance of the error-free and error-prone processes. If the *exr* system were less efficient at low temperatures one would expect this also to be reflected in an enhanced sensitivity to the lethal effect of ultraviolet and this is in fact observed (93, 111). The enhancing effect of ionizing radiation on ultraviolet mutagenesis (112, 113) might be interpretable in terms of an inactivation of the error-free component (which would also account for the increase in ultraviolet lethality), thus increasing the proportion of the total postreplication repair carried out by the error-prone component.

Recently caffeine has been found to have an antimutagenic effect after high doses in Hcr$^+$ strains and after low doses in Hcr$^-$ strains (83, 114). This has also been interpreted (114) in terms of an effect on postreplication recombinational repair.

It can be seen that recombination is likely to assume increasing importance in discussions of mutagenesis and we shall consider it from a more general viewpoint in a subsequent section.

One factor which has received little attention is the state of the DNA at the time of irradiation. Single-stranded DNA might be expected to occur near the replication point,f or instance, and might be more susceptible to certain types of lesion, e.g. cytosine hydrates. There would appear to be some profit in experimenting with synchronous cultures of Hcr$^-$ bacteria (cf. 115, 116).

Eukaryotic cells.—There has been an increase in recent years in the number of useful mutation systems available with eukaryotic cells, including both forward and reverse mutations. Induction of mutations by ultraviolet has been reported in *Aspergillus* (117, 118), *Neurospora* (117, 119, 120), *Chlamydomonas* (121, 122), *Saccharomyces* (e.g. 122–125), *Ophiostoma* (126), and *Ustilago* (127, 128). Although many of the loci involved have been well characterized genetically, it is not yet generally possible to state with certainty the nature of the mutations at the molecular level (transitions, frameshifts, etc.). Steps are, however, being made in this direction (123, 129).

The only information available concerning the nature of the premutational lesion comes from photoreactivation experiments where it is assumed that photoreactivation is a reflection of the splitting of pyrimidine dimers by visible light and a photoreactivation enzyme. This assumption has been justified for *Neurospora* by Kilbey & De Serres (120) on the following grounds: 1. an enzyme has been obtained from *Neurospora* which has the ability to repair ultraviolet-damaged DNA *in vitro* in the presence of light; 2. the action spectra for *in vitro* (i.e. direct) photoreactivation and *in vivo* photoreactivation are similar. Kilbey & De Serres (120) found that photoreactivation reduced the frequency of all types of *ad 3B* mutants induced by ultraviolet, including those suspected of being base-pair substitutions and those possibly deletions or additions (frameshifts). It therefore seems likely that pyrimidine dimers can give rise to all the mutational types. They also observed that the nonphotoreactivable mutations, those remaining after maximum photoreactivation ($\sim$30–40 per cent), were of the same types as those removed by photoreactivation. One may conclude that they arise either from pyrimidine dimers which for some reason are not susceptible to enzymic splitting, or from some other lesion. If the latter, it is necessary to assume that they are equivalent to pyrimidine dimers in terms of the mutations they can give rise to.

A further study (119) showed that ultraviolet-induced damage leading to reversion of 15 out of 16 auxotrophic loci in *Neurospora* was photoreactivated to the same extent as lethal damage. One, the inositol allele 37401, was photoreactivated much less than lethal damage. Other inositol alleles behaved normally. It would appear that photoreactivation can reveal some specificity at certain sites, either because damage other than pyrimidine dimers is formed there, or because dimers formed there are less amenable to the photoreactivation enzyme. Study of action spectra might, as Kilbey suggests, throw some light on the problem.

Davies & Levin (121) have also interpreted their data on the assumption that photoreversal of induced mutation is attributable to splitting of pyrimidine dimers. They found that the photoreversible sector for reversions from acetate dependence (0.9) was higher than for killing (0.8) in wild-type (radiation-resistant) strains of *Chlamydomonas reinhardi*. An appreciable fraction of the photoreversible damage was also dark-reparable as judged by the mutability of ultraviolet sensitive strains. In these strains photoreactivation resulted in mutation frequencies comparable with those of the wild type; in UVS-1 a photoreactivable sector in excess of 0.99 was observed.

An unusual dose-response curve was exhibited by UVS-1 which was much more mutable than the wild type at low doses ($\sim$30 ergs mm^{-2}) but much less so at high doses. Davies & Levin suggest an explanation in terms of loci analogous to *exr* and *hcr* in bacteria. They postulate involvement of UVS-1 in both types of repair; at high doses the *exr*-like activity is reduced so that few mistakes or mutations are produced. It must be conceded that such an interpretation is highly speculative.

In diploid yeast, pyrimidine dimers appear to be involved in about 80 per cent of mutations to ad^- in a strain heterozygous at this locus, as deduced from action spectra for induction and photoreversal (130, 131). It is likely, however, that many of the mutants arose from recombinational events such as mitotic recombination and allelic conversion; the authors suggest a figure of 20 to 30 per cent but this may well be an underestimate. It is not known whether the recombinational mutations share the same action spectra and photoreversal characteristics as the total mutations.

There seems little doubt that premutational lesions induced by ultraviolet are reparable by dark processes in many genera. Both approaches which have been used with bacteria, isolation of sensitive and resistant strains and inhibition or stimulation of repair, have been successfully used with eukaryotes. Arlett (118), for example, reports an ultraviolet-resistant strain of *Aspergillus nidulans* which was not mutable by ultraviolet (the decrease in mutation rate was at least 1000-fold compared to the normal strain). One possible explanation for this property (which was found to have a cytoplasmic determinant) is a repair system with an extremely high efficiency for repairing premutational lesions.

A sensitive strain of *Aspergillus* was studied by Chang, Lennox & Tuveson (117) and found to be less mutable in general than the wild type, whether the comparison was made on the basis of equivalent dose or of equivalent survival. Only at high doses, with mutations to methionine independence, was the sensitive strain more mutable than the wild type. Similar experiments with a sensitive strain of *Neurospora* gave even more clear-cut results for mutation to acriflavine or caffeine resistance. With UVS-1 no induced mutants to caffeine resistance were detectable. In general, then, these mutants behave rather similarly to Exr⁻ strains of *E. coli* (110). One must not forget that acriflavine and caffeine are well known as inhibitors of dark-repair processes in bacteria and their action in the plating medium may not be simply that of a selective agent for resistant mutants.

There are certain to be many other types of radiation-sensitive mutants in fungi which would be useful in understanding the mechanism. An ultraviolet-sensitive mutant of *Aspergillus rugulosus* has, for example, been reported which has the same ultraviolet mutability as its parent (117), and one of *Schizosaccharomyces pombe* has a greatly reduced ultraviolet mutability (132).

The ascomycete *Ophiostoma multiannulatum* has been developed as a mutation system by Zetterberg (126). He has concentrated on revealing the existence of postirradiation processes by a postirradiation incubation in liquid medium during which the frequency of prototrophic mutants decreases. The decrease appears to be dependent upon an enzymic process, requiring energy and dependent upon temperature, and it ceases when the cells are starved of nitrogen or glucose, or treated with sodium azide. Unlike the superficially similar process observed with nonsense suppressor mutations in bacteria, that in *Ophiostoma* occurs in both minimal and rich media.

It is possible in diploid cells to construct mutation systems in which mutation is almost certainly due to a recombination event. Such a system is the gene conversion observed in diploid yeast (*S. cerevisiae*) heteroallelic at the *his-1* locus. Snow (125) has isolated six ultraviolet-sensitive strains and found that their sensitivity to the lethal effect of ultraviolet was parallelled by their sensitivity to the production of prototrophs by ultraviolet. He concludes that the inability to repair properly DNA damaged by ultraviolet results in enhanced probability for a mitotic recombination event. Snow's ultraviolet-sensitive mutants (not sensitive to gamma rays) would appear to be entirely analogous to the Hcr⁻ mutants of *E. coli*. A similar mutant in *Ustilago* has been reported by Holliday (128). In addition, he isolated ultraviolet-sensitive mutants resembling the Rec⁻ mutants of *E. coli*. They are X-ray sensitive and have a strong influence on recombination, particularly in preventing ultraviolet-induced gene conversion and abolishing spontaneous gene conversion.

It should perhaps be noted before passing on to other matters that while it may be interesting to speculate on the similarities between ultraviolet-sensitive mutants of eukaryotes and bacteria, their biochemical characterization is rudimentary and we should not be surprised if greater complexities are revealed in eukaryotic cells, particularly where enzymes involved in recombination are concerned.

There have been a number of recent attempts to establish the strandedness of the newly induced mutation, i.e., is it transmitted to one or both daughter duplexes at the first replication? This is of considerable interest in view of its relevance to the mechanism of mutagenesis and in view of the unexpected results sometimes found with other mutagens and other systems. Nasim & Auerbach (133) included ultraviolet among the mutagens they examined in the purple ad-7 strain of *Schizosaccharomyces pombe*. Forward mutations at five antecedent loci in the adenine pathway yield white cells and these were scored as complete and mosaic colonies. The former were assumed to come from cells whose DNA was mutated on both strands, the latter from cells whose DNA was mutated on only one strand. With ultraviolet the proportion of mosaic colonies fell from 48 to 13 per cent with increasing dose. Qualitatively this trend would be expected if complete colonies arose from mosaic cells in which the complementary (unmutated) strand had suffered an inactivating event, but quantitatively the decrease was much too small to be compatible with this hypothesis. The authors concluded that their data were compatible with two hypotheses, one postulating that complete and mosaic mutations arise by different molecular mechanisms, the other assuming that mosaics are transformed into completes through repair of mismatched bases.

The paper of Nasim & Auerbach (133) brought forth a comment by Haefner (134) who pointed out that their results were likely to be severely biased by lethal sectoring, that is the production of a nonviable cells at the first postirradiation division. By means of pedigree analyses of cultures grown and treated in the same way as those of Nasim & Auerbach, Haefner found that up to 69 per cent of treated cells had only one viable first-generation descendent. He concludes that one may only interpret segregation patterns for mutants if it is shown, by individual pedigree analyses, that they are derived from that fraction of the population which gives rise to 100 per cent viable progeny.

This laborious task he subsequently undertook (135) and demonstrated that 5 out of 23 pedigrees containing auxotrophic or slow growth mutations were complete (all mutant cells at the fourth generation). Within such pedigrees all branches carried the same homoallelic mutation. Ten out of the 23 clones were half mutant, half wild type. Haefner's conclusions concerning the origin of the complete clones, however, were essentially the same as those of Nasim & Auerbach. With the recently available data concerning the time of DNA synthesis in *S. pombe* (136), Haefner suggested that at least some of the mosaic clones arose from cells in which DNA duplication had already occurred at the gene concerned, i.e. there were two DNA duplexes at that locus.

Nasim (132) attempted to test the hypothesis that complete mutant clones arise from cells in which a mutation on one strand of the DNA is extended to the other strand by virtue of its being excised and repolymerized to match the mutated strand. He observed a greater proportion of mosaic colonies in two strains of *S. pombe* sensitive to both the lethal and the mutagenic action of ultraviolet and concluded that this supported the hypothesis. This conclusion must be regarded with some caution in view of the evidence that pyrimidine dimers may persist for more than one generation time in bacteria defective in excision repair (109, 69). It is conceivable that mosaic colonies arise from cells in which a pyrimidine dimer gives rise to a mutation at the second or third replication cycle after ultraviolet. In a strain deficient in excision repair, more dimers would persist to the second or third cycles than in a wild-type strain and would thus generate a higher proportion of mosaic colonies.

Cross (122) has investigated the segregation of ultraviolet-induced mutations in *Chlamydomonas* and concludes that the patterns are consistent with the presence of polyneme chromosomes (having two or more parallel molecules of DNA). The abstract which the reviewer has seen does not, however, give the evidence on which this conclusion is based.

While steady progress is being made with eukaryotes particularly in relation to repair phenomena, there are very few clues to the mechanisms ininvolved in converting unrepaired damage into mutations. Apart from systems such as gene conversion in heteroallelic diploids there is little evidence for or against the proposition that recombination is involved in ultraviolet mutagenesis.

IONIZING RADIATION

Viruses.—The chief problem with viruses has been to demonstrate convincingly any mutagenic action of ionizing radiation on free viruses. Many years ago Gowen (137) reported an increase in mutation frequency after exposure of tobacco mosaic virus to X rays, but doubt was cast on this after Mundry (138) showed that the increase may have reflected an improved efficiency in detecting mutants pre-existing in the untreated population. In 1950 Kaplan and his colleagues reported the induction of plaque-type mutations in phage κ of *Serratia* by X rays (139). The c locus of κ appears to have a rather exceptional mutability, however, and mutations have been detected in other systems only with some difficulty, for example in coliphages T4 (140, 141), ϕX174 (142), T2 (143), s_d (144–146), and λ (quoted in 145).

So little information is available that it is not possible to generalize about the initial damage. Some loci show a linear dose-response curve, others a cumulative or multihit curve (147). There seems to be some measure of agreement that free radicals formed in the liquid outside the phage are not instrumental in mutagenesis since changing the concentration of radical scavenger (nutrient broth) does not alter the frequency of induced mutants (142, 140–147). There is evidence, however, that if phage DNA is stripped of its protein coat then mutations in ϕX174 bacteriophage may be produced by the action of aqueous free radicals (142) but only in the absence of oxygen. The authors suggest that although oxygen has an overall protective effect against ionizing radiation under such conditions, such damage as does occur could well be so extensive (involving perhaps a reaction with oxygen) as to be lethal. No mutations would then be seen in the presence of oxygen. They point out, however, that the data do not exclude the possibility of a specific action of anoxic radiation at the mutable sites. Brown (141) has questioned whether these workers have rigorously excluded the possibility that their "induced" mutations are not the result of selecting for pre-existing spontaneous mutations.

The possible direct conversion of one base (or base pair) to another appears to have been eliminated for gamma-ray-induced reversion of an amber mutation in intracellular T4 phage E51 (148). The amber codon is specified by a mutation in gene 56 which is essential for DNA synthesis. In a host which is Su⁻, only those newly induced mutations which can be fully established in the absence of DNA synthesis will grow. In an Su⁻ host this limitation is removed. The experiments showed that at least 90 and possibly 100 per cent required DNA synthesis and could not therefore be due to direct base conversion.

Reuger & Kaplan (147) found that both premutational and lethal lesions induced in *Serriatia* phage κ were stable to heat, light, and peroxidase, thus differing from those induced by ultraviolet. They suggest that thyminehydroperoxide might be an important lesion produced by ionizing radiation.

From these workers also comes the only observation relevant to the possible reparability of premutational lesions induced by ionizing radiation. They observed (147) that no mutants were produced when their phage was assayed on a particular strain CN. They incline to the view that this strain has an unusually effective repair system, able to nullify all the premutational lesions. There are, however, other possibilities. One is that strain CN is like Exr⁻ strains of *E. coli* which are believed to be deficient in a repair system which makes mistakes (i.e. mutations) when coping with both ultraviolet (58) and ionizing radiation (37) damage. Such strains repair somewhat less damage but repair it accurately. A further possibility is that in strain CN, mutations can be induced but are unable to express themselves for some physiological reason (like *rII* mutants of T4 in the absence of λ repressor).

The mechanism by which premutational lesions are established as functional mutations (if this is in fact needed) is still a matter for speculation. It has long been known that multiplicity reactivation can occur when more than one damaged phage infects a cell. That is to say, the genes of the damaged phages may become rearranged (or undergo a process of recombination) so that a viable phage genome is formed which can then multiply. The possibility that mutations may arise during such recombination was considered by Brown (141) who rejected it on the grounds that his *r* mutants of phage T4 were produced under conditions where bacteria outnumber phage by 1000:1 (i.e. there was a multiplicity of infection of 0.001). It should, however, be pointed out that the induction of *r* and *r* mottled mutants in his experiments was not statistically significant at 125 krad and only just significant (assuming only statistical errors $p < 0.05$) at 250 krad. In his second, and more successful series in which reversion of *rII* mutants by ionizing radiation was clearly demonstrated, the multiplicity of infection was 10. The occurrence of recombination was thus overwhelmingly probable.

There is obviously a need for more information on the involvement of recombination in the mutation of bacteriophage.

Bacteria.—The location of the initial damage by ionizing radiation which leads to mutations is believed to be the DNA, largely because of evidence that X-ray-induced mutations in Hfr male *E. coli* are transferable to the female by conjugation immediately after irradiation (149).

A number of workers have sought to show whether mutations arise from the same types of lesions as lethal events, and the most common approach has been to compare the effect on both mutation induction and killing of modifying agents. Chemical protective agents, in general, have been shown to affect both parameters to the same extent. Early work (150, 151) not carried out under controlled gas conditions is difficult to interpret because the effect of oxygen depletion was also involved.

Bridges (152) described the effect of chemicals representative of four different classes of chemical protectors on the effect of gamma radiation given anoxically to *E. coli* WP2. Table I shows that the protective factors against mutational and lethal damage were very similar for cysteine, glycerol, and thiourea. Additional data of Stern (152) extend this correlation to cysteamine, S,2-aminoethylisothiourea, β-mercaptoethanol, and cysteine. Only with dimethyl sulphoxide did Bridges (152) obtain a difference; mutation induction was less affected than was lethality. There is, however, a possibility that dimethyl sulphoxide carried over onto the plating medium might have stimulated growth and/or expression of mutations and thus have given a falsely low value for protection. An opposite effect of a chemical in the plating medium was found by Stern (153) for cysteine and only by washing the bacteria free of cysteine was it possible to demonstrate that the presence of the compound during irradiation affected lethality and mutation equally.

TABLE I

PROTECTION RATIOS FOR THE EFFECTS OF VARIOUS COMPOUNDS ON *Escherichia coli* WP2 EXPOSED TO GAMMA RADIATION UNDER ANOXIA[a]

Compound	Protection ratio	
	Inactivation	Mutation
Cysteine (0.1 M)	1.7	1.7
Glycerol (1 M)	1.9	1.8
Thiourea (0.2 M)	1.9	2.0
Dimethyl sulphoxide (2 M)	1.9	1.5

[a] From Bridges, B. A., *J. Gen Microbiol.*, **31**, 405–12 (1963).

An interesting observation of Bridges was that no protection, either of lethality or of mutation induction, was obtained below a dose of ⌐10 krad. Stern also observed this with cells grown in minimal medium and suggested that radiation might increase the permeability of the cell to the chemicals. The physiological state of the cell, however, is obviously of some importance because cells grown in a rich medium are protected from zero dose upwards (153).

In general, then, there is no evidence that chemical protectors can distinguish between damage which ultimately gives rise to mutations and damage which is lethal. The situation with sensitizing chemicals is rather different. Oxygen is known to be more effective on lethal damage (dose-enhancing factor ⌐3) than on mutagenic damage (dose-enhancing factor ~1.5–2.0) in *E. coli* WP2 (152, 154). Indeed, if the oxygen effect obtained for mutagenesis at very low doses is extrapolated back to zero dose, a modifying factor of only ~1.3 is obtained (155). A significant, though smaller, difference was found with the chemical sensitizer N-ethylmaleimide where a modifying factor of 1.8 was found for mutation induction compared with a factor of 2.1 for lethality (152). At least insofar as oxygen is concerned, the results are consistent with the hypothesis that only a fraction of potentially lethal damage can give rise to mutations and that this fraction is affected by oxygen to a relatively small extent.

The nature of this damage is still uncertain. Calculations of the target size using the track segment method yield a value of $\sim 5 \times 10^{-8}$ cm for the sensitive region within which one ion cluster induces a mutation with a probability of approximately unity in *E. coli* WP2 (156). This would represent a rather small part of a nucleotide, certainly not likely to be an event involving both strands of the chromosome.

An analysis of more extensive data obtained with radiations of differing LET was interpreted as indicating that potentially lethal damage in *E. coli* could be divided into two classes: double-stranded DNA damage which is largely irreparable, and single-stranded DNA damage which could be repaired to varying degrees in different strains (157). A similar analysis for mutagenic damage (158) indicated that it is largely single stranded. Circumstantial evidence suggests that the single-stranded damage might be scission of the sugar-phosphate backbone. Estimates of the oxygen enhancement ratio for single-strand scissions vary around the value two (159, 160) but may even be as low as unity when all enzymic repair is eliminated (161). Unfortunately, this information is not of great assistance in comparing the mutagenic lesion with the single-strand scission, as it is clear that the majority of single-strand damage is reparable, even in sensitive strains, and we have no reliable knowledge of the oxygen-enhancement factors for those components of single-strand damage which remain *in vivo* after repair is complete.

There is no photoreversal of X-ray-induced mutational damage in *E. coli* WP2 (162), which indicates that, as expected, pyrimidine dimers are not involved.

If premutational lesions are indeed single stranded, then one would expect that the kinetics of induction should be single-hit, i.e. linear with dose. Linear dose-response curves have been reported for logarithmic phase *E. coli* WP2 continuously irradiated in continuous culture (163) and flash irradiated, both for true revertants and for ochre suppressor mutations (37). On the other hand, there are reports of response curves showing an upward curvature with increasing dose. In three instances these have been for stationary (nongrowing) cells (152, 154, 162), and a further instance has been with bacteria growing at 16° C (163a). In one report where a detailed analysis was performed (152) the curve was not well fitted to a multihit model. The best fit was obtained with an exponential model (of the type mutant frequency $= ae^{bx}$]) which has no foundation in any theoretical model. Further investigation of this question is desirable. It should be possible to determine whether the departure from linearity is connected with the expression rather than the induction of mutations. It is also possible that an upward curvature of a dose-response curve can be explained in terms of (*a*) a dose-dependent impairment of repair capacity, analogous to that postulated for the cumulative ultraviolet curve (63) or (*b*) a dose-dependent change in the relative proportions of repair carried out by the mistake-proof and mistake-prone systems postulated by Witkin (110) for ultraviolet and extended by Bridges, Law & Munson (37), to ionizing-radiation damage.

It will be apparent by now that even in a relatively simple system such as a bacterium so many factors can be anticipated to affect dose-response curves that very great caution must be exercised in their interpretation. The difficulties with eukaryotes must be expected to be even greater.

The evidence for repair in bacteria of premutational lesions induced by ionizing radiation is scanty. Certainly single-strand lesions in general are readily reparable in most cells but we do not yet know whether there is a particular fraction of single-strand damage which alone is mutagenic and if so, what processes may operate to remove it before mechanisms are set in train which lead to mutations.

It is known that premutational damage produced in *E. coli* WP2 by ionizing radiation is not subject to excision repair as a strain of WP2 deficient in the latter (Hcr⁻) has the same mutability by ionizing radiation as the wild-type WP2 (93, 164). This in itself would seem to indicate that the damage has not the nature of an alkylated base or pyrimidine dimer which one can envisage as readily cut out by an excision process. Repair of a single-strand DNA scission, if such be the premutational lesion, would presumably not require any further excision.

As there is no excision repair, it is reassuring to find (93) that there is no mutation frequency decline (believed to involve excision repair of ultraviolet damage at suppressor loci). Nor is there any enhancing effect of adding nutrient broth to the plating medium, apart from that due to the content of the required amino acid necessary if the mutation is to be expressed (93).

Acriflavine is frequently held to inhibit dark repair but is without effect in X-ray mutagenesis (93).

Almost the only postirradiation factor which influences the induced mutation rate is temperature. Munson & Jeffery (163) reported that the rate of mutation induction in continuously gamma-irradiated turbidostat cultures of *E. coli* WP2 was lower, the lower the temperature of growth from 37° down to 16° C. This was subsequently shown to be due in large measure to a process of low-temperature mutation loss (155). This process had the characteristics of a metabolic process and did not persist beyond the first nuclear doubling. A similar phenomenon was observed with ultraviolet mutagenesis (93) and a possible explanation has already been offered in the appropriate section above.

Also in that section the reader will have read of current hypotheses for the mechanism by which pyrimidine dimers give rise to mutations. That of Witkin (110) supposes a mistake-prone system and a mistake-proof system that enables cells to cope with DNA damage (believed to be a single-strand gap) which occurs whenever the replication enzymes attempt to replicate a region of DNA containing a pyrimidine dimer. If, as is suggested (37), a similar situation exists with regard to ionizing-radiation damage, then low-temperature mutation loss could be explained in terms of differing temperature kinetics for the two systems, the mistake-prone being favoured at high, and the mistake-proof at low temperatures.

The suggestion that there might be a common pathway for ultraviolet and ionizing-radiation mutagenesis was made by Bridges, Law & Munson (37) who envisaged the single-strand gap in DNA as the common mutagenic precursor. Gaps are known to occur in newly synthesized DNA when DNA containing ultraviolet-induced pyrimidine dimers is replicated (59); they also appear to arise as a result of direct breakage of single strands by ionizing radiation followed by the digestion of one of the ends by exonuclease action (165). The locus specifying the mistake-prone system is *exr* (110), and Bridges et al. predicted on the basis of their hypothesis that mutant strains deficient at this locus (*Exr*⁻), besides being unmutable by ultraviolet, should also be unmutable by ionizing radiation and by thymine starvation, which is also believed to give rise to single-strand breaks (166). These predictions were in general confirmed. Exr⁻ strains of *E. coli* were unmutable by thymine starvation, and had less than one-tenth the mutability by gamma radiation of Exr⁺ strains. A similar prediction concerning thymine starvation has been arrived at independently by Witkin (72).

Even if some 10 per cent of mutations induced by gamma radiation occur by some unknown mechanism, it is clear that most occur via a mechanism for which the Exr⁺ phenotype is essential. Since single-strand breaks are present immediately after gamma irradiation but only as replication continues after ultraviolet, it would be expected that gamma-induced mutations would be expressed (that is, able to produce colonies on plates in the complete absence of the required amino acid) rather earlier then ultraviolet-induced mutations in an Hcr⁻ strain. Indeed they might be expressed even in the absence of further DNA replication, or at least 50 per cent—those not on the transcription strand of DNA might have to wait until replication occurred before the information was transferred to a transcription strand.

Kada, Doudney & Haas (167) have in fact reported some years ago that some 50 per cent of X-ray-induced mutations in *E. coli* WP2 could be expressed in the absence of DNA synthesis. This agreement with expectation might, however, be fortuitous as one may dispute the rigour of their exclusion of DNA synthesis on two grounds. 1. At the dose used (20 krad) a great deal of breakdown of DNA is known to occur (165) and yet the DNA content did not change for some time after irradiation. Therefore there must have been DNA synthesis to maintain the net DNA balance. 2. Bacteria plated on minimal plates and suddenly deprived of the required amino acid may still complete the DNA replication cycles on which they are engaged, as they would if the same operation were carried out in liquid media (168, 169). This could amount to further average replication of some 40 per cent of the DNA after plating. It is also possible, I suppose, to argue that even if all the mutations were expressible in the absence of DNA replication there might be insufficient tryptophan in the bacteria on minimal plates to permit the synthesis of the deficient enzyme in the tryptophan biosynthesis pathway. We must agree with Kada et al. (167) when they remark that "it is very difficult to establish definitive proof which eliminates the last possibility that DNA turnover is involved in the X-ray-induced mutation mechanism."

In the author's laboratory with doses some 10 or 20 times lower than those used by Kada et al., we still find a proportion ($\sim$10–40 per cent) of mutations which appear to express themselves on minimal medium immediately after irradiation but we are still unable to rule out residual DNA replication on the plates.

In the experiments of Kada et al. the 50 per cent which appeared to need DNA synthesis in order that expression might occur also became very sensitive to chloramphenicol during the first 10 min after irradiation. Presumably there is some critical step which is sensitive to the rate of protein synthesis, though whether it is a repair step inhibited by protein synthesis or a step in the mutagenic process requiring protein synthesis remains to be determined.

Further discussion of the *exr*-dependent mutation process is given in a subsequent section.

A number of workers have attempted to determine the state of the newly induced mutation at the moment it is first replicated after irradiation. Is it double stranded (i.e. passed on to both daughter genes and their progeny) or single stranded (passed on to only one daughter duplex and its progeny)? Initially it appeared that in *E. coli* WP2 there was a mixture of both single-stranded and double-stranded mutations (170) but soon the apparent proportions were found to be temperature dependent, more single-stranded mutations occurring at lower incubation temperatures (156). Subsequently the apparent single strandedness at lower temperatures (e.g. 16° C) was shown to depend upon the temperature of growth *before* irradiation which appeared to cause a G2 resting stage in the DNA replication cycle not present at 37° C (155). After recently reviewing some of the difficulties involved in this type of experiment, Bridges & Munson concluded that "when a mutational event is induced by ionizing radiation, both daughter chromatids produced at the first replication are mutant" and that this is true for both true revertants and suppressor mutants (14). This conclusion implies: mutations are established on both strands of DNA before the first replication after irradiation; or if they are confined to one strand, that strand is responsible for specifying the information at that point in both new daughter strands at the first replication. The second proposition is, of course, inconsistent with the simple Watson-Crick model for DNA replication.

Eukaryotic cells.—Eukaryotic cells have been used in mutation work for many years and as might be expected from the greater complexity of their genetic apparatus, rather more types of mutation are available for study and the mechanisms of mutagenesis are less well understood at the molecular level than those operating in prokaryotes. Indeed, there do not appear to be any systems where the molecular change involved in a mutation is known for certain.

Biochemical mutations such as those from auxotrophy to prototrophy are fairly simple to operate and genetic analysis usually makes it relatively easy to distinguish supersuppressors from true revertants. Induction of biochemical mutants by ionizing radiation has been studied in yeasts (109, 171–173), fungi (174), and mammalian cells (175); induction of supersuppressors has been studied in yeasts (176, 177).

In diploid cells possibilities for mechanisms involving mitotic recombination are increased and a number of systems have been studied which are known to involve reciprocal and nonreciprocal recombination and gene conversion (e.g. 171). The phenotypic change is from auxotrophy to prototrophy.

There are also a number of systems which do not involve automatic selection against nonmutants. Their sensitivity is much less than that of biochemical reversions and they are therefore only suitable when there is a very high mutability. Recessive lethal and slow growth mutations in *Paramecium* have been studied for many years by Kimball and his colleagues (178, 179), and recessive lethals in yeast by a number of groups (e.g. 180, 181). More recently temperature-sensitive mutations in *Paramecium* have been reported by Igarashi (182, 183) who presents evidence that they may be related to the lethal type.

An unusual mutation to X-ray resistance has been reported by Moustacchi (124). These mutations are produced with such a high frequency that their target size has been estimated to be ~4 per cent of the total genome.

In all these nonselective mutation systems the number of loci involved is unknown although likely to be large. Magni (181), for example, has estimated that there are between 50 and 100 "sites" for recessive lethal mutation in yeast and that their average sensitivity is 30–50 times greater than that of mutations to biochemical requirements induced in the same organism by X rays. Obviously, gross chromosomal changes are likely to be involved.

The induction of pigment mutations by ionizing radiation in *Euglena* (184) and *Chlorella* (185) has been reported to be linear with dose.

A careful study of the kinetics of induction of forward mutants at the *ad-3* locus in *Neurospora* revealed two different classes (186). One type (ad-3^R) is the predominant type at low doses and is induced linearly with dose. The other type (ad-3^{1R}) differs from the ad-3^R in being unable to grow as a homokaryon on adenine-supplemented media. Induction is proportional to the square of the dose and these mutants predominate at higher doses. The ad-3^{1R} are believed to involve genetic regions adjacent to the ad locus but it is clear that they do not arise from ad-3^R events by a single further event as might be expected.

The events which cooperate to give rise to ad-3^{1R} mutants are in fact induced at a much higher rate ($>4.01 \times 10^{-7}$ per rad per survivor) than those giving rise to the ad-3^R point mutations (1.75×10^{-9} per rad per survivor for ad-$3A^R$; 3.64×10^{-9} per rad per survivor for ad-$3B^R$). Even the latter rates are considerably higher than those found for X-ray-induced base-pair transitions in *E. coli* WP2 [$\sim 1 \times 10^{-11}$ per rad per survivor (156, 187)] and mutations to prototrophy in cultured Chinese hamster cells [7×10^{-10} per rad per survivor (174)]. Values of 10^{-7} per rad per survivor (or more) are considerably higher than one would theoretically expect for a mutation at a specific base pair. For mutations inducible at one of many base pairs (e.g. most forward mutations), induced mutation rates of 10^{-8} to 10^{-10} would be expected.

Webber & De Serres (186) also make the point that the two-hit type are only detectable as mutations in a diploid organism. Since they cannot grow as a homokaryon they presumably would be lethal in a haploid organism. Thus, if the behaviour of the ad locus were in any way typical, one would tend to isolate mainly one-hit point mutations in haploid lines and mainly two-hit [multilocus deletion (188)] mutations in diploid lines, particularly at the high doses normally used for the isolation of mutants. Moreover, since the two-hit type are not revertible, comparison of the revertibility of X-ray-induced mutants should only be undertaken if low doses were used in inducing the forward mutations, so that the majority of mutants would be one-hit in origin.

More recently de Serres et al. (188) have studied the effect of varying the dose rate on the induction of one-hit point mutations and two-hit multilocus deletions. As was found in *Paramecium*, there was a large dose-rate effect for the two-hit mutations. With a reduction in the rate of exposure from 1000 to 10 R per min, the forward mutation rate for the one-hit events that cooperate in pairs was reduced by a factor of 3.2. Such an effect was taken as evidence for repair of the initial events. The slower the rate of irradiation, the smaller the number of events remaining unrepaired at any time to interact with one another.

The absence of a dose-rate effect was taken by these authors as further evidence that lesions leading to point mutations are qualitatively different from those which interact to form multilocus deletions. It seems to me, however, that this is not a valid interpretation since if there is no interaction between lesions to give point mutations there would be no expectation of a dose-rate effect even if the lesions were susceptible to repair, and even if they were the same lesions which interact to give multilocus deletions. The distinction between the two types of mutation must for the moment rest on the other evidence reported above.

Dose-response curves with a multihit component are often found for the induction of biochemical mutations in yeast (e.g. 189, 190) but their cause is not yet understood.

Several workers have investigated the effect of chemical protective agents on mutation induction but their conclusions are not always valid because of their failure to take account of the effect of oxygen depletion. It has been known for some years that relatively low concentrations of sulphydryl comcompounds rapidly deplete the medium of oxygen upon irradiation and cause apparent protection due to abolition of the sensitizing action of oxygen (191). Genuine protection can only be considered as established if the presence or absence of oxygen is controlled. Cysteine has been reported to protect *Saccharomyces* by a factor of ~2 (171, 192) but in neither case is it possible to eliminate anoxia as the cause. Even in the experiments of Kølmark with *Neurospora crassa* (193), where the author was aware of the development of endogenous anoxia by respiration at 20° and 30° C, it is not possible to eliminate oxygen depletion by radiation chemical action in the experiments at 0° C. In a companion paper (194) Kølmark reports the effect of endogenous anoxia on lethality and mutation induction. The dose-modifying factor for both was in the range 1.9 to 1.7. It is not clear in these experiments how effective were the precautions taken to prevent oxygen depletion in the supposed "oxic" series at 0° C. Development of anoxia at doses above ~50 krad probably occurred as a result of radiation chemical reactions and may well explain the increased radioresistance at higher doses and the concomitant loss of "anoxic" protection against mutation induction.

Rather more adequate control of oxic and anoxic conditions appears to have been achieved in the experiments of Mortimer, Brustad & Cormack (177) who observed the effect of oxygen on lethality and mutation induction in diploid yeast for a range of fast charged particles (summarized in Table II). It is interesting that the induced mutation rates are comparable to those obtained for base-change and suppressor mutations in *E. coli* (156) if allowance is made for the diploid nature of the yeast and an unknown proportion of yeast cells which have completed DNA synthesis. The oxygen effect for lethality was not large and tended to decrease at higher LET. That for mutation induction was larger but decreased similarly with increasing LET. In contrast to the most recent results of Bridges & Munson (158) with *E. coli* WP2, Mortimer et al. found that there was a peak LET for mutagenesis as well as for killing at a little over 10^3 MeV g^{-1} cm². For a given

TABLE II

Mutation Induction and Lethality of Diploid Yeast in Relation to LET and Oxygen[a]

Radiation	Lethality		His+ suppressor mutation		Try+ mutation		Try+ suppressor mutation	
	LD$_{50}$ (krad)	M (O$_2$)[b]	β[c]	M (O$_2$)	β	M (O$_2$)	β	M (O$_2$)
X ray	47	1.6	2.96	2.77	1.65	1.92	2.70	2.57
D$^+$	39	1.9	3.38	2.25	1.64	2.22	4.01	2.71
He^{2+}	36	1.6	3.82	2.11	2.24	2.57	3.53	2.39
Li^{3+}	28	1.6	4.79	1.85	2.43	1.96	5.37	1.83
B^{5+}	15	1.6	5.61	1.75	3.44	2.22	6.78	1.93
C^{6+}	15	1.3	5.49	1.61	2.84	1.48	7.12	1.83
Ne^{10+}	20	1.2	2.16	2.35	1.22	1.72	2.79	1.63

[a] From Mortimer et al., *Rad. Res.*, **26**, 465–82 (1965).

[b] M (O$_2$) = ratio of aerobic to anaerobic values of β or anaerobic to aerobic values of LD$_{50}$.

[c] β = induced mutation rate per survivor per rad ($\times 10^{-10}$).

amount of inactivation, the probability of recovering mutants among the survivors decreased with increasing LET, as was also observed with *E. coli* WP2. The authors suggest that for the induction of lethality the track core of neon ions is always effective in air or nitrogen (oxygen effect of 1.2), whereas for mutation the entire effect can be accounted for by the action of the associated delta rays (oxygen effect of up to 2.35).

Information on the repair of premutational damage induced by ionizing radiation is almost as scarce for eukaryote cells as for bacteria. The dose-rate experiments of De Serres et al. (188) have already been cited as evidence for repair of those lesions which interact to form multilocus deletions. To detect repair of single-hit premutational lesions one must be able either to inhibit repair or to isolate mutants which are deficient in repair. Radiation-sensitive mutants of bacteria have been a fruitful source of strains possessing abnormal mutability (77, 110) and it is likely that the radiation-sensitive mutants of eukaryotes currently being reported will prove to be equally useful. Another useful approach may involve the use of synchronized cultures as is done with the *Paramecium* system.

The well-established *Paramecium* system has provided good evidence for repair and has been reviewed by Kimball (178). The essential points are: (*a*) a number of postirradiation treatments can decrease the amount of mutation; (*b*) DNA synthesis in the micronucleus is probably the endpoint for the mutation process since the maximum amount of mutation is produced by irradiation just before the DNA synthetic phase and postirradiation treatments are effective when started before but not after DNA synthesis; (c) the rate of repair is greatest in normal log phase, least in stationary phase and can be decreased by a variety of metabolic inhibitors; (d) the damage which interacts in a two-hit manner is more rapidly repaired than that inducing point mutations [cf. the *Neurospora* results (188)].

More recently Igarashi (182, 183) has obtained very similar results for the induction of temperature-sensitive mutations in *Paramecium*.

Our knowledge of the processes involved in ionizing-radiation mutagenesis of eukaryotic cells is fragmentary, to say the least. It is clear that most mutations are irrevocably established at the time of DNA synthesis in *Paramecium* (178, 183) but there is a recent report (179) that a small fraction of the mutations are fixed independently of DNA replication, a phenomenon that is also suspected to occur in bacteria.

The reader will now be aware of the current speculations concerning the role of recombination processes in mutagenesis. The genetic situation in diploid cells allows further scope for recombination processes which is denied haploid cells. It is possible to set up mutation systems where recombination is inevitably involved. An example is the heteroallelic diploid where two noncomplementing alleles at the same locus are involved. Manney & Mortimer (195) studied many such heteroallelic diploids of *Saccharomyces* at both the *ar4* and the *tr5* locus. They found that at sublethal doses, induction was linearly related to dose and the results were consistent with the simple hypothesis that an X-ray lesion anywhere between the two mutations can lead to a recombination, which may be reciprocal or nonreciprocal. Mortimer (quoted in 180) has also shown that these lesions are stable for several hours since recombination can occur following the formation of a zygote by irradiated haploids.

A related phenomenon termed mitotic recombination can also occur when diploids are heterozygous at different loci. Haefner (196) obtained a nonlinear response for nine such combinations in *Saccharomyces* (as judged by frequencies after three doses of ionizing radiation). Recombination tended to be higher nearest the centromere. Detailed pedigrees showed that the recombinations were always reciprocal. Moreover, some recombinations occurred during generations after the immediate postirradiation generation indicating that the damage can persist for more than one cell cycle. Information up to 1965 on both heteroallelic and heterozygous diploid recombination is given by James & Werner (180). Schwaier (171) has recently reported that X rays are highly efficient at inducing mitotic recombination and gene conversion but only weak at inducing forward mutations in *Saccharomyces*.

Obviously eukaryotes, particularly the yeasts and fungi, are likely to be exceedingly useful for the detailed study of the relation of recombination to mutation.

A number of workers have attempted to answer the question as to whether newly induced mutations are established on both DNA strands by the time of the first DNA replication after radiation. Yamasaki, Ito & Matsudaira (172) found that the dose-response curve for mosaic mutations at the *ad* locus was concave downwards whereas that for whole-colony mutations was concave upwards. The results were consistent with the conversion of "mosaics" to "wholes" by inactivation of the unmutated part of the genome within the cell by a second 'hit.' It is obviously attractive to ascribe the mutated and unmutated parts of the genome to the separate strands of the DNA duplex, but the results do not exclude the possibility that the mosaic colonies arise from cells in which DNA duplication has already occurred.

In a more comprehensive study, James & Werner (197) made pedigree analyses for three generations of recessive lethal mutations in *Saccharomyces*. Pedigree analyses enabled them to take account of lethal sectoring (the presence of subclones of dead cells) which, they found, grossly distorted the estimate of genetic mosaicism. With the use of pedigrees which were either free of lethal sectors or in which the lethal sectors did not interfere with the determinations, only 4 out of 14 mutations were found to be mosaic, and most if not all of this mosaicism was exposed at the first postirradiation cell division. Very similar data have been obtained with recessive lethal and slow-growth mutations in *Paramecium* (198) where only 4 out of 19 were mosaic. It seems clear that the majority of recessive lethal mutations in both yeast and *Paramecium* are such as to appear in both original strands of DNA as well as in both daughter strands at the first replication. The situation is thus very similar to that found with bacteria (see above).

Although unicellular organisms have been the chief eukaryotic material for studies at the cellular level, there are a few reports of X-ray mutagenesis with mammalian cell lines including both presumed point mutations (174, 199) and heritable changes of a much less well-defined character (200). We may expect mutagenesis of mammalian cells to be a very profitable field during the next decade.

Recombination as a Mutagenic Process

While it is unlikely that all spontaneous mutations arise as mistakes during recombination, there is now a good deal of evidence that recombination can give rise to mutations in the absence of any applied mutagen. Magni & Von Borstel, for example, showed that the spontaneous mutation rate for certain alleles in yeast was much higher during meiosis than during mitosis (201). A large proportion of meiotic revertants were associated with a exchange in the region of the specific locus (202). Moreover, a meiotic reversion of thr_{4-1} was abolished when one of the two homologous chromosomes carried a long deletion covering thr_4 locus. Similar results have been reported for *Ascobolus immersus* (203) and *Neurospora crassa* (204, 205) where again meiotic reversions were associated with outside marker recombination.

Magni believes that meiotic mutations are largely frameshifts due to unequal crossing over and it is quite possible that these may predominate among mutations induced during recombination. Strigini (206), for example, found that frameshift mutants of T4 phage reverted spontaneously during growth and recombination with 40 per cent recombination of outside markers whereas for base-substitution mutations the figure was 13 per cent. Streisinger et al. (207) also reports frameshift mutants arising during recombination in T4 phage. Most bacterial systems, however, are not designed for the detection of frameshift mutations, so that their contribution may have been underestimated.

Another system where there is direct evidence for the association of mutation and recombination is the transformation of *Bacillus subtilis*. Yoshikawa (208) found that the frequency of auxotrophic mutants among transformants was ten to twenty times higher than that among control bacteria. The effect was specifically induced by homologous DNA at the site of integration. A high frequency of *ind⁻* mutations associated with recombination during transformation was also reported by Chernik & Krivisky (209).

In bacteria Demerec's (210) phenomenon of selfing is probably attributable to mutation associated with recombination. He observed that the *Arg-A162* allele transduced into *Salmonella* with the same allele reverted with a higher frequency than in transduced control cells.

It is clear, then, that recombination can give rise to mutations. Can the mutagenic effect of radiations be attributed to their ability to enhance recombination? In one or two instances already mentioned the evidence is quite strong, for example allelic conversion and mitotic recombination in fungi and yeasts and possibly ultraviolet-induced mutation in temperate bacteriophages.

As for bacteria, the evidence is largely circumstantial. The *exr* locus appears to be involved both in recombination and in mutation induction by ultraviolet (58), ionizing radiation, and thymine starvation (37). All these treatments have in common the ability to produce single-strand gaps in DNA and to stimulate recombination (211–213). Howard-Flanders and his colleagues have given evidence that recombination is involved in repair of postreplication gaps caused by pyrimidine dimers (e.g. 214) and it is useful as a working hypothesis to suppose that mutation is associated with this recombination.

An experiment by Howard-Flanders, Wilkins, and Rupp [quoted by Witkin (72)] strongly supports the expectation that both daughter genomes are necessary in the same cell for recombination and thus repair of ultraviolet-irradiated DNA. If this should be generally true for recombination repair and *exr*-dependent mutation there are implications for ionizing-radiation mutagenesis which have yet to be explored. Ionizing radiation, unlike ultraviolet, produces single-strand gaps without the need for DNA replication. One would therefore predict that those regions of the genome which are not duplicated within the cell (or the nucleus?) should not be mutable by ionizing radiation.

It should also be noted that whereas it is easy to account for frameshift mutations simply in terms of unequal crossing-over [as Grigg (215) suggested to explain X-ray mutagenesis in bacteria], the production of base-change mutations requires a mistake-producing step in which the specificity of at least one base is changed. Whereas such a step may normally be involved in mutagenesis, it is important to realize that it may well operate during other processes involving DNA which are, strictly speaking, not recombination

It is thus quite possible that the *exr* system is involved in mutagenesis as part of a nonrecombination repair process acting on X- or γ-ray-induced gaps, as well as being involved in recombination repair of postreplication gaps.

This is about the extent of our knowledge of the involvement of recombination in mutagenesis. Any further speculation should probably await further experimental results.

Conclusions

One fact which has emerged quite clearly is that expressible mutations do not arise to a significant extent as a direct action of radiation upon genetic material. If such an action exists it has yet to be adequately demonstrated in a living system. In general it would appear that the initial damage is located in the DNA and that it is relatively nonspecific in that many different molecular changes (for example transitions, transversions, frameshifts, recombinational events) may arise from one particular type of damage (this is especially true of pyrimidine dimers produced by ultraviolet). One of the common precursors of a mutation could be a single-stranded gap in DNA but although such gaps could well be both mutagenic and recombinogenic, definite evidence is lacking.

There is a strong suspicion that recombination between two DNA molecules may be involved in mutagenesis but we still lack two important pieces of information. 1. Although we know that recombination can be mutagenic, we do not know whether it may give rise to all the various types of molecular mutation that are found after irradiation. 2. Although ultraviolet and ionizing radiations are known to induce both recombination and mutation, it is not known whether one can lead to the other. The fact that Exr⁻ bacteria are deficient both in recombination ability and in mutability does not in itself establish a cause and effect relationship. Fortunately many experimental approaches are available to test this hypothesis so that one may expect steady progress in the understanding of ultraviolet and ionizing-radiation mutagenesis.

As far as visible and near-visible radiations are concerned, the evidence for the involvement of recombination is weak but this may merely reflect the relative lack of attention paid to these radiations.

In my opinion mutagenesis in cellular and subcellular systems (by chemicals, radiations, and viruslike particles) has reached a stage where the foundations have been laid. Clean systems are available, and molecular biologists, geneticists, and cell biologists can at last begin to link together their various approaches towards an understanding of the mechanisms involved. The knowledge we gain may be expected to throw light on matters of immediate concern to mankind such as carcinogenesis and ageing (where mutation of somatic cells is likely to be a contributory factor), mutation in the population by environmental radiations and chemicals (the latter presenting an almost totally unknown risk in our present world of pharmaceuticals, preservatives, and pesticides), and genetic engineering (including plant, animal, and microorganism breeding).

LITERATURE CITED

1. Kimball, R. F., *Ann. Rev. Nucl. Sci.*, **1**, 479–94 (1952)
2. Ponnamperuma, C. A., Lemmon, R. M., Calvin, M., *Science*, **137**, 605–7 (1962)
3. Ponnamperuma, C. A., Lemmon, R. M., Bennett, E. L., Calvin, M., *Science*, **134**, 113 (1961)
4. Grossman, L., *Photochem. Photobiol.*, **7**, 727–35 (1968)
5. Witkin, E. M., Sicurella, N. A., *J. Mol. Biol.*, **8**, 610–13 (1964)
6. Kubitschek, H. E., *Proc. Natl. Acad. Sci. U.S.*, **52**, 1374–81 (1964)
7. Watson, J. D., Crick, F. H. C., *Nature*, **171**, 964–67 (1953)
8. Löwdin, P.-O., *Rev. Mod. Phys.*, **35**, 724–32 (1963)
9. Ladik, J., *J. Theoret. Biol.*, **6**, 201–7 (1964)
10. Rein, R., Ladik, J., *J. Chem. Phys.*, **40**, 2466–70 (1964)
11. Rein, R., Harris, F. E., *Science*, **146**, 649–50 (1964)
12. Löwdin, P.-O., *Electronic Aspects of Biochemistry*, 167–201 (Academic Press, New York, 1964)
13. Pullman, A., *Electronic Aspects of Biochemistry*, 135–52 (Academic Press, New York, 1964)
13a. Holliday, R., *Genet. Res.*, **3**, 472–86 (1962)
14. Bridges, B. A., Munson, R. J., *Current Topics in Radiation Research*, **4**, 95–188 (North-Holland, Amsterdam, 1968)
15. Kaplan, R. W., Kaplan, C., *Exptl. Cell Res.*, **11**, 378–92 (1956)
16. Ritchie, D. A., *Genet. Res.*, **5**, 168–69 (1964)
17. Ritchie, D. A., *Genet. Res.*, **6**, 474–78 (1965)
18. Kubitschek, H. E., *Science*, **155**, 1545–46 (1967)
19. Webb, R. B., Malina, M. M., *Science*, **156**, 1104–5 (1967)
20. Leff, J., Krinsky, N. I., *Science*, **158**, 1332–35 (1967)
21. Ashwood-Smith, M. J., Copeland, J., Wilcockson, J., *Nature*, **214**, 33–35 (1967)
22. Kaplan, R. W., *Nature*, **163**, 573–74 (1949)
23. Ito, T., Hieda, K., *Mutation Res.*, **5**, 184–86 (1968)
24. Webb, R. B., Kubitschek, H. E., *Biochem. Biophys. Res. Commun.*, **13**, 90–94 (1963)
25. Nakai, S., Saeki, T., *Genet. Res.*, **5**, 158–61 (1964)
26. Stewart, C. R., *Genetics*, **59**, 23–31 (1968)
27. Böhme, H., Wacker, A., *Biochem. Biophys. Res. Commun.*, **12**, 137–39 (1963)
28. Brendel, M., Kaplan, R. W., *Mol. Gen. Genet.*, **99**, 181–90 (1967)
29. Drake, J. W., McGuire, J., *J. Virol.*, **1**, 260–67 (1967)
30. Brendel, M., *Mol. Gen. Genet.*, **101**, 111–15 (1968)
31. Barricelli, N. A., Del Zoppo, G., *Mol. Gen. Genet.*, **101**, 51–58 (1968)
32. Bellin, J. S., Grossman, L. I., *Photochem. Photobiol.*, **4**, 45–53 (1965)
33. Simon, M. I., Van Vunakis, H., *J. Mol. Biol.*, **4**, 488–99 (1962)
34. Simon, M. I., Van Vunakis, H., *Arch. Biochem. Biophys.*, **105**, 197–206 (1964)
35. Sussenbach, J. S., Berends, W., *Biochim. Biophys. Acta*, **95**, 184–85 (1965)
35a. Freifelder, D., Davison, P. F., Geiduschek, F. P., *Biophys. J.*, **1**, 389–400 (1961)
36. Bellin, J. S., Yankus, C. A., *Biochim. Biophys. Acta*, **112**, 363–71 (1966)
37. Bridges, B. A., Law, J., Munson, R. J., *Mol. Gen. Genet.*, **103**, 266–73 (1968)
38. Uretz, R. B. (Personal communication, 1967)
39. Kubitschek, H. E., Henderson, T. R., *Proc. Natl. Acad. Sci. U.S.*, **55**, 512–19 (1966)
40. Jacob, F., *Compt. Rend.*, **238**, 732–34 (1954)
41. Weigle, J. J., *Proc. Natl. Acad. Sci. U.S.*, **39**, 628–36 (1953)
42. Tessman, E. S., Ogaki, T., *Virology*, **12**, 431–49 (1960)
43. Weigle, J. J., Dulbecco, R., *Experientia*, **9**, 372–73 (1953)
44. Tessman, E. S., *Virology*, **2**, 679–88 (1956)
45. Drake, J. W., *J. Mol. Biol.*, **6**, 268–83 (1963)
46. Krieg, D. R., *Virology*, **9**, 215–27 (1959)
47. Drake, J. W., *J. Bacteriol.*, **91**, 1775–80 (1966)
48. Folsom, C. E., *Genetics*, **47**, 611–22 (1962)
49. Kaplan, R. W., Winkler, U., Wolf-Ellmauer, H., *Nature*, **186**, 330–31 (1960)

50. Zavil'gel'skii, G. B., Kriviskii, A. S., *Genetika*, **1**, 12–19 (1965)
51. Drake, J. W., *J. Cellular Comp. Physiol.*, **64**, *Suppl. 1*, 19–31 (1964)
52. Belykh, R. A., Kriviskii, A. S., Chernik, T. P., *Genetika*, **4**, 62–70 (1968)
53. Bresler, S. E., Perumov, D. A., *Dokl. Akad. Nauk SSSR*, **158**, 967–69 (1964)
54. Drake, J. W., *J. Bacteriol.*, **92**, 144–47 (1966)
55. Winkler, U., *Z. Vererbungsl.*, **96**, 75–78 (1965)
56. Winkler, U., *Z. Vererbungsl.*, **97**, 18–28 (1965)
57. Winkler, U., *Z. Vererbungsl.*, **97**, 29–39 (1965)
58. Witkin, E. M., *Brookhaven Symp. Biol.*, **20**, 17–55 (1967)
59. Rupp, W. D., Howard-Flanders, P., *J. Mol. Biol.*, **31**, 291–304 (1968)
60. Stent, G. S., *Advan. Virus Res.*, **5**, 95–149 (1958)
61. Devoret, R., *Compt. Rend.*, **260**, 1510–13 (1965)
61a. Devoret, R., in discussion after paper by H. E. Kubitschek in *Mutation as Cellular Process* (Churchill, London, in press 1969)
62. Witkin, E. M., *Rad. Res. Suppl. 6*, 30–53 (1966)
63. Bridges, B. A., *Mutation Res.*, **3**, 273–79 (1966)
64. Doudney, C. O., *Mutation Res.*, **4**, 280–97 (1966)
65. Witkin, E. M., *Genetics*, **48**, 916 (1963)
66. Hill, R. F., *J. Gen. Microbiol.*, **30**, 289–97 (1963)
67. Bridges, B. A., Dennis, R. E., Munson, R. J., *Mutation Res.*, **4**, 502–4 (1967)
68. Osborn, M., Person, S., *Mutation Res.*, **4**, 504–7 (1967)
69. Bridges, B. A., Munson, R. J., *Proc. Roy. Soc.*, *B*, **171**, 213–26 (1968)
70. Person, S., Osborn, M., *Proc. Natl. Acad. Sci. U.S.*, **60**, 1030–37 (1968); (Personal communication)
71. Bridges, B. A., Dennis, R. E., Munson, R. J., *Genetics*, **57**, 897–908 (1967)
72. Witkin, E. M., *Proc. 12th Intern. Congr. Genet.*, 111 (In press) (1969)
73. Zampieri, A., Greenberg, J., *Genetics*, **57**, 41–51 (1967)
74. Schwartz, N. M., *Genetics*, **57**, 495–503 (1967)
75. Dawson, G. W. P., Smith-Keary, P. F., *Heredity*, **18**, 1–20 (1963)
76. Kondo, S., Kato, T., *Photochem. Photobiol.*, **5**, 827–37 (1966)
77. Hill, R. F., *Photochem. Photobiol.*, **4**, 563–68 (1965)
78. Witkin, E. M., Sicurella, N. A., Bennett, G. M., *Proc. Natl. Acad. Sci. U.S.*, **50**, 1055–58 (1963)
79. Witkin, E. M., *Mutation Res.*, **1**, 22–36 (1964)
80. Kondo, S., Jagger, J., *Photochem. Photobiol.*, **5**, 189–200 (1966)
81. Witkin, E. M., *Proc. Natl. Acad. Sci. U.S.*, **50**, 425–30 (1963)
82. Kaplan, R. W., Witt, G., *Z. Vererbungsl.*, **97**, 209–17 (1965)
83. Horneck-Witt, G., Kaplan, R. W., *Mol. Gen. Genet.*, **101**, 123–30 (1968)
84. Witkin, E. M., *Science*, **152**, 1345–53 (1966)
85. Hainz, H., Kaplan, R. W., *Z. Allgem. Mikrobiol.*, **3**, 113–25 (1963)
86. Ashwood-Smith, M. J., Bridges, B. A., *Mutation Res.*, **3**, 135–44 (1966)
87. Bridges, B. A., Ashwood-Smith, M. J., Munson, R. J., *Proc. Roy. Soc. B*, **168**, 203–15 (1967)
88. Smith, K. C., Yoshikawa, H., *Photochem. Photobiol.*, **5**, 777–86 (1966)
89. Smith, K. C., O'Leary, M. E., *Science*, **155**, 1024–26 (1967)
90. Donnellan, J. E., Hosszu, J. L., Rahn, R. O., Stafford, R. S., *Nature*, **219**, 964–65 (1968)
91. Rahn, R. O., Hosszu, J. L., *Photochem. Photobiol.*, **8**, 53–63 (1968)
92. Zamenhof, S., Reddy, T. K. R., *Rad. Res.*, **31**, 112–20 (1967)
93. Munson, R. J., Bridges, B. A., *Mutation Res.*, **3**, 461–69 (1966)
94. Igali, S., Bridges, B. A. (Unpublished observations)
95. Witkin, E. M., *Proc. 10th Intern. Congr. Genet.*, **1**, 280–89 (1958)
96. Witkin, E. M., *J. Cellular Comp. Physiol.*, **58**, *Suppl. 1*, 135–44 (1961)
97. Clarke, C. H., *Mol. Gen. Genet.*, **100**, 225–41 (1968)
98. Paribok, V. P., Kassinova, G. V., Bandas, E. L., *Tsitologiya*, **9**, 1496–1502 (1967)
99. Vechet, V., *Folia Microbiol.*, **13**, 279–390 (1968)
100. Shankel, D. M., Kleinberg, J. A., *Genetics*, **56**, 589 (1967)
101. Setlow, R. B., *J. Cellular Comp. Physiol.*, **61**, *Suppl. 1*, 51–68 (1964)
102. Sideropoulos, A. S., Shankel, D. M., *J. Bacteriol.*, **96**, 198–204 (1968)

103. Clarke, C. H., *Mol. Gen. Genet.*, **99**, 97–108 (1967)
104. Doudney, C. O., White, B. F., Bruce, B. J., *Biochem. Biophys. Res. Commun.*, **15**, 70–75 (1964)
105. Sideropoulos, A. S., Johnson, R. C., Shankel, D. M., *J. Bacteriol.*, **95**, 1486–88 (1968)
106. Berrie, A. M. M., *Proc. Natl. Acad. Sci. U.S.*, **39**, 1125–33 (1953)
107. Witkin, E. M., *Proc. Natl. Acad. Sci. U.S.*, **39**, 427–33 (1953)
108. Doudney, C. O., Nishioka, M. (Personal communication)
109. Bridges, B. A., Munson, R. J., *Biochem. Biophys. Res. Commun.*, **30**, 620–24 (1968)
110. Witkin, E. M., *Brookhaven Symp. Biol.*, **20**, 17–55 (1967)
111. Alper, T., *Phys. Med. Biol.*, **8**, 365–85 (1963)
112. Davies, D. R., Arlett, C. F., Munson, R. J., Bridges, B. A., *J. Gen. Microbiol.*, **46**, 329–38 (1967)
113. Bridges, B. A., Munson, R. J., Arlett, C. F., Davies, D. R., *J. Gen. Microbiol.*, **46**, 339–46 (1967)
114. Witkin, E. M., Farquharson, E. L., *Mutation as Cellular Process* (Churchill, London, in press, 1969)
115. Kunicki-Goldfiner, W. J. H., Mycielski, R., *Acta Microbiol. Polon.*, **15**, 113–18 (1966)
116. Stonehill, E. H., Hutchison, D. J., *J. Bacteriol.*, **92**, 136–43 (1966)
117. Chang, L.-T., Lennox, J. E., Tuveson, R. W., *Mutation Res.*, **5**, 217–24 (1968)
118. Arlett, C. F., *Mutation Res.*, **3**, 410–19 (1966)
119. Kilbey, B. J., *Mol. Gen. Genet.*, **100**, 159–65 (1967)
120. Kilbey, B. J., De Serres, F. J., *Mutation Res.*, **4**, 21–29 (1967)
121. Davies, D. R., Levin, S., *Mutation Res.*, **5**, 231–36 (1968)
122. Cross, R. A. (Ph.D. thesis, Univ. Calif., 1965) (*Nucl. Sci. Abstr.*, **20**, 16294)
123. Gilmore, R. A. (Thesis, *UCRL 16851*, 1965) (*Nucl. Sci. Abstr.*, **20**, 45364)
124. Moustacchi, E., *Mutation Res.*, **2**, 403–12 (1965)
125. Snow, R., *Genetics*, **56**, 591–92 (1967)
126. Zetterberg, G., *Mutation Res.*, **3**, 393–409 (1966)
127. Holliday, R., *Genet. Res.*, **8**, 323–37 (1966)
128. Holliday, R., *Mutation Res.*, **4**, 275–88 (1967)
129. Malling, H. V., De Serres, F. J., *Rad. Res.*, **31**, 637–38 (1967)
130. Ito, T., Yamasaki, T., Domon, M., Ishizaka, S., Matsudaira, Y., *Japan. J. Genet.*, **39**, 136–46 (1964)
131. Ito, T., Domon, M., Yamasaki, T., *Japan. J. Genet.*, **41**, 233–40 (1966)
132. Nasim, A., *Genetics*, **59**, 327–33 (1968)
133. Nasim, A., Auerbach, C., *Mutation Res.*, **4**, 1–14 (1967)
134. Haefner, K., *Mutation Res.*, **4**, 514–16 (1967)
135. Haefner, K., *Genetics*, **57**, 169–78 (1967)
136. Bostock, C., Donachie, W. D., Masters, M., Mitchison, J. M., *Nature*, **210**, 808–10 (1966)
137. Gowen, J. N., *Symp. Quant. Biol.*, **9**, 187–93 (1941)
138. Mundry, K. W., *Z. Induktive Abstammungs-, Vererbungsl.*, **88**, 115–27 (1957)
139. Kaplan, R. W., Winkler, U., Wolf-Ellmauer, H., *Nature*, **186**, 330–31 (1960)
140. Brown, D. F., *Nature*, **212**, 1595–96 (1966)
141. Brown, D. F., *Mutation Res.*, **3**, 365–73 (1966)
142. van der Ent, G. M., Blok, J., Linckens, E. M., *Mutation Res.*, **2**, 197–204 (1965)
143. Ardashnikov, S. N., Soyfer, V. N., Goldfarb, D. M., *Biochem. Biophys. Res. Commun.*, **16**, 455–59 (1964)
144. Krivisky, A. S., *Genetics Today* (*Proc. 11th. Congr. Genet.*), **1**, 56 (Pergamon, London, 1963)
145. Krivisky, A. S., *Physiology of Gene and Mutation Expression*, 75–85 (Academia, Prague, 1966)
146. Kriviskii, A. S., *Dokl. Akad. Nauk SSSR*, **161**, 707–10 (1965) (*Nucl. Sci. Abstr.*, **20**, 80)
147. Rueger, W., Kaplan, R. W., *Z. Allgem. Mikrobiol.*, **6**, 253–69 (1966)
148. Bridges, B. A., Dennis, R. E., Munson, R. J. (In preparation)
149. Kada, T., Marcovich, H., *Genetics Today* (*Proc. 11th. Congr. Genet.*), **1**, 62–63 (Pergamon, London, 1963)
150. Künkel, H. A., Kamm, P., Höhne, G., *Strahlentherapie*, **114**, 94–102 (1961)
151. Künkel, H. A., Rodegra, H., *Naturwissenschaften*, **50**, 594 (1963)
152. Bridges, B. A., *J. Gen. Microbiol.*, **31**, 405–12 (1963)
153. Stern, M. E., van Dillewijn, J., Blok, J., *Mutation Res.*, **5**, 349–57 (1968)
154. Deering, R. A., *Rad. Res.*, **19**, 169–78 (1963)
155. Bridges, B. A., Munson, R. J., *Mutation Res.*, **1**, 362–72 (1964)

156. Munson, R. J., Bridges, B. A., *Nature*, **203**, 270–72 (1964)
157. Munson, R. J., Neary, G. J., Bridges, B. A., Preston, R. J., *Intern. J. Rad. Biol.*, **13**, 205–24 (1968)
158. Munson, R. J., Bridges, B. A., *Biophysik* (In press) (1969)
159. Lett, J. T., Caldwell, I., Dean, C. J., Alexander, P., *Nature*, **214**, 790–92 (1967)
160. Boyce, R. P., Tepper, M., *Virology*, **34**, 344–51 (1968)
161. Dean, C. J., Ormerod, M. (Personal communication, 1969)
162. Kada, T., Brun, E., Marcovich, H., *Ann. Inst. Pasteur*, **99**, 547–66 (1960)
163. Munson, R. J., Jeffery, A., *J. Gen. Microbiol.*, **35**, 191–203 (1964)
163a. Munson, R. J., Bridges, B. A. *Nature*, **210**, 922–25 (1966)
164. Bridges, B. A., Munson, R. J., *Biochem. Biophys. Res. Commun.*, **22**, 268–73 (1966)
165. Emmerson, P. T., Howard-Flanders, P., *Biochem. Biophys. Res. Commun.*, **18**, 24–29 (1965)
166. Mennigmann, H. D., Szybalski, W., *Biochem. Biophys. Res. Commun.*, **9**, 398–404 (1962)
167. Kada, T., Doudney, C. O., Haas, F. L., *Genetics*, **46**, 683–701 (1961)
168. Maaløe, O., Hanawalt, P. C., *J. Mol. Biol.*, **3**, 144–55 (1961)
169. Lark, K. G., Repko, T., Hoffman, E. J., *Biochem. Biophys. Acta*, **76**, 9–24 (1963)
170. Bridges, B. A., Munson, R. J., *J. Mol. Biol.*, **8**, 768–69 (1964)
171. Schwaier, R., *Mol. Gen. Genet.*, **101**, 203–11 (1968)
172. Yamasaki, T., Ito, T., Matsudaira, Y., *Japan. J. Genet.*, **39**, 147–50 (1964)
173. Haefner, K., *Biophysik*, **1**, 413–17 (1964)
174. De Serres, F. J., *Natl. Cancer Inst. Monogr. No. 18*, 33–52 (1965)
175. Kao, F.-T., Puck, T. T., *Biophys. J. Soc. Abstr. 8*, A-144 (1968)
176. Magni, G. E., Puglisi, P. P., *Symp. Quant. Biol.*, **31**, 699–704 (1966)
177. Mortimer, R., Brustad, T., Cormack, D. V., *Rad. Res.*, **26**, 465–82 (1965)
178. Kimball, R. F., *Repair from Genetic Radiation Damage*, 167–76 (Pergamon, London, 1963)
179. Kimball, R. F., Perdue, S. W., *Mutation Res.*, **4**, 37–50 (1967)
180. James, A. P., Werner, M. M., *Rad. Bot.*, **5**, 359–82 (1965)
181. Magni, G. E., *2nd Intern. Congr. Rad. Res. Abstr. 19* (1962)
182. Igarashi, S., *Mutation Res.*, **3**, 13–24 (1966)
183. Igarashi, S., *Mutation Res.*, **3**, 25–33 (1966)
184. Matsuoka, T., *Bull. Inst. Chem. Res. Kyoto Univ.*, **42**, 1–9 (1964)
185. Khropova, V. I., Kvitko, K. V., Zakharov, I. A., *Issled. Genet., No. 2*, 69–76 (1964) (*Nucl. Sci. Abstr.*, **20**, 3716)
186. Webber, B. B., De Serres, F. J., *Proc. Natl. Acad. Sci. U.S.*, **53**, 430–37 (1965)
187. Munson, R. J., Bridges, B. A. (Unpublished observations)
188. De Serres, F. J., Malling, H. V., Webber, B. B., *Brookhaven Symp. Biol.*, **20**, 56–75 (1967)
189. Heslot, H., *Abhl. Deut. Akad. Wiss. Berlin Kl. Med.*, **1**, 193–228 (1963)
190. Hrishi, N., James, A. P., *Can. J. Genet. Cytol.*, **6**, 357–63 (1964)
191. Bridges, B. A., Koch, R., *Intern. J. Rad. Biol.*, **3**, 49–58 (1961)
192. Schaedel, U., Lochmann, E.-R., Laskowski, W., *Nature*, **211**, 431–32 (1966)
193. Kølmark, H. G., *Mutation Res.*, **2**, 229–35 (1965)
194. Kølmark, H. G., *Mutation Res.*, **2**, 222–28 (1965)
195. Manney, T. R., Mortimer, R. K., *Science*, **143**, 581–83 (1964)
196. Haefner, K., *Z. Vererbungsl.*, **98**, 82–90 (1966)
197. James, A. P., Werner, M. M., *Mutation Res.*, **3**, 477–85 (1966)
198. Kimball, R. F., *Mutation Res.*, **1**, 129–38 (1964)
199. Szybalski, W., *Symp. Intern. Soc. Cell Biol.*, **3**, 209–221 (1964) (*Nucl. Sci. Abstr.*, **19**, 43844)
200. Todd, P., *Mutation Res.*, **5**, 173–83 (1968)
201. Magni, G. E., Von Borstel, R. C., *Genetics*, **47**, 1097–1108 (1962)
202. Magni, G. E., *Proc. Natl. Acad. Sci. U.S.*, **50**, 975–80 (1963)
203. Paszewski, A., Surzycki, S., *Nature*, **204**, 809 (1964)
204. Kiritani, K., *Japan. J. Genet.*, **37**, 42–56 (1962)
205. Bausum, H. T., Wagner, R. T., *Genetics*, **51**, 815–30 (1965)
206. Strigini, P., *Genetics*, **52**, 759–76 (1965)
207. Streisinger, G., Okada, Y., Emrich, J., Newton, J., Tsugita, A., Terzaghi, E., Inouye, M., *Symp. Quant. Biol.*, **31**, 77–84 (1966)
208. Yoshikawa, J., *Genetics*, **54**, 1201–14 (1966)

209. Chernik, T. P., Krivisky, A. S., *Genetika*, **4**, 75–86 (1968)
210. Demerec, M., *Proc. Natl. Acad. Sci. U.S.*, **48**, 1696–1704 (1962)
211. Curtiss, R., *Genetics*, **58**, 9–54 (1968)
212. Holliday, R., *Replication and Recombination of Genetic Material*, 157–74 (Australian Acad. Sci., Canberra, 1968)
213. Gallant, J., Spotswood, T., *Genetics*, **52**, 107–18 (1965)
214. Howard-Flanders, P., *Ann. Rev. Biochem.*, **37**, 175–200 (1968)
215. Grigg, G. W., *Australian J. Biol. Sci.*, **17**, 907–20 (1964)

BOSON RESONANCES

By I. BUTTERWORTH

Rutherford High Energy Laboratory
Chilton, Didcot, Berkshire, England
and
Physics Department
Imperial College of Science and Technology
London, England

CONTENTS

1. THEORETICAL BACKGROUND

1.1 INTRODUCTION

The strongly interacting particles, or hadrons, fall into two classes: the half-integral spin fermions or baryons, and the integral spin bosons or mesons. This article is concerned with the latter. Since for bosons the baryonic number B equals zero, it follows that the Gell-Mann–Nishijima formula that relates the additive quantum numbers possessed by hadrons, namely:

$$Q = I_Z + \frac{B+S}{2} \quad \text{(hadrons)} \qquad 1.$$

simplifies to:

$$Q = I_Z + S/2 \quad \text{(mesons)} \qquad 2.$$

where Q, S, and I_Z are respectively the electric charge, strangeness, and Z component of i spin.

For a meson the hypercharge $Y=B+S$ is identical with the strangeness. Since Q is integral it follows from Equation 2 that mesons of even strangeness, and in particular nonstrange mesons, have integral i spin; those of odd strangeness, in particular $S = \pm 1$, have half-integral i spin.

With a proton or neutron as the only practical target, mesons may be studied in *formation* experiments only by use of antibaryon beams. Since, to date, the only available monochromatic antibaryon beam is one of antiprotons, it follows that formation experiments are limited to $\bar{p}p$ and $\bar{p}n$ interactions. Only nonstrange $I = 0$ or 1 systems of mass greater than 2 m_p, i.e. $M > 1880$ MeV, can possibly be studied in this way. Most information on mesons has come, therefore, from *production* experiments where the meson is seen as an enhancement in the effective mass of some group of particles in a multibody final state. This enforced complexity of final states has guaranteed that, to date, meson studies have been made predominantly using the bubble-chamber technique. The difficulties of analysis for multibody final states are legion. They will be illustrated by specific examples in the following sections.

1.2 Meson Quantum Numbers

1.2.1 Parity.—For bosons, the parity of particle and antiparticle is the same in contrast to fermions where the parity reverses between antiparticles. If the parity is related to spin by $P = (-1)^J$, the boson is said to be of *natural* parity; if by $P = (-1)^{J+1}$ of *unnatural* parity. Thus the pseudoscalar particles (π, K, η etc.) are of unnatural parity.

Any boson decaying into two pseudoscalar mesons must be of natural parity with spin equalling the relative orbital angular momentum l of the products. Moreover, since any system must be symmetric to the interchange of two identical bosons, it follows that a particle decaying into two identical 0^- mesons (e.g. $\pi^0\pi^0$, $\eta\eta$, K^+K^+ etc.) must not only be in the natural parity sequence but must have even spin, i.e. $J^P = 0^+2^+4^+$ etc. Another important parity constraint is that a 0^+ meson cannot decay into a 0^- and 1^- meson.

$$0^+ \nrightarrow 0^- + 1^- \hspace{5em} 3.$$

1.2.2 Symmetry within an i multiplet.—Two members of the same i multiplet become 'identical' when i spin is included in the formulation of the state. The overall wavefunction $\psi_I\psi_{\text{space}}$ has then to be symmetric to the interchange of two such mesons. The rules for the symmetry of ψ_I are those of ordinary spin: if I_0 is the i spin of the two separate mesons, then the combined state ψ_I is symmetric for $I = 2I_0$, $2I_0 - 2$, $\cdots$ and antisymmetric for $2I_0 - 1$, $2I_0 - 3$, $\cdots$. Thus for a system of two π's, the total $I = 2$ or 0 states are symmetric in ψ_I and hence in the space part, the $I = 1$ state is antisymmetric. With the statements given in Section 1.2.1, it follows that for an object decaying into two π's by strong (i-conserving) decay, I, J, P are all even or odd together,

$$\text{e.g. } IJ^P = 01^- \nrightarrow \pi\pi$$

Similarly for an $S = 2$ meson that decays strongly to KK, the $I = 1$ state and hence the space part is symmetric; the $I=0$ state is antisymmetric,

$$\text{e.g. } IJ^P = 11^- \nrightarrow KK \qquad (S = 2)$$

I spin also inhibits certain strong decays because of the required symmetry of Clebsch-Gordan coefficients. The coupling $|I, I_Z\rangle \rightarrow |I_1, I_{1Z}\rangle + |I_2, I_{2Z}\rangle$ is zero whenever $I_{1Z}=I_{2Z}=I_Z=0$ *and* I_1+I_2+I is odd. Thus $|1, 0\rangle \nrightarrow |1, 0\rangle + |1, 0\rangle$. For example $\rho^0 \nrightarrow \pi^0 + \pi^0$, by this rule. (It is also separately forbidden by the rule of Section 1.2.1 that two identical 0^- mesons must have J even.)

1.2.3 C *parity.*—Any system for which all the additive quantum numbers I_Z, Q, S, and B are zero transforms into itself under the action of the charge-conjugation operator which transforms particles into antiparticles, i.e. $C|\Psi\rangle = C|\psi\rangle = \pm|\psi\rangle$. The evenness or oddness of C is the C parity of the system. It is conserved in strong and probably electromagnetic interactions.

The photon has odd C. This follows since for the electromagnetic interaction of a charged particle described by:

$$\mathcal{H} = p^2/2m + A_\mu j_\mu$$

to remain unchanged under charge conjugation when the charge current j_μ reverses sign, A_μ, the photon field, must also change sign. It follows that observation of $\pi^0 \rightarrow \gamma\gamma$, $\eta \rightarrow \gamma\gamma$ etc. shows these mesons to be of even C.

A $\pi^+\pi^-$ system is in an eigenstate of C. Since the π is spinless the actions of C and P are identical: they both interchange the particles. Thus for a $\pi^+\pi^-$ system I, J, P, C are all even or odd together. Similarly a K^-K^+ or $K^0\overline{K}^0$ system has $C=P$.

Special interest attaches to the case $M \rightarrow K^0\overline{K}^0$. Since $C=P$, it follows that CP is even. If decay is to $K_1^0K_1^0$ or $K_2^0K_2^0$ the CP of the final state $= +1.(-1)^l = (-1)^J$ [intrinsic CP even, CP of relative motion equals P of relative motion $=(-1)^l$]. If decay is to $K_1^0K_2^0$, then $CP=(-1)(-1)^l =(-1)^{J+1}$. Since CP is even, it follows that an object of even spin ($J^{PC}=0^{++}$, 2^{++} etc.) must decay to $K_1^0K_1^0$ or $K_2^0K_2^0$ and an object of odd spin ($J^{PC}=1^{--}$, 3^{--} etc.) must decay to $K_1^0K_2^0$.

1.2.4 G *parity.*—If a system has $B=S=0$ but Q (and hence I_Z via Equation 1) $\neq 0$, it is not an eigenstate of C. However, operation of C serves solely to reverse Q (and hence I_Z), which can be corrected by rotating the z axis of i space through 180° (operation R). Thus, though not an eigenstate of C, it *is* an eigenstate of $G=RC$. Since the strong interaction is the only one invariant to rotations in i space, the corresponding G parity will be conserved *only* in strong interactions.

Since G has been introduced to extend, in some way, the concept of C to a whole i multiplet, it is clearly the same for all the members of a multiplet. Thus to define G for a multiplet, with $B=S=0$, we can find its value for the neutral ($B=S=Q=0$) member for which C is defined. Since this member has $I_Z=0$ we can carry out the 180° rotation in i space about the

I vector which lies in the x, y plane. A rotation of ϑ about $\mathbf{n}$ is described by the operation $e^{i\vartheta\mathbf{n}.\mathbf{I}}$. Thus $R\big| I,\ 0\rangle = e^{i\pi I}\big| I,\ 0\rangle = (-1)^I\big| I,\ 0\rangle$. Thus for this member of the multiplet $G = C(-1)^I$, and this same value will apply to the other members of the multiplet.

Using this rule the π is of odd G. Thus an object decaying *strongly* into an even number of pions is of even g parity.

Similarly, an object decaying strongly into K^+K^-, $K^0\overline{K}^0$, $K^+\overline{K}^0$ has defined G since these systems have $B = S = 0$. They all have the same G as K^+K^- for which $C = P = (-1)^J$. Thus any object decaying into $K\overline{K}$ strongly has $G = (-1)^{I+J}$.

1.3 $SU(3)$

1.3.1 SU(3) *multiplets*.—All hadrons may be attributed to an $SU(3)$ multiplet whose members all have the same external quantum numbers of J and P. Figure 1 shows the pattern of $Y(=S$ for mesons) against I_Z for the octet, decuplet, antidecuplet, and 27-plet. For the (8) and (27) G parity is defined along the central line, $Y = Q$; C only at the centre. Since the (10) and ($\overline{10}$) transform into each other under C, and not into themselves, neither contains members of defined C. As may be seen, particles can be considered as being in v-spin and u-spin multiplets as well as being in i-spin multiplets.

$SU(3)$ is badly broken and members of the same multiplet do not have the same mass. However, formulae for the mass breaking can be derived. Since field equations invariably involve mass-squared, in the case of mesons it is assumed that all mass formulae apply to mass-squared. Thus in the octet we may derive the Gell-Mann–Okubo formula:

$$m_{1/2}{}^2 = \tfrac{3}{4}m_0{}^2 + \tfrac{1}{4}m_1{}^2 \qquad\qquad 4.$$

where suffices refer to i spin.

This is a special case of the general formula:

$$m^2 = m_0{}^2 + aY + b[I(I+1) - Y^2/4] \qquad\qquad 4a.$$

[Hereafter (mass)^2 will be written as m.]

A second consequence of $SU(3)$ breaking is that finite transition rates between two multiplets, say an (8) and a (1), with the same external quantum numbers are possible. The situation is then akin to that for $K^0\overline{K}^0$ where the S-violating weak interaction mixes the two states to produce different physically observable states $K_1{}^0$ and $K_2{}^0$. In this case the octet and singlet would mix to form a nonet of mesons. However, since I and Y *are* conserved, only the $I = Y = 0$ member of the octet ϕ_8 would mix with the singlet ($I = Y = 0$), ϕ_0. The observed particles would not be ϕ_8 and ϕ_0, but two different objects ϕ and ϕ'. Since we want to keep states normalized, i.e. $\big|\phi_8\big|^2 + \big|\phi_0\big|^2 = \big|\phi\big|^2 + \big|\phi'\big|^2$, we may write:

$$\begin{aligned} \big|\,\phi\rangle &= \cos\vartheta\,\big|\,\phi_0\rangle + \sin\vartheta\,\big|\,\phi_8\rangle \\ \big|\,\phi'\rangle &= -\sin\vartheta\,\big|\,\phi_0\rangle + \cos\vartheta\,\big|\,\phi_8\rangle \end{aligned} \qquad\qquad 5.$$

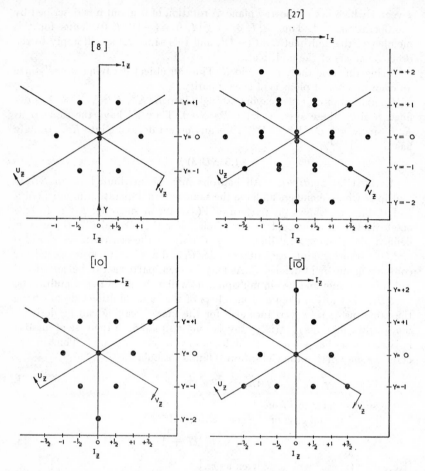

Fig. 1. The $SU(3)$ (8), (10), ($\overline{10}$), and (27) multiplets.

where ϑ is called the mixing angle.

Without mixing, ϕ_8 and ϕ_0 would have masses m_8 and m_1.

$$m_1 = \langle \phi_0 | \, \mathfrak{IC} \, | \phi_0 \rangle \qquad m_8 = \langle \phi_8 | \, \mathfrak{IC} \, | \phi_8 \rangle \qquad\qquad 6.$$

With mixing, we have finite cross terms and in terms of the $SU(3)$ states instead of defined masses we have the mass matrix:

$$\begin{pmatrix} \langle \phi_8 | \, \mathfrak{IC} \, | \phi_8 \rangle & \langle \phi_8 | \, \mathfrak{IC} \, | \phi_0 \rangle \\ \langle \phi_0 | \, \mathfrak{IC} \, | \phi_8 \rangle & \langle \phi_0 | \, \mathfrak{IC} \, | \phi_0 \rangle \end{pmatrix} = \begin{pmatrix} m_8 & a \\ a^* & m_1 \end{pmatrix} \qquad\qquad 7.$$

Here ϑ is the angle of rotation which diagonalizes this matrix. By transforming the mass matrix according to RMR^{-1} where R is the rotation matrix:

$$R = \begin{pmatrix} \cos \vartheta & -\sin \vartheta \\ \sin \vartheta & \cos \vartheta \end{pmatrix} \qquad 8.$$

it can readily be confirmed that the mixing angle is given by:

$$\tan 2\vartheta = \frac{-\lambda}{(m_8 - m_1)} \qquad 9.$$

where $\lambda = a + a^*$.

The squared masses of the physical states are found as follows:

$$m_\phi = \langle \phi | \mathfrak{IC} | \phi \rangle = \langle \cos \vartheta \cdot \phi_0 + \sin \vartheta \cdot \phi_8 | \mathfrak{IC} | \cos \vartheta \cdot \phi_0 + \sin \vartheta \cdot \phi_8 \rangle$$

i.e. $m_\phi = m_1 \cos^2 \vartheta + m_8 \sin^2 \vartheta + \lambda \sin \vartheta \cos \vartheta$ 10a.

$$m_{\phi'} = m_1 \sin^2 \vartheta + m_8 \cos^2 \vartheta - \lambda \sin \vartheta \cos \vartheta \qquad 10b.$$

Substituting for ϑ from Equation (9),

$$m_\phi = \tfrac{1}{2}(m_8 + m_1) - \tfrac{1}{2}[(m_8 - m_1)^2 + \lambda^2]^{1/2} \qquad 11a.$$

$$m_{\phi'} = \tfrac{1}{2}(m_8 + m_1) + \tfrac{1}{2}[(m_8 - m_1)^2 + \lambda^2]^{1/2} \qquad 11b.$$

Thus:

$$m_\phi + m_{\phi'} = m_1 + m_8 \qquad 12.$$

Also substituting in Equation 11 for λ^2 and subtracting 11b from 11a we find:

$$\tan^2 \vartheta = \frac{m_{\phi'} - m_8}{m_8 - m_\phi} \qquad 13.$$

It can be seen that from a study only of masses the sign of the mixing angle cannot be determined. Interchanging the physical particles ϕ and ϕ' in Equation 13 changes ϑ to $\pi/2 - \vartheta$. If ϕ is defined as the physical particle which is more singlet than octet, the mixing angle will be less than 45°. Methods by which the mixing angle might be checked independently of the masses are discussed below.

1.3.2 SU(3) *constraints on meson decays.*—$SU(3)$ rules put constraints on allowed meson decays which are valid to the degree that $SU(3)$ is a good symmetry. Firstly there are the coupling rules. For example, since the combination of an octet and singlet can only give an octet, $(8) \times (1) = (8)$; it follows that a meson seen decaying into octet plus singlet mesons would have, itself, to come from an octet. Or, as a second example, since the (35) is contained in $(8) \times (8) \times (8)$ but not in $(8) \times (8)$, a meson in a (35) would not decay into two octet mesons but could into three.

Secondly, a state of two mesons from the same $SU(3)$ multiplet becomes a state of two identical bosons if $SU(3)$ is included in a description of the state, i.e. $\psi_{SU(3)} \cdot \psi_{\text{spin}} \cdot \psi_{\text{orbit}}$ is symmetric to the interchange of the mesons. The $SU(3)$ states obtained by combining two octets have the interchange symmetries for $\psi_{SU(3)}$ indicated below:

$$(8) \times (8) \to (1) + (8) + (8) + (10) + (\overline{10}) + (27)$$
$$\quad\quad\quad\quad S \quad\; S \quad\; A \quad\; A \quad\;\; A \quad\; S$$

14.

Consider for example two 0^- mesons from the same octet: ψ_{spin} is absent, ψ_{orbit} is symmetric for $J = 0, 2, 4, \cdots$ and is antisymmetric for $J = 1, 3, 5$ etc.

It follows from Equation 14 that a $J^P = 0^+$, 2^+ meson which is in a (10) or $(\overline{10})$ multiplet cannot decay into two such 0^- mesons. A $J^P = 1^-$, 3^-, $\cdots$ meson from an $SU(3)$ singlet or (27) similarly cannot decay into two such mesons.

$SU(3)$ Clebsch-Gordan coefficients may of course be used to relate decays of particles in the same multiplet. Thus we see from Equation 14 that the coupling $(8) \times (8)$ to the (1), $(\overline{10})$, $(\overline{10})$ or (27) is unique and a single amplitude is involved in each case.

For the case of $(8) \to (8) \times (8)$ there are, in principle, two coupling constants usually described as D and F. However, C invariance will ensure that one or the other is zero. This may readily be seen by describing the octet as $q^i \bar{q}_j = A_j{}^i$, when the coupling amplitudes are $\sum_{ijk} A_j{}^i B_k{}^j C_i{}^k = \text{Tr } ABC$, or alternatively Tr BAC (cyclic permutations leaving the trace unchanged). The symmetric D coupling is Tr $A(BC+CB)$ and the antisymmetric F coupling is Tr $A(BC-CB)$. However, for a meson octet the action of the charge-conjugation operator C on the element $A_j{}^i$ is to turn it into the antiparticle $A_j{}^i$. Hence $C \cdot \sum_{ijk} A_j{}^i B_k{}^j C_i{}^k = \pm \sum A_k{}^i B_j{}^k C_k{}^i = \pm \text{Tr } BAC$, the $\pm$ sign depending on whether the product C parities are even or odd. Hence we note that if the product C parities of the three octets are odd, there will only be F coupling, if even, only D coupling.

Levinson, Lipkin & Meshkov (1) have pointed out that if a decay allowed by $SU(3)$ is forbidden by some other selection rule, this will imply selection rules for other members of the $SU(3)$ multiplet. Thus if the decay is one where a single $SU(3)$ coupling constant applies, e.g. $(27) \to (8) \times (8)$, all decays are uniquely related by the $SU(3)$ Clebsch-Gordan coefficients and suppression of the decay of one member implies suppression of the decays of the other members. For instance if the C-defined member of a (27) has odd CP it will not be able to decay into $\pi^+ \pi^-$ for which CP is even. It follows that no members of the (27) will be able to decay into two pseudoscalar mesons.

Since we have seen above that only D or F coupling will apply to any given case of $(8) \to (8) + (8)$, again single suppression of some $SU(3)$ allowed decay will prevent the whole octet so decaying. Thus for example no member of an octet whose neutral member has odd CP can decay into two 0^- mesons. One must ensure that the process is indeed $SU(3)$ allowed. Thus G-parity rules should not be invoked since they only consist of C plus i spin, i.e. C plus $SU(3)$, and these rules will already have been implicitly

TABLE I

PROPERTIES OF QUARKS

	I	I_z	S	B	Q
p	$\frac{1}{2}$	$+\frac{1}{2}$	0	$\frac{1}{3}$	$+\frac{2}{3}$
n	$\frac{1}{2}$	$-\frac{1}{2}$	0	$\frac{1}{3}$	$-\frac{1}{3}$
Λ	0	0	-1	$\frac{1}{3}$	$-\frac{1}{3}$

invoked in suppressing one or other of the D or F couplings. Similarly the fact that the ρ^0 is forbidden to decay to $\pi^0\pi^0$ because a $\pi^0\pi^0$ state must be of even spin does not forbid all 1^- mesons decaying to two pseudoscalars since this decay is already forbidden by i spin, i.e. by $SU(3)$.

1.3.3 SU(3) *properties in terms of quarks.*—The properties of $SU(3)$ multiplets can be derived by considering them as created by the addition of the basic quark triplets, whose properties are given in Table I. Thus the (8) and (1) multiplets can be built from $q\bar{q}$. The $qq\bar{q}\bar{q}$ combination would yield (10), $(\overline{10})$, and (27) in addition to (8) and (1). Let us consider in particular the (8) and (1) obtained by taking $q\bar{q}$. With $SU(3)$ breaking, the complete 3×3 matrix $q^i\bar{q}_j$ can be written:

$$q^i\bar{q}_j = \begin{pmatrix} p\bar{p} & p\bar{n} & p\bar{\Lambda} \\ n\bar{p} & n\bar{n} & n\bar{\Lambda} \\ \Lambda\bar{p} & \Lambda\bar{n} & \Lambda\bar{\Lambda} \end{pmatrix} \qquad 15.$$

where, as can be checked from the quark properties given in Table I, the off-diagonal terms correspond to unique meson states. Using the names of the 0^- mesons we have:

$$q^i\bar{q}_j = \begin{pmatrix} p\bar{p} & \pi^+ & K^+ \\ \pi^- & n\bar{n} & K^0 \\ K^- & \overline{K}^0 & \Lambda\bar{\Lambda} \end{pmatrix} \qquad 16.$$

The $SU(3)$ singlet state η_0, having no $SU(3)$ properties, must be unaffected by rotations in $SU(3)$ space, i.e. $R^{-1}MR\to M$. It must therefore behave like the unit matrix.

Hence

$$\eta^0 = \frac{p\bar{p} + n\bar{n} + \Lambda\bar{\Lambda}}{\sqrt{3}} \qquad 17.$$

The set of octet states are obtained by subtracting this term from the 3×3 matrix $q^i\bar{q}_j$ to give a traceless matrix:

$$\text{i.e. octet} = q^i\bar{q}_j - \tfrac{1}{3}\delta_j{}^i q^k\bar{q}_k \qquad 18.$$

Since $\pi^{\pm}$ are made purely from p, n-type quarks it follows that the π^0 must also be so made. Hence

$$\pi^0 = \frac{p\bar{p} - n\bar{n}}{\sqrt{2}} \qquad\qquad 19.$$

Only the $I = 0$ octet member η_8 remains to be defined. If only p, n-type quarks were involved it would be of the form $p\bar{p}+n\bar{n}$, i.e. behaving like a unit matrix for I-spin rotations. However we have to include $\Lambda\bar{\Lambda}$ quarks and guarantee orthogonality to η_0. This is achieved if we write $\eta_8 \propto p\bar{p}+n\bar{n}-2\Lambda\bar{\Lambda}$ since

$$\langle p\bar{p} + n\bar{n} - 2\Lambda\bar{\Lambda} \| p\bar{p} + n\bar{n} + \Lambda\bar{\Lambda}\rangle = 0. \quad \text{Normalizing:}$$

$$\eta_8 = \frac{p\bar{p} + n\bar{n} - 2\Lambda\bar{\Lambda}}{\sqrt{6}} \qquad\qquad 20.$$

we see that the complete octet matrix (Equation 18) is given by:

$$\begin{pmatrix} \dfrac{\eta_8}{\sqrt{6}} + \dfrac{\pi^0}{\sqrt{2}} & \pi^+ & K^+ \\[2mm] \pi^- & \dfrac{\eta_8}{\sqrt{6}} - \dfrac{\pi^0}{\sqrt{2}} & K^0 \\[2mm] K^- & \overline{K}^0 & -\sqrt{\tfrac{2}{3}}\,\eta_8 \end{pmatrix} \qquad 21.$$

If $SU(3)$ is broken, then we get mixing of the states η_0, η_8. An important case arises if the mixing angle (Equation 5) is given by $\tan\vartheta_Z = \frac{1}{2}$.

By Equation 5 the state which is more octet than singlet is then given by

$$\eta = -\tfrac{1}{3}\cdot\frac{p\bar{p} + n\bar{n} + \Lambda\bar{\Lambda}}{\sqrt{3}} + \tfrac{2}{3}\cdot\frac{p\bar{p} + n\bar{n} - 2\Lambda\bar{\Lambda}}{\sqrt{6}} = -\Lambda\bar{\Lambda}$$

The state which is more singlet than octet is then given by:

$$\eta' = \frac{p\bar{p} + n\bar{n}}{\sqrt{2}}$$

The mixing angle $\vartheta_Z = \tan^{-1}\frac{1}{2} = 35.3°$ is known as the ideal mixing angle. For this angle one of the isosinglet states is purely made of $\Lambda\bar{\Lambda}$ quarks, the other of $p\bar{p}$ and $n\bar{n}$ quarks.

Should this situation arise we can write the $q\bar{q}$ matrix as

$$q^i\bar{q}_j = \begin{pmatrix} \dfrac{\eta' + \pi^0}{\sqrt{2}} & \pi^+ & K^+ \\[2mm] \pi^- & \dfrac{\eta' - \pi^0}{\sqrt{2}} & K^0 \\[2mm] K^- & \overline{K}^0 & -\eta \end{pmatrix} \qquad 22.$$

Many simple $SU(3)$ breaking rules can be obtained by assuming that the Λ and $\bar\Lambda$ quarks when present add (each) Δ, a positive (mass)^2, to the (mass)^2 of the meson: i.e. $m_\pi = m_8{}^0$ [where $m_8{}^0$ is the octet squared mass without $SU(3)$ breaking].

$$m_K = m_8{}^0 + \Delta$$

$$m_{\eta_8} = \left\langle \frac{p\bar p + m\bar n - 2\Lambda\bar\Lambda}{\sqrt 6} \;\middle|\; M \;\middle|\; \frac{p\bar p + n\bar n - 2\Lambda\bar\Lambda}{\sqrt 6} \right\rangle$$

$$= m_8{}^0 + \tfrac{4}{3}\Delta$$

i.e. $\quad m_K = \tfrac{3}{4}m_8 + \tfrac{1}{4}m_\pi.$

23.

where all m's refer to (mass)^2. Thus the Gell-Mann–Okubo mass formula may be derived.

A similar substitution for the explicit singlet and octet isosinglet quark states in the mass matrix (Equation 7) gives for this matrix

$$\begin{bmatrix} \left\langle \frac{p\bar p+n\bar n-2\Lambda\bar\Lambda}{\sqrt6} \middle| M \middle| \frac{p\bar p+n\bar n-2\Lambda\bar\Lambda}{\sqrt6} \right\rangle & \left\langle \frac{p\bar p+n\bar n-2\Lambda\bar\Lambda}{\sqrt6} \middle| M \middle| \frac{p\bar p+n\bar n+\Lambda\bar\Lambda}{\sqrt3} \right\rangle \\[2mm] \left\langle \frac{p\bar p+n\bar n+\Lambda\bar\Lambda}{\sqrt3} \middle| M \middle| \frac{p\bar p+n\bar n-2\Lambda\bar\Lambda}{\sqrt6} \right\rangle & \left\langle \frac{p\bar p+n\bar n+\Lambda\bar\Lambda}{\sqrt3} \middle| M \middle| \frac{p\bar p+n\bar n+\Lambda\bar\Lambda}{\sqrt3} \right\rangle \end{bmatrix}$$

$$= \begin{bmatrix} m_8{}^0 + \dfrac{4}{3}\Delta & \dfrac{-2\sqrt2}{3}\Delta I \\[3mm] \dfrac{-2\sqrt2}{3}\Delta I & m_1{}^0 + \dfrac{2}{3}\Delta \end{bmatrix}$$

24.

In this expression I is the overlap integral between octet and singlet states. It would be unity for identical states and is otherwise less than unity. Applying Equation (9):

$$\tan 2\vartheta = \frac{\dfrac{4\sqrt2}{3}\Delta I}{(m_8{}^0 - m_1{}^0) + \tfrac{2}{3}\Delta}$$

25.

It will be seen that ideal mixing is obtained if $m_8{}^0 = m_1{}^0$ and $I = 1$.

On the same assumption that $SU(3)$ breaking arises from the term Δ, we may prove the Schwinger mass formula for the masses of the nonet of observed particles, i.e.

$$(m_\eta - m_\pi)(m_{\eta'} - m_\pi) - \tfrac{4}{3}(m_K - m_\pi)(m_\eta + m_{\eta'} - 2m_K)$$
$$= \tfrac{8}{9}(m_K - m_\pi)^2(1 - I^2) \geq 0$$

26.

where, as in all formulae, m stands for squared mass. (To prove, use Equation 11 to define the physical masses m_η and $m_{\eta'}$; substitute in this formula $m_8 = m_8{}^0 + \tfrac{4}{3}\Delta = m_\pi + \tfrac{4}{3}\Delta$ and similarly $m_1 = m_1{}^0 + \tfrac{2}{3}\Delta$; substitute

$$\lambda = -\frac{4\sqrt2}{3} \cdot \Delta I.)$$

1.4 $SU(6)$

In $SU(6)$ the basic triplet of $SU(3)$ is coupled with the basic doublet of spin to yield a six-component "quark." Multiplets built by adding these "quarks" will hopefully describe both spin and $SU(3)$ properties of particles.

The simplest $q\bar{q}$ combination yields (1) and (35) multiplets. With quarks considered as fermions so that q and $\bar{q}$ have opposite parity, the generated mesons are of negative parity. The singlet consists of an $SU(3)$ and spin singlet $J^P = 0^-$ meson. The (35) breaks into an $SU(3)$ (8) of 0^-, an $SU(3)$ (1) of 1^-, and an $SU(3)$ (8) of 1^-. To the extent that $SU(6)$ were unbroken, the particles in the (35) would have identical masses. In particular m_8 and m_1 for the (1) and (8) of the 1^- mesons would be the same. The nature of the wavefunctions for the (1) and (8) would be similar. If then $SU(3)$ breaking is considered with $m_8{}^0 = m_1{}^0$ for the 1^- mesons and plausibly $I = 1$ in view of the similarity of the mixing states, we would then find (Equation 25) ideal mixing for the 1^- mesons. Since the nonet of 0^- mesons involves mixing both $SU(3)$ multiplets and $SU(6)$ multiplets one would not expect mixing to be ideal; one would expect less mixing.

In $SU(6)$ one may generate higher spin states only by adding more basic quarks.

Positive parity states could be obtained from $q\bar{q}q\bar{q}$. The simplest $SU(6)$ multiplets would then be the (189) and (405)—whose $SU(3)$-spin substructure would be given by:

$$(189) = (27, 1)^+ + (10, 3 + \overline{10}, 3)^+ + (10, 3 - \overline{10}, 3)^- + (8, 5)^+$$
$$+ (8, 3)^+ + (8, 3)^- + (8, 1)^+ + (1, 5)^+ + (1, 1)^+ \qquad \text{27a.}$$

$$(405) = (27, 5)^+ + (27, 3)^- + (27, 1)^+ + (10, 3 + \overline{10}, 3)^-$$
$$+ (10, 3 - \overline{10}, 3)^+ + (8, 5)^+ + (8, 3)^+ + (8, 3)^- \qquad \text{27b.}$$
$$+ (8, 1)^+ + (1, 5)^+ + (1, 1)^+$$

In these expressions the first number gives the $SU(3)$ multiplicity, the second the multiplicity of the spin state. The superscript gives C parity. The (10) and $(\overline{10})$ have no C-defined members and C is defined only for $(10) \pm (\overline{10})$. Parity is even throughout.

Using $SU(6)$, one can only obtain high-spin mesons at the cost of introducing higher $SU(3)$ multiplets. Thus the existence of the 2^+ mesons would, in $SU(6)$, imply the existence of (27) (10) and $(\overline{10})$ multiplets.

Mass formulae are not simply derived for broken $SU(6)$. Gursey & Radicati (2) semiempirically suggest an extension of the Gell-Mann–Okubo formula (1.4a), simply inserting a spin term identical to the i-spin term.

$$m = m_0 + aY + bJ(J + 1) + C[T(T + 1) - Y^2/4] \text{ (mass squared)} \qquad 28.$$

It is only suggested that this applies to the (35). It predicts for the 0^- and 1^- mesons that:

$$m_{K*} - m_\rho = m_K - m_\pi \qquad 29.$$

Experimentally we have 0.22 and 0.21 (GeV)^2 respectively.

The difficulty with the $SU(6)$ approach is to make it respectable relativistically. We shall see that the predictions of the quark model and $SU(6)$ are very similar for the 0^- and 1^- mesons. For higher mesons, at the present time, the quark model is the more successful in accounting for observed states. An extension of $SU(6)$, the so-called $SU(6)_W$ group, might be applicable to relativistic processes which are colinear, i.e. two-body decays or simple three-particle vertices. (See Section 2.)

1.5 Nonrelativistic Quark Model

In this model a meson is envisaged as made up of a real $q\bar{q}$ pair rotating with relative orbital angular momentum l. The model is unsatisfactory in many regards. It is somewhat illogical to describe as $q\bar{q}$ systems both the π and the ρ when the latter, by strong interaction theory, spends much of its time as two π's. If the nonobservation of quarks is attributed to their high mass ($M_q > 5M_N$ say), then the nonrelativistic ability to separately quantize l seems inappropriate in view of the large loss of rest mass involved in forming a meson. It has been argued by Dalitz (3), however, that it is the large quark mass that permits nonrelativistic treatment of angular momentum. Thus, if one states that $p_q \ll M_q c$ is a condition for nonrelativistic motion, and that $p_q \sim \hbar/r$ where r represents the size of the system, then $\hbar/r \ll M_q c$ can still be satisfied for $M_q = 5M_N$ down to $r \sim 10^{-14}$ cm.

On this model all mesons fall into $SU(3)$ octets and singlets, in general mixing to form nonets. The discussion of Section 1.3.1 will be appropriate for any value of l.

Since the quarks are fermions, the parity of a meson is given by $(-1)^{l+1}$. The C parity, when defined, will be given by $(-1)^{l+s}$. Since (Section 1.3.2) C can in some sense be used to describe a complete multiplet, nonets will be described hereafter by reference to J^{PC}. In general four nonets will be generated for a given value of l, three from the spin-triplet state of the $q\bar{q}$ system and one from the spin-singlet state. Their properties will be as listed in Table II.

We note for example that on this model all natural-parity states must have $C = P$. For the special case $l = 0$ we generate only two nonets with respectively $J^{PC} = 0^{-+}$ and 1^{--}. Since in this case the orbital angular momentum that typifies the model is zero, the results derived are similar to those from $SU(6)$. For $l > 0$ the results of the model differ from those of pure $SU(6)$.

In this model, one may consider the possibility of radial excitations of the quark "molecule" with radial quantum number n. Dalitz (3) has used a potential of the form $V(r) = V(0) + \frac{1}{2}M\omega^2 r^2$ in the Blankenbecler-Sugar equation, obtained from the Bethe-Salpeter equation, to find that the energy eigenvalues are given by:

$$E^2 = E_0^2 + 4\sqrt{3}M_q\omega(2n + L - 2) \qquad 30.$$

In this model, higher spins are obtained by increasing l rather than the number of quarks. However, some of the advantages of $SU(6)$ could be

TABLE II

Properties of Meson States Generated in the Quark Model

	S=1			S=0
J	$l+1$	l	$l-1$	l
P	$(-1)^J$	$(-1)^{J+1}$	$(-1)^J$	$(-1)^{J+1}$
C	$(-1)^J$	$(-1)^{J+1}$	$(-1)^J$	$(-1)^J$
CP	$+$	$+$	$+$	$-$

maintained by stating that the spin and $SU(3)$ properties (independent of l) show approximate $SU(6)$ symmetry. On this assumption we could expect near-ideal mixing of all the nonets arising from the spin-triplet state and less mixing for the spin-singlet nonets. The group generated is $SU(6) \times O(3)$.

Within this model there are many ways in which $SU(3)$ or $SU(6)$ symmetry might be broken. One might, for example, have spin-orbit forces proportional to $l.S$. Since

$$<l.S> = \left\langle \frac{J^2 - l^2 - S^2}{2} \right\rangle = \frac{j(j+1) - l(l+1) - S(S+1)}{2}$$

one would then get for corresponding members of the four nonets associated with a given value of l the sequence of mass shifts:

$$^3L_{l+1} \quad ^1L_l \quad ^3L_l \quad ^3L_{l-1}$$
$$a \quad\quad 0 \quad -a \quad -(l+1)a$$

31.

Tensor forces would give a different mass-shift sequence. The mass shift between the 0^- and 1^- mesons suggests a spin-spin force, i.e. one dependent on $\mathbf{\delta}_q \cdot \mathbf{\delta}_{\bar{q}}$.

Such forces do not in themselves break $SU(3)$; they simply break $SU(6)$ for the $q\bar{q}$ system. Of course these terms could readily also be made $SU(3)$ breaking.

In Section 1.3.2, the possibility of breaking $SU(3)$ by associating an extra contribution Δ to squared mass from the Λ quark was considered. If the interquark force is itself $SU(3)$ breaking, Δ could be a function of whether one was considering octet or singlet $SU(3)$ states—the mass matrix (1.2.4) would then become:

$$\begin{vmatrix} m_8 + \frac{4}{3}\Delta_8 & \dfrac{-2\sqrt{2}}{3}\Delta_{18}I \\ \dfrac{-2\sqrt{2}}{3}\Delta_{18}I & m_1 + \frac{2}{3}\Delta_1 \end{vmatrix}$$

32.

This can be made identical to the matrix 24 by putting $\Delta = \Delta_8$, $m_1' = m_1 + \frac{2}{3}(\Delta_1 - \Delta_8)$ and $I' = \Delta_{18}/\Delta_8 I$ to give:

$$\begin{vmatrix} m_8 + \frac{4}{3}\Delta & \dfrac{-2\sqrt{2}}{3}\Delta I' \\[2ex] \dfrac{-2\sqrt{2}}{3}\Delta I' & m_1' + \frac{2}{3}\Delta \end{vmatrix}$$

One could still derive the same formulae for mixing angles, the Schwinger mass formula etc.—but now I' would be freed from the constraint of being less than one. Moreover since Δ_8 and Δ_1 would be functions of the particular space wavefunction of the potential, Δ and I' would vary from nonet to nonet.

The simplicity of the model rapidly disappears as such breaking terms are included.

If it is assumed that Equation 30 describes to first approximation the masses of corresponding nonet members—say the $I = 1$ members—then a linear relationship between (mass)^2 and l is predicted, e.g. the 1^-, 2^+, 3^-, $\cdots$ mesons that come from the $S = 1$, $l = 0$, 1, 2 states will lie on a linear spin vs. (mass)^2 trajectory. Linear Regge trajectories for these states would be predicted with degenerate trajectories for states of odd and even signature. It may be shown (4) that such degeneracy results from the absence of exchange forces. In the quark model this is understandable since such forces would involve exchange of qq or $\bar{q}\bar{q}$ systems and there are certainly no double quark states of low mass.

The experimental data presented in this article will be discussed in the light of the quark model as the simplest, and so far most successful, description of the hadron states.

We shall see that there is no evidence for mesons with $SU(3)$ properties or J^{PC} assignments that are incompatible with the model.

An examination of Regge amplitudes for production processes with unequal mass kinematics shows that singularities occur which must in practice be cancelled. Though other methods have been proposed (4), Freedman & Wang (5) have shown that correct analyticity properties can be assured by assuming that Regge trajectories occur in families, each major trajectory with intercept $\alpha(0)$ leading to a set of daughters with the same Y, I, and C but intercepts $\alpha(0) - n$, where n is an integer. This condition only applies to the intercept at $t = 0$; however, if the odd daughters had rising trajectories so that they materialized as particles, states incompatible with the quark model would appear. Thus the first daughter of the $J^{PC} = 2^{++} A_2$ would have $J^{PC} = 1^{-+}$, breaking the rule from the quark model that no natural-parity state shall have odd CP. Gell-Mann & Zweig (6) have sought to make the quark model relativistic. They then predict that any state of natural parity and those unnatural-parity states that arise from the spin-singlet quark state will have daughters (even and odd) down to $J = 0$. The unnatural-parity mesons with spin J from the spin-triplet state will be replaced by a

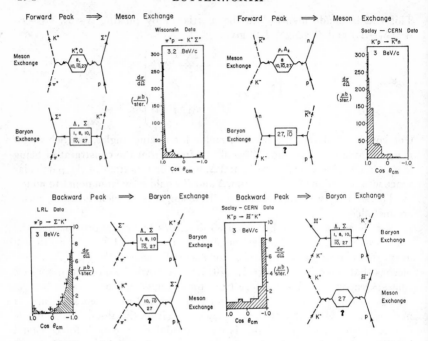

Fig. 2. Evidence for the existence or suppression of various exchange processes (9).

parity doublet $J^{\pm}$ of the same C with daughters down to $J = 1$. There is no experimental evidence for these states which are in conflict with the simple quark model (7).

2. THE EXISTENCE OF HIGHER REPRESENTATION MESONS

Of crucial importance to an understanding of meson states is the existence or otherwise of mesons which would have to be attributed to the (27) (10) or $(\overline{10})$ representations of $SU(3)$. Such states do not occur in the simple quark model. They would imply the existence of $I = 2$, $S = 0$ states (27), $I = \frac{3}{2}$, $S = \pm 1$ states (27, 10, $\overline{10}$), $I = 1$, $S = \pm 2$ states (27), and $I = 0$, $S = \pm 2$ states (10, $\overline{10}$). There is no convincing direct evidence for such states. Indirect evidence for their nonexistence has been pointed out by Goldhaber (8) and Barger (9) in the strong suppression of peripheral processes which would involve their exchange, as in forward meson production for $\pi^- p \rightarrow \Sigma^- K^+$ or $K^- p \rightarrow \Xi^- K^+$. Figure 2 shows various data. We see that processes involving exchange of higher representation mesons have cross sections smaller by more than 2 orders of magnitude than those involving exchange of known mesons. However, Figure 3 shows results from the large (52 events/μb) study by Abolins et al. (10) of $\pi N \rightarrow Y^*_{1385} K$ in the range 2 to 4 GeV/c. The cosine of the angle between the incident π and produced K is shown in the figure for the charge-symmetric reactions $\pi^+ n \rightarrow Y^* + K^0$ and $\pi^- p \rightarrow Y^* + K^+$. A for-

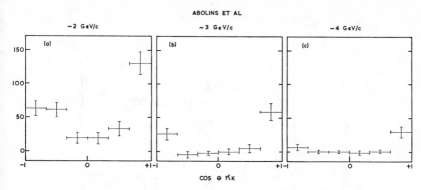

FIG. 3. Angular distribution for the charge-symmetric reactions
$\pi^+ n \to Y^{*+} K^0$ and $\pi^- p \to Y^{*-} K^+$ (10).

ward peak is seen, which if interpreted as meson exchange implies the existence of an $I = \frac{3}{2} K^*$. The presence of the forward peak throughout the momentum range argues against s-channel N^* formation. The evidence that there may be a $K^*_{3/2}$ is clearly very indirect and might indicate Regge cut exchange.

As to direct evidence, the $\pi^+ \pi^+$ and $K^+ \pi^+$ systems have been exhaustively studied in $\pi^+ p \to n \pi^+ \pi^+$ and $K^+ p \to n K^+ \pi^+$. No enhancements have been reported. Any resonances present are produced with cross sections well below 10 μb.

A $K^+ K^{0+}$ enhancement at 1280 MeV was suggested by Ferro-Luzzi et al. (11) in $K^+ p \to K^+ K^+ \Lambda$, $K^+ K^+ \Sigma^0$, $K^+ K^0 \Sigma^+$, $K^+ K^+ \Lambda \pi^0$, $K^+ K^0 \Lambda \pi^+$. However, a compilation by Dodd et al. (12) which tripled the statistics washed out the effect.

In the reaction $p \bar{p} \to K_1^0 K^{\pm} \pi^{\pm} \pi^+ \pi^- \pi^0$, and in no other channels, at 3.0 GeV/c, Bock et al. (13) reported an enhancement in the $K^* \pi$ system with $I_z = \pm \frac{3}{2}$. Subsequently further data at 3.6 and 4 GeV/c were added (14) and Figure 4 shows all the data and those at 3 GeV/c alone; the enhancement is at 1265 ± 10 MeV with width 50 ± 20 MeV. Baltay et al. (15) in the same reaction at 3.7 GeV/c saw nothing. There are four doubly charged $K \pi \pi$ combinations per event so the background is large in the figures. The effect is not statistically too significant and in view of the conflict between the two experiments requires confirmation. Incident-K^+ experiments do not report an $I = \frac{3}{2}$ peak in this region (8).

An $I = \frac{3}{2} K \pi \pi$ state has also been suggested at 1170 MeV. Positive claims of a $K \pi \pi$ state at this mass produced in $\pi^- p \to (\Lambda, \Sigma) K \pi \pi$ were made by Wangler et al. (16) (Wisconsin) at 3 GeV/c (164-event sample) and Miller et al. (17) at 2.7 GeV/c (242-event sample). Nothing was seen in the larger experiment of Dahl et al. (18) (96 events at 2.8 GeV/c, 141 at 3 GeV/c, 1100 at 3.2 GeV/c). The bump is thus not convincing. In this reaction it could, anyway, be $I = \frac{1}{2}$. The Wisconsin group (19), however, have supported their claim by

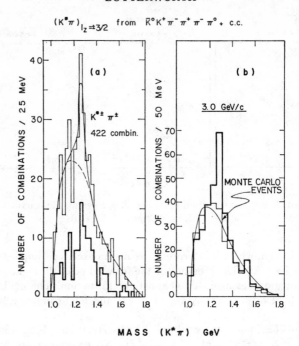

$(K^*\pi)_{|I_z|=\pm 3/2}$ from $\bar{K}^0 K^+ \pi^- \pi^+ \pi^- \pi^0$ + c.c.

FIG. 4. (a) Effective mass of $K^*\pi$ with $|I_z| = \frac{3}{2}$ in $p\bar{p} \to K_1^0 K^{\pm}\pi^{\mp}\pi^+\pi^-\pi^0$ at 3.0, 3.6, and 4.0 GeV/c (14). (b) The data at 3.0 GeV/c shown separately.

further evidence in $K^+p \to p\pi^-\pi^+\pi^+K^0$, the bump being $(K^*\pi)^{++}$ and so $I = \frac{3}{2}$. Figure 5a shows all $(K\pi\pi)^{++}$ events, 5b the $(K^*\pi)^{++}$ selected events. On a purely statistical basis the enhancement is unimpressive [the dot-dash curve is a background estimate by Rosenfeld (20) allowing for peripheralism and the finite K^* width]. Moreover one is worried about kinematic effects arising because both π^+ can give $\Delta(1238)$ with the proton. Figure 5c shows what happens if you eliminate events with $1.17 < p\pi^+ < 1.3$ GeV. Though the peak at 1170 stays, the statistical significance drops away completely. Indeed, the authors themselves state that alternative explanations are a $K\pi\pi$ resonance or a symmetric Δ^{++} decay. [The $\Delta^{++} \to p\pi^+$ in $K^+p \to K^*p\pi^+\pi^-$ decays with π^+ preferentially parallel to the K^*. This could be the *result* of a $(K\pi\pi)^{++}$ resonance overlapping the Δ^{++} events or the *source* of the apparent $K\pi\pi$ enhancement.] No other authors report this resonance.

Vanderhaghen et al. (21) have claimed in π^-d interactions at 5 GeV/c a $\rho^-\pi^-$ $I = 2$ enhancement at the A_2 mass. Their results and those of two earlier experiments are shown in Figure 6. One does not find the bump convincing, particularly as for two thirds of the events the fitting programme inserted an unseen spectator in the reaction $\pi^-d \to p\pi^-\pi^-\pi^0(p)$, which could be dangerous when looking for small effects. The cross section is only 15 ± 5

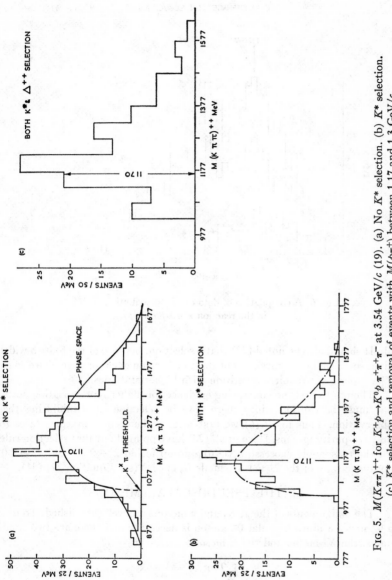

FIG. 5. $M(K\pi\pi)^{++}$ for $K^+p \to K^0p\,\pi^+\pi^+\pi^-$ at 3.54 GeV/c (19). (a) No K^* selection. (b) K^* selection. (c) K^* selection and removal of events with $M(p\pi^+)$ between 1.17 and 1.3 GeV/c.

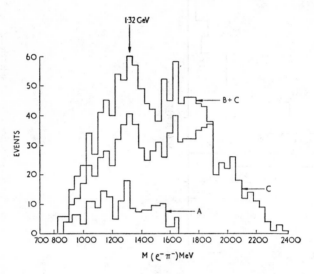

$\pi^-\ d \longrightarrow p\ \ \varrho^-\pi^-\ (p)$

A. SEIDLITZ ET AL. 3·2 GeV/C $\Delta^2 < 7\ (\text{GeV})^2$; N* OUT: 165 EVENTS.

B. ABOLINS ET AL. 3·7 GeV/C   A II Δ^2;   N* OUT: 342 EVENTS.

C. VANDERHAGEN ET AL. 5 GeV/C   A II Δ^2;   N* IN;   892 EVENTS.

FIG. 6. A compilation of data on the distribution of $M(\rho^-\pi^-)$ in the reaction $\pi^-d \rightarrow pp\rho^-\pi^-(p)$.

μb. It should also be noted (22) that ρ selection, because of the finite ρ width, gives some slight enhancement of this mass region even for pure phase space. The experiment requires repetition with higher statistics.

Though there is no convincing evidence for higher representation states, one must be cautious since there may be selection rules inhibiting their observation. Thus in $SU(6)$ we first meet higher representation states for positive-parity mesons. Horn et al. (23) have pointed out that selection rules for the reasonably successful $SU(6)_W$ group forbid $2^+ \rightarrow 0^-0^-$ and $2^+ \rightarrow 1^-0^-$ decays for any of the 2^+ mesons made from $qq\bar{q}\bar{q}$ (see Equation 27) (24).

3. THE PSEUDOSCALAR MESONS

The assignments of the π, K, and η mesons are well established. To form a 0^- nonet a ninth isoscalar 0^- meson is necessary and there are two candidates, the X^0 meson and the E meson.

3.1 THE X^0 MESON

The X^0 is a neutral meson, $M = 958 \pm 1$ MeV, $\Gamma < 4$ MeV, with $\eta\pi\pi$ and $\pi\pi\gamma$ modes. Its mass degeneracy with the charged δ reported in missing mass-spectrometer experiments ($M = 962 \pm 5$ MeV, $\Gamma < 5$ MeV) raises the question as to whether either or both of those decay modes are due to the δ^0.

TABLE III

Cross-Section Limits for Production of a Negative Meson at 960 MeV[a]

Reaction $K^-n \to \Lambda X^-$	All events (μb)	$\Delta^2 \leq 0.7$ (GeV/c)^2 (μb)
1. $X^- \to \pi^- \pi^0$	<26	<19
2. $X^- \to \pi^- MM$	<37	<25
3. $X^- \to \pi^- \pi^+ \pi^-$	<16	<10
4. $X^- \to \pi^- \rho$	<11	<8
5. $X^- \to \pi^- \pi^+ \pi^- \pi^0$	<14	<9
6. $X^- \to \pi^- \eta$	<13	<9
7. $X^- \to \pi^- \omega$	<12	<8
8. $X^- \to \pi^- \pi^+ \pi^- MM$	<5	<4
9. $X^- \to \pi^- \pi^0 \eta$	<16	<12
10. $X^- \to$ one charged	<42	<29
11. $X^- \to$ three charged	<20	<12

[a] Reported by Barbaro-Galtieri et al. (26), for data at 2.11 and 2.65 GeV/c combined.

Since X_0 is seen in $K^-p \to \Lambda X^0$ the i spin is 0 or 1; if it is the neutral δ then $I = 1$.

That $I_{X^0} = 0$ and thus that the δ and X^0 are not the same object has been established in several deuterium experiments reporting the absence of $K^-n \to \Lambda X^-$ which would have twice the cross section of $K^-p \to \Lambda X^0$ for $I = 1$ (25–27). Barbaro-Galtieri et al. (26), studying K^-d at 2.11 and 2.65 GeV/c, would expect to see 378 ± 32 events; they see less than 42. They check separately the two decay modes, thus $X^- \to \pi\pi\eta$ would be expected to give 158 ± 18 events and less than 21 are seen. They, in fact, look for decay modes other than those known for X^0. Table III gives their 99 per cent confidence level upper limits on cross sections (averaged at the two momenta) at all Δ^2 and low Δ^2. To these limits they thus also see no δ. The $SABRE$ collaboration studying K^-d interactions at 3 GeV/c also separately search for $X^- \to \eta\pi\pi$ and $\gamma\pi\pi$ (27). The former mode for $I = 1$ would be seen at 56 ± 10 μb, the latter at 44 ± 2 μb. They see 2 ± 2 μb for the former. For the latter in $K^-n \to \Lambda\pi^- MM$ there is a finite signal of 13 ± 5 μb, but not enough to be X^-. This small peak is at 990 ± 10 MeV, rather high for the δ. (We shall see below that the probable decay of δ is to $\eta\pi$, so this peak could be due to δ going into $\eta\pi$ with neutral η decay.) Thus the X^0 is certainly $I = 0$.

J^P is less well established. Let us consider the two modes in turn. The Dalitz plot of the $\pi\pi\eta$ decay is approximately uniformly populated. This eliminates natural parity, since such a state decaying into three 0^- mesons has zero density at the boundaries. [At the boundary the products are collinear and the state is describable by only one direction. The orientation of this direction can only be given by $Y_m{}^J$, there being no other parameter, when $P = (-1)^J \times (-1)$,[3] i.e. boundary occupation is only possible for $P = (-1)^{J+1}$.] Let p_η be the η momentum in the X^0 c.m. and ϑ the angle between the $\pi\pi$

direction and the η in the $\pi\pi$ c.m. Since $I_{X^0}=I_\eta=0$, $I_{\pi\pi}=0$, hence $l_{\pi\pi}$ is even (Section 1.2.2). If $l_\eta=l_{\pi\pi}=0$, then $J^P(X^0)=0^-$ and there is uniform population of the plot. This is the normally accepted solution. If $l_{\pi\pi}=0$ and $l_\eta=1$, then 1^+ is obtained. Since the π's are in an s wave there is no ϑ dependence, but the angular momentum barrier $(p_\eta)^{2l_\eta}$ will yield a p_η^2 variation. Such a variation is inconsistent with the data. However $p_\eta^2\simeq2M_\eta T_\eta$ and $m_{\pi\pi}^2$ $=m^2 x^0+m_\eta^2-2m_X^0(T_\eta+m_\eta)$; thus p_η^2 is linearly related to $m^2_{\pi\pi}$. The $\pi\pi$ system is in the s-wave isoscalar state which might resonate at a mass ~700 MeV (Section 7.2). Such a resonance at higher $m_{\pi\pi}$ values would enhance low p_η values, cancelling the normal p_η^2 dependence. A fit to the observed uniform plot could again be obtained. A J^P assignment of 2^- can be obtained with $l_{\pi\pi}=2$ and $l_\eta=0$ or $l_{\pi\pi}=0$ and $l_\eta=2$. Two comparable amplitudes being present, by adjusting their ratio approximate uniformity of the plot can again be obtained. Hence this mode does not separate 0^+, 1^-, 2^+ assignments.

Turning to the $\pi\pi\gamma$ mode we use the same variables replacing p_η by p_γ. This electromagnetic decay is C-conserving. With C_X even and C_γ odd it follows $C_{\pi\pi}$ and hence $l_{\pi\pi}$ is odd. Assume $l_{\pi\pi}=1$. Let $\mathbf{a}$ be a unit *axial* vector along the magnetic-field direction of the γ, when $\mathbf{k}_\epsilon=\mathbf{a}\wedge\mathbf{p}_\gamma$ is a vector along the electric field and $\mathbf{k}_m=(\mathbf{a}\wedge\mathbf{p}_\gamma)\wedge\mathbf{p}_\gamma$ is an axial vector along the magnetic field. To make the X^0 pseudoscalar the transition must be $M1$ (i.e. the photon behaves like a 1^+ state) with $|ME|^2=|\mathbf{p}_\pi\cdot\mathbf{k}_m|^2$. If $\mathbf{k}_m$ has an azimuthal angle ϕ about p_γ and is resolved into $k_m{}^{||}$ and $k_m{}^\perp$, components parallel and perpendicular to the $\gamma\pi\pi$ plane, then $|ME|^2=|\mathbf{p}\cdot\mathbf{k}_m{}^{||}|^2\propto\sin^2\vartheta$. The data of Kalbfleisch et al. (28) fit such a distribution with an 80 per cent confidence level (29). If the X^0 is axial vector one has an $E1$ transition (the photon behaves like a 1^- state) with $|ME|^2=|\mathbf{p}_\pi\wedge\mathbf{k}_\epsilon|^2=|\mathbf{p}_\pi\wedge\mathbf{k}^{||}|^2$ $+|\mathbf{p}_\pi\wedge\mathbf{k}^\perp|^2=k^2\cos^2\phi\,\cos^2\vartheta+k^2\sin^2\vartheta\propto1+\cos^2\vartheta$ on averaging over ϕ. The confidence level of a fit to this distribution is 8 per cent. However, the Q value of the decay is high and thus an $M2$ transition contribution cannot be ruled out. A small admixture of such a transition could raise the confidence level of the 1^+ assignment to quite acceptable values. If $J^P(X^0)=2^-$, the transition can again be $M1$ leading to a distribution of the form $(6+\sin^2\vartheta)$ and this can be fitted to a 40 per cent confidence level.

We thus see that neither three-body decay mode permits a decision between the 0^-, 1^+, and 2^- assignments.

Bollini et al. (30) have observed what is probably the decay of X^0 to two photons in a study of $\pi^-p\to n\gamma\gamma$ at 1.93 GeV/c. A peak of five events over a background of one event was observed. Since a spin 1 particle cannot decay into two photons this observation eliminates the 1^+ assignment.

It has been pointed out by Foster et al. (31) that the ratio of the cross sections for $\bar{p}p\to\eta\pi^+\pi^-$ and $\bar{p}p\to X^0\pi^+\pi^-$ at rest are those expected from phase space and that the detailed behaviour of η and X^0 in the final-state Dalitz plot is similar. This suggests but does not prove that η and X^0 have the same quantum numbers.

3.2 Use of X^0 as a 0^- Nonet Member

Experimentally the X^0 could have J^P of 0^- or 2^-. In view of its low mass the former seems the more likely. If it is assumed that π, K, η, and X^0 form a nonet, the mass formula gives for the magnitude of the mixing angle, $|\vartheta| = 10.4° \pm 0.2°$. On the quark model we would expect this angle to be negative (Equation 25). The overlap function I in the Schwinger mass formula (Equation 26) is ~ 0.5.

Since the total width of the X^0 is only known to be < 4 MeV, the observation of Bollini et al. (30) only gives an upper limit for $\Gamma_{\gamma\gamma}(X^0)$ of (220_{-200}^{+140}) keV. If $\Gamma_{\gamma\gamma}$ were actually known, the 0^- and 2^- assignments for the X^0 could be distinguished in two ways. Firstly, the amplitude for X^0 production by the Primakoff effect $\gamma\gamma \to X^0$ would be known and the $(2J+1)$ term in the density of final states would yield a factor of 5 difference in the cross section for the 0^- and 2^- assignments (32). Secondly, the mixing angle could be checked using:

$$\Gamma(X^0 \to \pi\pi) = \frac{m^3_{X^0}}{\sin^2 \vartheta}\left[\left(\frac{\Gamma(\pi^0 \to \gamma\gamma)}{3m_\pi^3}\right)^{1/2} - \cos\vartheta\left(\frac{\Gamma(\eta \to \gamma\gamma)}{m_\eta^3}\right)^{1/2}\right]^2$$

This formula arises because γ is equivalent to a $U=0$ octet member, hence $\gamma\gamma$ couples to the $SU(3)$ singlet with amplitude A_1, and to the $U=0$ octet member, i.e. $(\sqrt{3}\pi^0 - \eta^8)$ with amplitude A_8. Hence:

$$A(\eta) = -\sin\vartheta A_1 + \cos\vartheta\left[-\frac{1}{\sqrt{3}}A(\pi^0)\right]$$

$$A(X^0) = \cos\vartheta A_1 + \sin\vartheta\left[-\frac{1}{\sqrt{3}}A(\pi^0)\right]$$

$$= \cos\vartheta\left[\frac{A(\eta)}{\sin\vartheta} - \frac{1}{\sqrt{3}}A(\pi^0)\frac{\cos\vartheta}{\sin\vartheta}\right] - \frac{\sin\vartheta}{\sqrt{3}}A(\pi^0)$$

i.e. $$\Gamma(X^0) = \frac{1}{\sin^2\vartheta}\left[\frac{\Gamma^{1/2}(\pi^0)}{\sqrt{3}} - \cos\vartheta\cdot\Gamma^{1/2}(\eta)\right]^2$$

The m^3 terms represent adjustments for the varying phase space due to the different particle masses. The present upper limit on $\Gamma_{\gamma\gamma}(X^0)$ does not permit a useful check.

A less reliable check of the mixing angle has been made by comparing X^0 and η production in πN collisions. Since πN is made exclusively of n and p-type quarks, we might expect in such collisions that only the X^0, η combination which contains no $\Lambda\bar{\Lambda}$ quarks, $(p\bar{p}+n\bar{n})/2$, would be produced. Since

$$X^0 = \cos\vartheta\cdot\eta_0 + \sin\vartheta\cdot\eta_8 = \cos\vartheta\cdot\frac{p\bar{p}+n\bar{n}+\Lambda\bar{\Lambda}}{\sqrt{3}}$$

$$+ \sin\vartheta\frac{p\bar{p}+n\bar{n}-2\Lambda\bar{\Lambda}}{\sqrt{6}}$$

and

$$\eta = -\sin\vartheta\cdot\eta_0 + \cos\vartheta\cdot\eta_8 = -\sin\vartheta\frac{p\bar{p} + n\bar{n} + \Lambda\Lambda}{\sqrt{3}}$$
$$+ \cos\vartheta\frac{p\bar{p} + n\bar{n} - 2\Lambda\bar{\Lambda}}{\sqrt{6}}$$

it follows that the amplitude for X^0 production is:

$$A(X^0) \propto \sqrt{\tfrac{2}{3}}\cos\vartheta + \sqrt{\tfrac{2}{3}}\sin\vartheta$$

and that for η production is:

$$A(\eta) \propto \sqrt{\tfrac{2}{3}}\cdot(-\sin\vartheta) + \sqrt{\tfrac{1}{3}}\cos\vartheta$$

Hence:

$$\frac{\sigma(\pi N \to N X^0)}{\sigma(\pi N \to N\eta)} = \frac{\sigma(\pi N \to N^* X^0)}{\sigma(\pi N \to N^*\eta)} \propto \cot^2(\vartheta - \vartheta_V) \qquad 33.$$

where $\vartheta_V = \tan^{-1} 1/\sqrt{2}$, is the ideal mixing angle. The factor of proportionality has to allow for the effect of mass differences. It is this that makes the technique unreliable. Figure 7 shows checks at various Q values (33). The solution that makes the mixing angle large has been ignored. The agreement

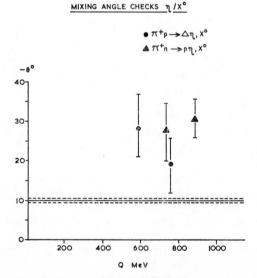

FIG. 7. Values obtained for the pseudoscalar mixing angle by comparing η and X^0 production in πN collisions.

with a quadratic mass formula is not good, but the sign is that expected from the quark model. A linear mass formula would give $\vartheta = \pm(24 \pm 1)°$ which fits these results better. However it will be seen below that only by using quadratic formulae is Δ, the strange quark mass excess, reasonably constant between different nonets.

3.3 The E Meson

The E meson is seen as a $K\bar{K}\pi$ state with $M = 1424 \pm 6$ MeV and $\Gamma = 71 \pm 10$ MeV. Since $K_1^0 K_1^0 \pi$ decay is seen but not $K_1^0 K_2^0 \pi$, its C parity is even. No E^{++} of E^{--} has been seen, hence $I \neq 2$. That $I = 0$ and $\neq 1$ is shown in two antiproton experiments. In $p\bar{p} \to K\bar{K}3\pi$ at rest studied by Baillon et al. (34) (Figure 8), no $E^{\pm}$ is seen in the experimentally observable $K_1^0 K_1^0 \pi^{\pm}$ and $K^+ K^- \pi^{\pm}$ distributions. This suggests $I = 0$. However, an unlikely cancellation could allow $I = 1$. If $I = 1$, $G(E) = -1$, hence $G(p\bar{p})$ in the reaction $p\bar{p} \to E\pi\pi$ is also -1 which is possible for $(C_{p\bar{p}}, I_{p\bar{p}}) = (1, 1)$ or $(-1, 0)$. There are thus three orthogonal amplitudes for $E^{\pm}(\pi\pi)^{\mp}$ production:

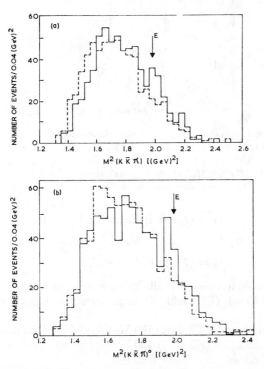

FIG. 8. (a) $(K\bar{K}\pi)^0$ (solid line) and $(K\bar{K}\pi)^{\pm}$ (broken line) spectra in the reaction $p\bar{p} \to K_1^0 K_1^0 \pi^+\pi^-\pi^0$ at rest. (b) $(K\bar{K}\pi)^0$ (solid line) and $(K\bar{K}\pi)^{\pm}$ (broken line) spectra in $p\bar{p} \to K^+ K^- \pi^+\pi^-\pi^0$ at rest (34).

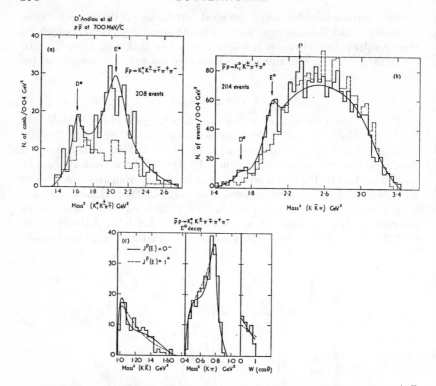

FIG. 9. Observation of E meson in $p\bar{p}$ annihilations at 0.7 GeV/c. (a) $M^2(K_1K^{\pm}\pi^{\mp})$ shown solid and $M^2(K_1K^{\pm}\pi^{\pm})$ shown dashed for $p\bar{p}\to K_1K^{\pm}\pi^{\mp}\pi^+\pi^-$. (b) $M^2(K_1K^{\pm}\pi^{\mp})$ shown solid and $M^2(K_1K^{\pm}\pi^0)$ shown dashed for $p\bar{p}\to K_1K^{\pm}\pi^{\mp}\pi^0$. (c) Decay properties of the E, namely $M(K\bar{K})$, $M(K\pi)$ and $W(\cos\theta)$, the angular distribution of K mesons in the $K\bar{K}$ CM. The curves correspond to the 0^- and 1^+ hypotheses (35).

$$A_1 = \langle I_{p\bar{p}} = 1 \| E\pi\pi \text{ with } I_{\pi\pi} = 2 \rangle$$

$$A_2 = \langle I_{p\bar{p}} = 1 \| E\pi\pi \text{ with } I_{\pi\pi} = 1 \rangle$$

$$A_3 = \langle I_{p\bar{p}} = 0 \| E\pi\pi \text{ with } I_{\pi\pi} = 1 \rangle$$

Only in the unlikely event that all three are zero will $E^{\pm}$ production be suppressed for $I(E) = 1$. (E^0 would still be produced through $A_4 = \langle I_{p\bar{p}}=1 \| E\pi\pi$ with $I_{\pi\pi}=0 \rangle$.)

In the reaction $p\bar{p}\to K\bar{K} 2\pi$ at 700 MeV/c, though E^0 is produced neither E^+ nor E^- is seen (35) (Figure 9) and in this case no cancellation could suppress both of E^+ and E^-. Hence $I(E) = 0$.

If $I = 0$, then the G parity is even and the $K\bar{K}$ subsystem of the $K\bar{K}\pi$ decay has $G = -1$. The mode $K_1^0K^{\pm}\pi^{\mp}$ is seen for which $I(K\bar{K}) = 1$, hence $J^P(K\bar{K}) = 0^+2^+$ etc. (Section 1.2.3). The $K\bar{K}$ system peaks at low masses so

0^+ is plausible for that system. Hence $J^P(E) = 0^- 1^+ 2^-$ etc. To distinguish the alternatives requires detailed data fitting. The most complete is that by Baillon et al. (34) who use two methods.

Firstly the decay is considered as made up of three amplitudes, $K^*\overline{K}$, $\overline{K}^*K$, and $(K\overline{K})\pi$ where $K\overline{K}$ is considered either as 0^+BW (see Section 7.1) or as a 3F scattering length. Appropriate matrix elements are readily found,

$$\text{e.g.} \quad 0^- \rightarrow K^*\overline{K} \quad \text{has} \quad ME \propto p_{K*} \cdot p_K$$

$$1^+ \rightarrow K^*\overline{K} \quad \text{has} \quad ME \propto p_{K*}$$

$$0^- \rightarrow (K\overline{K})\pi \quad \text{has} \quad ME \propto \text{constant}$$

$$1^+ \rightarrow (K\overline{K})\pi \quad \text{has} \quad ME \propto p_\pi$$

where p_{K*} is the π and K momentum in the K^*CM, p_K is the CM momentum of the odd K, p_π is the CM momentum of the odd π. With allowance for the odd intrinsic parities of the three products, it is evident that these matrix elements are of the correct form. If we fit the data for the three amplitudes for any given assignment, the probabilities of 0^- 1^+ and 2^- are respectively 2 per cent, 0.2 per cent, and 0 per cent for the reaction $p\bar{p} \rightarrow E^0\pi^+\pi^-$, and 30 per cent, 5 per cent, and 3 per cent in the reaction $p\bar{p} \rightarrow E^0\pi^0\pi^0$. It is assumed that the low probabilities in the first reaction result from neglecting interference between the two possible E^0's.

The second method is to fit, not the decay, but the production of E^0. The $p\bar{p}$ system can be in the $^1S_0$ or $^3S_1$ states which have $C = (-1)^{l+S}$, i.e. $+1$ and -1 respectively. Since $C_E = +1$, in the case of $^1S_0$ annihilation the C parity of the dipion is even and the fit is limited to $l = 0$ and 2. In the case of $^3S_1$ annihilation $C_{\pi\pi}$ and $l_{\pi\pi}$ must be odd and only $l_{\pi\pi} = 1$ is considered. Fitting for 0^-, 1^+, and 2^- respectively, χ^2 values of 44, 40, and 52 are obtained for 37 degrees of freedom. Hence 2^- gives a poor fit and 0^- and 1^+ are acceptable. However, the 1^+ fit demands that some 85 per cent of the $\pi\pi$ system be in the d wave, which is unlikely as its energy is always less than 500 MeV.

Hence for $\bar{p}p$ annihilations at rest 0^- is the preferred assignment.

D'Andlau et al. (35) similarly fit the data for E^0 decay in the reaction $p\bar{p} \rightarrow K_1^0 K^\pm \pi^\mp \pi^+ \pi^-$ at 700 MeV/c. [Only the scattering-length description of the $K\overline{K}$ system is used for the $(K\overline{K})\pi$ amplitude.] The consequences for the Dalitz plot variables and for the angular distribution of the K meson in the $K\overline{K}$ c.m. are shown in Figure 9. The 0^- assignment is preferred but 1^+ cannot be ruled out.

Dahl et al. (36) observe E^0 in $\pi^- p \rightarrow n(K\overline{K}\pi)^0$ between 2.9 and 4.2 GeV/c but the large background makes fitting difficult. Figure 10 shows the results of the fit. In this case 1^+ seems preferred but 0^- cannot be ruled out.

Since only one group claims to have found $J^P(E)$ unambiguously, it is desirable to await further results before the assignment is considered finally established.

3.4 Classification of the E Meson

If the E meson is used instead of the X^0 as the ninth nonet member, the mixing angle drops to $6.2 \pm 0.1°$ with a quadratic formula. There are no independent mixing-angle checks on this assumption. If the E meson is 0^- but not a member of the lowest 0^- nonet, then on the quark model it could represent a radially excited state of the $l=0$ spin-singlet $q\bar{q}$ system with

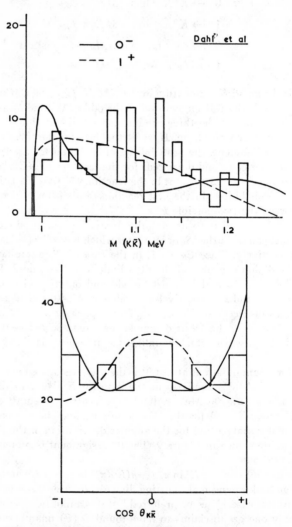

FIG. 10. Fits to the decay distributions for E^0 produced in $\pi^- p \to n \, (K\overline{K}\pi)^0$ at 2.9 and 4.2 GeV/c (36).

radial quantum number, $n=2$. Using Equation 30 we would expect the mass of such a meson to be approximately the same as that of states with $n=1$ and $l=2$. Such mesons are commonly held to be those in the 1650–1750 mass region. One would then expect to produce the $I=1$ and $I=\frac{1}{2}$ members of the radially excited nonet by diffractive dissociation of incident π and K. However, from the density of states term in the cross section $(2J+1)$, they would be produced with cross sections only $\frac{1}{5}-\frac{1}{7}$ of those for the A_3 and L mesons if these are indeed diffractively produced states of spin 2 or 3. At such a low cross section they could easily have evaded experimental detection.

Regardless of which of E or X^0 are assigned to the nonet, the mass-squared difference Δ associated with the nonet is 0.22 GeV/c^2.

In conclusion there is a well-established 0^- nonet, but a final decision as to whether X^0 or E is the ninth member must await further experimental work.

4. THE VECTOR MESONS

4.1 THE 1^- NONET

The 1^- nonet consisting of ρ, $K^*(890)$, ω, and ϕ is well established. With a quadratic formula the mixing angle is $39.9 \pm 1.1°$ [using $m(\rho) = 765 \pm 15$], the ϕ being somewhat more octet than singlet. The mixing is thus close to ideal and this is also reflected in the fact that $m_\omega \simeq m_\rho$ (Section 1.2.3) and that the overlap function in the Schwinger mass formula is approximately unity. The ϕ is thus almost purely $\lambda\bar{\lambda}$-type quarks explaining its suppressed decay to the $\rho\pi$ system which consists of p, n-type quarks (3π decay represents only 20 ± 2 per cent of the ϕ decay, the rest is to $K\bar{K}$). In fact, by an argument similar to that leading to Equation 33:

$$\frac{\omega \to 3\pi}{\phi \to 3\pi} \propto \cot^2 (\vartheta - \vartheta_V)$$

Again allowance for mass differences must be made, and Glashow & Socolow (37) have shown that

$$\frac{\omega \to 3\pi}{\phi \to 3\pi} = \frac{1}{17} \cot^2 (\vartheta - \vartheta_V)$$

This formula leads to a mixing angle of $39 \pm 1°$, in excellent agreement with the value from the mass formula.

Comparison of ω and ϕ production in incident-pion collisions permits a check to be made on the mixing angle by the argument leading to Equation 33. Comparing $\pi^+ n \to p\phi$ and $\pi^+ n \to p\omega$, Benson et al. (33) deduce that the mixing angle is either $(27 \pm 4)°$ or $(43 \pm 4)°$; comparing $\pi^+ p \to N^{*++}\phi$ and $\pi^+ p \to N^{*++}\omega$, Lai & Schumann (33) find that the angle is $(35 \pm 7)°$.

If it is assumed that the photon transforms like the $U=0$ member of the octet, then the decay of ϕ and ω to $e^+ e^-$, assumed to be mediated by a virtual photon, is an indicator of the octet component. Neglecting symmetry-breaking effects, $\Gamma(\omega \to e^+ e^-)/\Gamma(\phi \to e^+ e^-) = \tan^2 \vartheta$. Observation of leptonic decays is difficult and branching ratios obtained vary from experiment to experiment. Producing ϕ from incident pions, Binnie et al. (38) and Bollini et al. (39) respectively report that $\Gamma_{ee}/\Gamma_{total}$ equals $(7.2 \pm 3.9) \times 10^{-4}$ and $(6.1 \pm 2.6) \times 10^{-4}$. Using photoproduced ϕ, Becker et al. (40) find $\Gamma_{ee}/\Gamma_{K^+K^-}$ which, using the best value of the K^+K^- branching ratio, gives $\Gamma_{ee}/\Gamma_{total}$ $= (2.9 \pm .4) \times 10^{-4}$. Augustin et al. (41) in a colliding electron-beam experiment find $(3.96 \pm .62) \times 10^{-4}$, using their own $3\pi/K\bar{K}$ ratio. If they use the previous value of Lindsey et al. (42), which has twice the error, they obtain $(3.53 \pm .67) \times 10^{-4}$. For the ω meson, Bollini et al. (43) report $\Gamma_{ee}/\Gamma_{total}$ $= (.4 \pm .15) \times 10^{-4}$ and Augustin et al. (44) report $(.76 \pm .14) \times 10^{-4}$. The mean ϕ branching ratio from these experiments is $(3.97 \pm .5) \times 10^{-4}$ and the mean ω branching ratio is $(.58 \pm .11) \times 10^{-4}$. Combining with the accepted values for the total widths of the ϕ and ω, respectively 3.7 ± 0.6 MeV and 12.6 ± 1.1 MeV (45), use of $\tan^2 \vartheta = \Gamma(\omega \to ee)/\Gamma(\phi \to ee)$ yields a value for $|\vartheta|$ of $(35.2_{-4.5}^{+3.6})°$. The above determinations of the leptonic decay rate of the ω ignore $\omega - \rho$ interference and the calculation ignores symmetry breaking effects. A number of authors (46) have put forward explicit models for symmetry breaking in the leptonic decays using, in general, the algebra of currents, but the present errors on the experimental data prevent distinction between these models.

$SU(3)$ checks on two-body decays are only possible for $K^* \to K\pi$, $\phi \to K\bar{K}$, and $\rho \to \pi\pi$ which should be in the ratio $\frac{1}{4}: \frac{1}{2} \cos^2 \vartheta : \frac{1}{3}$. Goldberg (47), using the formula:

$$\Gamma \propto C^2 \cdot p/M^2 \left(\frac{p^2 X^2}{p^2 + X^2} \right)^L \qquad\qquad 34.$$

to correct for the different phase spaces available, where C^2 is the $SU(3)$ Clebsch-Gordan coefficient squared given above, p is the momentum of the decay products, M is the particle mass, and X is some inverse interaction radius (not critical), has carried out a fit to the three experimental widths and obtains a χ^2 probability of 31 per cent. However, an equally good fit is obtained from phase space alone, and thus the check is not significant.

There is still considerable uncertainty as to the best mass and width of the ρ meson. Many authors have used the simple Breit-Wigner form to fit their data, neglecting the energy dependence of the width. All methods involve some theoretical assumptions. One would expect that for ρ^0, the most reliable method of determining the width would be to use colliding electron beams, since there are no other strongly interacting particles in the final state, and in principle the background can be explicitly calculated. Expressing the pionic electromagnetic form factor in terms of the Omnes function, and using an energy-dependent width, Roos (48) fits the colliding-beam data of Auslender et al. (49) and Augustin et al. (50) to obtain $M = 770 \pm 4$ MeV and $\Gamma = 122 \pm 6$ MeV. However, the Particle Data group (45) only feel able to give values of $M = 765 \pm 10$ MeV and $\Gamma = 125 \pm 20$ MeV for both ρ^0 and $\rho^\pm$.

4.2 RELATIONSHIP OF 0^- AND 1^- MESONS

On the quark model, the 0^- and 1^- mesons are the two nonets obtained with zero relative orbital angular momentum. In $SU(6)$ they represent components of the two lowest multiplets, the (1) and the (35). The near-ideal mixing of the vector mesons compared with the small mixing of the pseudoscalar nonet is as expected from $SU(6)$, though in this connection the large mass difference between corresponding members of the two nonets, for example m_π and m_ρ, would indicate violent $SU(6)$ breaking. The mass-squared difference Δ between ρ and K^* is 0.21 GeV^2, almost identical to the value for the 0^- mesons as expected for a simple quark model. The simplest assumption for the quark model would be that large spin-spin forces [$SU(6)$ breaking] separate the $S=0$ and $S=1$ nonets, and that $S=1$ nonets show near $SU(6)$ symmetry in the $q\bar{q}$ forces, so that near-ideal mixing results for $S=1$ states (we shall see that this is indeed true for the 2^+ mesons) and small mixing for $S=0$ states.

5. THE 2^+ NONET

The normal 2^+ nonet attribution is the A_2, $K^*(1400)$, f and f'. The situation has been complicated by the observation of splitting of the A_2 peak.

5.1 A_2 SPLITTING

The splitting of the A_2 was first reported by the CERN Missing-Mass Spectrometer group. In their initial experiment (51, 52) the reaction $\pi^- p \rightarrow p M^-$ was studied at 6 and 7 GeV/c. $M_M{}^2$ is given by:

$$M_M{}^2 = (E_1 + m_p - E_3)^2 - p_1{}^2 - p_3{}^2 + 2p_1 p_3 \cos \vartheta$$

where p_1 and E_1 refer to the incident pion, and ϑ, p_3, and E_3 to the recoil proton, all quantities being in the laboratory system. The form of the relationship between recoil angle and momentum for different masses is shown in Figure 11. In their first experiment, data were taken at the minimum of the appropriate curve in Figure 11 so that only the recoil angle, and not the recoil momentum, needed accurate measurement. The experiment was subsequently repeated (53) at an incident momentum of 2.6 GeV/c with the recoil proton selected to be at 0^0 when it will be seen from Figure 11 that only the recoil momentum need be measured accurately. By coincidence the four-momentum transfer in the two experiments was in the same region ($.21 \leq t \leq .39$ GeV^2 in the first experiment, and $t \sim .22$ GeV^2 in the second). The mass resolution in the two experiments was ~ 16 MeV and 10 MeV respectively.

Figure 12 shows the separate and combined missing-mass distributions. The background is large but the splitting is very significant. The number of events in the two peaks is compatible with equality at both momenta. If there is a momentum dependence, it is such that the lower peak increases relative to the upper with increasing momentum. A fit of the combined data to two incoherent Breit-Wigner (BW) resonances on an incoherent back-

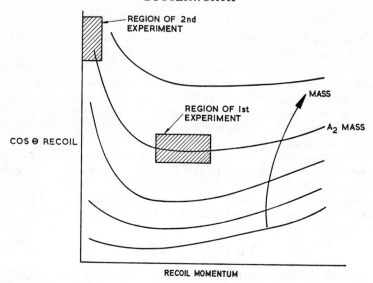

FIG. 11. Form of the relationship between the cosine of the laboratory angle and the momentum of the recoiling proton in the reaction $\pi^-p \rightarrow pM^-$.

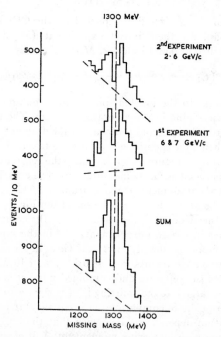

FIG. 12. The splitting of the A_2 meson observed by the CERN Missing-Mass Spectrometer Group (51, 52).

ground gives a confidence level of <0.2 per cent. (If the two resonances separately have destructive interference with coherent background amplitudes, each making up to 16 per cent of the total background, a good fit can be obtained by assuming different values of J^P for the two peaks.) Three alternative good fits can be obtained by assuming that the peaks correspond to the same J^P:

(a) Two adjacent BW resonances $A_2{}^L$ and $A_2{}^H$ at 1289 and 1309 MeV, each with widths $\sim$22 MeV.

(b) A destructively interfering narrow A_2' at approximately the same mass of 1298 MeV as a broad A_2, the widths being $\sim$12 MeV and $\sim$90 MeV respectively.

(c) A special case of (a) when the BW resonances merge to give a dipole resonance formula,

$$N \propto \left| \frac{(M - M_0)\Gamma}{(M - M_0)^2 + \Gamma^2/4} \right|^2$$

with $M_0 = 1298$ MeV and $\Gamma = 28$ MeV.

Confirmation of the splitting of the $\pi\rho$ decay mode of the A_2 has been reported in bubble-chamber experiments on π^-p interactions at 6 GeV/c (55) and 4 GeV/c (56) and on $p\bar{p}$ interactions at 1.2 GeV/c (57). None are of high statistical significance.

For a part of the first CERN MM experiment, wide-gap wire chambers measured the laboratory angles of the decay pions from the A_2 (54). The A_2 momentum and hence the combined momentum of the three pions is known from the proton recoil. Except when the pions are coplanar, measurement of their angles then allows their individual momenta to be deduced, so yielding a point on the decay Dalitz plot (six-way folded since the charges are not distinguished). Moreover, the effective mass of the pions can be checked against the missing mass to eliminate events involving π^0 production. Energy balance to ±15 MeV was demanded. Coplanar events will, in general, fail this test, and this loss plus that resulting from the finite solid angle of counters was corrected for, by Monte Carlo calculations. Fits for 675 events in the overall A_2 region are shown in Table IV assuming resonance plus incoherent 3π and $\rho\pi$ phase-space backgrounds, together with fits for the $A_2{}^H$ and $A_2{}^L$ regions separately. (The sample used did not, in itself, show splitting.)

Both from the mass fitting and from the Dalitz plot, the CERN experiment thus favours both halves of the A_2 peak being of the same J^P, namely 2^+. (Should one peak have to be of different J^P, $A_2{}^H$ is the one that might not be 2^+.)

This conclusion is supported by Aguilar-Benitez et al. (58) who study $p\bar{p} \to K_1{}^0 K^{\pm} \pi^{\mp}$ at 0, 0.7, and 1.2 GeV/c. At the two higher momenta, K^* events are removed to improve the A_2 signal; at rest this is not done since there is evidence that A_2 and K^* interfere and such a selection does not enhance the A_2 signal. Figure 13 shows the $K_1{}^0 K^{\pm}$ mass distribution for 3217

TABLE IV

Confidence Levels for Different J^P Assignments in the A_2 Region[a]

Mass interval	J^P	% resonance	% $\rho\pi$ background	% 3π background	$P(\chi^2)$
Total A_2	2^+	40	20	40	38
	1^-	30	40	30	0.1
1260–	$2^-(p)$	30	40	30	0.1
1360 MeV	$1^+(s)$	40	30	30	0.1
	$1^+(d)$	30	40	30	0.1
$A_2{}^L$	2^+	40	10	50	38
1254–					
1307 MeV	1^-	30	40	30	0.3
$A_2{}^H$	2^+	40	0	60	54
1307–					
1360 MeV	1^-	30	40	30	10

[a] Chikovani et al. (54).

selected events. A fit to a single BW has only 4 per cent probability, to two incoherent BW's 28 per cent probability ($M_L = 1281 \pm 3$, $\Gamma_L = 22_{-7}^{+10}$; $M_H = 1325 \pm 3$, $\Gamma_H = 22_{-7}^{+10}$), and to the dipole 65 per cent probability ($M = 1303 \pm 2$, $\Gamma = 31 \pm 4$); other fits are not performed. Even if the splitting is ignored, the 124_{-27}^{+41} MeV width found for the overall peak is compatible with the normal unresolved A_2 width.

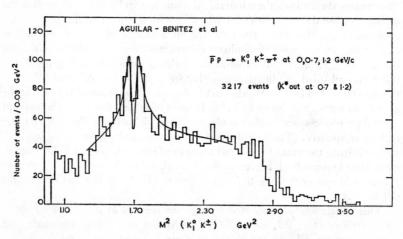

Fig. 13. $M(K^0 K^\pm)$ from the reaction $p\bar{p} \rightarrow K_1^0 K^\pm \pi^\mp$ at 0, 0.7, and 1.2 GeV/c (58). At the two higher momenta, K^* events have been removed.

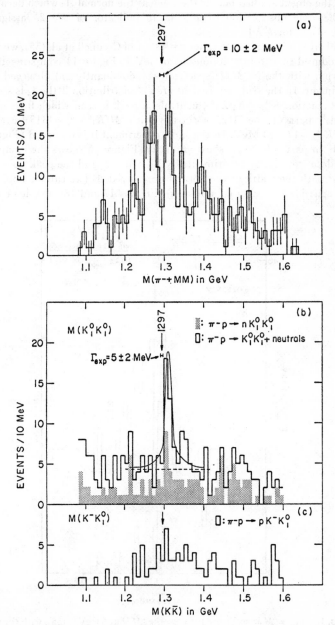

FIG. 14. A comparison of the $K\overline{K}$ spectra in $\pi^-p\rightarrow nK_1^0K_1^0$, $\pi^-p\rightarrow K_1^0K_1^0$ neutrals and $\pi^-p\rightarrow pK^-K_1^0$ with the X^- spectrum in $\pi^-p\rightarrow pX^-$ at 6 GeV/c (55).

If the object studied here is the same as the normal A_2 which decays into $\rho\pi$, then the observation of $K\overline{K}$ is an indicator of the 2^+ assignment, $[G = (-1)^{I+J}$ for $K\overline{K}]$.

This result is in direct conflict with that of Crennell et al. (55), who study A_2 produced in $\pi^- p$ interactions at 6 GeV/c. Figure 14 compares the $K\overline{K}$ spectrum with the $(\pi^- MM)$ spectrum (predominantly $\pi\rho$) observed in this experiment. In the truly equivalent $(K\overline{K})^-$ distribution little A_2 is seen, but in the reaction $K^- p \rightarrow K_1^0 K_1^0$+neutrals, a peak is seen which lines up with the $A_2{}^H$ part of the MM spectrum with $M(K_1^0 K_1^0) = 1315 \pm$ MeV and $\Gamma(K_1^0 K_1^0) = 12 \pm 10$ MeV. In the same experiment it is observed (Figure 15) that both parts of the A_2 show $\eta\pi$ decay. Figure 16 shows the compilation of $\eta\pi$ data from many experiments by French (59) and the peak does indeed appear with the width of the normal unresolved A_2 (we cannot hope to observe splitting in such a compilation). Figures 17 and 18 show his compila-

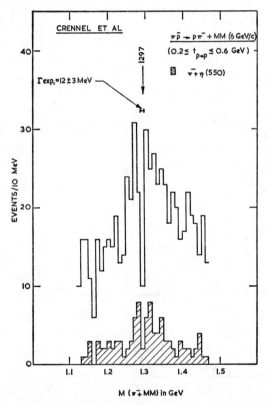

FIG. 15. The $\pi^- MM$ spectrum for $\pi^- p \rightarrow p\pi^- MM$ at 6 GeV/c (55), with the requirement that $0.2 \leq t_{pp} \leq 0.6$ (GeV/c)². The inner histogram is for those events where the missing mass is compatible with the η mass.

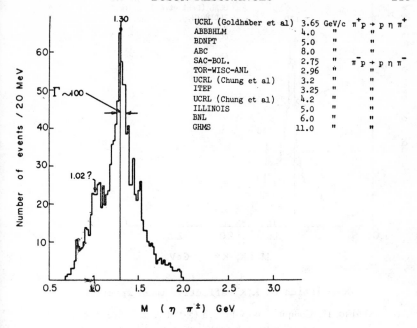

Compilation of $\eta^0\pi^\pm$ effective spectra.

FIG. 16. Compilation by French (59) of $\eta\pi^\pm$ effective-mass
spectra from the reactions $\pi^\pm p \rightarrow p\pi^\pm\eta$.

tions of $K\overline{K}$ distributions. The $K_1^0 K_1^0$ distribution might be affected by the
f' and is rather broad, but the $K_1^0 K^\pm$ distribution, though it appears shifted
upwards in mass compared with the normal value for the A_2 mass, appears
to be of the standard width. Of course there are clear dangers in making such
compilations. Lipkin & Meshkov have pointed out (60) that to the degree
that η is nearly pure $SU(3)$ octet, there is no J^P assignment for an $I = 1$
meson that permits $\rho\pi$ and $\eta\pi$ decay and forbids $K\overline{K}$ decay. J^P and G-parity
conservation for an $I = 1$ object with $\rho\pi$ and $\eta\pi$ decay forbid all but the
sequences $J^{PC} = 1^{-+}$, 3^{-+}, $\cdots$ or 2^{++}, $4^{++} \cdots$. The former sequence is
forbidden by the arguments of Section 1.3.2, namely that $SU(3)$ prevents
any particle from a multiplet whose neutral members have odd CP from decaying into $\eta\pi$. Only the 2^{++}, 4^{++} sequence remains, for which there is a
definite ratio of $K\overline{K}$ to $\pi\eta$ decay of $\frac{2}{3}$ from $SU(3)$ (mass differences are
ignored). (It should be stressed that these arguments prevent association of
either peak with an odd Regge daughter.) To understand the observation of
Crennell et al. one would thus have to suppose that there are three "A_2's"
so that the $\eta\pi$, $\rho\pi$, and $K\overline{K}$ modes need not be related to the A_2^H and A_2^L
positions to the same particles, or that $SU(3)$ is very badly broken, or that

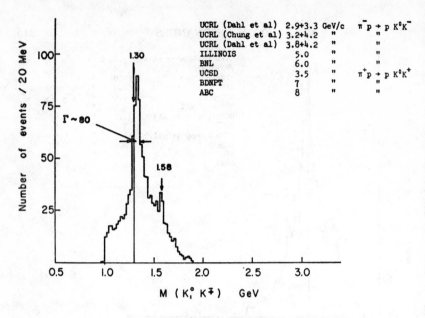

Compilation of $K_1^0 K^\pm$ effective mass spectra.

FIG. 17. Compilation by French (59) of $K^\pm K_1^0$ effective-mass spectra from the reactions $\pi^\pm p \to K^\pm K_1^0$.

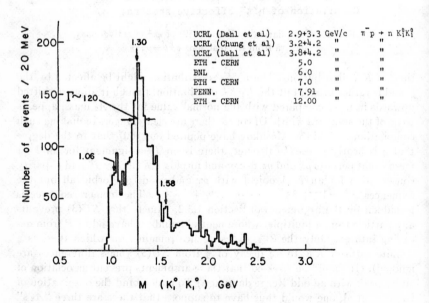

Compilation of $K_1^0 K_1^0$ effective mass spectra.

FIG. 18. Compilation by French (59) of $K_1^0 K_1^0$ effective-mass spectra from the reaction $\pi^- p \to n K_1^0 K_1^0$.

the η mixing angle is grossly wrong. In view of this and of the fact that the observation is in conflict with the experiment of Aguilar-Benitez et al., it will be assumed here that the result is spurious and that both halves of the A_2 are $J^{PC} = 2^{++}$.

The solution of a narrow A_2' destructively interfering with a broad A_2 has a number of attractions:

    (*i*) It preserves the normal A_2 width for $SU(3)$ checks (Section 5.3).

    (*ii*) It gives a natural explanation of why the two observed peaks have the same height at different momenta without invoking a dipole fit. (Of course it might be argued that two neighbouring resonances of the same J^P will be of similar cross sections anyway.)

    (*iii*) French (59) has reported that at high momentum transfer (>0.5 GeV/c) the combined results of experiments to study $\pi^- p \to p \pi^+ \pi^- \pi^-$ at 4 GeV/c and $\pi^+ p \to p \pi^+ \pi^+ \pi^+$ at 8 GeV/c show a narrow, 20 MeV wide, peak in the $(3\pi)^\pm$ spectrum at exactly the position where, at lower momentum transfers, the dip in the A_2 structure appears (Figure 19). This might conceivably be the A_2' if one supposed that for some reason its differential cross section has a weaker dependence on momentum transfer that that of the A_2.

(It should be pointed out that two resonances in the same partial wave must destructively interfere at their overlap. The presence of two resonances implies that the phase shift rises through $\pi/2$ and then through $3\pi/2$. Between, it must pass through π, thus giving a minimum in the cross section.)

5.2 Observations on Other Members of the 2^+ Nonet

Ascoli et al. (61) have reported in the reaction $\pi^- p \to n \pi^+ \pi^- \pi^+ \pi^-$ at 5 GeV/c an 11.5 ± 4 μb enhancement in the $\pi^+ \pi^+ \pi^- \pi^-$ mass distribution at

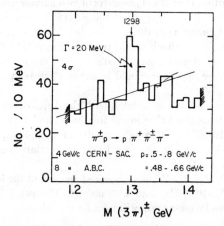

FIG. 19. Combined 3π spectra from two experiments on $\pi^\pm p \to p \pi^\pm \pi^+ \pi^-$ at 4 and 8 GeV/c in the higher momentum-transfer region (59).

the f^0 mass. The angular distributions fit the assumption that the 4π system is made up of $\rho\rho$ in a relative s wave (though m_f is less than twice the central ρ mass). With this assumption, the ~ 10 per cent branching ratio for $\pi^+\pi^-\pi^+\pi^-/\pi^+\pi^-$ which they observe would correspond to a total $4\pi/\pi^+\pi^-$ branching ratio of 30 per cent, which is greater than has normally been assumed.

No splitting of other 2^+ mesons has been reported.

5.3 ASSIGNMENTS IN THE 2^+ NONET

Use of the central masses for A_2, $K^*(1400)$, f and f' (i.e. adopting the broad A_2 rather than either $A_2{}^H$ or $A_2{}^L$) leads to a mixing angle of $(27.6 \pm 2.3)°$. The near-ideal value of this angle is compatible with the f' decay being predominantly to $K\overline{K}$. (The fact that the $\pi\pi$ branching ratio for the f' is less than 14 per cent shows that the mixing angle is within $13°$ of ideal.) Comparison of f and f' production cross sections by Lai & Schumann (33) gives a mixing angle of $(26 \pm 2.5)°$ or $(47 \pm 3)°$.

The large number of decays for 2^+ mesons allows $SU(3)$ checks to be made. There are six observed $2^+ \rightarrow 0^- 0^-$ decays and four others compatible with zero. The $SU(3)$ predictions fit well, but the $\pi\eta$ and $K\overline{K}$ decays of the A_2 are sufficiently small and unreliable to provide no test of the correct A_2 width. The situation is very different for the $0^- 1^-$ decays. A check on the five observed decays fits well if a broad A_2 is used, and is very critically dependent on the relative width of the A_2 and $K^*(1400)$. The final-state momenta in $A_2 \rightarrow \pi\rho$ and $K^*(1400) \rightarrow K^*\pi$ are respectively 410 and 413 MeV. Thus phase-space corrections (p^5) affect the relative widths only by about 10 per cent. The $SU(3)$ factor for $\Gamma(A_2 \rightarrow \rho\pi)/\Gamma(K^*1400 \rightarrow K^*\pi) = \frac{8}{3} = 2.67$. Use of a broad A_2 gives 2.6 ± 0.4 for this ratio. Use of either of the narrow $A_2{}^H$ or $A_2{}^L$ states gives 0.83 ± 0.4.

If then the solution that the A_2 consists of two narrow and adjacent states (or a dipole) is adopted, it follows that the $K^*(1400)$ must be similarly split. It could be true of course that the splitting is a fundamental property of all 2^+ states, but the $SU(3)$ check would only require $K^*(1400)$ splitting. The conjunction of a broad and a narrow state is the more attractive solution, and we will see below that there could be other 2^+ states which might be associated with the A_2'.

The excess mass-squared associated with the Λ quarks is $.34$ GeV^2, which differs somewhat from the value found for the 0^- and 1^- mesons. The near-ideal mixing is reflected by the rough equality of m_f and m_{A_2} (1264 ± 10 MeV and 1297 ± 10 MeV), and by the fact that the overlap function I approximates to unity ($I \sim 0.92$).

On the quark model, the 2^+ mesons arise from one of the four $l_{q\bar{q}} = 1$ states together with $J^{PC} = 0^{++}$ 1^{++} and 1^{+-} nonets. The possible assignment of mesons to the three latter nonets will next be considered.

6. AXIAL-VECTOR MESONS

In the quark model one expects two axial-vector nonets with $J^{PC} = 1^{+\pm}$.

Since C is not defined for the strange members, these can be expected to mix. The only 1^+ meson whose existence and quantum numbers are both reasonably well established is the isovector B meson.

6.1 THE B MESON

The B meson ($M = 1221 \pm 16$ MeV, $\Gamma = 123 \pm 16$ MeV) is seen as a $\pi\omega$ enhancement in incident-pion interactions and in antiproton annihilations at rest (45). The $\pi\omega$ decay fixes $I = 1$, $C = -1$ and rules out $J^P = 0^+$. The absence of $\pi\pi$ and $K\overline{K}$ decay suggests $J^{PC} \neq 1^{--}$, 3^{--} etc. In all incident-pion experiments, the ω decay normal has a $\sin^2$ distribution relative to the ω direction in the B centre of mass. This eliminates the 0^- assignment which must give a $\cos^2$ distribution. Ascoli et al. (62), performing a Jackson analysis of the direction of ω alignment in the reaction $\pi^- p \rightarrow pB$, $B \rightarrow \pi\omega$ at 5 GeV/c, find that the only acceptable assignments for J^P are either 1^+ or 2^+, 3^-, 4^+ etc. In the latter case a fit is only obtainable if the B is produced fully aligned along the direction of the incident pion. It seems implausible that in a peripheral interaction a high-spin object can be aligned in this way. (For ω exchange the maximum value allowed for $|J_Z|$ is of course one.) Thus the assignment $J^{PC} = 1^{+-}$ is reasonably well established. Regardless of any model for meson states, the existence of the $I = 1$ B meson implies at least the existence of a 1^{+-} octet.

6.2 THE D MESON

The D meson ($M = 1285 \pm 4$ MeV, $\Gamma = 31 \pm 4$ MeV) is seen as a $K\overline{K}$ enhancement in incident-pion interactions and in antiproton annihilations (45). (Unfortunately the latter reaction is not seen at rest, so the initial state is not as well defined as for E production and this will explain the uncertainty in J^P assignment.) I spin and C parity are certainly 0 and $+1$ respectively. The I spin can be particularly well decided in the $\bar{p}$ experiments (63) where D can be sought in the reactions:

$$p\bar{p} \rightarrow (K_1^0 K^{\pm}\pi^{\mp})^0\omega$$
$$(K_1^0 K^{\pm}\pi^{\mp})^0\pi^0$$
$$(K^{\pm}K_1^0\pi^0)^{\pm}\pi^{\mp}$$
$$(K_1^0 K_1^0\pi^{\pm})^{\pm}\pi^{\mp}$$
$$(K_1^0 K_2^0\pi^{\pm})^{\pm}\pi^{\mp}$$
$$(K_1^0 K^{\pm}\pi^{\mp})^0\rho^0$$
$$(K_1^0 K_1^0\pi^0)^0\rho^0$$
$$(K_1^0 K_2^0\pi^0)^0\rho^0$$
$$(K_1^0 K_1^0\pi^{\pm})^{\pm}\rho^{\mp}$$
$$(K_1^0 K_2^0\pi^{\pm})^{\pm}\rho^{\mp}$$

since all of $p\bar{p} \to D\pi$, $D\rho$, and $D\omega$ are seen. This highly overdetermined situation makes the I-spin assignment certain. The $C = +1$ assignment follows from the observation of $K_1^0 K_1^0 \pi^0$ decay and from the fact that the $I = 1$ $K^{\pm} K_1^0$ subsystem peaks at low masses implying $l_{K\bar{K}} = 0$ when $G_{K\bar{K}} = -1$ and hence $G_D = C_D = +1$.

It is this $I = 1$ $K\bar{K}$ threshold enhancement (possibly associated with the δ meson) that prevents a unique J^P assignment. If we assume the $K\bar{K}$ system is s-wave dominated, the $K\bar{K}\pi$ D system must be of unnatural parity and

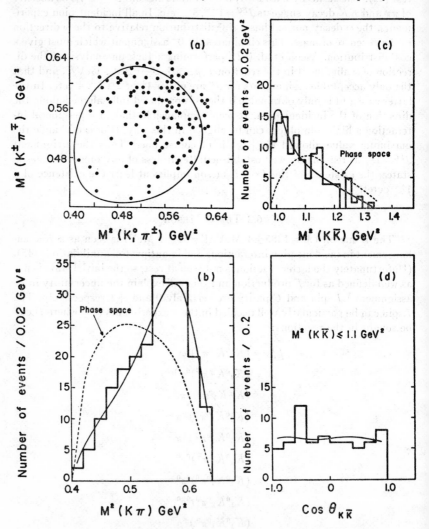

FIG. 20. Decay properties of D meson (63). (a) Dalitz plot.
(b) and (c) Dalitz projections. (d) Cos $\theta_{K\bar{K}}$ in the $K\bar{K}$ centre of mass.

this is confirmed by the fact that the Dalitz-plot density does not go to zero at the boundaries. If there were no final-state interactions, the l_π values of 0, 1, and 2 corresponding to the assignments 0^-, 1^+, 2^- would lead respectively to Dalitz-plot densities which were constant, proportional to $p_\pi{}^2$ and to $p_\pi{}^4$. Figure 20 shows the Dalitz plot, the $K\overline{K}$ system peaks at low masses. Since there is an inverse linear relationship between $M_{K\overline{K}}$ and p_π this peaking can be contributed to by the D-decay matrix element (if $J^P = 1^+$ or 2^-), a fundamental $K\overline{K}$ enhancement or the effect of K^* production (since we see that this region corresponds to the overlap of the two K^* bands). Depending on the relative amount of these contributions, any of the assignments 0^-, 1^+ 2^- is acceptable.

Thus J^P is unknown for the D meson. If it is indeed 1^+, it is, as yet, the only well-established isoscalar meson which could be used as one of the four expected from the quark model with this spin and parity.

6.3 THE H MESON

A second isoscalar state with $J^P = 1^+$, 2^- etc., the H meson, decaying into $\pi^+\pi^-\pi^0$, has been claimed in three experiments, one studying $\pi^+p\rightarrow\Delta^{++}$ $\pi^+\pi^-\pi^0$ at 4 GeV/c (64), the others $\pi^+d\rightarrow pp\pi^+\pi^-\pi^0$ at 3.65 and 3.3 GeV/c (65, 66). No H was seen in an experiment on π^+p interactions at five momenta between 2.95 and 4.1 GeV/c (22) nor in π^+d interactions at 5.1 GeV/c (67) and between threshold and 2.3 GeV/c (68). It has been suggested that the effect is due to ρ selection (22), but in fact for all three of the positive claims the enhancement was present before background was reduced in this way. This is illustrated in Figure 21 for the 4 GeV/c π^+p data. It has also been claimed that the effect is due to misfitting of $X^0\rightarrow\pi\pi\gamma$ (69). At least for the data of Figure 21, this does not seem to be true. The authors claim a cross section of 150 μb (70); Barbaro-Galtieri & Soding (69) find 70 ± 25 μb for the same data, by recalculating background to allow for ρ selection and peripheralism. At this energy the X^0 production determined from the $\pi\pi\eta$ mode seen in the same experiment is 71 ± 32 μb (71); with a 22 ± 3 per cent branching ratio for $\pi\pi\gamma$ this corresponds to a cross section of only 16 ± 7 μb for the $\pi\pi\gamma$ mode. Moreover the authors (70) refitted the events in the region of the enhancement to $\pi\pi\gamma$ and did not obtain a sharp X^0 peak. The situation on the H remains the same as for some time, that its observation in so few experiments is suspicious and that it cannot be regarded as an established state.

6.4 THE A_1 MESON

The A_1 is claimed as a $\pi\rho$ enhancement in the reaction $\pi^\pm p\rightarrow p\rho^0\pi^\pm$ with a mass of about 1070 MeV and width of about 80 MeV. The situation is somewhat confused since some authors claim to see the enhancement strongly (e.g. Figures 22 and 23), yet experiments differ sufficiently that a compilation of data shows very little enhancement, particularly for incident π^- (Figures 24, 25, 26). (Figure 26 shows that the often-made statement that the A_1 is never produced with charge exchange may not be correct.) Either a large

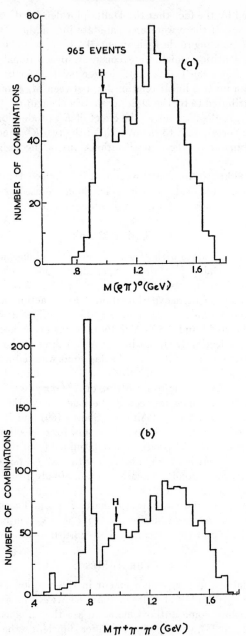

FIG. 21. Evidence for H meson in $\pi^+ p \rightarrow \Delta^{++}\pi^+\pi^-\pi^0$ at 4 GeV/c (64).
(a) $M(\rho\pi)$. (b) Unselected $M(\pi^+\pi^-\pi^0)$.

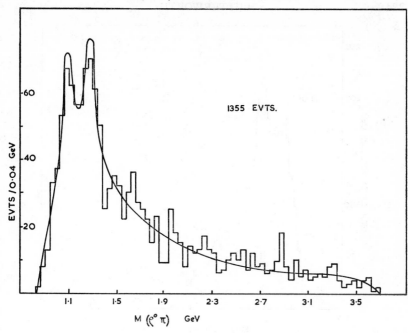

FIG. 22. $M(\pi^+\pi^-\pi^-)$ for the reaction $\pi^-p\to p\pi^+\pi^-\pi^-$ at 11 GeV/c studied by the *GHMS* collaboration.

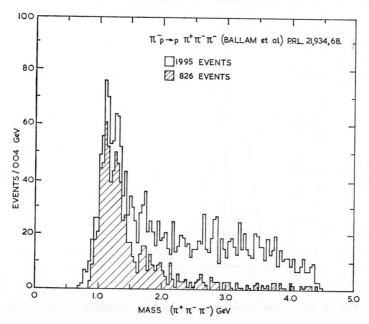

FIG. 23. $M(\pi^+\pi^-\pi^-)$ for the reaction $\pi^-p\to p\pi^+\pi^-\pi^-$ at 16 GeV/c. The shaded histogram shows events associated with ρ^0 and excluding Δ^{++}.

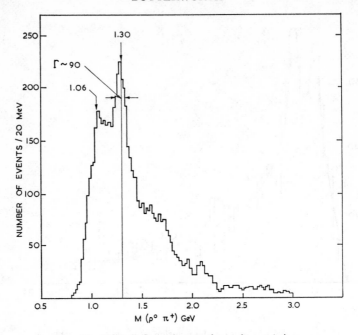

FIG. 24. $M(\rho^0\pi^+)$ from the reactions $\pi^+p \rightarrow p\rho^0\pi^+$ or $\pi^+n \rightarrow n\rho^0\pi^+$, compiled by French (59).

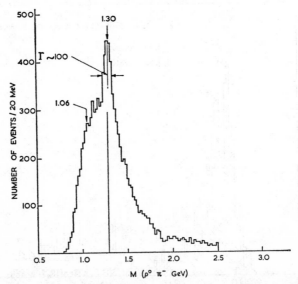

FIG. 25. $M(\rho^0\pi^-)$ from the reaction $\pi^-p \rightarrow p\rho^0\pi^-$, compiled by French (59).

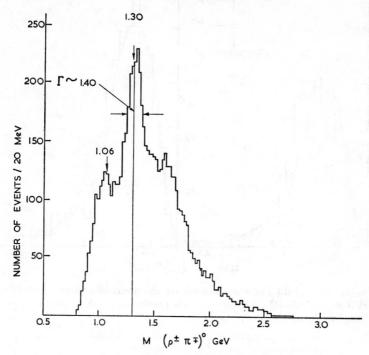

FIG. 26. $M(\rho^{\pm}\pi^{\mp})$ from the reactions $\pi^{-}n \rightarrow p\rho^{\pm}\pi^{\mp}$ and
$\pi^{-}p \rightarrow \Delta^{++}\rho^{\pm}\pi^{\mp}$, compiled by French (59).

number of experiments have been the subject of wild statistical fluctuations, or, more reasonably, a complex phenomenon is taking place.

There has been much discussion as to whether the enhancement is kinematic in origin. The effect proposed by Deck (72) (Figure 27a) in which the incident π dissociates into $\rho + \pi$ with the π scattering from the nucleon, even when extended to include the diagrams of Figures 27 b and c, has not proved able to produce a peak sufficiently narrow. However, by using Regge amplitudes (73) for exchange diagrams, the broad features of the threshold enhancement are explained. Thus Figure 28 shows data for $\pi^{-}p \rightarrow ^{0}\rho\pi^{-}p$ at 13

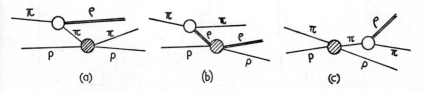

FIG. 27. Possible Deck diagram, for the reaction $\pi p \rightarrow p\pi\rho$.

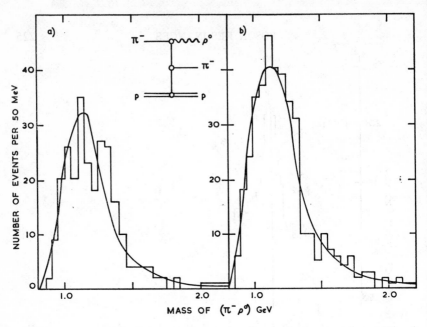

FIG. 28. (a) and (b) The $\pi^-\rho^0$ spectrum for the reactions $\pi^-p \to p\rho^0\pi^-$ at 13 and 20 GeV/c respectively (74). The curves are the predictions of the double Regge model normalized to the number of events.

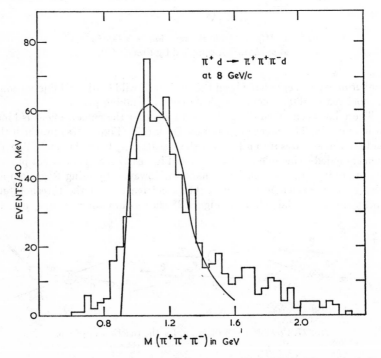

FIG. 29. $M(\pi^+\pi^+\pi^-)$ for the reaction $\pi^+d \to \pi^+\pi^+\pi^-d$ at 8 GeV/c. The curve is from a Reggeized Deck calculation normalized in the interval 1.0–1.2 GeV.

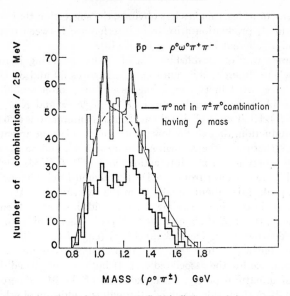

FIG. 30. $M(\rho\pi)$ in $p\bar{p}\to3\pi^+ 3\pi^- \pi^0$ at 3 GeV/c (79) containing noninterfering combinations of the $\pi^+\pi^-$ and $\pi^+\pi^-\pi^0$ in the ρ^0 and ω^0 regions respectively. The inner histogram is for events where the π^0 is not in a $\pi^{\pm}\pi^0$ combination having a mass in the $\rho^{\pm}$ region.

and 20 GeV/c (74) fitted by the double Regge diagram shown in the insert; Figure 29 shows data for $\pi^+d\to\pi^+\pi^+\pi^-d$ at 8 GeV/c (75). In each case the calculated curve is normalized to the data but absolute calculations are correct to 65 per cent and 60 per cent respectively. (It is not shown that all angular correlations are correctly predicted.) Moreover, the distribution of four-momentum transfer to the $\pi\rho$ system varies monotonically with $\pi\rho$ effective mass, the coefficient α in the relation $\partial N/\partial t \propto e^{-\alpha t}$ gradually falling with increase of mass, with no discontinuity at the A_1 mass (76).

Possibly one is witnessing an example of the duality of describing a system by t-channel or s-channel processes (77): the t-channel double Regge calculation gives the average behaviour of the cross section, but this is not incompatible with the diffraction production of resonances which would be observed as fine structure in the cross section. Certainly if the A_1 is assumed real, the sharp t distribution, the suppressed charge exchange cross section, and the observed decay angles suggest diffractive production. (It has also been argued that the A_1 production cross section is energy independent as expected from the Pomeron trajectory, but such an argument must be considered weak when the very existence of the A_1 is disputable.) The duality concept, attractive as it is, leaves experimental problems, for it means that the A_1 must be established by the study of fine structure, which is statistically difficult, and it makes difficult the question of background corrections. It also weakens, or at least makes less obvious, the suggestion that the observed variations in A_1 production arise from interference between resonance formation and a coherent Deck background (78).

The surest way of establishing the A_1 is by observing it in unrelated channels where different dynamics obtain. Figures 30 and 31 show A_1 enhancements reported in two experiments on antiproton annihilations (79, 80). Figure 32 shows A_1 production in $K^+p \rightarrow K^0p\ 3\pi$ and $K^+p \rightarrow K^0p\ 4\pi$ at 9 GeV/c (92c). High-multiplicity channels in π^-p interactions at 6 GeV/c (81) show A_1 production, though the low resolution of this experiment prevented a study of structure in the A_1 region. Figures 33, 34, and 35 show mass distribution from studies of π^-p interactions at 5 (82), 6.7 (83), and 16 GeV/c (84). (In Figure 33, data from the Deck and non-Deck channels are combined; the peaks arise about equally from the two samples.) We note that it is a feature of the distributions shown that in addition to the A_1 enhancement a second peak, the $A_{1.5}$, at about 1200 MeV is also seen. This second peak has also been reported in the Deck channel (85) (e.g. Figure 36).

One concludes therefore that the A_1 probably exists and that there is growing evidence for the associated $A_{1.5}$. If both are produced diffractively they are of unnatural parity, since exchange of the 0^+ Pomeron can only change the J^P of the incident particle through the addition of orbital angular momentum. Direct J^P determinations of the A_1 enhancement are compatible

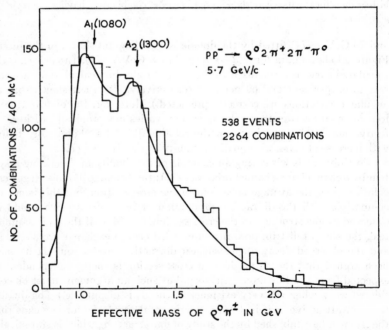

FIG. 31. $M(\rho^0\pi^\pm)$ from $p\bar{p} \rightarrow 3\pi^+\ 3\pi^-\ \pi^0$ at 5.7 GeV/c (80).

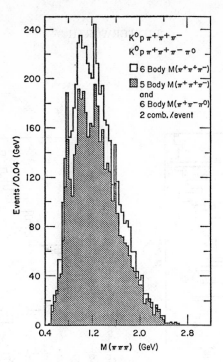

FIG. 32. Combined neutral and singly charged tripion effective masses for the reactions $K^+p \to K^0p \ \pi^+\pi^+\pi^-$ and $K^+p \to K^0p \ \pi^+\pi^+\pi^-\pi^0$ at 9 GeV/c (92c).

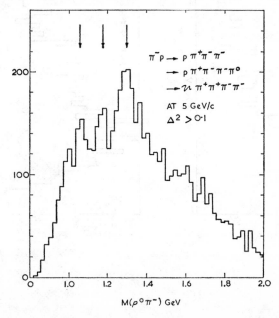

FIG. 33. $M(\rho^0\pi^-)$ from the reactions $\pi^-p \to p \ \pi^-\pi^+\pi^-$, $\pi^-p \to p\pi^-\pi^+\pi^-\pi^0$ and $\pi^-p \to n\pi^+\pi^+\pi^-\pi^-$ at 5 GeV/c (82).

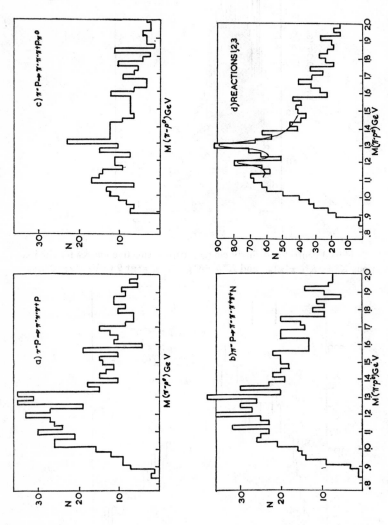

FIG. 34. $M(\pi^-\rho^0)$ from the reactions: (a) $\pi^-p \to p\pi^-\pi^+\pi^-$. (b) $\pi^-p \to n\pi^-\pi^-\pi^+\pi^+$. (c) $\pi^-p \to p\pi^-\pi^+\pi^-\pi^0$. (d) Combinations of (a), (b), and (c). The data are obtained from a study of π^-p interactions at 6.7 GeV/c (83).

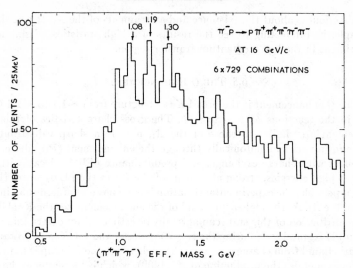

FIG. 35. $M(\pi^+\pi^-\pi^-)$ from $\pi^-p\rightarrow p\pi^+\pi^+\pi^-\pi^-\pi^-$ at 16 GeV/c (84).

with the assignment 1^+ although 2^- cannot be excluded (86, 87). The non-isotropy of the $\pi\rho$ decay indicates for the 1^+ assignment the presence of a significant d-wave contribution to the decay (86). If the $A_{1.5}$ is real and of the same J^P as the A_1 some of the variations observed from experiment to experiment could be explained by their interference.

More experimental work is needed to establish the $A_{1.5}$ and to remove re-

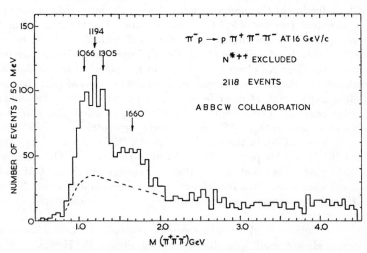

FIG. 36. $M(\pi^+\pi^-\pi^-)$ from the reaction $\pi^-p\rightarrow p\pi^+\pi^-\pi^-$ at 16 GeV/c when events with the $(p\pi^+)$ system in the N^* band are removed (84).

maining doubts about the existence and assignments of the A_1. It would, for example, be interesting to see the results of a high statistics missing-mass experiment in the low momentum-transfer region.

6.5 THE Q ENHANCEMENT

The Q enhancement is the broad $K\pi\pi$ structure from $\sim$1.1 to $\sim$1.4 GeV seen in the reactions $K^{\pm}p\rightarrow(K\pi\pi)^{\pm}p$. The gross characteristics of this enhancement are similar to those of the A_1, namely a sharp t distribution, whose form varies monotonically through the enhancement (76), absence of the effect with charge exchange, and predominance of 0^-1^- decay with, at least at high energies, strong alignment of the K^* in the $|1, 0\rangle$ state. However, if possible, the experimental situation is even more confused than in the case of the A_1. At the higher-mass end of the enhancement lies the $K^*(1400)$. The contribution of this state can normally be estimated by determining the amount of its production using the two-body decay and the $K\pi\pi/K\pi$ branching ratio found from charge exchange channels where the Q bump is not seen. At higher energies the production of $K^*(1400)$, which is by meson exchange, falls sufficiently that this correction is not large. Many experimenters find no evidence for structure within the Q enhancement and treating it as a single phenomenon with $\Gamma\sim$200 MeV, find 1^+ the only acceptable J^P assignment (88).

The broad enhancement does not have a simple Breit-Wigner form and is probably again explicable in gross by double Regge exchange processes. Figure 37 shows the consequence of such a calculation for $K^-p\rightarrow K^{*0}\pi^-p$ at 12.6 GeV/c (88g). The only double Regge diagram considered is that involving K^* production at the meson vertex and π^- at the intermediate vertex. A four-momentum transfer of <1 GeV^2 was demanded at the K to K^* and p to p vertices. Shown are the fits to the mass distributions, the K to $K\pi\pi$ momentum-transfer distribution, and the Treiman-Yang distribution (correlation of the K to K^* plane with the p_{in} to p_{out} plane). The calculation is normalized, but the calculated cross section is 80 per cent of that observed.

Before a discussion of possible resonant structure in the enhancement, the rather confusing experimental facts will be assembled. One should remember that many authors (88) fail to observe structure.

(*a*) Shen et al. (89), in the reaction $K^+p\rightarrow(K^{*0}\pi^+)p$ at 4.6 GeV/c, found an (80 ± 20) MeV wide peak at 1320 ± 10 MeV (Figure 38). The effect was enhanced by selecting equatorial K^* decays ($|\cos\vartheta_{KK}|$ in the K^* CM <0.8). This is *not* what is expected for diffractive production of a 1^+ state. No certain J^P assignment was possible, but the $K\pi/K^*\pi$ branching ratio was $<15\pm15$ per cent.

(*b*) Bassompierre et al. (90), in the same reaction at 5 GeV/c, obtained the structureless overall $K^*\pi$ distribution of Figure 39. However, when events were divided into those with equatorial and those with polar decays (Figure 39 upper histograms) (always removing events with $\Delta^2<0.1$ GeV^2), three peaks were claimed at 1230 ± 15, 1280 ± 10, and 1310 ± 15 MeV with

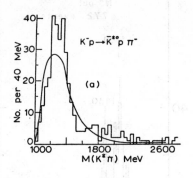

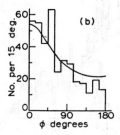

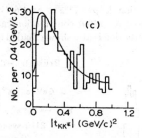

Fɪɢ. 37. The reaction $K^- p \to \overline{K}^{*0} p \pi^-$ at 12.6 GeV/c. (a) $M(K^* \pi)$. (b) Treiman-Yang angle (correlation of $K - K^*$ and p_{in}-p_{out} planes). (c) Four-momentum transfer between K and K^*. The curves are the results of a double Regge-model calculation normalized to the data.

widths 80 ± 20, 60 ± 20, and 60 ± 20 MeV respectively. The two outer peaks appeared upon polar selection and so it was suggested that these were diffractively produced unnatural-parity states; the 1280 peak appeared together with $K^*(1400)$ upon equatorial selection as had the 1320 peak of Shen et al.

However at 5.5 GeV/c the same selections by the Hopkins group (91) (Figure 39b lower histograms) gave peaks in completely different places.

The above three experiments are very close in E^*. If the differences reported are due to real interference effects, these are strongly E^* dependent and compilation of data from different experiments is virtually impossible.

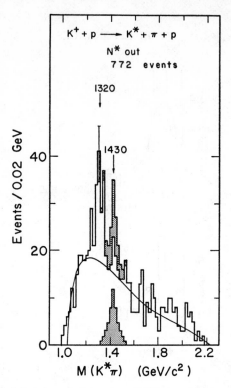

FIG. 38. $M(K^*\pi)$ for the reaction $K^+p \to K^{*0}\pi^+p$ at 4.6 GeV/c (89). The small inner histogram shows the expected $K^*(1400)$ contribution calculated from the observed two-body decay. This contribution is also shown shaded in the main histogram.

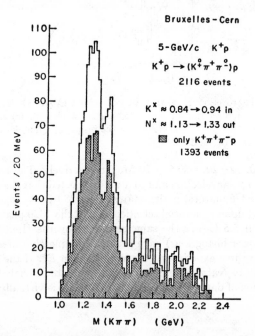

Fig. 39a

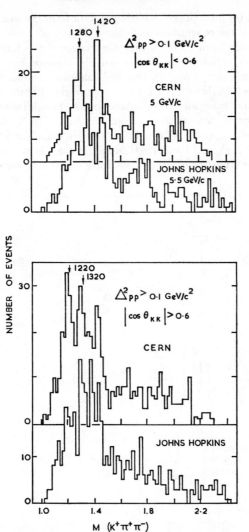

Fig. 39b

M (K⁺π⁺π⁻)

FIG. 39. (a) Unselected $K^*\pi$ spectrum from the reaction $K^+p \rightarrow K^+\pi^+\pi^-p$ at 5 GeV/c (90). (b) (Upper histograms) the same data requiring $\Delta^2 > 0.1$ (GeV/c)² and dividing into equatorial, $|\cos\theta_{KK}| < 0.6$, and polar, $|\cos\theta_{KK}| > 0.6$, decays. (Lower histograms) the corresponding data at 5.5 GeV/c (91).

(c) In the same reaction at 9 GeV/c, Goldhaber et al. (92a) found in addition to the $K^*(1400)$ two peaks at 1250 ± 10 MeV ($\Gamma = 50 \pm 20$ MeV) and 1360 ± 10 MeV ($\Gamma = 80 \pm 20$ MeV). The decay $K^*(890)$ were aligned in the $|1, 0\rangle$ state as expected for the diffractive production, and the 1^+ assignment was preferred for both peaks. Addition of data from the same reaction at 10 GeV/c (93) left the conclusions unchanged (Figure 40).

Turning away from the diffractive dissociation channel, relevant data are:

(*a*) In the reaction $p\bar{p}\to K\bar{K}\pi\pi$ at rest, the C meson has been reported as an $I=\frac{1}{2}$ $K\pi\pi$ enhancement (94). Detailed analysis is reported by Astier et al. (95). In $p\bar{p}\to K_1^0 K_1^0 \pi^+\pi^-$ the C is seen strongly (Figure 41a). Maximum likelihood fitting using all mass combinations and angles simultaneously and assuming channel contribution from $K^*\bar{K}\pi$, $K^*\bar{K}^*$, $K\bar{K}\rho$, $(K\pi)_{s\text{-wave}}$ $(K\pi)_{s\text{-wave}}$, and $C\bar{K}$ (with $K^*\pi$ and $K\rho$ decay) demands a 1^+ C meson with $M=1242_{-10}^{+9}$ MeV and $\Gamma=127_{-25}^{+7}$ MeV. (If the C is of unnatural parity it can only come from the $J^{PC}=1^{--}$ $p\bar{p}$ state and 0^-, 2^- are explicitly elimi-

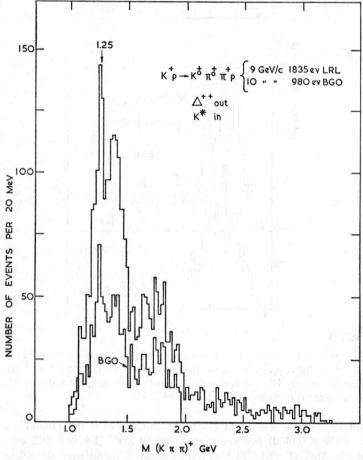

FIG. 40. $M(K^*\pi)$ from $K^+p\to K^{+0}\pi^{-0}\pi^+\pi^-$ at 9 and 10 GeV/c (92a, 93).

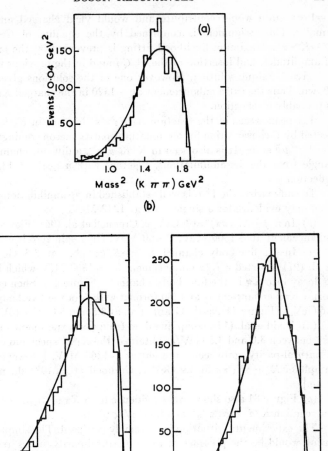

FIG. 41. The reaction $p\bar{p} \rightarrow K\bar{K}\pi\pi$ at rest (95). (a) Effective mass squared of $K_1^0\pi^+\pi^-$ from $p\bar{p} \rightarrow K_1^0K_1^0\pi^+\pi^-$. (b) The charged and neutral $(K\pi\pi)$ effective mass squared from $p\bar{p} \rightarrow K^0K^\pm\pi_\mp\pi^0$. The curves correspond to the introduction of a single resonance at $M = 1242$ MeV and $\Gamma = 127$ MeV.

nated; for natural parity, addition of the 0^{-+} $p\bar{p}$ state introduces extra amplitudes and only a separate fit to masses and angles is performed, but the fit is much worse than for the 1^+ assignment.) Inclusion of a second C at 1320 (suggested in earlier data) does not improve the fit.

In $p\bar{p} \rightarrow K_1^0K^\pm\pi^\mp\pi^0$ both charged and neutral C is seen (Figure 41b). The neutral mode is seen more strongly showing $I_C = \frac{1}{2}$, since for $I = \frac{3}{2}$ only the

$I=1$ $p\bar{p}$ state would contribute and would yield charged and neutral C equally. This assignment is confirmed by the equality of $C^0{\rightarrow}K^{*0}\pi^0$ and $C^0{\rightarrow}K^{*\pm}\pi^{\mp}$. Maximum likelihood fitting is prevented by the large number of amplitudes, and insertion of the 1^+ C found in the previous channel does not give a unique solution. However one of the solutions gives the curves shown. Thus the extra enhancement at $\sim$1320 in the charged $K\pi\pi$ spectrum is probably a reflection.

The peak is seen in the reaction $K_1^0 K_2^0 \pi^0 \pi^0$ but not in $K_1^0 K_1^0 \pi^0 \pi^0$ as expected by C conservation for an unnatural-parity meson coming purely from the 1^{--} $p\bar{p}$ state. It is also seen in $K^+ K^- \pi^+ \pi^-$, a difficult channel to disentangle from the 4π annihilations, together with a peak at 1320 due to ϕ reflection.

To summarize, the 1^+ C is well established in $\bar{p}p$ annihilations but there is then only evidence for a single state at 1242 MeV.

(*b*) In $\pi^- p{\rightarrow}\Lambda(K\pi\pi)^0$ at 6 GeV/c, Crennell et al. (96) (Figure 42) find an enhancement at $\sim$1300 MeV, $\Gamma\sim$60 MeV. Only spin zero is excluded.

(*c*) In the five-body channel $K^+ p{\rightarrow}K^0 p\pi^+ \pi^+ \pi^-$ at 3.5 GeV/c, Bishop et al. (97) reported a $K^*\pi$ enhancement at $\sim$1300 MeV which appears also to decay into $K\pi$ in the four-body channels (Figure 43). Shen et al. (89) report a $K\pi\pi$ enhancement at 1280 formed with Δ in the five-body channel at 4.6 GeV/c (Figure 44), as do Grard et al. at 3.5 and 5 GeV/c (8).

(*d*) Dodd et al. (12), compiling data from four experiments on $K^+ p{\rightarrow}K_1^0 p\pi^+$ between 3.0 and 3.5 GeV/c, obtained the $K_1^0 \pi^+$ spectrum of Figure 45. A natural-parity state seems present at $\sim$1260 MeV. [However, in a large sample of $K^- n{\rightarrow}K_1^0 \pi n$ at 3.9 GeV/c, Crennell et al. (98) do not find this state.]

(*e*) Figure 81 also shows weak evidence for a $K\pi$ state at $\sim$1280 MeV in the reaction $K^- d{\rightarrow}dK^- \pi^+ \pi^-$ at 12.6 GeV/c (88g).

The experimental situation is extremely confused. The simplest interpretation would be the presence of two unnatural-parity states at $\sim$1250 and $\sim$1340 MeV and a natural-parity state at about 1280–1300 MeV. The lower of the unnatural-parity states (C-meson) would be 1^+, the upper undetermined; the natural-parity state would be preferably 2^+. Such a firm conclusion is not justified from the present data. Much more information is required; it is particularly important to confirm the nature of the mesons reported in nondiffractive channels. All that seems definite is that there is one 1^+ state in the C meson. If there are two 1^+ mesons in the same mass region they will interfere and variations in the observed spectrum from experiment to experiment are to be expected. Goldhaber (92b) has indicated how drastic such variations could be.

It has been pointed out above that strange mesons from the 1^{+-} and 1^{++} octets can mix. Indeed, since there is a single octet-octet coupling to the $SU(3)$ singlet, if two K^* states are seen to be produced diffractively, i.e., by the exchange of the singlet Pomeron, then this already implies that they are $SU(3)$ mixed. If they were not mixed, the octet from which they derive could be determined by checking whether the amplitudes of $K^*\pi$ and $K\rho$ decay are

Crennell et al., Brookhaven

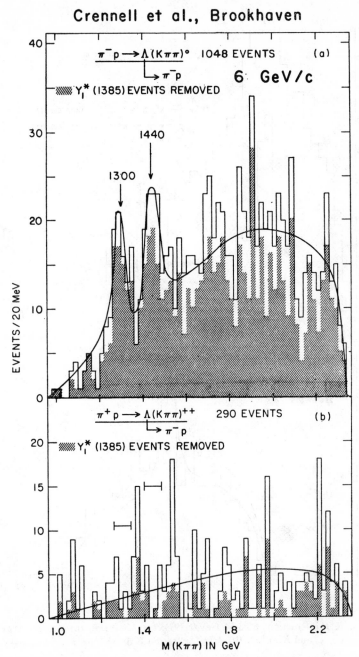

FIG. 42. (a) $M(K\pi\pi)^0$ for $\pi^-p \to \Lambda(K\pi\pi)^0$ at 6 GeV/c. (b) $M(K\pi\pi)^{++}$ for $\pi^+p \to \Lambda(K\pi\pi)^{++}$ at 6 GeV/c (96).

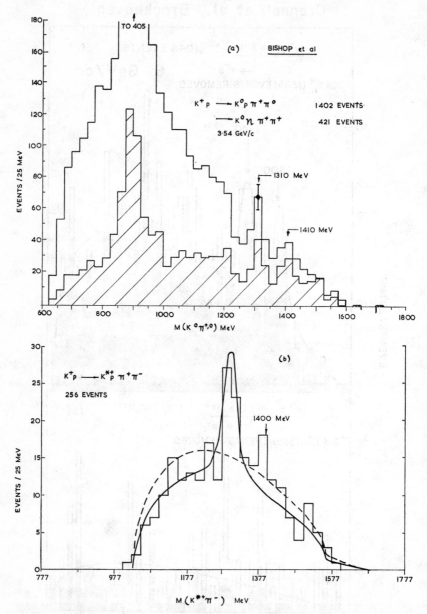

FIG. 43. K^+p interactions at 3.54 GeV/c (97). (a) $M(K^0\pi^{+,0})$ from $K^+p \rightarrow K^0p\pi^+\pi^0$ and $K^0n\pi^+\pi^+$. (b) $M(K^{*+}\pi^-)$ from $K^+p \rightarrow K^{*+}p\pi^+\pi^-$.

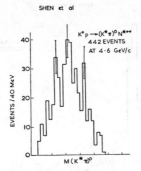

FIG. 44. $M(K\pi\pi)$ in the reaction $K^+p \to \Delta^{++}(K^*\pi)^0$ at 4.6 GeV/c (89).

constructive or destructive. This follows from the argument of Section 1.3.2 that the decay of a K^* from the 1^{+-} octet into 1^{--} and 0^{-+} mesons can only involve D coupling for which $K^*\pi$ and $K\rho$ amplitudes are in antiphase, whereas the decay of a 1^{++} member involves only F coupling for which they are in phase. Using this method Astier et al. (95) find the C meson more compatible with 1^{++}. Bomse et al. (88f) find that the lower-mass half of the Q enhancement is consistent with 1^{++} and not with 1^{+-}, whereas the upper half is consistent with 1^{+-} but gives a poor fit to 1^{++} [Chien et al. (88c), however, find the entire Q enhancement fits 1^{++}].

Further comments on the Q enhancement are made in Section 10.9.

6.6 THE 1^{++} AND 1^{+-} NONETS

Even if all states are identified it is not possible to deduce the parameters m_1, m_8, ϑ, and Δ for the two nonets together with the internonet mixing angle ϕ, since this represents nine unknowns to be found from eight physical masses.

Let it be assumed that the A_1 and B are the $I = 1$ members of 1^{++} and 1^{+-} nonets. Then, since Δ is approximately the same for the spin-singlet and spin-triplet $q\bar{q}$ states for $l_{q\bar{q}} = 0$, one might expect that the K^*'s from these two $l = 1$ nonets would, if unmixed, have squared masses higher by $\sim.34$ GeV^2, the Δ value found for the 2^{++} nonet. This predicts masses of ~ 1220 and 1340 MeV, suprisingly close to the masses of possible 1^+ K^* mesons. This suggests that if the K^* states are indeed confirmed at these masses, their $SU(3)$ mixing is small. (Such mixing would cause mass shifts completely analogous to octet-singlet mixing—i.e. would lead to increased separation in their masses.)

Such $SU(3)$ checks of $K^*\pi$ and $K\rho$ decay as have been done tend to support this picture.

The identification of the $I = 1$ and $I = \frac{1}{2}$ states is thus encouraging. The situation for the $I = 0$ states, however, is most unsatisfactory since only the D meson (1^{++}) looks in the least convincing. If one had unmixed octets, then

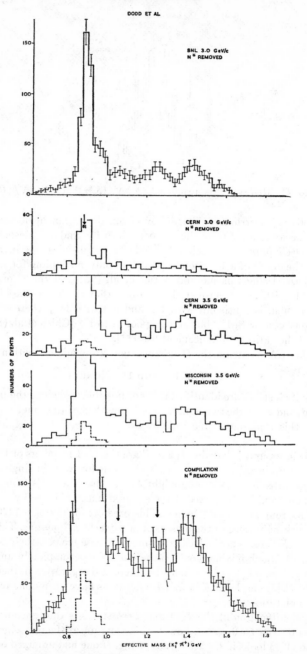

FIG. 45. The separate and combined $K_1^0\pi^+$ spectra for four experiments studying the reaction $K^+p\to K_1^0\pi^+p$. In each case events where the $p\pi^+$ combination is in the N^* region have been removed (12).

the Gell-Mann–Okubo mass formula would predict $I=0$ states at ~1300 MeV (1^{++}) and ~1380 MeV. (1^{+-}). The D thus lies close to the Gell-Mann–Okubo mass even though we might expect to have near-ideal mixing in this $S_{q\bar{q}}=1$ state (as for the 1^{--} and 2^{++} nonets). The as yet unfound D' could have $K\bar{K}\pi$. $\pi\pi\eta$, or 4π decay modes. For the 1^{+-} states only the extremely doubtful H meson is known.

7. SCALAR MESONS

On the quark model we expect a nonet of $J^{PC}=0^{++}$ states from $l_{q\bar{q}}=1$.

7.1 $I=1$ SCALAR CANDIDATE, THE δ MESON

The allowed strong decays of an $I\ J^{PC}=10^{++}$ state are $K\bar{K}$ and $\pi\eta$. With these couplings we would not expect such a state to be produced with a large cross section in incident-pion collisions. However, a threshold s-wave enhancement of the $K\bar{K}$ system has been reported in $p\bar{p}$ annihilations. Thus in the reaction $p\bar{p}\rightarrow K_1^0 K^{\pm}\pi^{\mp}$ studied at rest by Astier et al. (99), the threshold enhancement of Figure 46 is seen, which cannot be fitted by a coherent addition of amplitudes for production of K^*, A_2, or $K\pi$ in the s wave (the spread of theoretical curves in Figure 46 shows the effect of varying the $K\pi$ scattering length between -0.4 and $+0.4$ F). The state is so near to threshold that it is surely s-wave. (Indeed the $K\bar{K}$ decay angle with respect to the bachelor π is isotropic.) This is not, however, entirely conclusive since if J^{PC} ($K\bar{K}$) $=1^{--}$, the particle would come from the $^3S_1$ and $^1S_0$ $p\bar{p}$ states with the π in a p wave with respect to $K\bar{K}$. This would lead respectively to $\sin^2\vartheta$ and $\cos^2\vartheta$ distribution, and a mixture could be isotropic. Detailed fitting does not prove that the enhancement is a resonance. Good fits are obtained to a BW at 1016 ± 6 MeV, i.e. above threshold, or to a positive real scattering length of $2.0_{-0.5}^{+1.0}$ F, or

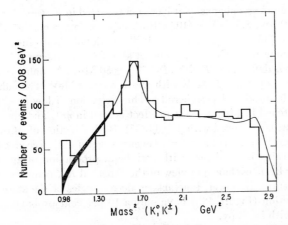

FIG. 46. $M(K_1^0 K^{\pm})$ for the reaction $p\bar{p}\rightarrow K_1^0 K^{\pm}\pi^{\mp}$ at rest (99). The curve shows a fit allowing for coherent production of K^*, A_2, and $K\pi$ in the s wave (the spread in the curve in the low-mass region shows the effect of varying the $K\pi$ scattering length between -0.4 and $+0.4$ F).

to a complex "bound state" scattering length with an imaginary part of $0.0_{-0.0}^{+0.5}$ F and a real part of $-2.3_{-4.0}^{+0.3}$ F. This last possibility is particularly interesting since it would correspond to a narrow bound state of mass 975_{-10}^{+15} MeV.

If the last assumption were correct, it would be tempting to identify the state with the $I = 1$ or 2 δ^- reported as a narrow resonance at 962.0 ± 5.0 MeV with $\Gamma \leq 5$ MeV by Kienzle et al. in a missing-mass study of $\pi^- p \rightarrow p M^-$ at 3–5 GeV/c (100). Such identification would make $J^{PC} = 0^{++}$ and would suggest a predominantly $\eta\pi$ decay.

The $D \rightarrow K \overline{K} \pi$ decay shows the $I = 1$ $K \overline{K}$ enhancement referred to above (it is this enhancement that makes it difficult to find the spin and parity of the D). Hence, if the above identification is correct we might expect to find $\eta\pi\pi$ decay of the D with the $\eta\pi$ system resonating at the δ mass. Defoix et al. (101) have reported exactly this effect. In $\bar{p}$ annihilations at 1.2 GeV/c, where D production is known to take place, 3500 events of the type $p\bar{p} \rightarrow 3\pi^+ 3\pi^- \pi^0$ were studied. The $\pi^+\pi^-\pi^0$ spectrum suggests η production Figure 47a, though the large number of combinations prevent the observation of a clear η signal. However, if $\pi^+\pi^-\pi^0$ combinations in the η region ("η" = 528–568 MeV) are chosen, then the "η" $\pi^\pm$ and "η" $\pi^+\pi^-$ plots of Figures 47b and c are obtained. The presence of X^0 in the latter distribution confirms the presence of η. One notes that $\eta\pi^\pm$ shows a peak at ~970 MeV (i.e. the δ mass), and $\eta\pi^+\pi^-$ at 1310 MeV (i.e. close to the D mass, though a little higher than the standard mass). Selection of the $\pi^+\pi^-\pi^0$ outside the η region attenuates these peaks. To prove that these peaks are correlated, Figure 47 shows the effect of selecting events so as to combine $\delta^\pm \pi^\mp$ rather than any $\eta \pi^+\pi^-$ combination. The D enhancement is more evident. Fits to the data yield $M(\delta) = 975$ MeV, $\Gamma(\delta) \leq 25$ MeV, $M(D) = 1310$ MeV, $\Gamma(D) = 40$ MeV, and $(D \rightarrow K \overline{K} \pi)$ $/(D \rightarrow \eta\pi\pi) = 0.124 \pm 0.035$.

Moreover an $\eta\pi$ enhancement has been reported in $K^- p \rightarrow \Lambda \pi^+ \pi^- M$ at 4.6 GeV/c (102) and 5.5 GeV/c (103). Both experiments report $\pi^\pm M$ enhancements at the δ mass but disagree markedly on details. Thus in a seven-event /μb experiment at 5.5 GeV/c, Ammar et al. (103) find the $\eta\pi^-$ plot of Figure 48a where η is defined as M with $M = M(\eta) \pm 50$ MeV. An enhancement of 40 events is seen at 980 ± 10 MeV with width 80 ± 30 MeV. The enhancement appears to be produced peripherally. Thus selection of low t reduces background (hatched in Figure 48a), no effect is seen in $\eta\pi^+$, and about half the enhancement is produced with Y^{*+} (1385). Both selection of M out of the η region and examination of $\eta \rightarrow \pi^+\pi^-\pi^0$ events suggest exclusive $\eta\pi$ association. Thus everything is consistent with peripheral production of an $\eta\pi$ system, presumably by K exchange in view of the expected $K \overline{K}$ coupling of the suggested 0^{++} state. However, the Jackson decay angles of the state assuming such production (Figure 48b) have only a 10 per cent probability of being consistent with isotropy.

Figure 49 shows the results obtained by Barnes et al. (102) for the same

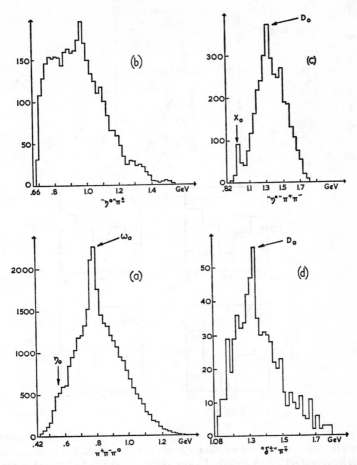

FIG. 47. Evidence for δ production in $p\bar{p} \rightarrow 3\pi^+ 3\pi^- \pi^0$ at 1.2 GeV/c (101). (a) $M(\pi^+\pi^-\pi^0)$. (b) and (c) "η" $\pi^\pm$ and "η" $\pi^+\pi^-$ spectra where "η" denotes a $(3\pi)^0$ system with $528 < M(3\pi) < 568$ MeV. (d) "$\delta^\pm$" $\pi^\mp$ distribution where "δ" denotes an "η"π system with $955 < M < 995$ MeV.

reaction (3.5 events/μb) at 4.6 GeV/c. The $M\pi^-$ distribution (not now η selected) is shown. A peak at the δ mass is seen, but in detail the results disagree with those of Ammar et al. Thus though $M\pi^-$ shows a peak and not $M\pi^+$, there seems no other evidence for peripheralism in the 4.6 GeV/c data. Selection of low momentum transfer does not enhance the signal, and Y^{*+} is *antiselected* together with Y^{*-} to reduce background in histograms B and C of Figure 49. Moreover these authors claim that not more than 30 per cent of the events are η^0 associated.

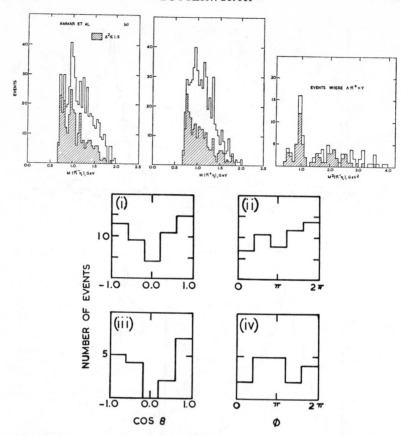

Fig. 48. (a) $\pi^-\eta$ and $\pi^+\eta$ mass distributions for $K^-p \rightarrow \Lambda\pi^- M$, at 5.5 GeV/$c$ (103) where η denotes a missing mass within ± 50 MeV of the η mass. The hatched distribution is for $\Delta^2(\pi\eta) \lesssim 1.5 (\text{GeV}/c)^2$. (b) Jackson decay angles θ and ϕ for the δ region (940 to 1040 MeV). The events have the peripheral selection $\Delta \lesssim 1.5$ (GeV/c)^2. The distributions (iii) and (iv) differ from those of (i) and (ii) in that the extra requirement that the $\Lambda\pi^+$ mass be between 1335 and 1435 MeV is applied.

The difficulty in reconciling these data lies in the observed widths of the enhancement. The $\eta\pi$ and $K\overline{K}$ coupling object seen in $\overline{p}$ annihilations would be compatible with the missing-mass δ or the $\eta\pi$ object produced by incident $\overline{K}$, but it is hard to see how the latter can be made compatible with each other. (This difficulty of correlating missing-mass spectrometer widths with those of objects of known decay modes will reappear in the high-mass region.) The simplest interpretation would be the existence of a strongly decaying $IJ^{PC} = 10^{++}$ meson with $K\overline{K}$ and $\pi\eta$ coupling but which is different from the δ. However, more data are needed.

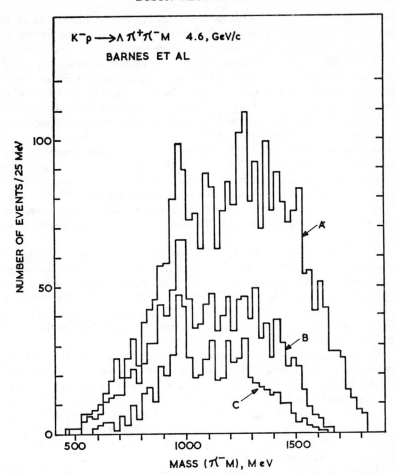

FIG. 49. π^-M spectrum for the reaction $K^-p \rightarrow \pi^+\pi^-M$ at 4.6 GeV/c (102) with $M^2 > 0.05$ GeV^2. (a) 2736 events X^0 removed. (b) 1295 events X^0 removed and $\Lambda\pi^+$ and $\Lambda\pi^- > 1430$ MeV. (c) 648 events X^0 removed, $\Lambda\pi^+$ and $\Lambda\pi^- > 1430$ and $M^2 < 0.9$ GeV^2, $E_{\text{lab}}(\pi^+) < 1$ GeV.

Less convincing claims for a narrow isovector object decaying into 3π with a somewhat lower mass than the δ have been made by Juhala et al. (104) studying $K^-p \rightarrow K^-\pi^-\pi^+\pi^0 p$ and $K^-\pi^-\pi^+\pi^+ n$ at 4.6 and 5 GeV and by Allison et al. (105) studying $K^-p \rightarrow \Sigma 4\pi$ at 6 GeV/c. There are unconvincing claims in incident-pion experiments of an $\pi\eta$ object at ~ 1020 MeV (106, 107).

7.2 $I = \frac{1}{2}$ SCALAR CANDIDATE $K\pi$ (1100)

A scalar strange meson can be expected to decay mainly to $K\pi$. Trippe et al. (108) have sought to investigate $K\pi$ scattering by studying the reactions $K^+p \to K^0\pi^0\Delta^{++}$ and $K^+p \to K^+\pi^-\Delta^{++}$ at 7.3 GeV/c. Figure 50a shows the energy dependence of the spherical harmonic moments of the $K\pi$ angular distribution relative to the incident K^+. Since the $K^+\pi^-$ and $K^0\pi^0$ moments are compatible it can be assumed that the $I = \frac{3}{2}$ interaction is not important. Since the dominant feature of the low-energy region is the existence of the p-wave $K^*(890)$, the fact that $\langle Y_1^0 \rangle$ goes to zero at just below 1 GeV when the p-wave phase is $\sim 150^0$ suggests that the s-wave has a phase of $\sim 60^0$ here and the fact that Y_1^0 becomes positive again at higher energies suggests that δ_s is rising. Thus one concludes that the s-wave $K\pi$ system is at least approaching resonance in this region. On the assumption that the only contributions to $K\pi$ scattering are the $K^*(890)$, the $K^*(1400)$, and the s wave, the s-wave cross section shown inset in Figure 50 is obtained, using the Durr-Pilkuhn form factors for off-mass-shell correction. It appears that the s-wave cross section reaches the unitarity limit. These two pieces of evidence suggest the presence of a broad ($\Gamma \sim 400$ MeV) $K\pi$ scalar resonance at ~ 1.1 GeV.

There have also been claims for narrow $K\pi$ states in this region. In the reaction $K^+p \to K^0\pi^+p$ at 3.5 GeV/c, de Baere et al. (109) reported an enhancement at 1080 MeV, seen only in those events where the π^+ was also associated with the p to form an N^*. Dodd et al (12), compiling data from this and other experiments at 3 and 3.5 GeV/c, obtained the histograms of Figure 45. The compilation showed the enhancement at ~ 1080 MeV for the overall sample on *removing* N^*.

On the other hand, in the reaction $K^-d \to K_1^0 \pi^- n p$ at 3.9 GeV/c, Crennell et al. (98) failed to see this enhancement but saw one at 1160 ± 10 MeV, $\Gamma = 90 \pm 30$ MeV (Figure 51). The authors point out that they see strong interference with the $N^*(1238)$ and $Y^*(1815)$ so that it is conceivable that interference effects in this and in the K^+ experiment alter the apparent mass of the state. They also point out, however, that the effect is at the same mass in the N^* and Y^* bands which would be a very fortuitous coincidence if its mass were being affected by interference.

In none of the experiments claiming to see a narrow state is a direct J^P determination possible.

Clearly more experimental data is necessary to clarify the situation. The existence of a shoulder to the $K^*(890)$ has been a feature of several experiments and there is little doubt that the s wave is large here. The existence of a resonance and certainly its width cannot however be regarded as fully established by the present data.

7.3 $I = 0$ SCALAR CANDIDATES

7.3.1 *The* $\epsilon^0(700)$ *meson.*—The marked asymmetry in the decay of the ρ^0 meson has indicated for some time the existence of a large s-wave contribution to $\pi\pi$ scattering in this mass region, and there has been considerable study of the reaction $\pi^-p \to n\pi^+\pi^-$ in order to abstract the $I = 0$ $\pi\pi$ phase

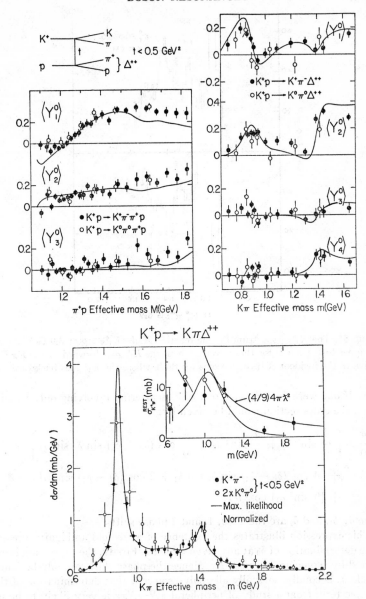

FIG. 50. The study of $K\pi$ scattering in the reaction $K^+p\to K^0\pi^0\Delta^{++}$ and $K^+p\to K^+\pi^-\Delta^{++}$ at 7.3 GeV/c (108). (a) Energy dependence of the spherical harmonic moments of the $K\pi$ angular distribution relative to the incident K^+. (b) One-pion exchange fit to the experimental data and (inset) the s-wave $K\pi$ cross section.

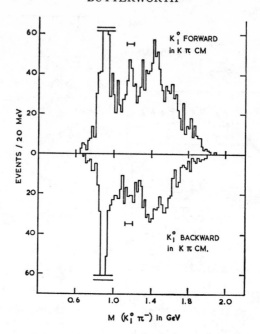

FIG. 51. The $K_1{}^0\pi^-$ spectrum for the reaction $K^-d \to K_1{}^0\pi^-np$ at 3.9 GeV/c (98). The upper histogram is for those events where the $K_1{}^0$ goes forward in the $K\pi$ CM relative to the incident K^-; the lower is for those where the $K_1{}^0$ goes backward.

shifts. If one were dealing with real $\pi^+\pi^-$ scattering involving only s and p waves, the cross section would be given by:

$$\frac{\partial\sigma}{\partial\Omega} = \left[\frac{4}{9}\sin^2\delta_0 + \frac{1}{9}\sin^2\delta_2 + \frac{4}{9}\cos(\delta_0 - \delta_2)\sin\delta_0\sin\delta_2\right]$$
$$+ \left[4\cos(\delta_0 - \delta_1)\sin\delta_0\sin\delta_1 + 2\cos(\delta_2 - \delta_1)\sin\delta_2\right]\cos\vartheta$$
$$+ \left[9\sin^2\delta_1\right]\cos^2\vartheta$$

where δ_0, δ_1, and δ_2 are the $I = 0$, 1, and 2 phase shifts.

This expression illustrates the existence of a twofold ambiguity present in the determination of δ_0 at any given $\pi\pi$ mass. Firstly, the expression is unaffected by replacing δ_0 by $\delta_0 - n\pi$. Hence the phase shift can only be found modulo π. Secondly, virtually all analyses accept that determination of the isotropic term from a study of peripheral $\pi N \to N\pi\pi$ is very likely to be influenced by absorptive corrections. Thus determination of δ_0 is made using the term linear in $\cos\vartheta$ and assuming δ_1 is dominated by the ρ. This term is ambiguous to the interchange $\delta_0 \to \pi/2 - (\delta_0 - \delta_1)$. The situation that then arises is shown in Figure 52 from the analysis of Marateck et al. (110), which shows that at each $m_{\pi\pi}$ value two solutions are possible (even if δ is defined to be between 0 and 180°). Moreover, since the two solutions became degenerate

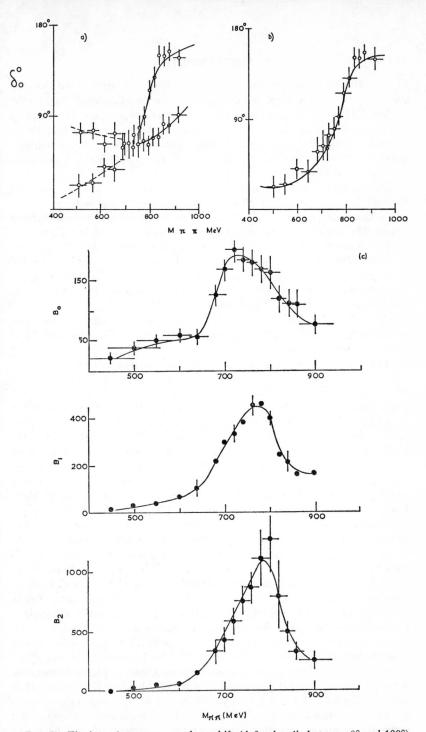

FIG. 52. The isoscalar s-wave $\pi\pi$ phase shift (defined to lie between $0°$ and $180°$) obtained by Marateck et al. (110). (a) Showing the ambiguous "up" and "down" assignments at each value of $M\pi\pi$. (b) The preferred down-up solution. (c) The coefficients B_0, B_1, and B_2 which are proportional to the pole extrapolated coefficients of an expansion of the $\pi\pi$ scattering angle distribution in terms of a series in $\cos \vartheta_{\pi\pi}$.

at $m_{\pi\pi} \sim 700$ MeV, four solutions become possible generally referred to as the up-up, up-down, down-up, and down-down solutions following Malamud & Schlein (111). All solutions give a large s wave in the ρ region. The down-up solution would give a fairly narrow resonance, and the up-up a broad resonance in the 700 MeV region, whereas the up-down and down-down solutions *if* resonant would have the resonant mass above ~ 1000 MeV. A resonance at ~ 700 MeV will be referred to as the ϵ^0 meson.

Mareteck et al. have assembled a large sample of data (15 000 events) for the reaction $\pi^- p \to n\,\pi^+\pi^-$ with $\Delta^2 < 12\,\mu^2$, and are thus able to attempt a Chew-Low extrapolation to the pion pole. In determining the phase shifts of Figure 52a, only the extrapolated terms in $\cos\vartheta$ are used but Figure 52c shows the extrapolated coefficients (isotropic, B_0; linear in $\cos\vartheta$, B_1; quadratic in $\cos\vartheta$, B_2). The sharp peak in B_0, which is not at the ρ mass but at ~ 720 MeV and so is not felt to be an absorptive consequence of ρ production, encourages the authors to accept the down-up solution of Figure 52b, i.e. to claim a resonance at ~ 720 MeV with $\Gamma \sim 140$ MeV. This choice clearly depends on the validity of the extrapolation procedure, but this analysis represents the first large enough to attempt such a choice from the peripheral $\pi^+\pi^-$ production alone.

The most direct way of confirming the choice is to examine $\pi^+ n \to p\pi^0\pi^0$ or $\pi^- p \to n\pi^0\pi^0$, but unfortunately the data here are still not as full as one would like. Several experiments (112) have shown that there is no very narrow $\pi^0\pi^0$ enhancement at the ρ mass. Feldman et al. (113) study $\pi^- p \to n\pi^0\pi^0$ at 1.53 and 1.27 GeV/c using neutron time of flight and γ detection in lead spark chambers to identify the reaction kinematically. Only the data at the higher momentum are fitted in detail. At this low energy the reaction is dominated by N^* formation, and analysis hinges on the inability to fit the $\pi^0\pi^0$ spectrum by Monte Carlo calculations assuming only N^* formation and $n\pi^0\pi^0$ phase space. It is argued that an s-wave state, $M = 715 \pm 35$ MeV, $\Gamma = 150 \pm 30$ MeV, is also needed. This would be compatible with the peripheral down-up solution adopted by Marateck et al., but at such a low energy the analysis is clearly very complicated. At higher energies the observed $\pi^0\pi^0$ spectrum does not support so narrow a resonance. Thus Figure 53 shows the spectrum for $\pi^+ n \to p\pi^0\pi^0$ at 2.15 BeV/c studied by Braun et al. (114), together with an earlier spectrum published by Corbett et al. (115), of the same shape but unaccountably lower in magnitude by a factor of 3. The dotted curve shows the absolute prediction of the peripherally determined s-wave phase shift using the Malamud & Schlein up-up solution, i.e. with the assumption of a broad $\pi\pi$ scalar resonance. This agrees very well in magnitude, but it will be seen that even this broad resonance solution drops too quickly at the higher $\pi^0\pi^0$ masses. The data shown in Figure 54 from the missing-mass experiment performed in a bubble chamber for $\pi^+ d \to ppM$ at momenta between 1.1 and 2.3 GeV/c by Smith & Manning (116) show almost identical characteristics. Neither experiment supports the presence of a state as narrow as 140 MeV. However two comments should be made.

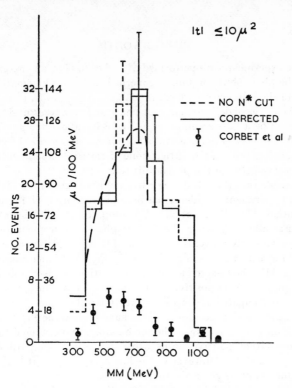

FIG. 53. The dipion effective mass for the reaction $\pi^+ n \rightarrow p\pi^0\pi^0$ at 2.15 GeV/c
(114). For description of curve see text.

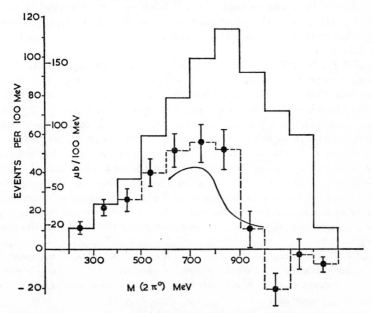

FIG. 54. $M(2\pi^0)$ (dotted) obtained from the missing mass (solid) in the reaction
$\pi^+ d \rightarrow ppMM$ between 1.1 and 2.3 GeV/c (116) by subtracting out $3\pi^0$, η, ω, and N^*.
The curve is the Malamud & Schlein prediction.

Firstly, the experiments are performed at rather too low an energy, thus the falloff of the phase-shift solution is governed rather by kinematics than by the width of the resonance (indeed, though the solution fits, phase space fits about as well). Secondly, virtually all phase-shift analyses neglect d waves and this could be a false assumption in the region above the ρ. We shall see below that there may be evidence for a d-wave resonance at ~ 1 GeV. In view of these facts it is probably true that none of the $\pi^0\pi^0$ experiments are decisive in choosing between the phase-shift solutions.

It is possible that the solution is π less than that shown in Figure 58, i.e. negative δ. Experimental evidence has been put forward both for and against such a solution. Thus Braun et al. (114) compare the ratio of $\pi^0\pi^0$ and $\pi^+\pi^-$ production at 500 MeV when the p-wave phase has dropped to $\leq 8°$. The ratio is approximately unity. If it is assumed that $|\delta_0| > |\delta_2|$, then this result is obtained for real $\pi\pi$ scattering if $\delta_0 \simeq -4\delta_2$. Peripheral analysis of $\pi^-\pi^0$ production (e.g. 117) shows δ_2 to be small and negative in this region. Hence this argument suggests δ_0 is positive. It is suggested that since $\pi^+\pi^-$ and $\pi^0\pi^0$ production are compared, absorptive correction will not affect this argument.

On the other hand, Biswas et al. (118) have studied the effect of Coulomb interference on $\pi^+\pi^-$ scattering. The same ambiguity between δ and $\delta-\pi$ results so that either positive or negative values are compatible with the peripheral phase shifts. However in the low-energy region it is the negative solution which appears correctly to go to zero.

Clearly one requires further analysis of the behaviour of the phase shift near threshold, and further data on the $\pi^0\pi^0$ spectrum, before the existence and width of the ϵ^0 can be established. It has been strongly argued by Lovelace et al. (119) that an interpretation of backward πp elastic scattering using dispersion relations demands the existence of a scalar $\pi\pi$ resonance in this general region.

7.3.2 *The* S*(1070).— An $I=0$ scalar meson should, apart from $SU(3)$ mixing arguments, show both $\pi\pi$ and $K\overline{K}$ couplings. It has been known for some time that the $I=0$ $K\overline{K}$ system shows a threshold enhancement ($J^{PC} = 0^{++}$) which is evident in $K\overline{K}$ systems produced in the reaction $\pi^-p \rightarrow nK\overline{K}$ and in $\overline{p}$ annihilations. What is not clear is whether this enhancement is due to a resonant state or to a large positive scattering length. Figure 55 shows a compilation of the low-mass $K_1^0 K_1^0$ spectrum found in various experiments. One notes considerable differences between experiments. For some (120–123) the enhancement is better fitted by a resonance standing some distance above threshold, for some it is better fitted by a scattering-length solution (124), and for some it is fittable in either way (125, 126). A sufficiently large number of experiments are suggestive of a resonance above threshold that the possibility must be taken very seriously, but with such variations between experiments one cannot regard the state as well established. Also if accepted, its mass and width are not well determined. Generally speaking, bubble-chamber experiments claim a much narrower width than do spark-chamber experiments (Figure 56).

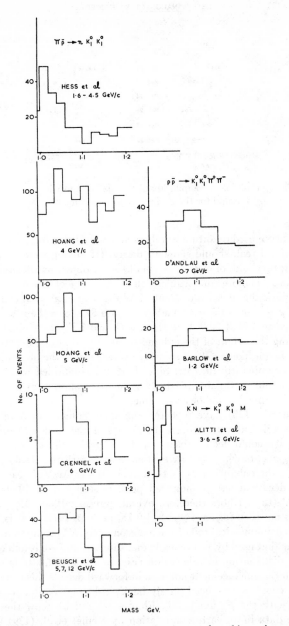

FIG. 55. Compilation of $K_1^0 K_1^0$ spectrum at low masses found in various experiments.

DETERMINATION OF S* PROPERTIES.

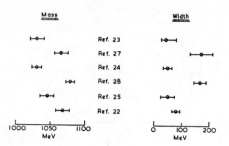

FIG. 56. A compilation of the masses and widths found for the S^* in various experiments.

With its supposed quantum numbers and since it is produced by incident pions, the S^*, if real, should show $\pi\pi$ decay. (If mixing were near ideal the branching ratio could, of course, be small.) A $\pi\pi$ object has been seen at this mass but its J^P is more compatible with 2^+. The object was first reported in the sonic spark-chamber experiment of Whitehead et al. (127) to study the reaction $\pi^- p \to n\pi^+\pi^-$ at 3.1–3.6 GeV/c. Figure 57 shows the $\pi^+\pi^-$ spectrum. The enhancement is at 1085 ± 10 MeV, with width $\Gamma \leq 25$ MeV. It is argued by comparing the height of the enhancement with that of the f^0 that the enhancement is not likely to be a spin-zero state even if the object in question decayed predominantly into $\pi\pi$. Since if it were identified with the S^* the $K\overline{K}$ mode would be dominant, this argument to show that it is not S^* is strengthened. Spin 2 would be acceptable.

Miller et al. (128) have reported the same enhancement in a bubble-chamber experiment containing 7916 events of the type $\pi^- p \to n\pi^+\pi^-$ at 4 GeV/c. ($I = 0$ is confirmed by absence of a $\pi^-\pi^0$ peak.) The peak is seen clearly by selecting events where the $\pi\pi$ system decays with the $\cos\vartheta_{\pi^-\pi^-} < -0.75$ (Figure 58); the enhancement is then at 1050 ± 15 MeV, $\Gamma < 40$ MeV. For equatorial decays a less pronounced peak is seen but at 1080 ± 10 MeV, $\Gamma < 40$ MeV (the authors therefore overall put $M = 1060 \pm 15$ MeV, $\Gamma < 70$ MeV). It is not seen for $\cos\vartheta_{\pi^-\pi^-} > +0.75$, perhaps because it is hidden by the forward asymmetrical tail of the ρ meson or by N^* production. The same argument as that used by Whitehead et al. shows that $J = 0$ is unlikely. Moreover decay distribution in this region (Figure 59) is compatible with $J = 2$, though with the enhancement not seen in forward decays this is not convincing since it could be produced purely backward as is possible for an s wave interfering with the p-wave ρ tail. Figures 60 and 61 show the backward hemisphere data in an early compilation by Veillet et al. (129) for experiments between 4 and 8 GeV/c and for an experiment at 6 GeV/c (130).

This $\pi\pi$ enhancement thus, most probably, does not represent confirmation of the S^*.

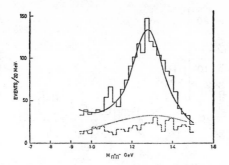

FIG. 57. Mass spectrum of $\pi^+\pi^-$ in $\pi^-p \to n\pi^+\pi^-$ at $3.1-3.69$ GeV/c (127). The broken histogram shows the estimated background.

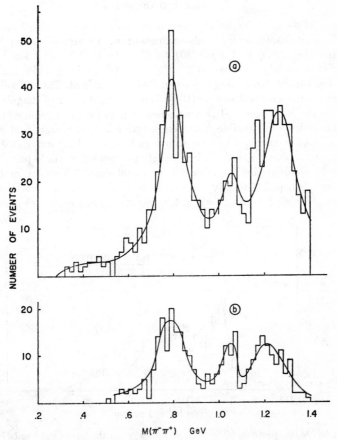

FIG. 58. $M(\pi^+\pi^-)$ for $\pi^-p \to n\pi^+\pi^-$ at 4 GeV/c (128). (a) For $\cos\vartheta_{\pi^-\pi^-} < -0.75$. (b) As (a) but with $\Delta^2 < 0.1$ GeV^2.

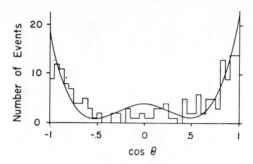

FIG. 59. The $\pi\pi$ decay distribution for the region $1.02 < M_{\pi\pi} < 1.1$ GeV and $\Delta^2 < 0.1$ GeV^2 for the reaction $\pi^- p \to n\pi^+\pi^-$ at 4 GeV/c (128).

7.4 THE SCALAR NONET

No candidate for the scalar nonet, as we have seen, is trouble free. Moreover the nine candidates listed above cannot form a nonet. This may be seen by using the $\delta(960)$ and $K\pi(1080)$ to give the Gell-Mann–Okubo octet-isosinglet mass, when a mass ~ 1200 MeV is obtained, i.e. heavier than both the ϵ^0 and the S^*. No mixing angle can then be obtained. The mass-squared difference between the δ and $K\pi(1080)$ is of the right order of magnitude. If these are therefore accepted as nonet members, and the ϵ^0 is retained but the S^* is discarded as a possible scattering-length effect, the near-ideal mixing, as expected for an $S_{q\bar{q}}=1$ state, would yield the real S^* mass at ~ 1280. Since such an object would not be strongly $\pi\pi$ coupled it would be produced weakly in incident-pion experiments. It would show $K\bar{K}$ decay and could easily have escaped notice, lying as it would in the A_2 region. [Bettini et al. (131) have claimed a scalar $I=0$ state decaying to $\rho\rho$ at 1410 MeV, but there

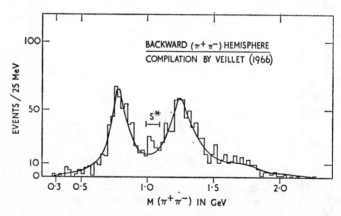

FIG. 60. Mass spectrum for $\pi^-\pi^-$ in $\pi^- p \to n\pi^+\pi^-$ in the backward hemisphere. From a compilation by Veillet (129).

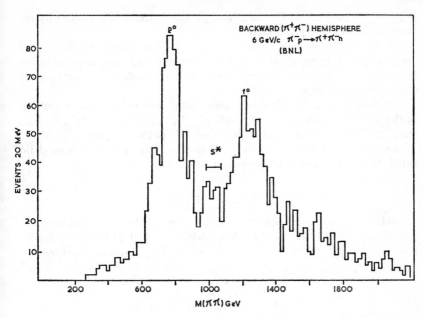

FIG. 61. $M(\pi^+\pi^-)$ for $\pi^-p\to n\pi^+\pi^-$ at 6 GeV/c in the backward hemisphere (130).

seems no reason why such a state should not preferentially show $\pi\pi$ decay. Beusch et al. (126) found a shoulder to the $K_1^0K_1^0$ decay of the A_2 at $\sim$1440 MeV.]

With width uncertainties for all candidates, $SU(3)$ checks are not possible.

8. SURVEY OF THE POSITIVE-PARITY MESONS

It is clear from what has been said above that there are still considerable uncertainties in the correct classification of the positive-parity mesons. Of the four nonets expected on the quark model for $l=1$, none is completely trouble free. It is worth pointing out, however, that the squared masses of the $I=1$ candidates, i.e. those members that can show no mixing, have values of $\sim$0.93 (0^{++}), $\sim$1.14 (1^{++}), $\sim$1.46 (1^{+-}), and $\sim$1.66 (2^{++}) with successive differences of 0.21, 0.3, and 0.2 GeV^2. This is the approximate equality of spacing expected from the quark model with spin-orbit forces dominating in the separation of nonets (Equation 31).

It is not inconceivable that a second 2$^+$ nonet of narrow states is appearing, the $I=1$ candidate being the A_2', the $I=\frac{1}{2}$ candidate being the possible natural-parity state in the Q region, and one $I=0$ member being the $\pi\pi$ state at the S^* mass. An ideally mixed fourth state would have a mass in the g-meson region and we shall see below that there is a possible disagreement between the parameters for the $g^{\pm}$ and g^0, which would be explained by the presence of an isoscalar $\pi\pi$ state.

In connection with the rather low-mass $J=2$ $\pi\pi$ candidate, the normal statement from Regge analyses that the Pomeron trajectory must be rather flat can be avoided if cut-exchange corrections are made to the normal pole-exchange calculations.

On the other hand it might be argued that there is evidence for a doubling of the positive-parity mesons, e.g. split A_2; A_1, $A_{1.5}$; the two $K\pi\pi$ Q mesons; the $K^*(1400)$ and the natural-parity possibility in the Q region.

Even in this low-mass region considerably more data are required for satisfactory classification of the meson state.

9. THE F' MESON

Before we turn to the higher-mass region, one further state should be discussed: the $F'(1550)$ reported by Aguilar-Benitez et al. (132) as a $K\overline{K}\pi$ enhancement in $p\bar{p}\to K\overline{K}\pi\pi$ at 700 MeV/c where the channels studied were:

$$p\bar{p} \to K_1^0 K_1^0 \pi^+\pi^- \quad \text{(484 events)}$$

$$K_1^0(K^0)\pi^+\pi^- \quad \text{(1305 events)}$$

$$K_1^0 K^{\pm}\pi^{\mp}\pi^0 \quad \text{(2114 events)}$$

Figure 62 shows the results. The width of the state is $\sim$40 MeV. The presence of the enhancement in $K_1^0 K^{\pm}$ π^0 shows $I>0$. The decay is predominantly into K^*K and the relative enhancements seen in these channels are compatible with $I=1$ K^*K decay.

Mode	Prediction	Experiment
$K_1^0 K_2^0 \pi^{\pm}$	a	66 ± 17
$K_1^0 K_1^0 \pi^{\pm}$	$a/3$	10 ± 10
$K_1^0 K^{\pm} \pi^0$	a	66 ± 14

To find the G parity one would like to study the $K\overline{K}$ subsystem of the $K\overline{K}\pi$ state, though if decay is exclusively to K^*K this is not easy, since $I_{(K\overline{K})}$ need not then be unique. Absence of ϕ in the odd C, $K_1^0 K_2^0$ substructure *might* imply $I_{K\overline{K}}=1$, when $G_{K\overline{K}}=+1$, i.e. $G_F=-1$.

Should this assumption be correct, the state is clearly a candidate for the radially excited π, formed in a situation analogous to the E which might be the radially excited η. However, the authors tentatively disfavour $J^P=0^-$ though they believe the state to be in the unnatural sequence (private communication). On the other hand if one examines the $K\overline{K}$ compilations of Figures 29 and 30 there may be an indication of a $K\overline{K}$ mode of decay. If this is indeed the case, the object is of natural parity and if the above G-parity assignment were taken seriously it would indicate another 2^+ state.

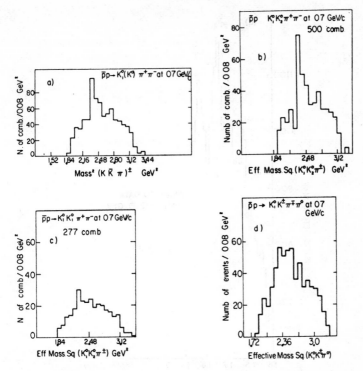

Fig. 62. $M^2(K\overline{K}\pi)$ distributions for $p\bar{p}\rightarrow K_1^0 K_1^0 \pi^+\pi^-$, $p\bar{p}\rightarrow K_1^0(K^0)\pi^+\pi^-$ and $p\bar{p}\rightarrow K_1^0 K^{\pm}\pi^{\mp}\pi^0$ at 700 MeV/c reported by Aguilar-Benitez et al. (132), where $K\pi$ or $\overline{K}\pi$ is in the K^* region.

10. HEAVY NONSTRANGE MESONS BELOW THE $N\overline{N}$ THRESHOLD (i.e. 1600–1880 MeV)

10.1 MISSING-MASS EXPERIMENTS

It is convenient to divide the heavier mesons into those which have masses below that of two nucleons, i.e. 1880 MeV, and those above, which can thus decay into $N\overline{N}$. Three, and possibly four, nonstrange narrow states have been found in the lower region in the CERN Missing-Mass Spectrometer Experiment (133), as yet unconfirmed independently. Figure 63a shows combined data from runs at 7, 11.5, and 12 GeV/c, and Figure 63b the effect of removing the very large background. The peaks R_1, R_2, R_3 at 1630 ± 15, 1700 ± 15, and 1748 ± 16 MeV with widths compatible with the 30 MeV resolution were found independently at the different momenta, and are statistically very significant. The peak R_4 at 1830 ± 15 MeV is much less convincing. The resonance production is studied at comparatively high momentum transfer $.2 < t < .3$ GeV^2. The I spin may be 1 or 2 and little can be said

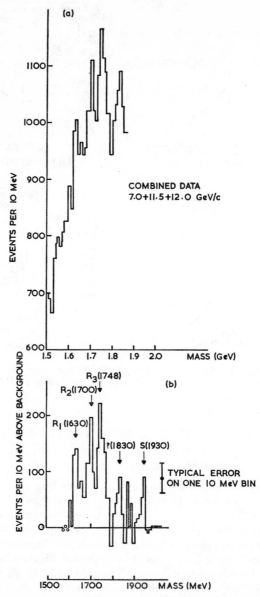

FIG. 63. Missing-mass spectrum in the R-meson region from the CERN experiment (133). (a) Combined data from runs at 7, 11.5, and 12 GeV/c. (b) Number of events above background.

of the decay modes except that all of R_1, R_2, R_3 appear to have decays involving one and three charge products.

We shall find that few bubble-chamber experiments find objects as narrow as those reported in the missing-mass experiment. There is no evidence that this is due to bad resolution. It should always be remembered that the two techniques may be sensitive to different states. Firstly, in the MM technique with no selection on decay modes, background is exceedingly large and the method tends to be preferentially sensitive to narrow peaks standing clearly above background. Secondly, the bubble-chamber technique invariably gets most data from the peripheral low memontum-transfer region, whereas we have seen that the MM technique has been used with $.2 < t < .3$. Finally, narrow states are, by definition, weakly coupled, and will be harder to observe in the statistically limited bubble-chamber technique.

The different states observed in this region using the bubble-chamber technique will next be described: firstly, even G-parity states; secondly, odd G-parity states.

10.2 THE g MESON

The g meson which exhibits $\pi\pi$ decay is a well-established state. It is seen decaying into $\pi^+\pi^-$ and $\pi^\pm\pi^0$ in $\pi N \to N\pi\pi$ at higher energies but is not seen in $\pi^+ p \to n\pi^+\pi^+$ or in $\pi^+ n \to p\ \pi^0\pi^0$. It is thus is vector in nature and must be in the J^P sequence 1^-, 3^-, 5^- etc.

Figures 64a and 64e show the $\pi\pi$ spectra reported by Crennell et al. (134), for the reactions $\pi^- p \to \pi^-\pi^+ n$ and $\pi^- p \to \pi^-\pi^0 p$ at 6 GeV/c. Neutral and charged enhancements are seen, each with a width of 200 ± 100 MeV, but with masses given respectively by $M_{g^0} = 1720 \pm 20$ MeV and $M_{g^-} = 1640 \pm 25$ MeV. This incompatibility in mass will be further discussed below. To determine the spin of the state, it is assumed that g production takes place via pion exchange. The $\pi\pi$ scattering-angle distribution for events with $t < 1.0$ GeV^2 is fitted to the expansion $\Sigma_n\ A_n\ P_n\ (\cos\vartheta)$ and the even moments obtained are shown in Figure 64. The decay of the g^0 is very asymmetric, which indicates strong interference of different partial waves, and the authors feel unable to use this decay to determine J^P. A similar statement has been made by Armenise et al. (135) who study the reaction $\pi^+ n \to p\ \pi^+\pi^-$ at 5.1 GeV/c. Figure 65 shows the marked asymmetry of $\pi\pi$ scattering in this experiment and that this does not vary significantly through the g^0 region. The authors find no structure in the coefficients A_n normalized to A_0 up to tenth order. However, for g^- Crennell et al. feel that analysis is possible. Peaks are seen in A_2, A_4, and possibly A_6. The A_4 term probably eliminates $J^P = 1^-$ and the data suggest $J^P = 3^-$, though the authors themselves point out that a 1^- state interfering with a spin 3 background could produce the same distributions. Remember that such semiqualitative statements are not to be compared with the detailed phase-shift studies carried out in formation experiments.

The angular distributions published by Johnston et al. (136) (Figure 66) for the same reactions at 7 GeV/c do not suggest that g^- behaves very differ-

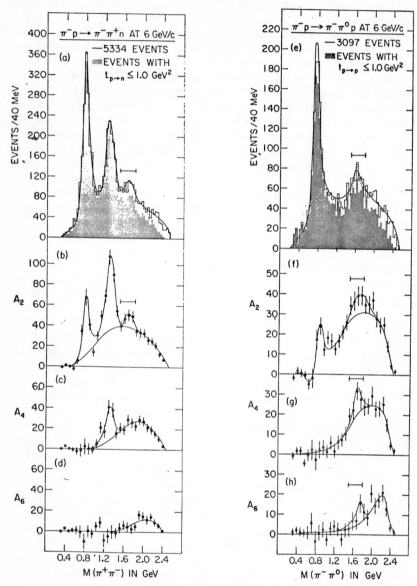

FIG. 64. Production of g meson at 6 GeV/c observed by Crennell et al. (134). The figures show the $M(\pi\pi)$ spectra and the even moments of a Legendre polynomial analysis of the $\pi\pi$ scattering angle distribution in the g region.

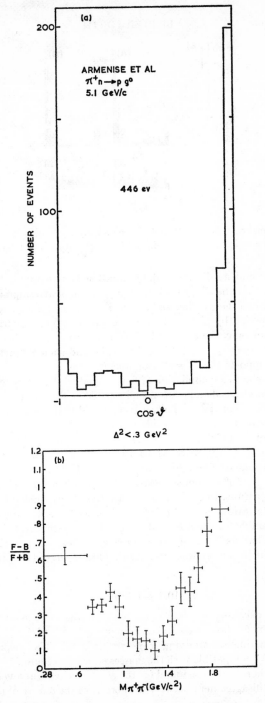

Fɪɢ. 65. (a) Angular distribution of g^0 decay in the reaction $\pi^+ n \rightarrow p g^0$ at 5.1 GeV/c (135). (b) Forward-backward asymmetry of $\pi^+ \pi^-$ scattering through the g region in the same experiment.

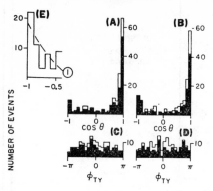

FIG. 66. Dipion scattering angle and Treiman-Yang angle distributions for events in the g region (136). (a) and (c) refer to g^0. (b) and (d) refer to g^-.

ently from g^0 and no direct J^P determination is attempted. However, the authors state that the slight backward peak in the scattering-angle distribution is peculiar to the g region and is thus indicative of a high-spin state. Assuming production by pion exchange, the Treiman-Yang distribution being reasonably isotropic, the authors by a comparison of the f^0 and g^0 cross sections conclude that $J_g = 3 \pm 1$. Since J must be odd such a conclusion, if correct, again supports the $J^P = 3^-$ assignment.

Figure 67 shows a compilation of masses and widths reported by various authors for the charged and neutral g meson. There may be a tendency for the g^0 to be heavier than the $g^\pm$. This, if true, would indicate the presence of an isoscalar state on the high-mass side of the g^0.

Crennell et al. (134) find a $K^\pm K_1^0$ enhancement in $\pi^\pm p \rightarrow K^\pm K_1^0 p$ at a mass of $1640_{-25}{}^{+20}$ MeV with width $79_{-25}{}^{+70}$ MeV, with no corresponding $K_1^0 K_1^0$ enhancement. The $K^\pm K_1^0$ decay has $I = 1$. The absence of the effect in the even C $K_1^0 K_1^0$ system suggests that C (and hence P) is odd. It is thus reasonable to assume that this is an alternative decay mode of the g meson. The branching ratio $g \rightarrow K^- K^0 / g \rightarrow \pi^- \pi^0 = 0.08_{-0.08}{}^{+0.08}$. No direct spin-parity analysis is possible for the $K\bar{K}$ mode at this level of statistics. The same $K\bar{K}$ enhancement has also been reported in two earlier experiments. Ehrlich et al. (137) saw it in $(K\bar{K})^\pm$ and not in $K_1^0 K_1^0$, Abrams et al. (138) in $K_1^0 K_2^0$ and not in $K_1^0 K_1^0$.

10.3 $\rho(1700) \rightarrow 4\pi$

An object decaying into four pions is seen in the same mass region as the g meson. Figure 68 shows a compilation by French (59) of the $(4\pi)^\pm$ spectrum from five experiments studying $\pi^\pm p \rightarrow p\pi^\pm \ \pi^+ \pi^- \pi^0$. The peak appears at 1700 ± 15 MeV, compatible with the masses shown in Figure 74 for the g^0, but somewhat higher than those found for the $g^\pm$. At the present level of statistics there is slight disagreement between authors as to the nature of the 4π decay.

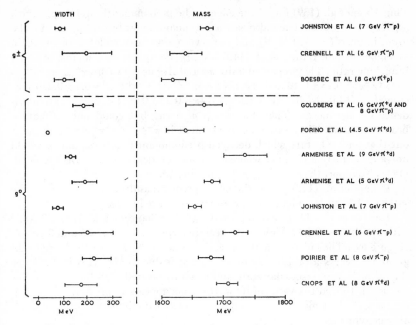

FIG. 67. Values found by various authors for the widths and masses of the $g^{\pm}$ and g^0.

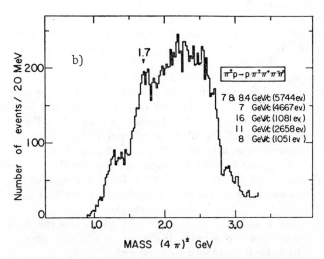

FIG. 68. Compilation of $(4\pi)^{\pm}$ mass spectra (59).

Thus Caso et al. (139) found the effect to be predominantly due to $\rho 2\pi$ with $\rho\rho/\rho 2\pi = 0.48 \pm 0.16$. They also saw a $\pi\omega$ enhancement but it appeared to be lower in mass ($\sim$1630 MeV) and narrower ($\sim$60 MeV wide) than the main 4π enhancement. Baltay et al. (140) saw $\rho^+\rho^0$ decay but were unable to find a branching ratio. They reported finite $\pi\omega$ and $A_2\pi$ decay with $\omega\pi^+(\omega\to\pi^+\pi^-\pi^0)$ $/\pi^+\pi^+\pi^-\pi^0 = 0.25 \pm 0.10$ and $A_2{}^0\ \pi^+(A_2\to\pi^+\pi^-\pi^0)/\pi^+\pi^+\pi^-\pi^0 = 0.4 \pm 0.2$. Johnston et al. (136) also found $\omega\pi$ ($\omega\to 3\pi$)/$4\pi = 0.25 \pm 0.10$, but had negative evidence for $A_2\pi$ decay. They too saw $\rho^-\rho^0$ decay but could not estimate a branching ratio. They saw no enhancement in $\pi^-p\to n\ \pi^+\pi^-\pi^+\pi^-$ and point out that an $I = 1$ state which decayed predominantly into $\pi\omega$ and $\rho\rho$ would not appear in this channel whereas an $A_2{}^\pm\ \pi^\pm$ decay would. Biswas et al. (141) also report the state in $\rho^-\rho^0$ but not in $\rho^0\rho^0$.

Ballam et al. (142) found the effect predominantly due to $\rho^-\rho^0$ ($\rho^-\rho^0/4\pi$ $= 0.83 \pm 0.3$) with evidence for $\omega\pi$ and no evidence for $A_2\pi$.

Danysz et al. (143) have reported a $(4\pi)^0$ enhancemen at 1717 ± 7 MeV with a width of 40 ± 17 MeV in the reaction $\bar{p}p\to 3\pi^+\ 3\pi^-$ at 2.5 and 3 GeV/c (Figure 69). The enhancement is certainly due to $\rho^0\pi^+\pi^-$ but the large background prevents a decision as to whether it is due to $\rho^0\rho^0$. If it is due to $\rho\rho$ then it must be an isoscalar state and is not associated with the states so far discussed. Indeed the value of its width alone appears to prevent its identification with the g meson. Seen in data from two experiments it seems reasonably convincing.

Figure 70 shows the $\pi\omega$ spectrum obtained by Johnston et al. (136) together with the variation with mass of four angles: ϑ the angle between incident π and bachelor π in the $\pi\omega$ CM, ϕ the Treiman-Yang angle, θ' the angle between the ω decay normal and the ω direction in the ω CM, and ϕ' the corresponding azimuth angle taken from an x axis in the $\pi\omega$ production plane. If the $\pi\omega$ decay is due to the $J^P = 3^-$ g meson produced by π exchange, then it will be produced in the $|3, 0\rangle$ state. In the $\pi\omega$ decay, to obtain negative parity l must be odd and can only equal three. Moreover the orbital state $|3, 0\rangle$ is forbidden since $|3, 0\rangle\nleftrightarrow|0, 0\rangle + |1, 0\rangle + |3, 0\rangle$. Hence the orbital state would have to be $|3, \pm 1\rangle$. The curve given on the $\cos\vartheta$ histogram corresponds to this alignment and fits the data with a probability of 75 per cent. Curves corresponding to $J^P = 1^-$ or 5^- fit with only 5 per cent and 10^{-4} per cent probabilities respectively. The $|1, \pm 1\rangle$ alignment forced on the ω by π exchange results in a $\sin^2\vartheta'\ \sin^2\phi'$ distribution of the ω decay normal. The data, as is shown, are compatible with such a distribution.

Clearly the simplest hypothesis is to assume that the object decaying into 4π is the g meson and that this has $J^P = 3^-$. However, the present discrepancies in masses and in decay modes are highly suggestive that there is more than one even G-parity state in this mass range.

The g meson is too wide to associate with any of one the R mesons. It could be a mixture of two or more. If the narrow $\rho 2\pi$ state seen by Danysz et al. is not isoscalar it could be associated with R_2 or R_3.

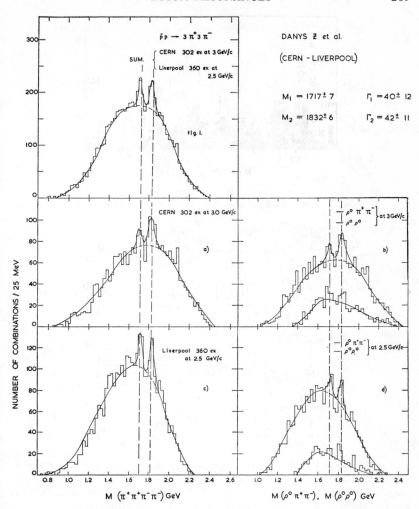

FIG. 69. Evidence presented by Danysz et al. (143) for two $(4\pi)^0$ states in $p\bar{p}\rightarrow3\pi^+3\pi^-$ at 2.5 and 3.0 GeV/c. Data from the two momenta are shown separately and combined. The effect of $\rho^0\pi^+\pi^-$ and $\rho^0\rho^0$ selection is also shown.

10.4 OTHER EVEN G-PARITY STATES

There have been unconvincing claims for 4π states at 1592, 1610, and 1630 MeV, all in $\bar{p}$ annihilations (112). Danysz et al. (143), in $p\bar{p}\rightarrow3\pi^+3\pi^-$ at 2.5 and 3 GeV/c, see a state at 1832 ± 6 MeV, with a width of 42 ± 11 MeV (Figure 69). Its properties are identical to those of the 1717 enhancement,

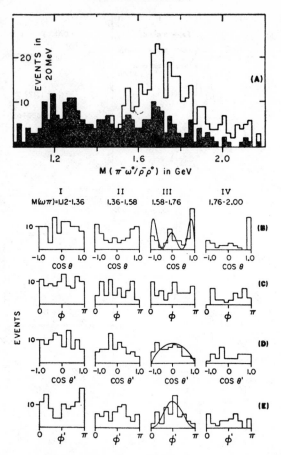

FIG. 70. (a) $M(\pi^-\omega)$ (shaded) $M(\rho^-\rho^0)$ (open) in the reaction $\pi^-p\rightarrow p\pi^-\pi^-\pi^+\pi^0$ at 7 GeV/c (136). For events having both $\pi\omega$ and $\rho\rho$ configurations, $\pi\omega$ is plotted. (b)–(e) Angular distributions ϑ ϕ ϑ' and ϕ' as described in the text. The solid curves show the expected distribution for production of a 3^- state by pion exchange.

namely that it is $\rho 2\pi$ associated but one cannot be sure if it is $\rho\rho$ associated. If it is not isoscalar, which it would be if due to $\rho\rho$, then it might be the same as the statistically weak R_4. Again the evidence for its existence is reasonably convincing. Supporting evidence (Figure 71) has been claimed (144) in the reaction $p\bar{p}\rightarrow 2\pi^+ 2\pi^-\pi^0$ at 1.2 GeV/c.

10.5 ODD G-PARITY STATES—THE A_3

The existence of an $I=1$ 3π enhancement at 1640 MeV (Figure 72), the A_3, is well established in the channels $\pi^\pm p\rightarrow p$ $\pi^\pm\pi^+\pi^-$. However, as in the case of the A_1 the question whether this state is a resonance or a kinematic

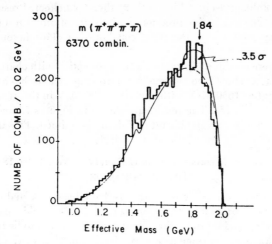

FIG. 71. $M(\pi^+\pi^+\pi^-\pi^-)$ in the reaction $p\bar{p}\to2\pi^+2\pi^-\pi^0$ at 1.2 GeV/c (144).

enhancement arises since it is close to the πf threshold and the decay is predominantly to πf. A compilation by Ferbel (145) shows that $\rho\pi$ constitutes <40 per cent of the 3π decay. A πf branching ratio of 35 ± 20 per cent is obtained. The 3π decay has not been reported other than in the "Deck" channel. Since, however, πf does not constitute 100 per cent of the decay, the en-

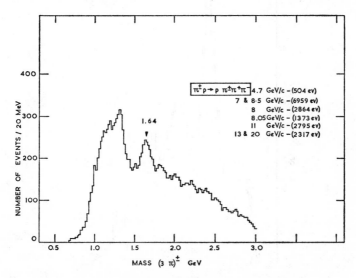

FIG. 72. Compilation of $3\pi^\pm$ spectra from the reactions $\pi^\pm p\to p\pi^\pm\pi^+\pi^-$ which show the A_3 (59).

hancement is unlikely to be a kinematic πf threshold effect. Presumably one is again witnessing diffractive production of a state of unnatural parity, and the spin-parity analysis by the ABC collaboration (146) favours such an assignment.

If the A_3 were produced diffractively, its production with charge exchange in $\pi^+ n \to p \ \pi^+ \pi^- \pi^0$ should be suppressed. Two experiments have reported a 3π enhancement at 1650 MeV with a width ~ 100 MeV in this reaction at 5.1 and at 8 GeV/c (67, 75). The results are shown in Figures 73 and 74. However, in both experiments significant $\rho^0 \pi^0$ decay is claimed. This implies that the decaying meson is isoscalar and not the A_3.

10.6 OTHER MESONS OF ODD G PARITY IN THE 1600–1880 MEV MASS REGION

Two 5π states, too narrow to be the A_3, have been reported in the reaction $\bar{p}p \to 3\pi^+ \ 3\pi^- \pi^0$ at 3 GeV/c by Danysz et al. (147). In this reaction, they find evidence for ρ and ω production. If events are selected to lie in these two regions (which is not the same as a ρ, ω selection in view of the large background), two "ρ" "ω" enhancements are found, one at 1689 ± 10 MeV ($\Gamma = 38 \pm 18$ MeV) and one at 1848 ± 11 MeV ($\Gamma = 67 \pm 27$ MeV) (Figure 75). It cannot be stated if the decay is to $\omega \ 2\pi$ or to $\omega \rho$ and so the i spin is un-

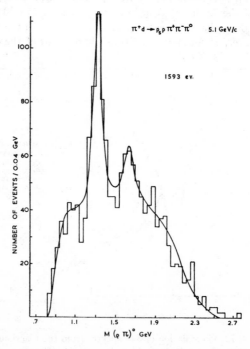

FIG. 73. $M(\rho\pi)^0$ for the reaction $\pi^+ n \to p \pi^+ \pi^- \pi^0$ at 5.1 GeV/c (67).

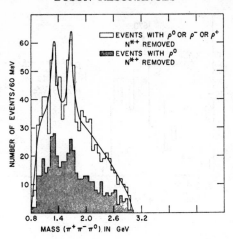

FIG. 74. $M(\rho\pi)^0$ for the reaction $\pi^+ n \rightarrow p\pi^+\pi^-\pi^0$ at 8 GeV/c (75).

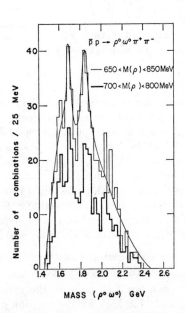

FIG. 75. $M(2\pi^+2\pi^-\pi^0)$ from the reaction $p\bar{p} \rightarrow 3\pi^+3\pi^-\pi^0$ when $\pi_a^+\pi_b^-\pi^0$ lies in the ω^0 region and $\pi_c^+\pi_d^-$ in the ρ^0 region (147).

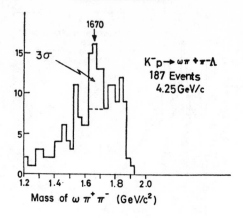

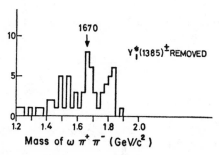

FIG. 76. $M(\omega^0\pi^+\pi^-)$ from the reaction $K^-p \to \Lambda\omega\pi^+\pi^-$ at 4.25 GeV/c (148).

known. The enhancements are respectively only 2.5 and 3 S.D., and are not seen at 2.5 or 5.7 BeV/c. The evidence is thus not compelling. However Yost et al. (148) also report an $\omega\,\pi^+\pi^-$ state of unknown i spin at 1670 ± 18 MeV with a width of 50 ± 15 MeV in the reaction $K^-p \to \Lambda\,\omega\,\pi^+\pi^-$ at 4.25 GeV/c. The enhancement is a three S.D. peak (Figure 76) and in this experiment the $\pi^+\pi^-$ system does not appear to form a ρ^0.

The higher of the two Danysz et al. peaks might be the same as a peak reported in the $(3\pi)^-$ system for events in the reaction $\pi^-p \to p\,\pi^-\pi^-\pi^+\pi^-$ at 13 and 20 GeV/c (149). Figure 77 shows the enhancement in question at 1840 MeV on the high-mass side of the A_3, with a width of $\sim$80 MeV.

10.7 STATES OF UNKNOWN G PARITY IN THE 1660–1880 MEV MASS REGION

French et al. (150) see a *two* S.D. peak in the mass combination $(K^*\overline{K} + \overline{K}^*K)$ between 1675 and 1725 MeV for the reaction $p\bar{p} \to K^0K^\mp\,\pi^\pm\pi^+\pi^-$ at 3.0, 3.6, and 4.0 GeV/c and point out that Baltay et al. at 3.7 GeV/c (151) also see a *two* S.D. peak in the same place. No quantum numbers are known

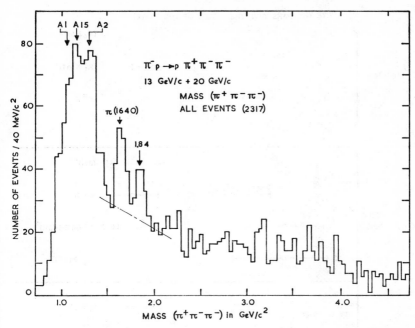

FIG. 77. $M(3\pi)$ from the reaction $\pi^-p\to p\pi^+\pi^-\pi^-$ at 13 and 20 GeV/c (149).

and if real it could be tied to either the 4π or the 5π enhancements which are seen in essentially the same channel of $p\bar{p}\to X\pi^+\pi^-$ (143, 147). Likewise they see a four S.D. ($K_1^0K^0$ $m\pi^0$) peak in $p\bar{p}\to\pi^+\pi^-K_1(MM)$ at 1820 ± 12 MeV with $\Gamma=50\pm20$ of unknown quantum numbers that could be tied to either the 4π or the 5π enhancement seen at this second mass.

Cooper et al. (152) in the reaction $p\bar{p}\to\pi^+\pi^-(MM)$ between 1.2 and 1.6 GeV/c report an enhancement in the $(\pi MM)^\pm$ system at 1720 MeV. It is not evident that MM corresponds to the neutral decay of a single particle.

10.8 The Relationships of the Nonstrange State Between 1600 and 1880 MeV

Figure 78 shows the mass and width (plotted as horizontal bars about the mass value) for the states discussed above. Clearly much more work is needed before a satisfactory identification of states in this region can be made.

The only state for which the spin and parity is at least experimentally suggested is the g meson. Figure 79a shows a plot of spin against mass-squared for the $1^-\rho$ meson, the 2^+A_2, and the 3^-g meson. The masses lie remarkably well on a common linear trajectory, which suggests exchange degeneracy with states of opposite signature on the same trajectory. We have seen in Section 1.5 that this fact has a simple explanation on the nonrelativ-

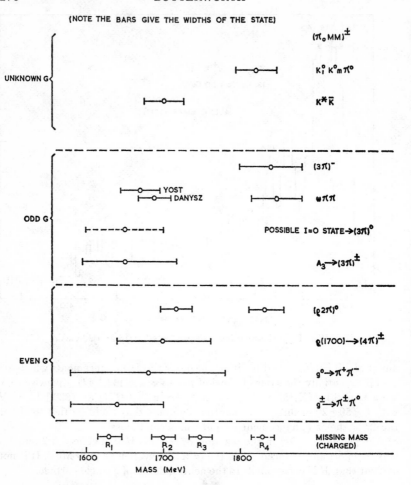

FIG. 78. The masses and widths (given by the horizontal bars) of states reported in the R-meson region.

istic quark model. On this model the 3^{--} state arises from $l = 2$ together with three other nonets (1^{--}, 2^{--}, 2^{-+}). If spin-orbit forces dominated the mass splitting, then by Equation 1.31, the splittings expected would be:

3^{--}	2^{-+}	2^{--}	1^{--}
$2a$	0	$-a$	$-3a$

If the splitting was in the same sense as for the $l = 1$ nonets, a would be positive and the higher-spin states would be heavier. It will be noted from Figure 78 that the R mesons, including the doubtful R_4 of unknown quantum

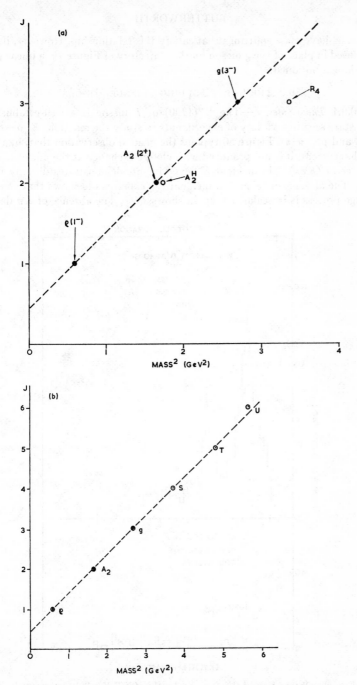

FIG. 79. (a) J versus mass-squared for the ρ, A_2, and g mesons. (b) The masses of the S, T, and U mesons shown on the same trajectory.

numbers, have mass splittings that satisfy this relationship. However, if the R_4 is used in place of the g meson on the trajectory of Figure 79, a linear plot is no longer obtained.

10.9 THE CORRESPONDING STRANGE MESONS

10.9.1 *The L meson.*—The $K^*(1790)$ or L meson is a well-established $K\pi\pi$ state seen in a variety of experiments to study the reactions $K^\pm p \to p K^\pm$ $\pi^+\pi^-$ and $pK^0 \pi^\pm\pi^0$. Figure 80 typifies the general observation that, like the Q enhancement, it is not produced with charge exchange at the nucleon. The absence of $(K\pi\pi)^{++}$ in incident-K^+ experiments demonstrates directly that $I = \frac{1}{2}$. The absence of $(K\pi\pi)^0$ in incident-K^- experiments shows that the exchange process is isoscalar, i.e. again shows $I = \frac{1}{2}$. The absence of $K\pi$ decay

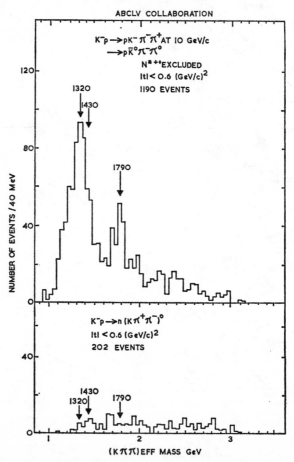

FIG. 80. Observation of the L meson by the $ABCLV$ collaboration (88d).

suggests unnatural parity, which, together with the evidence for isoscalar exchange, suggests once more a diffractive production process. However, again the question must be raised as to whether the effect is due to a kinematic threshold effect of the double Regge exchange type, since here one is just above the $K^*(1400)\pi$ threshold. Fortunately two experiments have clearly demonstrated that the L meson has decay modes other than $K^*(1400)\pi$ as shown in Table V. If the state is produced diffractively, one would also expect it to be of unnatural parity. The $ABCLV$ collaboration (88d), carrying out a detailed fit to $K\pi\pi$, $K\rho$, $K^*\pi$, and $K^*(1400)\pi$, concluded that $J^P = 1^+$, 2^-, or 3^+. The Hopkins group (88g) see coherent production of the L meson in the reaction $K^-d \to K\pi\pi d$ at 12.6 GeV/c (Figure 81). Absence of production of $K^*(890)$ or $K^*(1400)$ in $K^-d \to K\pi d$ suggests that only states of unnatural parity are produced with a deuteron. (Any exchange object must be isoscalar and extreme peripherability is ensured by the deuteron form factor.) This is what one would expect for such a coherent process. Moreover one expects that Pomeron exchange will leave the L messn in the state $|J, 0\rangle$ relative to the incoming K. The K^*_{1400} from the L meson shows the $|2, 0\rangle$ alignment expected if it arises from a 2^- meson in the $|2, 0\rangle$ state decaying into $K^*(1400)$ and π in a relative s wave. However, the $K\pi\pi$ decay normal, though distributed relative to the incident K with a distribution containing the $\sin^4$ term expected for a spin 2 state, shows some abnormalities. There is a marked asymmetry in its distribution relative to the normal to the production plane. This asymmetry could be indicative of finite exchange of nonzero-spin mesons or the interference of a 2^- L meson with other resonances or background.

The $ABCLV$ collaboration have pointed out that the upper limit for $K^*\eta$ decay when compared with the $K^*\pi$ branching ratio allows one to decide

TABLE V

DECAY MODES OF THE L MESON[a]

Mode	$ABCLV$ collaboration (88d)	Hopkins group (88g)
$K\pi$	<10	<10
$K\rho$	11 ± 9	20^{+15}_{-20}
$K^*(890)\pi$	34 ± 12	20^{+9}_{-14}
$K\omega$	8 ± 5	—
$K^*(1400)\pi$	19 ± 15	20^{+9}_{-15}
$K\pi\pi$	28 ± 13	31^{+30}_{-15}
$K^*\eta$	<5	—

[a] Branching ratios in percentages.

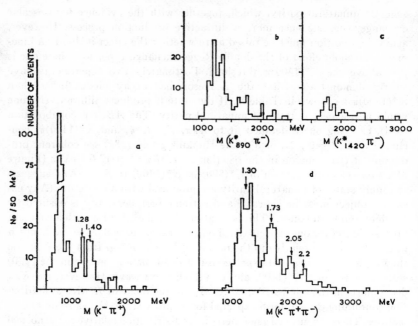

FIG. 81. Invariant mass distributions from the reaction $K^-d \rightarrow K^-\pi^+\pi^-d$ at 12.6 GeV/c. (a) $K^-\pi^+$ mass. (b) $K^-\pi^+\pi^-$ mass for $0.85 < M(K^-\pi^+) < 0.94$ GeV. (c) $K^-\pi^+\pi^-$ mass for $1.33 < M(K^-\pi^+) < 1.44$ GeV. (d) Overall $K^-\pi^+\pi^-$ mass.

whether D or F coupling is involved in the $L \rightarrow 1^-0^-$ decay. The positive C-parity assignment for the nonet from which the L comes is strongly ruled out.

It would thus appear that the best assignment for the L meson is that it comes from a nonet with $J^{PC} = 2^{--}$. However, in several experiments the enhancement does not show the Breit-Wigner form and one suspects that there is more than one resonance in this region. Indeed evidence for a second state, the $K^*(1660)$, has been presented.

10.9.2 $K^*(1660)$.—Evidence for this state has been presented by the Birmingham-CERN-Bruxelles collaboration (153) in K^+p studies at 5 GeV/c. It is claimed that it shows both $K\pi$, $K^*(890)\pi$, and $K^*(1400)\pi$ modes and so is of natural parity. Figure 82a shows the $K^*(890)\pi$ mass distribution in the channel $K^+\pi^+\pi^-p$ with N^* antiselected to reduce background. However the $K\pi$ and $K^*(1400)\pi$ modes are reported respectively in the channels $K^0\pi^+p$ and $K^{0+}\pi^{0-}\pi^+p$ but only in interference with the N^* (Figures 82b and c). In the channel $K^0p\ \pi^+\pi^+\pi^-$ the $(K3\pi)^+$ mass distribution shows the same enhancement with a suggestion (Figure 83) that it might decay into $Q^0\pi^0$ with Q subsequently decaying to $K^0\pi^+\pi^-$. The mass of the state is 1660 ± 10

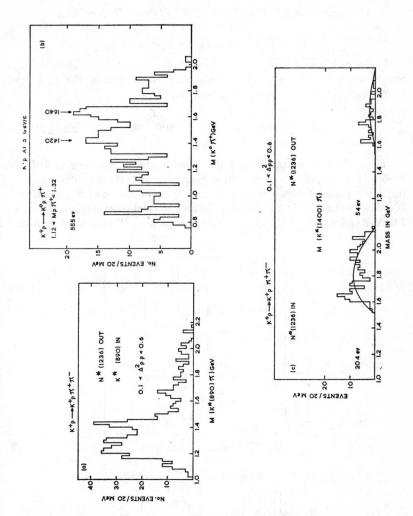

FIG. 82. Evidence for $K^*(1660)$ from the Birmingham-CERN-Bruxelles collaboration (153). (a) $M(K^*\pi)$ in the reaction $K^+p \to pK^+\pi^+\pi^-$ (N^* removed). (b) $M(K^0\pi^+)$ in the reaction $K^+p \to K^0p\pi^+$ in the N^* region. (c) $M(K^*_{1400}\pi)$ in the reaction $K^+p \to pK^+\pi^+\pi^-$ both in and outside the N^* region.

MeV, the width 60 ± 20 MeV. There has also been a claim of a $K\omega$ state at this mass (154).

11. MESONS ABOVE THE NUCLEON-ANTINUCLEON THRESHOLD

11.1 NONSTRANGE STATES

Figure 84 shows a compilation of results from the CERN Missing-Mass Spectrometer experiment with enhancements referred to as the S, T, and U mesons with the following parameters:

	S	T	U
Mass	1929 ± 14	2195 ± 15	2382 ± 24
Width	≤ 35	≤ 13	≤ 30

These $I = 1$ or 2 states have unknown quantum numbers and decay modes. However, it may be seen from Figure 79b that the degenerate ρ, A_2, g trajectory passes through these masses at spins of 4, 5, and 6 respectively.

Abrams et al. (155) have studied $\bar{p}p$ and $\bar{p}d$ total cross sections from 1 to 3 GeV/c and deduce the $I = 1$ and $I = 0$ cross sections (Figure 85) to find that

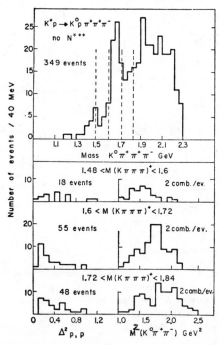

FIG. 83. $M(K^0\pi^+\pi^+\pi^-)$ in the reaction $K^+p \rightarrow pK^0\pi^+\pi^+\pi^-$ at 5 GeV/c, together with $M^2(K^0\pi^+\pi^-)$ from the varying regions of $K3\pi$ effective mass (153).

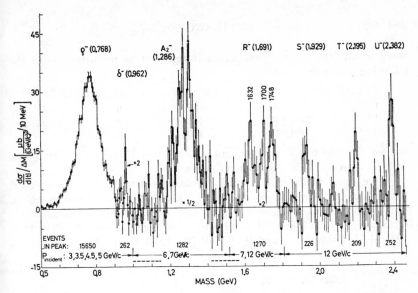

FIG. 84. Boson mass spectrum from the missing-mass spectrometer at CERN.

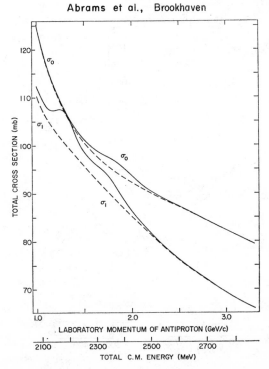

Abrams et al., Brookhaven

FIG. 85. $I=0$ and $I=1N\overline{N}$ total cross sections deduced by Abrams et al. (155).

three broad structures appear which if regarded as resonances have the following parameters:

I	E^*	Full width	mb	$\frac{1}{2}(J+\frac{1}{2})x$
1	2190 ± 5	85	6	.4
1	2345 ± 10	140	3	.3
0	2380 ± 10	140	2	.2

We see that the two $I=1$ states are in the T and U regions but are much broader than the MM spectrometer peaks.

Cline et al. (156) have studied $p\bar{p}$ elastic scattering in the backward hemisphere between 300 and 700 MeV/c, i.e. through the S-meson region. The behaviour of the differential cross section for various scattering-angle regions is shown in Figure 86. Evidence for narrow resonances ($I=0$ or 1) at 1925 MeV ($\Gamma\sim10$ MeV) and 1945 MeV ($\Gamma\sim22$ MeV) is seen. (The parame-

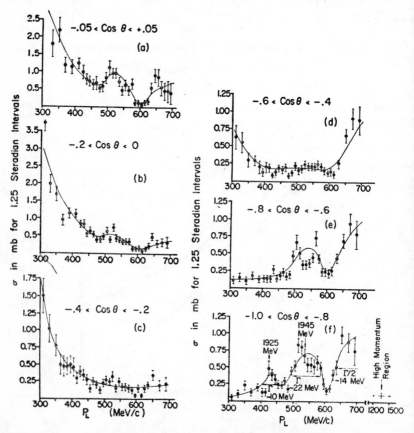

FIG. 86. Energy dependence of $p\bar{p}$ differential cross section for various $\cos\vartheta$ cuts (156).

ters assume no background interference.) The behaviour of the backward peak at the high-energy limit of the experiment suggests a third state with mass ≥ 1975 MeV and width ≥ 20 MeV. The 1925 peak is seen to occur only in the most backward direction, which suggests a high-spin state (the absence of a peak at $\vartheta \sim 0°$ might suggest resonance in an odd orbital angular momentum state). The two higher-energy backward peaks are broader, which suggests lower angular momentum, and the appearance of the enhancements at $\vartheta \sim 0°$ might suggest even orbital angular momentum states. Of course detailed study requires a knowledge of background behaviour but the background between the peaks is encouragingly small. It is clearly tempting to identify one or other of the two lower-mass resonances with the S meson.

Several experiments have studied $\bar{p}p$ and $\bar{p}n$ cross sections in the vicinity of the T meson (157). The bubble-chamber topological cross sections (157) show no significant structure (Figure 87c) nor does the single-pion production cross section (Figure 87b). On the other hand the $\bar{p}p$ backward elastic scattering cross section (Figure 87a) shows a broad enhancement which is probably not shown by $\bar{p}n$ (though the data are poorer) and is thus from the $I=1$ state. It appears to be at a higher mass than the T meson or the peak in the total cross section.

Evidence has been put forward in the reaction $\pi^+ p \rightarrow p\pi^+ \pi^0$ at 8 GeV/c for a $\pi^+\pi^0$ resonance on the high-mass side of the g meson (158) (Figure 88), with a mass of 1900 ± 40 MeV and a width of 216 ± 105 MeV. The same state may have been seen in $\pi^- p$ interactions at 7 GeV/c (136).

Many high-energy experiments show suggestive enhancements in the mass region above 1900 MeV. In general, however, such enhancements are as yet unconfirmed in other experiments. For examples, the reader is referred to (112), and to (59).

11.2 Heavy Strange Mesons

If there are heavy nonstrange bosons, then from $SU(3)$ there must be corresponding heavy strange bosons. Those heavier than 2060 MeV could couple to the $Y\overline{N}$, $\overline{Y}N$ systems. Alexander et al. (159) have reported a possible $K^* \rightarrow \overline{Y}N$ decay.

In 90K photographs of K^+p interactions at 9 GeV/c the following antihyperon production was found amongst events with seen V^0's.

State	No. of events	μb
1. $\overline{\Lambda}pp + \overline{\Sigma}^0 pp$	25	11.4 ± 2.3
2. $\overline{\Lambda}pp\ \pi^0$	21	9.0 ± 2.0
3. $\overline{\Lambda}pn\ \pi^+$	76	35.2 ± 4.0
4. $\overline{\Lambda}pp\ \pi^+\pi^-$	6	2.7 ± 1.1

In 1. the 25 events consist of 8 sure $\overline{\Lambda}$, 4 sure $\overline{\Sigma}^0$, and 13 ambiguous treated as $\overline{\Lambda}$ in the analysis in view of the nonisotropic γ-decay distribution they give if treated as $\overline{\Sigma}$. In 2. and 3. $\overline{\Sigma}^0$ cannot be fitted and some of the events will be

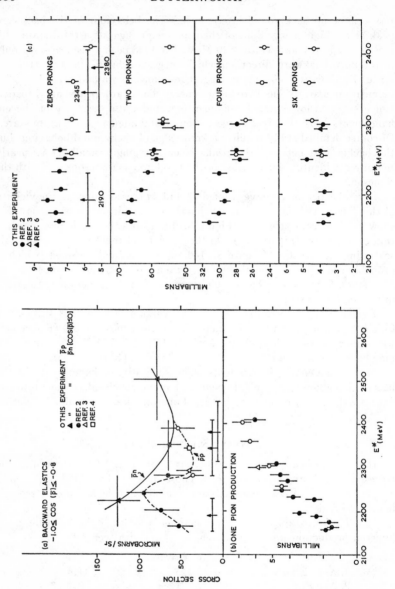

FIG. 87. (a) $p\bar{p}$ and $\bar{p}n$ elastic cross sections in the backward direction. Proton target data for cos $\vartheta < -0.8$, neutron target data for cos $\vartheta < 0$. (b) The single-pion production cross section in $p\bar{p}$ interactions. (c) The topological cross sections for $p\bar{p}$ interactions (157a).

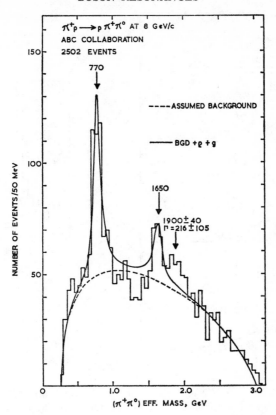

FIG. 88. Dipion effective mass in $\pi^+ p \rightarrow p\pi^+\pi^0$ at 8 GeV/c (158).

of this type. It is checked that this ambiguity is not the source of the claimed enhancement. Figure 89a shows the $\overline{Y}N$ mass plot for both four- and three-body (shown alone shaded) channels. Two combinations are plotted per event. A peak is seen at 2240 ± 20 MeV with width $\sim$70 MeV. Figure 89 shows the peripheral nature of the reaction, baryons always peaking forward or back. Figure 89b is a peripheral selection with one N forward ($\cos\vartheta_{Nf} > 0$) and one back ($\cos\vartheta_{Nb} < 0$)—only $\overline{Y}N_f$ being plotted. This selection leaves 95 of the 122 original events and makes the enhancement more pronounced, which suggests its interpretation as K^* produced by nonstrange exchange. The $\overline{\Lambda}$ polarization is 0.7 ± 0.49 in the enhancement and $.07 \pm 0.28$ outside. All four unique $\overline{\Sigma}pp$ events lie out of the peak area. One can safely say that the enhancement is due to $\overline{\Lambda}N$ and so $I = \frac{1}{2}$. No $K\pi$, $K^*\pi$, $K\omega$, or $K\phi$ modes could be found.

The Birmingham, Glasgow, Oxford collaboration report the same en-

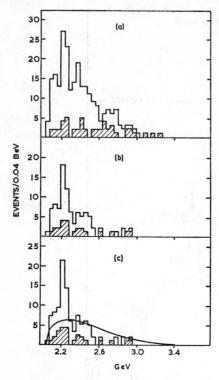

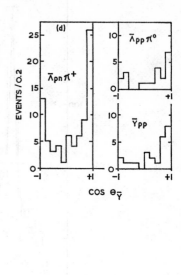

FIG. 89. Evidence for $\overline{Y}N$ enhancement in K^+p interactions at 9 GeV/c. (a) Mass distribution for all $\overline{Y}N$ combinations from both the three- and four-body reactions listed in the text. (Inner histogram shows three-body alone.) There are two combinations per event. (b) Peripheral events. (c) Peripheral events weighted for decay probability. (d) Production centre-of-mass angular distribution of the $\overline{Y}$ for the reactions listed.

hancement in K^+p interactions at 10 GeV/c (160), but negative evidence has been presented at 8.25 GeV/c and at 12.7 GeV/c (161).

Alexander et al. also obtain the $K^0(4\pi)^-$ spectrum of Figure 90 for the reaction $K^+p \rightarrow K^0p\ \pi^+\pi^+\pi^-\pi^0$ at 9 GeV/c. There is an enhancement of four S.D. at 2640 MeV with width $\sim$80 MeV. The effect is shown of demanding that $\pi^+\pi^-\pi^0$ lie in the "A" region or that either $\pi^+\pi^-\pi^0$ be in the "A" region or $K^0\pi^+$ in the $K^*(890)$ region. All histograms show the effect.

11.3 THE HIGH-MASS REGION AND THE QUARK MODEL

It is clear that the knowledge of the higher-mass states is still too rudimentary to permit any clear identification of states with those expected on the quark model. Moreover, for each value of orbital angular momentum l

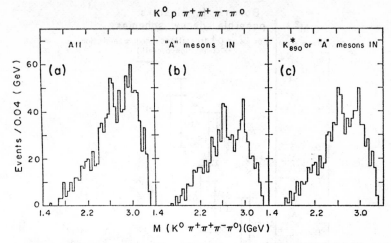

FIG. 90. $M(K^0\pi^+\pi^+\pi^-\pi^0)$ for $K^+p \to K^0p\pi^+\pi^+\pi^-\pi^0$ at 9 GeV/c (92c). (a) All events. (b) With $M(\pi^+\pi^-\pi^0)$ in the A region. (c) With $M(\pi^+\pi^-\pi^0)$ in the A region or $M(K^0\pi^+)$ in the $K^*(890)$ region.

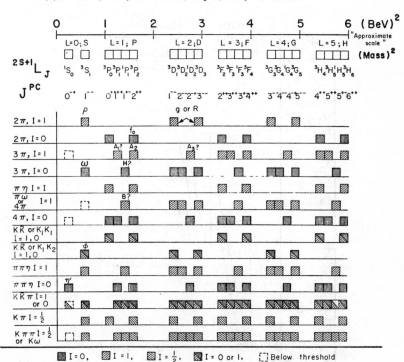

FIG. 91. Allowed decays into the better-known mesons for nonets generated by the quark model (162).

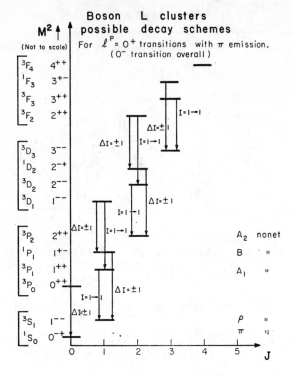

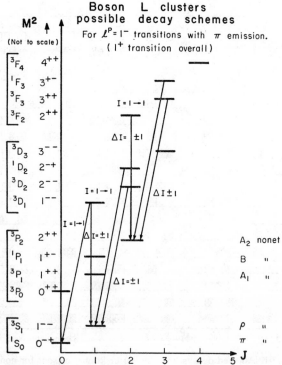

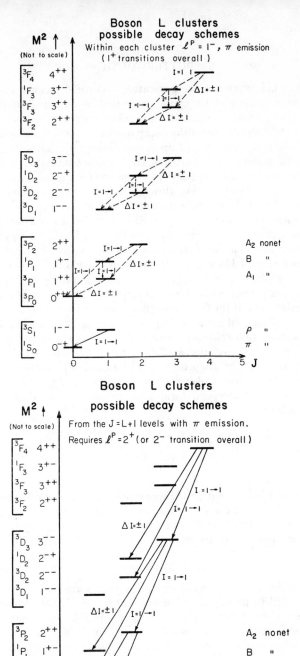

FIG. 92. Allowed cascade decays via π emission for nonets generated by the quark model (162). (a) $l_\pi = 0$. (b) $l_\pi = 1$ with change in $l_{q\bar{q}}$. (c) $l_\pi = 1$ with no change in $l_{q\bar{q}}$. (d) $l_\pi = 2$ from the $J = l+1$ nonets.

on this model, four nonets are generated; this could mean a considerable overlapping of states with the consequence that resolution of unique mesons would prove extremely difficult. To guide experimental investigations Goldhaber (162) has discussed the decay modes expected for mesons describable by the quark model with l up to five. Figure 91 shows the permitted decays to the better-known particles. He has also pointed out that in such decays, the member of the cluster of four nonets for a given value of l which has $J = l+1$ consistently decays through higher angular momentum states and may be expected to have a narrower width. Of course the preferential decay of such states will be by cascading down to a lower-mass but still high-spin state, say by π emission, since this will involve a much reduced angular momentum barrier. Figure 92 shows the allowed decays. Since those decays which stay within a given cluster of nonets will be highly inhibited or disallowed by phase-space considerations, we again note that the $J = l+1$ member can only decay with $\Delta l = 2$ and hence may be expected to be narrow. It is interesting that if the (mass)^2 of the states continues to increase linearly with l (or J), the actual mass spacing between states becomes less, leading to reduced phase space for cascade decay. Goldberg (163) has calculated that, in the limit of large J, decay into two light or two heavy particles gives rise to a width varying as $\Gamma \alpha J^{-aJ}$ where $a > 0$ ($a = 1$ for two light particles). The growing number of such partial-decay channels still cannot give rise to a width which competes with the partial width for the emission of a single-pion mass μ, which varies as:

$$\Gamma \propto \left[\frac{1}{J \log {}^J/C} \right]^{2\mu J^{1/2}}$$

where C is a constant.

Such conclusions, based on the assumption that trajectories continue to rise linearly, show that narrow states should continue to be found in the high-mass region. If taken literally they would imply that eventually mesons would become stable against strong decay at $M \sim 20$ GeV.

12. CONCLUSIONS

The experimental data in the high-mass region are clearly not yet sufficiently detailed to permit a serious attempt to classify meson states. Thus, for example, the validity of the quark model in this region cannot yet be tested. However, almost any model of bosons indicates that there are still many states to be discovered. We have seen, moreover, that even in the region below 1600 MeV many unresolved problems remain. Missing-mass experiments over a wider range of momentum transfer, with recoil neutron and recoil proton detection using both incident π and K beams, are desperately needed. More sophisticated decay-product analysis in such experiments is wanted. The operation or imminent operation of semiautomatic measuring devices at many laboratories will lead to bubble-chamber experiments of improved statistics. It would be very desirable to see these backed up by the construction of bubble chambers capable of greater resolution. The use of electronic triggering for fast cycling chambers could see the detailed analysis of the bubble-chamber technique allied to the statistical accuracy of the counter technique. One suspects that the growing complexity of the field of boson resonances badly needs such improvements in experimental methods if the underlying structure is to be clearly understood. Despite this, it is remarkable that though the definite establishment of a single state with the appropriate quantum number would destroy it, the nonrelativistic quark model remains the most satisfactory scheme for the classification and discussion of hadron states.

ACKNOWLEDGMENTS

I should like to express my gratitude to Dr. T. C. Bacon and Dr. G. A. Ringland for a number of helpful suggestions in the preparation of this review. I am especially grateful to my wife for her constant support.

REFERENCES AND ANNOTATIONS

1. Levinson, C. A., Lipkin, H. J., Meshkov, S. *Nuovo Cimento*, **32**, 1376 (1964)
2. Gursey, F., Radicati, L. *Phys. Rev. Letters*, **13**, 173 (1964)
3. R. H. Dalitz, *Proc. Philadelphia Topical Conf. Meson Spectroscopy, Philadelphia 1968* (Benjamin, New York, 1968). This article contains a large bibliography on the quark model
4. For a general discussion see Collins, P. D. B., Squires, E. H., *Regge Poles in Particle Physics, Springer Tracts in Modern Physics, No. 45* (Berlin, 1968)
5. Freedman, D. Z., Wang, J. M., *Phys. Rev.*, **153**, 153 (1967)
6. Gell-Mann, M., Zweig, G., cited in Harari, H. *Proc. 14th Intern. Conf. High Energy Phys., Vienna, 1968*
7. For a detailed bibliography on the quark model see Ref. 3
8. Goldhaber, G., *Proc. 13th Intern. Conf. High Energy Phys., Berkeley, 1967* (Univ. California Press, 1967)
9. Barger, V., *Rev. Mod. Phys.*, **40**, 129 (1968)
10. Abolins, M. A., Dahl, O. I., Danburg, J., Davies, R., Hoch, P., Miller, D. H., Rader, R., Kirz, J., *Phys. Rev. Letters*, **22**, 427 (1969)
11. Ferro-Luzzi, M., George, R., Gold-schmidt-Clermont, Y., Henri, V. P., Jongejans, B., Leith, D. W. G., Lynch, G. R., Muller, F., Perreau, J. M., *Phys. Letters*, **17**, 155 (1965)
12. Dodd, W. P., Joldersma, T., Palmer, R. B., Samios, N. P., *Phys. Rev.*, **177**, 1991 (1969)
13. Bock, R., French, B. R., Kinson, J. B., Simak, V., Badier, J., Bazin, M., Equer, B., Rouge, A., *Phys. Letters*, **12**, 65 (1964)
14. French, B. R., Kinson, J. B., Rigopoulos, R., Simak, V., McDonald, F., Petmezas, G., Riddiford, L., *Nuovo Cimento*, **52**, 438 (1967)
15. Baltay, C., Lach, J., Sandweiss, J., Taft, H. D., Yeh, N., Stonehill, D. L., Stump, R., *Phys. Rev.*, **142**, 932 (1966)
16. Wangler, T. P., Erwin, A. R., Walker, W. D., *Phys. Letters*, **9**, 71 (1964)
17. Miller, D. H., Kovacs, A. Z., McIlwain, R. L., Palfrey, T. R., Tautfest, G. W., *Phys. Letters*, **15**, 74 (1965)

18. Dahl, O. I., Hardy, L. M., Hess, R. I., Kirz, J., Miller, D. H., *Phys. Rev.*, **163**, 1377 (1967)
19(a). Bishop, J. M., Goshaw, A. T., Erwin, A. R., Thompson, M. A., Walker, W. D., Weinberg, A., *Phys. Rev. Letters*, **16**, 1069 (1966)
   (b) Goshaw, A. T., Erwin, A. R., Walker, W. D., Weinberg, A., *Wisconsin Preprint 1967*
20. Rosenfeld, A., *Proc. Univ. Pennsylvania Conf. Meson Spectroscopy, Philadelphia, 1968* (Benjamin, New York, 1968)
21. Vanderhaghen, R., Huc, J., Fleury, P., Duboc, J., George, R., Goldberg, M., Makowski, B., Armenise, N., Ghidini, B., Picciarelli, V., Romano, A., Forino, A., Gessaroli, R., Quareni, G., Quareni-Vignudelli, A., *Phys. Letters*, **24B**, 493 (1967)
22. Fung, S. Y., Jackson, W., Pu, R. T., Brown, D., Gidal, G., *Phys. Rev. Letters*, **21**, 50 (1968)
23. Horn, D., Lipkin, H. J., Meshkov, S., *Phys. Rev. Letters*, **17**, 1200 (1966)
24. The footnote given in Ref. 23 stating that a 1^+ meson made of $q\,q\,\bar{q}\,\bar{q}$ could not decay into 0^-1^- is in error
25. Martin, H. J., Crittenden, R. R., Schroeder, L. S., *Phys. Letters*, **22**, 352 (1966)
26. Barbaro-Galtieri, A., Matison, M., Rittenberg, A., Shively, F. T., *Phys. Rev. Letters*, **20**, 349 (1968)
27. SABRE Collaboration, *Phys. Letters*, **26B**, 674 (1968)
28. Kalbfleisch, G. R., Dahl, O. I., Rittenberg, A., *Phys. Rev. Letters.* **13**, 349 (1964)
29. Rittenberg, A., quoted in *UCRL 8030*, Jan. 1969 edition
30. Bollini, D., Buhler-Broglin, A., Dalpiaz, P., Massam, T., Navach, F., Navarria, F. L., Schneegans, M. A., Zichichi, A., *Nuovo Cimento*, **58A**, 289 (1968)
31. Foster, M., Gavillet, Ph., Labrosse, G., Montanet, L., Salmeron, R. A., Villemoes, P., *Nucl. Phys.*, **B8**, 174 (1968)
32. Zaslavsky, A., Ogievetsky, V., Tybor, W., *Dubna Preprint No. E2-4064* (1968)
33. Benson, G., Lovell, L., Murphy, C., Roe, B., Sinclair, D., Van der

Velde, J. (Univ. Michigan preprint); Lai, K., Schumann, T. (Brookhaven Natl. Lab. preprint), cited by Dalitz, R. H., *Proc. 13th Intern. Conf. High Energy Phys.*, *Berkeley, 1966* (Univ. California Press, 1967)

34. Baillon, P., Edwards, D., Marechal, B., Montanet, L., Tomas, M., D'Andlau, C., Astier, A., Cohen-Ganouna, J., Della-Negra, M., Wojcicki, S., Baubillier, M., Duboc, J., James, F., Levy, F., *Nuovo Cimento*, **50A**, 393 (1967)

35. D'Andlau, C., Astier, A., Cohen-Ganouna, J., Della-Negra, M., Lorstad, B., Aguilar-Benitez, M., Barlow, J., Jacobs, L. D., Malecki, M., Montanet, L., Contrib. *14th Intern. Conf. High Energy Phys.*, *Vienna, 1968* (Unpublished)

36. See also Ref. 18

37. Glashow, S., Socolow, R., *Phys. Rev. Letters*, **15**, 329 (1965)

38. Binnie, D. M., Duane, A., Faruqui, A. R., Horsey, P. J., Jones, W. G., Kay, M. E., Mason, D. C., Nicholson, P. J., Rahman, I. U., Walters, J., Wilson, J. G., *Phys. Letters*, **27B**, 106 (1968)

39. Bollini, D., Buhler-Broglin, A., Dalpiaz, P., Massam, T., Navach, F., Navarria, F. L., Schneegans, M. A., Zichichi, A., *Nuovo Cimento*, **56A**, 1173 (1968)

40. Becker, U., Bertram, W. M., Binkley, M., Jordan, C. L., Knasel, T. M., Marshall, R., Quinn, D. J., Rohde, M., Smith, A. J. S., Ting, S. C. C., *Phys. Rev. Letters*, **21**, 1506 (1968)

41. Augustin, J. E., Bizot, J. C., Buon, J., Delcourt, B., Haissinski, J., Jeanjean, J., Lalanne, D., Marin, P. C., Nguyen-Ngoc, H., Perez-y-Jorba, J., Richard, F., Rumpf, F., Treille, D., *Phys. Letters*, **28B**, 517 (1969)

42. Lindsey, J. S., Smith, G. A., *Phys. Rev.*, **147**, 913 (1966)

43. Bollini, D., Buhler-Broglin, A., Dalpiaz, P., Massam, T., Navach, F., Navarria, F. L., Schneegans, M. A., Zichichi, A., *Nuovo Cimento*, **57A**, 404 (1968)

44. Augustin, J. E., Beneksas, D., Buon, J., Gracco, V., Haissinski, J., Lalanne, D., Leplanche, F., Lefracois, J., Lehmann, P., Marin, P., Rumpf, F., Silva, E., *Phys. Letters*, **28B**, 513 (1969)

45. Particle Data Group, *UCRL-8030.* Jan. 1969 edition

46. For example, Kroll, N., Lee, T. D., Zumino, B., *Phys. Rev.*, **157**, 1376 (1967); Oakes, R. J., Sakurai, J. J., *Phys. Rev. Letters*, **19**, 1266 (1967); Das, T., Mathur, V. S., Okubo, S., *Phys. Rev. Letters*, **19**, 470 (1967)

47. Goldberg, M., *Proc. Symp. Present Status of SU(3). Argonne Natl. Lab., 1967,* 17

48. Roos, M., cited in Ref. 45

49. Auslender, V. L., Budker, G. I., Pestov, Yu. N., Sidorov, V. A., Skrinsky, A. N., Khabakhpashev, A. G., *Phys. Letters*, **26B**, 433 (1967) and *Novosibirsk Preprint No. 243* (1968)

50. Augustin, J. E., Bizot, J. C., Buon, J., Haissinski, J., Lalanne, D., Marin, P., Nguyen-Ngoc, H., Perez-y-Jorba, J., Rumpf, F., Silva, E., Tavernier, S., *Phys. Letters*, **28B**, 508 (1969)

51. Levrat, B., Tolstrup, C. A., Schubelin, P., Nef, C., Martin, M., Maglic, B. C., Keinzle, W., Focacci, M. N., Dubal, L., Chikovani, G., *Phys. Letters*, **22**, 714 (1966)

52. Chikovani, G. E., Focacci, M. N., Lechanoine, C., Levrat, B., Maglic, B. C., Martin, M., Schubelin, P., Dubal, L., Fischer, M., Grieder, P., Nef, C., *Phys. Letters*, **25B**, 44 (1967)

53. Benz, H., Chikovani, G. E., Damgaard, G., Focacci, M. N., Kienzle, W., Lechanoine, C., Martin, M., Nef, C., Schubelin, P., Baud, R., Bosnjakovic, B., Cotteron, J., Klanner, R., Weitsch, A., *Phys. Letters*, **28B**, 233 (1968)

54. Chikovani, G. E., Focacci, M. N., Kienzle, W., Kruse, U., Lechanoine, C., Martin, M., Schubelin, P., *Phys. Letters*, **28B**, 526 (1969)

55. Crennell, D. J., Karshon, U., Lai, K. W., Scarr, J. M., Skillicorn, I. O. *Phys. Rev. Letters*, **20**, 1318 (1968)

56. CERN 4 GeV/c /π⁻p Experiment (Private communication)

57. Donald, R. A., Edwards, D. N., Foster, M., Moore, R. S., *Univ. Liverpool Preprint, 1969*

58. Aguilar-Benitez, M., Barlow, J., Jacobs, L. D., Malecki, P., Montanet, L., Tomas, M., Lorstad, B., West, N. (Submitted to *Phys. Letters*)

59. French, B. R., *Proc. 14th Intern. Conf. High Energy Phys., Vienna, 1969*

60. Lipkin, H. J., Meshkov, S., *Phys. Rev. Letters*, **22**, 212 (1969)

61. Ascoli, G., Crawley, H. B., Mortara, D. W., Shapiro, A., *Phys. Rev. Letters*, 21, 1712 (1968)
62. Ascoli, G., Crawley, H. B., Mortara, D. W., Shapiro, A., *Phys. Rev. Letters*, 20, 1411 (1968)
63. D'Andlau, C., Astier, A., Dobrzynski, L., Siaud, J., Barlow, J., Montanet, L., Tallone-Lombardi, L., Adamson, A. M., Duboc, J., Goldberg, M., Donald, R. A., Edwards, D. N., Lys, J. E. A., *Nucl. Phys.*, B5, 693 (1968)
64. Aachen-Berlin-Birmingham-Bonn-Hamburg-London (I.C.)-Munich Collaboration, *Phys. Letters*, 11, 167 (1964)
65. Benson, G., Marquit, E., Roe, B., Sinclair, D., Van der Velde, J., *Phys. Rev. Letters*, 17, 1234 (1966)
66. Cohn, H. O., McCulloch, R. D., Bugg, W. M., Condo, G. T., *Nucl. Phys.*, B1, 57 (1967)
67. Armenise, N., Ghidini, B., Picciarelli, V., Romano, A., Forino, A., Gessaroli, R., Leninara, L., Quareni, G., Quareni-Vignudelli, A., Cartacci, A., Dagliana, M. G., Di Caporiacco, G., Parrini, G., Barrier, M., Laberrique-Frolow, J., Quinquard, J., *Phys. Letters*, 26B, 336 (1968)
68. Abolins, M., Dahl, O. I., Danburg, J. S., Davies, D., Hoch, P., Kirz, J., Miller, D. H., Rader, R., cited Butterworth, I., *Proc. Heidelberg Intern. Conf. Elementary Particles*, 24 (North-Holland, Amsterdam, 1968)
69. Barbaro-Galtieri, A., Soding, P., *Proc. Univ. Pennsylvania Conf. Meson Spectroscopy, Philadelphia, 1968*
70. Aachen-Berlin-Birmingham-Bonn-Hamburg-London(I.C.)-Munich Collaboration, *Phys. Rev.*, 138B, 897 (1965)
71. Aachen-Berlin-Bonn-Hamburg Munich Collaboration, *Nuovo Cimento*, 44, 530 (1966)
72. Deck, R. T., *Phys. Rev. Letters*, 13, 169 (1964)
73. Berger, E. L., *Phys. Rev.*, 166, 1525 (1968)
74. Ioffredo, M. L., Brandenburg, G. W., Brenner, A. E., Eisenstein, B., Eisenstein, L., Johnson, W. H., Kim, J. K., Law, M. E., Salzberg, B. M., Scharenguivel, J. H., Sisterson, L. K., Szymanski, J. J., *Phys. Rev. Letters*, 21, 1212 (1968)
75. Cnops, A. M., Hough, P. V. C., Huson, F. R., Kenyon, I. R., Scarr, J. M., Skillicorn, I. O., Cohn, H. O., McCulloch, R. D., Bugg, W. M., Condo, G. T., Nussbaum, M. M., *Phys. Rev. Letters*, 21, 1609 (1968)
76. ABC and ABCLV Collaborations, *Phys. Letters*, 27B, 336 (1968)
77. Chew, G. F., Pignotti, A., *Phys. Rev. Letters*, 20, 1078 (1968)
78. Goldhaber, G., *Phys. Rev. Letters*, 19, 976 (1967)
79. Danysz, J. A., French, B. R., Simak, V., *Nuovo Cimento*, 51A, 801 (1967)
80. Fridman, A., Maurer, G., Michalon, A., Oudet, J., Schiby, B., Strub, R., Voltolini, C., Cuer, P., *Phys. Rev.*, 168, 1268 (1968)
81. Alyea, E. D., Crittenden, R. R., Galloway, K. F., Lee, K. Y., Martin, H. J., Suen, K. F., *Phys. Rev. Letters*, 21, 1421 (1968)
82. Ascoli, G., Crawley, H. B., Kruse, U., Mortara, D. W., Schafer, E., Shapiro, A., Terrault, B., *Phys. Rev. Letters*, 21, 113 (1968)
83. Von Krogh, J., Miyashita, S., Kopelman, J. B., Libby, L. M., *Phys. Letters*, 27B, 253 (1968)
84. Aachen-Berlin-Bonn-CERN-Warsaw Collaboration (Submitted to *Nucl. Phys.*, 1968); Contrib. *14th Intern. Conf. High Energy Phys., Vienna, 1968*
85. For a compilation of earlier data see Butterworth, I., *Proc. Heidelberg Intern. Conf. Elementary Particles*, 11 (North-Holland, Amsterdam, 1968)
86. Ballam, J., Brody, A. D., Chadwick, G. B., Fries, D., Guiragossian, Z. G. T., Johnson, W. B., Larsen, R. R., Leith, D. W. G. S., Martin, F., Perl, M., Pickup, E., Tan, T. H., *Phys. Rev. Letters*, 21, 934 (1968)
87. Slattery, P., Kraybill, H., Forman, B., Ferbel, T. *Nuovo Cimento*, 50A, 377 (1967)
88. For example,
    (a) Park, J. C., Kim, S., *Phys. Rev.*, 174, 2165 (1968)
    (b) Berlinghieri, J., Farber, M. S., Ferbel, T., Forman, B., Melissinos, A. C., Yamanouchi, T., Yuta, H., *Phys. Rev. Letters*, 18, 1087 (1967)
    (c) Chien, C. Y., Dauber, P. M., Malamud, E. I., Mellema, D. J., Schlein, P. E., Schreiner, P. A., Slater, W. E., Stork, D. H., Ticho,

H. K., Trippe, T. G., *Phys. Letters*, **28B**, 143 (1968)

(d) Aachen - Berlin - CERN - London (I.C.)-Vienna Collaboration, *Nucl. Phys.*, **8B**, 9 (1968)

(e) Denegri, D., Callahan, A., Ettlinger, L., Gillespie, D., Goodman, G., Luste, G., Mercer, R., Moses, E., Pevsner, A., Zdanis, R., *Phys. Rev. Letters*, **20**, 1194 (1968)

(f) Bomse, F., Borenstein, S., Callahan, A., Cole, J., Cox, B., Ellis, D., Ettlinger, L., Gillespie, D., Luste, G., Mercer, R., Moses, E., Pevsner, A., Zdanis, R., *Phys. Rev. Letters*, **20**, 1519 (1968)

(g) Denegri, D., Callahan, A., Ettlinger, L., Gillespie, D., Goodman, G., Luste, G., Mercer, R., Moses, E., Pevsner, A., Zdanis, R., *Phys. Rev. Letters*, **20**, 1194 (1968); Contrib. *14th Intern. Conf. High Energy Phys.*, *Vienna*, *1968*

89. Shen, B. C., Butterworth, I., Fu, C., Goldhaber, G., Goldhaber, S., Trilling, G. H., *Phys. Rev. Letters*, **17**, 726 (1966)

90. Bassompierre, G., Goldschmidt-Clermont, Y., Grant, A., Henri, V. P., Hughes, I., Jongejans, B., Lander, R. L., Linglin, D., Muller, F., Perreau, J. M., Saitov, I., Sekulin, R. L., Wolf, G., de Beare, W., Debaisieux, J., Dufour, P., Grard, F., Heughebaert, J., Pape, L., Peeters, P., Verbeure, F., Windmolders, R., Jobes, M., Matt, W., *Phys. Letters*, **26B**, 30 (1967)

91. Johns Hopkins group cited French, B. R., *Proc. 14th Intern. Conf. High Energy Phys.*, *Vienna*, *1968*

92(a) Goldhaber, G., Firestone, A., Shen, B. C., *Phys. Rev. Letters*, **19**, 972 (1967)

(b) Goldhaber, G., *Phys. Rev. Letters*, **19**, 976 (1967)

(c) Alexander, G., Firestone, A., Goldhaber, G. (Submitted to *Phys. Rev.*, 1969)

93. Birmingham-Glasgow-Oxford Collaboration cited French, B. R., *Proc. 14th Intern. Conf. High Energy Phys.*, *Vienna*, *1968*

94(a) Armenteros, R., Edwards, D. N., Jacobson, T., Montanet, L., Shapira, A., Vandermeulen, J., d'Andlau, Ch., Astier, A., Baillon, P., Cohen-Ganouna, J., Defoix, C., Siaud, J., Ghesquiere, C., Rivet, P., *Phys. Letters*, **9**, 207 (1964)

(b) Barash, N., Kirsch, L., Miller,

D., Tan, T. H., *Phys. Rev.* **145**, 1095 (1966)

95. Astier, A., Cohen-Ganouna, J., Della-Negra, M., Marechal, B., Montanet, L., Zoll, J., Baubillier, M., Duboc, J., Levy, F., James, R., Edwards, D. N., Donald, R., *CERN/D.Ph.II/68-32*

96. Crennell, D. J., Kalbfleisch, G. R., Lai, K. W., Scarr, J. M., Schumann, T. G., *Phys. Rev. Letters*, **19**, 44 (1967)

97. Bishop, J. M., Goshaw, A. T., Erwin, A. R., Thompson, M. A., Walker, W. D., Weinberg, A., *Phys. Rev. Letters*, **16**, 1069 (1966)

98. Crennell, D. J., Karshon, U., Lai, K. W., O'Neall, J. S., Scarr, J. M., *Phys. Rev. Letters*, **22**, 487 (1969)

99. Astier, A., Cohen-Ganouna, J., Della-Negra, M., Marechal, B., Montanet, L., Tomas, M., Baubillier, M., Duboc, J., *Phys. Letters*, **25B**, 294 (1967)

100. Kienzle, W., Maglic, B. C., Levrat, B., Lefebvres, F., Freytag, D., Blieden, H., *Phys. Letters*, **19B**, 438 (1965)

101. Defoix, C., Rivet, P., Siaud, J., Conforto, B., Widgoff, M., Shively, F., *Phys. Letters*, **28B**, 353 (1968)

102. Barnes, V., Dornan, P., Guidoni, P., Samois, N., Goldberg, M., Leitner, J., cited by B. R. French, *Proc. 14th Intern. Conf. High Energy Phys.*, *Vienna*, *1968*

103. Ammar, R., Davis, R., Kropac, W., Mott, J., Slate, D., Werner, B., Derrick, M., Fields, T., Schweingruber, F., *Phys. Rev. Letters*, **21**, 1832 (1968)

104. Juhala, R. E., Leacock, R. A., Rhode, J. I., Kopelman, J. B., Libby L. M., Urvater, E., *Phys. Letters*, **27B**, 257 (1968)

105. Allison, W. W. M., Cruz, A., Schankel, W., Haque, M. M., Tuli, S. K., Finney, P. J., Fisher, C. M., Gordon, J. D., Turnbull, R. M., Erskine, R., Sisterson, K., Paler, K., Chaudhuri, P., Eskreys, A., Goldsack, S. J., *Phys. Letters*, **25B**, 619 (1967)

106. Alitti, J., Baton, J. P., Deler, B., Nevue-Rene, M., Crussard, J., Ginestet, J., Tran, A. H., Gessaroli, R., Romano, A., *Phys. Letters*, **15**, 69 (1965)

107. Weickowicz, R. P., Reynolds, B. G., Albright, J. R., Bradley, R. H., Harms, B. C., Harrison, W. C.,

Lannutti, J. E., Sims, W. H., *Phys. Letters*, **28B**, 199 (1968)

108. Trippe, T. G., Chien, C. Y., Malamud, E., Mellema, J., Schlein, P. E., Slater, W. E., Stork, D. H., Ticho, H. K., *Phys. Letters*, **28B**, 203 (1968)

109. de Baere, W., Debaisieux, J., Dufour, P., Grard, F., Heughebaert, J., Pape, L., Peeters, P., Verbeure, F., Windmolders, R., Goldsmidt-Clermont, Y., Henri, V. P., Jongejans, B., Moiseev, A., Muller, F., Perreau, J. M., Prokes, A., Yarba, V., *Nuovo Cimento*, **51A**, 401 (1967)

110. Marateck, S., Hagopian, V., Selove, W., Jacobs, L., Oppenheimer, F., Schultz, W., Gutay, L. J., Miller, D. H., Prentice, J., West, E., Walker, W. D., *Phys. Rev. Letters*, **21**, 1613 (1968)

111. Malamud, E., Schlein, P. E., *Phys. Rev. Letters*, **19**, 1056 (1967)

112. See I. Butterworth, *Proc. Heidelberg Intern. Conf. Particle Physics*, 11 (North-Holland, Amsterdam, 1968)

113. Feldman, M., Frati, W., Gleeson, R., Halpern, J., Nussbaum, M., Richert, S., *Phys. Rev. Letters*, **22**, 316 (1969)

114. Braun, K. J., Cline, D., Scherer, V., *Phys. Rev. Letters*, **21**, 1275 (1968)

115. Corbett, I. F., Damerell, C. J. S., Middlemas, N., Newton, D., Clegg, A. B., Williams, W. S. C., Carroll, A. S., *Phys. Rev.*, **156**, 1451 (1967)

116. Smith, G. A., Manning, R. H., *Phys. Rev.*, **171**, 1399 (1968)

117. Baton, J. P., Laurens, G., Reignier, J., *Nucl. Phys.*, **B3**, 349 (1967)

118. Biswas, N. N., Cason, N. M., Johnson, P. B., Kenney, V. P., Poirier, J., Shephard, W. D., Torgerson, R., *Phys. Letters*, **27B**, 513 (1968)

119. Lovelace, C., Heinz, R. M., Donnachie, A., *Phys. Letters*, **22**, 332 (1966)

120. Crennell, D. J., Kalbfleisch, G. R., Lai, K. W., Scarr, J. M., Schumann, T. G., Skillicorn, I. O., Webster, M. S., *Phys. Rev. Letters*, **16**, 1025 (1966)

121. Alitti, J., Barnes, V. E., Crennell, D. J., Flaminio, E., Goldberg, M., Karshon, U., Lai, K. W., Metzger, W. J., O'Neall, J. S., Samios, N. P., Scarr, J. M., Schumann, T. G., *Phys. Rev. Letters*, **21**, 1705 (1968)

122. Aguilar-Benitez, M., Barlow, J., Jacobs, L. D., Malecki, P., Montanet, L., D'Andlau, C., Astier, A., Cohen-Ganouna, J., Della-Negra, M., Lorstad, B. (Submitted to *Phys. Letters*)

123. Barlow, J., Lillestol, E., Montanet, L., Tallone-Lombardi, L., D'Anglau, C., Astier, A., Dobrzynski, L., Wojcicki, S., Adamson, A. M., Duboc, J., James, F., Goldberg, M., Donald, R. A., James, R., Lys, J. E., Nisar, T., *Nuovo Cimento*, **50A**, 701 (1967)

124. Hess, R. I., Dahl, O. I., Hardy, L. M., Kirz, J., Miller, D. H., *Phys. Rev. Letters*, **17**, 1109 (1966)

125. Hoang, T. F., Earthy, D. P., Phelan, J. J., Roberts, A., Sandler, C. L., Bernstein, S., Marguilies, S., McLeod, D. W., Groves, T. H., Biswas, N. N., Cason, N. M., Kenney, V. P., Marrafino, J. M., McGahan, J. T., Poirier, J. A., Shephard, W. D., *Phys. Rev. Letters*, **21**, 316 (1969)

126. Beusch, W., Fischer, W. F., Gobbi, B., Pepin, M., Polgar, E., Astbury, P., Branti, C., Finocchiaro, G., Lassalle, J. C., Michelini, A., Terwilliger, K. W., Websdale, D., West, C. M., *Phys. Letters*, **25B**, 357 (1967)

Note: These authors only find the resonance fit acceptable, but the data are similar to those in Ref. 27 and a scattering-length fit is possible (Phelan, J. J., private communication)

127. Whitehead, C., McEwen, J. G., Ott, R. J., Aitken, D. K., Bennett, G. W., Jennings, R. E., *Nuovo Cimento*, **53A**, 817 (1968)

128. Miller, D. H., Gutay, L. J., Johnson, P. B., Kenney, V. P., Guiragossian, Z. G. T., *Phys. Rev. Letters*, **21**, 1489 (1968)

129. Cited Goldhaber, G., *Proc. 13th Intern. Conf. High Energy Phys. Berkeley, 1966*, 123 (Univ. California Press, Berkeley, 1967)

130. Lai, K. W., cited by French, B. R., *Proc. 14th Intern. Conf. High Energy Phys., Vienna, 1968*

131. Bettini, A., Cresti, M., Limentani, S., Loria, A., Peruzzo, L., Santangelo, R., Bertanza, L., Bigi, A., Carrara, R., Casali, R., Hart, E., Lariccia, P., *Nuovo Cimento*, **42**, 695 (1966)

132. Aguilar-Benitez, M., Barlow, J., Jacobs, L. D., Malecki, P., Montanet, L., D'Andlau, C., Astier, A., Cohen-Ganouna, J., Della-Negra,

M., Lorstad, B., cited French, B. R., *Proc. 14th Intern. Conf. High Energy Phys., Vienna, 1968*

133(a) Levrat, B., Tolstrup, C. A., Schubelin, P., Nef, C., Martin, M., Maglic, B. C., Kienzle, W., Focacci, M. N., Dubal, L., Chikovani, G., *Phys. Letters*, **22**, 714 (1966)

(b) Focacci, M. N., Kienzle, W., Levrat, B., Maglic, B. C., Martin, M., *Phys. Rev. Letters*, **17**, 890 (1966)

(c) Dubal, L., Focacci, M. N., Kienzle, W., Lechanoine, C., Levrat, B., Maglic, B. C., Martin, M., Schubelin, P., Chikovani, G., Fischer, M., Grieder, P., Neal, H. A., Nef, C., *Nucl. Phys.*, **B3**, 435 (1967)

134. Crennell, D. J., Hough, P. V. C., Kalbfleisch, G. R., Lai, K. W., Scarr, J. M., Schumann, T. G., Skillicorn, I. O., Strand, R. C., Webster, M. S., Baumel, P., Bachman, A. H., Lea, R. M., *Phys. Rev. Letters*, **18**, 323 (1967); Contrib. *14th Intern. Conf. High Energy Phys., Vienna, 1968*

135. Armenise, N., Ghidini, B., Picciarelli, V., Romano, A., Silvestri, A., Forino, A., Gessaroli, R., Leninara, L., Quareni, G., Quareni-Vignudelli, A., Cartacci, A. M., Dagliana, M. G., Di Caporiacco, G., Barrier, M., Laberrique-Frolow, J., Quinquard, J., Sene, M., Loskiewicz, J., *Nuovo Cimento*, **54A**, 999 (1968)

136. Johnston, T. F., Prentice, J. D., Steenberg, N. R., Yoon, T. S., Garfinkel, A. F., Morse, R., Oh, B. Y., Walker, W. D., *Phys. Rev. Letters*, **20**, 1414 (1968)

137. Ehrlich, R., Selove, W., Yuta, H., *Phys. Rev.*, **152**, 1194 (1966)

138. Abrams, G. S., Kehoe, B., Glasser, R. G., Sechi-Zorn, B., Wolsky, G., *Phys. Rev. Letters*, **18**, 620 (1967)

139. Caso, C., Conte, F., Tomasini, G., Cords, D., Diaz, J., von Handel, P., Mandelli, L., Ratti, S., Vegni, G., Daronian, P., Daudin, A., Gandois, B., Kochowski, C., Mosca, L., *Nuovo Cimento*, **54A**, 983 (1968)

140. Baltay, C., Kung, H. H., Yeh, N., Ferbel, T., Slattery, P. F., Rabin, M., Kraybill, H. L., *Phys. Rev. Letters*, **20**, 887 (1968)

141. Biswas, N. N., Cason, N. M., Groves, T. H., Kenney, V. P., Poirier, J. A., Shephard, W. D., *Phys. Rev. Letters*, **21**, 50 (1968)

142. Ballam, J., Brody, A. D., Chadwick, G. B., Guiragossian, Z. G. T.,

Johnson, W. B., Leith, D. W. G. S., Pickup, E., Contrib. *14th Intern. Conf. High Energy Phys., Vienna, 1968*

143. Danysz, J. A., French, B. R., Kinson, J. B., Simak, V., Clayton, J., Mason, P., Muirhead, H., Renton, P., *Phys. Letters*, **24B**, 309 (1967)

144. Donald, R., et al., cited in Ref. 59

145. Ferbel, T., *Proc. Philadelphia Conf. Meson Spectroscopy, Philadelphia, 1968*

146. Aachen-Berlin-CERN Collaboration, *Nucl. Phys.*, **B4**, 501 (1968)

147. Danysz, J. A., French, B. R., Simak, V., *Nuovo Cimento*, **51A**, 801 (1967)

148. Yost, G. P., et al. cited in Ref. 59

149. Ioffredo, M. L., et al. cited in Ref. 59

150. French, B. R., Kinson, J. B., Rigopoulos, R., Simak, V., McDonald, F., Petmezas, G., Riddiford, L., *Nuovo Cimento*, **52A**, 442 (1967)

151. Baltay, C., Lach, J., Sandweiss, J., Taft, H. D., Yeh, N., Stonehill, D. L., Stump, R., *Phys. Rev.*, **142**, 932 (1966)

152. Cooper, W. A., Hyman, L. G., Manner, W. E., Musgrave, B., Voyvodic, L., Contrib. *14th Intern. Conf. High Energy Phys., Vienna, 1968*

153. Jobes, M., Matt, W., Bassompierre, G., Goldschmidt-Clermont, Y., Grant, A., Henri, V. P., Hughes, I., Jongejans, B., Lander, R. L., Linglin, D., Muller, F., Perreau, J. M., Siatov, I., Sukelin, R., Wolf, G., de Baere, W., Debaisieux, J., Dufour, P., Grard, F., Heughebaert, J., Pape, L., Peeters, P., Verbeure, F., Windmolders, R., *Phys. Letters*, **26B**, 49 (1967); Contrib. *14th Intern. Conf. High Energy Phys., Vienna, 1968*

154. Carmony, D. D., Hendricks, T., Lander, R. L., *Phys. Rev. Letters*, **18**, 615 (1967)

155. Abrams, R. J., Cool, R. L., Giacomelli, G., Kycia, T. F., Leontic, B. A., Lai, K. K., Michael, D. N., *Phys. Rev. Letters*, **18**, 1209 (1967)

156. Cline, D. English, D. J., Reeder, D. D., Terrell, R., Twitty, J., *Phys. Rev. Letters*, **21**, 1268 (1968)

157(a) Ma, Z. M., Parker, D. L., Smith, G. A., Sprafka, R. J., Abolins, M. A., Rittenberg, A., Contrib. *14th Intern. Conf. High Energy Phys., Vienna, 1968*

(b) Cooper, W. A., Hyman, L. G., Manner, W., Musgrave, B., Voyvodic, L., *Phys. Rev. Letters*, **20**, 1059 (1968); Contrib. *14th Intern. Conf. High Energy Phys., Vienna, 1968*

(c) Lynch, G. R., Foulks, R. E., Kalbfleisch, G. R., Limentani, S., Shafer, J. B., Stevenson, M. L., Xuong, N. H., *Phys. Rev.*, **131**, 1276 (1963)

158. See Ref. 146. This represents an updating of the previous analysis published by Deutschmann et al., *Phys. Letters*, **18**, 351 (1965)

159. Alexander, G., Firestone, A., Gold-

haber, G., Shen, B. C., *Phys. Rev Letters*, **20**, 755 (1968)

160. Birmingham-Glasgow-Oxford Collaboration, Contrib. *14th Intern. Conf. High Energy Phys., Vienna, 1968*

161(a) Bassompiere, G., et al.; (b) Berlinghieri, J., et al. both cited in Ref. 59

162. Goldhaber, G., *Proc. 1967 CERN Sch. of Phys., CERN Report 67–24* (1967)

163. Goldberg, H., *Phys. Rev. Letters*, **21**, 778 (1968)

PION-NUCLEON INTERACTIONS

By R. G. Moorhouse

Glasgow University, Glasgow, Scotland

CONTENTS

1. LOW-ENERGY CONSTANTS, DISPERSION RELATIONS, AND SUM RULES

1.1 TOTAL CROSS SECTIONS

To take a first look at the pion-nucleon system we plot, in Figures 1–4, the total cross sections for $\pi^{\pm}$-p elastic scattering. At the lower energies these curves exhibit marked peaks and dips (and shoulders) corresponding to resonances in states of particular angular momentum and parity J^P, as discussed in Section 2. At higher energies the behaviour, within the limits of present experimental error, becomes smooth. This does not necessarily signify lack of resonances, or that high-energy resonances have a large width. Even with resonances at higher energy similar to those at lower energy, smoother behaviour is to be expected because of the decreasing importance of any single state of particular J^P among the increasing number of angular momentum states contributing to the total cross section.

The total cross sections for π^+p and π^-p both appear to be tending to constancy at high energy, thus fulfilling the condition on the total cross section of the original Pomeranchuk theorem (3, 4) leading to particle and antiparticle cross sections being asymptotically equal. Certainly the π^+p and π^-p cross sections appear to be approaching each other and to form the best indication among elementary-particle cross sections of this possible asymptopia.

1.2 THE A, B, AND PARTIAL-WAVE AMPLITUDES

We will first fix some necessary notation and conventions for pion-nucleon elastic scattering, at the same time reminding the reader of some well-known features of the pion-nucleon interaction.

As shown in Figure 5 we assign 4-momenta p_1 and q_1 to the initial nucleon and meson respectively and p_2 and q_2 to the final nucleon and meson respectively. We also use the notation $E = \sqrt{M^2 + \mathbf{p}^2}$ and $\omega = \sqrt{m^2 + \mathbf{q}^2}$ for the energies of a nucleon and meson, where M is the nucleon, m the meson mass. The relation between the S matrix and the invariant scattering amplitude T is defined by [we use the metric and spinor normalization of Bjorken & Drell (5)]

$$S_{fi} = \delta_{fi} + i(2\pi)^4 \delta(p_1 + q_1 - p_2 - q_2)\left(\frac{M^2}{4E_1E_2\omega_1\omega_2}\right)^{1/2} \bar{u}(p_2)Tu(p_1) \quad 1.1$$

and from Lorentz covariance and parity conservation

$$T = A + \tfrac{1}{2}(\not{q}_1 + \not{q}_2)B \quad 1.2$$

where A and B are scalar functions of the kinematic invariants

$$s = (p_1 + q_1)^2 = M^2 + m^2 + 2M\omega_L,$$
$$t = (q_1 - q_2)^2 = -2q^2(1 - \cos\theta) \quad 1.3$$

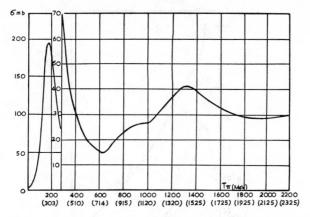

FIG. 1. Total cross sections for π^+p scattering from 0 to 2200 MeV pion kinetic energy T_π. The bracketed scale on the abscissa is the pion laboratory momentum in MeV/c.

where ω_L is the energy of the incident pion in the laboratory system, q is the centre-of-mass momentum of the pion, and θ is the scattering angle in the centre of mass.

The pion-nucleon interaction is nearly isospin invariant (see Section 3.1) and there are two sets of invariant amplitudes A, B corresponding to total isospin $\frac{1}{2}$ and $\frac{3}{2}$ for the pion-nucleon system. We use the notation $A^{1/2}$, $B^{1/2}$, $A^{3/2}$, $B^{3/2}$ and define $(+)$ and $(-)$ amplitudes by

$$A^{(+)} = \tfrac{1}{3}A^{1/2} + \tfrac{2}{3}A^{3/2}, \qquad A^{(-)} = \tfrac{1}{3}A^{1/2} - \tfrac{1}{3}A^{3/2} \qquad \text{1.4a}$$

$$A^{1/2} = A^{(+)} + 2A^{(-)}, \qquad A^{3/2} = A^{(+)} - A^{(-)} \qquad \text{1.4b}$$

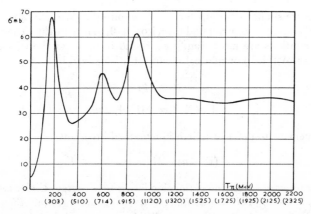

FIG. 2. Total cross section for π^-p scattering from 0 to 2200 MeV pion kinetic energy T_π. The bracketed scale on the abscissa is the pion laboratory energy in MeV/c.

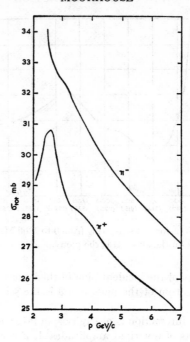

FIG. 3. Total cross sections for $\pi^{\pm}p$ scattering from 2 to 7 GeV/c (pion laboratory momentum) (1).

and if A_+ corresponds to the process $\pi^+ + p \rightarrow \pi^+ + p$ and A_- to $\pi^- + p \rightarrow \pi^- + p$

$$A_+ = (A^{(+)} - A^{(-)}), \qquad A_- = (A^{(-)} + A^{(+)}) \qquad \text{1.4c}$$

There are similar relations for the B amplitudes. We may note that the $A^{(\pm)}, B^{(\pm)}$ amplitudes correspond to an isospin 0,1 state in the t channel. The crossing relations are most simply expressed in terms of the $(+)$ and $(-)$ amplitudes: A third kinematical invariant u is defined by

$$u = (p_1 - q_2)^2 \qquad \text{1.5a}$$

$$s + u + t = 2M^2 + 2m^2 \qquad \text{1.5b}$$

It is convenient to define a fourth kinematical invariant

$$\nu = (s - u)/4M = \omega_L + t/4M \qquad \text{1.6}$$

which for forward scattering is the pion laboratory energy. Under crossing of the pions ($q_1 \leftrightarrow -q_2$, $\pi^+ \leftrightarrow \pi^-$) we find $s \leftrightarrow u$, $\nu \leftrightarrow -\nu$ and we obtain the following relations between the invariant amplitudes, expressed as functions of the kinematical invariants ν and t

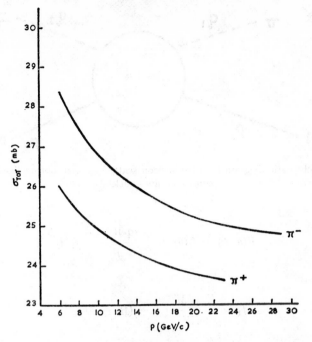

FIG. 4. Total cross sections for $\pi^{\pm}p$ scattering from 6 GeV/c pion laboratory momentum (2).

$$A^{(+)}(-\nu, t) = A^{(+)}(\nu, t), \qquad A^{(-)}(-\nu, t) = - A^{(-)}(\nu, t)$$
$$B^{(+)}(-\nu, t) = - B^{(+)}(\nu, t), \qquad B^{(-)}(-\nu, t) = B^{(-)}(\nu, t) \qquad 1.7$$

To define partial-wave amplitudes let $|m_1\rangle$ and $|m_2\rangle$ be the Pauli spinor states for the initial and final nucleon states with components of spin m_1 and m_2 along the z direction. Then we define a Pauli scattering matrix F in the centre-of-mass system by

$$\langle m_2 | F | m_1 \rangle = \frac{M}{4\pi W} \bar{u}(p_2) T u(p_1) \qquad 1.8$$

giving for a differential cross section in the centre-of-mass system

$$\frac{d\sigma}{d\Omega} = \left(\frac{M}{4\pi W}\right)^2 \sum | \bar{u}(p_2) T u(p_1) |^2 = \sum | \langle m_2 | F | m_1 \rangle |^2 \qquad 1.9$$

where $W = +\sqrt{s}$ and where Σ denotes sum and/or average over nucleon spins according to which differential cross section is being measured. Transforming the right-hand side of 1.8 by expressing u as a projection operator acting on a Pauli spinor

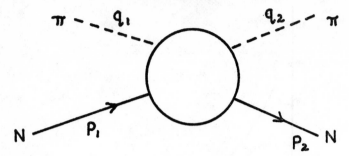

FIG. 5. Schematic diagram for pion-nucleon scattering, with dotted lines denoting pions, solid lines nucleons.

$$F(\theta, \phi) = f_1(\theta) + \frac{(\delta \cdot \mathbf{q}_2)(\delta \cdot \mathbf{q}_1)}{q^2} f_2(\theta) \qquad \text{1.10}$$

where

$$f_1 = \frac{E + M}{8\pi W} \left[A + (W - M)B \right]$$

$$f_2 = \frac{E - M}{8\pi W} \left[-A + (W + M)B \right] \qquad \text{1.11a}$$

$$\frac{1}{4\pi} A = \frac{W + M}{E + M} f_1 - \frac{W - M}{E - M} f_2$$

$$\frac{1}{4\pi} B = \frac{1}{E + M} f_1 + \frac{1}{E - M} f_2 \qquad \text{1.11b}$$

The form 1.10 is a necessary consequence of rotational invariance. The explicit dependence on energy of $f_1(\theta)$ and $f_2(\theta)$ shown by 1.11a has been omitted. The partial waves corresponding to total angular momentum $j = \ell \pm \frac{1}{2}$ are denoted by $f_{\ell\pm}$ where

$$\left.\begin{aligned} f_1 &= \sum_{\ell=0}^{\infty} (f_{\ell+} P_{\ell+1}'(z) - f_{\ell-} P_{\ell-1}'(z)) \\ f_2 &= \sum_{\ell=1}^{\infty} (f_{\ell-} - f_{\ell+}) P_{\ell}'(z) \end{aligned}\right\} \qquad \text{1.12}$$

where $z = \cos\theta$, $P_\ell(z)$ is the Legendre polynomial of order ℓ, and $P_\ell'(z) = (d/dz)\, P_\ell(z)$.

$$f_{\ell\pm} = \frac{1}{2} \int_{-1}^{1} dz (P_\ell(z) f_1(z) + P_{\ell\pm1} f_2(z)) \qquad 1.13$$

Derivations and further details of the pion-nucleon scattering formalism may be found in the article of Hamilton & Woolcock (6) and in the original paper of Chew, Goldberger, Low & Nambu (7).

1.3 FORWARD DISPERSION RELATIONS AND THE PION-NUCLEON COUPLING CONSTANT

The fixed momentum transfer dispersion relations are

$$\text{Re } A^{(\pm)}(\nu, t) = \frac{1}{\pi} P \int_{m+t/4M}^{\infty} d\nu' \text{ Im } A^{(\pm)}(\nu', t) \left(\frac{1}{\nu' - \nu} \pm \frac{1}{\nu' + \nu} \right) \qquad 1.14$$

$$\text{Re } B^{\pm}(\nu, t) = \frac{g^2}{2M} \left(\frac{1}{\nu_B - \nu} \mp \frac{1}{\nu_B + \nu} \right)$$
$$+ \frac{1}{\pi} P \int_{m+t/4M}^{\infty} d\nu' \text{ Im } B^{(\pm)}(\nu', t) \left(\frac{1}{\nu' - \nu} \mp \frac{1}{\nu' + \nu} \right) \qquad 1.15$$

where $\nu_B = -m^2/2M + t/4M$ and g^2 is the rationalized pseudoscalar Watson-Lepore (8) coupling constant.[1] The first term on the right-hand side of 1.15 is called the Born term and corresponds to the renormalized perturbation theory Feynman graphs of Figure 6. These Born terms only appear in the B amplitude because the fact that the pion is pseudoscalar means that the Born terms must involve nucleon spin flip. In deriving 1.14 and 1.15 the crossing relations 1.7 are used. Under the usual field theoretical assumptions (of microcausality and the asymptotic conditions) the analyticity leading, if no subtractions are necessary, to these dispersion relations can be proved (9) for fixed t such that

$$-\frac{32}{3} \left(\frac{2M + m}{2M - m} \right) m^2 \leq t \leq 0 \qquad 1.16$$

From 1.10 the forward scattering amplitude in the centre-of-mass system is

$$f(q, 0) \equiv f_1(q, 0) + f_2(q, 0) \qquad 1.17$$

Using 1.11a and the kinematic relations 1.3, 1.6 and

$$E \pm M = \frac{(W \pm M) - m^2}{2W} \qquad 1.18$$

[1] This coupling constant has value (see below) $g^2/4\pi \sim 15$. It is defined by $<p_2|j_i(0)|p_2> = g(s)\bar{u}(p_2)\gamma_5\tau_i u(p_i)(M^2/E_1E_2)^{1/2}$, $g = g(m^2)$, where $j_i(x)$ is the source of the pion field $\phi_i(x)$, $(\Box + m^2)\phi_i = j_i$, $s = (p_1 - p_2)^2$.

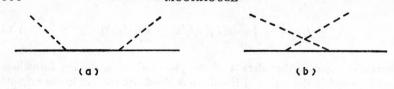

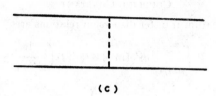

FIG. 6. Feynman diagrams for (a) the direct Born term in πN scattering, (b) the crossed Born term in πN scattering, (c) the Born (single-pion exchange) term in $N-N$ scattering.

we find

$$f(q, 0) = \frac{M}{4\pi W} \left\{ A(\omega_L, 0) + \omega_L B(\omega_L, 0) \right\} \qquad 1.19$$

Let $f_L(q_L, 0)$ be the forward scattering amplitude in the laboratory system and q_L the incident pion momentum in that system, corresponding to q in the centre-of-mass system, so that

$$q_L/q = W/M \qquad 1.20$$

The optical theorem takes the same form in either the laboratory or the centre-of-mass system:

$$\text{Im} f(q, 0) = \frac{q}{4\pi} \sigma_{\text{tot}}(\omega_L); \qquad \text{Im} f_L(q_L, 0) = \frac{q_L}{4\pi} \sigma_{\text{tot}}(\omega_L) \qquad 1.21$$

and also

$$f_L(q_L, 0)/q_L = f(q, 0)/q \qquad 1.22$$

From 1.19, 1.20, and 1.22 we find the simple form

$$f^L(\omega_L) \equiv f_L(q_L, 0) = \frac{1}{4\pi} \left\{ A(\omega_L, 0) + \omega_L B(\omega_L, 0) \right\} \qquad 1.23$$

Since A and B obey dispersion relations in ω_L (which equals ν in the forward direction), so does $f^L(\omega_L)$. We can write this down using the optical theorem 1.21 and from 1.15 evaluating the residues of $f^L(\omega_L)$ at the Born poles $\omega_L = \pm m^2/2M$. With one subtraction[2] we find if $f_\pm{}^L(\omega_L)$ are the forward amplitudes for $\pi^\pm p$ scattering

$$\mathrm{Re}\, f_\pm{}^L(\omega_L) = C \mp \frac{f^2}{\omega_L \mp m^2/2M}$$

$$+ \frac{\omega_L}{4\pi^2} P \int_m^\infty \frac{d\omega_L'}{\omega_L'} q_L' \left[\frac{\sigma_\pm(\omega_L')}{\omega_L' - \omega_L} - \frac{\sigma_\mp(\omega_L')}{\omega_L' + \omega_L} \right] \qquad 1.24$$

where C is a subtraction constant, $\sigma_\pm$ are the total cross sections for $\pi^\pm p$ scattering, and the equivalent pseudovector coupling constant f^2 is defined as

$$f^2 = \frac{g^2}{4\pi} \left(\frac{m}{2M} \right)^2 \qquad 1.25$$

Using a forward dispersion relation with data available in 1959, Spearman (11) found f^2 to be in the range $0.075 < f^2 < 0.085$. Samaranayake & Woolcock (12) have used 1.24 with the assumption, consistent with the high-energy data, that above 10 GeV (13)[3]

$$\tfrac{1}{2}(\sigma_- - \sigma_+) = 0.425/q_L^{0.5} \qquad 1.26$$

They used values of $\mathrm{Re}\, f_\pm(w)$ from phase-shift analysis (see Section 2 below) or by extrapolation of differential cross sections to the forward direction, and found the optimum values of C and f^2 to satisfy 1.24. Their result is

$$f^2 = 0.0822 \pm 0.0018 \qquad 1.27$$

Woolcock (6, 10) had earlier evaluated f^2 from the dispersion relation

$$\frac{1}{4\pi M} \mathrm{Re}\, B_+(\omega_L, 0) = \frac{-4f^2}{\omega_L - m^2/2M}$$

$$+ \frac{P}{\pi} \int_m^\infty \frac{d\omega_L'}{4\pi M} \left[\frac{\mathrm{Im}\, B_+(\omega_L', 0)}{\omega_L' - \omega_L} - \frac{\mathrm{Im}\, B_-(\omega_L', 0)}{\omega_L' + \omega_L} \right] \qquad 1.28$$

The left-hand side was evaluated using results of phase-shift analysis and the dispersion integral using phase-shift analysis at low energies, a resonance dominance approximation at intermediate energies up to 2 GeV and above 2

---

[2] On the assumption of finite range of the π-N interaction as $\nu \to \infty$ then $|A_\pm(\nu,t)|/\nu$, $|B_\pm(\nu,t)| < K$ where t is fixed and K is a constant; so $f^L(\omega_L)/\omega_L < K$. This assumption of finite range gives σ_{tot} bounded as $\omega_L \to \infty$. If σ_{tot} were to increase as log s, then a finite range R should be replaced by R ln q, and K by K ln ν (10). The dispersion relation 1.24 is evidently convergent if the total cross sections obey the Pomeranchuk theorem.

[3] The units in Equations 1.26 and 1.33 are $\hbar = c = m = 1$.

GeV cross-section information [the integral depending, nearly, only on σ_- $-\sigma_+$ (6)]. A fit of $-4f^2/(\omega_L - m^2/2M)$ to these evaluations gave

$$f^2 = 0.081 \pm 0.003 \qquad\qquad 1.29$$

[On the other hand Hohler, Baacke & Steiner (14) report from forward dispersion relations on $B_\pm$ that $f^2 = 0.077 \pm 0.004$ and from dispersion relations on $f^{(-)}$ that $f^2 = 0.076 \pm 0.004$. These analyses include high-energy charge exchange data and more sophisticated phase-shift analysis results than those used by Woolcock.]

The longest-range part of the nucleon-nucleon force is that due to simple pion exchange, illustrated in Figure 6(c). This one-pion exchange contribution (OPEC) can be introduced for the higher partial waves, with f^2 as a parameter, into phase-shift analyses of nucleon-nucleon scattering. Mac-Gregor, Arndt & Wright (15) in an analysis from 0 to 400 MeV (nucleon laboratory kinetic energy) found $g^2 = 14.72 \pm 0.83$ corresponding to

$$f^2 = 0.0762 \pm 0.0043 \qquad\qquad 1.30$$

In view of the conflicting evidence from various groups, if we use various methods and data, and bear in mind the possibility of systematic data error, it seems only safe to state that $f^2 = 0.08$ with a 6 per cent error.

1.4 Pion-Nucleon s-Wave and p-Wave Scattering Lengths

As $\omega_L \to m$, Re $f_\pm{}^L(\omega_L)$ tends to linear combinations of the $T = \frac{1}{2}$ and $T = \frac{3}{2}$ s-wave pion-nucleon scattering lengths a_1 and a_3 (defined in Section 2 below). From 1.24

$$\operatorname{Re} f_+(m) + \operatorname{Re} f_-(m) = \left(1 + \frac{m}{M}\right)\frac{2}{3}(a_1 + 2a_3)$$

$$= 2C - \frac{1}{M}\frac{2f^2}{(1 - m^2/4M^2)} + \frac{m^2}{2\pi^2}\int_m^\infty \frac{d\omega_L{}'}{\omega_L{}'q_L{}'}[\sigma_+(\omega_L{}') + \sigma_-(\omega_L{}')] \quad 1.31$$

$$\operatorname{Re} f_-(m) - \operatorname{Re} f_+(m) = \left(1 + \frac{m}{M}\right)\frac{2}{3}(a_1 - a_3)$$

$$= \frac{1}{m}\frac{4f^2}{(1 - m^2/4M^2)} + \frac{m}{2\pi^2}\int_m^\infty \frac{d\omega_L{}'}{q_L{}'}[\sigma_-(\omega_L{}') - \sigma_+(\omega_L{}')] \quad 1.32$$

Having determined f^2 and $C(=-0.116 \pm 0.002)$ using 1.24, Samaranayake & Woolcock (12) evaluate the integrals 1.31 and 1.32 to determine $(a_1 - a_3)$ and $(a_1 + 2a_3)$. Above 10 GeV they use the form 1.26 for $\sigma_- - \sigma_+$ and the following fit[3] to the high-energy data for $\sigma_+ + \sigma_-$:

$$\tfrac{1}{2}(\sigma_+ + \sigma_-) = 1.104 + 8.30/q_L{}^{0.9} \qquad\qquad 1.33$$

The results are, in units of the pion Compton wavelength,

$$(a_1 - a_3) = 0.292 \pm 0.020, \quad (a_1 + 2a_3) = -0.035 \pm 0.012 \quad 1.34$$

An earlier calculation by Hamilton & Woolcock (6) of the best fit of low-energy parameters to similar sum rules to 1.31 and 1.32 and sum rules from the evaluation at $\omega_L = m$ of dispersion relations for $B^{(\pm)}$ $(\omega_L, 0)$ and $f_1^{(-)}(\omega_L 0)$ and also some other information [such as the result (6), from the Panofsky ratio, $a_1 - a_3 = 0.254 \pm 0.012$] gave

$$a_1 = \quad 0.171 \pm 0.005 \qquad a_3 = -0.088 \pm 0.004 \qquad 1.35$$

$$a_{31} = -0.038 \pm 0.005 \qquad a_{31} = -0.029 \pm 0.005$$
$$\qquad\qquad\qquad\qquad\qquad\qquad\qquad\qquad\qquad\qquad\qquad 1.36$$
$$a_{11} = -0.101 \pm 0.007 \qquad a_{33} = \quad 0.215 \pm 0.005$$

The quantities $a_{2T,2J}$ of 1.36 are the p-wave π-N scattering lengths as defined in Section 2 below. There is a disagreement with the results 1.34 of Samaranayake & Woolcock (12), which Hamilton (16) has criticized as being dependent on uncertainties in the high-energy total cross-section behaviour owing to the relatively slowly convergent integrals 1.31 and 1.32. Hamilton (16) argues powerfully that

$$a_1 - a_3 = 0.271 \pm 0.007, \quad a_1 + 2a_3 = -0.002 \pm 0.008 \quad 1.37$$

There is a region of controversy about the important s-wave scattering lengths. However, it seems rather certain that, for example, $a_1 - a_3 = 0.275$ to within 8 per cent.

PCAC and current algebra.—Tomozama, Weinberg and others (17) have shown that one can derive the s-wave scattering lengths by using PCAC, current algebra, and 'soft pion' techniques. One may consider (17) the elastic scattering of a π^+ on a nucleon (proton or neutron). The reduction formula for the pion-nucleon scattering S matrix is used, so that two pion field operators $\phi^+(x)$, $\phi^-(x)$ appear. These are eliminated in favour of the axial vector current by using the PCAC relation:

$$\phi^{\pm}(x) = \frac{g(0)}{\sqrt{2}\, g_A M m^2} \frac{\partial J_\mu^{A\pm}(x)}{\partial x_\mu} \qquad 1.38$$

where g_A (~ 1.18) is the axial-vector coupling constant, $g(0)$ is the nucleon/pionic form factor, normalized to g at the pion mass squared. $J^{A\pm}$ is the axial-vector current for isospin $1 \pm i2$. One finds that a term in the S-matrix element contains the commutator, specified by the once integrated current algebra

$$\left[\int J_0^{A+}(\mathbf{x}, y_0) d^3x, J_\mu^{A-}(\mathbf{y}, y_0) \right] = 2\delta_{\mu 0} J_0^{V3}(\mathbf{y}, y_0) \qquad 1.39$$

where $J_0^{V3}(\mathbf{y}, y_0)$ is the time component of the vector current so that

$$\int J_0^{V_3}(\mathbf{y}, y_0) d^3x = T_3$$

where T_3 is the third component of isospin. Other terms enter in the S-matrix element but these give zero contribution in the soft pion limit, q_1, q_2 $\to 0$ (17). Assuming that the variation in extrapolation from the zero pion mass limit to the physical pion mass is small, one obtains

$$4\pi \frac{a}{m}(1 + m/M) = -\left(\frac{g(0)}{g_A M}\right)^2 \langle N \mid T_3 \mid N \rangle \qquad 1.40$$

where a is the $\pi^+ N$ scattering length. Equation 1.40 gives

$$2a_3 + a_1 \equiv \frac{3}{2}(a(\pi^+ p) + a(\pi^+ n)) = 0 \qquad 1.41$$

$$a_1 - a_3 = \frac{3}{2}(a(\pi^+ n) - a(\pi^+ p))$$

$$= \frac{3m}{2}\left(\frac{g(0)}{g_A M}\right)^2 / 4\pi(1 + m/M) \qquad 1.42$$

$$\simeq 0.30\ m^{-1}$$

As one certainly does not expect PCAC to be good to better than 10 per cent, this constitutes a very satisfactory agreement with the results 1.34 or 1.37. If one substitutes the expression 1.42 for $\frac{2}{3}((a_1 - a_3)/m)(1 + (m/M))$ into 1.32, and uses 1.25 one obtains

$$\frac{1}{1 - m^2/4M^2} - \frac{1}{g_A^2}\frac{g^2(0)}{g^2(m^2)} = \frac{2M^2}{g^2(m^2)\pi}\int \frac{d\omega_L'}{q_L'}(\sigma_+(\omega_L') - \sigma_-(\omega_L')) \quad 1.43$$

which in zero pion mass limit, $m \to 0$, reduces to the Adler-Weisberger relation.

Equation 1.41 is related to the Adler consistency condition.

$$A^{(+)}(\nu = 0, \nu_B = 0, q_1^2 = 0) = \frac{g(0)g(m^2)}{M} \qquad 1.44$$

which is a consequence of PCAC when the initial pion is soft, $q_1 \to 0$, and the final is on the mass shell, $q_2^2 = m^2$ (see e.g., 17). The consistency condition is verified by dispersion relations to 10 per cent. The total forward amplitude $f_L^{(+)}$ (Equations 1.19, 1.23) at threshold is proportional to $(2a_3 + a_1)$, and the Adler consistency condition is just the condition for the constant term from $A^{(+)}$ to cancel with the contribution from the pole term $B^{(+)}$ leaving the dispersion term to vanish as $m^2 \to 0$ as evident from Equation 1.31.

2. PARTIAL-WAVE AMPLITUDE ANALYSIS AND RESONANCES

The analysis of pion-nucleon elastic scattering data has been one of the principal sources of information on the pion-nucleon system. For example, the s- and p-wave phase shifts at low energies are related, through continuation in the energy, to the zero (kinetic) energy parameters, that is the s- and p-wave scattering lengths which appear in sum rules and dispersion relations like those of Section 1.4. The p_{33} resonance was identified with certainty by partial-wave analysis in 1954 (18). Of recent years pion-nucleon partial-wave analysis has been one of the most fruitful means of discovering new resonances. Partly because of lack of data, analysis of elastic kaon-nucleon scattering has not been so active.

2.1 Elastic Scattering Amplitudes and Experiments

It is usual to write the centre-of-mass scattering amplitude 1.10 as

$$F_{m_2 m_1}(\theta, \phi) = f(\theta)\delta_{m_2 m_1} + ig(\theta)\langle m_2 | \, \mathbf{d} \, | m_1 \rangle \cdot \mathbf{n} \qquad 2.1$$

$$\mathbf{n} = \mathbf{q}_1 \times \mathbf{q}_2 / | \, \mathbf{q}_1 \times \mathbf{q}_2 \, | \qquad 2.2$$

Here $\mathbf{n}$ is the normal to the production plane and $f(\theta)$, $g(\theta)$ are given in terms of the amplitudes $f_1(\theta)$, $f_2(\theta)$ of 1.10 and in terms of the partial-wave amplitudes, by

$$f = f_1 + \cos\theta f_2, \qquad g = f_2 \sin\theta \qquad 2.3$$

$$f(\theta) = \sum_{\ell=0}^{\infty} \left\{ (\ell+1)f_{\ell+} + \ell f_{\ell-} \right\} P_\ell(\cos\theta),$$

$$\qquad 2.4$$

$$g(\theta) = \sum_{\ell=1}^{\infty} \left\{ f_{\ell+} - f_{\ell-} \right\} \sin\theta P_\ell'(\cos\theta)$$

$$f_{\ell\pm} = (\eta_{\ell\pm} e^{2i\delta_{\ell\pm}} - 1)/2iq \qquad 2.5$$

where $\delta_{\ell\pm}$ is the real part of the phase shift,[4] and $\eta_{\ell\pm}$ is the inelasticity parameter. It is these two numbers for each partial wave, which it is the object of partial-wave analysis, otherwise phase-shift analysis, to determine. Where no particular partial wave is referred to we may denote these numbers by δ and η. The latter is subject to the unitarity restriction

$$| \, \eta \, | \leq 1 \qquad 2.6$$

where equality means that the scattering is purely elastic. We can summarize the information to be obtained from experiment by considering

---

[4] The threshold behaviour is given by $\delta_{\ell\pm} = aq^{2\ell+1} + \cdots$, where a is the 'scattering length' for that particular partial wave.

scattering from polarized protons where the polarization[5] of the protons is P_1. Setting $P_1 = 0$ gives the corresponding quantities for scattering from an unpolarized target. The differential cross section and the polarization of the recoil nucleon are given by

$$\frac{d\sigma}{d\Omega} = |f(\theta)|^2 + |g(\theta)|^2 + 2\,\mathrm{Im}[f(\theta)g^*(\theta)](\mathbf{n}\cdot\mathbf{P}_1)$$

2.7a

$$\frac{d\sigma}{d\Omega}\mathbf{P}_2 = 2\,\mathrm{Im}[f(\theta)g^*(\theta)]\mathbf{n} - 2\,\mathrm{Re}[f(\theta)g^*(\theta)](\mathbf{n}\times\mathbf{P}_1)$$

$$+ (|f(\theta)|^2 + |g(\theta)|^2)(\mathbf{n}\cdot\mathbf{P}_1)\mathbf{n}$$

$$- (|f(\theta)|^2 - |g(\theta)|^2)(\mathbf{n}\times(\mathbf{n}\times\mathbf{P}_1))$$

2.7b

We see that at a given angle θ there is the differential cross section $|f|^2 + |g|^2$ and three polarization quantities: $\mathrm{Im}[fg^*]$, $\mathrm{Re}[fg^*]$ and $|f|^2 - |g|^2$. To measure all these will determine f and g apart from a common phase, indeterminable from 2.7. To measure $|f|^2 + |g|^2$, $\mathrm{Im}[fg^*]$, $\mathrm{Re}[fg^*]$ will determine f, g (always apart from the common phase) but without distinction as to which is f and which is g. To measure $|f|^2 + |g|^2$, $\mathrm{Im}[fg^*]$, and $|f|^2 - |g|^2$ will leave the relative phase of f, g undetermined by π. The situation is further complicated by the charge or isospin dependence where if we denote by F_+, F_-, and F_{ex} amplitudes for $\pi^+p\to\pi^+p$, $\pi^-p\to\pi^-p$, and $\pi^-p\to\pi^0n$ scattering, and pure isospin amplitudes by $F^{1/2}$, $F^{3/2}$ then (see Equations 1.4 above)

$$F_+ = F^{3/2}, \qquad F_- = \frac{2}{3}F^{1/2} + \frac{1}{3}F^{3/2},$$

$$F_{ex} = \frac{\sqrt{2}}{3}(F^{3/2} - F^{1/2}) = \frac{1}{\sqrt{2}}(F_+ - F_-)$$

2.8

Bilenky & Ryndin (19) have shown that, at a given angle θ, the amplitudes $f^{3/2}$, $g^{3/2}$, $f^{1/2}$ and $g^{1/2}$ can be determined to within one overall phase by measuring nine experimental quantities, namely

$\pi^+p \to \pi^+p$: $|f|^2 + |g|^2$, three polarization quantities

(one only as to sign)

$\pi^-p \to \pi^-p$: $|f|^2 + |g|^2$, two polarization quantities

$\pi^-p \to \pi^0n$: $|f|^2 + |g|^2$, one polarization quantity

The present experimental situation is that about five of these nine quan-

[5] The proton polarization $\mathbf{P}$, defined in the rest system of the protons, is given by $\rho_{m'm} = \frac{1}{2}(\delta_{m'm} + \mathbf{P}\cdot<m'|\mathfrak{d}|m>)$ where $\rho_{m'm}$ is the 2×2 density matrix of the protons; $|\mathbf{P}|\leq 1$.

tities may have been measured at a given angle. It is probably fortunate that some further experimental information, together with some assumptions used in partial-wave analysis, make the measurement of all the nine quantities not necessary. The reasons for this are as follows.

Suppose one considers the determination of the f,g amplitudes for, say, $\pi^- p \rightarrow \pi^- p$. Then in principle from single scattering and double scattering (recoil nucleon polarization detection) experiments one can determine $|f(\theta)|, |g(\theta)|$ and their relative phase. One can see on inspection of 2.4, with the very important assumption of a finite number of nonzero partial-wave amplitudes, that this knowledge is sufficient to determine $|f_{\ell\pm}|$ and their relative phases.[6] The absolute phases can be determined by use of the optical theorem, $\operatorname{Im} f(0) = (q/4\pi)\,\sigma_{tot}$, and forward dispersion relations to determine $\operatorname{Re} f(0)$. Alternatively, one could obtain the absolute phase from the Coulomb scattering which is important at small enough angles. The information to be obtained from total cross sections can be explicitly expressed in terms of partial waves by

$$\sigma_{tot} = \frac{4\pi}{q} \sum_{\ell=0}^{\infty} \left\{ (\ell + 1)\,\operatorname{Im} f_{\ell+} + \ell\,\operatorname{Im} f_{\ell-} \right\} \qquad 2.9$$

Resonances often occur shown as peaks or shoulders in the total cross section (Figures 1–4) and it should be noted that because of unitarity (Equation 2.6) the maximum contribution from any partial wave of angular momentum j is $(4\pi/q^2)(j+1/2)$.

From a resonance point of view the most important quantities are partial-wave amplitudes in pure isospin states, which are given in terms of the partial-wave amplitudes for $\pi^\pm p$ elastic scattering by the corresponding equations to 2.8. We see that in principle we do not need to measure charge exchange scattering, $\pi^- + p \rightarrow \pi_0 + n$. However, the inevitable experimental errors would alone make the measurement of all three quantities desirable.

In practice, with the present state of experimental technique, there are, at or near a given energy, many gaps in the data. These may be missing points in the angular range and information on recoil nucleon polarization from polarized targets is currently not available. This leaves only an incomplete knowledge of $|f|^2 + |g|^2$ and $\operatorname{Im}(f^*g)$, which is not sufficient to determine the partial-wave amplitudes without further assumptions. It now seems that remarkable successes can be achieved by using the heuristic principle of smooth behaviour of the partial wave amplitudes as a function of energy. How this principle is used and the resulting limitations and uncertainties will be described in Sections 2.2–2.5.

2.2 POLYNOMIAL FITTING

Given an angular distribution such as a differential cross section or

---

[6] The principle is readily seen by taking max $(\ell) = 1$ in Equation 2.4.

polarization at a given energy, a useful first step is to analyze into powers of $\cos \theta$ or orthogonal polynomials such as $P_n (\cos \theta)$. For example

$$\frac{d\sigma}{d\Omega}(\theta) = \sum_{n=0}^{N} a_n(\cos \theta)^n \qquad \qquad 2.10a$$

$$\frac{d\sigma}{d\Omega}(\theta) = \sum_{m=0}^{M} c_m P_m(\cos \theta) \qquad \qquad 2.10b$$

For a given experimental angular distribution $E_i = d\sigma/d\Omega(\theta_i)$ at K angles θ_i, with errors Δ_i on E_i, the a_n for example can be determined by the usual procedures of curve fitting. Series in orthogonal polynomials such as P_m ($\cos \theta$) have the advantage that for K values of $\cos \theta_i$ uniformly distributed between -1 and 1 with uniform errors Δ_i, there are small (vanishing in the limit $K \to \infty$, $\Delta_i \to 0$) correlations between the C_m. This advantage is decreased if the $\cos \theta_i$ are markedly nonuniformly distributed, owing, for example, to gaps in the angular range of measurement.

Curve fitting of this type can show up bad data values, as points which polynomials of no reasonable degree can fit. These are often obvious by eye. More valuably it is a first step towards a complete analysis into partial waves.

More explicitly we expand

$$\left| f(\theta) \right|^2 + \left| g(\theta) \right|^2 = \sum_{m=0} c_m P_m(\cos \theta) \qquad \qquad 2.11a$$

$$2 \, \mathrm{Im}[f(\theta)g^*(\theta)] = \sum_{m=1} d_m P_m(\cos \theta) \qquad \qquad 2.11b$$

$$2 \, \mathrm{Re}[f(\theta)g^*(\theta)] = \sum_{m=1} e_m P_m(\cos \theta) \qquad \qquad 2.11c$$

The expansion coefficients c_m, d_m, e_m are bilinear in the partial-wave amplitudes:

$$c_m = \sum_{\ell\pm,\ell'\pm} \gamma_m(\ell\pm, \ell'\pm) \, \mathrm{Re}(f_{\ell\pm}f_{\ell'\pm}{}^*) \qquad \qquad 2.12a$$

$$d_m = \sum_{\ell\pm,\ell'\pm} \delta_m(\ell\pm, \ell'\pm) \, \mathrm{Im}(f_{\ell\pm}f_{\ell'\pm}{}^*) \qquad \qquad 2.12b$$

$$e_m = \sum_{\ell\pm,\ell'\pm} \epsilon_m(\ell\pm, \ell'\pm) \, \mathrm{Re}(f_{\ell\pm}f_{\ell'\pm}{}^*) \qquad \qquad 2.12c$$

The γ_m ($\ell\pm\ell\pm$) and $\delta_p(\ell\pm, \ell'\pm)$ have been given by Olson & Trower (20) up to $m=11$ and $p=8$. Knowledge of c_m, d_m (e_m) obviously gives the upper limit of the angular momentum of nonnegligible partial-wave amplitudes, and it may give a subset of $f_{\ell\pm}$ containing the subset of important $f_{\ell\pm}$. Inspection of the c_m, d_m, (e_m) at various energies may give a first glimpse of what partial-wave amplitudes could resonate.

2.3 PARTIAL-WAVE ANALYSIS

Ambiguities.—We have seen that because data on polarization of recoil nucleons from scattering of pions by polarized targets is not available, it is not possible to determine the partial-wave amplitudes at a single energy without some other principle. Some ambiguities due to lack of data are named. Suppose that at some energy we only have the scattering differential cross section from an unpolarized target and that $f_{\ell\pm}$ are a set of partial-wave amplitudes that fit the data, then the *Minami ambiguity* (21) is the existence of another set of partial-wave amplitudes, $f_{\ell\pm}{}^M$, such that $f_{(\ell+1)-}{}^M = f_{\ell+}$, $f_{(\ell-1)+}{}^M = f_{\ell-}$ which give exactly the same differential cross section. The two solutions in the Minami ambiguity are obtained one from the other by parity interchange. For scattering experiments from a polarized target (Equation 2.7a with $\mathbf{P}_1 \neq 0$) there is the *generalized Minami ambiguity* between solutions $f_{\ell\pm}$ and $f_{\ell\pm}{}^M$ which are such that $f_{(\ell+1)-}{}^M = f_{\ell+}{}^*$ and $f_{(\ell-1)+}{}^M = -f_{\ell-}{}^*$. These two solutions give exactly the same value for $|f|^2 + |g|^2$ and for Im (fg^*). One can immediately prove these ambiguities by using Equation 2.3 and noting from Equation 1.13 that the Minami ambiguity corresponds to the interchange $f_1 \leftrightarrow f_2$, and the generalized ambiguity to $f_1 \leftrightarrow -f_2{}^*$.

One eliminates these ambiguities by resorting to the hypothesis of continuous and not too rapid variation with energy. One looks at partial-wave amplitudes nearby (generally lower) in energy and the demand of similarity in the solutions at neighbouring energies is sufficient to resolve the ambiguities. It is plain that this is an iterative process, which descends to threshold, where ambiguities may be resolved by observation of threshold behaviour, $f_{\ell\pm} \to q^{2\ell}$. Interference with Coulomb scattering, particularly at lower energies where it is more easily observed, also resolves ambiguities.[7]

Single-energy analysis.—The range of interaction between pion and nucleon is $\sim 1/2m$ and so partial waves with $\ell > L_q = q/2m$ are expected to be small and this is confirmed by polynomial analysis as described in Section 2.2. So one first limits the number of unknown partial waves by taking some *ansatz* (for example, zero) for the higher ones. One then finds all the solutions (that is, satisfactory fits of the partial-wave amplitudes to the data) at each energy by minimizing

$$F = \sum_i \left(\frac{E_i - T_i}{\Delta_i} \right)^2 \qquad 2.13$$

[7] There are sometimes other guides to the elimination of ambiguities.

If one took the generalized Minami ambiguous solution in the neighbourhood of the 33 resonance, one would obtain a partial-wave amplitude with a rapidly descending phase shift. This would, however, be very suspicious for this almost purely elastic phase shift because of the Wigner condition (22) limiting the rate of descent of phase shift for potentials of a given range.

For a potential of finite range R, $(d\delta/dq) \geq -R$. If it is possible to apply this theorem to pion-nucleon scattering in the purely elastic case, it allows the phase shift to decrease by no more than $20°$ in 100 MeV/c.

In 2.13 E_i, Δ_i are the experimentally measured quantities (differential cross sections etc.) and errors, and T_i is the corresponding calculated quantity containing as the variable parameters the partial-wave amplitudes $f_{l\pm}$ of the given energy. Large errors Δ_i lead to large numbers of minima, each with corresponding solutions for $f_{l\pm}$. Taking one solution at the lowest energy one selects the solutions continuous with it at the next energy, and so on till the highest energy. One thus generates a number of possible paths from lowest to highest energy, each path leading through one of the solutions at each energy. Bareyre et al. (23) and Bareyre, Bricman & Villet (24) find the paths by inspection while the work done at Berkeley by Johnson et al. and reported at the Heidelberg Conference (25) and by Johnson & Steiner (26) and the work of Lovelace & Wagner reported at the Vienna Conference (27) used a minimum pathlength computer program. These analyses used the *ansatz* that higher partial waves are zero.

The work of Auvril et al. (28) and Donnachie et al. (25, 29, 30) is more complicated. A division of partial waves into three types is made: (*a*) lower l partial waves, $0 \leq l < L_i$; (*b*) intermediate l partial waves, $L_i \leq l < L_2$ small and nonresonating, taken from partial-wave dispersion relations with assumed knowledge of the longer-range forces; (*c*) higher partial waves, $L_2 \leq l$, set equal to zero. Solutions at each energy are found for the lower partial waves and paths are found from the lowest to the highest energy, each path close to one of the solutions at each energy. Each path is found from partial-wave dispersion relations with the left-hand cut discontinuity (or forces) parametrized, the parameters being determined by a best fit to the solutions. The smooth paths define new partial-wave amplitudes at each energy, and these values are used as input to the fitting (to data) program and new solutions found. The procedure is then iterated. At some stage the previously determined amplitudes of type (*b*) above are allowed to vary freely.

We delay comment on these methods till Section 2.5.

Energy-dependent analysis.—We have seen the complications that arise in the single-energy type of partial-wave analysis, in imposing smooth variation from energy to energy. In the energy-dependent type of analysis one seeks to overcome this difficulty by parametrizing each partial-wave amplitude as a function of energy and fitting the data at many energies simultaneously. The technical price that has to be paid for this builtin smoothness with energy is a large increase in the number of parameters that have to be simultaneously fitted to the data. The energy range that can be thus parametrized is limited by the speed and capacity of existent computers, since the larger the energy range the greater the number of parameters necessary for the description of the energy variation of each partial-wave amplitude.

A simple but effective parametrization is to develop the real phase shifts $\delta_{l\pm}$ and the absorption parameters $\eta_{l\pm}$, as power series in the centre-of-mass momentum q. This method was used by Roper (31), Roper & Wright (32), and Roper, Wright & Feld (33), with some refinement for partial waves known to be resonant. Another method used (34–36) was to take dispersion relations for the inverse partial-wave amplitudes parametrizing the inelasticity and the left-hand cut, the latter as a series of poles with variable residues, and thus not assuming knowledge of the forces driving the pion-nucleon system. Both these analyses assumed the higher partial waves to be zero.

Obviously there are a great many reasonable parametrizations. For example, one may parametrize each partial-wave amplitude as possible Breit-Wigner resonances and variable background (37). Or one could use dispersion relations in cos θ. Or one might parametrize in terms of crossed-channel and direct-channel Regge poles (Section 3).

2.4 Partial-Wave Amplitudes to 300 MeV

In pion-nucleon scattering there is a natural boundary in the energy variable at pion kinetic energy $\neq 300$ MeV. Up to that energy the scattering is, to a very close approximation, purely elastic (including charge exchange). Shortly above that energy, inelastic scattering processes of pion production appear, at first predominantly in the p_{11} wave, but by 500 MeV inelasticity has appeared in most waves. Below 300 MeV, there is only one parameter characterizing each partial wave, namely the real phase shift δ, and there is enough data available for partial-wave analysis to be, for the larger waves over most of that region, rather unambiguous. In an analysis of all the data in the energy range 0 to 350 MeV Roper & Wright (32) parametrized the nonresonant phase shifts by polynomials in the momenta, writing

$$\tan \delta_{l\pm} = q^{2l+1} \sum_{n=0}^{N} \lambda_n q^n \qquad 2.14$$

where the λ_i are parameters to be varied in the minimization: 2.14 carries the correct threshold behaviour. The resonant partial-wave amplitude for the p_{33} state, $f_{1+}{}^{3/2}(q)$, was represented as the sum of a Breit-Wigner resonance and a background:

$$qf_{1+}{}^{3/2}(q) = \frac{\frac{1}{2}\Gamma(q)}{\omega_R - \omega - \frac{1}{2}i\Gamma(q)} + \frac{1}{2i} (\exp (2i\bar{\delta}_{1+}{}^{3/2}) - 1) \qquad 2.15$$

where ω_R is the value of the centre-of-mass energy of the pion at resonance. The width $\Gamma(q)$ was allowed an energy variation suggested by Layson (38) which ensured that $\Gamma(q) \propto q^3$ for small q

$$\Gamma(q) = \frac{4\gamma^2 M}{\omega + \omega_R} [(qR)j_1{}^2(qR) + n_1{}^2(qR)] \qquad 2.16$$

where γ and R are constants. The form 2.15 is not automatically unitary for arbitrary values of the parameters, but the parameter ranges may be restricted so that unitarity is preserved.

Below pion laboratory kinetic energy of 200 MeV only s and p waves are large, but near this energy d waves become important and to obtain the best fits to the data it is necessary to take into account f waves as well. Even when all the data is included, a number of solutions can be found, between which it is difficult to distinguish. However, these differ only in details (39) and all are close to the particular solution shown in Table 1. The choice of solution at 300 MeV does tend to be important (though probably not critical) as a starting point of higher-energy analyses. Dynamical predictions based on longer-range pion-nucleon forces (40, 41) are in reasonable agreement with the values at $\sim$300 MeV of Table I; detailed comments and comparisons on this and other points are given by Hull & Lin (39).

Apart from the p_{33} phase shift, one of the most interesting features of the solution is that the p_{11} phase shift is negative at low energies, but passes through zero near 170 MeV to become positive and rapidly increasing at the end of our energy range. This is particularly important because many dynamical models, constructed to account for the nucleon state (which has the same quantum numbers as the p_{11} partial wave), are inconsistent with this behaviour and must be rejected.

2.5 RESONANCES

Interpretation of strong interactions.—According to the analyses described in Section 2.3 most partial waves (of angular momentum and energy satisfying $\ell < q/2m$) exhibit strong interactions insofar as there is at least one energy in whose neighbourhood the amplitude varies relatively quickly as a function of energy. Characteristically at these energies the partial-wave amplitude describes a counterclockwise circle in the Argand diagram (Figure 7) corresponding to the T-matrix element (42) for a transition from channel i to channel f of the same J, P, T.

$$\langle i | T' | f \rangle = \frac{e^{2i\delta_\infty}}{2} \frac{(\gamma_i \gamma_f)^{1/2}}{W_R - W - i\gamma/2}$$

$$+ (c_{ri}c_{rf}e^{i\delta_\infty} \sin \delta_\infty + \sum_\alpha{}' c_{\alpha i}c_{\alpha f}e^{i\delta_\alpha} \sin \delta_\alpha) \qquad 2.17a$$

where γ, γ_i, and γ_f are the total width, partial width to channel i, and partial width to channel f; the c's are the elements of an orthogonal matrix transforming from the physical channels to the eigenchannels, δ_α are eigenphases, and δ_∞ is the displacement of the resonant eigenphase from 90°; the prime in 2.17a indicates that the resonant eigenphase is omitted from the sum; $T' = (q)^{1/2}T(q)^{1/2}$ where q is the diagonal matrix of centre-of-mass momenta. Here T is the usual T matrix normalized so that $<i|T|i> = T_{ii} = (\eta e^{2i\delta} - 1)/2iq$. Equation 2.17, with all elements slowly varying compared to $1/(W_R - W - i\gamma/2)$, is valid in a neighbourhood of the resonance such that all eigenphases vary slowly except the resonance eigenphase (43). Generally more than one eigenphase varies in the neighbourhood of a resonance and then the appropriate form is (43):

TABLE I

VALUES OF THE REAL PHASE SHIFTS 0–310 MeV[a]

T_π (MeV)	s_{11}	s_{31}	p_{11}	p_{31}	p_{13}	p_{33}	d_{13}	d_{33}	d_{15}	d_{35}	f_{15}	f_{35}	f_{17}	f_{37}
6	+ 2.758	− 0.761	− 0.150	−0.030	−0.015	+ 0.228	0	0	0	0	0	0	0	0
31	+ 5.483	− 2.934	− 1.220	−0.336	−0.168	+ 2.818	+0.005	+0.003	+0.003	−0.006	0	0	0	0
58	+ 6.920	− 5.135	− 2.045	−0.847	−0.412	+ 7.810	+0.029	+0.015	+0.018	−0.028	+0.001	0	0	0
98	+ 8.293	− 8.305	− 2.060	−1.818	−0.863	+ 20.475	+0.122	+0.057	+0.071	−0.110	+0.004	0	0	0
120	+ 8.880	− 9.996	− 1.458	−2.422	−1.139	+ 31.526	+0.214	+0.096	+0.123	−0.187	+0.009	0	0	0
144	+ 9.458	− 11.785	− 0.344	−3.113	−1.454	+ 48.152	+0.357	+0.154	+0.200	−0.300	+0.019	+0.002	+0.004	+0.001
170	+10.055	− 13.642	+ 1.375	−3.879	−1.805	+ 70.594	+0.659	+0.236	+0.313	−0.459	+0.037	+0.003	+0.007	+0.001
195	+10.632	− 15.334	+ 3.522	−4.610	−2.146	+ 91.699	+0.840	+0.334	+0.452	−0.651	+0.064	+0.006	+0.012	+0.002
220	+11.237	− 16.920	+ 6.166	−5.317	−2.486	+108.122	+1.186	+0.452	+0.622	−0.882	+0.105	+0.009	+0.019	+0.003
240	+11.756	− 18.104	+ 8.659	−5.853	−2.755	+117.680	+1.527	+0.560	+0.782	−1.095	+0.150	+0.014	+0.026	+0.005
270	+12.616	− 19.723	+13.090	−6.586	−3.151	+127.720	+2.161	+0.744	+1.062	−1.459	+0.243	+0.023	+0.039	+0.008
310	+13.959	− 21.563	+20.416	−7.389	−3.658	+136.274	+3.297	+1.025	+1.509	−2.023	+0.430	+0.042	+0.063	+0.014

[a] The energy variable is the pion laboratory kinetic energy. The phase shifts are from (32).

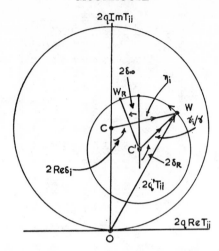

FIG. 7. A plot of $2qT_{ii}(W)$ where T_{ii} is the elastic scattering amplitude. With increasing W the amplitude follows an anticlockwise path on the circle with centre C' and with radius equal to the elasticity γ_i/γ of the resonant eigenstate with respect to channel i; δ_∞ denotes the background phase in the resonant eigenstate.

$$\langle i\,|\,T'\,|\,f\rangle = C_{if} + \frac{1}{2}\,\frac{t_i t_f}{W_R - W - i\Gamma/2} \qquad\qquad 2.17\text{b}$$

where C_{if} is a slowly varying background and t_i, t_f are complex parameters such that $|t_i| = \sqrt{\gamma_i}, |t_f| = \sqrt{\gamma_f}$. In 2.17b unitarity has to be imposed as a condition on the parameters whereas in 2.17a it is automatic, provided $c_{\alpha i}$ is an orthogonal matrix. If we limit the term resonance[8] to T- or K-matrix poles, then we shall *presume* that where major portions of such arcs exist there is a resonance (some relevant questions are discussed in Section 4). In just a comparatively few cases where the energy variation is both well known and characteristic, or where there is good evidence from other channels (see below), can we be confident that there is a resonance.

We wish to distinguish clearly between the interpretation of a strong interaction in a partial wave as a resonance, and the question of the existence of that strong interaction. We now discuss the latter point.

Reliability of partial-wave analyses.—We have seen how at a given energy, lack of some experimental measurements together with errors on existing ones give rise to a multiplicity of possible partial-wave solutions, and how the principle of energy continuity reduces this number. In the whole energy range from (N^*) mass of $\sim$1 GeV/c^2 to $\sim$2 GeV/c^2 the partial-wave analyses

[8] A general discussion of resonances particularly relevant to this section is given by Dalitz (42). See also Goebel & McVoy (43).

are in broad agreement on the strong interactions or resonances to be found therein. One might inquire whether some types of solution may have been missed owing to insufficient search in multidimensional parameter space, and for associated latent theoretical or other prejudice. In this connection, the analyzers are working in a climate of opinion which favours established resonances. Secondly there is theoretical prejudice from partial-wave dispersion relations, used with some assumptions on the long-range forces, overt in the analyses of Donnachie, Lovelace et al. (25, 27–30) but which also exists in others (35). This is because the solution at higher energy depends on the solution at lower energy and in particular that at the pion laboratory kinetic energy of 300 MeV corresponding to a total mass $\sim$1300 MeV/c^2. As mentioned in Section 2.4 there is some uncertainty in the solutions at this energy and the type of solution indicated by partial-wave dispersion relation has been favoured (35, 39). [The analyses of Bareyre et al. (23, 24) do not contain this prejudice.] An additional criticism is that the existing analysis set higher partial waves equal to zero, and there might well be a nonnegligible combined effect from these waves. This type of situation can be illustrated at lower energies, for example at 300 MeV, where f waves cannot be determined from the available data alone, but if f waves are calculated (40) from the theory of long-range pion forces then the d waves change considerably from their values when f waves are zero.

Despite these latent reservations, we consider that the balance of probabilities is in favour of the correctness of the strong interactions or resonances (though not necessarily of the less important waves) found by the present partial-wave analyses. This consideration is partly from the great amount of searching for solutions performed, and partly from the coincidence of the results with observations of strong interaction or resonance in other channels, which we will now discuss.

Resonance-like photoproduction has been observed at the mass of the $N_{3/2}*(1236)$, $N_{1/2}*(1520)$, $N_{1/2}*(1690)$, $N_{3/2}*(1920)$, and higher resonances. There is no question on the $N_{3/2}*(1236)$. However, one must be careful of adducing the other photoproduction phenomena as evidence for the correctness of the pion-nucleon scattering partial-wave analysis; strong interactions are evident at these energies from π-N differential and total cross section without analysis, and from final-state interactions one would expect enhancement of photoproduction at these energies. To support the pion-nucleon scattering partial-wave analysis assignments of resonances to particular J, P, T states we require an independent analysis of the pion photoproduction experiments for which not enough data yet exists, though there are strong indications (44, 45) for the assignment $N_{1/2}*(1520)$, $J^P = 3/2^-$.

It is in channels other than γN that two special cases support the partial-wave analysis, and particularly strongly because the states concerned are of low angular momentum which makes them more difficult to resolve.

Firstly, there is the $T = \frac{1}{2}$, $J^P = \frac{1}{2}^-$, s_{11} pion-nucleon system, with a resonance mass, according to pion-nucleon partial-wave analysis, in the range

1500–1560 MeV/c^2; the pion-nucleon amplitude shows a particularly strong energy dependence in the absorption parameter. Now experiments also show nearly isotropic η production in the reaction

$$\pi^- + p \to \eta + n \qquad\qquad 2.18$$

particularly strong just above threshold and which can be shown to be s-wave from the threshold excitation. This is sufficient to confirm the correctness of the s_{11} wave given by the partial-wave analysis in this energy region. Moreover, detailed investigation of the η production supports a resonance hypothesis for this wave (46–48).

Secondly there is the $T = \frac{1}{2}$, $J^P = \frac{1}{2}^+$, p_{11} pion-nucleon system with a strong interaction centered in the mass range 1400–1500 MeV/c^2; the partial-wave analysis reveals also in this case a rather strong energy dependence on the absorption parameter. But there is evidence (49–51) from missing-mass experiments in p-p and π-p collisions of the production of a system,[9] say (R) with invariant energy centred around 1420 MeV:

$$p + p \to p + (R); \qquad \pi + p \to \pi + (R)$$

The system (R) is produced much more preferentially forward than any of the resonances such as $N_{1/2}^*$ (1520), which are also observed, thus strongly suggesting the mechanism of diffraction dissociation of the nucleon so that (R) has the quantum numbers $T = \frac{1}{2}$, $J^P = \frac{1}{2}^+$. Though there is thus evidence for a system (R) with the same quantum numbers as the p_{11} strong interaction of the partial-wave analysis, we cannot adduce independent evidence that the strong interaction is a resonance. There is also a body of more tentative supporting evidence such as Morgan's (53) isobar model analysis of the $T = \frac{1}{2}$ $\pi + N \to 2\pi + N$ processes for a second p_{11} resonance (Table II).

Resonance results.—We show in Figure 8 the Argand diagram of the s_{11} wave from the analysis of Davies (37) and in Figure 9 from that of Donnachie, Kirsopp & Lovelace (25, 30). These two figures illustrate a not untypical difference in the results of partial-wave analysis by different groups, and a rather difficult case of separating the resonance loops from the background (Equations 2.17). In Table II we list the resonance parameters of Donnachie, Kirsopp & Lovelace (25, 27, 30) where the problem of background is resolved by taking the energy of minimum η, corresponding to maximum absorption in the partial wave, as the resonance energy for the inelastic resonances; also listed are the resonance parameters of Davies & Moorhouse (27, 37). Here the partial-wave amplitudes are parametrized by 2.17b, with possibly more than one resonance and a slowly varying background term, so that the position with corresponding energy W_R comes out

---

[9] Gellert et al. (52), from an analysis of the momenta in the events $p + p \to p + p + \pi^- + \pi^+$, present a case for the phenomenon (R) in p-p collisions being merely a reflection of strong $N^*(1236)$ production. However, one is left with the problem of finding a mechanism in the π-p collisions.

TABLE II
Resonance Results of Different Partial-Wave Analyses

Partial wave $L_{2T\,2J}\ J^P,\ T$	1[a] Mass MeV/c^2	1[a] Γ_{tot} MeV	1[a] Γ_{el}/Γ_{tot}	2[b] Mass MeV/c^2	2[b] Γ_{tot} MeV	2[b] Γ_{el}/Γ_{tot}	3[c] Mass MeV/c^2	3[c] Γ_{tot} MeV	3[c] Γ_{el}/Γ_{tot}
$p_{33}(1)$ 3/2$^+$, 3/2	1236	125	1.00						
$s_{31}(1)$ 1/2$^-$, 3/2	1640	177	0.28	1617	141	0.28	1650	130	?
$d_{33}(1)$ 3/2$^-$, 3/2	1690	270	0.14	1650	188	0.12	—	—	—
$p_{33}(2)$ 3/2$^+$, 3/2	1690	280	0.10	—	—	—	—	—	—
$f_{35}(1)$ 5/2$^+$, 3/2	1910	350	0.16	1840	136	0.20	—	—	—
$p_{31}(1)$ 1/2$^+$, 3/2	1930	340	0.30	1914	290	0.18	—	—	—
$f_{37}(1)$ 7/2$^+$, 3/2	1950	220	0.40	1935	196	0.50	1980	140	0.57
$p(1)$ 1/2$^+$, 1/2	1470	210	0.66	1460	390	0.50	1505	205	0.68
$d_{13}(1)$ 3/2$^-$, 1/2	1520	114	0.57	1512	106	0.45	1515	110	0.54
$s_{11}(1)$ 1/2$^-$, 1/2	1550	116	0.33	1502	36(?)	0.36	1515	105	?
$d_{15}(1)$ 5/2$^-$, 1/2	1680	173	0.40	1670	115	0.50	1655	105	0.40
$f_{15}(1)$ 5/2$^+$, 1/2	1690	132	0.68	1685	104	0.54	1680	105	0.64
$s_{11}(2)$ 1/2$^-$, 1/2	1710	300	0.80	1766	404	0.56	1665	110	?
$d_{13}(2)$ 3/2$^-$, 1/2	1730?	?	?	—	—	—	—	—	—
$p_{13}(1)$ 3/2$^+$, 1/2	1860	300	0.20	1844	450	0.40	—	—	—
$p_{11}(2)$ 1/2$^+$, 1/2	1750	327	0.32	1770	445	0.43	—	—	—
d_{13} 3/2$^-$, 1/2	2030?	290	0.26	—	—	—			
g_{17} 7/2$^-$, 1/2	2190	300	0.35	—	—	—			
f_{17} 7/2$^+$, 1/2	1983	225	0.128	—	—	—			
g_{17} 7/2$^-$, 1/2	—	—	—	1906	319	.06			

[a] Donnachie, Kirsopp & Lovelace (25, 27, 30).
[b] Davies (27, 37).
[c] Bareyre, Bricman & Villet (24).

according to the Wigner condition as the point of fastest energy variation on the loop (Figure 7). Though this latter criterion certainly gives the real part of the pole position correctly in the limit of a narrow resonance, it is not clear that it gives the real part of the pole position more correctly than the minimum η criterion in the case of a broad resonance, particularly when there is considerable background. The third set of resonance parameters in

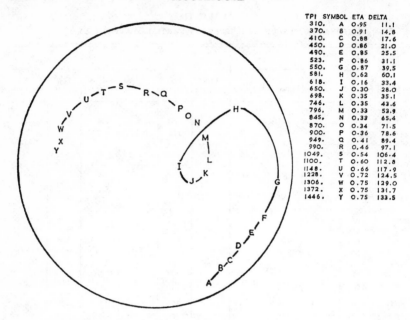

TPI	SYMBOL	ETA	DELTA
310.	A	0.95	11.1
370.	B	0.91	14.8
410.	C	0.88	17.6
450.	D	0.86	21.0
490.	E	0.85	25.5
523.	F	0.86	31.1
550.	G	0.87	39.5
581.	H	0.62	60.1
618.	I	0.16	33.4
650.	J	0.30	28.0
698.	K	0.35	35.1
746.	L	0.35	43.6
796.	M	0.33	53.9
845.	N	0.33	65.4
870.	O	0.34	71.5
900.	P	0.36	78.6
949.	Q	0.41	89.4
990.	R	0.46	97.1
1049.	S	0.54	106.4
1100.	T	0.60	112.8
1148.	U	0.66	117.9
1228.	V	0.72	124.5
1306.	W	0.75	129.0
1372.	X	0.75	131.7
1446.	Y	0.75	133.5

FIG. 8. Argand diagram for the s_{11} partial-wave amplitude from (37). The legend shows the pion laboratory kinetic energy T_π and the corresponding δ, η for various points on the graph.

Table II is that of Bareyre, Bricman & Villet (24), with the resonance position and width again estimated from the rate of variation with energy. In Table II a dash denotes that though the energy region has been analyzed, the resonance indicated has not been observed or only observed very dubiously. A blank denotes that the energy region has not been analyzed.

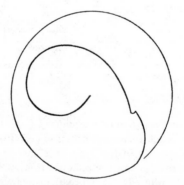

FIG. 9. Argand diagram for the s_{11} partial-wave amplitude from (30).

Notation.—As shown in Table II we use the notation $L_{2T,2J}(1)$, $L_{2T,2J}(2)$, for the lowest energy resonance, second lowest energy resonance, in the $L_{2T,2J}$ state. For example $p_{11}(1)$, $p_{11}(2)$, ... The motivation for this notation is the present variability in quoted masses, though obviously the notation is only useful when the existence and quantum numbers of resonances up to a common mass are rather certain.

2.6 RESONANCE CLASSIFICATION IN THE QUARK MODEL

The most successful attempt at classifying the N^* resonances (and other baryon resonances since it subsumes SU_3) is the L-excitation quark model. The nonrelativistic quark model (54) in which heavy quarks sit in a deep potential well, with three quarks forming a $\{56\}$ of SU_6 for baryons and quark-antiquark forming a $\{35\}+\{1\}$ of SU_6 for mesons, can form additional states by orbital motion (L excitation) of the quarks. This leads to classification of the baryons in $[\{SU_6\}, L^P]$ supermultiplets where, for three quarks, $\{SU_6\} = \{56\}$, $\{70\}$, or $\{20\}$ containing $\{1\}$, $\{8\}$, and $\{10\}$ representations of SU_3. There is no firm evidence for the existence of other multiplets such as $\{\overline{10}\}$ or $\{27\}$ and in particular all the N^* resonances of Table II are consistent with singlet, octet, or decuplet assignments.[10] The symmetric quark model, first proposed by Greenberg (57), takes the quark state to be completely symmetric under interchange of quark coordinates, implying para-Fermi statistics for the quarks. With symmetric radial wavefunctions, avoiding nucleon form factor zeros, and as seems natural from the quark-force point of view, this implies that the orbital $\times SU_6$ wavefunction is completely symmetric.[11] Among such states is the supermultiplet $[\{70\}, 1^-]$ which can accommodate the low-lying negative parity resonances shown in Table II. To illustrate the quark model we show in Table III the sequence of zero, single, and double quantum excitations of harmonic oscillator (shell-model) wavefunctions for three quarks (with the spurious centre-of-mass motions removed) (59). We note, for example, that the natural place for the lowest p_{11} (Roper) resonance, having the quantum number of the nucleon, is in the second $\{56\}$ of Table III. This then would be a low-lying supermultiplet, about equal in mass to the $[\{70\}, 1^-]$ supermultiplet.

Dalitz (60–63) has studied the assignments of baryon and meson resonances in the L-excitation quark model. In the baryon part of that work, perhaps the most important role has been played by the N^* resonances of Table

---

[10] It has been suggested (55) that the lowest p_{11} resonance may belong to a $\{\overline{10}\}$ representation. Suggested tests for this have proved negative. In particular if this resonance belonged to a $\{\overline{10}\}$, the transition $\gamma+p \to N^{*+}$ would be forbidden and $\gamma+n \to N^{*0}$ would be allowed; present experimental evidence is that neither of these transitions is strong (56).

[11] This might be the significant part of the theory. Gell-Mann has emphasized that the results of the L-excitation quark model can be used without commitment to 'concrete' quarks. See for example (58).

TABLE III

BARYON CONFIGURATIONS IN THE L-EXCITATION QUARK MODEL

Harmonic oscillator states	$\{SU_6\}$	L	P
$(1s)^3$	$\{56\}$	0	$+$
$(1s)^2(1p)$	$\{70\}$	1	$-$
$(1s)^2(2s)$ $(1s)^2(1d)$ $(1s)\ (1p)^2$	$\{56\}$ $\{56\}$ $\{70\}$ $\{70\}$ $\{20\}$	0 2 0 2 1	$+$ $+$ $+$ $+$ $+$

II. The fact that the ℓ-excitation model fits most of the lower-mass N^* resonances into the $[\{56\},\ 0^+]$ and $[\{70\},\ 1^-]$ supermultiplets is one of the strongest arguments for the ℓ-excitation quark model. There is corresponding success with the rather less well-known strange baryon resonances.[12]

The $\{56\}$, $\{70\}$, and $\{20\}$ of SU_6 have the following $(SU_3,\ 2S+1)$ structure, where S is the quark spin:

$$\{56\} = (10, 4) \oplus (8, 2)$$
$$\{70\} = (10, 2) \oplus (8, 4) \oplus (8, 2) \oplus (1, 2) \qquad 2.19$$
$$\{20\} = (8, 2) \oplus (1, 4)$$

The quark orbital angular momentum L is combined with S to form the total resonance angular momentum J and multiplets $(SU_3,\ 2S+1)^{J,P}$. In Table IV we show how the N^* resonances fit the multiplets. This nearly fills the first four multiplets of Table IV, leaving only one or two resonances from Table II. These presumably fit into higher multiplets. Within the quark model, assignments somewhat different from these are possible (65) and we may investigate details of the quark-quark forces to attain the desired level structure (59–67).

Another way of classifying the N^*, and other baryon resonances of higher spin, is by using higher SU_6 groups such as $\{700\}$ and $\{1134\}$, corresponding to extra quark-antiquark pairs added to the original three-quark nucleon. Such multiplets certainly provide places for all the resonances of

---

[12] Harari (64) has graphically illustrated how well the Λ and Σ resonances fit with the N^* resonances into octets and decuplets of SU_3. This observation can be used as an argument (i) for $SU_3|$, or (ii) for the resonances of Table II, or (iii) for the correctness of the Λ and Σ resonance assignments, or (iv) for any combination of (i) (ii) and (iii).

TABLE IV

N^* RESONANCES IN L-EXCITATION QUARK-MODEL CONFIGURATIONS

Mass MeV/c^2	$\{56\}, L=0^+$	$\{70\}, L=1^-$	$\{56\}, L=0^+$	$\{56\}, L=2^+$
900	(8, 2) N			
1000				
1100				
1200	(10, 4) p_{33} (1)			
1300				
1400			(8, 2) p_{11} (1)	
1500		(8, 2) d_{13} (1) (8, 2) s_{11} (1)		
1600		(10, 4) s_{31} (1) (10, 4) d_{33} (1) (8, 4) d_{15} (1)		
1700		(8, 4) d_{13} (2)? (8, 4) s_{11} (2)	(10, 4) p_{33} (2)	(8, 2) f_{15} (1)
1800				(8, 2) p_{13} (1) (10, 4) f_{35} (1)
1900				(10, 4) p_{31} (1) (10, 4) f_{37} (1) (10, 4) ?

Table II but they also imply the existence of higher isospin and SU_3 multiplets, though the observation of these may be inhibited by selection rules (68).

2.7 N^* RESONANCE BRANCHING RATIOS

There is generally little information on the N^* branching ratios for strong decay, other than that to the pion-nucleon channel. In particular there is little evidence bearing on SU_3 predictions. The branching ratios of the $T=\frac{3}{2}$, decuplet, resonances into channels like $\Delta\pi$, $N\rho$, $K\Sigma$ etc. are not known.[13] In the $T=\frac{1}{2}$, octet, resonances there is the ratio $D/F+D=\alpha$ to be determined before SU_3 tests can be applied. Davies & Moorhouse (48) in an analysis of

$$\pi^- + p \rightarrow \eta + n$$

just above η threshold found a value of $\alpha = -.5$ or 2. for the $s_{11}(1)$ and $\alpha = -.1$ or 1.6 for the $d_{13}(1)$, both disagreeing with the SU_6 prediction. It appears (27) that the branching ratios of the $d_{15}(1)$ and $f_{15}(1)$ into the $K\Lambda$ channel are very small; this is not surprising because of the closeness of threshold at 1608 MeV, which would in any case make the interpretation of branching ratios in such high angular momentum channels difficult.

Many quark-model calculations have been made (59, 65–67) of the partial widths $\Gamma_{el} = \Gamma(N^* \rightarrow \pi N)$ of Table II. The authors claim success for the calculations though it is difficult to assess the significance in view of the large uncertainty evident from Table II.

Electromagnetic decays.—The photoexcitation of nucleons to form N^* resonances has much of interest which awaits theoretical and experimental elucidation.

(i) $p_{33}(1)$, $f_{33}(1)$, ... The quark model (60, 70) and SU_{6_w} predict that in the transition $p_{33}(1) \rightarrow N+\gamma$ the $E2$ amplitude vanishes, leaving a purely $M1$ transition amplitude. From analysis of the data the magnitude appears to be given by $|E2/M1| \sim 0.04$ compared with an expected value of 0.18 (67, 70).[14] This prediction is due to the assignment of the $p_{33}(1)$ to a $\{56\}$, $L=0^+$ configuration. Similarly the assignment of $f_{37}(1)$ to a $\{56\}$, $L=2^+$ configuration leads to the prediction $|E4/M3| = 0$ and so on for all the Δ Regge trajectory resonances, though there may be correction factors in quark-model calculations (67).

The $M1$ matrix element calculated (72) in the quark model is smaller by a factor 1.28 than the experimental amplitude (67).

(ii) $p_{11}(1)$. In the photoproduction of pions from protons

$$\gamma + p \rightarrow \pi^0 + p, \qquad \gamma + p \rightarrow \pi^+ + n \qquad\qquad 2.20$$

there is no evident bump corresponding to this resonance and any evidence for its photoproduction rests upon partial-wave (multipole) analysis. Estimates for the partial width from such analyses range from $(\Gamma\gamma/\Gamma)^{1/2} = 4$. $\times 10^{-2}$ (44) to 0 (45, 73). It is perhaps significant that the p_{11} has not yet shown in inelastic electron scattering (74, 75), whereas it shows markedly in inelastic p-p and π-p collisions (59–42). If the p_{11} is assigned to a $\{56\}$ $L=0^+$ radially excited state in the quark model (Tables III and IV), then calculations neglecting recoil give a zero partial width though detailed calculations

---

[13] Though Walker (69) deduces $[f_{37}(1) \rightarrow \rho + N]$ from photoproduction data and gains agreement with the SU_{6_w} result $\Gamma\rho = 2\Gamma\pi$ for magnetic excitation of Δ resonances.

[14] Also dispersion relation fits to data tend to give a minimum of $|E2/M1|$ at the resonance position (71).

by Faiman & Hendry (76) give $\Gamma(p_{11}(1) \to p\gamma) = 0.13$ MeV. Dispersion relation calculations of the $p_{11}(1)$ photoproduction (55, 71) find a small branching ratio for 2.20 but a large one (55) for

$$\gamma + n \to \pi^0 + n, \qquad \gamma + n \to \pi^- + p \qquad 2.21$$

The latter prediction does not seem to accord with experiment (56) as mentioned above in connection with the suggested $\{\overline{10}\}$ assignment of the $p_{11}(1)$.

(iii) $d_{13}(1)$ and $f_{15}(1)$. Each of these has two possible helicity amplitudes for decay into $\gamma + N$, which we denote by $A(J_z = \frac{3}{2})$ and $A(J_z = \frac{1}{2})$ where J_z is the component of total spin along the centre-of-mass direction of decay. It might be an empirical rule (44, 45) that for both resonances $A(J_z = \frac{1}{2}) = 0$. Bietti (77) has attempted an explanation: it is also possible that the rule is related to the classification of particle states in the infinite momentum frame into representations of chiral $SU_2 \times SU_2$ (78).

(iv) The $\{70\}$ $L^P = 1^-$ configuration contains an (8, 2) with $J = \frac{3}{2}, \frac{1}{2}$ and an (8, 4) with $J = \frac{5}{2}, \frac{3}{2}, \frac{1}{2}$. The quark model predicts (79) that the coupling of the (8.4) $J = \frac{5}{2}, \frac{3}{2}, \frac{1}{2}$ to $\gamma + N$ vanishes. In the simple assignments of Table IV these are the $d_{15}(1)$, $d_{13}(2)$?, and $s_{11}(2)$. However, there may be mixing of (8,2) and (8,4) for $J = \frac{5}{2}, \frac{3}{2}$ which modifies (76) the expected result for $d_{13}(2)$ and $s_{11}(2)$.

3. HIGH-ENERGY SCATTERING

3.1 ELASTIC SCATTERING

Like all elementary-particle elastic scattering at high energies, pion-nucleon elastic scattering (80–84) exhibits a marked forward peak. Figure 10 shows that at a pion laboratory momentum of 2.0 GeV/c the $\eta^- p$ cross section has fallen through more than 2 decades at $\cos\theta = 0.5$; at 8 GeV/c it falls 6 decades in the same angular interval. Foley et al. (84) have investigated the extreme forward direction so as to observe the interference with Coulomb scattering and thus measure the ratios $\alpha_\pm = \mathrm{Re} f_\pm(0)/\mathrm{Im}\ f_\pm(0)$ where (Equation 1.17) $f_\pm(0)$ are $\pi^\pm p$ forward scattering amplitudes. These are obtained by fitting the experimental cross section, in and near the interference region, to the formula

$$\frac{d\sigma}{dt}(\pi^\pm p) = \left| \mp \frac{2\sqrt{\pi}}{137\beta|t|} \exp(\pm 2i\delta) + \frac{\pi}{q}(\alpha_\pm + i)\ \mathrm{Im} f_\pm(t) \right|^2 \qquad 3.1$$

where δ, the average phase difference due to the Coulomb interaction is given by

$$\delta = \frac{1}{137\beta} \ln \frac{1.06}{qa\theta} \qquad 3.2$$

where a is the nucleon radius. In the small forward interval used, $\alpha_\pm$ is as-

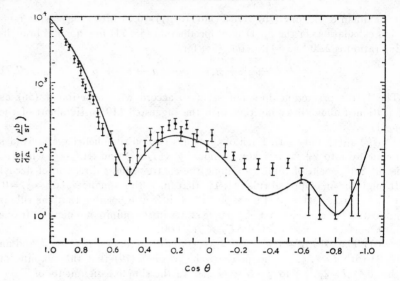

FIG. 10. $\pi^- p$ elastic scattering at pion laboratory momentum of 2.0 GeV/c (85).
The curve is the optical-model fit of Kozlowsky & Dar (86).

sumed to be constant [the effective (84) interval in the fits is $0.001 < -t$
< 0.004 (GeV/$c)^2$] with the form Im $f(t) = (q/4\pi)\ \sigma_{\text{tot}}\ \exp(\tfrac{1}{2}bt + \tfrac{1}{2}ct^2)$. The results are shown in Figure 11, where the errors shown do not include a sys-
tematic scale error of ± 0.02 equal and of opposite sign for $\pi^- p$ and $\pi^+ p$
scattering. These results may be used to check the predictions of forward
dispersion relations, which we write in a subtracted form of 1.24 (see also
Equations 1.31, 1.32) as

$$\operatorname{Re} f^{L(+)}(\omega_L) = \tfrac{1}{2}(\operatorname{Re} f_-^L(\omega_L) + \operatorname{Re} f_+^L(\omega_L))$$

$$= \operatorname{Re} f^{L(+)}(m) + \frac{f^2 q_L^2}{M(1 - m^2/4M^2)} \frac{1}{(\omega_L^2 - (m^2/2M)^2)}$$

$$+ \frac{q_L^2}{4\pi^2} P \int_m^\infty \frac{d\omega_L'}{q_L'} \frac{(\sigma_+(\omega_L') + \sigma_-(\omega_L'))}{(\omega_L'^2 - \omega_L^2)} \omega_L' \qquad 3.3$$

$$\operatorname{Re} f^{L(-)}(\omega_L) = \tfrac{1}{2}(\operatorname{Re} f_-^L(\omega_L) - \operatorname{Re} f_+^L(\omega_L))$$

$$= \frac{2f^2 \omega_L}{\omega_L^2 - (m^2/2M)^2} + \frac{\omega_L}{4\pi^2} P \int_m^\infty \frac{d\omega_L' q_L'}{\omega_L'^2 - \omega_L^2}$$

$$\cdot (\sigma(\omega_L') - \sigma_+(\omega_L')) \qquad 3.4$$

Using $f^2 = 0.081$ and Re $f^+(m) = -0.002$ (the results being insensitive to
reasonable variations of these parameters such as $\pm .004$ in f^2) and evaluating
Im $f^\pm(\omega_L)$ from the optical theorem a good agreement with $\alpha_\pm$ as evaluated

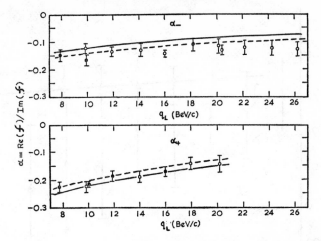

FIG. 11. $\alpha_\pm = \mathrm{Re}\, f_\pm(q_L,0)/\mathrm{Im}\, f_\pm(q_L,\,0)$ evaluated by the Coulomb interference method. The solid curve is $\alpha_\pm$ as evaluated from forward dispersion relations, while the dashed curve is a result of displacing the solid curve by 0.02 being the systematic scale error which cancels in $\alpha_- + \alpha_+$.

from Coulomb interference is found (84) as illustrated in Figure 11. The result verifies the πp forward dispersion relations up to 20 GeV and this may be expressed by remarking (84, 88) that if there exists a small acausal region or fundamental length ℓ, the forward dispersion relations would break down at an energy corresponding to $1/\ell$. These results verify that $\ell < 10^{-15}$ cm.

We can also check isospin invariance which gives

$$\mathrm{Re}\, f^{L(-)}(\omega_L,\,0)$$

$$= \left[\frac{1}{2} \frac{d\sigma_{\mathrm{CE}}}{d\Omega}(0°) - \frac{q_L^2}{64\pi^2}\left(\sigma_{\mathrm{tot}}(\pi^- p) - \sigma_{\mathrm{tot}}(\pi^+ p)\right)^2 \right]^{1/2} \qquad 3.5$$

where $(d\sigma_{\mathrm{CE}}/d\Omega)\,(0°)$ is the laboratory differential cross section at $0°$ for charge exchange scattering (89, 90). The result is shown in Figure 12.

Diffraction scattering.—We have seen that the forward amplitude is predominantly imaginary, nearly 90 per cent at 20 GeV, and that as shown in Figure 10 and Figure 13 the differential cross section displays maxima and minima, as do the cross sections for $\bar{p}p$ and $K^- p$ scattering. This suggests that a first picture of elastic scattering is that of diffraction scattering. One may take an optical model in which the interaction region is represented by a black or grey sphere. The curve shown in Figure 10 is that of Kozlowsky & Dar (86) who take a four-parameter optical model. Their sphere is characterized by a radius, edge diffusivity, transparency, and a parameter giving a real part of the amplitude. Their fit to $\pi^- p$ elastic scattering between 2 and

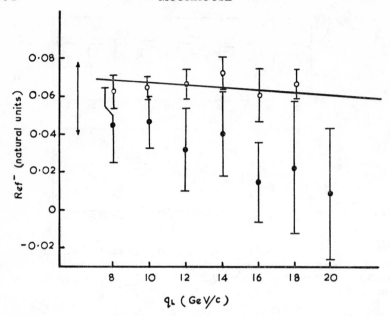

FIG. 12. Values of Re $f^{(-)}(\omega_L)$ obtained from the Coulomb interference method (full circles) and from the charge exchange data (open circles). The solid curve is the value obtained using forward dispersion relations (84).

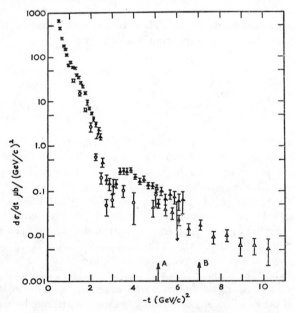

FIG. 13. $\pi^- p$ elastic scattering at ~ 6 GeV/c (full circles, 5.8 GeV/c; full triangles, 5.9 GeV/c) and ~ 8 GeV/c (open triangles, 7.9 GeV/c).

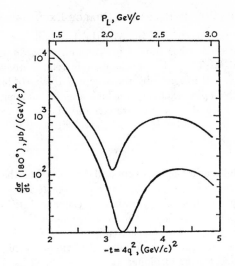

FIG. 14. Differential cross sections at 180° for π^+p scattering (upper curve) and π^-p scattering (lower curve).

12 GeV is qualitatively good for $0 < -t \lesssim 3 (\text{GeV}/c)^2$ with rather constant values of the parameters except for the diffusivity which increases with energy from 0.08 to 0.24 F. The radius of the sphere decreases from 0.84 to 0.81 F.

In both π^+p and π^-p elastic scattering there are dips at constant values of t, namely $t \sim -1$ $(\text{GeV}/c)^2$ and $t \sim -3$ $(\text{GeV}/c)^2$. The dip at $t \sim -1$, corresponding on a diffraction picture to the first diffraction minimum, is very evident at a pion laboratory momentum of 2.0 GeV/c and at higher energies, as in Figure 4, becomes a kink or shoulder. For scattering with pion laboratory energy >2 GeV there is an evident dip (Figure 13) in π^-p and π^+p scattering at $t \sim -3$ $(\text{GeV}/c)^2$. Below 2 GeV incident energy, $t = -3$ is in an unphysical region of the scattering amplitude corresponding to $\cos \theta < -1$. Booth (91) has pointed out how the movement of this dip in the angular distribution from $\cos \theta < -1$ for $q_L < 2$ GeV/c to $\cos \theta > -1$ for $q_L > 2$ GeV/c corresponds to the minimum in the 180° differential cross section at $q_L = 2$ GeV/c shown in Figure 14. Since the dip moves through 180° where the spin-flip amplitude $g(\theta)$ vanishes, it must be associated with the spin-nonflip amplitude $f(\theta)$. We may picture the high-energy scattering as predominantly spin-nonflip diffraction scattering, with diffraction minima at $t \sim -1$, -3, $(\text{GeV}/c)^2$. It may be possible to construct such a picture using a generalized optical model (92) with a suitable impact parameter function $\chi(s, b)$. (From this viewpoint the optical-model curve of Figure 10 may have too much structure, and we might associate the experimental structure round $\cos \theta = -0.4$ with resonances, although of course the average effect of such resonances may be included in a suitable optical potential.) Booth (91) associates the dip at $t \sim -3$ with a t-channel Regge pole.

3.2 t-CHANNEL REGGE POLES AND CHARGE EXCHANGE SCATTERING

While optical-type models parametrize the data and describe how energetic particles view each other during interaction, they do not explain the viewing mechanism. That this mechanism is related to allowed particle exchanges is suggested not only by the accepted theory of highly peripheral interactions (such as pion exchange for high angular momentum in nucleon-nuclear interaction) and interaction ranges but also by the empirical observation of Morrison that cross sections behave with energy like

$$\sigma = \sigma_0 q_L^{-n} \qquad 3.6$$

where q_L is the laboratory momentum and n depends on the quantum numbers allowed to be exchanged in the particular scattering process: (i) $n \sim 0$, vacuum exchange; (ii) $n \sim 2$ charge or isospin exchange; (iii) $n \sim 2.5$, strangeness exchange; (iv) $n \sim 3\text{--}4$, baryon number exchange.

Regge pole exchange (93) currently appears to be the best way of giving quantitative form to quantum number exchanges at high energies and charge exchange scattering of pions:

$$\pi^- + p \to \pi^0 + n \qquad 3.7$$

is one of the most successful examples. It is helpful to define

$$A' = A + \frac{\omega_L + t/4M}{1 - t/4M^2} B \qquad 3.8$$

Then the differential cross section, analogous to 2.7a but using t, A', and B instead of $\cos \theta$, f, and g is

$$\frac{d\sigma}{dt} = \frac{1}{\pi s}\left(\frac{M}{4q}\right)^2 \left\{ (1 - t/4M^2) \mid A' \mid^2 + \frac{t}{4M^2}\left(\frac{q_L^2 - st/4M^2}{1 - t/4M^2}\right) \mid B \mid^2 \right\}$$

$$- \frac{\sin \theta}{16\pi\sqrt{s}} \operatorname{Im} (A'B^*)(\mathbf{n} \cdot \mathbf{P_1}) \qquad 3.9$$

and the optical theorem (see Equations 1.21 and 1.23) is

$$\sigma_{\text{tot}} = \frac{1}{q_L} \operatorname{Im} A'(s, t = 0) \qquad 3.10$$

The t-channel Regge pole contribution to A' and B is a sum of terms, one for each Regge pole, like

$$A'(s, t) = C(t) \frac{1 \pm \exp (-i\pi\alpha)}{\sin \pi\alpha}\left(\frac{s}{s_0}\right)^{\alpha(t)} \qquad 3.11a$$

$$B(s, t) = D(t) \frac{1 \pm \exp (-i\pi\alpha)}{\sin \pi\alpha} \alpha(t) \left(\frac{s}{s_0}\right)^{\alpha(t)-1} \qquad 3.11b$$

for large s. A', B come from terms involving P_α ($-\cos\theta_t$), P_α' ($-\cos\theta_t$) respectively, for large $-\cos\theta_t$. The positive signature trajectories (corresponding to the $+$sign in 3.11) if continued to $t > 0$ generate mesons of spin parity 0^+, 2^+, 4^+, . . . , and the negative signature trajectories (corresponding to the $-$sign in 3.11) generate mesons of spin parity 1^-, 3^-, 5^-, ... The isospin has not been stated explicitly in 3.11 but $A'^{(+)}$, $B^{(+)}$ amplitudes correspond to pure t-channel isospin 0, and $A'^{(-)}$, $B^{(-)}$ to pure t-channel isospin 1. Bose statistics require that $A'^{(+)}$, $B^{(-)}$ amplitudes have positive-signature Regge poles and $A^{(-)}$, $B^{(-)}$ amplitudes have negative-signature Regge poles.

In the πp charge exchange reaction 3.7 the exchanged quantum number has zero strangeness and isospin $T = 1$, and having negative signature it must be in the spin-parity class 1^-, 3^-, 5^-, . . . The only such particle known is the ρ meson, and one thus tries to fit the data with the negative-signature ρ trajectory alone, with remarkable first-order success (94). The charge exchange amplitudes are given by $-\sqrt{2}\,A'^{(-)}$, $-\sqrt{2}\,B^{(-)}$ (see Equation 2.8) and the following results are obtained:

(i) s-Dependence at fixed t ('shrinkage') is consistent with the formula:

$$\frac{d\sigma}{dt}(s, t) \sim F(t)\left(\frac{s}{s_0}\right)^{2\alpha_\rho(t)-2} \qquad 3.12$$

and the ρ trajectory for $t < 0$ can be found from the data. A linear best fit (95) in the range $|t| < (1\ \mathrm{GeV}/c)^2$ is

$$\alpha(t) = 0.57 + 0.91t \qquad 3.13$$

which extrapolates close to the ρ-meson position at $\alpha = 1$, $t = 0.56\ (\mathrm{GeV}/c)^2$; this strongly confirms the connection with particles.

(ii) Phase rule. α_ρ (0) according to (i) is roughly $\frac{1}{2}$ and so the real and imaginary parts of the forward amplitudes are roughly equal. The imaginary part can be found from the optical theorem together with isospin invariance (Equations 2.8 and 1.4).

$$\frac{1}{q_L}\,\mathrm{Im}\,A'^{(-)}(t = 0) = \frac{1}{\sqrt{2}}\,(\sigma_{\mathrm{tot}}(\pi^+ p) - \sigma_{\mathrm{tot}}(\pi^- p)) \qquad 3.14$$

The real part to within a sign can then be found from the observed charge exchange forward cross section, and this agrees with the phase rule (96).

(iii) Dip at $t = -0.6$. The contribution of B to the cross section must vanish at $t = 0$, and a strong B term thus explains the rise near $t = 0$. However, from 3.11b, B vanishes at $\alpha(t) = 0$ corresponding to $t = -0.6$, which is just where a dip is observed in the differential cross section. However, the amplitude and thus A does not vanish at this point, which implies that the Regge pole residue function $C(t)$ does not vanish. The factorization[15] property of

---

[15] A review has been given by Bertocchi (96) of factorization mechanisms at integer values of α. The success of the simple Regge pole picture seems to show that in this case anyway, the effect of Mandelstam-Wang fixed poles (97) and of cuts (93) can be neglected.

Regge pole residues is then satisfied by a sense-choosing mechanism at the (wrong signature) zero of $\alpha_p(t)$.

The points (i) (ii) and (iii) just mentioned support the description of high-energy (2–18 GeV/c) πp charge exchange by a single Regge ρ exchange. However, this model has A' and B in phase, according to 3.11, and consequently zero polarization (Equation 3.9), and experiment shows appreciable polarization at 6 and 11 GeV/c (98). This polarization can be produced by not too large additions to the amplitude, consistent with the maintenance of (i) (ii) and (iii) above, and the Regge ρ exchange as the dominant effect. Beaupre, Logan & Sertorio (99) and Barger & Phillips (100) have fitted all the data using a second ρ trajectory, ρ'. As this implies a physical ρ' meson of mass 1–1.2 GeV/c^2 which has not been observed, the ρ' trajectory used might be taken to represent an effective cut contribution. Among other possibilities Michael (101) has calculated the effect of Regge ρ exchange plus absorption.

White (102) has successfully applied the Gottfried-Jackson absorption prescription (103), whose original purpose was to modify and correct single-particle exchange, to modify the Regge ρ exchange, and to produce the observed polarization without variable parameters.

3.3 STRAIGHT-LINE TRAJECTORIES, u-CHANNEL POLES, AND π-p BACKWARD SCATTERING

As illustrated in Figure 3 the bumps and shoulders in the $\pi^+ p$ total cross section (1) implies on the simplest interpretation $T = \frac{3}{2}$ resonances at $\sim$2450, 2840, and (possibly) 3220 MeV/c^2. If the resonances are assigned spin 11/2, 15/2, and 19/2 respectively they lie, to within the mass errors, on the straight-line spin versus (mass)^2 Chew-Frautschi plot (Figure 15) together with the $p_{33}(1)$ and $f_{37}(1)$ of Table II. Whether this Δ Regge trajectory, like the ρ trajectory, is a straight line awaits final confirmation in experimental findings on the spins and parities of the $N_{3/2}{}^*$ (2450), $N_{3/2}{}^*$ (2840), and $N_{3/2}{}^*$ (3220). The trajectory has positive parity and negative signature, $\tau = (-1)^{J-1/2}$ and if we denote the mass by $\sqrt{u}/c$, where u is measured in (GeV/c)^2, the trajectory of Figure 15 is given (105) by:

$$\Delta: \quad \text{Re } \alpha(\sqrt{u}) = 0.15 + 0.90u \qquad 3.15$$

Shown in Figure 16 are two more conjectural trajectories. The N_α trajectory has $P = +1$, $\tau = +1$; the only known resonances it contains are the nucleon and the $f_{15}(1)$; it has trajectory equation (105);

$$N_\alpha: \quad \text{Re } \alpha(\sqrt{u}) = -0.39 + 1.01u \qquad 3.16$$

From the $\pi^- p$ and $\pi^+ p$ total cross sections one can extract the $T = \frac{1}{2}$ total cross section using the relation (derived from Equations 1.4 and 1.21)

$$\sigma_{\text{tot}}(T = \tfrac{1}{2}) = \tfrac{3}{2}\sigma(\pi^- p) - \tfrac{1}{2}\sigma(\pi^+ p) \qquad 3.17$$

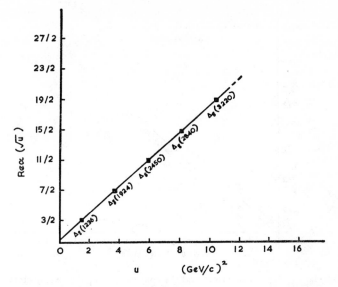

FIG. 15. The conjectured Regge trajectory of $T = 3/2$ resonances.

Resonances are indicated at $\sim$2210, 2640, and 3020 MeV/c^2 which if assigned a spin difference of 2 lie on a straight line. The N_γ Regge trajectory
is obtained by assigning spins of 7/2, 11/2, and 15/2 resulting in the trajectory, with $\tau = -1$,

$$N_\gamma: \quad \text{Re } \alpha(\sqrt{u}) = -0.90 + 0.92u \qquad 3.18$$

The $d_{13}(1)$ lies near this line (Figure 16) and if this also belongs to the trajectory, then the trajectory has parity $P = -1$. Since

$$u = -2q^2(1 + \cos \theta) + (M^2 - m^2)^2/s \qquad 3.19$$

u is small in the backward direction, for large s, and u-channel, necessarily
baryon, exchanges have long been conjectured to play an important role in
ηp backward scattering. Barger & Cline (105) have speculated that the $\pi^\pm p$
scattering amplitude in the backward hemisphere, and particularly at 180°,
is given, in the region of laboratory momentum $\sim 2 - \sim 9$ GeV/c by the sum
of direct-channel resonances and u-channel Regge poles.

This is shown schematically in Figure 17 for $\pi^- p$ scattering, in which case
only isospin $T = \frac{3}{2}$ can be exchanged in the backward direction, and this u-
channel exchange is assumed dominated by the Δ Regge trajectory. This
model, known as the interference model, has been subject to criticism, as
discussed in Section 4 below, on the grounds that perhaps the crossed-channel amplitude already contains a considerable portion of the direct-channel
resonant amplitudes. This possible error is known as double counting. We
will describe the Barger & Cline results from a heuristic point of view, and

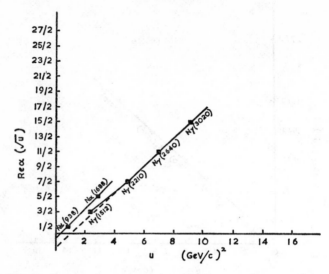

FIG. 16. The N_α Regge trajectory, and the conjectured N_γ Regge trajectory of $T=\frac{1}{2}$ resonances.

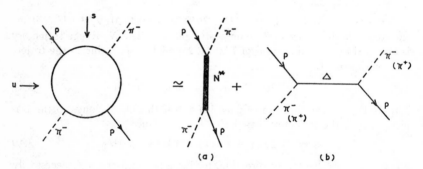

FIG. 17. Schematic illustration of the interference model of Barger & Cline (105) for backward $\pi^- p$ elastic scattering. The amplitude is the sum of direct-channel resonances amplitudes (a) and the crossed-channel, with Δ Regge trajectory exchange, amplitude (b).

assume that double counting has not occurred. Their prime result is a verification of the parity assignments of the resonances in the Δ and N_γ trajectories. The amplitudes used, and illustrated schematically in Figure 17, are constructed as follows.

In 1.11a A and B may be regarded as functions of the two independent variables s and u—$A(s,u)$ and $B(s,u)$ The nucleon energy in the s-channel centre-of-mass system is given by

$$E = (s + M^2 - m^2)/2W \qquad 3.20$$

and is an odd function of $W = \sqrt{s}$ so that from 1.11a

$$f_1{}^s(\sqrt{s}, u) = \frac{E^s + M}{8\pi\sqrt{s}} [A(s, u) + (\sqrt{s} - M)B(s, u)] \qquad 3.21$$

$$f_2{}^s(\sqrt{s}, u) = f_1{}^s(-\sqrt{s}, u) \qquad 3.22$$

where a superfix $s(u)$ is added to denote that the quantity refers to the $s(u)$ channel. The corresponding u-channel amplitude to 3.21 is

$$f_1{}^u(\sqrt{u}, s) = \frac{E^u + M}{8\pi\sqrt{u}} [A(u, s) + (\sqrt{u} - M)B(u, s)] \qquad 3.23$$

Considering as in Figure 17 $\pi^- p$ scattering in the s channel (denoted by the suffix -), then under crossing of the pions $\pi^- \leftrightarrow \pi^+$, leading to isospin $\frac{3}{2}$ in the u channel and the crossing relations (Section 1.2):

$$A_-(s, u) = A^{3/2}(u, s), \qquad B_-(s, u) = -B^{3/2}(u, s)$$

and from 1.11, 3.21–3.23

$$f_1{}^s(\sqrt{s}, u)_- = \frac{E^s + M}{2\sqrt{s}} \left[(\sqrt{u} - \sqrt{s} + 2M)\frac{f_1{}^{u3/2}(\sqrt{u}, s)}{E^u + M} \right.$$
$$\left. + (\sqrt{u} + \sqrt{s} - 2M)\frac{f_1{}^{u3/2}(-\sqrt{u}, s)}{E^u - M} \right] \qquad 3.24$$

For $\pi^+ p$ and charge exchange scattering the relationship is, symbolically,

$$f_1{}^s_+ = \frac{1}{3} f^{u3/2} + \frac{2}{3} f^{u1/2} \qquad 3.25$$

$$f_1{}^s(CE) = \frac{\sqrt{2}}{3} (f^{u3/2} - f^{u1/2}) \qquad 3.26$$

For the Regge pole contribution[16] of Figure 17(b), Barger & Cline (105) use the approximation

$$f_1(\sqrt{u}, s) = \frac{E^u + M}{\sqrt{u}} (\alpha + \tfrac{1}{2})(\alpha + \tfrac{3}{2})\gamma \frac{1 + i\tau e^{i\pi\alpha}}{\cos \pi\alpha}$$
$$\cdot \left(\frac{s - M^2 - m^2}{s_0}\right)^{\alpha - 1/2} \qquad 3.27$$

where α, γ are $\alpha(+\sqrt{u})$, $\gamma(+\sqrt{u})$ or $\alpha(-\sqrt{u})$, $\gamma(-\sqrt{u})$ for $\tau P = -1$ or $\tau P = +1$ respectively. The additional assumptions, probably valid for small

[16] Note that the Regge trajectory function $\alpha(\sqrt{u})$ is real below the physical u threshold, $u < (M+m)^2$.

$|u|$ and thus for backward scattering between $\sim$2 GeV/c and $\sim$8 GeVc, are introduced:[17]

$$\gamma = \text{const}, \qquad \alpha(\sqrt{u}) = a + bu \qquad\qquad 3.28$$

In $\pi^- p$ scattering 3.27 involves only the Δ trajectory, Equation 3.15 with $\tau p = +1$. The direct-channel poles are taken to be those of the Δ, $N\alpha$, and $N\gamma$ trajectories making a contribution to $f(\sqrt{s}, \cos\theta)$ at 180° of

$$f_-^{\text{res}}(180°) = \frac{1}{q}\left[\frac{1}{3}\sum_\Delta \frac{(-1)^\ell(J+\frac{1}{2})x_\Delta}{\epsilon - i} + \frac{2}{3}\sum_{N\alpha,N\gamma} \frac{(-1)^\ell(J+\frac{1}{2})x_N}{\epsilon - i}\right]$$

$$\epsilon = (M_R{}^2 - s)/M_R\Gamma \qquad\qquad 3.29$$

where M_R, Γ are the resonance mass and width, x is the elasticity and $(-1)^\ell$ is $+1$ or -1 according as the parity P of the resonance is -1 or $+1$ respectively. Consequently when this amplitude 3.29 is added to the crossed-channel amplitude obtained from 3.27, 3.25, 3.22 and 2.3, a sensitive test of the resonance parities is provided by the interference. M_R, Γ are estimated from total cross sections or phase-shift analysis; x, where not known, is a parameter of the fit to experiment, as is γ; and s_0 is fixed at 0.4 (GeV/c)^2 (103).

It is found that a good fit to $\pi^- p$ 180° scattering can be obtained when the parity assignments of the resonances in 3.29 are made according to their Regge trajectories, but not otherwise. This is shown in Figure 18. With the parameters determined by this fit a good qualitative but not detailed quantitative agreement near 180° is obtained. For the $\pi^+ p$ scattering near 180° Barger & Cline report a satisfactory fit to experiment. Chiu & Stirling (106) reach somewhat more negative conclusions in the $\pi^+ p$ case and make a general examination of the interference model including forward and charge exchange scattering when the Regge backround involves the t channel. They conclude that the interference model works well either when there are large cancellations between nearby resonance amplitudes or when the resonance contributions are much smaller than the Regge backround. However, their criticisms assume that the intrinsic phase of the resonance (δ_∞ in 2.17a) is zero, as indeed do Barger & Cline. This is not necessarily so, and the point is discussed further in Section 4.3.

We may note the marked dip in $\pi^- p$ backward scattering near $q_L = 2$ GeV/c which is well fitted by the interference model (Figure 18). This is the same dip or kink described in Section 3.1 that occurs at all energies at $t = -3$ (GeV/c)^2 as perhaps a second diffraction-type minimum, which moves in to the physical region though 180° at $q \sim$2 GeV/c. The interference

---

[17] The naive derivation of the Regge asymptotic formula, as a limit for example of $P_\ell(-\cos\theta^u)$ for $\cos\theta^u \to \infty$, breaks down owing to the finiteness of $\cos\theta^u$ at $\theta^s = 180°$. The problem has been resolved by correct analytic continuations involving the existence of Regge daughters. A review is given in (96).

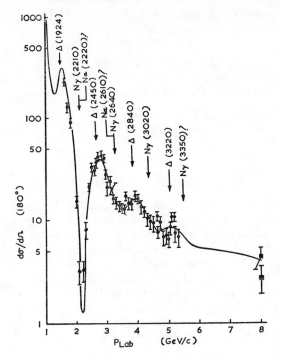

FIG. 18. Interference model fit to $\pi^- p$ elastic differential cross section at 180°.

model thus implies that the resonances are such as to produce a dip at the required position. Finite-energy sum rules may aid in understanding this situation.

In favour of the Regge exchange interpretation of backward scattering, though not necessarily of the interference model, may be quoted the well-observed dip near the backward direction, at a constant value $u = -0.2$ $(GeV/c)^2$ over a large range of u energy. This is satisfactorily near the value of u where the nucleon Regge trajectory (Equation 3.16) goes through the value $\alpha = -\frac{1}{2}$. At this point on the Regge trajectory, factorization theorems on Regge residues show that the residue vanishes (at $\alpha = -\frac{1}{2}$ there is a non-sense-nonsense transition at a wrong signature point (96)). According to this theory only the Δ Regge trajectory should contribute, at this point $u = 0.2$, to the $\pi^+ p$ scattering and thus it should bear a ratio $1:9$ to $\pi^- p$ scattering at the same angle and energy. This ratio is satisfied.

McDowell symmetry and parity doublets.—An interesting point arises connected with the forms 3.15, 3.16, and 3.18 and their dependence on $(\sqrt{u})^2$ rather than $\sqrt{u}$. We wish to discuss these trajectories in the s channel so we now write the Δ trajectory function, for example,

$$\text{Re } \alpha_\Delta(W) = 0.15 + 0.90W^2 \qquad\qquad 3.30$$

McDowell symmetry, obvious from 1.13 and 1.11, is the statement

$$f_{l+}(W) = f_{(l+1)-}(-W) \qquad 3.31$$

This symmetry between partial waves of opposite parity requires a Regge pole in partial-wave amplitudes with parity P, to necessitate a Regge pole of the same signature in the partial waves of opposite parity:

$$f_P{}^J(W) = \beta(W)/(J - \alpha(W)) \qquad 3.32$$

$$f_{-P}{}^J(W) = \tilde{\beta}(W)/(J - \tilde{\alpha}W) \qquad 3.33$$

where

$$\tilde{\alpha}(W) = \alpha(-W), \qquad \tilde{\beta}(W) = \beta(-W) \qquad 3.34$$

So if these were even functions of W, we would expect parity doubling of resonances. Now 3.30, for example, being firstly approximate and secondly only formulated for positive W, does not necessarily imply parity doubling. However, it may be suggestive. Even if Regge trajectories were even functions of W, the residue functions need not be and Regge poles may be missing due to vanishing of $\tilde{\beta}(W)$ at the pole energy. Some possible McDowell parity doublets may be found on examination of Table II, such as $f_{15}(1)$, $d_{15}(1)$. The nucleon and the $f_{15}(1)$ are on the N_α trajectory. However, there is no parity doublet of the nucleon on the $d_{15}(1)$ trajectory or any other trajectory. This could be due to vanishing of the residue function (107).

3.4 $\pi^{\pm}p$ ELASTIC SCATTERING AND REGGE POLES

Philips & Rarita (94) showed that high-energy meson-nucleon scattering can be fit using a number of Regge poles with trajectory and residue parameters determined by the data. In particular they used the P, P', and ρ trajectories to fit $\pi^{\pm}p$ elastic scattering for 6 GeV/$c^2 \lesssim q_L < 16$ GeV/c and -0.9 (GeV/$c)^2 \lesssim t < 0$ and also $\pi^- p \rightarrow \pi^0 n$ differential cross sections. Incidentally, though Philips & Rarita made no explicit fit to elastic $\pi^{\pm}p$ polarizations, their predicted polarizations are quite a good fit to the data—though not of course in the case of charge exchange polarization as discussed in Section 3.2.

However, as discussed in Section 3.1 the elastic $\pi^{\pm}p$ differential cross sections have a diffraction-type structure with dips or kinks at $t \sim -0.8$ and $t \sim -2.8$. The Philips-Rarita analysis does not extend to large enough $|t|$ to include these. (The $t \sim -0.8$ 'kink,' as it has become at high energy, is at the end of the range of fit and so the characteristic behaviour for $|t| > \sim 0.8$ is not included.) Similar dips or inflection points also occur in $K^- p$ and $\bar{p}p$, though not so markedly in pp and not in K^+p. [These last two systems apparently do not have strong resonance formation and we may recollect that the dips at $t \sim -2.8$ (GeV/$c)^2$ can apparently be reproduced by resonance interference with backround as described in Section 3.3.]

Barger & Philips (108, 109) have attempted to describe this diffraction-type phenomenon purely in terms of Regge poles by the cyclic residue

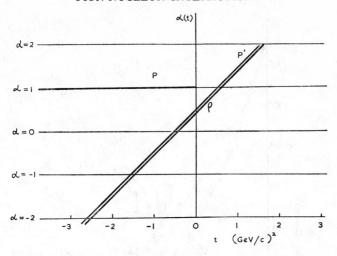

FIG. 19. Schematic diagram of P, P', and ρ Regge trajectories.

ansatz. Besides the dips the *ansatz* explains the zeros of the polarization (110) and provides a semiqualitative fit to the differential cross sections and polarization in the region -4.0 $(GeV/c)^2 \lesssim t \leq 0$.

The Regge trajectories used in the *ansatz* for the $\pi^{\pm}p$ data are the P, P', and ρ as shown in Figure 19. The contributions to the A' and B amplitudes are taken to be (compare Equations 3.11, 3.12)

$$A_{\pm}' = -\gamma_P \exp(-i\pi\alpha_P/2)\nu^{\alpha_P} - \gamma_{P'} \exp(-i\pi\alpha_{P'}/2)\nu^{\alpha_{P'}}$$
$$\mp i\gamma_\rho \exp(-i\pi\alpha_\rho/2)\nu^{\alpha_\rho} \qquad \qquad 3.35$$

$$B_{\pm} = -\beta_P \exp(-i\pi\alpha_P/2)\nu^{\alpha_P} - \beta_{P'} \exp(-i\pi\alpha_{P'}/2)\nu^{\alpha_P}$$
$$\mp i\beta_\rho \exp(-i\pi\alpha_\rho/2)\nu^{\alpha_\rho} \qquad \qquad 3.36$$

where the $\pm$ suffix corresponds to $\pi^{\pm}p$ scattering and where the γ and β are essentially the residues C and D of 3.11, 3.12 divided by $\sin(\pi\alpha/2)$ for P and P' and $\cos(\pi\alpha/2)$ for ρ. The cyclic residue *ansatz* is that the P' and ρ residues have recurring zeros, the important residues being given by

$$\gamma_{P'} = \lambda(t) \sin(\pi\alpha_{P'}(t)/2) \qquad \qquad 3.37$$

$$\beta_\rho = \xi(t) \sin(\pi\alpha_\rho(t)/2) \qquad \qquad 3.38$$

In $\pi^{\pm}p$ differential cross sections the A' (nonspinflip) amplitude is dominant, and in it the P and P' contributions. Now 3.37 has zeros at $\alpha_{P'}=0$ and $\alpha_{P'}=-2$, that is (Figure 19) at $t\sim-0.5$ and $t\sim-2.5$. Through these zeros and the interference between the P and P' trajectories points of inflection or dips occur near $t\sim-0.5$ and $t\sim-2.5$ and a good fit to the differential cross sections can be achieved as shown in Figure 20.

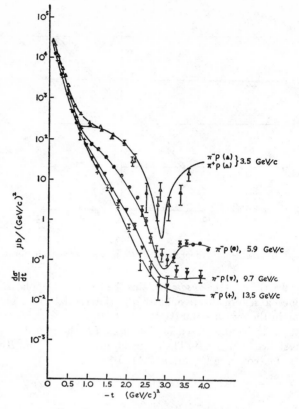

Fig. 20. Cyclic residue *ansatz* fit to $\pi^{\pm}p$ differential cross sections.

Secondly consider the polarization. We neglect the ρ contribution to A', known to be small from analyses of $\pi^- + p \rightarrow \pi^0 + n$, and we find from 3.9, 3.35, 3.36 that

$$P(\pi^{\pm}p)\frac{d\sigma}{dt}(\pi^{\pm}p) = \pm\frac{\sin\theta}{16\pi\sqrt{s}}\sum_{K=P,P'}\beta_{\rho}\gamma_K\cos\frac{\pi}{2}(\alpha_K - \alpha_{\rho})\nu^{\alpha K + \alpha_{\rho} - 1} \qquad 3.39$$

$$+ (P, P' \text{ interference terms})$$

The P-P' interference terms have the same sign for π^+p and π^-p, so the observed mirror symmetry of polarization (108) shows that this term is small. [This implies that $\beta_{P'}$ must have the same zeros as $\gamma_{P'}$ in Equation 3.37 (109).] In the P-ρ term both β_{ρ} and the cosine factor vanish near $t = -0.5$, giving an approximate double zero. In the P'-ρ term both β_{ρ} and $\gamma_{P'}$ vanish at $t = -0.5$ by the *ansatz* equations 3.37, 3.38, giving then an approximate double zero in the polarization 3.39 at $t \sim -0.5$. For similar reasons a double

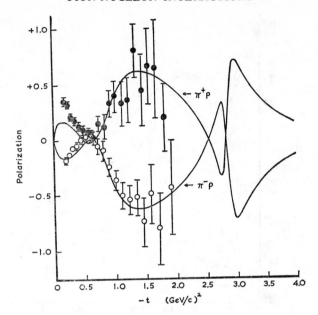

FIG. 21. Cyclic residue *ansatz* fit to $\pi^\pm p$ polarization at GeV/c.

zero also occurs at $t \sim -2.5$. The fit to the polarization is shown in Figure 21 and to the differential cross section in $\pi^- + p \rightarrow \pi^0 + n$ in Figure 22.

Among sophisticated optical-type treatments of high-energy scattering the Chou-Yang model (111) is important. This treats each elementary particle in the collision as an extended object, and relates, for example, high-energy scattering cross sections of protons to the proton form factor. The possibility of combining Regge pole exchange with absorption has already been noted in Section 3.2 in connection with $\pi^- + p \rightarrow \pi^0 + n$. It is of course likely to be particularly relevant in obtaining the diffraction-type minima in, for example, $\pi^\pm p$ scattering at $t \sim -0.8$ and $t \sim -2.8$. Such treatments automatically put in the equivalent of Regge cuts, so care must be taken that successful features of the Regge theory that may be spoiled by cuts (such as the factorization theorems with the occurrence of dips before noted in $\pi^- p$ charge exchange and $\pi^+ p$ backward scattering) are not in fact too much changed. Arnold & Blackmon (112) have fitted elastic and charge exchange scattering in such a hybrid model with Regge poles and Chou-Yang absorption.

4. DYNAMICS

4.1 NUCLEON RESONANCES FROM PARTICLE EXCHANGE

The successful prediction of the p_{33} resonance, whose exact position and width depend on adjustable parameters, is the achievement of many theories

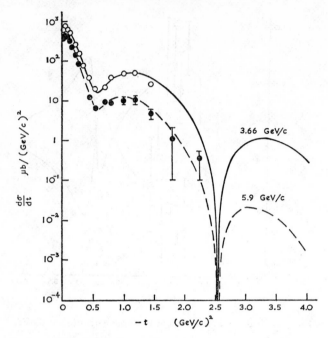

FIG. 22. Cyclic residue *ansatz* fit to the $\pi^- + p \to \pi^0 + n$ differential cross sections.

from strong coupling through the Tamm-Dancoff method to partial-wave dispersion relations. The reason lies in the diagram of Figure 6(b) representing nucleon exchange in pion-nucleon scattering. The force corresponding to this diagram, projected into the p_{33} channel, is strongly attractive enough to give resonance, and the successful theories contain this force or its equivalent as a consequence of the nature of πNN coupling with pseudoscalar pions. While this is highly encouraging for the proponents of a more general dynamical theory with forces generated by the exchange of known particles, it is not of course conclusive. For example the existence of this strong N-exchange force of the right sign might reflect an underlying symmetry and the resonance forces might lie deeper, as in the quark model.

The study of the dynamics of particle exchange through single variable dispersion relations is exemplified in the classic paper of Chew, Goldberger, Low & Nambu (7) and the Mandelstam representation (113) gave hope of describing strong interactions by finding the double spectral function [cf. Chew (114)]. Chew & Mandelstam (115) initiated the bootstrap idea for the ρ meson, in which the exchange of the ρ meson between two pions itself provides the force which makes the two pions resonate in a ρ-meson state, thus perhaps making possible a self-consistent calculation to find the mass and width of the ρ. Chew & Frautschi (116) extended the bootstrap idea to encompass a self-consistent calculation of all particle (resonance) masses and widths (117). It has not seemed possible to produce a very convincing success with this program. However, one should probably distinguish between the basic idea of the bootstrap and its execution. The basic idea, that a set of particles or resonances is generated by the strong forces produced by the exchange of the same set of particles or resonances, may well be correct; it is supported by the qualitative observation that in most cases strong forces in the correct, resonance states are produced by particle exchange, as in N exchange and ρ exchange mentioned above. But finding the masses and widths, by an iterative or similar process, may well be subject to inherent instabilities due to lack of knowledge of the forces caused by interchange of more massive particles. In any case the program seems beyond the grasp of present calculational techniques.

In this situation a much less ambitious approach is open. One may consider the exchange of known particles or resonances of mass very roughly ~ 1 GeV$/c^2$, and ask whether the forces so obtained are attractive and big enough in the correct channels to generate the observed resonances at about the right mass. In this approach there is no attempt at an iterative bootstrap and we must be careful not to try to apply the method where shorter-range forces (than those due to the exchange of the particles in question) are important. Donnachie & Hamilton (118) use such an approach to the N^* resonances, and in the remainder of this section we will give their results.

The partial-wave amplitudes $f_{\ell\pm}(s)$ (Equations 1.13 and 2.5) are assumed to obey a dispersion relation with a right-hand physical cut from $s = (M\ m)^2$ to $s = \infty$ and a left-hand cut structure with Re$s < M^2 + 2m^2$ (10). The particle interchanges contribute to the left-hand or unphysical cut, and the more massive the particle the further to the left does its contribution to the cut occur. To diminish the importance of very massive particle exchanges, Donnachie & Hamilton consider a dispersion relation in

$$F_{\ell\pm}(s) \equiv \frac{1}{q^{2\ell}} f_{\ell\pm}(s) \qquad 4.1$$

This amplitude is finite at threshold, $q = 0$ (since $\delta_\ell \sim q^{2\ell+1}$), and the $1/q^{2\ell+1}$ factor diminishes the contribution from distant cuts on the left and on the right. The dispersion relation is

$$\mathrm{Re}\, F_{\ell\pm}(s) = \frac{1}{\pi} P \int_{(M+m)^2}^{\infty} ds' \frac{\mathrm{Im}\, F_{\ell\pm}(s')}{s' - s}$$
$$+ \frac{1}{\pi} \int_{\text{unphysical cuts}} ds' \frac{\mathrm{Im}\, F_{\ell\pm}(s^2)}{s' - s} \qquad 4.2$$

The unitarity condition

$$\mathrm{Im}\, f_{\ell\pm}^{-1} = -q \frac{\sigma_{\mathrm{el}}(\ell\pm)}{\sigma_{\mathrm{tot}}(\ell\pm)} \qquad 4.3$$

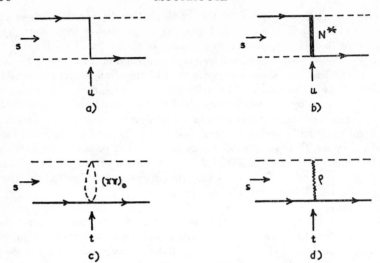

FIG. 23. The four processes which give the longer-range parts of the pion-nucleon interaction. They are (a) N exchange; (b) N^* exchange; (c) exchange of the $T=0$, $J=0$ pion pair: $(\pi\pi)_0$; (d) ρ exchange.

where $\sigma_{\mathrm{el}}(\ell\pm)$, $\sigma_{\mathrm{tot}}(\ell\pm)$ are the elastic, total cross sections for scattering in the $j=\ell\pm\frac{1}{2}$ partial wave, connects Re $F_{\ell\pm}(s)$ with Im $F_{\ell\pm}(s)$ on the right-hand cut. If one knew $R_{\ell\pm}=\sigma_{\mathrm{el}}(\ell\pm)/\sigma_{\mathrm{tot}}(\ell\pm)$ everywhere and also Im $F_{\ell\pm}(s)$ on the left-hand cut everywhere, one could in principle find $F_{\ell\pm}(s)$ on the physical cut.

Exchanges of N, N^*, ρ, and π-π s waves, shown in Figure 23, give contributions to Im $F_{\ell\pm}(s^1)$ on unphysical cuts nearest to the physical region. It is assumed that other contributions to the left-hand cuts will be sufficiently diminished by the $q^{2\ell}$ factor, the higher ℓ the greater the effect. Far left-hand cut contributions, corresponding to short-range forces, are comparatively more important at higher energies as can be seen from 4.2. It has been estimated (41) that the diagrams of Figure 23 are dominant up to pion laboratory kinetic energy of 900 MeV in the case of d waves and 1500 MeV for f waves. The factor $q^{2\ell}$ also sharply suppresses the high-energy rescattering integral, from the first term on the right-hand side of 4.2. This would leave the contribution of the diagrams of 4.2 (that is the near left cut, or comparatively long-range force contributions) as the principal driving terms in the resonance region of energy. Donnachie & Hamilton assume that the likelihood of a resonance occurring is to be measured by the sign and magnitude of the contribution of N, N^*, ρ, and $(\pi$-$\pi)_0$ (s-wave) exchange to

$$F_{\ell\pm}'(s) \equiv \frac{1}{\pi} \int_{\mathrm{unphysical\ cuts}} ds' \frac{\mathrm{Im}\ F_{\ell\pm}(s')}{s'-s} \qquad 4.4$$

TABLE V

SOME TYPICAL VALUES OF THE CONTRIBUTIONS TO $F_{l\pm}'(s)$ FROM THE
FOUR EXCHANGE PROCESSES IN THE CASE OF THE
STRONGLY ATTRACTIVE AMPLITUDES[a]

	p_{33} 200 MeV	d_{13} 600 MeV	f_{15} 600 MeV	f_{15} 900 MeV	f_{37} 1.3 GeV
N	$+4.2 \cdot 10^{-2}$	$-6.0 \cdot 10^{-5}$	$+6.3 \cdot 10^{-7}$	$+2.8 \cdot 10^{-7}$	$+1.4 \cdot 10^{-6}$
N^*	$+1.0 \cdot 10^{-3}$	$-2.7 \cdot 10^{-4}$	$+3.9 \cdot 10^{-6}$	$+2.7 \cdot 10^{-6}$	$+8.8 \cdot 10^{-8}$
$(\pi\pi)_0$	$+1.2 \cdot 10^{-2}$	$+4.9 \cdot 10^{-5}$	$+1.8 \cdot 10^{-6}$	$+7.2 \cdot 10^{-7}$	$+4.0 \cdot 10^{-7}$
ρ	$+3.0 \cdot 10^{-4}$	$+1.02 \cdot 10^{-3}$	$+1.87 \cdot 10^{-5}$	$+1.18 \cdot 10^{-5}$	$+3.07 \cdot 10^{-6}$
$F_{l\pm}'(s)$	$5.5 \cdot 10^{-2}$	$0.74 \cdot 10^{-3}$	$2.50 \cdot 10^{-5}$	$1.55 \cdot 10^{-5}$	$4.96 \cdot 10^{-6}$
$1/2q^{2l+1}$	$1.1 \cdot 10^{-1}$	$1.43 \cdot 10^{-3}$	$1.38 \cdot 10^{-4}$	$2.57 \cdot 10^{-5}$	$5.78 \cdot 10^{-6}$

[a] The energy values are laboratory pion energies. The last row gives the unitary limit for comparison.

The resultant isospin and spin properties of 4.4 can be summarized by classifying the pion-nucleon states into four sets: (i) $T = \frac{1}{2}$, $j = l - \frac{1}{2}$; (ii) $T = \frac{3}{2}$, $j = l + \frac{1}{2}$; (iii) $T = \frac{1}{2}$, $j = l + \frac{1}{2}$; (iv) $T = \frac{3}{2}$, $j = l - \frac{1}{2}$. The important features of the forces in these sets are as follows:

(i) ρ Exchange is strong and attractive and generally dominates the other interactions at the relevant energies,

(ii) N exchange is strong and attractive for odd l, strong and repulsive for even l.

(iii), (iv) N, N^* forces are weak, ρ-exchange force repulsive.

The detailed conclusions are:

(i) $T = \frac{1}{2}$, $j = l - \frac{1}{2}$: The partial waves in the energy range are the p_{11}, d_{13}, and f_{15}. It is very satisfactory that resonances or particles occur in all these waves. The d_{13} and f_{15} resonate at pion laboratory kinetic energies of $\sim$600 and 900 MeV. In Table V we show the contributions of the individual exchange processes to $F_{l\pm}'(s)$ at resonance, and the total. A positive contribution corresponds an attractive force. From 2.5 and 4.1,

$$\text{Re } F_{l\pm}(s) = \eta_{l\pm} \sin 2\delta_{l\pm}/2q^{2l+1} \qquad 4.5$$

giving an upper limit from unitarity of $1/2q^{2l+1}$ on Re $F_{l\pm}(s)$. From 4.2 a comparison of $F_{l\pm}'(s)$ with this unitary limit is a good measure of the strength of the forces due to the particle exchanges. We see from Table V that the combined contribution of N, N^*, ρ and $(\pi-\pi)_0$ exchange approaches the unitary limit at resonance. (At high enough energies this contribution exceeds the unitary limit—but this is in a range of energy where the peripheral approach is not valid.) One can say that this very strong force is consistent with the occurrence of resonance, and perhaps that the approach to the unitary limit gives the approximate resonance energy, as seen from the f_{15} figures of Table V.

The baryon exchange forces contain a factor $(-1)^\ell$, and this is also illustrated in Table V. The baryon exchange forces are repulsive for the d_{13} and attractive for the f_{15}. The N_γ Regge trajectory, containing the d_{13}, lies lower than the N_α Regge trajectory, containing the f_{15} (Figure 16), corresponding to more 'binding' in the N_α trajectory, which is indeed provided by the N, N^* forces. Of course, when one considers Regge trajectories generated from a Schrödinger equation with an ordinary and a space exchange potential, it is the space exchange potential which is responsible for signature and the splitting of a trajectory into two trajectories of odd and even signature respectively (93). For the baryons space exchange forces are generated by baryon exchange and what is interesting is that baryon exchange appears to be of the right sign and a reasonable magnitude to provide the observed splitting.

The p_{11} wave is much more complicated in the energy range considered, since it contains the nucleon and the $p_{11}(1)$, with the phase shift changing sign at 170 MeV (Section 2.4). While it is satisfactory that the forces of Figure 23 are strongly attractive in the p_{11} case, a dynamical explanation of the observed situation may be complicated. One of the factors is the inelasticity of the p_{11} wave which increases very rapidly above a pion laboratory energy of 300 MeV ($W \sim 1300$ MeV), (24, 31, 35). In the case of scattering with no inelasticity, resonances due to CDD poles (119) (characterized by a zero in the scattering amplitude and a breakdown of Levinson's theorem)[18] are not implied by the left-hand cut singularities but are consistent with them. Thus, one may say that CDD poles are not dynamically generated. Where there is inelasticity (say two open channels) Bander, Coulter & Shaw (122) have shown [see also Atkinson, Dietz & Morgan (123) and Hartle & Jones (124)] that a resonance dynamically generated when the problem is considered, correctly, as one of the two coupled channels is not implied when the problem is considered in one channel only, and from the one-channel point of view its existence is as a CDD pole. This is probably relevant to the p_{11} partial-wave amplitude where there is strong inelasticity (125). A realistic multichannel dynamical theory is very difficult, because of uncertainty of the important channels, anomalous thresholds, unstable particle channels such as πN^* and possible three-particle channels such as $\pi\pi N$.

(ii) $T=\frac{3}{2}$, $j=\ell+\frac{1}{2}$: There is excellent agreement since resonances are observed (Table II) in p_{33} and f_{37} where the forces are strongly attractive (Table V) but probably not in d_{35}, where if there were one it would be very

---

[18] Levinson's theorem, proved for a nonrelativistic potential theory, in its original form (120) states that the phase shift $\delta_\ell(q)$ changes by an integral multiple of π as q varies from 0 to ∞: $\delta_\ell(0) - \delta_\ell(\infty) = n_b\pi$, where n_b is the number of bound states in the partial wave. When CDD poles are present the theorem must be modified (121) to $\delta_\ell(0) - \delta_\ell(\infty) = (n_b - n_p)\pi$, where n_p is the number of CDD poles.

inelastic and thus subject to CDD considerations. While there is a resonance in s_{31} (Table II) there is certainly no disagreement since this resonance is, being in an s wave, far too high in energy for the peripheral approach to apply.

(*iii*) $T=\frac{1}{2}$, $j=\ell+\frac{3}{2}$; (*iv*) $T=\frac{1}{2}$, $j=\ell-\frac{1}{2}$: According to the peripheral approach there should be no resonances. Consulting Table II, $s_{11}(1)$, $s_{11}(2)$, and $p_{31}(1)$ are too high in energy (having $\ell=0$ and 1) for the peripheral approach to apply. To $f_{35}(1)$ the peripheral approach should apply but there is strong inelasticity so CDD pole considerations apply; $d_{15}(1)$ is marginal in energy and again CDD considerations may apply.

In conclusion, there seems to be no inconsistency in considering resonances as being generated by particle exchange.[19] However, owing to inelastic effects and short-range forces (high mass exchange) the predictive power of the method is extremely limited. If the resonances of Table II are correct the L-excitation quark model has much more predictive power. If one considers the quarks to be actual, bound particles then the nucleon resonances, from the one-channel or multichannel (nonquark) viewpoint, would presumably all be CDD poles.

4.2 FINITE-ENERGY SUM RULES

If an amplitude satisfies a dispersion relation

$$\frac{1}{\pi} \int_{-\infty}^{\infty} d\nu' \, \frac{\mathrm{Im}\, A(\nu', t)}{\nu' - \nu} = A(\nu, t)$$

and if $A(\nu,t)$ falls off faster than $\nu^{-1/2}$ as $\nu \to \infty$ then the *superconvergence relation*,

$$\int_{-\infty}^{\infty} d\nu' \, \mathrm{Im}\, A(\nu', t) = 0,$$

results.

Consider a pion-nucleon scattering amplitude, odd under crossing (Equation 1.7), for example $A'^{(-)}(\nu,t)$ asymptotically equal to a sum, $R(\nu,t)$, of Regge poles. Since $A'^{(-)}(\nu,t) - R(\nu,t)$ converges suitably at infinity we may use Cauchy's theorem and crossing to write a superconvergence relation (127) of the type first introduced by Igi (128–130):

$$\int_{0}^{\infty} \mathrm{Im}\left\{ A'^{(-)}(\nu, t) - R(\nu, t) \right\} = 0 \qquad 4.6a$$

If $A'^{(-)}(\nu, t) = R(\nu, t)$ for $\nu \geq \nu_1$:

$$\int_{0}^{\nu_1} \mathrm{Im}\left\{ A'^{(-)}(\nu, t) - R(\nu, t) \right\} = 0 \qquad 4.6b$$

[19] Carruthers (126) has also considered the families of nucleon resonances with similar results.

and if the asymptotic form of R

$$R \sim \sum \frac{\beta(\pm 1 - \exp(-i\pi\alpha))}{\sin \pi\alpha \Gamma(\alpha + 1)} \nu^\alpha \qquad 4.7$$

is valid at $\nu = \nu_1$, then evaluation of 4.6b gives (128–133):

$$-4\pi f^2 \left(\frac{1 - t/2m^2}{1 - t/4M^2}\right) + \int_{m+t/4M}^{\nu_1} \operatorname{Im} A'^{(-)}(\nu, t)d\nu$$

$$= \sum \frac{\beta\nu_1^{\alpha+1}}{(\alpha + 1)\Gamma(\alpha + 1)} \qquad 4.8$$

The first term on the left-hand side is the Born pole contribution which was not explicitly included in 4.6.

As emphasized by Dolen, Horn & Schmid (131, 132), the importance of a Regge pole in the sum on the right-hand side of 4.8 is proportional to its importance in fitting the high-energy data. So the sum rules like 4.8 connect high-energy data with low-energy data or parameters, for example pion-nucleon phase shifts. Equation 4.8 can be generalized so long as there are enough Regge poles in the sum 4.7 that $A'^{(-)}(\nu,t) - R(\nu,t)$ falls off faster than $1/\nu^{n+1}$, to:

$$S_n \equiv \int_0^{\nu_1} \nu^n d\nu \operatorname{Im} A'^{(-)}(\nu, t) = \sum \frac{\beta\nu^{\alpha+n+1}}{(\alpha + n + 1)\Gamma(\alpha + 1)} \qquad 4.9$$

where n is integral and even to preserve the crossing. The integral in 4.9 includes the term coming from the Born pole. This type of sum rule makes greater demands on the excellence of the high-energy fit. One can, by involving both the real and imaginary parts of the amplitude, further generalize the sum rule to nonintegral n (134, 135).

For $A'^{(-)}(\nu,t)$ the only known t-channel Regge trajectory is the ρ, so for small $|t|$ the sum on the right-hand side of 4.8 contains only one term. Igi & Matsuda (130) evaluate the sum rule for $t = 0$, which enables the use of the optical theorem:

$$\operatorname{Im} A'^{(-)}(\nu, 0) = \frac{q_L}{2} \left(\sigma_{tot}(\pi^- p) - \sigma_{tot}(\pi^+ p)\right)$$

The cutoff ν_1 was chosen as 6 GeV/c, with an asymptotic form slightly different from 4.7 and 4.8 and with values of $\alpha_\rho(0)$ (~ 0.55) and $\beta\rho(0)$ (~ 5.72) that fitted the high-energy data, the sum rule was found to be well satisfied.

Dolen, Horn & Schmid (131, 132) have investigated the sum rules 4.8 and 4.9 with $n = 2$, and the corresponding sum rules for the crossing odd amplitude $\nu B^{(-)}(\nu,t)$. The principle of the investigation is to evaluate the integral on the left-hand side of 4.8 by using the results of partial-wave analysis

(Section 2) to evaluate Im $A'^{(-)}(\nu,t)$ and Im $\nu B^{(-)}(\nu,t)$ with Equations 1.11, 1.12, and 3.8. Unfortunately the uncertainty in the partial-wave analyses at high energies means that a low value for ν_1 is necessary; the value chosen is 1.13 GeV/c. At such a value the parameters of 4.7, when chosen to fit the high-energy data, will not necessarily give a good fit. One comparison, not involving the residue function β, is to evaluate different moments for $\nu B^{(-)}$. We define

$$B_n(t) = \frac{1}{\nu_1^{n+1}} \int_0^{\nu_1} \nu^n d\nu \; \mathrm{Im}\,(\nu B^{(-)}(\nu,\, t)) = \sum \frac{\beta \alpha \nu_1^\alpha}{(\alpha + n + 1)\Gamma(\alpha + 1)} \qquad 4.10$$

If we approximate the right-hand side by a single pole, good agreement of $(\alpha+3)B_2$ with $(\alpha+1)B_0$ is obtained, when α is determined from high-energy fits. Alternatively from 4.10

$$\alpha(t) = \frac{3B_2(t) - B_0(t)}{B_0(t) - B_2(t)} \qquad 4.11$$

giving, for example, $\alpha(0) = 0.4 \pm 0.2$ to be compared with 0.57 from the high-energy data described by 3.13. Also B_0, B_2 have zeros at $t = -0.43 \pm 0.1$, -0.52 ± 0.1 respectively, corresponding to a zero of $\alpha(t)$ at $t = -0.55$ from 3.13. (The dip phenomenon in πp charge exchange scattering which this result predicts is discussed in Section 3.2.) Thus the effective ρ trajectory determined from the sum rule seems to lie somewhat lower than the presumably more accurately determined ρ trajectory from high-energy data. This deviation could be due to the effective presence of a ρ' trajectory or a cut, or a form different from 4.10 for the Regge pole term could be better (this latter point also raises the question of the backround integral). However, the residue function β as determined from 4.10 is also in rather good agreement with the determination from high-energy data. So one may conclude that the results of extrapolating the ρ trajectory down to $\nu = 1.13$ GeV/c are, roughly, good though perhaps not surprisingly so in view of previous indications in favour of this extrapolation. It also seems that the sum rule for $\nu B^{(-)}$ is not too sensitive to the value of ν_1 since for $t = 0$ the prominent resonances enter with opposite sign and the continuum contribution gets rapidly smaller as t decreases, as illustrated in Figure 24.

Perhaps the most important connection between low- and high-energy data established by the finite-energy sum rule for $\nu B^{(-)}$ is the prediction of the dip phenomenon in πp charge exchange scattering, at approximately the correct value of t, as stated above in noting the vanishing of B_0 and B_2. This result is qualitatively independent of uncertainties in the partial-wave analysis because the nucleon Born term gets very large and negative as t decreases from zero and the continuum decreases rapidly, giving rise to a zero[20] in the

[20] The cyclic residue *ansatz* (Section 3.4 above) necessitates a zero of $B_0, B_2 \ldots$ at $t \sim -2.5$ also.

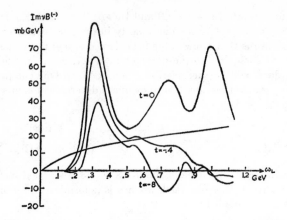

Fig. 24. The imaginary part of $\nu B^{(-)}$ for various values of t, plotted as a function of $\omega_L = \nu - t/4M$. The partial-wave amplitudes used in the evaluation are those of (24). The Born term contribution to $\int \mathrm{Im}\, \nu B^{(-)}$ at $t = 0$, $-.4$, $-.8$ is -1.25 ± 0.05, -14.4 ± 0.6, -27.6 ± 1.4 mb GeV^2 respectively. The smooth line is the Regge ρ contribution at $t = 0$, extrapolated from high-energy data.

range $0.4 < -t < 0.6$. As remarked by Dolen, Horn & Schmid, B is mainly a sum of terms (Equations 1.11, 1.12) containing $P_\ell{}'(\cos \theta) = P_\ell{}'(1 + t/2q^2)$ where ℓ is the angular momentum of the partial wave contributing, and ℓ, q^2 for the prominent resonances are such that the $P_\ell{}'(1 + t/2q^2)$ are zero for $0.4 < -t < 0.6$.

The integral for $\mathrm{Im}\, A^{(-)}(\nu, t)$, apart from the special case $t = 0$, is less certainly evaluated and depends more severely on ν, because of alterations in sign of the contributions of prominent resonances as shown in Figure 25. However $\int^{\nu_1}\mathrm{Im}\, A'^{(-)}$ is much smaller than $\int^{\nu_1}\mathrm{Im}\, B^{(-)}$ near $t = 0$, which predicts, from the low-energy data, the near forward peak in high-energy charge exchange scattering (Figure 22) discussed in Section 3.2. Also it seems that $\int^{\nu_1}\mathrm{Im}\, A^{1(-)}(\nu, t)$ changes sign between $t = 0$ and $t = -0.2$, which implies a zero between these t values in $A^{1(-)}(\nu, t)$ in high-energy scattering. This result is consistent with the crossover of the magnitudes of the $\pi^+ p$ and $\pi^- p$ differential cross sections near $t = 0$ (94).

It seems from the above discussion and Section 3.2 that ρ exchange occupies a rather special position in high-energy physics because of the existence of the process $\pi^- + p \to \pi^0 + n$ where the quantum numbers only allow $\rho(\text{or } \rho')$ Regge exchange in the t channel, and because good results are obtained by studying this process on the assumption of Reggeized ρ exchange. One should note the study of $\pi + \pi \to \pi + \omega$ where there is only one invariant amplitude and only the ρ (or ρ') trajectory is allowed in the s, t, or u channels. In this case the ρ bootstrap appears to achieve some success (136–138).

Barger & Phillips (139) have examined a family of continuous-moment sum rules for $A'^{(+)}(\nu, t)$ and $B^{(+)}(\nu, t)$ to find conditions on the P, P' Regge parameters. They find solutions $B^{(+)}$, of a different character from some suggested by fits (140, 141) to high-energy data, where it is not well determined. The approximate relation $B^{(+)} \sim A'^{(+)}/\nu$ is suggested.

4.3 DUALITY AND INTERFERENCE

Dolen, Horn & Schmid (132) make the following statement of duality: "There are two complete representations of any scattering amplitude: one is the partial-wave series which can be dominated by direct-channel resonances, or might have a large nonresonating backround, and the other is the Regge asymptotic series consisting of pole (and cut) terms ν^{α} plus a background integral in the j plane."

The finite-energy sum rules tell us not only that the sum of Regge terms ν^{α} is an average of the experimental curve from 0 to ν, but also, because of the various-moment finite-energy sum rules, that this average should roughly hold semilocally over regions say $\Delta\nu \sim 1$ GeV (The value $\nu_1 = 1.13$ GeV, which had to be used in evaluation, is too low for this to be very evident in Figures 24 and 25 because of the presence of a considerable Born term. In Figure 26 we plot Im $A'^{(-)}(\nu,0)$ obtained using the optical theorem and the sum of Regge amplitudes, showing the semilocal averaging effect.) It is the background integral that provides the wiggles superposed on the Regge curve.

The interference model, described in Section 3.3 above, represents the amplitude as

$$F = F_{\text{Regge}} + F_{\text{res}} \qquad 4.12$$

whereas the duality representation is

$$F = F_{\text{Regge}} + F_{\text{res}} - \langle F_{\text{res}} \rangle \qquad 4.13$$

where $\langle F_{\text{res}} \rangle$ is the smoothed-out semilocal averaged resonance amplitude. If $\langle F_{\text{res}} \rangle$ is small, then duality and the interference are the same. But if, for example, all the resonances enter with the same sign, then $\langle F_{\text{res}} \rangle$ is considerable and according to the finite-energy sum-rule argument given about the interference model would not be valid. In this circumstance, the interference model would have committed the fault of double counting insofar as some of the F_{regge} amplitude is already contained in the F_{res} amplitude and vice-versa.

In this situation the intrinsic sign of the resonance contribution, as shown in Equations 2.17a,b, is of primary importance. Dolen, Horn & Schmid (132) give a number of arguments against the interference model on the ground of double counting. However, it appears that they assume an intrinsic positive sign for all resonance contributions (corresponding to $\delta_{\infty} = 0$ in Equation 2.17a) and that the force of these arguments is considerably reduced (though they might apply to some papers on the interference model) when one considers that the resonance may enter with intrinsic negative sign ($\delta_{\infty} = \pi/2$) in Equation 2.17a (142). So one might, for example, have resonances alternat-

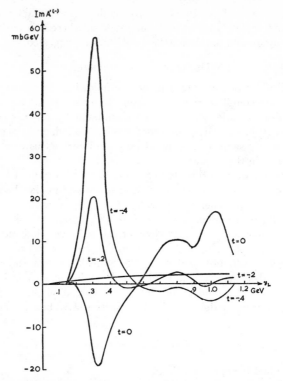

FIG. 25. The imaginary part of $A'^{(-)}$, for various values of t, plotted as a function of $\omega_L = \nu - t/4M$. The partial-wave amplitudes used in the evaluation are those of (24). The Born term contribution to $\int \mathrm{Im}\, A'^{(-)}$ at $t=0$, $-.2$, $-.4$ is -1.25 ± 0.5, -7.5 ± 0.4, -13.0 ± 0.6 mb GeV^2 respectively. The smooth line is the Regge ρ contribution at $t=0$, extrapolated from high-energy data.

ing in sign to produce $\langle F_{\mathrm{res}} \rangle = 0$ with a resultant coincidence of duality and the interference model.

In support of the interference model, Alessandrini, Amati & Squires (143) have proposed the following *ansatz* for the scattering amplitudes:

$$F = F_s + F_t + F_u \qquad\qquad 4.14$$

In 4.14 F_s, F_t, F_u, amplitudes are due to Regge poles in s, t, u, respectively. So for physical s, F will contain direct-channel poles from F_s and crossed-channel poles from F_t and F_u and for physical t, F will contain direct-channel poles from F_t and crossed-channel poles from F_s and F_u— and similarly for physical u.

Such an *ansatz* implies finite-energy sum rules and the consequent duality with $\langle F_{\mathrm{res}} \rangle = 0$ coexisting with the interference model. It is proposed to

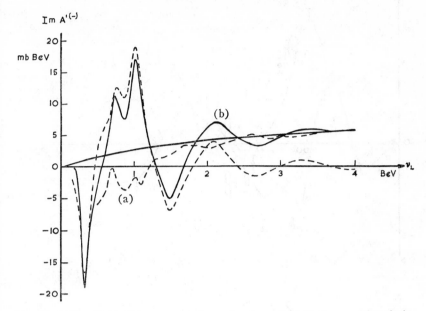

FIG. 26. (a) Im $A'^{(-)}(\nu,0)$ from experiment using the optical theorem and isospin invariance. (b) The extrapolated Regge approximation to Im $A'^{(-)}(\nu,0)$.

determine the resonances by taking the crossed-channel Regge poles (extrapolated if necessary) in F_t and F_u as background, determining F from experiment and partial-wave analysis and so finding $F_s = F - F_t - F_u$, and determining the s-channel resonances. Such a procedure can be applied for example to pion-nucleon scattering (see below) where it will produce resonances which may or may not be the same as those of Table II, but which probably cannot be quarrelled with as an interpretation of the pion-nucleon scattering data alone. The idea gives a more definite prescription for determining the background to the resonances, subject however to uncertainties as to exactly what should be taken as the Regge pole extrapolation to smaller energies. The separation of the background is one of the chief sources of uncertainty in the resonance parameters listed in Table II. Whether the *ansatz* be true or not is more difficult to determine, though there are a number of conditions which it must certainly satisfy. For example, resonance parameters determined using different physical channels must be the same. The reactions $\overline{K}N \to \overline{K}N$, $\overline{K}N \to \pi\Lambda$, $\overline{K}N \to \pi\Sigma$ should be particularly useful here; unfortunately the data from associated production reactions such as $\pi^- + p \to K^\circ + \Lambda$, $\pi^- + p \to K^\circ + \Sigma^\circ$ seems to be not yet sufficient or accurate enough to make a meaningful test for the N^* resonances.

Donnachie & Kirsopp (144) have investigated the results of applying the *ansatz* to pion-nucleon elastic scattering. The partial-wave projections of the

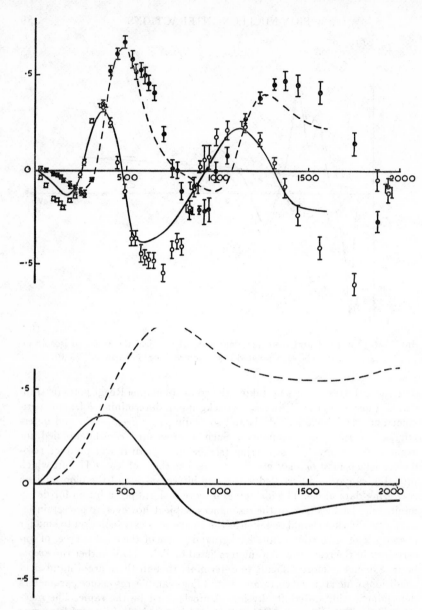

FIG. 27. The p_{11} amplitude plotted as a function of energy after subtraction of the t-channel Regge poles and the direct-channel nucleon pole. The real part is denoted by solid circles, and the imaginary part by open circles with the bars representing the errors from the partial-wave analysis. The curves approximately through the points are real and imaginary parts of the fitted superposition of Breit-Wigner resonances. The lower curves are the real and imaginary parts of the subtracted amplitude $(F_i)_{1-}$. From (144).

t-channel Regge poles, P, P', and ρ, are subtracted from the partial-wave amplitudes $f_{l\pm}$ determined from partial-wave analyses (Section 2). The resultant amplitude can then be fitted by a set of Breit-Wigner formulae (with not necessarily zero phase, δ_∞):

$$f_{l\pm} - (F_t)_{l\pm} = \sum \text{Breit-Wigner amplitudes}$$

The set is almost the same as the set of resonances of Table II, except of course for the parameter details. Figure 27 shows the resonance fit in the case of the p_{11} wave. This particular wave is probably the most remarkable result since the $p_{11}(1)$ has in this fit a mass of 1410 MeV/c^2 and a width of 140 MeV in close agreement with the values always suggested in production processes (Section 2.5). The superposition of Breit-Wigner amplitudes is generally a better fit in the $T = \frac{1}{2}$ than in the $T = \frac{3}{2}$ case, which perhaps implies the necessity of some more exchanged Regge poles in the $T = \frac{3}{2}$ amplitudes.

Reggeons or resonances?—Schmid (145) has pointed out that a partial-wave amplitude, projected from a crossed-channel Regge pole, describes circles in the Argand diagram. This behaviour is reminiscent of the circles found by partial-wave analysis, which have been interpreted as resonances (Section 2). It has been suggested (143, 146) that some of the circles found by partial-wave analysis (particularly those which do not belong to the $\Delta, N\alpha$ or $N\gamma$ trajectories) are the projection of crossed-channel Regge poles. Such 'reggeons' would lack second-sheet poles (145) and would not be interpreted as resonances. But the original Schmid circles (145) seem to have the wrong energy dependence to be the circles seen in partial-wave analysis and interpreted as resonances. Schmid (147) interprets these slow (as a function of energy) reggeon circles as showing that the Regge poles in crossed channels, like the antecedent particle exchange, provide a possible force which generates the resonances. We also see from Figure 27 that the variation of the reggeon amplitude with energy seems too slow to be itself a resonance, and that resonances and reggeons can coexist. However, at this time no one has tried to fit the pion-nucleon scattering data directly by, say, the Δ, N_α, N_γ Regge trajectories and the crossed-channel Regge trajectories.

Analytic models (148) for scattering amplitudes which have second-sheet poles and Regge asymptotic behaviour have been constructed. Igi (149) for example has attempted such a model for the πN system.

Modified duality.—Harari (150) has suggested, and Gilman, Harari & Zarmi (151) have developed a modified duality in which (in the sense of the finite-energy sum rules resembling 4.8) the Pomeron (on the right-hand side) is built from the background in Im A (on the left-hand side) while other trajectories are built from the resonance. Harari duality can be represented schematically as:

$$\text{Pomeron exchange} \quad \longleftrightarrow \quad \text{direct-channel background} \qquad 4.15a$$

$$\text{Other Regge exchange} \longleftrightarrow \text{direct-channel resonance} \qquad 4.15b$$

implying that Pomeron exchange or background are equally valid descriptions of one part of the Im A while other Regge exchanges or direct-channel resonances are equally valid descriptions of the other part. So the amplitude is composed of either side of 4.15a plus either side of 4.15b. The Harari duality explains immediately a number of phenomena; for example the constancy at high energy of K^+N and NN total cross sections—since these systems appear to have no strong direct-channel resonances the cross sections should be given by Pomeron exchange alone, unmixed with other Regge exchanges which would give a decreasing contribution to the total cross section.

Harari & Zarmi (152) have recently projected the pion-nucleon partial-wave amplitudes (discussed in Section 2 above) on to pure isospin $I_t = 0$ and $I_t = 1$ in the crossed, or t channel. Plotted on an Argand diagram the $I_t = 0$ amplitudes show considerable imaginary background to the resonance circles while the $I_t = 1$ amplitudes do not. Since Pomeron exchange is present in the $I_t = 0$ amplitude but not in the $I_t = 1$ amplitude, this observation of Harari & Zarmi constitutes a verified prediction of the Harari model.

LITERATURE CITED

1. Citron, A., Galbraith, W., Kycia, T. F., Leontic, B. A., Phillips, R. H., Rousset, A., Sharp, P. H., *Phys. Rev.*, **144**, 1101 (1966)
2. Foley, K. J., Jones, R. S., Lindenbaum, S. J., Love, W. A., Ozaki, S., Platner, E. D., Quarles, C. A., Willen, E. H., *Phys. Rev. Letters*, **19**, 330 (1967)
3. Pomeranchuk, I. Ya., *Zh. Eksperim. Teor. Fiz.*, **34**, 725 (1958) [Transl.: *JETP*, **34**, 499 (1958)]
4. Weinberg, S., *Phys. Rev.*, **124**, 2049 (1961)
5. Bjorken, J., Drell, S. D., *Relativistic Quantum Fields* (McGraw-Hill, New York, 396 pp., 1965)
6. Hamilton, J., Woolcock, W. S., *Rev. Mod. Phys.*, **35**, 737 (1963)
7. Chew, G. F., Goldberger, M. L., Low, F. E., Nambu, Y., *Phys. Rev.*, **106**, 1337 (1957)
8. Watson, K. M., Lepore, J. V., *Phys. Rev.*, **76**, 1157 (1949)
9. Bremermann, H. J., Oehme, R., Taylor, J. G., *Phys. Rev.*, **109**, 2178 (1958)
10. Hamilton, J., *Strong Interactions and High Energy Physics*, 281–369 (Moorehouse, R. G., Ed., Oliver & Boyd, Edinburgh, 475 pp., 1964)
11. Spearman, T. D., *Nuovo Cimento*, **15**, 147 (1960)
12. Samaranayake, V. K., Woolcock, W. S., *Phys. Rev. Letters*, **15**, 936 (1965)
13. Hohler, G., Baacke, *J. Phys. Letters*, **18**, 181 (1965)
14. Hohler, G., Baacke, J., Steiner, F. (Private communication)
15. MacGregor, M. H., Arndt, R. A., Wright, R. M., *Phys. Rev.*, **169**, 1128 (1968)
16. Hamilton, J., *Phys. Letters*, **20**, 687 (1966)
17. Adler, S. L., Dashen, R. F., *Current Algebras* (Benjamin, New York, 394 pp., 1968)
18. De Hoffmann, F., Metropolis, N., Alei, E. F., Bethe, H. A., *Phys. Rev.*, **95**, 1586 (1954)
19. Bilenky, S. M., Ryndin, R. M., *Dubna Preprint E-2521* (1965)
20. Olson, L. E., Trower, W. P., *Univ. Illinois, Urbana, Phys. Dept. Tech. Rept. 146*
21. Minami, S., *Progr. Theoret. Phys.* (*Kyoto*), **11**, 213

22. Wigner, E. P., *Phys. Rev.*, **98**, 145 (1958)
23. Bareyre, P. Bricman, C., Stirling, A. V., Villet, G., *Phys. Letters*, **18**, 342 (1965)
24. Bareyre, P., Bricman, C., Villet, G., *Phys. Rev.*, **165**, 1730 (1967)
25. Lovelace, C., *Proc. Heidelberg Intern. Conf. Elementary Particles*, 79–116 (Filtuth, H., Ed., North-Holland, Amsterdam, 550 pp., 1968)
26. Johnson, C. H., Steiner, H. M., *Univ. Calif. Radiation Lab. Rept. UCRL 18001* (1967)
27. Donnachie, A., *Proc. Vienna Intern. Conf. High Energy Phys.* 139–58 (Prentki, J., Steinberger, J., Eds., CERN, Geneva, 527 pp., 1968)
28. Auvril, P., Donnachie, A., Lea, A. T., Lovelace, C., *Phys. Letters*, **12**, 76 (1964)
29. Donnachie, A., *Proc. Scottish Univ. Summer Sch.* (Preist, T. W., Vick, L. L. J., Eds., Oliver & Boyd, Edinburgh, 1966)
30. Donnachie, A., Kirsopp, R. G., Lovelace, C., *Phys. Letters*, **26B**, 161 (1968)
31. Roper, L. D., *Phys. Rev. Letters*, **12**, 342 (1964)
32. Roper, L. D., Wright, R. M., *Phys. Rev.*, **138**, B921 (1965)
33. Roper, L. D., Wright, R. M., Feld, B. T., *Phys. Rev.*, **138**, B190 (1965)
34. Bransden, B. H., Moorhouse, R. G., O'Donnell, P. J., *Phys. Letters*, **11**, 339 (1964)
35. Bransden, B. H., Moorhouse, R. G., O'Donnell, P. J., *Phys. Rev.*, **139**, B1566 (1965)
36. Bransden, B. H., Moorhouse, R. G., O'Donnell, P. J., *Phys. Letters*, **19**, 420 (1965)
37. Davies, A. T. (To be published)
38. Layson, W. M., *Nuovo Cimento*, **27**, 724 (1963)
39. Hull, M., Lin, F., *Phys. Rev.*, **139**, B630 (1965)
40. Kane, G. L., Spearman, T. D., *Phys. Rev. Letters*, **11**, 45 (1963)
41. Donnachie, A., Hamilton, J., Lea, A. T., *Phys. Rev.*, **135**, B515 (1964)
42. Dalitz, R. H., *Ann. Rev. Nucl. Sci.*, **13**, 339 (1963)
43. Goebel, C. J., McVoy, K. W., *Phys. Rev.*, **164**, 1932 (1967)
44. Chau, Y. C., Dombey, N., Moorhouse,

R. G., *Phys. Rev.*, **163**, 1632 (1967)
45. Walker, R., *Calif. Inst. Technol. Preprint CALT 68-159*
46. Hendry, A., Moorhouse, R. G., *Phys. Letters*, **18**, 171 (1965)
47. Logan, R. K., Uchiyama-Campbell, F., *Phys. Rev.*, **149**, 1220 (1966)
48. Davies, A. T., Moorhouse, R. G., *Nuovo Cimento*, **52A**, 1112 (1967)
49. Chadwick, G. B., Collins, G. B., Duke, P. J., Fujii, T., Hien, N. C., Kemp, M. A. R., Turkot, F., *Phys. Rev.*, **128**, 1823 (1962)
50. Anderson, E. W., Bleser, E. J., Collins, G. B., Fujii, T., Menes, J., Turkot, F., Carrigan, R. A., Edelstein, R. M., Hien, N. C., McMahon, T. J., Nadelhaft, I., *Phys. Rev. Letters*, **16**, 855 (1966)
51. Foley, K. T., Jones, R. S., Lindenbaum, S. J., Love, W. A., Ozaki, S., Platner, E. D., Quarles, C. A., Willen, E. H., *Phys. Rev. Letters*, **19**, 397 (1967)
52. Gellert, E., Smith, G. A., Wojcicki, S., Colton, E., Schlein, P. E., Ticho, H. K. *Phys. Rev. Letters*, **17**, 884 (1966)
53. Morgan, D., *Phys. Rev.*, **166**, 1731 (1968)
54. Morpurgo, G., *Physics*, **2**, 95 (1965)
55. Donnachie, A., *Phys. Letters*, **24B**, 420 (1967)
56. Hilpert, H. G., Lauscher, P., Matziolis, M., Schnuckers, H., Weber, H., Meyer, A., Pose, A., Böckmann, K., Idschok, U., Müller, K., Paul, E., Propach, E., Butenschön, H., Kübeck, H., Lüke, D., Seebeck, H., Spitzer, H., Storim, C., Brandt, S., Braun, O., Steffen, P., Stiewe, J., Schlamp, P., Weigl, J., Wilkinson, K., *Nucl. Phys.*, **88**, 535 (1968)
57. Greenberg, O. W., *Phys. Rev. Letters*, **13**, 598 (1964)
58. Gell-Mann, M., *Proc. 1965 Oxford Intern. Conf. Elementary Particles*, 183 (Rutherford Lab., 1966)
59. Faiman, D., Hendry, A. W., *Phys. Rev.*, **173**, 1720 (1968)
60. Dalitz, R. H., *High Energy Physics*, 253 (Gordon & Breach, New York, 1966)
61. Dalitz, R. H., *Proc. 1965 Oxford Intern. Conf. Elementary Particles*, 157 (Rutherford Lab., Chilton, Jan. 1966)
62. Dalitz, R. H., *Proc. 13th Intern. Conf. High Energy Phys., Berkeley, 1966*, 215 (Univ. Calif. Press, 1967)
63. Dalitz, R. H., *Elementary Particle Physics*, 56 (Takeda, G., Fujii, A., Eds., Benjamin, New York, 1967)
64. Harari, H., *Proc. 14th Intern. Conf. High Energy Phys., Vienna, 1968*, 195 (CERN, Geneva, 1968)
65. Mitra, A., Ross, M., *Phys. Rev.*, **158**, 1630 (1967)
66. Mitra, A., *Rutherford Lab. Preprint RPP/A44* (1968)
67. Morpurgo, G., *Proc. 14th Intern. Conf. High Energy Phys., Vienna, 1968*, 225 (CERN, Geneva, 1968)
68. Horn, D., Lipkin, H. J., Meshkov, S., *Phys. Rev. Letters*, **17**, 1200 (1966)
69. Walker, J., *Phys. Rev. Letters* (In press)
70. Becchi, C., Morpurgo, G., *Phys. Letters*, **17**, 352 (1965)
71. Rollnik, H., *Proc. 1967 Heidelberg Intern. Conf. Elementary Particles*, 400 (Filtath, H., Ed., North-Holland, Amsterdam, 1968)
72. Dalitz, R. H., Sutherland, D., *Phys. Rev.*, **146**, 1180 (1966)
73. Betourne, C., Bizot, J. C., Perez-y-Jorba, J., Treille, D., Schmidt, W., *Phys. Rev.*, **172**, 1343 (1968)
74. Cone, A. A., Chen, K. W., Dunning, J. R., Jr., Hartwig, G., Ramsey, N. F., Walker, J. K., Wilson, R., *Phys. Rev.*, **156**, 1490 (1967)
75. Cone, A. A., Chen, K. W., Dunning, J. R., Jr., Hartwig, G., Ramsey, N. F., Walker, J. K., Wilson, R., *Phys. Rev. Letters*, **14**, 326 (1965)
76. Faiman, D., Hendry, A. W., *Phys. Rev.* (To be published)
77. Bietti, A., *Phys. Rev.*, **142**, 1258 (1966)
78. Love, A., Moorhouse, R. G. (To be published)
79. Moorhouse, R. G., *Phys. Rev. Letters*, **16**, 772, 968 (1966)
80. Coffin, C. T., Dikmen, N., Ettlinger, L., Meyer, D., Saulys, A., Terwilliger, K., Williams, D., *Phys. Rev.*, **159**, 1169 (1967)
81. Orear, J., Rubinstein, R., Scarl, D. B., White, D. H., Krisch, A. D., Frisken, W. R., Read, A. L., Ruderman, H., *Phys. Rev.*, **152**, 1162 (1966)
82. Foley, K. J., Gilmore, R. S., Lindenbaum, S. J., Love, W. A., Ozaki, S., Willen, E. H., Yamada, R., Yuan, L. C. L., *Phys. Rev. Letters*, **15**, 45 (1965)
83. Ashmore, A., Damerell, C. J. S., Frisken, W. R., Rubinstein, R., Orear, J., Owen, D. P., Peterson, F. C., Read, A. L., Ryan, D. G., White, D. H., *Phys. Rev. Letters*, **21**, 387 (1968)

84. Foley, K. J., Jones, R. S., Lindenbaum, S. J., Love, W. A., Ozaki, S., Platner, E. D., Quarles, C. A., Willen, E. H., *Phys. Rev. Letters*, **19**, 193 (1967)

85. Damouth, D. E., Jones, L. W., Perl, M. L., *Phys. Rev. Letters*, **11**, 287 (1963)

86. Kozlowsky, B., Dar, A., *Phys. Letters*, **20**, 311 (1966)

87. Bethe, H. A., *Ann. Phys. (N.Y.)*, **3**, 190 (1958)

88. Oehme, R., *Phys. Rev.*, **100**, 1503 (1955)

89. Mannelli, I., Bigi, A., Carrara, R., Wahlig, M., Sodickson, L., *Phys. Rev. Letters*, **14**, 408 (1965)

90. Stirling, A. V., Sonderegger, P., Kirz, J., Falk-Vairant, P., Guisan, O., Bruneton, C., Borgeaud, P., Yvert, M., Guillaud, J. P., Caversazio, C., Amblard, B., *Phys. Rev. Letters*, **14**, 763 (1965)

91. Booth, N. E., *Phys. Rev. Letters*, **21**, 465 (1968)

92. Arnold, R. C., *Phys. Rev.*, **153**, 1523 (1967)

93. Omnès, R., *Ann. Rev. Nucl. Sci.*, **16**, 263 (1966)

94. Phillips, R. J. N., Rarita, W., *Phys. Rev.*, **139**, B1336 (1965)

95. Hohler, G., Baacke, J., Schaile, H., Sonderegger, P., *Phys. Letters*, **2P**, 79 (1966)

96. Bertocchi, L., *Proc. 1967 Heidelberg Intern. Conf. Elementary Particles*, 197 (Filtuth, H., North-Holland, Amsterdam, 1968)

97. Mandelstam, S., Wang, L. L., *Phys. Rev.*, **160**, 1490 (1967)

98. Bonamy, P., Borgeaud, P., Falk-Vairant, P., Guisan, O., Sonderegger, P., Caversazio, C., Guillaud, J. P., Schneider, J., Yvert, M., Mannelli, I., Sergiampetri, F., Vincelli, L., *Phys. Letters*, **23**, 501 (1968)

99. Beaupre, J., Logan, R. K., Sertorio, L., *Phys. Rev. Letters*, **18**, 259 (1967)

100. Barger, V., Phillips, R. J. N., *Phys. Rev. Letters*, **21**, 865 (1968)

101. Michael, C., *Rutherford Lab. Rept. RPPA45*

102. White, J. N. J., *Phys. Letters*, **27B**, 92 (1968)

103. Gottfried, K., Jackson, J. D., *Nuovo Cimento*, **34**, 735 (1964)

104. Phillips, R. J. N., Rarita, W., *Phys. Letters*, **14**, 598 (1965)

105. Barger, V. D., Cline, D., *Phys. Rev.*, **155**, 1792 (1967)

106. Chiu, C. B., Stirling, A. V., *CERN Preprint TH840* (1967)

107. Barger, V. D., *Proc. 1961 Intern. Conf. Particles Fields at Rochester*, 655 (Hagen, C. R., Guralnik, G., Mathur, V. S., Eds., Interscience, New York, 1967)

108. Barger, V., Phillips, R. J. N., *Phys. Rev. Letters*, **20**, 564 (1968)

109. Barger, V., Phillips, R. J. N., *Phys. Rev.* (To be published)

110. Belletini, G., *Proc. 14th Intern. Conf. High Energy Phys., Vienna, 1968*, 330 (CERN, Geneva, 1968)

111. Chou, T. T., Yang, C. N., *Phys. Rev.*, **170**, 1591 (1968)

112. Arnold, R. C., Blackmon, L. M., *Phys. Rev.*, **176**, 2082 (1968)

113. Mandelstam, S., *Phys. Rev.*, **115**, 1741 (1959)

114. Chew, G. F., *Ann. Rev. Nucl. Sci.*, **9**, 29 (1959)

115. Chew, G. F., Mandelstam, S., *Phys. Rev.*, **119**, 467 (1960)

116. Chew, G. F., Frautschi, S. C., *Phys. Rev. Letters*, **7**, 394 (1961)

117. Zachariasen, F., *Strong Interactions and High Energy Physics*, 370–409 (Moorhouse, R. G., Ed., Oliver & Boyd, Edinburgh, 475 pp., 1964)

118. Donnachie, A., Hamilton, J., *Ann. Phys. (N.Y.)*, **31**, 410 (1965)

119. Castillejo, L., Dalitz, R. H., Dyson, F. J., *Phys. Rev.*, **101**, 453 (1965)

120. Levinson, N., *Dan. Mat. Fys. Medd. No. 9*, 25 (1949)

121. Vaughn, H. T., Aaron, R., Amado, R. D., *Phys. Rev.*, **124**, 1258 (1961)

122. Bander, M., Coulter, P. W., Shaw, G. L., *Phys. Rev. Letters*, **14**, 270 (1968)

123. Atkinson, D., Dietz, K., Morgan, D., *Ann. Phys. (N.Y.)*, **37**, 77 (1966)

124. Hartle, J. B., Jones, C. E., *Phys. Rev.*, **140**, 390 (1965)

125. Bart, G. R., Warnock, R. L., *Argonne Natl. Lab., High Energy Phys. Div. Preprint*

126. Carruthers, P., *Phys. Rev.*, **133**, B497 (1964)

127. de Alfaro, V., Fubini, S., Furlan, G., Rossetti, C., *Phys. Letters*, **21**, 576 (1966)

128. Igi, K., *Phys. Rev. Letters*, **9**, 76 (1962)

129. Igi, K., *Phys. Rev.*, **130**, 820 (1963)

130. Igi, K., Mutsuda, S., *Phys. Rev. Letters*, **18**, 625 (1967)

131. Dolen, R., Horn, D., Schmid, C., *Phys. Rev. Letters*, **19**, 402 (1967)

132. Dolen, R., Horn, D., Schmid, C., *Phys. Rev.*, **166**, 1768 (1968)

133. Soloviev, L. D., Logemov, A. R., Tankhelidze, A. N., *Phys. Letters,* **24B,** 18 (1967)
134. Barger, V., Philips, R. J. N., *Phys. Letters,* **26B,** 730 (1968)
135. Della Selva, A., Masperi, L., Odorico, R., *Nuovo Cimento,* **54A,** 979 (1968)
136. Mandelstam, S., *Phys. Rev.,* **166,** 1539 (1968)
137. Ademollo, M., Rubinstein, H., Veneziano, G., Virasono, M., *Phys. Rev. Letters,* **19,** 1402 (1967)
138. Schmid, C., *Phys. Rev. Letters,* **20,** 628 (1968)
139. Barger, V., Phillips, R. J. N., *Phys. Rev. Letters,* **26B,** 730 (1968)
140. Chiu, C. B., Chu, S. Y., Wang, L. L., *Phys. Rev.,* **161,** 1563 (1967)
141. Rarita, W., Riddell, R. J., Jr., Chiu, C. B., Phillips, R. J. N., *Phys. Rev.,* **165,** 1615-39 (1968)
142. Bando, M., Hattori, T., *Kyoto Univ. Preprint KUNS-1139* (1968)

143. Alessandrini, V. A., Amati, D., Squires, E. J., *Phys. Letters,* **27B,** 463 (1968)
144. Donnachie, A., Kirsopp, R. G., *Glasgow Univ. Preprint* (To be published)
145. Schmid, C., *Phys. Rev. Letters,* **20,** 689 (1968)
146. Collins, P. D. B., Johnson, R. C., Squires, E. J., *Phys. Letters,* **27B,** 23 (1968)
147. Schmid, C., *CERN Preprint TH958* (1968)
148. Veneziano, G., *Nuovo Cimento,* **57A,** 190 (1968)
149. Igi, K., *Phys. Letters,* **28B,** 264 (1968)
150. Harari, H., *Phys. Rev. Letters,* **20,** 1395 (1968)
151. Gilman, F. J., Harari, H., Zarmi, Y., *Phys. Rev. Letters,* **21,** 323 (1968)
152. Harari, H., Zarmi, Y. (To be published)

PARITY AND TIME-REVERSAL INVARIANCE IN NUCLEAR PHYSICS

By Ernest M. Henley

Department of Physics, University of Washington, Seattle, Washington

CONTENTS

1. INTRODUCTION

Symmetries have always played a central role in the development of physics. Whenever dynamical laws of motion are not well established, symmetries and their associated conservation laws give useful restrictions on the nature of the forces involved.

Among the symmetries of physical laws, those associated with space-time have been crucial in classical as well as in quantum theories. In addition to the invariances under continuous transformations, there are those related to

367

discrete transformations. Of particular importance are parity or space inversion and time reversal or time inversion.

Parity is often likened to reflection in a mirror or plane, e.g., the xy plane $(x{\rightarrow}x, y{\rightarrow}y, z{\rightarrow}-z)$. Although this operation is not identical to full inversion, it is directly related to it through a subsequent rotation of 180° about the z axis $(x{\rightarrow}-x, y{\rightarrow}-y, -z{\rightarrow}-z)$. Time inversion, on the other hand, has been likened to running a movie forwards and backwards. Although this analogy is useful, it can lead to misconceptions in quantum mechanics (1).

Classically, both Newton's equations of motion (for a conserved force) and Maxwell's equations are invariant under both space and time inversions. Classical examples of space and time inversion for systems of particular interest to subatomic physics studies have been given by Schiff (2). In statistical mechanics the notion of the "arrow of time" has attracted considerable philosophical and physical interest (3, 4). Thus it appears that the approach to equilibrium defines a direction of time.

Parity and time-reversal transformations and invariances were introduced into quantum mechanics by Wigner (5, 6). A further transformation that was ushered in by the advent of relativistic quantum theory was charge-conjugation invariance, later generalized to particle-antiparticle conjugation (7). These symmetries, their development, and their applications to nuclear and particle physics have been described in earlier volumes of the *Annual Review of Nuclear Science* (7–9).

Before 1956 the evidence for invariance under parity, charge-conjugation, and time-reversal transformations was thought to be extremely strong. All tests in atomic, nuclear, and particle physics indicated validity of the symmetries. However, as Lee & Yang (10) forcefully pointed out, there were no tests of these invariances for the weak interactions responsible for β decay and the decays of the strange particles. Indeed, experiments carried out at their suggestion showed that both parity (P) and charge-conjugation (C) invariances were violated in these interactions. Landau argued (11) that not all was lost if a combined invariance, namely CP, the product of a space inversion and particle-antiparticle conjugation, remained valid. The combined CP transformation is related to time reversal (T) through the TCP theorem (12). This theorem, which can be shown to follow from invariance under relativistic transformations in local field theory, is very powerful, since it relates properties of particles and antiparticles (9) as well as the various space-time symmetries. The best experimental evidence for its validity comes from the measured mass difference of the K^0 and $\overline{K}^0$ (9). This validity means that CP conservation is equivalent to time-reversal invariance.

The second blow was struck in 1964 by Christenson et al. (13) who demonstrated that a violation of CP invariance occurred in the decay of the long-lived neutral K meson. Although this is a small violation, the relative amplitude of the CP-odd to the CP-even decay mode being some $2{\times}10^{-3}$, it has been verified by many subsequent experiments (9). The violation of CP

invariance implies that T invariance is not valid if the TCP theorem holds. These results have led to a revival of interest in all of the space-time invariance properties of the strong, electromagnetic, and weak interactions.

The CP noninvariance observed in the decay of the neutral kaon appears, at first, to have no observable consequences for nuclear physics, since it seems to be related to a superweak interaction (14). However, other theories have been put forward to explain the small violation (15–18). Of particular interest are the proposals by Bernstein, Feinberg & Lee (16), by Barshay (17), and by Prentki & Veltman (18) that the small violation of CP invariance could be due to the interference of a T-conserving weak interaction and a much stronger T-violating interaction, the latter being roughly comparable to the electromagnetic one in strength. Indeed, Bernstein et al. showed that the evidence for C conservation of the electromagnetic current of the strongly interacting particles, i.e. hadrons, was weak. It is particularly the suggestion of Bernstein, Feinberg & Lee (16) which has stimulated physicists to reexamine the evidence for T invariance in the strong and electromagnetic interactions of the hadrons. The theory will be developed in Section 2 and the experimental status of T invariance in nuclear physics will be reviewed in Section 3.4.

Contrary to the case of time-reversal invariance, the evidence for parity conservation in the strong and electromagnetic interactions has increased since 1956. The interest in nuclear-parity experiments thus stems from two aspects: first, sufficiently accurate experiments can be carried out to detect a parity violation caused by the weak interactions (i.e. roughly 1 ppm); second, measurements of parity violation in nuclear systems offer a unique tool to gain insight into a part of the weak interactions. We will detail the theoretical framework for understanding parity tests in nuclear physics in Section 2 and will review the measurements and their meaning in Sections 3.2 and 3.3.

Charge-conjugation invariance is not of direct interest in low- and medium-energy nuclear physics because the creation of a particle-antiparticle pair requires a center-of-mass (c.m.) energy of at least $2m_\pi \approx 280$ MeV for bosons or $2M_N \approx 1876$ MeV for baryons. In the absence of boson creation or absorption reactions, the role of C is thus considerably reduced. However, this does not imply that the effects of C invariance are negligible. Indeed, the TCP theorem makes it impossible to discuss P and T invariances without considering the implications of C. The properties of C will be used in the theoretical developments of Section 2. Furthermore, if P and T are *both* violated, then, as we shall show, nondegenerate states may possess an electric dipole moment. The search for such moments is being avidly pursued, and will be discussed in Section 3.6.

2. THEORETICAL FRAMEWORK

2.1 BASIC THEORY OF P

To introduce the notation which will be used throughout this article and to discuss the consequences of P and T,[1] we briefly review the fundamentals of these operations here.

The operation of parity or space inversion is defined by

$$\mathbf{r} \to \mathbf{r}_P = -\mathbf{r}; \qquad t \to t_P = t \qquad\qquad 1.$$

whereas the time-reversal operation is

$$\mathbf{r} \to \mathbf{r}_T = \mathbf{r}; \qquad t \to t_T = -t \qquad\qquad 2.$$

These operations have significance in classical as well as quantum mechanics. For simplicity, we assume that the dynamics of the system can be described by a Hamiltonian operator. Although such a description is not relativistic, this restriction is not of primary importance for the phenomena to be discussed. Furthermore, the derivations sketched below can be generalized for relativistic applications.

If the Hamiltonian is invariant under the space reflection transformation P

$$PHP^{-1} = H, \quad \text{or} \quad [P, H] = 0 \qquad\qquad 3.$$

then the parity is an observable constant of the motion. The reason is that the transformation operation is a discrete one with $P^\dagger P = PP^\dagger = P^2 = 1$. The operator P is Hermitian as well as unitary and thus is an observable with eigenvalues ± 1 (the choice of phase is taken to be the usual one). It follows that there exists an eigenstate ψ_P corresponding to ψ, with $\psi_P = P\psi$, which satisfies the same (Schroedinger) equation as ψ. Since both states have the same eigenvalue, they must be degenerate or differ at most by a multiplicative factor of magnitude unity, i.e. a phase. The latter is the normal case in nuclear physics, and we have $P\psi = \eta\psi = \pm\psi$. The plus and minus signs correspond to positive and negative parities, respectively. For degenerate eigenfunctions, we can always choose linear combinations which have definite parities.

If the Hamiltonian is not invariant under the parity transformation, it can be broken up into two parts,

$$H = H^{(e)} + H^{(o)}$$
$$PH^{(e)}P^{-1} = H^{(e)} \qquad\qquad 4.$$
$$PH^{(o)}P^{-1} = -H^{(o)}$$

In that case the nondegenerate eigenfunctions do not have a definite parity. However, if $H^{(o)} \ll H^{(e)}$ then the eigenfunctions consist of a dominant part with a definite parity, the normal part, and a small fraction with the opposite parity, χ

----

[1] We use P, T as an abbreviation for the parity and time-reversal operations and for the operators themselves.

$$\psi \approx (\psi_{\text{norm}} + \mathfrak{F}\chi) \qquad\qquad 5.$$

with $\mathfrak{F} \ll 1$. This applies to hadrons, since only their weak interactions do not conserve parity.

The constants of the motion corresponding to the parity are multiplicative, so that for a state of several particles, the parity is the product of the parities of the pieces. From the definition of the parity operator it follows that

$$P\mathbf{p}P^{-1} = -\mathbf{p}, \qquad P\mathbf{r}P^{-1} = -\mathbf{r} \qquad\qquad 6.$$

where $\mathbf{p}$ is the linear momentum operator. Since the orbital angular momentum is a pseudovector or axial vector, we have by direct generalization to spins and thus to an arbitrary angular momentum

$$P\mathbf{J}P^{-1} = \mathbf{J}, \quad \text{or} \quad [P, \mathbf{J}] = 0 \qquad\qquad 7.$$

By contrast, we have for the helicity (19), $\Lambda \equiv \mathbf{J} \cdot \hat{\mathbf{p}}$, with $\hat{\mathbf{p}}$ a unit vector in the direction of $\mathbf{p}$,

$$P\Lambda P^{-1} = -\Lambda \qquad\qquad 8.$$

The foregoing remarks have been restricted to nonrelativistic one-particle wave mechanics. For Dirac particles, the parity operator has eigenvalues that differ for positive and negative energy solutions and therefore no longer has a unique eigensolution (7). In addition, in a many-particle theory (e.g. field theory) where particles can be destroyed and created, it is necessary to define the "intrinsic" parity of the particles. These features will not be of direct concern to us, but are required in tests where, for instance, a pion is emitted or absorbed. In general, the parity of a given state function is the product of the intrinsic parities of the particles and those of the relative motions. In one-particle theories, where no particles are created or destroyed, the intrinsic parities of any physical process, e.g. a reaction such as (d, p) stripping, are identical in the initial and final states, and therefore need not be taken into account. Although the absolute parities of the proton and neutron and their relative parities cannot be measured (7), their parities are taken to be positive and identical, because they both belong to the same isospin $\frac{1}{2}$ multiplet, the nucleon. The parity of the neutral pion, on the other hand, can be measured and has been found to be negative.

Although the parity of the photon is meaningful, a photon cannot be brought to rest; it is therefore not possible to separate its angular momentum and spin. Thus, only the total parity (intrinsic times angular part) appears in the theory of particle interactions with the electromagnetic field. The transformation of the electromagnetic-field operators can be obtained as follows. The vector potential (Coulomb potential) transforms like a scalar. This follows because the classical interaction $j \cdot A \equiv j_0 A_0 - \mathbf{j} \cdot \mathbf{A}$ is invariant under the parity transformation. The charge density j_0 is a true scalar and $\mathbf{j}$, which is proportional to the velocity of the particle, is a true vector. It thus follows that

$$PAP^{-1} = -A, \qquad PA_0P^{-1} = A_0 \qquad\qquad 9.$$

The transformation of the electric- and magnetic-field intensities follows directly from their relationship to the vector potential. We note that for the electric field $\mathcal{E}$ and magnetic $\mathcal{B}$

$$P\mathcal{E}P^{-1} = -\mathcal{E}$$
$$P\mathbf{B}P^{-1} = \mathbf{B} \qquad\qquad 10.$$

The selection rules for electromagnetic transition moments follow from these definitions and will be discussed in Section 2.3.

2.2 Basic Theory of T

In quantum mechanics, the time development of a system is given by a diffusion-like equation. As was first shown by Wigner (6), this requires that the time-reversal operator be antiunitary. Thus, consider the Schroedinger equation ($\hbar = c = 1$)

$$i\partial\psi(t)/\partial t = H\psi(t) \qquad\qquad 11.$$

If the Hamiltonian is real, then the same equation is satisfied by $\psi^*(-t)$, so that we can identify the time-reversed wavefunction $\psi_T(t)$ with $\psi^*(-t)$.

More generally, we can express the antiunitary nature of T by writing the operator as[2]

$$T = U_T K \qquad\qquad 12.$$

where U_T is unitary and K is the operator of complex conjugation, with

$$K^2 = 1, \qquad K^{-1} = K$$

The operator T is antilinear in that

$$T(a\psi_1 + b\psi_2) = a^*T\psi_1 + b^*T\psi_2$$

However, T preserves the normalization of a state function; indeed we have

$$\langle T\phi(t) \mid T\psi(t) \rangle = \langle \phi(-t) \mid \psi(-t) \rangle^* = \langle \psi(-t) \mid \phi(-t) \rangle \qquad\qquad 13.$$

For the matrix element $\langle \phi | \Theta | \psi \rangle$ we have

$$\langle T\phi \mid \Theta_T \mid T\psi \rangle = \langle \phi(-t) \mid \Theta \mid \psi(-t) \rangle^* \qquad\qquad 14.$$

[2] The operator T depends on the representation chosen, since the complex nature of a function depends on the representation. However, this dependence is of no consequence for physical observables. Schwinger (20) introduced a further way of defining time-reversal invariance. We shall not use the Schwinger definition here (see, however, 7).

where

$$\Theta_T = T\Theta T^{-1} = U_T \Theta^* U_T^{-1} \qquad 15.$$

What are some of the consequences of the time-reversal operation? We note, first of all, that for a spinless particle which satisfies the Schroedinger equation, we can take $T = K$. If H is invariant under the time-reversal transformation,

$$KHK^{-1} = H^* = H \qquad 16.$$

and it follows that both ψ and $\psi_T = \psi^*$ have the same eigenvalue. If there is no degeneracy, then ψ and ψ^* can at most differ by a constant, which means that ψ can be chosen real. For instance, all nondegenerate bound states of such a Hamiltonian can be so chosen.

For the linear momentum, angular momentum and helicity operators, we have

$$T\mathbf{p}_{op}T^{-1} = -\mathbf{p}_{op} \qquad 17.$$

$$T\mathbf{J}T^{-1} = -\mathbf{J} \qquad 18.$$

$$T\Lambda T^{-1} = \Lambda \qquad 19.$$

These transformations show that T cannot be taken as simply K for particles of half-integral spin. Indeed, for spin $\frac{1}{2}$, for instance, the operators σ_x and σ_z are real, whereas σ_y is imaginary in the standard representation. Equation 18 then shows that U_T is $[\exp(i\alpha)]\sigma_y$ where α is an arbitrary phase. The choice $\alpha = -\pi/2$ leads to $KU_T = U_TK$, and is often used for this reason. We then have

$$T = U_TK = -i\sigma_yK = e^{-i\pi S_y}K \qquad 20.$$

In the last form, Equation 20 can be used for any spin and, by a choice of phase, for arbitrary spin j

$$T|j, m\rangle = (-1)^{j-m}|j, -m\rangle \qquad 21.$$

The difference between half-integral spin and integer spin systems under time reversal can be brought out from the above discussion. Application of double time reversal to a state function must lead back to the original system; if β refers to quantum numbers other than the angular momentum, we have

$$T^2|\beta; j, m\rangle = \eta_T|\beta; j, m\rangle \qquad 22.$$

with $\eta = \pm 1$ (because T is antiunitary). From Equation 21, we find

$$T^2|\beta; j, m\rangle = (-1)^{2j}|\beta; j, m\rangle \qquad 23.$$

and this can be shown to be independent of phase convention. The eigenvalue is thus -1 for states of half-integral spin and $+1$ for integral spin. For a system of N particles of half-integral spins, we have

$$T^2 \mid N\rangle = (-1)^N \mid N\rangle$$

This difference in sign between half-integral and integral spin systems shows that no matrix elements can connect the two cases even if T does not commute with the Hamiltonian. Furthermore, for an odd number of fermions (e. g. nucleons) an immediate degeneracy follows, the Kramers degeneracy (21). To wit, if $T^2\psi = -\psi$ the state $T\psi$ cannot be identical to the state ψ. The reason is that if $[T, H] = 0$, both states have the same energy E, and if they were the same state, then the application of T^2 leads to an inconsistency ($E = -E$). The states of energy E must thus be at least doubly degenerate for a system of an odd number of nucleons. This has immediate consequences for odd-A deformed nuclei, which are specified by quantum numbers j, m, Ω, where m is the projection of the angular momentum on the space-fixed axis and Ω is that on the body-fixed axis. In this case, it is known that the states of different Ω are not usually degenerate. However, the above tells us that the states with quantum numbers $\pm\Omega$ are required to be degenerate if the Hamiltonian commutes with the time-reversal operator.

For a system of bosons no such restriction applies, and, as we saw earlier in this subsection, the state $T\psi$ may be identical to the state ψ, this leading to the real basis of bound states. For states of nonintegral spins, a "real basis" can also be found.

In the case of continuum states there are obvious degeneracies. The integral equation for the scattering wavefunction in a potential V is given by

$$\psi^{(+)}(\beta; \mathbf{k}; s, \nu) = \phi + \frac{1}{E - H + i\epsilon} V\phi \qquad 24.$$

where ϕ is the (plane wave and bound states) solution in the absence of V and the superscript $(+)$ means that the solution at infinity corresponds to spherical outgoing waves. From the previous discussion, we find that the time-reversal solution is given

$$T\psi^{(+)}(\beta; \mathbf{k}; s, \nu) = (-1)^{s-\nu}\psi^{(-)}(\beta_T; -\mathbf{k}; s, -\nu)$$

$$= \phi_T + \frac{1}{E - H_T - i\epsilon} V_T\phi_T \qquad 25.$$

where the subscript T stands for time-reversed. If $[T, H] = 0$, then the time-reversed solution, Equation 25, is one with reversed momenta and spins and with spherical ingoing waves. This state is readily visualized in terms of running a movie backwards. Such a state is difficult to prepare,[3] but Equation 25 nevertheless leads to physically observable consequences. For a Hermitian potential which is invariant under T, we have $V_T = V$. In nuclear physics, however, the interaction of a nuclear projectile with a nucleus is often approximated by a one-channel optical potential which is complex. Some care is required in this case (22, 23), since the approximation is not invariant

[3] For a lucid discussion of the difficulties this raises, see (1).

under T; it is necessary to replace $V(r)$ by its complex conjugate quantity $V^*(r)$, or more generally, V by V_T in Equation 25.

Lastly we note that arguments similar to those made for parity show that the electromagnetic potentials $\mathbf{A}$ and A_0 are, respectively, odd and even under T

$$TAT^{-1} = -\mathbf{A}; \qquad TA_0T^{-1} = A_0 \qquad\qquad 26.$$

For the electric- and magnetic-field operators, we therefore have

$$T\mathcal{E}T^{-1} = \mathcal{E}; \qquad T\mathbf{B}T^{-1} = -\mathbf{B} \qquad\qquad 27.$$

2.3 Consequences of P and T Invariance in Nuclear Physics

2.3.1 *Static electric and magnetic moments.*—The various multipole-moment operators can be obtained from classical electromagnetic theory by means of the correspondence principle (24). The quantum-mechanical 2^L static moments of a distribution of charge and currents are defined as the expectation values of the corresponding operators in a state of angular momentum j and z component $m = j$

$$\langle \mathcal{O}(L) \rangle \equiv \langle \beta; j, m = j \, | \, \mathcal{O}(L, N = 0) \, | \, \beta; j, m = j \rangle \qquad\qquad 28.$$

There are clearly restrictions which arise from angular momentum conservation alone; thus, the 2^L pole can only occur for a system with $2j \geq L$. Angular momentum conservation is, however, not the only restriction. As will be shown in the following paragraphs, both parity and time-reversal invariance of the electromagnetic field's interaction with the nuclear particles allow only even electric and odd magnetic multipole moments if the state in question is nondegenerate.

If parity is conserved, then the expectation value of the multipole operator vanishes unless

$$P\mathcal{O}(L)P^{-1} = \mathcal{O}(L) \qquad\qquad 29.$$

If we designate electric and magnetic multipoles by the subscripts E and M, then because

$$P\mathcal{O}_E(L, N)P^{-1} = (-1)^L\mathcal{O}_E(L, N)$$
$$P\mathcal{O}_M(L, N)P^{-1} = (-1)^{L+1}\mathcal{O}_M(L, N) \qquad\qquad 30.$$

we obtain the selection rule enunciated above and summarized in Table I. Since it is primarily the low multipole moments which can be measured accurately, we consider briefly the electric and magnetic dipole moments. For a nucleus or other nonrelativistic system the corresponding operators are

$$\mathcal{O}_E(1) = \sum_i e_i\mathbf{r}_i$$
$$\mathcal{O}_M(1) = \sum_i (g_{iL}\mathbf{L}_i + g_{iS}\mathbf{S}_i) \qquad\qquad 31.$$

TABLE I

SELECTION RULES THAT FOLLOW FROM P AND T INVARIANCE[a]

Symmetry	Multipole moments		
	electric (Coulomb)	magnetic	transverse electric
Parity	even L	odd L	even L
Time reversal	even L	odd L	odd L

[a] The allowed static multipole moments are listed.

where the subscripts i refer to the particles which make up the system and L and S are the orbital and spin angular momentum operators with corresponding gyromagnetic ratios g. It is readily found that Equation 30 is satisfied by these operators. Thus, a nondegenerate state cannot support a static electric dipole moment. This conclusion can be understood classically through inspection of Figure 1a, where it is seen that the state obtained by a reflection about the center of mass differs from the initial one (2). The same trans-

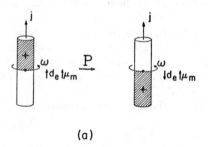

(a)

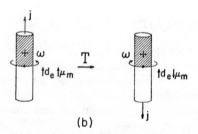

(b)

FIG. 1. Classical example of parity and time-reversal transformations for electric (d) and magnetic (μ) dipole moments: 1a shows the parity transformation and 1b the time-reversal transformation.

formation for a current loop (rotating charge), which corresponds to a magnetic dipole moment, leads to the same state.

The restriction imposed on the static moments by time-reversal invariance of the Hamiltonian for the system can be studied by means of the groundwork laid in Section 2.2. We note that the phase factor of Equation 21 is zero for the case $m = j$. However, any other phase convention for the state functions would lead to the same conclusion. If we use Hermiticity, and Equation 14, then we see that time-reversal invariance requires

$$\langle \beta; j, j \mid \Theta(L, 0) \mid \beta; j, j \rangle = \langle \beta; j, -j \mid T\Theta(L, 0)T^{-1} \mid \beta; j, -j \rangle \qquad 32.$$

By means of the Wigner-Eckart theorem and the reality of the Clebsch-Gordan coefficients, we note that Equation 32 implies

$$(jL - m0 \mid j - m)T\Theta(L, 0)T^{-1} = (jLm0 \mid jm)\Theta(L, 0) \qquad 33.$$

Thus, T invariance requires

$$T\Theta(L, 0)T^{-1} = (-1)^L\Theta(L, 0) \qquad 34.$$

However, for electric moments, which arise primarily from the charge distribution of the source, we have

$$T\Theta_E(L, N)T^{-1} = (-1)^N\Theta_E(L, -N) \qquad 35.$$

whereas for magnetic multipoles, which are due to currents and magnetizations, we have

$$T\Theta_M(L, N) = (-1)^{N+1}\Theta_M(L, N) \qquad 36.$$

From Equations 34–36, we note that the selection rules due to time-reversal invariance, summarized in Table I, are identical to those which follow from parity conservation. Only even electric and odd magnetic static multipoles may occur. Furthermore, a physical argument similar to that made for parity shows that the electric dipole moment must vanish. Thus, in the state with $m = j$, the system whose dipole moment is being considered is polarized along j since this is the only direction defined for the stationary system. Time reversal does not alter the charge distribution, but does reverse the spin direction, so that a new state is obtained. This is illustrated in Figure 1b. Hence if time-reversal invariance is valid, the dipole moment must be zero. Formally, this argument states that the dipole moment must be proportional to $\mathbf{J}$, since this is the only vector of the stationary system. Since $\mathbf{J}$ is odd under T, it follows that the electric dipole moment must vanish.

In summary, we note that static odd electric moments and even magnetic ones are forbidden by P and/or by T invariance. *Both* P and T must therefore be violated for such moments to occur in nature.

Elastic and inelastic electron scatterings from nuclei are valuable tools for the study of nuclear structure. To the extent that these scatterings can be treated as due to one-photon exchanges, an analysis in terms of multipole moments can be carried out, and leads to selection rules that can be derived along lines similar to those above (25). Indeed a multipole analysis shows that selection rules similar to those above hold. However, in addition to the Coulomb field, transverse electric-interaction multipoles contribute. The properties of these multipoles under P and T are summarized in the last column of Table I. It is interesting that the parity and time-reversal selection rules are not identical for these multipoles. Indeed, if *both* P and T hold, then no transverse electric multipoles can contribute at all to the elastic scattering of electrons (25).

2.3.2 *Transition electromagnetic moments.*—Space-time symmetries also restrict the transition electromagnetic moments. Such transitions may occur with the emission or absorption of photons and take place between bound states of the system, e.g., nucleus, or between bound and continuum states. The radiation field can be expanded in electric and magnetic multipoles, in terms of which the selection rules become clear. Thus, angular momentum limits the possible multipoles which can occur in a transition to those which satisfy $|j-j'| \leq L \leq |j+j'|$, where j and j' are the angular momenta of the initial and final states, respectively.

If the Hamiltonian for the system (including its interaction with the electromagnetic field) is invariant under the parity transformation, then the initial and final states have definite parities η and η'. Conservation of parity then only allows those multipoles which satisfy

$$\eta_L = \eta\eta' \qquad\qquad 37.$$

Together with Equation 30, this leads to the selection rules summarized in Table II. When more than a single multipole contributes to the radiative transition, it follows that even electric multipoles add coherently only to odd magnetic ones and vice versa.

There are also restrictions which follow from T invariance. In particular, to first order in the fine-structure constant, the electromagnetic-transition matrix elements between bound states can all be chosen to be real (or real times a constant common phase). The proof of this assertion makes use of Hermiticity as well as T invariance, and therefore requires lowest-order perturbation theory to be valid for the transition. Since the fine-structure

TABLE II

ELECTROMAGNETIC TRANSITIONS ALLOWED BY PARITY CONSERVATION

Parities	Allowed multipoles	
	electric	magnetic
$\eta' = \eta$	even L	odd L
$\eta' = -\eta$	odd L	even L

constant, which controls the rate of the transition, is small compared to unity, this assumption is justified. There are various ways of proving the above assertion; here we shall show that the interference to two parity-conserving multipoles is real if parity is conserved. From Equation 14 we have

$$\langle \beta'; j', m' \mid \Theta(L, N) \mid \beta; j, m \rangle = \langle \beta'; j', -m' \mid T\Theta(L, N)T^{-1} \mid \beta; j - m \rangle^*$$
$$\times (-1)^{j+j'-m-m'} \quad 38.$$

Together with the behavior of the multipoles under time reversal, Equations 35, 36, and the symmetry of the Clebsch-Gordan coefficients under reversal of the magnetic quantum numbers, we find for the reduced matrix elements

$$E(L) \equiv \langle j' \| \Theta_E(L) \| j \rangle = (-1)^L \langle j' \| \Theta_E(L) \| j \rangle^* = (-1)^L E(L)^* \quad 39.$$
$$M(L) \equiv \langle j' \| \Theta_M(L) \| j \rangle = (-1)^{L+1} \langle j' \| \Theta_M(L) \| j \rangle^* = (-1)^{L+1} M(L)^* \quad 40.$$

If we combine Equations 39 and 40 with the parity selection rules summarized in Table II, we see that those matrix elements that can interfere have a relative phase of either 0° or 180°. This selection rule was first pointed out by Lloyd (26). By a suitable choice of phase for the levels of angular momentum j and j' it is, in fact, possible to make all reduced matrix elements real.

2.3.3 *Nuclear scatterings.*—Nuclear scatterings include all processes which occur through the intervention of hadronic forces. An important distinction between these scatterings and the electromagnetic transitions is that perturbation theory is not generally applicable. The consequences of P and T invariances for hadronic events can be obtained from the discussion given in Sections 2.1 and 2.2.

The scattering matrix between states $|\phi_i\rangle$ and $|\phi_f\rangle$, which are solutions of $H_0 \equiv H - H'$ and may include bound states, is given by

$$\langle \phi_f \mid S \mid \phi_i \rangle = 1 - 2\pi i \delta(E - H_0)\langle \phi_f \mid R \mid \phi_i \rangle \quad 41.$$

where

$$\langle \phi_f \mid R \mid \phi_i \rangle = \langle \phi_f \mid H' + \frac{1}{E - H_0 - H' + i\epsilon} H' \mid \phi_i \rangle \quad 42.$$
$$= \langle \phi_f \mid H' \mid \psi_i^{(+)} \rangle$$

and $\psi_i^{(+)}$ is defined by Equation 24. Because of the unitary nature of the parity operator, it follows that

$$PSP^{-1} = S \quad 43.$$

Hence, only states ϕ_i and ϕ_f of the same parity are connected by the S matrix; that is, parity invariance requires

$$\eta_i = \eta_f \quad 44.$$

Because of the antiunitary nature of the time-reversal operator, the conclusions which follow from T invariance are different; one obtains the principles of reciprocity and detailed balance. For the non-Hermitian R and S matrices, we find from Equations 14 and 25

$$TST^{-1} = S\dagger, \qquad TRT^{-1} = R\dagger \tag{45.}$$

or

$$\langle \phi_f \mid S \mid \phi_i \rangle = \langle T\phi_i \mid S \mid T\phi_f \rangle \tag{46.}$$

and similarly for R.

If we consider the general processes $a+b \rightleftharpoons c+d$, we can rewrite the last equation as

$$\langle \mathbf{k}_f; s_c, \nu_c; s_d, \nu_d \mid S \mid \mathbf{k}_i; s_a, \nu_a; s_b, \nu_b \rangle$$
$$= (-1)^{s_a+s_b+s_c+s_d-\nu_a-\nu_b-\nu_c-\nu_d} \tag{47.}$$
$$\times \langle -\mathbf{k}_i; s_a, -\nu_a; s_b, -\nu_b \mid S \mid -\mathbf{k}_f; s_c, -\nu_c; s_d, -\nu_d \rangle$$

where $\mathbf{k}_{ab} \equiv \mathbf{k}_i$ and $\mathbf{k}_{cd} = \mathbf{k}_f$ are the relative momenta of a and b and of c and d. Since the cross section is proportional to the square of the scattering matrix element, we have a relationship between forward and backwards scattering, which can be expressed as

$$\mid \langle \mathbf{k}_f; s_c, \nu_c; s_d, \nu_d \mid S \mid k_i; s_a, \nu_a; s_b, \nu_b \rangle \mid^2$$
$$= \mid \langle -\mathbf{k}_i; s_a, -\nu_a; s_b, -\nu_b \mid S \mid - \mathbf{k}_f; s_c, - \nu_c; s_d, - \nu_d \rangle \mid^2 \tag{48.}$$

(Cross sections also involve fluxes and densities of final state factors.) This relation is known as the reciprocity relation. Thus, as emphasized by Lee (1), although the time-reversed state of a plane wave and outgoing spherical wave in a final state cannot be prepared as an initial one, the rate of the reaction from the state ϕ_i to ϕ_f is directly related to that between the reversed states, namely from $T\phi_f$ to $T\phi_i$. Finally, if spins are not measured, and the beam and targets are not polarized, then rotational invariance guarantees that

$$\sum_{\nu_a, \nu_b, \nu_c, \nu_d} \mid \langle \mathbf{k}_f; s_c, \nu_c; s_d, \nu_d \mid S \mid \mathbf{k}_i; s_a, \nu_a; s_b, \nu_b \rangle \mid^2$$
$$= \sum_{\nu_a, \nu_b, \nu_c, \nu_d} \mid \langle \mathbf{k}_i; s_a, \nu_a; s_b, \nu_b \mid S \mid \mathbf{k}_f; s_c, \nu_c; s_d, \nu_d \rangle \mid^2 \tag{49.}$$

This relation is known as the principle of detailed balance; it relates forward $(ab \rightarrow cd)$ and backward $(cd \rightarrow ab)$ reaction cross sections.

For elastic scattering measurements, without measurements of spins, neither P nor T symmetries give any restrictions. The reason is that the S matrix must be a scalar under rotations, and the only relevant scalars are k_i^2, k_f^2, and $\mathbf{k}_i \cdot \mathbf{k}_f$, all invariant under both P and T. If polarized beams or targets are used, or polarizations in the final state are detected, then both P and T invariances impose restrictions which can be tested. To treat these cases, we consider the S matrix as an operator in spin-space. For systems of

TABLE III

Terms in the Elastic Scattering S-Matrix Forbidden by P, T, or by P and T Invariances for Systems of Spins Zero Through Unity

$(\mathbf{k}_i + \mathbf{k}_f \equiv \mathbf{K},\ \mathbf{k}_i - \mathbf{k}_f \equiv \mathbf{q})$

s_a, s_b	Terms forbidden by		
	P conservation (allowed by T)	T invariance (allowed by P)	P and T invariances
0,0	—	—	—
1/2,0	$\mathbf{\sigma} \cdot \mathbf{K}$	—	$\mathbf{\sigma} \cdot \mathbf{p}$
1/2,1/2	$\mathbf{\sigma}_1 \cdot \mathbf{K},\ \mathbf{\sigma}_2 \cdot \mathbf{K}$ $\mathbf{\sigma}_1 \cdot \mathbf{\sigma}_2 \times \mathbf{q}$	$(\mathbf{\sigma}_1 \times \mathbf{\sigma}_2) \cdot (\mathbf{q} \times \mathbf{K})$ $(\mathbf{\sigma}_1 \cdot \mathbf{K} \mathbf{\sigma}_2 \cdot \mathbf{q} +$ $\mathbf{\sigma}_1 \cdot \mathbf{q} \mathbf{\sigma}_2 \cdot \mathbf{K})$	$\mathbf{\sigma}_1 \cdot \mathbf{q},\ \mathbf{\sigma}_2 \cdot \mathbf{q}$ $\mathbf{\sigma}_1 \cdot \mathbf{\sigma}_2 \times \mathbf{K}$
1,0	$\mathbf{S} \cdot \mathbf{K},\ (\mathbf{S} \cdot \mathbf{K} \mathbf{S} \cdot \mathbf{q} \times \mathbf{K})$	$\mathbf{S} \cdot \mathbf{K}\, \mathbf{S} \cdot \mathbf{q}$	$\mathbf{S} \cdot \mathbf{q},\ (\mathbf{S} \cdot \mathbf{q}\, \mathbf{S} \cdot \mathbf{q} \times \mathbf{K})$

total spin $\frac{1}{2}$, (i.e., $s_a = \frac{1}{2}$, $s_b = 0$), P conservation rules out terms such as $\mathbf{\sigma} \cdot \mathbf{k}_i$, $\mathbf{\sigma} \cdot \mathbf{k}_f$, or any linear combinations thereof. If $[P, H] = 0$, then no tests of T invariance are possible, as no terms of the S matrix can be constructed which are odd under T, but even under P. However, T alone forbids terms such as $\mathbf{\sigma} \cdot (\mathbf{k}_i - \mathbf{k}_f)$. We summarize the restrictions imposed by P, and T invariances in Table III for spins up to 1 (27). For higher spins, the restrictions have been investigated by Csonka, Moravcsik & Scadron (28).

In nuclear reactions, the non-Hermiticity of the S matrix (in the absence of the first-order Born approximation) means that no single experiment tests T invariance. Since T invariance connects $\langle f | S | i \rangle$ to $\langle i_T | S | f_T \rangle$, it is generally necessary to perform "reciprocal" experiments or those connected by detailed balance. This requirement arises because time-reversal noninvariance introduces a phase into the S matrix. However, other phases occurring in the S matrix are due to final- or initial-state rescattering effects, and it is necessary to distinguish these phases from those introduced by T noninvariance. The requirement of reciprocal experiments generally restricts one to reactions with only two particles in the final state.

Because P is a Hermitian operator Equation 44 requires that columns 1 and 3 of Table III also apply to reactions. Thus, P invariance alone tells us that for scattering systems of spin $\frac{1}{2}$, any polarization produced must be perpendicular to the scattering plane. When more than two particles are produced in a reaction, then further restrictions occur. For instance, in the reaction $a + b \rightarrow c + d + e$, P invariance forbids the appearance of terms proportional to $\mathbf{k}_a \cdot \mathbf{k}_b \times \mathbf{k}_c$ in the S matrix.

2.3.4 *P-odd and T-odd nucleon-nucleon potentials.*—Many nuclear phenomena can be interpreted in terms of a potential model between nucleons. Such a potential arises from meson exchanges between nucleons. To the extent that the description in terms of a potential is valid (e.g., in the nonrelativistic domain), one can expect P-odd or T-odd N-N potentials if the meson-nucleon or other interactions are not invariant under these symmetries. If we allow potentials that may depend on the (relative) momenta and angular momenta of the two particles, and require other symmetries (29) (e.g., Galilean invariance, symmetry under exchange) to hold, then some terms in the potential which are forbidden by P, T, and by P and T together (29–31) are summarized in Table IV. For clarity, we list only the simplest type of terms of a given variety.

2.4 THEORY OF PARITY-VIOLATING NUCLEAR FORCE

In Subsections 1 and 2 we presented general features of the parity and time-reversal transformations. In this part we outline the fundamental theory of the parity-violating nuclear force, based on our present understanding of the weak interactions. Comparisons of theory with experiments (Section 3) thus test the theory of weak interactions, which necessarily involves basic elementary-particle physics.

The theory of parity-violating nuclear forces makes two basic assumptions.

TABLE IV

POTENTIAL TERMS THAT ARE ODD UNDER P, T OR P AND T, BUT SATISFY OTHER SYMMETRY REQUIREMENTS[a]

$$\mathbf{s} = \tfrac{1}{2}(\mathbf{\sigma}_1 - \mathbf{\sigma}_2), \quad \mathbf{S} = \tfrac{1}{2}(\mathbf{\sigma}_1 + \mathbf{\sigma}_2), \quad \mathbf{L} = \mathbf{r} \times \mathbf{p}$$

Type	P-odd T-even	T-odd P-even	P-odd T-odd	Possible isospin dependence
Static	$\mathbf{\sigma}_1 \cdot \mathbf{\sigma}_2 \times \mathbf{r}$	—	$\mathbf{s} \cdot \mathbf{r}$	$1, \ \tau_1 \cdot \tau_2,$
Momentum dependent	$\mathbf{s} \cdot \mathbf{p}$	$\mathbf{r} \cdot \mathbf{p}$	$\mathbf{\sigma}_1 \cdot \mathbf{\sigma}_2 \times \mathbf{p}$	$\tau_1^{(3)} \tau_2^{(3)}$ $\tau_1^{(3)} + \tau_2^{(3)}$
Static	—	—	$\mathbf{s} \cdot \mathbf{r}$	
Momentum dependent	$\mathbf{S} \cdot \mathbf{p}$	$\mathbf{\sigma}_1 \cdot \mathbf{\sigma}_2 \times \mathbf{L}$	$\mathbf{\sigma}_1 \cdot \mathbf{r} \mathbf{\sigma}_2 \cdot \mathbf{L} + \mathbf{\sigma}_2 \cdot \mathbf{r} \mathbf{\sigma}_1 \cdot \mathbf{L}$	$\tau_1^{(3)} - \tau_2^{(3)}$
Static	—	—	—	
Momentum dependent	$\mathbf{\sigma}_1 \cdot \mathbf{r} \mathbf{\sigma}_2 \cdot \mathbf{L} + \mathbf{\sigma}_2 \cdot \mathbf{r} \mathbf{\sigma}_1 \cdot \mathbf{L}$	$\mathbf{s} \cdot \mathbf{L}$	$\mathbf{S} \cdot \mathbf{p}$	$(\tau_1 \times \tau_2)^{(3)}$

[a] In all cases $+$h.c. is to be understood.

(a) The first is that the weak interaction can be written as a current-current one, with a Hamiltonian density[4] ($\mu = 0, 1, 2, 3$)

$$\mathcal{3C} = (2)^{-1/2} G J_\mu{}^\dagger J^\mu \qquad 50.$$

The current J_μ is decomposed into a leptonic part $J_\mu{}^{(l)}$ and a nonleptonic part $J_\mu{}^{(h)}$,

$$J_\mu = J_\mu{}^{(l)} + J_\mu{}^{(h)} \qquad 51.$$

The reasonableness of this assumption rests primarily on experiments of semileptonic and leptonic decays (9, 32, 33), where the terms $2^{-1/2} G[J_\mu{}^{\dagger (h)} \cdot J_\mu{}^{(l)} + \text{h.c.}]$ and $2^{1/2} G J_\mu{}^{\dagger (l)} J_\mu{}^{(l)}$ contribute. Much less is known about the validity of the current-current form for the interaction of the purely non-leptonic processes. One of the strongest pieces of evidence for its correctness is, in fact, the observation of parity nonconservation in nuclear phenomena (see Section 3). If the parity-violating weak interaction were solely due to terms in the Hamiltonian, such as those responsible for beta decay, e.g.,

$$2^{-1/2} G[\bar{\psi}_p \gamma^\mu (1 - A\gamma_5)\psi_n][\bar{\psi}_e \gamma_\mu (1 - \gamma_5)\psi_\nu]$$

(this implies that one cannot break up the Hamiltonian into a product of two pieces), then the parity-violating nuclear force would arise from diagrams such as those shown in Figure 2a and would be $\sim G^2$. However, if the current-current interaction is correct, we may have contributions such as that shown in Figure 2b, which is $\sim G$. Since, for small momentum transfers, the strength G in dimensionless units $(GM_N{}^2)$ is $\sim 10^{-5}$, the fact that parity

---

[4] We use the metric $x^\mu = (t, \mathbf{r})$, $x^2 = t^2 - r^2$.

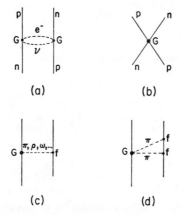

(a)   (b)

(c)   (d)

Fig. 2. Contributions to a parity-violating nuclear potential $V_{P.V.}$: 2a is of second order in the weak interactions; 2b is a contact term; 2c and 2d are meson exchange contributions.

violations of this order of magnitude have been observed is an argument for the correctness of the underlying assumption (*a*).

(*b*) The second basic assumption in the development of the theory is that *CP* nonconserving effects can be neglected. Although the observation of *CP* nonconservation in the K_2^0 system (9) shows that this assumption is not accurate, it must be remembered that the effects observed are $\sim 2 \times 10^{-3}$ of the *CP*-conserving ones. Hence, to an accuracy of roughly 1/1000 we can calculate parity-nonconserving effects due to the *CP*-conserving weak interactions alone. Some theories that have been proposed to explain the *CP* violation in the decays of the K_2^0 radically alter the basic structure of weak interactions (34–36). Assumptions (*a*) and (*b*) do not hold for these theories. Because they are highly speculative and the corresponding parity-violating nuclear forces have not been worked out, we do not consider these theories further.

The first theories of parity-violating nuclear forces were developed (36, 37) in the early 1960s. A potential model is assumed to be valid for low energies. A direct contact term arises naturally in the current-current inter-action model, and gives a potential (30, 38)

$$
V_{\text{P.V.}}^{(\text{cont})}(\mathbf{r}) = -\frac{G}{2(2)^{1/2}} \frac{1}{M} \left\{ (1 + \mu^V)(i\mathbf{\sigma}_1 \times \mathbf{\sigma}_2) \cdot \left[\mathbf{p}, v^{(\text{cont})}(\mathbf{r})\right] \right. \tag{52.}
$$
$$
\left. + (\mathbf{\sigma}_1 - \mathbf{\sigma}_2)\left\{\mathbf{p}, v^{(\text{cont})}(\mathbf{r})\right\} \right\} (\mathbf{\tau}_1 \cdot \mathbf{\tau}_2 - \tau_1^{(3)}\tau_2^{(3)}),
$$

with

$$
\mathbf{p} \equiv \tfrac{1}{2}(\mathbf{p}_1 - \mathbf{p}_2)
$$
$$
\mathbf{r} \equiv (\mathbf{r}_1 - \mathbf{r}_2)
$$
$$
v^{(\text{cont})}(\mathbf{r}) \equiv \delta(\mathbf{r})
$$

In Equation 52 $\{ \quad \}$ is an anticommutator, $[\quad]$ is a commutator, μ^V is the isovector anomalous moment, $\mathbf{p}$ is the momentum, and the subscripts refer to the two nucleons and the numerical superscripts to components. The source of the potential is the self-interaction of the nucleonic weak current, which for Equation 52 was taken to be

$$
J_\mu^{(h)} = \bar{\psi}_N \tau^+ \gamma_\mu (1 - \gamma_5)\psi_N \tag{53.}
$$

Higher-order terms in p/M have been dropped in Equation 52. Although $V_{\text{P.V.}}^{(\text{cont})}$ is obtained simply, its effects in the presence of the strong interactions are negligible because of the repulsive core which acts between nucleons.

However, the nucleons are surrounded by meson clouds, and the interaction can occur through these clouds. Thus, we can represent the parity-violating force by diagrams such as those of Figures 2c and 2d. In accord with assumption (*a*) the weak interaction *G*, shown by the blobs on one nucleon leg, arises from a current-current interaction; the other vertices f represent strong, parity-conserving interactions. The exchanged mesons are those present in the cloud. The contribution of the diagram represented by Figure 2d was computed by Blin-Stoyle (30) in perturbation theory. However, just as for the two-pion strong-interaction potential, the results are not trustworthy because both $\pi-\pi$ interactions (rescattering effects) and π-N rescatterings (e.g., leading to an N^* intermediate state) were not considered. Recently Lacaze (39) has included these effects and has found a static potential with a longer range than that deduced by Blin-Stoyle. To the extent that the strong N-N force can be represented by single-boson exchanges (40), it is perhaps possible to consider the important part of the two-pion exchange effects to be included in Figure 2c. Indeed, the results of Lacaze support this point of view. Since the single-boson exchanges are the only contributions which can be calculated at all reliably, we shall take this attitude here.

Because of the short-range repulsive character of the strong nucleon-nucleon forces, the major detectable contribution of the weak parity-violating force should come from the exchange of light mesons. Although the pion is the lightest of these, it was first shown by Barton (37) that there is no parity-violating force from single-pion exchange if only zero-strangeness, $\Delta S=0$, changing weak currents are included (and CP is conserved). The proof of this conclusion is not independent of theory, but assumes a current-current interaction; the dependence of the conclusion on these assumptions will be demonstrated below. In particular, the proof makes use of the isospin character of the weak nonleptonic ΔS or $\Delta Y=0$ currents. The proof which follows shows that the argument is equally valid for any pseudoscalar meson exchange.

For simplicity, we first show that (i) CP conservation, alone, is sufficient (independent of theory) to forbid the exchange of any neutral C-even pseudoscalar particle (e.g. π^0, η^0). Indeed, for such an exchange the most general "effective Hamiltonian" to represent the coupling at the weak blob in Figure 2c is

$$H_{\text{P.V.}}^{\text{eff}} = 2^{-1/2}G \int \{ c_1 \bar{\psi}_N \psi_N \phi + c_2 \bar{\psi}_N \gamma^\mu \psi_N \partial_\mu \phi \qquad\qquad\qquad 54.$$

$$+ c_3 \bar{\psi}_N \sigma^{\mu\nu} \psi_N \partial_\mu \partial_\nu \phi \} \, d^3r$$

where only parity-violating contributions are included, c_i are arbitrary constants, and ∂_μ stands for derivatives with respect to x^μ. The operators $\bar{\psi}$ and ψ create and destroy nucleons, respectively, and ϕ represents a pseudoscalar meson field with $C\phi C^{-1}=\phi$, where C is the charge-conjugation operator. The term in Equation 54 proportional to c_1 is odd under P, but even under C and therefore odd with respect to CP. The term proportional to c_2 involves a vector operator V^μ. The most general matrix element of this vector for nucleons on their mass shell is (41)

$$\langle N(p') \mid \bar{\psi}_N V^\mu \psi_N \mid N(p) \rangle = \bar{U}_N(p')[\gamma^\mu F_1(q^2) + i\sigma^{\mu\nu} q_\nu F_2(q^2) \\ + q^\mu F_3(q^2)]U_N(p) \qquad 55.$$

where $q_\mu = p_\mu' - p_\mu$, and $q^2 = q^\mu q_\mu$. The last term of Equation 55 has the CP character opposite to that of the first two and is therefore not allowed by CP conservation. The first two terms satisfy

$$\partial_\mu \langle N(p) \mid \bar{\psi}_N V^\mu \psi_N \mid N(p) \rangle = q_\mu \bar{U}_N(p')[\gamma^\mu F_1 + i\sigma^{\mu\nu} q_\nu F_2]U_N(P) = 0 \quad 56.$$

even though the current may not be a conserved one. Since, by a partial integration of the matrix element of the term in $H_{\mathrm{P.V.}}^{\mathrm{eff}}$ proportional to c_2 it can be brought to the form of Equation 56, we find that this term does not contribute. Partial integrations allow us to bring the last term in Equation 54 to a form where both derivatives operate on the nucleon matrix elements. The antisymmetry of $\sigma^{\mu\nu}$ then guarantees that this term also vanishes.

In generalizing the above argument to charged pseudoscalar mesons, it is simpler not to use C, but to use T or the properties under G conjugation (42), where

$$G = e^{i\pi I_2} C \qquad 57.$$

Since we need an interaction that is CP even and C odd, we note that if $H_{\mathrm{P.V.}}^{\mathrm{eff}} = H'$ has the character of $\Delta I = 1$, then $GH'G^{-1} = H'$ whereas for even values of ΔI (e.g. 0,2) $GH'G^{-1} = -H'$. Of the various possible combinations of the isospin operators together with the interactions given by Equation 54, the only possible one that is CP even can then be shown to be

$$H' = i \int \bar{\psi}_N(\tau^+ \phi^- - \tau^- \phi^+)\psi_N d^3r = \int \bar{\psi}_N(\tau \times \phi)^{(3)} \psi_N d^3r \qquad 58.$$

where $\phi^\pm = \phi^{(1)} \pm i\phi^{(2)}$, $\tau^\pm = \frac{1}{2}[\tau^{(1)} \pm i\tau^{(2)}]$. H' is the third component of an isovector and has the character of $\Delta I = 1$. We now make use of the fact that the strangeness-nonchanging weak nonleptonic currents are isovectors, e.g., the nucleon current is given by Equation 53. Since the product of two like currents of isospin 1 cannot give rise to an isospin 1, one cannot obtain the third component of an isovector. Indeed, we note an important selection rule here: (ii) Strangeness-nonchanging currents give rise to a parity-violating nuclear potential that carries isospin change of 0 or 2. The theorem rests on assumptions (a) and (b) together with isospin invariance of the hadronic forces and the isovector character of the weak, strangeness-nonchanging currents. It follows that if we write the nonleptonic currents in their presently most widely accepted form (referred to as the Cabibbo theory),

$$J_\mu = \left\{ [\mathfrak{J}_\mu{}^{(1)} + i\mathfrak{J}_\mu{}^{(2)}] \cos\theta + [\mathfrak{J}_\mu{}^{(4)} + i\mathfrak{J}_\mu{}^{(5)}] \sin\theta \right\} \qquad 59.$$

with

$$\mathfrak{J}_\mu = V_\mu + A_\mu \qquad 60.$$

then the above parity-violating force is proportional to $\cos^2 \theta$, where θ is the Cabibbo angle ($\sin \theta \approx 0.22$) (32). In Equation 59 the superscripts refer to SU (3) indices, with 1 and 2 corresponding to isospin components. The currents V_μ and A_μ are vector and axial-vector currents.

The above result (ii) does not restrict the potential contributions of Figures 2b–d to $\Delta I = 0, 2$ parts, but states that the $\Delta I = 1$ terms must arise from hypercharge (or strangeness)-changing currents and are therefore proportional to $\sin^2 \theta$.[5] The $\Delta I = 1$ parity-violating potential would therefore be expected to be considerably weaker ($\tan^2 \theta \approx 0.05$) than the $\Delta I = 0, 2$ potential. This applies, in particular, to the one-pion exchange potential, as we argued earlier. However, because of the long range of this contribution, the one-pion exchange cannot be neglected. Indeed, a measure of the contribution of this force may allow one to obtain information about the strangeness-changing nonleptonic currents and the validity of the current-current form of the interaction for the nonleptonic weak force. The isospin character of the parity-violating force is undetermined experimentally.

The contribution of the pion to the parity-violating force was first derived by McKellar (43) by means of current algebra and $SU(3)$. We shall not repeat his derivation here, but merely sketch the assumptions and the arguments. McKellar assumes current-algebra, partial-current conservation for the axial current, a smooth extrapolation from the pion pole to zero four-momentum (alternatively zero-mass pions), and Cabibbo currents, Equation 59. He makes use of the observation by Suzuki (44) and Sugawara (45) that the parity-violating nonleptonic decays of the hyperons can be understood with the above assumptions. The number of independent matrix elements is considerably reduced with the assumption of $SU(3)$ symmetry. For the S-state (parity-violating) decays of the hyperons there are but three independent matrix elements which can be classified according to their $SU(3)$ representation; they are a symmetric octet, an antisymmetric octet, and a 27-plet, A [27]. These amplitudes are determined from the pionic decays of the Λ^0, Σ, and Ξ hyperons; the experimental data require A [27] ≈ 0, which is consistent with the $\Delta I = \frac{1}{2}$ rule for the baryonic decays. $SU(3)$ symmetry completely determines the matrix $\langle N\pi | H_{\text{weak}}' | N \rangle$ in terms of these amplitudes. Since the latter are fixed by baryonic decays, there are no free parameters. The data give roughly

$$\langle p\pi^- | H_{\text{weak}}' | n \rangle = -\langle n\pi^+ | H_{\text{weak}}' | p \rangle = g$$
$$\approx 4.2 \pm 0.8 \times 10^{-8}$$
61.

The one-pion exchange potential is now simple to compute in terms of the amplitude g and the strong vertex f at the other nucleon leg. One finds (43)

$$V_{\text{P.V.}}{}^{(\pi)}(\mathbf{r}) = \frac{gf}{2^{3/2}M} (\mathbf{\delta}_1 + \mathbf{\delta}_2) \cdot \left(\mathbf{\nabla} \frac{e^{-m_\pi r}}{4\pi r} \right) (\mathbf{\tau}_1 \times \mathbf{\tau}_2)^{(3)}$$
62.

where $f^2/4\pi \approx 14.4$.

[5] There can be no $\sin \theta \cos \theta$ terms; these contribute to an overall change of strangeness interacon.

There are remaining uncertainties in the magnitude of the constant g in Equation 61. These are directly connected to the theoretical framework used in its evaluation. Thus, it is assumed that the weak currents that contribute are those given by Equation 59. However, it has been suggested that the vanishing of the 27-plet contribution to the nonleptonic decays of the strange baryons arises from the presence of additional $[SU(3)]$ currents (46, 47). These extra currents turn out to give a $\Delta I = 1$ potential proportional to $\cos^2 \theta$, and therefore enhance the value of g by roughly $\cot^2 \theta \sim 20!$ For the case of (47), the enhancement is -30 (48). A variety of current models are considered by Tadic (49) and Fischbach & Trabert (50). A different proposal is that dynamical enhancement of the octet (but not of the 27-plet) contributions in the decay of the baryons is due to the strong interactions (51, 52). This presumably would not greatly affect the constant g, since its numerical evaluation already uses experimental data. However, the enhancement for the matrix element of Equation 61 may be appreciably different from that for the matrix elements involved in the nonleptonic decays of the strange baryons. Experimental determinations of g are clearly important, since its numerical value is sensitive to the underlying weak-interaction theory.

In addition to pions, various vector mesons can be exchanged. The contributions from $\rho^{\pm}$ exchange were computed by Michel (38) by considering Figure 2b for physical nucleons, and thus with appropriate form factors. Michel took vector dominance to specify the form factor of the vector current, and neglected the momentum dependence of the axial-vector form factor. For small momentum transfers, the ratio of the axial vector to the vector coupling constant is $\lambda = 1.23 \pm 0.02$ (53). The potential is thus identical to that given by Equation 52 except that $v^{(\text{cont})}$ (r) is replaced by the ρ-meson exchange potential, and the result is multiplied by the constant $\lambda \cos^2 \theta$

$$V_{\text{P.V.}}{}^{(\rho)}(\mathbf{r}) = \lambda \cos^2 \theta \big\{ V_{\text{P.V.}}{}^{(\text{cont})} \big[\text{with } v^{(\text{cont})} \rightarrow v^{(\rho)} \big] \big\} \qquad 63a.$$

$$v^{(\rho)} \equiv m_\rho{}^2 \frac{e^{-m_\rho r}}{4\pi r} \qquad 63b.$$

A more general approach has been taken by Blin-Stoyle & Herczeg (54) and by Tadic (49), who use experimental form factors and note that an induced pseudoscalar and tensor form factors (54) may appear. Since the evidence for these contributions is questionable, we shall ignore them. However, these terms do contribute to a $\Delta I = 1$ potential (44). Neglecting them, but using form factors obtained from electron scattering experiments, one finds

$$V_{\text{P.V.}}{}^{(ff)}(\mathbf{r}) = - \frac{G\lambda}{2(2)^{1/2}} \frac{\cos^2 \theta}{M} \big\{ \mu i \boldsymbol{\delta}_1 \times \boldsymbol{\delta}_2 [\mathbf{p}, v(r)^{(ff)}]$$

$$+ (\boldsymbol{\delta}_1 - \boldsymbol{\delta}_2) \big\{ p, v_{(r)}{}^{(ff)} \big\}_+ \big\}$$

$$\times \big[\boldsymbol{\tau}_1 \cdot \boldsymbol{\tau}_2 - \tau_1{}^{(3)} \tau_2{}^{(3)} \big], \qquad 64.$$

$$\mu v^{(ff)} = \left(\frac{1}{2\pi}\right)^3 \int e^{i\mathbf{q}\cdot\mathbf{r}}d^3q[F_1(q^2) + F_2(q^2)]F_A(q^2),$$

$$v^{(ff)} = \left(\frac{1}{2\pi}\right)^3 \int e^{i\mathbf{q}\cdot\mathbf{r}}d^3q F_1(q^2)F_A(q^2)$$

where F_1, F_2 are defined by Equation 55 and F_A is the corresponding axial-vector form factor. This potential does not differ greatly from Equation 63 in the region outside the nucleon core, and both have been used in the analyses of experiments.

To obtain a feeling for the strengths of the parity-violating potentials, we compare the orders of magnitude of the static parts of $V_{\text{P.V.}}^{(\rho)}$ and $V_{\text{P.V.}}^{(\pi)}$ to each other and to the strong-interaction force at separations of the pion Compton wavelength ($r \approx 1.4$ F) and at the radius of the hard core ($r \approx 0.4$ F). At a pion Compton wavelength, the normal (hadronic) spin-spin term of the one-pion exchange potential is ≈ 1.4 MeV, whereas the strengths of the parity-violating potentials are roughly

$$V_{\text{P.V.}}^{(\pi)}(m_\pi^{-1}) \sim 2 \times 10^{-7} \text{ MeV}, \quad V_{\text{P.V.}}^{(\rho)}(m_\pi^{-1}) \sim 10^{-7} \text{ MeV} \quad \text{65a.}$$

At $r \approx 0.4$ F we have $V_{\text{hadr}} \approx 40$ MeV, and

$$V_{\text{P.V.}}^{(\pi)}(0.4 \text{ F}) \sim 4 \times 10^{-6} \text{ MeV}$$
$$V_{\text{P.V.}}^{(\rho)}(0.4 \text{ F}) \sim 3 \times 10^{-5} \text{ MeV} \qquad \text{65b.}$$

This comparison shows that $\overline{V}_{\text{P.V.}}/\overline{V}_{\text{hadr}} \sim 10^{-6} - 10^{-7}$ and makes it quite clear that $V_{\text{P.V.}}^{(\pi)}$ cannot be neglected despite its reduction by a factor of $\tan^2\theta \approx 20$ relative to $V_{\text{P.V.}}^{(\rho)}$.

In the above approach, the coupling of neutral vector mesons (e.g. ρ^0, ω^0, ϕ, $\cdots$) results from strong-interaction renormalization effects and is expected to be weaker; its effects are not included in Equations 63–64.

Our knowledge of the parity-violating nuclear forces is clearly far from complete. Not only can one question the replacement of continuum (e.g., two-pion) effects by resonances, but even the reliable evaluation of the resonance contributions is difficult. Although one may hope to obtain further information from nuclear physics parity-violation experiments, it must be remembered that the weak forces always occur in the presence of the strong ones. The effects of the latter are not trivial to include. To wit, although the electromagnetic interaction is known, the effects of electromagnetic corrections to the strong interactions have not yet been reliably computed (55).

2.5 A THEORY FOR T-ODD NUCLEAR FORCES

The basis for CP violation in the weak interactions is still poorly understood. However, a meaningful analysis of T violation in nuclear physics requires a theoretical framework. Without a theory it is impossible to estimate expected violations and to discuss experimental sensitivities meaningfully. Several alternatives have been proposed to explain the K_2^0 CP-odd decays. These include (*i*) superweak interactions (14), (*ii*) electromagnetic corrections (16, 17, 56, 57), (*iii*) weak-interaction changes (34–36). Most other suggestions have been ruled out by experimental findings (9). If the superweak-interaction hypothesis proves to be correct, then there is essentially no hope of finding T violation in nuclear experiments. Also some of the proposals (*iii*) lead to effects $\leq 10^{-9}$ of the strong force ones and are therefore also not testable in nuclear physics. Indeed, the only proposal which has interesting consequences in nuclear physics is that the violation arises from a medium strong interaction, and particularly from the electromagnetic interactions of hadrons (16, 17, 56). For this reason this theory is developed here and is adopted in the following section, where experiments are analyzed.

The suggestions for proposal (*ii*) rest on the observation that the CP violation in the $K_2{}^0$ decay is of the order of magnitude of α/π, where α is the fine-structure constant. To Arzubov & Filippov (57) this magnitude suggests a "milliweak" interaction, possibly of electromagnetic origin. Since the effect is then restricted to weak interactions, the effect on strong interactions is $\sim 10^{-9}$ and thus not presently detectable. However, as pointed out particularly by Bernstein et al. (16), experimental proof of charge-conjugation invariance for the electromagnetic interactions of the hadrons is weak. Although this invariance is identical to particle-antiparticle conjugation for the leptons, is this so for the hadrons? This hypothesis has not yet received clear-cut experimental verification. It has been shown that the annihilation of antiprotons by protons leads to charged pions for which C holds to $\lesssim 1$ per cent (58). For the η^0 decay into three pions, C invariance may or may not hold;[6] the violation, if any, is less than ~ 2 per cent (59). These experiments do not rule out the existence of a C-even electromagnetic current K_μ,

$$CK_\mu C^{-1} = K_\mu \qquad\qquad 66.$$

which is opposite to the normal electromagnetic current J, for which one has

$$CJ_\mu C^{-1} = -J_\mu \qquad\qquad 67.$$

Bernstein et al. (16) and Lee (56) suggest alternative seats for the C violation of the electromagnetic currents of the hadrons. They fall into two broad categories: (*a*) those for which $Q_K = nK_0 d^3r$ is zero for all physical states and (*b*) those for which this is not the case. In the latter instance there exist particles a^+ for which (56) $C|a^+\rangle = |a^+\rangle$. The consequences of the latter theory for nuclear phenomena have not been worked out, and we shall restrict ourselves to case (*a*). The isospin character for both theories is unknown; the current K_μ could be the third component of a pure isovector, might be an isoscalar [there is a slight theoretical preference for this case (60)], or more generally, may be an arbitrary combination of both. We shall take the latter point of view.

---

[6] The present evidence is not clear. See (59).

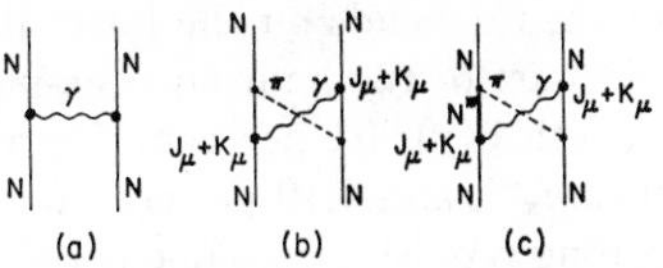

(a)   (b)   (c)

FIG. 3. Typical long-range electromagnetic corrections to nuclear forces: 3a, the Coulomb effect, does not contribute to a T-odd force; but 3b and 3c do.

For nucleons on the mass shell (free nucleons in both initial and final states), relativistic properties alone restrict the matrix element of the electromagnetic current to be of the form given by Equation 55. Current conservation, $\partial J_\mu / \partial x_\mu = 0$, is sufficient to quarantee that $F_3(q^2) = 0$.[7] These considerations show, for instance, that the Coulomb force (Figure 3a) has no contribution from K_μ or C-even terms. They also lead one to expect that the violation of C and consequently of T (assuming TCP) is reduced by at least m_π / M; contributions of this order of magnitude can arise when the nucleon is not on the mass shell, as in Figures 3b and 3c. In these cases, the restrictions above no longer apply. Current conservation is replaced by the more general Ward's identity (61). For Figure 3b the form factors $F_i(q^2)$, with $i = 1 \cdots 3$, are replaced by more general ones which can depend on the three independent relativistic scalars q^2, p^2, p'^2. (For Figure 3c, further generalizations occur if the spin of the N^* is $\geq \frac{3}{2}$.) If one of the nucleon legs is on the mass shell, then there are six form factors which may depend on p'^2 and q^2 (if $p^2 = m^2$) (64). The condition that the current be Hermitian gives restrictions on the momentum dependence, and Ward's identity leads to a relationship between the form factors (65). There is then no unique way of introducing T-odd terms in the generalized electromagnetic vertex. If one assumes that the influence of K_μ occurs through a form factor F_3 in Equation 55, which multiplies q_μ, it is still far from straightforward to compute the T-odd nucleon-nucleon force. Thus, although there are restrictions on the off-shell dependence of F_3 on q^2, p^2, and p'^2, there remains considerable freedom.

Huffman (66) has estimated the one-pion one-photon exchange T-odd force with the above as basis, but with the omission of N^* contributions. The radial dependence of the T-odd potential found depends on the assumed form of F_3 and will not be reproduced here. Of more interest are the general results obtained: (i) The strength of the T-odd force relative to the one-pion exchange potential is roughly $(2\alpha/\pi) (m_\pi/M_N)^3$. A factor $(m_\pi/M_N)^2$ comes from the off-the-mass-shell requirement for F_3 and another (m_π/M_N) comes from $\langle q_\mu \rangle / M_N$. The numerical coefficient depends to some extent on how the "strength" is defined; for the above, the integral of the potential was used.

[7] For particles of spin $> \frac{1}{2}$, the on-mass-shell restrictions from current conservation are not sufficient to rule out a T-odd current matrix element. This applies to the deuteron, considered as an elementary particle of spin 1 (62, 63).

The numerical strength then depends on the lower limit of the integration. Relative to the spin-spin part of the hadronic one-pion exchange potential, Huffman obtains a strength of 0.004 per cent for a lower cutoff $b = m_\pi^{-1}$, 0.012 per cent for $b = 0.75 m_\pi^{-1}$, and 0.10 per cent for $b = 0.5 m_\pi^{-1}$. Although, in the presence of the strong repulsive core one expects the pion exchange to be the most effective T-odd force because of its large range, this may be compensated by the loss of strength (e.g. m_π^2/M_N^2) due to off-the-mass-shell requirement. (ii) Huffman finds that only the isovector component of K_μ contributes and gives rise to a purely exchange force which acts only between neutrons and protons. (iii) All terms of the T-odd potential depend on the spin and have at least a linear dependence on the momentum; they all vanish for zero range and have an exponential behavior at large distances, e.g., $\delta_1 \cdot \delta_2(\mathbf{r}_1 - \mathbf{r}_2) \cdot (\mathbf{p}_1 - \mathbf{p}_2) f(|\mathbf{r}_1 - \mathbf{r}_2|)$, +h.c. It is these features that we shall use in Section 3.

3. EXPERIMENTS AND THEIR INTERPRETATION

3.1 CLASSIFICATION OF P AND T EXPERIMENTS

In Section 2 we laid the groundwork for the discussion of possible experiments to test P and T invariances. We saw that the parity operator is unitary whereas the time-reversal one is antiunitary. This difference leads to important distinctions in the experimental tests. In general, there are absolute selection rules for P invariance; that is, certain reaction amplitudes are forced to vanish by P invariance. On the other hand, time reversal does not generally lead to such selection rules, but rather predicts that certain pairs of matrix elements should be relatively real. It is only for elastic amplitudes that the antiunitary nature of the time-reversal operator leads to selection rules (e.g., the vanishing of certain electromagnetic moments). These distinctions are reflected in the types of experiments that can be carried out to test the two invariances. Thus, tests of T often use a comparison of two reactions or two transition amplitudes, whereas P-conservation tests seek forbidden processes or amplitudes.

A useful classification of tests of parity invariance was given by Wilkinson (67). This classification, which has been retained in the literature with minor modifications,[8] is:

1_P. experiments that violate the selection rule (Equation 44);

2_P. experiments that seek a nonvanishing expectation value of a helicity amplitude (see Equation 8) or of other polarization terms (see Section 2.3c and Table III);

3_P. experiments that look for asymmetries in the angular distributions of final-state particles with respect to an initial polarization.

The distinction between 2_P and 3_P is primarily operational, in that differ-

---

[8] Wilkinson's class 3_p experiments were sensitive to $\mathcal{F}^2$, because the initial polarization was assumed to be provided by parity-forbidden reactions.

ent detection techniques are used. These broad categories and the experiments belonging to them are discussed in detail in Section 3.2. There, we assume that T invariance is satisfied, since the seat of P violation is in the weak interactions for which CP invariance holds to better than a few parts in a thousand (see Section 2.4).

Time-reversal experiments can also be categorized, not only according to the type of experiment, but also according to the strength of the interaction:

1_T. weak-interaction experiments with nuclear particles;

2_T. electromagnetic tests;

3_T. polarization experiments in elastic scattering;

4_T. reciprocity or detailed-balance tests for reactions;

5_T. nuclear structure tests.

These categories will be used in the detailed discussion of experiments to date. P invariance will be assumed (see Subsection 2.5).

In addition to tests of P and T, separately, there are experiments that are sensitive to a violation of *both* P and T. Examples appear in Tables I and II. We shall discuss only one of these tests in detail, because the effects of P and T violation, together, are expected to be extremely small. Thus P is violated only by the weak interactions and CP violations are further reduced by 2–3 orders of magnitude. The extreme sensitivity required ($\sim 10^{-9}$) is difficult to achieve. It has been reached in measurements of static electric dipole moments, and it is these experiments that will concern us in Section 3.3.

3.2 Parity Experiments

Experiments of type 1_P, which violate the parity selection rule, are second order in the parity admixture coefficient $\mathfrak{F}$, Equation 5. In these experiments one seeks to measure the transition rate from (and to) the abnormal part of the state admixed by the parity-violating forces to (and from) $|\psi_{\text{norm}}\rangle$. With the notation of Equation 5 this rate is proportional to

$$\mathfrak{R} \propto \sum_{n_i} |\mathfrak{F}_{n_i}|^2 |\langle \psi_{f,\text{norm}}|S|\chi_{n_i}\rangle|^2 + \sum_{n_f} |\mathfrak{F}_{n_f}|^2 |\langle \chi_{n_f}|S|\psi_{i,\text{norm}}\rangle|^2 \qquad 68.$$

where $\mathfrak{F}_n$ in the initial or final state is given by

$$\mathfrak{F}_n = \frac{\langle \chi_n|V_{\text{P.V.}}|\psi_{\text{norm}}\rangle}{E_{\text{norm}} - E_n} \qquad 69.$$

Although $|\mathfrak{F}|^2$ is expected to be very small ($\lesssim 10^{-12}$), modern technology allows one to search for parity-forbidden transitions to roughly the required accuracy. It is clearly advantageous to seek cases where irregular states which lie close in energy, to $|\psi_{i,\text{norm}}\rangle$ or $|\psi_{f,\text{norm}}\rangle$, are connected to these states by sizable matrix elements of the parity-violating interaction $V_{\text{P.V.}}$; in such cases $\mathfrak{F}$ may be large compared to $\overline{V}_{\text{P.V.}}/\overline{V}_{\text{hadr}}$ (see Equation 65) which measures the relative strength of the parity-violating force.

The experiments of type 2_P and 3_P generally seek effects due to the interference of electric and magnetic radiations of the same multipole order between discrete states of a system (see Equation 37 and Table II). Circular polarization of a photon, which corresponds to a nonvanishing expectation of the helicity, Equation 8, arises from such an interference. For a parity admixture in the initial (decaying) state, the circular polarization measures the ratio

$$\overline{\mathfrak{F}R} \equiv \left| \frac{\sum_{n_i} \mathfrak{F}_{n_i} \langle \psi_{f,\text{norm}} | \mathcal{O}'(L) | \chi_{n_i} \rangle}{\langle \psi_{f,\text{norm}} | \mathcal{O}(L) | \psi_{i,\text{norm}} \rangle} \right| \qquad\qquad 70.$$

where $\overline{\mathfrak{F}}$ is the average amplitude of the states admixed by the parity-violating force and $\overline{R}$ is the average ratio of the matrix elements of $\mathcal{O}'(L)$ to $\mathcal{O}(L)$, which are multipole operators of opposite parity.

In the case of 3_P, generally the initial state is polarized and one seeks an asymmetry of the rate for emission of a particle with respect to this polarization. This asymmetry corresponds to the expectation of a pseudoscalar ($\langle \mathbf{J}_i \cdot \mathbf{k} \rangle$) and should vanish. If the emitted particle is a photon, then the asymmetry is approximately equal to $\overline{\mathfrak{F}}\,\overline{R}$, Equation 70. In both cases 2_P and 3_P the parity-violating effect can be enhanced by searching for cases where $\overline{R}$ and $\overline{\mathfrak{F}}$ are large. The value of $\overline{R}$ is much larger than unity if the regular parity-allowed matrix element is inhibited by dynamical effects or other selection rules and if the irregular transitions are enhanced. The value of $\overline{\mathfrak{F}}$ is large if states of the opposite parity and the same angular momentum as the normal state are close to each other in energy and are connected by large matrix elements of $V_{\text{P.V.}}$.

3.2.1 *Absolute selection rules.*—To test parity conservation by an absolute selection rule, one must seek a transition that is allowed except for parity conservation. One can use the S-state capture of a spinless particle (e.g. pion) or the emission of such a particle (e.g. alpha). It is advisable to use light nuclear systems with well-separated levels (67, 68). Since these tests require states in the continuum, care must be taken that the emission of an unobservable soft photon to the tails of neighboring levels of opposite parity does not spoil the test (67); good resolution is thus essential.

A particularly appropriate case, and one which has received considerable attention is the α decay of the 8.88 MeV 2^- state in ^{16}O. Since the only energetically allowed α decay is to the 0^+ ground state of ^{12}C, parity conservation can be tested. The 8.88 MeV state in ^{16}O is populated by a 1.1 per cent branch of the β decay of ^{16}N (see Figure 4), usually produced by a ^{15}N(d,p) reaction. The γ-decay branching ratios and the total radiative decay rate of the state have been measured (69). The branching ratio to, and the α-decay width of the 9.58 MeV 1^- state are both known (see Figure 4). This information, together with the number of alphas observed in the energy region of the 8.88 MeV level gives a value or sets an upper limit to Γ_α(8.88 MeV). Before 1965 (70, 72, 73) the best experimental value was (70) $\Gamma_\alpha \leq 6 \times 10^{-9}$ eV; a new

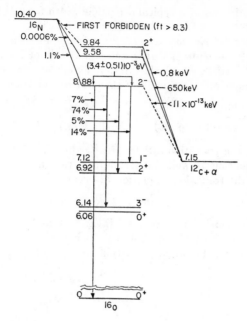

FIG. 4. Energy-level diagram, branching ratios, and approximate absolute widths of relevant $I=0$ states in ^{16}O produced in the β decay of ^{16}N.

measurement (71) gives $\Gamma_\alpha \lesssim 1.1 \times 10^{-9}$ eV. To convert this limit to a ratio $|\bar{\mathfrak{F}}_0|^2$, it is necessary to know a typical α-decay width for allowed transitions of $\Delta l = 2$ in this energy region; estimates (72, 73) set this at $\Gamma_\alpha \approx 6.7$ keV. Thus one finds

$$|\bar{\mathfrak{F}}_0|^2 \lesssim \frac{11 \times 10^{-13}\,\text{keV}}{6.7\,\text{keV}} \approx 3 \times 10^{-13} \qquad 71.$$

where the subscript on $\bar{\mathfrak{F}}$ means that it is a measure of the $\Delta I = 0$ part of $V_{\text{P.V.}}$. Since $\psi_{\text{norm}}(8.88$ MeV$)$ and the ground state of ^{12}C both have $I=0$, only the $\Delta I = 0$ part of $V_{\text{P.V.}}$ contributes to the decay and thus only $\bar{\mathfrak{F}}_0$ is relevant.

Rough theoretical estimates of the parity admixture $\bar{\mathfrak{F}}_0$ for this case have been made by Michel (38) with the potential $V_{\text{P.V.}}{}^{(p)}$, Equations 63, 64. For the levels shown in Figure 4, $\Delta E \equiv E_{\text{norm}} - E_n$ of Equation 69 is roughly $\Delta E \gtrsim 1$ MeV; this gives $|\bar{\mathfrak{F}}_0|^2 \approx 10^{-13}$. Experiments with improved resolution (71) and improved techniques (74) have been undertaken. If and when a parity violation is found, better theoretical calculations should prove valuable in estimating the influence of octet enhancement on the $\Delta I = 0$ component of $V_{\text{P.V.}}$, i.e., Equations 64–65. However, it is also necessary to make more reliable estimates of a typical α-particle width for a 2^+ level at the energy of the 8.88 MeV state.

3.2.2 *Circular-polarization experiments.*—In the absence of parity violation, the matrix elements of $\mathfrak{O}_M(L)$ and $\mathfrak{O}_E(L)$ cannot interfere. With the definitions of Equations 39–40, the interference of these two matrix elements produces a circular polarization $\mathcal{P}\delta$ of the emitted radiation

$$\mathcal{P}_\gamma = \frac{\mathfrak{R}_L(\theta) - \mathfrak{R}_R(\theta)}{\mathfrak{R}_L(\theta) + \mathfrak{R}_R(\theta)} = 2 \frac{\mathrm{Re}\sum_L M(L)^* E(L)}{\sum_L [\,|\,M(L)\,|^2 + \,|\,E(L)\,|^2]} \qquad 72.$$

where $R_{L,(R)}$ is the transition rate for left, (right) circularly polarized radiation. With Equation 70 and the assumption that a unique multipolarity contributes to the transition, Equation 72 becomes

$$\mathcal{P}_\gamma = 2\overline{\mathfrak{F}R} \qquad 73.$$

where we have assumed time-reversal invariance and $\overline{\mathfrak{F}} \ll 1$. For a parity-allowed transition of mixed multipolarity, such as $E(2) + M(1)$, with mixing ratio $\delta = E(2)/M(1)$, the primary parity-admixed multipolarity is $E(1)$ and Equation 72 becomes

$$\mathcal{P}_\gamma = \frac{2\overline{\mathfrak{F}R}}{1 + |\,\delta\,|^2} \qquad 74.$$

As indicated previously, it is advantageous to choose transitions for which the normal parity-allowed rate is severely inhibited and the irregular transition rate is enhanced. In light self-conjugate $(N=Z)$ nuclei this occurs, for instance, through isospin selection rules. The electric dipole matrix element $E(1)$ between two $I=0$ states is reduced 2 orders of magnitude (75). Such transitions are strictly forbidden in the absence of isospin mixing through the Coulomb interaction and other electromagnetic effects. Similarly, $M(1)$ matrix elements between two $I=0$ states are inhibited by roughly 1 order of magnitude (76) [i.e. by $(\mu_p + \mu_n)/(\mu_p - \mu_n) \approx 0.2$, where μ_N is the magnetic moment of the nucleon]. Other dynamical effects, which determine the structures of the states, may inhibit the transition.

Although the experiments are sensitive to first-order effects in $\overline{\mathfrak{F}}$, they remain extremely difficult to carry out. Even if $\overline{R}$ is as large as 100, it is necessary to measure the circular polarization of photons to better than ~ 1 part 10^5. Not only must a huge number of events be accumulated to obtain the necessary statistical accuracy, but it is also crucial to cancel out all possible systematic errors.

All the experiments to date giving more than an upper limit to $\overline{\mathfrak{F}}$ use heavy nuclei. A favorite transition is the 482 keV $\frac{5}{2}^+$ to $\frac{7}{2}^+$ transition to the ground state of ^{181}Ta. The 482 keV state is reached from the β decay of ^{181}Hf. Although the sensitivity is reduced by the large $E(2)$ admixture ($\delta \approx 7$), the $M(1)$ transition is strongly hindered (by about 3×10^6). The interference is with an $E(1)$ transition. Early experiments gave contradictory results, partially due to systematic errors of unknown origin. Boehm & Kankeleit (77) detected the polarization by means of plastic and liquid scintillators subsequent to transmission through magnetized iron. Later measurements by the same authors (78) and by Cruse & Hamilton (85) with improved equipment but similar detection techniques gave $\mathcal{P}_\gamma = -(0.1 \pm 0.4) \times 10^{-4}$, and $-(0.9 \pm 0.6) \times 10^{-4}$ respectively. The measurements for $M(1)$ transitions in ^{175}Lu and ^{203}Tl (78) gave no evidence for circular polarization (see Table V). In the interim, the original experiment was repeated by Bock & Schopper (79), who looked for the circular polarization with a NaI crystal after (forward) Compton scattering from magnetized iron. They obtained $\mathcal{P}_\gamma = (0.3 \pm 2.1) \times 10^{-4}$ and blamed mainly electronic instabilities for the early spurious result of (77). In both cases, effects from bremsstrahlung emitted in the β decay of ^{181}Hf must be removed; since the β-decay energy exceeds the γ-ray transition energy, undesired polarized photons can be obtained from the β transition. These effects are smaller in the NaI crystal than in the plastic scintillator used in (77), where they are of the same order of magnitude as the effect sought. More recently Grodzins & Genovese (80) used resonance absorption (Mössbauer effect) of the $\frac{3}{2}^-$ to $\frac{1}{2}^-$, $M(1)$, 14.4 keV γ ray in ^{57}Fe to test parity conservation. The circular polarization of the photons was detected by separating the $\Delta m = +1$ transitions from the $\Delta m = -1$ ones with a magnetic field. They found $|\mathcal{P}_\gamma| \leq 2 \times 10^{-3}$. The experiment was repeated by Kankeleit (81) who obtained $\mathcal{P}_\gamma = (2 \pm 6) \times 10^{-5}$.

Perhaps the most ingenious experiment to date is that of Lobashov and colleagues (82, 83), who find $\mathcal{P}_\gamma = -(6 \pm 1) \times 10^{-6}$ for ^{181}Ta. Because of the huge number of counts required to obtain a statistical accuracy of 1 part in 10^6, they chose not to detect individual photons, but rather to measure the integrated flux of photons. In such a method no pile-up problems occur and therefore a very intense source ($\sim$500 Ci) of ^{181}Hf can be used. Photon polarization was detected by Compton scattering from iron placed in a magnetic field, whose direction was reversed periodically. This continuous reversal induces an alternating current (frequency $= \frac{1}{2}$ cps) in the detector from the circularly polarized photons. The electric current from the detector was amplified and transformed into a mechanical force that drove a high-Q astronomical pendulum which was tuned to $\frac{1}{2}$ cps. The sign of the polarization is related to the phase of the pendulum's oscillations. Control experiments were carried out with the same setup for γ rays emitted in unhindered transitions [^{46}Sc and ^{82}Br, $E(2)$ and $E(1)$ transitions]. A sketch of the experimental arrangement is shown in Figure 5. This apparatus was also used to investigate the circular polarization of photons emitted in the $\frac{9}{2}^- \rightarrow \frac{7}{2}^+$, 396 keV, transition to the ground state of ^{175}Lu. Lobashov et al. (84) found $\mathcal{P}_\gamma = (4 \pm 1) \times 10^{-5}$ for this case.

The positive results obtained in these experiments are one of the few indications that the current-current interaction, Equation 50, holds for the nonleptonic part of the weak interactions. To determine $\bar{\mathfrak{F}}$ and thus to relate the results to $V_{\text{P.V.}}$, it is necessary to calculate $\bar{R}$. Both $\bar{\mathfrak{F}}$ and $\bar{R}$ are depen-

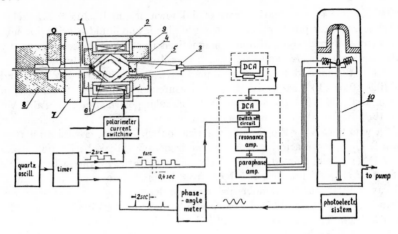

1 - Source.
2 - Polarimeter scattering core.
3 - Photodiode
4 - CsI (Tl) crystal.
5 - Light pipe.

6 - Lead-baffled collimators.
7 - Lead shield.
8 - Shipping container.
9 - Permalloy shield.
10 - Pendulum.

FIG. 5. Experimental arrangement used to detect the circular polarization of the 482 keV γ rays emitted from ^{181}Ta. (From 83.)

dent on nuclear structure effects. Indeed, it is generally not possible to separate $\overline{\mathscr{F}}$ from $\overline{R}$ because several levels with different $E(1)$ matrix elements may contribute. Most theoretical papers (30, 38, 86–91) have concerned themselves with ^{181}Ta, although some have also considered ^{175}Lu and ^{203}Tl (91). In most cases, the contributions of the parity-unfavored transitions were estimated by using Nilsson model wavefunctions (single-particle ones in a deformed potential) for the states and an effective single-particle parity-violating potential. Such a potential is computed from Equations 62–64 by averaging over all core particles,

$$\langle n' \mid V_{\text{P.V.}}{}^{(\text{s.p.})} \mid n \rangle = \sum_i \int d(1)d(2)\phi_{n'}{}^*(1)\phi_i{}^*(2)V_{\text{P.V.}}(1, 2)$$

$$\times \left[\phi_n(1)\phi_i(2) - \phi_n(2)\phi_i(1) \right]$$

75.

where the indicated integrals include appropriate spin and isospin summations and the summation $\sum_i$ is to be taken over all core states. Because of the exchange nature of the parity-violating force, a valence proton only interacts with core neutrons. For this reason, only the last (exchange) term contributes in Equation 75 and no truly single-particle potential can be given, except for a $V_{\text{P.V.}}(\mathbf{r})$ of zero range. The authors of (38, 87, and 89) make this approximation to obtain for the Cabibbo theory

$$V_{\text{P.V.}}{}^{(\text{s.p.})} = G' \mathfrak{d} \cdot \mathbf{p} \left(1 + \tau^{(3)} \frac{N - Z}{A} \right) \qquad 76a.$$

with

$$G' = \frac{\zeta}{2^{1/2} M_N} \left[G\lambda(\mu^v + 1) W_\rho(p) - \frac{fg}{m_\pi{}^2} \tau^{(3)} W_\pi(p) \right] \qquad 76b.$$

where ζ is the nucleon density and W_ρ, W_π are reduction factors which account for hard-core effects. McKellar (89) includes hard-core correlations to obtain average values $\overline{W}_\rho \approx 0.4$, $\overline{W}_\pi \approx 0.1$. Michel (38) and Wahlborn (87) do not include the pion contribution (i.e. they set $g = 0$). Maqueda & Blin-Stoyle (88) do not make the single-particle approximation, but only use a static two-body parity-violating potential. Recently Vinh Mau & Bruneau (90) calculated $\overline{\mathfrak{F}}\overline{R}$ due to the two-particle pion exchange potential $V_{\text{P.V.}}{}^{(\pi)}$ directly from Equations 62 to 70. Their result for $\mathcal{P}_\gamma$, shown in Table V, is considerably smaller than that obtained by McKellar (89) with an effective single-particle parity-violating potential. Although the latter author finds that roughly 80 per cent of $\mathcal{P}_\gamma$ arises from the ρ-exchange contribution and only 20 per cent from the pion (in the Cabibbo theory), there remains a discrepancy of a factor of ~ 6 between the two calculations. This shows the importance of treating the nuclear physics as well as possible.

McKellar (48), Tadic (49), and others have estimated the effects of added currents and other variants of the current-current interaction (see Section 2.4) on $\overline{R}\overline{\mathfrak{F}}$. Some of these findings are included in Table V.

3.2.3 *Asymmetry experiments.*—If the initial state has a polarization $\mathcal{P}_i$, then the correlation $\langle \mathcal{P}_i \cdot \mathbf{k}_f \rangle$, where $\mathbf{k}_f$ is some observed final-state momentum, cannot occur if parity is conserved. That is, there can be no asymmetry of the photon angular distribution relative to a plane perpendicular to $\mathcal{P}_i$. Among the methods used for producing the initial-state polarization are (i) magnetic fields, (ii) β decay, (iii) Bragg reflection of thermal neutrons from magnetized cobalt mirrors. In all of these cases, the subsequent emission of nuclear γ rays (momenta $\mathbf{k}$) serves as analyzer. The asymmetry a in the rate of photon emission

$$\mathcal{R} \propto 1 + a\mathcal{P}_i \cdot \hat{\mathbf{k}} \qquad 77.$$

is directly proportional to the weak-interaction mixing parameter $\overline{\mathfrak{F}}$, and arises from the interference of $E(L)$ and $M(L)$ radiation. The asymmetry is given by

$$a = 2\overline{R}\overline{\mathfrak{F}}\mathcal{Q} \qquad 78.$$

where $\mathcal{Q}$ is a geometrical constant of order unity determined by angular momentum considerations (86). As in the case of circular polarization, it is clearly advantageous to choose nuclei with highly retarded parity-allowed transitions.

TABLE V

EXPERIMENTAL MEASUREMENTS AND THEORETICAL CALCULATIONS OF THE CIRCULAR POLARIZATION OF PHOTONS EMITED IN NUCLEAR TRANSITIONS[a]

Nucleus	Transition	Exp. $\mathcal{P}_\gamma \times 10^5$	Exp. Ref.	Theoretical potential	Theor. $\mathcal{P}_\gamma \times 10^5$	Theor. Ref.
^{57}Fe	$3/2- \xrightarrow[\text{Irreg. } E(1)]{\substack{M(1) \\ 14.4 \text{ keV}}} 1/2-$	2 ± 6	81	Eq. 76 $W_\rho = 1,\, g = 0$	$\lesssim 2.6$	87
^{175}Lu	$9/2- \xrightarrow[\text{Irreg. } M(1)]{\substack{E(1) + M(2) \\ 396 \text{ keV}}} 7/2+$	4 ± 1	84	Eq. 76 $W_\rho = 1,\, g = 0$ Eq. 76, $W_\rho = 0.4,\, W_\pi = 0.1$ $\left.\right\}$ $g = 4.2 \times 10^{-8}$	$\pm(9 \pm 6)$ $\pm(3 \pm 2)$ $\left.\right\}$ $\pm(4.5 \pm 3)$	38 89
^{175}Lu	$5/2+ \xrightarrow[\text{Irreg. } E(1)]{\substack{E(2) + M(1) \\ 343 \text{ keV}}} 7/2+$	2 ± 3	78	Eq. 76, $W_\rho = 0.4,\, W_\pi = 0.1$ $\left.\right\}$ $g = 4.2 \times 10^{-8}$	-1 ± 0.5 $\left.\right\}$ -1.5 ± 1	89
^{181}Ta	$5/2+ \xrightarrow[\text{Irreg. } E(1)]{\substack{E(2) + M(1) \\ 482 \text{ keV}}} 7/2+$	3 ± 21 $-(1 \pm 4)$ $-(0.6 \pm 0.1)$ $-(9 \pm 6)$	79 78 82, 83 85	Eq. 76, $W_\rho = 1,\, g = 0$ Eq. 76, $W_\rho = 0.4,\, W_\pi = 0.1$ $\left.\right\}$ $g = 4.2 \times 10^{-8}$ Eq. 76, $W_\rho = 0.4,\, W_\pi \approx 0.1$ ($g \approx 8 \times 10^{-7}$, extra currents) $\left.\right\}$ Eq. 62, $g = 4.2 \times 10^{-8}$ $\left.\right\}$ Eq. 62, with extra currents, $g \approx 8 \times 10^{-7}$	-6 ± 3 -2 ± 1 $\left.\right\}$ -3 ± 1 8 ± 1 $-(10 \pm 5)$ -0.08 $\left.\right\}$ -1.55	87 89 90
^{203}Tl	$3/2+ \xrightarrow[\text{Irreg. } E(1)]{\substack{E(2) + M(1) \\ 273 \text{ keV}}} 1/2+$	-2 ± 3	78	Eq. 76, $W_\rho = 0.4,\, W_\pi = 0.1$ $\left.\right\}$ $g = 4.2 \times 10^{-8}$	$-(3 \pm 1)$ $\left.\right\}$ $-(4.5 \pm 2)$	89

Early experiments (92, 93) only were able to set upper limits on $\bar{\mathcal{F}}$. Boehm & Hauser (93) used β decays to populate excited nuclear states. They found $a = 0$ to within the accuracy of their measurements.

Of most interest are the experiments carried out by Abov and his colleagues (94, 95) and by Warming et al. (96) with the 9.05 MeV $1^+ \to 0^+$ transition to the ground state of ^{114}Cd. The initial state is polarized by the resonant capture of thermal neutrons, which are magnetized ($\mathcal{P}_n \geq 0.75$) by scattering from magnetized Co mirrors. The 9.05 MeV state decays primarily to highly excited states but has a small measured branch ($\sim 1.4 \times 10^{-3}$ per capture) to the ground state. The parity-allowed multipolarity is pure $M(1)$, and the parity-admixed transition is $E(1)$. Because the 1^+ state lies above the neutron separation energy there are many nearby 1^- states. Estimates of $\bar{\mathcal{F}}$ made by Haas et al. (97), who were the first to perform the experiment, gave $\bar{\mathcal{F}} \sim 10^3 \langle V_{\text{P.V.}} \rangle / \langle V_{\text{hadr}} \rangle$ where $\langle V_{\text{P.V.}} \rangle$ and $\langle V_{\text{hadr}} \rangle$ are average strengths of the parity-violating and parity-conserving matrix elements. Blin-Stoyle (30) used a harmonic-oscillator single-particle model to obtain the more reasonable estimate $\bar{\mathcal{F}} \approx 50 \langle V_{\text{P.V.}} \rangle / \langle V_{\text{hadr}} \rangle$. Together with an enhancement factor $\bar{R} \approx 10$ for the ratio of single-particle $E(1)/M(1)$ reduced matrix elements, this gives

$$a(\text{Cd}) \approx 10^3 \langle V_{\text{P.V.}} \rangle / \langle V_{\text{hadr}} \rangle \sim 10^{-4} \qquad 79.$$

The magnitude of the asymmetry is thus considerably enhanced for this transition, which explains why it is chosen. The results of various experiments and theories are summarized in Table VI.

The experimental setup used by Haas et al. (97) is shown in Figure 6; Abov et al. (94, 95) and Warming (96) used similar arrangements. Abov et al. used a rotating magnet and current driving foil to change the direction of polarization of the neutrons periodically. The asymmetry was detected by

TABLE VI

Measurements and Theoretical Calculations of the Asymmetry for the $1^+ \to 0^+$, 9.05 MeV, $M(1)$ Transition from Polarized ^{114}Cd[a]

Es accepted (MeV)	$a_{\text{exp}} \times 10l$	Ref.	Theoretical potential	a_{th} Ref. $\times 10^4$
8.3 – 9.3	1.2 ± 7.8	97	Eq. 76, $W_\rho = 1$, $g = 0$	± 6 38, 97
8.1 – 9.4	$-(3.7 \pm 0.9)$	94	Eq. 76, $W_\rho = 0.4$, $W_\pi = 0.1$ $\left.\begin{array}{l} \\ \\ \end{array}\right\}$ 89	$\left.\begin{array}{l} \pm 2 \\ \pm 3 \end{array}\right\}$
			$g = 4.2 \times 10^{-8}$	
8.5 – 9.5	$-(3.5 \pm 1.2)$	95	Eq. 76, $W_\rho = 0.4$, $W_\pi = 0.1$	± 8
			(extra currents, $\left.\begin{array}{l} \\ \\ \end{array}\right\}$	$\left.\begin{array}{l} \\ \pm 10 \end{array}\right\}$
8.8 – 9.5	$-(2.5 \pm 2.2)$	96	$g \approx 8 \times 10^{-7}$)	

[a] The two values for a_{th} in the second and third lines are for different signs of the interference between the π- and ρ-meson contributions to $V_{\text{P.V.}}$.

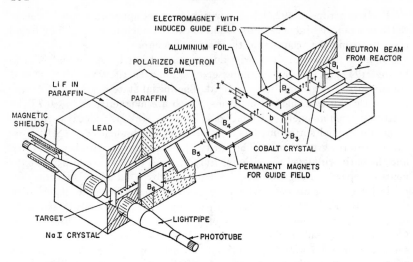

FIG. 6. Experimental arrangement used to measure the asymmetry of γ rays emitted
in the capture of polarized neutrons. (From 97.)

counting photons, in the energy interval of 8.1–9.4 MeV bracketing the
transition energy, in two scintillation counters at 0° and 180° relative to the
neutron spin direction. Instrumental instabilities were checked by alternating runs with polarized and unpolarized neutrons. Violations of the expected
order of magnitude were found for the ^{114}Cd energy region around 9 MeV
and are summarized in Table VI. Many control experiments were run, including ones with lower-energy transitions in ^{114}Cd dominated by $E(1)$
matrix elements and with other nuclei. These gave no asymmetry.

Theoretical calculations of the asymmetry in the ^{114}Cd transition are
similar to those for the circular polarization in ^{181}Ta and are summarized in
Table VI. All authors use an effective single-particle potential for $V_{\text{P.V.}}$ and
a single-particle model to estimate $\bar{R}$. As shown by the last entry in a_{th} the
results are not as sensitive to $V_{\text{P.V.}}^{(\pi)}$ and thus to various current-current
interaction models (43, 46–52) as the circular polarization of the γ rays
from ^{181}Ta.

3.3 PRESENT STATUS OF P VIOLATION; PROPOSED EXPERIMENTS

The experiments carried out in the last few years seem to show consistent
positive results for a parity-violating weak force of the order of magnitude
expected from the normal Cabibbo current-current interaction. If the nuclear
theory is assumed to be reliable, then additional currents (46) and most other
modifications of the weak currents (47, 98–99) would be ruled out [an exception is the γ_5-invariant theory of (99)]. However, Vinh Mau & Bruneau (90)
recently showed that this reliability has not yet been reached, so that the
above conclusion is still premature.

Proposals for different experiments to study parity noninvariance have been made from time to time (100). Light nuclei offer several advantages. In general, the structure of these nuclei is better known. Furthermore, with these targets one can separate the $\Delta I = 1$ part of $V_{\text{P.V.}}$ (i.e., $V^{(\pi)}$, Equation 62) from the $\Delta I = 0$, 2 parts [e.g., $V^{(\rho)}$, Equation 63].

Although such experiments have been attempted, no positive results have yet been obtained. The thermal capture of polarized neutrons by deuterons was attempted (101) and gave an asymmetry, $a = (0.78 \pm 1.55) \times 10^{-4}$. The experiment is being repeated by R. Wilson et al. (102) and by Lobashov. A theoretical analysis of this reaction was recently carried out by Moskalev (103), who predicts a circular polarization $\mathcal{P}_\gamma \approx 10^{-6} - 10^{-7}$ for photons from unpolarized neutrons, and an asymmetry $a \sim 10^{-6}$ from polarized-neutron capture (for the Cabibbo theory). Both $\Delta I = 0$ and $\Delta I = 1$ parts of the parity-violating force contribute to both $\mathcal{P}_\gamma$ and a. Among other light nuclear reactions studied is $^3\text{H}(d,n)^4\text{He}$ with deuterons polarized by an atomic-beam magnetic-resonance method (104). The lack of observed asymmetry ($\propto \mathcal{P}_d \cdot \mathbf{k}_f$) only limits $\underset{\sim}{\mathcal{F}}$ to be $\lesssim 3.8 \times 10^{-3}$.

The most fundamental nuclear experiment is a study of neutron capture by protons

$$n + p \rightarrow d + \gamma$$

or its inverse (105–108). It was shown by Danilov (107) that here $\mathcal{P}_\gamma$ is only sensitive to the $\Delta I = 0$ part of $V_{\text{P.V.}}$ and that the asymmetry a of photons obtained from the capture of polarized neutrons only depends on the $\Delta I = 1$ part of $V_{\text{P.V.}}$. The reason for this conclusion can be understood by considering the relevant matrix elements. At very low neutron energies, capture occurs only from S states. The three possible matrix elements are then

$$\mathfrak{M} = A\left[{}^1S_0(I = 1) \xrightarrow{M(1)} {}^3S_1(I = 0)\right]$$

$$+ \mathcal{F}_1\alpha_1\left[{}^3S_1(I = 0) \xrightarrow{E(1)} {}^3P_1(I = 1)\right]$$

$$+ \mathcal{F}_0\alpha_0\left[{}^1S_0(I = 1) \xrightarrow{E(1)} {}^1P_1(I = 0)\right]$$

where A is the amplitude of the parity-favored $M(1)$ transition. The amplitudes for the $E(1)$ transitions are labeled α_0, α_1, where the subscripts refer to the isospin of the final state admixed to the deuteron. Since the isospin of the normal deuteron is $I = 0$, this subscript is also the change of isospin induced by $V_{\text{P.V.}}$. The circular polarization arises from coherence among possible final states. Since the initial $^3S_1$ and $^1S_0$ state contributions add incoherently, the interference responsible for $\mathcal{P}_\gamma$ occurs between the final $^1S_1$ and $^1P_1$ states, so that $\mathcal{P}_\gamma \approx \mathcal{F}_0\alpha_0/A$, and thus measures $V_{\text{P.V.}}(\Delta I = 0)$. On the other hand, for the capture of polarized neutrons, the final $^3S_1$ and $^1P_1$ states add incoherently and it is the interference of α_1 and A which is responsible for the asymmetry $a(\approx \mathcal{F}_1\alpha_1/A)$. The estimates made by Danilov (107) are $\mathcal{P}_\gamma \approx 10^{-6} - 10^{-7}$ and $a \approx 10^{-8}$ for the Cabibbo theory. These numbers are smaller than those in Tables V and VI because the parity-favored $M(1)$ transition is not hindered. Even with extra currents, the asymmetry experiments are probably beyond present experimental techniques, since a substantial flux of polarized neutrons is required. The circular polarization might be somewhat larger than the above estimate through octet enhancement. Lobashov (108) has suggested that this polarization might be detectable by using neutrons from a reactor and extending the techniques for ^{181}Ta. The experiment would require roughly 1 order of magnitude higher sensitivity than that for ^{181}Ta. Dal'karov (109) has recently considered the inverse reaction $\gamma + d \rightarrow n + p$ with incident polarized radiation, but this is even more difficult experimentally.

A different experiment to isolate the $\Delta I = 1$ part of $V_{P.V.}$ has been proposed recently (110). The suggestion is to search for the circular polarization in retarded electromagnetic transitions of self-conjugate nuclei (51). As discussed earlier, $E(1)$ is hindered by 2 orders of magnitude (e.g. by $\alpha = 1/137$) and $M(1)$ by 1 order of magnitude for transitions among $I = 0$ levels of these nuclei. Two particularly suitable cases appear to be the $E(1)$ 1.080 MeV $0^- \rightarrow 1^+$ transition to the ground state of ^{18}F and the 4.388 MeV $E(1)$ $2^- \rightarrow 1^+$ transition from the 5.105 MeV level to the 0.717 state in ^{10}B. In both cases the admixture of other $I = 0$ states is shown to make negligible contribution to the circular polarizations so that these experiments are mainly sensitive to $V_{P.V.}^{(\pi)}$. In ^{18}F the $\Delta I = 1$ part of $V_{P.V.}$ is especially effective because there is a 0^+, $I = 1$, state at 1.043 MeV which is readily admixed to the 0^-, $I = 0$ state at 1.080 MeV. Rough estimates give $\mathcal{P}_\gamma = \mathcal{F}_1 R_1 \approx 10^{-5}$ for the Cabibbo theory in this instance ($\mathcal{P}_\gamma \approx 2 \times 10^{-4}$ with extra currents). Improved calculations are being carried out.

3.4 TIME-REVERSAL EXPERIMENTS

The classification of possible T experiments in nuclear physics has been given in Subsection 1. Here, we shall discuss these tests in more detail, describe experiments, and analyze the results in terms of both the general framework of Section 2 and the specific model presented in 2.5. Because tests of T invariance search for possible violations of <1 per cent, extreme care is required in the control of the experiments and in the theoretical analyses thereof. It is considerably more difficult to eliminate systematic uncertainties in these experiments than in those which test P conservation. Consequently, the upper limit of T-invariance violation obtained ($\lesssim 3 \times 10^{-3}$) is considerably higher than actual parity violations observed.

3.4.1 *Weak-interaction tests of* T.—It is in the weak decay of the K^0 system that CP invariance was first noted to fail. Do the decays of other hadrons show this asymmetry or a violation of T? T violation has been sought, for instance, in the decay $\Lambda^0 \rightarrow p + \pi^-$ (111). In the absence of final-state interactions between the proton and pion, one can use the first-order Born approximation. The Hermiticity of the interaction Hamiltonian together with T invariance shows that the polarization of the proton cannot have a component along $\mathbf{k}_p \times \boldsymbol{\mathcal{P}}_\Lambda$, where $\mathbf{k}_p$ is the momentum of the proton in the rest system of the Λ^0 and $\boldsymbol{\mathcal{P}}_\Lambda$ is the latter's polarization; that is, there can be no term in the S matrix for the decay proportional to $\langle \boldsymbol{\mathfrak{d}}_p \rangle \cdot \langle \boldsymbol{\mathfrak{d}}_\Lambda \rangle \times \mathbf{k}_p$ (see e.g. Table IV, which shows the absence of such a term for a Hermitian potential). Hermiticity is spoiled by the final-state interaction, but a correction for this effect can be applied if the $I = \frac{1}{2}$, S- and P-wave pion-nucleon phase shifts δ_S and δ_P are known; it is these interfering partial waves which permit T invariance to be tested in the Λ^0-decay amplitude. From elastic π-N scattering experiments, the difference of the phase shifts is deduced to be $(\delta_S - \delta_P) = 6.4 \pm 1.7°$. The measured coefficient of the term proportional to $\mathbf{k}_p \times \boldsymbol{\mathcal{P}}_\Lambda$ gives (111) $(\delta_S - \delta_P) = 9.0 \pm 3.8°$, which corresponds to a time-reversal phase $\phi = 2.6 \pm 4.2°$. Thus, within the errors, there is no evidence for a violation of T invariance.

More accurate measurements are possible in nonstrangeness-changing weak interactions, such as β decay. Since β decay occurs in the presence of both electromagnetic and strong interactions, one may be able to detect an effect no matter where the source of the T violation resides. If a violation is found, further experiments are required to sort out its origin.

Possible β-decay experiments were first discussed by Jackson, Treiman & Wyld in 1957 (112). Hermiticity of the weak-interaction Hamiltonian allows T invariance to be tested by[9]:

(*i*) Observations of recoil and electron momenta from a polarized nucleus. If T invariance holds, there must be as many electrons and neutrinos correlated so that $\mathbf{k}_e \times \mathbf{k}_\nu$ is parallel to the polarization of the nucleus $\boldsymbol{\mathcal{P}}_i = \langle \mathbf{J} \rangle$, as there are with $\mathbf{k}_e \times \mathbf{k}_\nu$ antiparallel to $\boldsymbol{\mathcal{P}}_i$. Thus, to search for a violation one searches for an asymmetry in the decay rate proportional to $\mathbf{J} \cdot \mathbf{k}_e \times \mathbf{k}_\nu$.

(*ii*) Measurements of the polarization of the electrons emitted by an unoriented nucleus together with the momenta of the electron and recoil nucleus, i.e., detection of a term proportional to $\boldsymbol{\mathfrak{d}} \cdot \mathbf{k}_e \times \mathbf{k}_\nu$ in the decay rate.

(*iii*) Observations of the electron polarization from an oriented parent nucleus, together with the measurement of the electron's momentum, i.e., an asymmetry $\mathbf{J} \cdot \boldsymbol{\mathfrak{d}} \times \mathbf{k}_e$ (since parity is not conserved in the weak decay, the fact that this term does not conserve parity is not detrimental).

(*iv*) Measurements of higher correlations, such as

$$(\boldsymbol{\mathfrak{d}} \cdot \mathbf{k}_e)\,(\mathbf{J} \cdot \mathbf{k}_e \times \mathbf{k}_\nu) \quad (113)$$

All of the experiments are designed to detect the nonreality of the matrix elements of a part of the weak-interaction Hamiltonian relative to other parts. Thus, as in all tests of T invariance an interference between two com-

---

[9] We omit expectation value symbols, $\langle \rangle$, henceforth, unless they are required to avoid confusion.

peting matrix elements is required. These occur between the possible types of couplings (e.g., vector, axial vector) which contribute to the β-decay Hamiltonian. If the interaction is purely of the V-A form, as appears to be the case, the interference terms (ii) and (iii) vanish and only (i) is sensitive to T-odd terms in the weak interaction. This conclusion remains valid even if the violation of T occurs in the electromagnetic or strong interactions, since the scalar and tensor couplings appear to be absent.

The rate $\mathfrak{R}$ of a transition from a state of angular momentum j to one of i' is

$$\mathfrak{R} \propto 1 + \frac{D\langle\mathbf{J}\rangle}{j} \cdot \mathbf{v}_e \times \hat{\mathbf{k}}_\nu + \text{further terms} \qquad 80.$$

where $\mathbf{v}_e$ is the velocity of the emitted electrons, $\mathbf{J}$ is the angular momentum operator whose expectation value is to be calculated for the state of angular momentum j, and the asymmetry coefficient D is given by

$$D = -4\delta_{j,j'}\left(\frac{j}{j+1}\right)^{1/2} \text{Im} \left(C_V C_A{}^* M_F M_{GT}{}^*\right)$$

$$= -4\delta_{j,j'}\left(\frac{j}{j+1}\right)^{1/2} C_V C_A M_F M_{GT} \sin\phi$$

A nonvanishing coefficient D requires a mixed Fermi-Gamow-Teller transition, the coupling constants for which are C_V and C_A, with nuclear matrix elements M_F and M_{GT}, respectively. The phase angle ϕ is the sum of that between C_V and C_A and between M_F and M_{GT}; $\sin\phi$ is zero if T invariance holds in both weak and hadronic interactions.

The first experiments consisted of studies of the β decay of polarized neutrons (114) and gave $\phi = 175° \pm 6°$. A more refined measurement of the angular correlation has recently been done by Erozolimsky et al. (115). An intense neutron beam is polarized by reflection from magnetized cobalt mirrors (see Section 3.2b). The spin direction is changed by nonadiabatic inversion of the magnetic field; the recoil protons are detected in two CsI scintillation counters, and the electrons by two sets of plastic scintillators, both sets of counters being placed symmetrically with respect to the target. The use of pairs of counters minimizes instrumental asymmetries. The asymmetry coefficient D was found to be $D = 0.01 \pm 0.01$, corresponding to a phase $\phi = 178.7° \pm 1.3°$.

Calaprice et al. (116) used the positron decay of ^{19}Ne, ^{19}Ne$\rightarrow^{19}$F$+e^+ +\nu$ to search for the asymmetry implied by a violation of T. The choice of ^{19}Ne was dictated by the requirement of a gaseous source in order to be able to measure the recoil and by the large mixture of Gamow-Teller and Fermi transitions ($|M_{GT}|/|M_F| = 1.5 \pm 0.1$), as well as by a reasonable lifetime ($\tau_{1/2} = 18.5$ sec). The ^{19}Ne was produced by the reaction ^{19}F$(p,n)^{19}$Ne, and was polarized in a Stern-Gerlach apparatus. The polarization and beam intensity were continuously monitored. The observed correlation $D = 0.002 \pm 0.014$ (the error is purely statistical) limits the phase angle ϕ of Equation 80 to $\phi = 180.2° \pm 1.6°$ consistent with T conservation.

In the absence of any definite theoretical framework, it should simply be noted that the maximum T-odd phase angle ϕ is $90°$. Thus the limit on time-reversal odd forces set by the above experiments is roughly $\mathfrak{F}_T \lesssim 0.01$, where $\mathfrak{F}_T$ is given by

$$\mathfrak{F}_T \equiv \left| \frac{\overline{V}_{T-\mathrm{odd}}}{\overline{V}_{T-\mathrm{even}}} \right| \approx \frac{\phi}{\pi/2} \qquad 81.$$

Although the experimental findings show no T violation, it is important to realize that a nonvanishing value of $\sin \phi$ could arise, in principle, as a result of the final-state scattering of the final electrons or positrons by the Coulomb field of the daughter nucleus. However, as shown by Jackson, Treiman & Wyld (112), with the neglect of recoil effects, the Coulomb potential corrections of order $Z\alpha m_e/k_e$ vanish if the weak interaction is purely of the V-A form. Recoil corrections are expected to be negligible, being of order $Z\alpha k_e/A\,M_N \lesssim 10^{-5}$, where $A\,M_N$ is the mass of the nucleus. However, there are non-Coulombic recoil corrections (117): the most important is an electromagnetic weak-magnetism type of correction, which is of order $(Z\alpha k_e/M_N)(\mu_i - \mu_f)$ for self-conjugate nuclei, where μ_i and μ_f are the magnetic moments of the initial and final nuclear states in nucleon magnetons. For the positron decay of ^{19}Ne this correction has been calculated (117), and gives $D_{\mathrm{recoil}} = 2.6 \times 10^{-4})\,(k_e/k_{e,\mathrm{max}})$; for the β decay of the neutron, we find that D_{recoil} is ~ 10 times smaller. These weak magnetism corrections are roughly 50 times (^{19}Ne) and 500 times (n) smaller than the present experimental errors. They may, nevertheless, have to be considered in the future, if significant improvements are made in the measurements.

3.4.2 *Electromagnetic tests.*—Electromagnetic tests of T include all experiments that make use of the electromagnetic interaction, be it through (i) the emission or absorption of photons, (ii) the scattering of photons, or (iii) the virtual exchange of photons as in electron scattering. Because the present evidence for charge-conjugation invariance of the electromagnetic interaction (current) of the hadrons is poor (58–59), experiments on T invariance are crucial.

Possible high-energy tests of class (i), above, include detailed-balance experiments, such as

$$\gamma + d \rightleftarrows n + p$$
$$\gamma + {}^3\mathrm{He} \rightleftarrows p + d \qquad 82.$$
$$\gamma + {}^3\mathrm{He} \rightleftarrows \pi^+ + {}^3\mathrm{H}$$

The first of these, suggested by Barshay (118), has been performed during the past few years. The tests are made at photon energies around 300 MeV where the contribution from the intermediate $\frac{3}{2}$, $\frac{3}{2}N^*$ resonance is large. This contribution ensures that part of the matrix element for the hadronic current is far off the mass shell, as required by the theory of Section 2.5 for sensitivity to T-odd effects. The photodisintegration of the deuteron has been examined most recently by a group at Stanford (119) in the energy region of 220 to 340 MeV with the detailed-balance experiment in mind. The inverse neutron capture experiment has been carried out independently at the Princeton-Pennsylvania accelerator (120) and at Berkeley (121). Detection of the photon-vector direction and deuteron momentum in the latter experiment determines the energy of the incident neutron despite the spread (± 50 MeV) of the incident beam. Care must be taken to distinguish directly produced photons from those emitted in the decay of the π^0 because of the large cross section for the background reaction $n+p \rightarrow \pi^0 + d$. Coplanarity and the different masses of the photon and π^0 are used by both groups to eliminate this background. Preliminary evidence reported (120) by the Princeton group indicates a possible deviation from detailed balance; the violation of T invariance, if any, found in the Berkeley experiment is smaller. The important results of these tests must await the full analysis of the experiments.

An example of a high-energy test of class (ii) is that suggested by Lipshutz (65); it makes use of Compton scattering from polarized protons to search for a correlation $\mathbf{\sigma}_p \cdot \mathbf{k}_i \times \mathbf{k}_f$. Such tests have not yet been carried out. Class (iii) tests seek similar correlations in reactions induced by charged leptons from or to polarized states. Reactions which have been proposed (62, 122) and used (123, 124) for this purpose are

$$e + p \rightarrow N^* + e \qquad\qquad 83a.$$
$$e + d \rightarrow e + d \qquad\qquad 83b.$$

from polarized targets or to polarized final states. In the study of reaction 83a, electrons were inelastically scattered from a (alcohol-water) target with protons polarized perpendicular to the reaction plane. Inelastically scattered electrons corresponding to excitations of the 1236 MeV ($j^P=\frac{3}{2}^+$, $I=\frac{3}{2}$); 1512 MeV ($j^P=\frac{3}{2}^-$, $I=\frac{1}{2}$); and 1688 MeV ($j^P=\frac{5}{2}^+$, $I=\frac{1}{2}$) resonances were detected. On reversal of the polarization no change in intensity for large four-momentum transfers was observed to within the experimental accuracy of 5–10 per cent. In the test which used reaction 83b, a vector deuteron polarization perpendicular to the scattering plane was sought. Incident electrons of 1 GeV were scattered from a liquid deuterium target at an angle corresponding to a deuteron four-momentum of 721 MeV/c; the deuteron polarization was analyzed by a measurement of the asymmetry after scattering from carbon. The polarization measured was $|\mathcal{P}_d| = 0.075$ ± 0.088, consistent with T invariance.

A somewhat different test of type (iii), involving virtual photons, uses the reaction $\pi^- + p \rightarrow n + e^+ + e^-$. The expectation value of $\mathbf{\sigma}_n \cdot (\mathbf{p}_+ \times \mathbf{p}_-)$ $\cdot (E_+ - E_-)$, with $+$ and $-$ subscripts referring to positrons and electrons, vanishes if T invariance holds. The asymmetry corresponding to this expectation value was recently found (125) to be -0.11 ± 0.13, consistent with zero.

The experiments listed above are all high-energy tests of T. As such, they may be sensitive to the matrix elements of the electromagnetic current far off the mass shell. However, this advantage must be balanced against the difficulty of achieving high precision, which can be achieved more readily by low-energy electromagnetic transitions between nuclear states. Suggested tests (126–128) are to employ the interference between two competing multipoles [usually $E(2)$ and $M(1)$] to measure the relative phase ϕ of the two

TABLE VII

TESTS OF TIME-REVERSAL INVARIANCE USING A MIXED-GAMMA-RAY TRANSITION[a]

γ-Ray polarization	Quantity measured	Degree of orientation required	
		Ω_a	Ω_b
None	$(\mathbf{k}\cdot\mathbf{j}_b)(\mathbf{k}\cdot\mathbf{j}_b\times\mathbf{j}_a)$	1	2
	$(\mathbf{k}\cdot\mathbf{j}_a)(\mathbf{k}\cdot\mathbf{j}_a\times\mathbf{j}_b)$	2	1
Circular	$(\mathbf{k}\cdot\boldsymbol{\sigma}\times\mathbf{k}\cdot\mathbf{j}_a\times\mathbf{j}_b)$	1	1
	$(\mathbf{k}\cdot\boldsymbol{\sigma})(\mathbf{k}\cdot\mathbf{j}_a\times\mathbf{j}_b)(\mathbf{j}_a\cdot\mathbf{j}_b)$	2	2
Linear	$(\mathbf{k}\cdot\mathbf{j}\times\boldsymbol{\epsilon})(\mathbf{k}\cdot\mathbf{j})(\boldsymbol{\epsilon}\cdot\mathbf{j})$,	3	0
	with a and b in various	2	1
	arrangements	1	2
		0	3

[a] From Reference (128).

matrix elements. As shown in Section 2, this phase must be 0° or 180° if T invariance is valid. Although measurements of $\cos\phi$ can be made (128), we shall restrict ourselves to first-order effects, proportional to $\sin\phi \approx \phi$, because the expected violation is small ($\sim 10^{-3}$). Unlike the situation for electromagnetic transition tests of parity conservation, the factor R (Equation 70) is of the order of unity in T-conservation tests, since both multipole transitions are allowed ones.

To detect the sine of the phase angle between two competing multipoles, it is necessary to measure correlations between the photon momentum and other observables. Because parity is known to hold to roughly 1 part in 10^5 of the strong interactions, it is generally assumed that the correlations should be invariant under P. Possible terms are listed in Table VII (128). The orientations of the state a can be produced by a magnetic field or by a preceding β decay. The analyzer of the polarization or more general orientation of the final state b can be a succeeding radiative transition to a state of lower energy. Other possible polarizers and analyzers are given in (128).

Early experiments used a β decay to polarize the initial state a and a second γ-ray transition to detect the polarization of the state b; attempts were made to observe a term such as that appearing in the first line of Table VII. The experiment which gave the smallest uncertainty in phase was that of Fuschini et al. (129) with $\sin \phi = (3 \pm 4) \times 10^{-2}$; previous experiments are listed in this reference. Subsequent to the observation of CP breakdown in the K_2^0, tests were undertaken with renewed vigor. These new experiments and their determination of the phase ϕ are summarized in Table VIII. The experiment by Garrell et al. compares the angular correlations in two mixed $2^+ \rightarrow 2^+(0^+)$ transitions of the same nucleus. These transitions have mixing ratios of opposite sign, so that a comparison reduces instrumental asymmetries. The difference of the phases listed in the first lines of Table VIII gives $\Delta \left| \delta \sin \phi \right| / \left| \delta_{\text{average}} \right| \approx (0.4 \pm 2.6) \times 10^{-2}$.

As in the parity experiments one can use the capture of reactor neutrons (see Figure 6), which are polarized by reflection from cobalt mirrors, to serve as the source of polarization. Recent accurate experiments (130, 131) with this technique have managed to lower the limit of the phase angle between an $E(2)$ and $M(1)$ radiation in ^{36}Cl to $(0.08 \pm 0.23) \times 10^{-2}$ (131), an order of magnitude improvement over the earlier experiments.

Other accurate tests of T in nuclear γ-ray transitions make use of the Mössbauer effect and set an upper limit of roughly 3×10^{-3} on the T-odd phase ϕ. These experiments measure the angular correlation of linearly (in fact elliptically) polarized photons with a nuclear orientation, i.e., the term shown on line 5 of Table VII. The schematic arrangement of Kistner's experiment (132) is shown in Figure 7. In it, he seeks an angular correlation from the absorption of elliptically polarized γ rays. The source is ^{99}Rh in a Ru powder. The elliptic polarization of the γ rays is achieved by interposing a stationary Ru-Fe plate identical to another one in which the photons are absorbed. Both plates are placed in a magnetic field sufficiently strong to fully resolve the hyperfine structure. The $M(1) + E(2)$ transitions observed were those from the 90 keV, $j = \frac{3}{2}$, $m = \pm \frac{3}{2}$ states to the ground state of $j' = \frac{5}{2}$, $m' = \pm \frac{5}{2}$. The reversal of m and m' or that of the magnetic field at the absorber causes a reversal in sign of the term $(\varepsilon \cdot \mathbf{k} \times \mathbf{j})\,(\mathbf{k} \cdot \mathbf{j})\,(\varepsilon \cdot \mathbf{j})$. It is this reversal which is sought. A Faraday rotation of the polarization is introduced because the magnetic field on the second absorber is not perpendicular to the direction of propagation of the photons (see Figure 7). This Faraday rotation mimics the T-odd term when the magnetic field is reversed but not when $m = \frac{3}{2}$, $m' = \frac{5}{2}$ is changed to $m = -\frac{3}{2}$ and $m' = -\frac{5}{2}$. The results quoted in Table VIII are those obtained after this Faraday rotation effect has been subtracted.

In the emission experiment of Atac et al. (133), the Mössbauer transition was the mixed $E(2) + M(1)$, 73 keV, $\frac{1}{2}^+ \rightarrow \frac{3}{2}^+$ transition in ^{193}Ir. In this case, the source was ^{193}Os; this source and the absorber were both embedded in iron and placed in a magnetic field. The γ rays are emitted with elliptic

TABLE VIII

RECENT TESTS OF T INVARIANCE IN NUCLEAR GAMMA-RAY TRANSITIONS[a]

Polarizer	Analyzer	Correlation sought	Nucleus	Transition	$\sin \phi \times 10^2$	Ref.
Beta decay	Subsequent $E(2)$ transition	$(\mathbf{k}\cdot\mathbf{j}_b)$ $(\mathbf{k}\cdot\mathbf{j}_b \times \mathbf{j}_a)$	^{56}Mn	$2^+ \xrightarrow[2.12\ \text{MeV}]{\delta=-0.28} 2^+$ $2^+ \xrightarrow[1.81\ \text{MeV}]{\delta=0.18} 2^+$	2.6 ± 1.4 -4.5 ± 2.7	134 134
			^{106}Pd	$2^+ \xrightarrow[1.050\ \text{MeV}]{\delta=-0.2} 2^+$	(0.4 ± 1.8)	135
Capture of polarized thermal neutrons	Subsequent $E(2)$ transition	$(\mathbf{k}\cdot\mathbf{j}_b)$	^{49}Ti	$1/2^- \xrightarrow[0.34\ \text{MeV}]{\delta=2.2} 3/2^-$	1.7 ± 2.5	130
	Subsequent $E(2)+M(1)$ transition	$(\mathbf{k}\cdot\mathbf{j}_b \times \mathbf{j}_a)$	^{36}Cl	$2^+ \xrightarrow[7.79\ \text{MeV}]{\delta=0.21} 3^+$	0.08 ± 0.23	131
Ru absorber	Resonant (Mössbauer) absorption	$(\mathbf{k}\cdot\mathbf{j} \times \boldsymbol{\epsilon})$	^{99}Ru	$3/2^+ \xrightarrow[90\ \text{keV}]{\delta=-1.64} 5/2^+$	0.10 ± 17 $[0 \pm 0.17]$	132
IR absorber	Resonant emission	$(\mathbf{k}\cdot\mathbf{j})(\boldsymbol{\epsilon}\cdot\mathbf{j})$	^{193}Ir	$1/2^+ \xrightarrow[73\ \text{keV}]{\delta=0.556} 3/2^+$	0.11 ± 0.38 $[0.02 \pm 0.38]$	133

[a] All transitions are mixed $E(2)+M(1)$, with δ the mixing ratio $E(2)/M(1)$ The numbers for $\sin \phi$, corrected for internal conversion effects, are shown in square brackets.

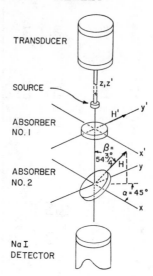

FIG. 7. Arrangement of the apparatus of the Mössbauer absorption experimental test
of time-reversal invariance. (From 132.)

polarization, which is detected in the absorber. Because the magnetic field
on the absorber was perpendicular to the direction of propagation of the
photon, no Faraday rotation was introduced. The experiment is thus complementary to that of Kistner and gives $\phi = (1.1 \pm 3.8)10^{-3}$.

With measurements of precision as high as those of Kistner and of Atac
et al., it is necessary to consider possible time reversal—even phases introduced by a final- or initial-state interaction which would spoil the tests. Such
phases are due to nuclear (136) or atomic (137, 138) radiative corrections
(see Figure 8). The latter ones are more important, as we shall see below.

Typical graphs for radiative corrections (136) are shown in Figure 8b.
Because the imaginary part of the propagator for the intermediate state
$(E - H_0 + i\epsilon)^{-1}$ is proportional to $\delta(E - H_0)$, the correction terms which spoil
time-reversal tests have at least one intermediate state on the energy shell.
The principal part of $(E - H_0 + i\epsilon)^{-1}$ does not introduce a phase. Thus the
spoiling factors (phases introduced by T-even final- or initial-state interactions) are reduced not only by e^2, but also by $(k/M)^L$ or $(kR)^L$, where k is
the momentum of the emitted photon, M is the mass of the nucleon, R is
a measure of the source dimension, and L is the multipolarity. Since all tests
use interfering $E(2) + M(1)$ transitions, these retardation factors are considerable, and imply that the phase introduced by radiative corrections is
roughly $< 10^{-6}$ (136) and thus negligible.

The atomic radiative corrections occur because the K and L inner electron shells lie within a wavelength of the emitted radiation. As for the nuclear
corrections, the "spoiling factors" are those which arise from the energy-

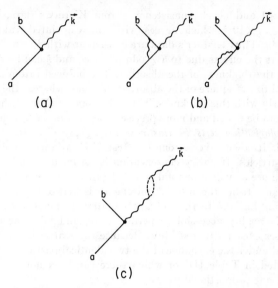

FIG. 8. Diagram for a radiative transition $a \rightarrow b$: 8a is the basic diagram; 8b shows some nuclear radiative corrections; 8c represents an atomic (internal conversion) correction (the dashed loop is an electron-hole pair).

conserving or pole terms. The most important intermediate states are those corresponding to internal conversion, shown in Figure 8c. Because the $M(1)$ and $E(2)$ nuclear current sources induce different conversion currents, there arises a phase difference between the $E(2)$ and $M(1)$ nuclear-transition matrix elements which has nothing to do with time-reversal noninvariance. Since the physical dimension of the atomic system is larger than the nuclear one, the retardation effects are smaller and the phase difference $\Delta\xi \equiv \xi(E2) - \xi(M1)$ has been computed to be (137, 138)

$$\Delta\xi(\text{Ru}) = -6.5 \times 10^{-3} \qquad\qquad 84a.$$

$$\Delta\xi(\text{Ir}) = 0.9 \times 10^{-3} \qquad\qquad 84b.$$

The effect is reduced in ^{193}Ir because K-shell conversion is forbidden energetically and only L-shell conversion occurs. The phase difference due to internal-conversion corrections is thus of the same order of magnitude as the accuracy of the measurements.

For the emission experiment, the time-reversal phase ϕ_T can be computed from the measured phase and Equation 84b

$$\phi_T = \phi_{\text{meas}} - \Delta\xi = (0.2 \pm 3.8) \times 10^{-3}$$

In the absorption experiment of Kistner, there is not sufficient information to remove both the Faraday and atomic effects. Hannon & Trammell (138) assume $\phi_T = 0$ and check consistency. From Kistner's data, they obtain $\Delta\xi = (-4.3 \pm 5) \times 10^{-3}$ which agrees with the calculated value given by Equation 84a. Thus Kistner's data are consistent with $\phi_T = 0$.

Note that the effects due to a T-odd phase ϕ, and ξ can be distinguished by varying the thickness of the absorber. The induced (atomic) effects are proportional to the square of the absorber thickness whereas the T-odd effect varies linearly with this thickness. Thus, the computation of the theoretical correction can be tested and more precise experiments are not ruled out.

3.4.3 *Polarization tests in elastic scattering experiments.*—Since elastic scattering is its own inverse, one can test T invariance directly by using polarized particles. If parity conservation is assumed, then to test T, it is necessary to use a system of total spin $> \frac{1}{2}$ (see Section 2.3.3). The simplest system which satisfies the above requirements is that of two protons. Elastic p-p scattering has the further advantages that no nuclear structure is involved, it is readily accessible experimentally, and fairly accurate phase-shift analyses exist up to 400 MeV. Because the scattering matrix must be symmetrical under the exchange of the two identical particles, the only term of those listed in Table III for which there can be a nonzero expectation value is the one proportional to

$$(\mathbf{\sigma}_1 \cdot \mathbf{q} \, \mathbf{\sigma}_2 \cdot \mathbf{K} + \mathbf{\sigma}_1 \cdot \mathbf{K} \, \mathbf{\sigma}_2 \cdot \mathbf{q})F(\mathbf{k}_i \cdot \mathbf{k}_f) \qquad 85.$$

Here $\mathbf{q} = \mathbf{k}_i - \mathbf{k}_f$, $\mathbf{k} = \mathbf{k}_i \mathbf{k}_f$, and F is a polynomial in $\cos \theta$, whose maximum power depends on the maximum relative angular momentum which is phase shifted. The term Equation 85 is even under P, but odd under T. Experimental requirements include a polarized target, and either a polarized beam or detection of the polarization of the scattered protons. It is also possible to compare the polarization of the scattered beam from an unpolarized incident beam with the asymmetry produced by a polarized beam (22, 139). Such experiments were done in 1958 and gave results consistent with T invariance to within the errors of a few per cent (140). More recently, a careful comparison of forward and reverse experiment was carried out by Handler et al. (141). The basic idea of the experiment is sketched in Figure 9 in the c.m. system, for ease of visualization. The experiment requires a triple scattering (polarizer and analyzer) but no measurements of absolute magnitudes; a direct comparison of the counting rates for the two geometrical configurations sketched in the figure gave the results (standard deviations

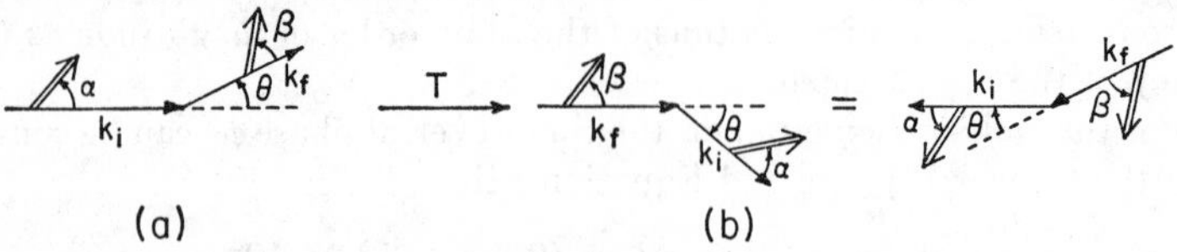

(a)　　　　　　　　　(b)

FIG. 9. Outline of the polarization test for T invariance in p-p scattering (141). The momenta and spins of one of the collision partners are shown in the c.m. system before and after the collision.

are used) for the difference in polarization of the *scattered* proton in Figures
9a and 9b.

$$\mathcal{P}_{(a)} - \mathcal{P}_{(b)} = 0.0019 \pm 0.009 \qquad 86.$$

Because the phase shifts are known for the scattering of two protons at 430
MeV, the result Equation 86 can be cast in terms of a phase due to a T-odd
interaction. This phase is found to be

$$\sin \phi \lesssim 0.06 \qquad 87.$$

The uncertainty of Equation 87 may be as large as 50 per cent because small
phase shifts with large errors are involved in the analysis. This is also the
cause of the disappointingly high limit set on ϕ. The interference terms which
contribute to the T-odd expression Equation 85 occur between P, F, and H
waves, all of which are small.

A comparison of the polarization and asymmetry in the elastic scattering
of 32.9 MeV protons by ^{13}C (142) gave (polarization/asymmetry) $=0.992$
± 0.025. In the experiment the need for absolute measurements of the polarization or asymmetry was circumvented by comparing results to those obtained with a ^{12}C target, where no polarization asymmetry can occur (see
2.3.3). No theoretical analysis of the result of the experiment has been carried out to extract a T-odd phase ϕ.

3.4.4 *Reciprocity or detailed-balance experiments.*—Aside from electromagnetic decays, the most precise (though not necessarily the most sensitive) experimental tests of T have been made by using detailed balance in
various reactions. From Equation 49, we find for the reactions $a+A \rightleftharpoons b+B$

$$1 + 2\chi \equiv \frac{(2s_a + 1)(2s_A + 1)p_a{}^2 d\sigma_{a\to b}/d\Omega}{(2s_b + 1)(2s_B + 1)p_b{}^2 d\sigma_{b\to a}/d\Omega} \qquad 88.$$

The ratio on the right-hand side is equal to unity, or $\chi=0$ if T invariance is
satisfied. Because it is only the absolute magnitude squared of (spin averaged) matrix elements that is compared in these experiments, there are
cases when no test of T is obtained. One well-known example is that when
the Born approximation is valid, the Hermiticity of the Hamiltonian is
sufficient to guarantee detailed balance (22, 24).

A further limitation of detailed-balance tests of T arises through the
unitarity of the S matrix (22). If only two states are relevant, or if the reaction proceeds through an isolated resonance, then measurements of cross
sections backward and forward do not test T invariance. The reason is that
the most general 2×2 unitary S matrix can be written as

$$S = \begin{pmatrix} \cos \theta e^{i\alpha} & i \sin \theta e^{i\phi} \\ i \sin \theta e^{i\phi} & \cos \theta e^{-i\alpha} \end{pmatrix} \qquad 89.$$

The diagonal matrix elements represent elastic scattering in initial and final states and the off-diagonal matrix elements are proportional to the reaction amplitudes. Since the phase is not measured, no test of T occurs. This theorem can be extended (22), and applies, for instance, whenever other elements of the S matrix are nonvanishing but do not interfere with those of Equation 89. We shall not discuss these restrictions further because for most hadronic reactions there are many open channels which make the S matrix a much larger matrix than 2×2 (exceptions may arise close to threshold), and unitarity is not a severe restriction.

Early experiments by Rosen & Brolley (143) made use of $p+{}^3\mathrm{H}\rightleftarrows d+d$ and by Bodansky et al. (144) of ${}^4\mathrm{He}+{}^{12}\mathrm{C}\rightleftarrows d+{}^{14}\mathrm{N}$. Both experiments agreed with detailed balance to within roughly 5 per cent. More recently new careful measurements of detailed balance have been reported. In addition to the studies of reaction 82 there is the measurement by a University of Washington group (145) for the reactions $d+{}^{24}\mathrm{Mg}\rightleftarrows p+{}^{25}\mathrm{Mg}$. The measurement consisted of a comparison for the forward and backward reaction of the ratios, $R(E)$, of the differential cross sections at angles close to 30° and 120° as a function of energy over a small region. Within this energy range there is a maximum as well as a minimum of R. Because no strong energy variations are found, it is expected that the reaction is predominantly a direct one; the experimenters estimate the compound nuclear contribution as being <25 per cent.

The targets were chosen to be light nuclei in order to have the first excited and ground states well separated; similar detection equipment was used for all measurements. To reduce systematic errors, measurements were carried out at the two angles simultaneously. The angles chosen were maxima of the angular distribution. Although it might appear that minima are better regions to have the uncorrelated T-odd matrix element show up, Weitkamp et al. (145) argue that when statistical errors control the uncertainties, maxima and minima are equally suitable and that for certain systematic errors maxima are preferable.

The comparison of the cross-section ratios $R(E)$ for the (p,d) and (d,p) reactions (see Figure 10) at energies close to the maximum gives

$$\frac{R_{(d,p)}(E)}{R_{(p,d)}(E)} \equiv 1 + \Delta = 0.999 \pm 0.003 \qquad\qquad 90.$$

Agreement with detailed balance is thus obtained to within 0.3 per cent. In translating this finding to a theoretical limit on T-odd forces, Weitkamp et al. note that there may be as many as 18 spin channels that add incoherently in the ${}^{24}\mathrm{Mg}(d,p){}^{25}\mathrm{Mg}$ reaction cross section. They thus estimate that the T-odd amplitude probably lies between $\Delta/3$ and $(2)^{1/2}\Delta$. The former applies when but one spin channel is effective and the latter when all of them contribute. It is clearly advantageous to choose low-spin targets and projectiles.

The second recent experiment (146) does not suffer from the disadvan-

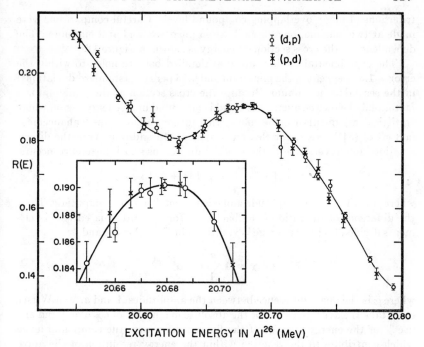

FIG. 10. Excitation function for $R(E) = [d\sigma(119.2°,E)/d\Omega]/[d\sigma(29.7°,E)/d\Omega]$ for the $^{24}Mg(d,p)^{25}Mg$ and $^{25}Mg(p,d)^{24}Mg$ reactions. The data near the peak at $E = 20.68$ MeV excitation energy in ^{26}Al are shown on an expanded scale in the inset. The solid curves are drawn to guide the eye. The standard errors shown include both statistical and systematic contributions. (From 145.)

tages of many independent spin channels. Von Witsch, Richter & von Brentano compare the backward and forward differential cross sections for the reaction

$$\alpha + {}^{24}Mg \rightleftarrows p + {}^{27}Al$$

In the experiment, scattered particles were detected at counters placed symmetrically about the beam axis at $\smile 172°$.

The advantage of measurements carried out to close to 180° is that, because the spins of ^4He and ^{24}Mg are zero, only a single spin channel contributes to the cross section at 0° and 180°. This is advantageous unless the T-odd matrix element vanishes at these angles because of other symmetry arguments. In any case there is then no incoherent sum over various spin channels with unknown T-odd contributions. In contrast to the Seattle test, this measurement was made in an energy region of strong fluctuations but was also carried out as a function of energy. This reaction is predominantly a compound nuclear one for which the fluctuations arise because of the contributions of many overlapping compound levels. Careful comparisons were made at two maxima, one for calibration purposes, and at a minimum. The dependence of the cross section on energy is shown in Figure 11.

The experimental findings are that detailed balance holds to within the error of 1.39 per cent in the minimum and 0.53 per cent (standard deviations) in the secondary maximum. Because the cross section at the minimum is a factor of 31 below the average one and is measured in an energy region where statistical arguments can be made, the authors assume an "enhancement factor" of $(31)^{1/2} = 5.57$ for this measurement. In addition, as in the direct-reaction time-reversal test, there is a gain because an interference occurs:

$$\sigma \propto \ \mid A + a \mid^2 \approx \ \mid A \mid^2 + 2Aa \cos \alpha$$

where A is the T-even amplitude and a is the smaller T-odd amplitude. Thus, the difference of the ratio of the backward for forward squared matrix elements divided by their average is χ, defined in Equation 88, and is

$$\chi = 4 \frac{a}{A} \cos \alpha$$

where α is the phase difference between the amplitudes A and a. Von Witsch et al. take the average value for the absolute magnitude of $\cos \alpha$ as $\frac{1}{2}$ and also use $\frac{1}{2}$ for the energy average of the fluctuating χ over the compound levels which contribute to the reaction within the energy resolution of the apparatus. Together with the statistical gain, they then find for the ratio of the T-odd to T-even reaction amplitudes, averaged in the above manner,

$$\left| \frac{a}{A} \right| \lesssim 3 \times 10^{-3}$$

with a confidence limit of 85 per cent.

Detailed-balance experiments were also carried out at a lower energy for the $^{16}O(d,\alpha)^{14}N$ reaction and its inverse (147). Here gas targets of high purity were used and absolute differential cross sections were compared at a peak, a valley, and a pleateau: detailed balance held to within ± 0.5 per cent at a confidence level of 68 per cent.

The experimental results of the low-energy detailed-balance tests of T invariance are summarized in Table IX.

Because the above tests are among the most precise to date, we shall discuss their sensitivity to T-odd forces in some detail. This sensitivity depends on a knowledge of the reaction mechanism responsible for the processes studied. Unfortunately, because strong interactions are involved, it is only possible to predict the cross sections in terms of models. To the extent that these models are able to reproduce the main features of the measurements, we can invoke them to analyze the T-odd force sensitivity of the tests.

We shall first discuss direct reactions. Robson (148) has argued that these processes may be quite insensitive to T-odd forces when the projectile and

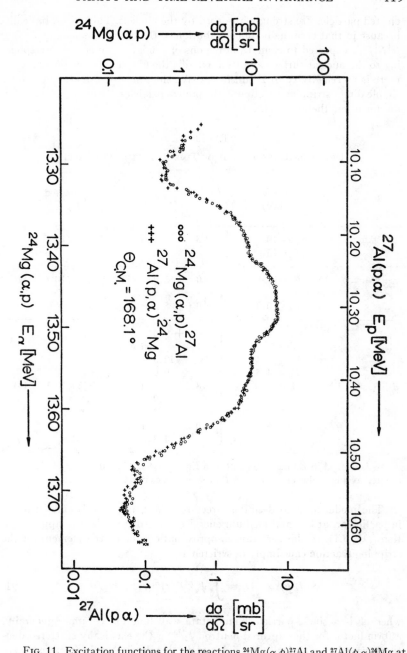

Fig. 11. Excitation functions for the reactions ^{24}Mg$(\alpha,p)^{27}$Al and ^{27}Al$(p,\alpha)^{24}$Mg at $\theta_{c.m.} = 168.1°$ in the minimum. The errors shown at the high-energy end are statistical. The curves are normalized at the peak of the maximum. (From 146.)

ejected particles are strongly absorbed by the nucleus. This arises, basically, because in that case the main contribution to the matrix element $\langle f|S|i\rangle$ or $\langle i|S|f\rangle$ is expected to come from regions of configuration space corresponding to the nuclear surface and from outside the nucleus. In the outer region, there is no sensitivity to T-odd forces. We shall present the argument for a simple direct stripping reaction with spinless particles. Spins complicate, but do not alter, the argument.

TABLE IX

DETAILED-BALANCE TESTS OF TIME-REVERSAL INVARIANCE[a]

Experiment	Energies (MeV)	$\left.\begin{array}{c}\chi\\\Delta\end{array}\right\}\times 10^3$		Estimated $|\bar{a}|/|\overline{A}|$	Ref.
$d+{}^{24}\mathrm{Mg}\rightleftharpoons p+{}^{25}\mathrm{Mg}$	$E_d\approx 10$ $E_p\approx 15$	$\Delta=-1\pm 3$		$\leq 3\times 10^{-3}$	145
$\alpha+{}^{24}\mathrm{Mg}\rightleftharpoons p+{}^{27}\mathrm{Al}$	$E_\alpha\approx 13.5$ $E_p\leq 10.3$	$\chi=\begin{cases}0 & \pm 13.9 \text{ at min.}\\ 0 & \pm 5.3 \text{ at max.}\end{cases}$		$\lesssim 3\times 10^{-3}$	146
$d+{}^{16}\mathrm{O}\rightleftharpoons \alpha+{}^{14}\mathrm{N}$	$E_d\approx 4.5$ $E_\alpha\approx 9$	$\chi=\begin{cases}+9.6\\ -11.4 \quad\text{at max.}\\ -12.3\\ +9.9\\ 8.2 \quad\text{at min.}\\ -7.9\\ +5.2\\ 0.8 \quad\text{at plateau}\\ -4.8\end{cases}$		$\leq 3\times 10^{-3}$	147

[a] χ is defined in Equation 88 and Δ in Equation 90. The ratio $|a|/|A|$ is the estimated average value of a T-odd to T-even matrix element.

The model used to describe direct reactions, in general, and stripping in particular, at low and medium energies is the distorted-wave approximation $(DWBA)$. In this zero-range approximation, the matrix element for the stripping reaction can simply be written as

$$S(d,p) \propto A_{fi} \equiv \int \psi_{\mathbf{k}}^{(-)*}(\mathbf{r})\Phi(\mathbf{r})\psi_{\mathbf{K}}^{(+)}(\mathbf{r}) \qquad 91.$$

where Φ is a single-particle bound-state wavefunction or, more generally, a form factor for the captured neutron $\psi_{\mathbf{K}}^{(+)}$ is the elastically scattered deuteron wave by an optical potential, and $\psi_{\mathbf{k}}^{(-)}$ is the elastically scattered wave for the proton by the final nucleus; the $(+)$ sign indicates outgoing waves at infinity, whereas the $(-)$ one means ingoing waves. If T invariance is valid, then (i) the bound-state wavefunction can be chosen to be real (see Section 2.2) and (ii) we have (see Equation 25)

$$\psi_{\mathbf{k}}^{(-)*} = \psi_{-\mathbf{k}}^{(+)} \qquad 92.$$

However, even if T is not valid, Equation 92 holds outside the nuclear inter-action volume where there are either no forces or only Coulomb forces. In this region of configuration space the asymptotic boundary conditions are sufficient to guarantee the equality of Equation 92. Furthermore, the wave-function or form factor $\Phi(r)$ is, at most, a real function times a constant phase factor, $e^{i\epsilon}$, in this region. Thus, we can write A_{fi} as

$$A_{fi} = \int_0^{R_0} \psi_{\mathbf{k}}^{(-)*}(\mathbf{r})\Phi(\mathbf{r})\psi_{\mathbf{K}}^{(+)}(\mathbf{r})d^3r$$

$$+ \int_{R_0}^{\infty} \psi_{\mathbf{k}}^{(-)*}(\mathbf{r})\Phi(\mathbf{r})\psi_{\mathbf{K}}^{(+)}(\mathbf{r})d^3r \qquad 93.$$

$$\equiv A_{fi}(\text{int}) + A_{fi}(\text{ext})$$

where the first term represents the contribution from the "internal" region, namely that volume where nuclear forces are present, and the second term is external contribution. Thus, whether T invariance is valid or violated, we find,

$$A_{if} = A_{if}(\text{int}) + A_{fi}(\text{ext})e^{-2i\epsilon} \qquad 94.$$

If $A(\text{int})$ can be neglected, then in such a "cutoff $DWBA$" a detailed-balance experiment does not test T. If the internal contribution is small, then, as Robson argues, direct reactions may not be sensitive to T-odd forces. How-ever, even a *small* internal contribution interferes with the external matrix element, so that a 10 per cent effect contributes 20 per cent to the matrix element. Furthermore, as shown in Table IV all T-odd forces are proportional to r [e.g., $r \cdot p + p \cdot r$], and vanish in zero range so that Equation 91 must be generalized.

The question raised by Robson was investigated in some detail (66, 149, 150). The authors of (149) add a T-odd potential of the simplest spin-inde-pendent type

$$v(\mathbf{r}) = \frac{\lambda}{2}\left[f(\mathbf{r})(\mathbf{r}\cdot\mathbf{p} + \mathbf{p}\cdot\mathbf{r}) + \text{h.c.}\right] \qquad 95.$$

to the interaction V_{np} and to the single-particle potential which binds the neutron to the target nucleus. For a strength v of roughly[10] 1 per cent of the

[10] The word "roughly" is used because the relative normalization depends on the average value of the operator p and on the form $f(r)$ of Equation 95, as well as on the definition of "strength."

T-even potential V_{np}, the violation of reciprocity averaged over angles is also ~1 per cent. The angle-averaged violation of reciprocity is defined by

$$\bar{\chi} = \frac{1}{4\pi} \int \; |\; \chi(\theta)\; |\; d\Omega \qquad\qquad 96.$$

The actual violation depends sensitively on the range of the force v, that is on the range of $f(r)$. For a neutron v of range equal to the nuclear radius (4.1 F) the violation is found to be 6.0 per cent. The angular dependence of the violation is shown in Figure 12; it is noted that no particular correlation of the violation of reciprocity is to be expected with the angular dependence of the differential cross section, except that extrema of the violation usually occur at the extrema of the differential cross section.

The reason that the usefulness of detailed-balance tests of T in direct reactions is not severely limited, as Robson's arguments imply, can be found in the interference of the external and internal (including the surface region)

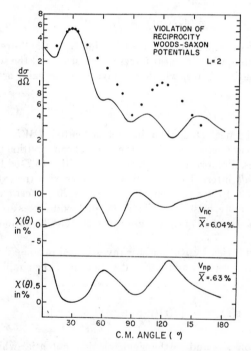

FIG. 12. Violation of reciprocity for the ^{24}Mg$(d,p)^{25}$Mg reaction for a 1 per cent parity-violating potential: 12a is the differential cross section computed for an $L=2$ capture, and is compared to the experimental data (solid dots); 12b shows the violation in reciprocity if the parity-violating interaction is purely in the neutron-core interaction and 12c in the neutron-proton interaction.

parts of the matrix elements. Whereas the T-even matrix element contributions from the internal region may be small, the same appears not to be true of the T-odd contributions, which are peaked close to the surface but inside the nuclear interaction region.

The effect of the T-odd nucleon-nucleon potential calculated by Huffman (66) (Section 2.5) was also used by him to estimate the T violation to be expected in the Seattle experiment. For this purpose, he calculates an effective single-particle potential, equivalent to the form of Equation 95. He finds roughly 0.2 per cent violation in the detailed-balance test. The potential $v(r)$ is found from a nucleon-nucleon T-violating force due to one-pion and one-photon exchange of magnitude 0.01 per cent of the normal one-pion exchange potential (lower cutoff $\approx 0.75 m_\pi^{-1}$). Because the strength of the n-p T-violating force depends sensitively on the lower radial cutoff (see Section 2.5), these results can only be taken as an order-of-magnitude estimate. The estimate assumes, furthermore, that the C-even current K_μ has an isovector component.

For compound nuclear reactions in an energy region of fluctuating cross sections, it has been argued by Ericson (151), Mahaux & Weidenmüller (152), and Moldauer (153) that a gain of sensitivity to T-odd effects may occur. The basic reason for this enhancement can be seen most easily in terms of the following simple argument given by Ericson. In the region of compound nuclear resonances, the S-matrix element from the state i (say $\alpha+A$) to the state f(e.g., $p+A'$) is proportional to

$$S_{fi} \propto A_{fi} \equiv \sum_c \gamma_{fc} \frac{1}{E - W_c} \gamma_{ci} \qquad 97.$$

where $\gamma_{ci}(\gamma_{cf})$ is the reduced width for forming the compound nuclear state from the state $i(f)$; the complex energy of this state is W_c. It is assumed that the region of the compound nucleus reached is one with many overlapping levels, namely those for which $\Gamma >> D$, where D represents the average spacing between levels and Γ the average width. If we break up the various contributions to Equation 97 into their T-even parts $A_{fi}^{(e)}$, and T-odd parts a_{fi}, we have, to first order in the T-odd parts

$$A_{fi}^{(e)} = \sum_c \gamma_{fc}^{(e)} \frac{1}{E - W_c^{(e)}} \gamma_{ci}^{(e)} \equiv \sum_c \gamma_{fc}^{(e)} \frac{1}{\Delta} \gamma_{ci}^{(e)} \qquad 98.$$

$$a_{fi} \approx \sum_c \gamma_{fc}^{(o)} \frac{1}{\Delta} \gamma_{fc}^{(e)} + \gamma_{fc}^{(e)} \frac{1}{\Delta} \gamma_{ci}^{(o)}$$

$$+ \gamma_{fc}^{(e)} \frac{1}{\Delta} v \frac{1}{\Delta} \gamma_{ci}^{(e)} \qquad 99.$$

where v is the T-odd interaction. Ericson neglects all but the last term of Equation 99 to estimate the enhancement to be expected for T-odd effects.

Moldauer examines the structure of the formula in detail, and finds that although the other two terms must be taken into account, there are cancellations in the sums such that the effective result is almost that obtained by Ericson. Although the average cross-section asymmetry may be zero, there are strong fluctuations which give rise to lack of detailed balance if T is violated. The estimate of the violation follows from Equations 98, 99. If we keep only the last term of Equation 99, drop the superscripts (e) in Equation 98, and carry out appropriate energy averages, we find

$$\frac{\overline{|a_{fi}|^2}}{\overline{|A_{fi}|^2}} \approx \overline{\langle c|v|c'\rangle^2} \sum_c \frac{1}{|E - W_c|^2}$$

$$= \overline{\langle c|v|c'|^2} \frac{2\pi}{\Gamma D} \approx \frac{w}{\Gamma} \left|\frac{v}{V}\right|^2$$

100.

Here V represents a T-even potential and w an energy over which the matrix elements of the T-odd force are thoroughly mixed. This quantity is hard to estimate. The "enhancement" of T-odd forces in the fluctuating region of cross sections is thus

$$R \sim (w/\Gamma)^{1/2}$$

101.

Ericson estimates w to be of the order of single-particle energy level spacings or a few MeV. With $\Gamma \approx 50$ keV, typical of the $A \approx 25$ region, he obtains $R \approx 6$. Mahaux & Weidenmüller (152) analyze the case of two interfering levels and conclude that w is more likely to be associated with the widths of doorway states; they show that $w < 1.5\Gamma_d$, where Γ_d is the spreading width of a doorway state. In that case R is of the order of unity rather than 6–10. Because the T-odd forces are momentum dependent and differ from the normal forces, all these estimates are only qualitative.

3.4.5 *Nuclear structure tests.*—Some effects of T-odd forces on nuclear structure were discussed in Section 2.2. In addition, Rosenzweig, Monahan & Mehta (154) and Favro & McDonald (155) have suggested that the statistical distribution of level widths and level spacings can be used to set limits on the T-odd forces present in nuclei. In a comparison of the distribution of widths with the Porter-Thomas one from $^{97}Au(n,\gamma)$ and $^{196}Pt(n,\gamma)$, Rosenzweig et al. conclude that the *probability* (squared amplitude) of the T-odd effects is $<25 \times 10^{-4}$. A comparison of the distribution of level spacings with that expected only when T-even forces are present shows maximum departures for small level spacings (156). Because, statistically, there are few levels with small spacings the test is difficult and no limit has been set from such an an analysis.

3.5 PRESENT STATUS OF T-VIOLATION SEARCHES

In summary of the previous sections, we can state that *no* violations of T invariance have yet been found. The present limits are roughly that the fractional violation of the strong-interactions strength $\mathfrak{F}_T$ is $\mathfrak{F}_T < 3 \times 10^{-3}$.

Although more accurate tests are clearly called for, such tests become extremely difficult and in many cases extraneous corrections which simulate T-violating effects must be taken into account. Up to now, no new techniques which would circumvent such effects have been proposed.

3.6. Searches for Simultaneous Violations of P and T Invariances

Some possible tests of the simultaneous breakdown of P and T symmetries were given in Section 2.3. Because the violation of P occurs in the weak interactions and CP or T violation is another 2×10^{-3} weaker, experiments which seek a simultaneous breakdown of P and T invariances of hadrons require a sensitivity of roughly one part in 10^9! A neutral atom of spin $\frac{1}{2}$ or the neutron, both of which can have no higher electric moments, are among the few possible systems in which such an accuracy can be reached. If the violation of T invariance occurs as a correction (e.g., electromagnetic or milliweak) to the weak interactions, then the electric dipole moment $d(n)$ for the neutron should be of order (157)

$$d(n) \backsim e\langle r \rangle GM_n{}^2 2 \times 10^{-3} \backsim 10^{-22} \, e\text{-cm} \qquad 102.$$

where $\langle r \rangle$ is a measure of the size of the neutron's charge distribution, and $GM_n{}^2$ is the dimensionless weak-interaction constant. On the other hand, for a direct violation of T invariance in the electromagnetic interactions of the hadrons (Section 2.5), an order-of-magnitude estimate gives $d(n) \sim e\langle r \rangle GM_n{}^2$ $\sim 10^{-19}$ e-cm (158).

Early experiments (159) gave

$$d(n) = (-0.1 \pm 2.4) \times 10^{-20} \, e\text{-cm}$$

Recently, attempts (160–162) have been made to detect a neutron dipole moment of the order of that given by Equation 102. In the experiment of (160) slow neutrons from a reactor were polarized by reflection from cobalt mirrors. These neutrons were deflected in a resonance spectrometer with radio-frequency coils, and an electrostatic field $\mathbf{\mathcal{E}}$ which is aligned parallel or antiparallel to a static magnetic field $\mathbf{B}$. The neutrons precess in these fields with a frequency proportional to their interaction,

$$H' = - \, \mathbf{\mu}_n \cdot \mathbf{B} - \mathbf{d} \cdot \mathbf{\mathcal{E}}$$

The radio frequency is set so that a maximum number of neutron spins are flipped if $d = 0$. The resonant frequency is shifted by $\pm d |\mathcal{E}|$ if the neutron possesses an electric dipole moment. The counting rate of detected neutrons is sensitive to this frequency. The absence of an effect gave $d(n) \lesssim 3 \times 10^{-23}$ e-cm. Shull & Nathans (161) find $d = (2.4 \pm 3.9) \times 10^{-22}$ e-cm. Recently by using a longer electric-field region and magnetic-field improvements, the authors of (160) managed to obtain an improved limit (163), $d(n) = (2 \pm 2)e$ $\times 10^{-23}$ cm. This result sets an upper limit somewhat lower than that estimated by simple-minded arguments, Equations 102. Thus, no simultaneous violation of P and T invariances has yet been found.

The theoretical interpretation of the experimental results is not clear. Taken at face value, they cast doubt on the theory of Section 2.5, in which the seat of C and therefore T violation is in the electromagnetic interactions of the hadrons. However, Lee (60) has suggested that the CP and strangeness-conserving parity-violating weak interaction may be suppressed, and thus may account for a very small neutron electric dipole moment. To rule out this explanation requires, at the very least, a separation of the $\Delta I = 0,2$ part of the parity-violating nuclear force in order to test the current-current basis for this interaction.

Barton & White (164) have used sideways dispersion relations to try to obtain a relatively reliable estimate for $d(n)$. They find 10^{-24} e-cm$\lesssim d(n)$ $\lesssim 10^{-23}$ e-cm for a milliweak (57) type theory and $d(n) \lesssim 6 \times 10^{-22}$ e-cm for an electromagnetic (16) violation of CP invariance. Alternatively, if the seat of the CP violation of the K_2^0 decay is solely in the strangeness-changing nonleptonic currents or is caused by a superweak interaction, then the electric dipole moment of the neutron would be $\sim 10^{-27}$ e-cm or less—far too small to be detectable by present experimental techniques.

4. SUMMARY

At the present time, violations of parity invariance of roughly 1 part in 10^2 have been observed in electromagnetic transitions of heavy nuclei. These violations correspond approximately to a parity-violating force one million times weaker than the hadronic forces. This magnitude lends support to the current-current form for the purely nonleptonic part of the weak interactions. Because of the importance of these results they need to be verified; such tests have been undertaken (165). Experiments on light nuclei are required to unravel the isospin dependence and thus obtain reliable information on the various contributions to the nonleptonic currents. Indeed some recent work (166) questions the validity of the $\Delta I = 0$ parity-violating potential $V_{P.V.}$ of Equations 63–64. The authors obtain a negligible value for the strength of the parity-violating potential due to vector meson exchanges. The new experiments on the α decay of the 8.88 MeV level of ^{16}O (71, 74) may be able to shed light on this question.

The lack of time-reversal violation found in various tests indicates that any interaction which violates T invariance is probably weaker than $\sim 3 \times 10^{-3}$ of the hadronic force strength. Together with the small upper limit found for the neutron electric dipole moment, these results do not favor the proposal that the electromagnetic interactions of the hadrons are responsible for CP violation. However, the experiments are not yet sufficiently accurate to rule out this proposal. If experiments on the K_1^0 and K_2^0 systems rule out the superweak and milliweak theories of CP violation, more accurate tests of T invariance in nuclear physics will become imperative.

ACKNOWLEDGMENTS

The author is indebted to Professor D. H. Wilkinson for discussions on the contents of this review, and to Professors H. Frauenfelder and J. Hannon for making available some of their work prior to publication. He is also grateful to Professors D. Bodansky, H. Frauenfelder, and I. Halpern for suggesting useful modifications of the manuscript.

LITERATURE CITED

1. Lee, T. D., in *Preludes in Theoretical Physics*, 5–16 (de Shalit, A., Feshbach, H., Van Hove, L., Eds., North-Holland, Amsterdam, 1966)
2. Schiff, L. I., *Physics*, **1**, 209 (1965)
3. Morrison, P., in *Preludes in Theoretical Physics*, 347-51 (de Shalit, A., Feshbach, H., Van Hove, L., Eds., North-Holland, Amsterdam, 1966)
4. Gold, T., *The Nature of Time* (Cornell Univ. Press, 1967). See especially Morrison, P., 121–48; Grünbaum, A., 149–86; Rosenfeld, L., 187–95.
5. Wigner, E. P., *Z. Physik*, **43**, 624 (1927)
6. Wigner, E. P., *Nachr. Akad. Wiss. Göttingen, II Math.-Phys. Kl.*, **31**, 546 (1932)
7. Wick, G. C., *Ann. Rev. Nucl. Sci.*, **8**, 1 (1958)
8. Tripp, R. D., *Ann. Rev. Nucl. Sci.*, **15**, 325 (1965)
9. Lee, T. D., Wu, C. S., *Ann. Rev. Nucl. Sci.*, **15**, 381 (1965)
10. Lee, T. D., Yang, C. N., *Phys. Rev.*, **104**, 254 (1956)
11. Landau, L. D., *Nucl. Phys.*, **3**, 127 (1957)
12. Pauli, W., in *Niels Bohr and the Development of Physics* (Pauli, W., Ed., McGraw-Hill, New York, 1955); Schwinger, J., *Phys. Rev.*, **91**, 720, 723 (1953) and **94**, 1366 (1953); Lüders, G., *Kgl. Danske Videnskab.Selskab Mat.-Fys. Medd.*, **28**, *No. 5* (1954)
13. Christenson, J. H., Cronin, J. W., Fitch, V. L., Turlay, R., *Phys. Rev. Letters*, **13**, 562 (1964)
14. Wolfenstein, L., *Phys. Rev. Letters*, **13**, 562 (1964); Lee, T. D., Wolfenstein, L., *Phys. Rev.*, **138**, B1490 (1965)
15. See Wolfenstein, L., Lectures at Intern.Sch.Phys."Ettore Majorana," July 1968 (To be published, Academic Press)
16. Bernstein, J., Feinberg, G., Lee, T. D., *Phys. Rev.*, **139**, B1650 (1965)
17. Barshay, S., *Phys. Rev. Letters*, **17**, 78 (1965)
18. Prentki, J., Veltman, M., *Phys. Rev. Letters*, **15**, 88 (1965)
19. Jacob, M., Wick, G. C., *Ann. Phys. (N.Y.)*, **7**, 404 (1959)
20. Schwinger, J., *Phys. Rev.*, **82**, 914 (1951)

21. Kramers, H. A., *Proc. Acad. Amsterdam*, **33**, 959 (1930)
22. Henley, E. M., Jacobsohn, B. A., *Phys. Rev.*, **113**, 225 (1959)
23. Biedenharn, L. C., *Nucl. Phys.*, **10**, 620 (1959)
24. Blatt, J. M., Weisskopf, V. F., *Theoretical Nuclear Physics*, (Wiley, New York, 1952) Appendix B; Chap. 10
25. de Forest, T., Jr., Walecka, J. D., *Advances in Physics (Phil. Mag. Suppl.)*, **15**, 1 (1966); Durand, L., de Celles, P., Marr, R. *Phys. Rev.*, **126**, 1882 (1962)
26. Lloyd, S. P., *Phys. Rev.*, **81**, 161 (1951)
27. Wolfenstein, L., Ashkin, J., *Phys. Rev.*, **85**, 947 (1958)
28. Csonka, P. L., Moravcsik, M. J., Scadron, M. D., *Phys. Rev. Letters*, **20**, 67 (1968), **14**, 861 (1965), *Nuovo Cimento*, **42A**, 743 (1966); Csonka, P., Moravcsik, M. J., *Phys. Rev.*, **152**, 1310 (1966)
29. Eisenbud, E., Wigner, E. P., *Proc. Natl. Acad. Sci. U.S.*, **27**, 281 (1941); Okubo, S., Marshak, R. E., *Ann. Phys. (N.Y.)*, **4**, 166 (1958)
30. Blin-Stoyle, R. J., *Phys. Rev.*, **118**, 1605 (1960)
31. Herczeg, P., *Nucl. Phys.*, **48**, 263 (1963); **75**, 655 (1966)
32. Adler, S. L., Dashen, R. F., *Current Algebras* (Benjamin, New York, 1968)
33. For the latest data, see *Rept. 14th Ann. High Energy Conf.* (Vienna, September 1968)
34. Nishijima, K., Swank, L. J., *Nucl. Phys.*, **B3**, 553, 565 (1967)
35. Das, T., *Phys. Rev. Letters*, **21**, 409 (1958)
36. Okubo, S., *Ann. Phys. (N.Y.)*, **49**, 219 (1968)
37. Barton, G., *Nuovo Cimento*, **19**, 512 (1961)
38. Michel, F. C., *Phys. Rev.*, **133**, B329 (1964)
39. Lacaze, R., *Nucl. Phys.*, **B4**, 657 (1968)
40. Sugawara, H., von Hippel, F., *Phys. Rev.*, **172**, 1764 (1968)
41. Drell, S. D., Zachariasen, F., *Electromagnetic Structure of Nucleons* (Oxford, 1961)
42. Lee, T. D., Yang, C. N., *Nuovo Cimento*, **3**, 749 (1956); Goebel, C., *Phys. Rev.*, **103**, 258 (1956)

43. McKellar, B. H. J., *Phys. Letters*, 26B, 107 (1967)
44. Suzuki, M., *Phys. Rev. Letters*, 15, 986 (1965)
45. Sugawara, H., *Phys. Rev. Letters*, 15, 870, 997 (1965)
46. d'Espagnat, B., *Phys. Letters*, 7, 209 (1963)
47. Oakes, R. J., *Phys. Rev. Letters*, 20, 1539 (1968)
48. McKellar, B. H. J., *Phys. Rev. Letters*, 21, 1822 (1968)
49. Tadic, D., *Phys. Rev.*, 174, 1694 (1968)
50. Fischbach, E., Trabert, K., *Phys. Rev.*, 174, 1843 (1968)
51. Dashen, R. F., Frautschi, S. C., Gell-Mann, M., Hara, Y., in Gell-Mann, M., Ne'eman, Y., *The Eightfold Way*, 254 (Benjamin, New York, 1964)
52. Chiu, Y. T., Schechter, J., *Phys. Rev. Letters*, 16, 1022 (1966)
53. Christensen, C. J., Nielsen, A., Bahnsen, A., Brown, W. K., Kustad, B. M., *Phys. Letters*, 26B, 11 (1967)
54. Blin-Stoyle, R. J., Herczeg, P., *Phys. Letters*, 23, 376 (1966); *Nucl. Phys.*, B5, 29 (1968)
55. Henley, E. M., in *Isospin in Nuclear Physics* (Wilkinson, D. H., Ed., North-Holland, to be published)
56. Lee, T. D., *Phys. Rev.*, 140, B959 and B967 (1965)
57. Arzubov, B. A., Filippov, A. T., *Zh. Eksp. Teor. Fiz. Pis'ma*, 8, 493 (1968) [Transl. *Soviet Phys. JETP Letters*, 8, 302 (1968)]
58. Baltay, C., Barash, N., Franzini, P., Gelfand, N., Kirsch, L., Lutjens, G., Severiens, J. C., Steinberger, J., Tycko, D., Zanello, D., *Phys. Rev. Letters*, 15, 591 (1965)
59. Cnops, A. M., Finocchiaro, G., Lassalle, J. C., Mittner, P., Zanella, P., Dufey, J. P., Gobbi, B., Pouchon, M. A., Muller, A., *Phys. Letters*, 22, 546 (1966); Gormley, M., Hyman, E., Lee, W., Nash, T., Peoples, J., Schultz, C., Stein, S., *Phys. Letters*, 21, 399, 402 (1968); Larribe, A., Leveque, A., Muller, A., Pauli, E., Revel, D., Tallini, T., Litchfield, P. J., Rangan, L. K., Segar, A. M., Smith, J. R., Finney, P. J., Fisher, C. M., Pickup, E., *Phys. Letters*, 23, 600 (1966)
60. Lee, T. D., in Proc. 1967 *Intern. Symp. Electron and Photon Interactions at High Energies*, 389–407

(NBS, Dept. Commerce, T(D-4500), 1967)
61. Ward, J., *Phys. Rev.*, 78, 1824 (1950); Takahashi, Y., *Nuovo Cimento*, 6, 370 (1957)
62. Kobzarev, I. Yu., Okun', L. B., Terent'ev, M. V., *Zh. Eksp. Teor. Fiz. Pis'ma*, 2, 466 (1965) [Transl. *Soviet Phys. JETP Letters*, 2, 289 (1965)]; Dubovik, V. M., Cheshkov, A. A., *Zh. Eksp. Teor. Fiz.*, 51, 169 (1966); 52, 706 (1967) [Transl. *Soviet Phys. JETP Letters*, 24, 111 (1967); 25, 464 (1967)]
63. Dubovik, V. M., Cheshkov, A. A., *Zh. Eksp. Teor. Fiz.*, 51, 165 (1965) [Transl. *Soviet Phys. JETP*, 24, 111 (1967)]
64. Bincer, A. M., *Phys. Rev.*, 118, 855 (1960)
65. Lipshutz, N. R., *Phys. Rev.*, 158, 1491 (1967)
66. Huffman, A. H. (Thesis, Univ. of Washington, 1968, unpublished); Huffman, A. H. (Submitted to *Phys. Rev.* for publication)
67. Wilkinson, D. H., *Phys. Rev.*, 109, 1603 (1958)
68. Tanner, N., *Phys. Rev.*, 107, 1203 (1959)
69. Pixley, R. E., Benenson, W., *Nucl. Phys.*, A91, 177 (1967)
70. Segel, R. E., Olness, J. W., Sprenkel, E. L., *Phil. Mag.*, 6, 163 (1961); *Phys. Rev.*, 123, 1382 (1961)
71. Boyd, D. P., Donovan, P. F., Marsh, B., Alburger, D. E., Wilkinson, D. H., Assimakopoulos, P., Beardsworth, E., *Bull. Am. Phys. Soc.*, 13, 1424 (1968); Alburger, D. E. (Private communication)
72. Alburger, D. E., Pixley, R. E., Wilkinson, D. H., Donovan, P. F., *Phil. Mag.*, 6, 171 (1961)
73. Donovan, P. F., Alburger, D. E., Wilkinson, D. H., *Proceedings of the Rutherford Jubilee*, 827 (Birks, J. B., Ed., Academic Press, New York, 1961)
74. Segel, E., Segel, R. E., Siemssen, R. (Private communication by R. E. Segel); Segel, R. E., *Bull. Am. Phys. Soc.*, 14, 565 (1969)
75. Radicati, L. A., *Phys. Rev.*, 87, 521 (1952); Gell-Mann, M., Telegdi, V. L., *Phys. Rev.*, 91, 196 (1953)
76. Morpurgo, G., *Phys. Rev.*, 110, 721 (1958)
77. Boehm, F., Kankeleit, E., *Phys. Rev. Letters*, 14, 312 (1965)

78. Boehm, F., Kankeleit, E., *Nucl. Phys.*, **A109**, 457 (1968)
79. Bock, P., Schopper, H., *Phys. Letters*, **6**, 284 (1965)
80. Grodzins, L., Genovese, F., *Phys. Rev.*, **121**, 228 (1961)
81. Kankeleit, E., *Congr. Intern. Phys. Nucl., Paris, 1964*, **2**, 1206 (CNRS, Paris, 1964)
82. Lobashov, V. M., Nazarenko, V. A., Saenko, L. F., Smotritskii, L. M., *Zh. Eksp. Teor. Fiz. Pis'ma*, **3**, 76 (1966) [Transl. *Soviet Phys. JETP Letters*, **3**, 47 (1966)]
83. Lobashov, V. M., Nazarenko, V. A., Saenko, L. F., Smotritskii, L. M., Kharkevich, G. I., *Zh. Eksp. Teor. Fiz. Pis'ma*, **5**, 73 (1967) [Transl. *Soviet Phys. JETP Letters*, **5**, 59 (1967)]; *Phys. Letters*, **25B**, 104 (1967)
84. Lobashov, V. M., Nazarenko, V. A., Saenko, L. F., Smotritskii, L. M., Kharkevich, G. I., *Zh. Eksp. Teor. Fiz. Pis'ma*, **3**, 268 (1966) [Transl. *Soviet Phys. JETP Letters*, **3**, 173 (1966)]
85. Cruse, D. W., Hamilton, W. D., *Nucl. Phys.*, **A125**, 241 (1969)
86. Blin-Stoyle, R. J., *Phys. Rev.*, **120**, 181 (1960)
87. Wahlborn, S., *Phys. Rev.*, **138**, B530 (1964)
88. Maqueda, E., Blin-Stoyle, R. J., *Nucl. Phys.*, **A91**, 460 (1967)
89. McKellar, B. H. J., *Phys. Rev. Letters*, **20**, 1542 (1968)
90. Vinh Mau, N., Bruneau, A. M. (To be published)
91. Szymanski, Z., *Nucl. Phys.*, **76**, 539 (1966)
92. Wilkinson, D. H., *Phys. Rev.*, **109**, 1614 (1958)
93. Boehm, F., Hauser, U., *Nucl. Phys.*, **14**, 615 (1960)
94. Abov, Yu. G., Krupchitsky, P. A., Oratovskii, Yu. A., *Phys. Letters*, **12**, 25 (1964); *Yad. Fiz.*, **1**, 479 (1965) [Transl. *Soviet J. Nucl. Phys.*, **1**, 341 (1965)]
95. Abov, Yu. G., Krupchitsky, P. A., Bulgakov, M. I., Ermakov, O. N., Karpikhin, I. L., *Phys. Letters*, **27B**, 16 (1968)
96. Warming, E., Stecher-Rasmussen, F., Ratynski, W., Kopecky, J., *Phys. Letters*, **25B**, 200 (1967)
97. Haas, R., Leipuner, L. B., Adair, R. K., *Phys. Rev.*, **116**, 1221 (1959)
98. Lee, T. D., *Phys. Rev.*, **171**, 1731 (1960)
99. Segrè, G., *Phys. Rev.*, **173**, 1730 (1968)
100. Lewis, R. R., *Phys. Rev. Letters*, **17**, 593 (1966); *Phys. Rev.*, **163**, 935 (1967); *Nucl. Phys.*, **113A**, 27 (1968). Lobov, G. A., Shabalin, E. P., *Nucl. Phys.*, **B9**, 75 (1969). Belyakov, V. A., *Zh. Eksp. Teor. Fiz.*, **54**, 1162 (1968) [Transl. *Soviet Phys. JETP*, **27**, 622 (1968)]
101. Forte, H., Saavedra, O., *Rept. Intern. Conf. Study Nucl. Struct. with Neutrons, Antwerp, 1965* (Reported in Ref. 103)
102. Wilson, R. (Private communication)
103. Moskalev, A. N., *Zh. Eksp. Teor. Fiz. Pis'ma*, **8**, 46 (1968) [Transl. *Soviet Phys. JETP Letters*, **8**, 27 (1968)]
104. Bonar, D. C., Drake, C. W., Headrick, R. D., Hughes, V. W., *Phys. Rev.*, **174**, 1200 (1968)
105. Blin-Stoyle, R. J., Feshbach, H., *Nucl. Phys.*, **27**, 395 (1961)
106. Partovi, F., *Ann. Phys.*, **27**, 114 (1968)
107. Danilov, G. S., *Phys. Letters*, **18**, 40 (1965)
108. Lobashov, V. M., *Yad. Fiz.*, **2**, 957 (1965) [Transl. *Soviet Phys. J. Nucl. Phys.*, **2**, 683 (1966)]
109. Dal'karov, O. D., *Zh. Eksp. Teor. Fiz. Pis'ma*, **2**, 197 (1968) [Transl. *Soviet Phys. JETP Letters*, **2**, 124 (1965)]
110. Henley, E. M., *Phys. Letters*, **28B**, 1 (1968)
111. Overseth, O. E., Roth, R., *Phys. Rev. Letters*, **19**, 391 (1967); Cleland, W. E., Bienlein, J. K., Conforto, G., Eaton, G. H., Gerber, H. J., Reinharz, M., Veltman, M., Gautschi, A., Heer, E., Renevey, J. Fr., Von Dardel, G., *Phys. Letters*, **26B**, 45 (1967)
112. Jackson, D., Treiman, S., Wyld, H., *Nucl. Phys.*, **4**, 206 (1957); *Phys. Rev.*, **106**, 517 (1957)
113. Ebel, M. E., Feldman, G., *Nucl. Phys.*, **4**, 213 (1957)
114. Burgy, M. T., Krohn, V. E., Novey, T. B., Ringo, G. R., Telegdi, V. L., *Phys. Rev.*, **120**, 1829 (1960); Clark, M. A., Robson, J. M., *Can. J. Phys.*, **38**, 693 (1960); **39**, 13 (1961)
115. Erozolimsky, B. G., Bondarenko, L. N., Mostovoy, Yu. A., Obinyakov, B. A., Zakharova, V. P., Titov, V. A., *Phys. Letters*, **27B**, 557 (1968)
116. Calaprice, F. P., Commins, E. D., Gibbs, H. M., Wick, G. L., Dob-

son, D. A., *Phys. Rev. Letters*, **18**, 918 (1967)

117. Callan, C. G., Jr., Treiman, S. B., *Phys. Rev.*, **162**, 1494 (1967)

118. Barshay, S., *Phys. Rev. Letters*. **17**, 49 (1966)

119. Anderson, R. L., Prepost, R., Wiik, B. H., *Phys. Rev. Letters*, **22**, 651 (1969)

120. Friedberg, C. E., Bartlett, D. F., Goulianos, K., Hammerman, I. S., Hutchinson, D. P., *Bull. Am. Phys. Soc.*, **14**, 76 (1969)

121. Young, K. (Private communication)

122. Christ, N., Lee, T. D., *Phys. Rev.*, **143**, 1310 (1966)

123. Chen, J. R., Sanderson, J., Appel, J. A., Gladding, G., Goiten, M., Hanson, K., Imrie, D. C., Kirk, T., Madaras, R., Pound, R. V., Price, L., Wilson, R., Zajde, C., *Phys. Rev. Letters*, **21**, 1279 (1968)

124. Prepost, R., Simonds, R. M., Wiik, B. H., *Phys. Rev. Letters*, **21**, 1271 (1968)

125. Ross, W. (Thesis submitted to Columbia Univ., 1969); *Nevis Rept. #167* (Unpublished)

126. Stichel, P., *Z. Phys.*, **150**, 264 (1958); de Sabbata, V., *Nuovo Cimento*, **21**, 659 (1961)

127. Lobov, G. A., *Soviet Phys. JETP Letters*, **1**, 157 (1965)

128. Jacobsohn, B. A., Henley, E. M., *Phys. Rev.*, **113**, 239 (1959)

129. Fuschini, E., Gadjokov, V., Maroni, C., Veronesi, P., *Nuovo Cimento*, **33**, 1309 (1964)

130. Kajfosz, J., Kopecky, J., Honzatko, J., *Phys. Letters*, **20**, 284 (1965); *Nucl. Phys.*, **A120**, 225 (1968)

131. Eichler, J., *Nucl. Phys.*, **A120**, 535 (1968)

132. Kistner, O. C., *Phys. Rev. Letters*, **19**, 872 (1967); Blume, M., Kistner, O. C., *Phys. Rev.*, **171**, 417 (1968)

133. Atac, M., Chrisman, B., Debrunner, P., Frauenfelder, H., *Phys. Rev. Letters*, **20**, 691 (1968); Atac, M., *Rept. #165, Univ. of Illinois* (Nov. 1967, unpublished)

134. Garrell, M. H., Frauenfelder, H., Ganek, D., Sutton, D. C. (To be published); Frauenfelder, H. (Private communication)

135. Perkins, R. B., Ritter, E. T., *Phys. Rev.*, **174**, 1426 (1968)

136. Henley, E. M., Jacobsohn, B. A., *Phys. Rev. Letters*, **16**, 706 (1966)

137. Hannon, J. P., Trammell, G. T., *Phys. Rev. Letters*, **21**, 726 (1968)

138. Hannon, J. P., Trammell, G. T. (To be published)

139. Bell, J. S., Mandl, F., *Proc. Phys. Soc. (London)*, **71**, 272 (1958)

140. Abashian, A., Hafner, E. M., *Phys. Rev. Letters*, **1**, 255 (1958); Hillman, P. Johansson, A., Tibell, G., *Phys. Rev.*, **110**, 1218 (1958)

141. Handler, R., Wright, S. C., Pondrom, L., Limon, P., Olsen, S., Kloeppel, P., *Phys. Rev. Letters*, **19**, 933 (1967)

142. Gross, E. E., Malanify, J. J., Van der Woude, A., Zucker, A., *Phys. Rev. Letters*, **21**, 1476 (1968)

143. Rosen, L., Brolley, J. E., Jr., *Phys. Rev. Letters*, **2**, 98 (1959)

144. Bodansky, D., Eccles, S. F., Farwell, G. W., Rickey, M. E., Robinson, P. C., *Phys. Rev. Letters*, **2**, 101 (1959)

145. Bodansky, D., Braithwaite, W. J., Shreve, D. C., Storm, D. W., Weitkamp, W. G., *Phys. Rev. Letters*, **17**, 589 (1966); Weitkamp, W. G., Storm, D. W., Shreve, D. C., Braithwaite, W. J., Bodansky, D., *Phys. Rev.*, **165**, 1233 (1968)

146. Von Witsch, W., Richter, A., von Brentano, P., *Phys. Letters*, **22**, 631 (1966); *Phys. Rev. Letters*, **19**, 524 (1967); *Phys. Rev.*, **169**, 923 (1968)

147. Thornton, S. T., Jones, C. M., Bair, J. K., Mancusi, M. D., Willard, H. B., *Phys. Rev. Letters*, **21**, 447 (1968)

148. Robson, D., *Phys. Letters*, **26B**, 117 (1968)

149. Henley, E. M., Huffman, A., *Phys. Rev. Letters*, **20**, 1191 (1968)

150. Moldauer, P., *Phys. Letters*, **26B**, 713 (1968)

151. Ericson, T. E. O., *Phys. Letters*, **23**, 97 (1966)

152. Mahaux, C., Weidenmüller, H. A., *Phys. Letters*, **23**, 100 (1966)

153. Moldauer, P. A., *Phys. Rev.*, **165**, 1136 (1968)

154. Rosenzweig, N., Monahan, J. E., Mehta, M. L., *Nucl. Phys.*, **A109**, 437 (1968)

155. Favro, L. D., McDonald, J. F., *Phys. Rev. Letters*, **19**, 1254 (1967)

156. Wigner, E. P., *SIAM Rev.*, **9**, 1 (1967)

157. Boulware, D. G., *Nuovo Cimento*, **40A**, 1041 (1965)

158. Feinberg, G., *Phys. Rev.*, **140**, B1402 (1965)

159. Smith, J. H., Purcell, E. M., Ramsey, N. F., *Phys. Rev.*, **108**, 120 (1957)

160. Miller, P. D., Dress, W. B., Baird, J. K., Ramsey, N. F., *Phys. Rev. Letters*, **19**, 381 (1967); Dress, W. B., Baird, J. K., Miller, P. D., Ramsey, N. F., *Phys. Rev.*, **170**, 1200 (1968)

161. Shull, C. G., Nathans, R., *Phys. Rev. Letters*, **19**, 384 (1967)

162. Cohen, V. W., Nathans, R., Silsbee, H. B., Lipworth, E., Ramsey, N. F., *Phys. Rev.*, **177**, 1942 (1969)

163. Miller, P. D., Dress, W. B., Baird, J. K., Ramsey, N. F., as reported by Cronin, J., in *Proc. 14th Intern. Conf. High Energy Physics, Vienna, 1968*, 281

164. Barton, G., White, E. D. (To be published)

165. Vanderleeden, J. C., Boehm, F., *Bull. Am. Phys. Soc.*, **14**, 587 (1969)

166. Feuer, M. (Thesis, Harvard Univ., 1969, unpublished); Olesen, P., Rao, J. S., *Phys. Letters*, **29B**, 233 (1969)

NUCLEAR MASS RELATIONS

By G. T. Garvey[1,2]

Department of Nuclear Physics, Oxford University, Oxford, England

Contents

GENERAL INTRODUCTION

This review is concerned with the present state of equations proposed to relate the observed energy differences either between ground states of neighbouring nuclei or between the levels in an isospin multiplet. The equations are cast in terms of mass differences rather than direct calculation of the masses themselves, because of the uncertainties encountered in any finite many-body calculation in nuclear physics. Even if the free nucleon-nucleon interaction were properly in hand, it is not all clear how to introduce it unambiguously into the calculation of the properties of a system of N neutrons and Z protons. Because of the difficulties involved in a direct and rigorous approach and the requirement to determine a "best" value for yet unmeasured masses, one is forced to adopt less fundamental procedures, some of which will be discussed in this review.

Basically one resorts to the use of mass relations because they provide more reliable estimates of masses than are available from a mass equation simply because they are less global in their approach and when properly employed bring to bear the most relevant data for the prediction of the desired mass. Sometimes the unknown mass can be calculated from a more formal theory of nuclear structure such as the shell model, but in general the nucleus involved is far from the valley of beta stability and either the required two-body matrix elements are not known or the number of configurations involved is so great that a shell-model calculation just is not realistic.

---

[1] A. P. Sloan Foundation Fellow 1967–1969.

[2] Permanent address—Physics Department, Princeton University, Princeton, New Jersey.

In addition to the knowledge about nuclear structure to be gained from success or failure of a particular mass relation amongst the already measured masses, there are other important aspects of these studies. As will be shown they can provide a useful tool in understanding the nature of isospin mixing, particularly in the light nuclei. Insofar as they predict the masses of nuclei far from the valley of beta stability in terms of the masses of the more stable nuclei, mass relations can reliably show changes in the nuclear binding energy for larger values of N-Z than have yet been encountered. Most directly they provide a useful guide to the experimentalists working on measuring the properties of nuclei far from the valley of stability (1–4) and to the astrophysicist interested in problems of nuclear synthesis, particularly as regards the "r" process (5–8).

Any equation which yields the mass of a group of nucleons contains in its structure a relation between the differences of these masses; however, this approach will not be stressed as it has had a long exposure and been refined and reviewed (9–11) several times, some of the reviews being quite recent. Mass equations as such will be introduced only for comparison or discussion and this review will concentrate instead on those equations whose content is expressed in terms of differences.

Mass relations quite naturally divide into two classes (a) those assuming charge independence or charge symmetry for the nuclear interactions and yielding expressions for the splittings of members of an isospin multiplet and (b) those relating the energies of states which differ in their specifically nuclear interactions. We begin with a review of the first class.

Mass Relations Following from the Assumption of Charge Independence

The basic idea underlying the relations to be treated in this section is that in first approximation the interactions determining the nuclear wavefunctions are assumed to be charge independent. Although the charge-dependent Coulomb force is large, it is not very effective in causing isospin mixing in the ground states of nuclei for a variety of reasons. The present status of the use of isospin for specifying nuclear wavefunctions can be drawn from the proceedings of recent conferences (12) and the entire topic has been reviewed in this series (13).

The splitting of the $(2T+1)$ members of a multiplet with isospin T will be assumed to arise solely from electromagnetic effects of which the Coulomb repulsion between protons is far and away the most important. The so-called isobaric-multiplet mass equation can be derived from quite general principles and shows much of the underlying physics in other mass relations based on the notion of charge independence. Thus we shall treat it first in this section.

The isobaric-multiplet mass equation.—The eigenvalues $E_{\alpha,T}$ of a charge-independent Hamiltonian H_{CI} are independent of the $2T+1$ values of T_z in a given isospin multiplet. This may be expressed formally as

$$H_{CI} \big| \alpha T T_z \big\rangle = E_{\alpha,T} \big| \alpha T T_z \big\rangle \qquad 1.$$

where α represents all the quantum numbers necessary for a unique specification of the eigenstate $|\alpha T T_z\rangle$, apart from the isospin T and its projection, $T_z = (N-Z)/2$, on the charge axis. If one assumes that the only charge-dependent perturbations are due to two-body interactions (certainly true for all electromagnetic interactions), the perturbations can be written as

$$H_{CD} = \sum_{i>j=1}^{A} V_0(i,j) + V_1(i,j)(t_{zi} + t_{zj}) + V_2(ij)t_{zi}t_{zj} \qquad 2.$$

where t_{zi} is a single-particle isospin operator ($T_z = \sum_{i=1}^{A} t_{zi}$) and where $V(i,j)$ denotes some two-body interaction dependent on the relative coordinates of particles i and j. This charge-dependent interaction is most usefully rewritten as

$$H_{CD} = \sum_{i>j=1}^{A} \left[V_0(ij) + \frac{V_2(ij)}{3} \mathbf{t}_i \cdot \mathbf{t}_j \right] + \sum_{i>j=1}^{A} [V_1(i,j)(t_{zi} + t_{zj})]$$
$$+ \sum_{i>j=1}^{A} \left[\frac{V_2(i,j)}{3} (3t_{zi}t_{zj} - \mathbf{t}_i \cdot \mathbf{t}_j) \right] \qquad 3.$$

The first sum in Equation 3 transforms as a scalar in isospin space, the second as a vector, and the last as a tensor of second rank. Treating H_{CD} as a perturbation and applying the Wigner-Eckart theorem yields

$$\langle \alpha T T_z \big| H_{CI} + H_{CD} \big| \alpha T T_z \rangle = E_{\alpha,T} + \langle \alpha T \| H_{CD}^{(0)} \| \alpha T \rangle$$
$$- \frac{T_z}{[T(T+1)]^{1/2}} \langle \alpha T \| H_{CD}^{(1)} \| \alpha T \rangle$$
$$+ \frac{3T_z^2 - T(T+1)}{[T(T+1)(2T+3)(2T-1)]^{1/2}} \langle \alpha T \| H_{CD}^{(2)} \| \alpha T \rangle \quad 4.$$

where the double-barred matrix elements are reduced with respect to T_z and the superscript in brackets on H_{CD} indicates it as the scalar (0), vector (1), or tensor (2) part of Equation 3. Thus Equation 4 can be simply written as

$$M(\alpha, T, T_z) = a(\alpha, T) + b(\alpha, T)T_z + c(\alpha, T)T_z^2 \qquad 5.$$

where $a(\alpha, T)$, $b(\alpha, T)$, and $c(\alpha, T)$ are constants associated with a particular multiplet. Equation 5 is referred to as the isobaric-multiplet mass equation and though it has been known for many years (14–16) it has only recently been a subject of considerable interest in nuclear physics (17–20).

Even if Equation 5 were to hold exactly, it would not establish the complete charge independence of nuclear forces, as a specifically nuclear two-body charge-dependent force has isospin transformation properties similar

TABLE I. The experimentally determined mass defects of the $T=3/2$ multiplets in which all four members are known. The last column lists the value of a possible cubic term to be added to Equation 5.

A	$T_z = 3/2$	$T_z = 1/2$	$T_z = -1/2$	$T_z = -3/2$	d(keV)
7	$26.11 \pm .030^a$	$26.19 \pm .040^a$	$26.75 \pm .030^a$	$27.94 \pm .100^a$	-25 ± 30
9	$24.965 \pm .005^b$	$25.743 \pm .005^a$	$27.075 \pm .005^a$	$28.916 \pm .005^a$	7.5 ± 2.3
13	$16.562 \pm .004^c$	$18.231 \pm .003^a$	$20.411 \pm .004^a$	$23.11 \pm .040^a$	1 ± 11
17	$7.871 \pm .015^c$	$10.267 \pm .006^a$	$13.149 \pm .004^a$	$16.47 \pm .250^a$	8 ± 42
21	$-.046 \pm .007^c$	$3.126 \pm .006^a$	$6.790 \pm .005^d$	$10.95 \pm .120^f$	-1 ± 20
23	$-5.148 \pm .003^c$	$-1.628 \pm .021^a$	$2.298 \pm .040^a$	$6.766 \pm .080^g$	-23 ± 26
25	$-9.356 \pm .009^c$	$-5.395 \pm .013^a$	$-1.017 \pm .005^a$	$3.86 \pm .120^f$	-14 ± 21
37	$-31.765 \pm .001^c$	$-25.941 \pm .030^a$	$-19.753 \pm .005^e$	$-13.23 \pm .050^{f,a}$	5 ± 17

[a] Cerny, J., *Ann. Rev. Nucl. Sci.*, **18**, 27 (1968)
[b] Nettles, P. H., et al., Paper VH, *Proc. 2nd Conf. Isospin in Nucl. Phys., March 1969*
[c] Mattauch, J., Thiele, W., Wapstra, A., *Nucl. Phys.*, **67**, 1 (1968)
[d] McDonald, et al., *Bull. Am. Phys. Soc.*, **13**, 635 (1968)
[e] Endt, P. M., Van der Leun, C., *Nucl. Phys.*, A105, 1 (1967)
[f] Cerny, J. (Private communication)
[g] Cerny, J., Mendelson, R., Wozniak, G., Esterl, J., Hardy, J., *Phys. Rev. Letters*, **22**, 612 (1969)

to those of the charge-dependent electromagnetic forces. Thus to examine the presumably small effects arising from two-body charge-dependent nuclear forces, the effects of the electromagnetic interaction between the nucleons must be accounted for with a degree of reliability difficult to achieve with present-day knowledge.

Equation 5 states that if the charge-dependent forces are two-body and are adequately treated by first-order perturbation theory, then three numbers are sufficient to specify the energies of all members of a isospin multiplet. Any multiplet with more than three members $(T>1)$ allows a direct test of this hypothesis. Only recently has one been able to measure all the members of sufficiently large multiplets and a recent review by Cerny (21) discusses the techniques and results encountered in these measurements. Table I lists the relevant masses in the $T = \frac{3}{2}$ multiplets that constitute the present body of data which tests Equation 5. Table II lists values for the coefficients determined from a χ^2 fit to the masses listed in Table I. In the case of the most accurately measured multiplet, $A = 9$, the value of $\chi^2 = 4.05$ corresponds to a probability of < 5 per cent and demonstrates the breakdown of Equation 5 at this level (~ 5 keV). This breakdown can be expressed by adding higher powers of T_z to Equation 5. If one assumes a cubic term of the form dT_z^3, the mass 9 multiplet yields $d = 7.5 \pm 2.3$ keV. It would be premature to claim that this indicates the existence of three-body charge-dependent forces. Equation 5 is not exact; charge-dependent two-body forces cause isospin mixing and the assumption of good T is only true in the first order. The effects of the electromagnetic interactions in the next order of perturbation theory have been looked at (20, 22, 23) and it is found that the largest

TABLE II. The coefficients in MeV of Equation 5 as extracted from a χ^2 fit to the $T=3/2$ multiplets presented in Table I. The error in each coefficient is listed in MeV in parenthesis. Column 5 lists the value of χ^2 for each fit while column 6 gives the maximum deviation in keV encountered with the fit and the corresponding T_z value.

A	a (MeV)	b (MeV)	c (MeV)	x^2	(Exp-Pred)$_{max}$(T_z)
7	26.398 (29)	$-.594$ (28)	.270 (27)	.67	45 $(-3/2)$
9	26.3425 (4)	-1.3185 (2.2)	.267 (2.5)	4.05	6.8 $(-1/2)$
13	19.257 (3.6)	-2.180 (4.9)	.256 (10)	.01	7 $(-3/2)$
17	11.647 (5)	-2.882 (7)	.243 (9)	.035	-46 $(-3/2)$
21	4.896 (5.9)	-3.664 (7.6)	.246 (7.3)	.001	3 $(-3/2)$
23	0.289 (21)	-3.960 (23)	.223 (17)	.75	35 $(-3/2)$
25	-3.260 (9)	-4.381 (13)	.211 (13)	.41	73 $(-3/2)$
37	-22.884 (11)	-6.178 (15)	.172 (16)	.18	-12 $(1/2)$

terms in second-order preserve the form of Equation 5. Smaller terms give rise to T_z^3 and T_z^4 dependency. While general arguments (24) indicate that the size of the coefficient of the cubic term is $\sim Zc(\alpha, T)/137$, it is not clear that the value observed in the mass 9 multiplet can be obtained in a second-order calculation while preserving the high isospin purity observed for the Be^9 (25) member of this multiplet.

A cubic term can arise from the following sources:

(a) the combined effect of the isovector and isotensor electromagnetic interaction calculated to second order in e^2;

(b) the isovector interaction in third order;

(c) an isospin mixing in the exterior region (13) in combination with the T_z dependence (20) of the boundary conditions;

(d) a three-body charge-dependent force.

Reference (23) examines the effect of (a) but does not properly account for the narrow width ($\sim.8$ keV) observed for the lowest $T=\frac{3}{2}$ state in Be^9. If the experimental evidence points to small-isospin mixing ($<1\%$) in the lowest $T=\frac{3}{2}$ states of Be^9 and B^9, then one will have to take explanation (d) much more seriously. Further experiments on the isospin mixing of these levels are required to clarify this situation.

From Tables I and II it is clear that for the purpose of predicting masses the isobaric-multiplet mass equation has an intrinsic accuracy ~ 20 keV.

The preceding discussion has been within a model-independent framework. No assumptions about nuclear wavefunctions were made except that they were eigenstates of T^2. Model-dependent assumptions about the nature of the wavefunctions have been made in an effort to calculate values of the coefficients of Equation 5. An extensive discussion of this problem would lead us into the details of Coluomb-energy calculations but these efforts need to be mentioned as they have direct bearing on mass relations. The most complete studies along these lines have been made by Hecht (26, 27) and Jänecke (22, 28, 29). The two-body Coulomb interaction between nucleons is written as

$$H_c = \sum_{i>j=1}^{A} \frac{e^2}{r_{ij}} (\tfrac{1}{2} - t_{zi})(\tfrac{1}{2} - t_{zj})$$

$$= \sum_{i>j=1}^{A} \left\{ \left(\frac{e^2}{r_{ij}}\right)\left(\frac{1}{4} + \frac{t_i \cdot t_j}{3} - \frac{(t_{zi} + t_{zj})}{2} + \frac{1}{3}(3t_{zi}t_{zj} - t_i \cdot t_j)\right)\right\} \qquad 6.$$

where r_{ij} is the separation distance between particles i and j.

Hecht has obtained (26, 27) analytic expressions for the isoscalar, isovector, and isotensor terms of the Coulomb energy in terms of two-body Coulomb matrix elements for two models of the many-body wavefunction. One of the schemes, the jj-coupling shell model in the low seniority limit ($v \leq 2$), should be a reasonable approximation for the lowest states of a specific T in, say, the $1f_{7/2}$ region. The other scheme studied by Hecht assumes the Wigner supermultiplet (14, 30) model to hold within a major oscillator shell and expressions are obtained for states with spatial symmetries appropriate to the lowest-lying configurations of a given T and A. The form of the expressions for the Coulomb energy is quite similar in both the j-j coupling and supermultiplet models. This is particularly true as regards the effects of Coulomb pairing energy (31–33) and appears rather difficult to distinguish between the two models if purely phenomenological fits to data are attempted. In addition to the assumption of a specific model for the single-particle configurations and their couplings, crucial assumptions regarding the appropriate radial wavefunctions must be made if calculations are to be carried out. The work of Wilkinson et al. (34) and Nolan et al. (35) has shown the importance of the effect of the proton-separation energy on Coulomb-energy differences. No theory of Coulomb-energy differences assuming radial wavefunctions generated from, say, a single infinite harmonic-oscillator potential can be trusted to give reliable results.

Of course, the observed value of the coefficients b and c contains small contributions from charge-dependent effects other than the Coulomb force such as the electromagnetic spin-orbit interaction (36, 37), the charge-dependent part of the relativistic Thomas effect (36), and the neutron-proton mass difference. Jänecke has applied Hecht's expressions to the observed energy splitting in the known isospin multiplets within the $1f_{7/2}$, $2s - 1d$, and $1p$ shells, making allowances for several of the above effects. His approach is not to calculate the two-body Coulomb matrix elements but to leave them as simple A-dependent parameters (to take account of changes in the nuclear size with A) which are fixed by least squares fit to the data. As there are many more data than free parameters a significant test can be made. The fits obtained by Jänecke are in general quite good, particularly in the $1f_{7/2}$ region where the resulting predictions are well within experimental error, one pathological case ($Sc^{42} - Ca^{42}$) having been dropped from the sample. This subject has been very completely reviewed by Jänecke in a recent article (22). The most serious criticism against this work is the treatment of the radial matrix elements and the related neglect of the effects of proton separation energy. Attempts to fit the experimental data generating the radial wavefunctions from a harmonic-oscillator potential have not been successful (22), in particular the value obtained for V_0 (the Coulomb interaction of a pair of protons coupled to $J = 0$) is much smaller than the corresponding value extracted from experiment. This is most likely due to neglecting the fact that the nuclear pairing increases the binding energy and thus causes these protons to have a larger overlap with other protons in the nucleus. More realistic calculations with finite wells should be employed to separate the effects of binding and pairing correctly.

Earlier (17, 18), hope had been expressed that one could extract significant information from Equation 5 regarding the charge independence of nuclear forces. Wilkinson (38) investigated the possibility of making the reasonable assumption that if nuclear force was not charge independent it was very likely charge symmetric ($V_{nn} = V_{pp} \neq V_{np}$). A charge-symmetric two-body interaction has no isovector part (20), so that only the coefficient c is different from what is to be expected from electromagnetic perturbations. In a uniformly charged sphere model for the nucleus the Coulomb contribution to the coefficients b and c of Equation 5 is

$$b_{Coul} = \frac{3e^2}{5R_c} (1 - A)$$

$$c_{Coul} = \frac{3e^2}{5R_c} \qquad\qquad 7.$$

where R_c is the charge radius appropriate to the multiplet. The Coulomb contributions to b and c are thus related

$$b_{Coul} = (1 - A)c_{Coul} \qquad\qquad 8.$$

This is really only rigorously true for the isospin multiplets with $T = A/2$. However, it is approximately true and in any instance more accurate expressions (26, 27) for b and c may be used, the important point being that in first order the Coulomb contributions to b and c are related to one another. To compare the values of the Coulomb contributions one must, of course, remove the effects of other charge-dependent terms. Accordingly one writes

$$b = b_{Coul} + b_{so} + b_{CD} + \Delta_{np}$$

and $c = c_{\text{Coul}} + c_{so} + c_{CD}$

where Δ_{np} is the neutron-hydrogen atom mass difference (782 keV) and b_{so} and c_{so} arise from the spin orbit and Thomas precision interactions. The terms b_{CD} and c_{CD} represent the contribution to b and c from nonelectromagnetic charge-dependent effects. For charge-symmetric interactions b_{CD} is equal to zero. In the uniform-charge model of the nucleus assumed above, corrections must be applied for the effects of Coulomb pairing before b_{Coul} and c_{Coul} $+c_{CD}$ can be compared to test for charge independence. Wilkinson (38) extracted the numbers b and c from isospin multiplets in the $1p$ shell and applied spin-orbit and pairing corrections to these numbers to obtain b_{Coul}. The values of b_{Coul} were then plotted against A and the slope of the best-fit straight line was then compared to $c_{\text{Coul}}+c_{CD}$. This comparison yields c_{CD} $= -.010 \pm .017$ MeV. Relating this value to the two-body nuclear force, it is claimed (38) that this charge-dependent effect is <1 per cent of the $^{31}S_0$ interaction, which is the largest of the $T=1$ two-body nuclear interactions. Jänecke (29) has also compared b_{Coul} to c_{Coul} using the equations of Hecht (27) and found that the term which characterizes the pairing interaction is different for the vector and tensor Coulomb energies by some 20 per cent; he suggests that this may be evidence for a small charge-dependent effect. However, it has been recently (20, 23) noted that the higher-order effects of the electromagnetic field, particularly the Coulomb isovector interactions in second order, have a strong effect on the coefficient c and until these effects are better understood it will be difficult to unambiguously show a positive effect of nuclear charge-dependent force in a nucleus via a comparison of b_{Coul} to c_{Coul}.

Approximate mass relations based on charge independence.—The isobaric-multiplet mass equation certainly represents an accurate format for predicting the masses of yet undiscovered nuclei. However, its implementation required a large amount of input data as one must know three members of the specific multiplet containing the nuclear level whose mass is to be determined; otherwise uncertain approximation procedures are involved. The relative insensitivity of the Coulomb energy to the details of the nuclear system allows one to assume in some approximation that it is state independent, but three important effects can cause appreciable variation in the Coulomb-energy differences. The comparison of analogue levels in mirror nuclei shows this to be an approximation which may be in error the order of several hundred kilovolts. For example lining up the ground states of C^{13} and N^{13} there is a difference in the Coulomb energies of the first three excited states in C^{13} and N^{13} of .72, .17, and .29 MeV relative to the Coulomb-energy difference of the ground states. The largest shift is in the first excited state whose wavefunction is predominantly a $2s^{1/2}$ particle added to a C^{12} core. This shift is understood to arise from the Thomas-Ehrman effect (39–41)

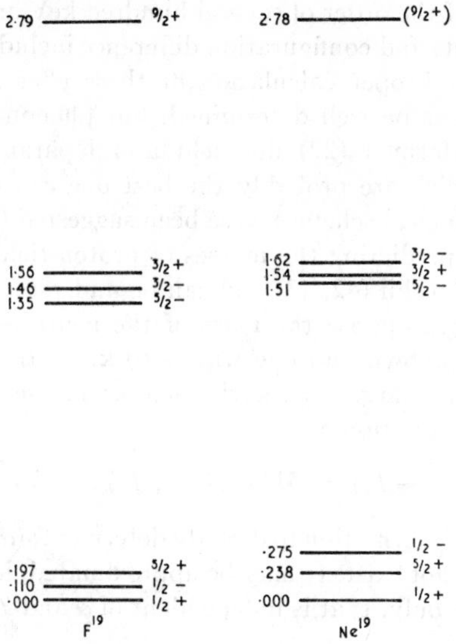

2·79 ——————— 9/2+   2·78 ——————(9/2+)

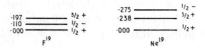

1·56 ——————— 3/2+   1·62 ——————— 3/2 −
1·46 ——————— 3/2 −   1·54 ——————— 3/2 +
1·35 ——————— 5/2 −   1·51 ——————— 5/2 −

　　　　　　　　　　　　　　　·275 ——————— 1/2 −
·197 ——————— 5/2 +   ·238 ——————— 5/2 +
·110 ——————— 1/2 −
·000 ——————— 1/2 +   ·000 ——————— 1/2 +

F^{19}　　　　　　　　　　　　Ne^{19}

FIG. 1. Level diagram showing the positions of the lowest-lying levels in the mirror nuclei F^{19} and Ne^{19}.

which is largest for s states and is particularly large in this case because the state is unbound in N^{13}.

Differences in Coulomb energy arising from different single-particle configurations can also be observed in the shifts between states in the mirror nuclei. Figure 1 shows the energies of the low-lying levels in F^{19} and Ne^{19}. The negative parity states in F^{19} all appear depressed by 160 ± 010 keV relative to their energy in Ne^{19}. This is readily understood as the negative parity states in F^{19} and Ne^{19} are predominantly due to the excitation of a particle out of the $1p$ shell into the $2s-1d$ shell such that the configuration of the particles in the $2s-1d$ shell possesses maximum spatial symmetry, hence have $T=0$. Thus in Ne^{19} the particle promoted from the p shell is a neutron, and the Coulomb energy of the negative and positive parity states in Ne^{19} is the same while in F^{19} the particle promoted from the $1p$ shell is a proton with the result that the Coulomb energy in these negative parity states is lower than for the lowest-lying positive-parity states.

The last important effect that an equation relating differences in Coulomb energies must take account of arises from pairing (31–33). A pair of protons coupled to total angular momentum zero $(J=0)$ has a higher Coulomb energy by as much as 150 keV than the same pair with $J \neq 0$. This effect has been known for a long time and is most readily seen by noting the second differences (32) in the Coulomb energies of mirror nuclei.

Hence if errors of the order of several hundred keV are to be avoided the binding-energy effects and configuration difference including pairing must be taken into account. Proper calculation of these effects requires that the nuclear wavefunctions be well determined, but phenomenological fits such as Jänecke has performed (22) do yield useful parameterizations of the Coulomb energy which are probably the best one can do at present. Two simpler phenomenological schemes have been suggested (42–44) which are of considerable use in predicting the masses of proton-rich nuclei. The first of these is due to Goldanskii (42, 43) and can be understood within the framework of Equation 5. Suppose the mass of the neutron-rich member of an isospin multiplet is known and one wishes to know the mass of the corresponding proton-rich charge-symmetric member of the multiplet. One may write directly from Equation 5

$$M(\alpha, A, T, -T_z) = M(\alpha, A, T, T_z) - 2b(\alpha, A, T)T_z \qquad 9.$$

Generally, sufficient information to directly determine $b(\alpha, A, T)$ for the multiplet of interest does not exist. It may be approximated, however, by treating b to be A dependent only, that is independent of α and T. Thus for odd A

$$M(\alpha, A, T, \tfrac{1}{2}) - M(\alpha, A, T, -\tfrac{1}{2}) = b(\alpha, A, T) = M(\beta, A, \tfrac{1}{2}, \tfrac{1}{2})$$
$$- M(\beta, A, \tfrac{1}{2}, -\tfrac{1}{2}) \qquad 10.$$

where β plays the same role as α, and $M(\beta, A, \tfrac{1}{2}, T_z)$ is the mass of a $T = \tfrac{1}{2}$ level usually taken as the ground state. For even A

$$M(\alpha, A, T, 1) - M(\alpha, A, T, -1) = 2b(\alpha, A, T) = M(\beta, A, 1, 1)$$
$$- M(\beta, A, 1, -1) \qquad 11.$$

where the mass differences on the right-hand side of Equations 10 and 11 are well known from $A = 6$ to $A \sim 50$. This procedure neglects the possibility that the single-particle configurations may be different if one is near magic numbers and even if the single-particle states are correct the important effects of proton pairing are not correctly treated, particularly for odd A. The expression obtained by Hecht for the isovector Coulomb energy (27) in a j^n, low-seniority approximation is

$$b(\alpha, A, T) = -3A_c - 3B(n - 1) - 12C(j + 1)$$
$$- \frac{3C(n - 2j - 1)}{2T(T + 1)} + \frac{3C(2T + 1)(2j + 3)(-1)^{(n/2)-T}}{2T(T + 1)} \qquad 12.$$

for A odd, seniority equal to one and reduced isospin equal to one half. The total isospin of the level is denoted by T; n, the number of particles in the j shell, is related to A via $n = A - A_0$ where A_0 is the number of nucleons in the doubly magic core. A_c represents the interaction of the protons in the j shell with the other protons in the nucleus and

$$B = \frac{2(j+1)\overline{V}_2 - V_0}{2(2j+1)} \qquad C = \frac{V_0 - \overline{V}_2}{4(2j+1)} \qquad 13.$$

The V_J's are the two-body matrix element of the Coulomb interaction of a pair of protons in the j shell coupled to angular momentum J.

$$V_J = \left\langle j^2 J \left| \frac{e^2}{3r_{ij}} \right| j^2 J \right\rangle \qquad 14.$$

V_2 is an average of V_J over all even values of J except $J=0$ and is given by

$$\overline{V}_2 = \frac{\displaystyle\sum_{\substack{J=2n \\ n \neq 0}}^{2j-1} (2J+1)V_J}{\displaystyle\sum_{\substack{J=2n \\ n \neq 0}}^{2j-1} (2J+1)}$$

Using Equation 12 one obtains

$$M(\alpha, A, T, \tfrac{1}{2}) - M(\alpha, A, T, -\tfrac{1}{2}) - M(\beta, A, \tfrac{1}{2}, \tfrac{1}{2}) + M(\beta, A, \tfrac{1}{2}, -\tfrac{1}{2})$$

$$= b(\alpha, A, T) - b(\alpha, A, \tfrac{1}{2}) = 3C(n - 2j - 1)\left[\frac{2}{3} - \frac{1}{2T(T+1)} \right]$$

$$+ 3C(2j+3)(-1)^{(n-1)/2}\left[\frac{2T+1}{2T(T+1)} (-1)^{(1/2)-T} - \frac{4}{3} \right] \qquad 15.$$

It is therefore evident that Equation 10 will not in general treat Coulomb pairing effects properly, as C is the difference in energy between a pair of protons coupled to $J=0$ and the average of the other couplings. In particular when $T=3/2$,

$$b\left(\alpha, A, \frac{3}{2}\right) - b\left(\beta, A, \frac{1}{2}\right) = \frac{8}{5} C(n - 2j - 1)$$

$$- 28 \frac{C}{5}(2j+3)(-1)^{(n-1)/2} \qquad 16.$$

the second term is the dominant one and so the difference is either positive or negative as $A = 4m+1$ or $4m-1$. The above equation has been written down by Jänecke who has shown (22) its effect in the difference in the excitation energies of the lowest $T=3/2$ states in $T_z = \pm\tfrac{1}{2}$ nuclei. The effect of Equation 14 is also seen in Table IV where the predictions for the masses of the known $T_z = -3/2$ nuclei using Equation 10 are shown in column 5. In $A = 9, 17, 21,$ 25, and 37 the predicted masses are smaller than observed, while in $A = 7, 13,$ and 23 they are larger. Of course the validity of the j^n assumption is not correct for $A = 17$ and 13 and indeed its use can be questioned throughout the $2s - 1d$ shell; however, the same alternation in the Coulomb shift is predicted in the supermultiplet model (27). The size of the alternation is larger than expected (22) from simple Coulomb pairing effects and as mentioned above additionally reflects the effect of the proton binding energy. This is certainly the case for $A = 7$ shown in Table IV where the prediction is incorrect by 40 per cent in the Coulomb-energy difference, largely because the $T = 3/2$ multiplet is in the continuum.

An alternative procedure which takes account of the single-particle configurations was proposed by Kelson & Garvey (44). Their formula can be written as

$$M(\alpha_0, A, T, T_z) - M(\alpha_z A, T, -T_z)$$
$$= \sum_{i=1}^{2T_z} M(\alpha_0{}^i, A_i, \tfrac{1}{2}, \tfrac{1}{2}) - M(\alpha_0{}^i, A_i, \tfrac{1}{2}, -\tfrac{1}{2}) \quad 17.$$

or in terms of the vector coefficients of the multiplets involved

$$2T_z b(\alpha_0, A, T) = \sum_{i=1}^{2T_z} b(\alpha_0{}^i, A_i, \tfrac{1}{2}) \quad\quad 18.$$

where $A_i = A - 2T_z - 1 + 2i$, and $\alpha_0{}^i$ simply designates the quantum numbers appropriate to the ground state of the $T = \tfrac{1}{2}$ system with A_i nucleons. This expression therefore replaces $2T_z b(\alpha_0 A, \tfrac{1}{2})$ in Equation 9 with the sum shown above. In addition to taking account of the single-particle configurations, it treats pairing to a better approximation than is true of the Goldanskii approach (Equations 9, 10). The degree of this approximation is easily worked out from Hecht's equations (23) but the number of special cases makes it difficult to write a general rule. The simplest case to which Equation 17 can be applied has $T = 1$ and yields:

$$M(\alpha_0 A, 1; 1) - M(\alpha_0 A, 1, -1) = M(\alpha_0{}^{(1)}, A - 1, \tfrac{1}{2}, \tfrac{1}{2})$$
$$- M(\alpha_0{}^{(1)}, A - 1, \tfrac{1}{2}, -\tfrac{1}{2}) + M(\alpha_0{}^{(2)}, A + 1, \tfrac{1}{2}, \tfrac{1}{2})$$
$$- M(\alpha_0{}^{(2)}, A + 1, \tfrac{1}{2}, -\tfrac{1}{2}) \quad\quad 19.$$

Table III presents the results of applying this equation over the range $8 \leq A \leq 42$ where appropriately accurate data exists. The two cases of poorest agreement occur for $A = 12$ and 16 where the proton separation energy in the $T_z = -1$ member is small, thus introducing the possibility of Thomas-Ehrman shifts mentioned earlier. In $A = 8$, even though the last proton in B^8 is only bound by 0.134 MeV which would cause a level shift, the effect is compensated by the small proton binding energy in B^9. In $A = 16$ where the largest discrepancy is noted an interesting case is presented. In comparing N^{16} and F^{16} a change in the level ordering has occurred (45) because of larger Thomas-Ehrman shifts for the low-spin states relative to higher-spin levels. The first four levels in N^{16} are at 0.00 (2−), .120 MeV (0−), .296 MeV (3−), and .396 MeV (1−), while in F^{16} the levels of corresponding spin and parity (45) oc-

TABLE III. A list of the deviations from zero encountered when using Equation 19. Column 1 lists the value of Z employed. Column 2 designates the nucleus with $Z+2$ protons and Z neutrons. This nucleus is therefore either odd-odd or even-even in N and Z. The next column gives the value obtained for Equation 19 in that particular case. The last column lists the proton binding energy of the nucleus in column 2. The experimental masses are taken from Mattauch et al., *Nucl. Phys.*, **67**, 1 (1967) and Endt & Van der Leun, *Nucl. Phys.*, **A105**, 1 (1967).

Z	Proton-rich $T=1$ member	Deviation (MeV) even-even	Deviation (MeV) odd-odd	$B(p)$ (MeV)
3	B^8		$-.047 \pm .006$	$-.134$
4	C^{10}	$-.003 \pm .014$		-4.05
5	N^{12}		$.234 \pm .010$	$-.370$
6	O^{14}	$-.008 \pm .004$		-4.64
7	F^{16}		$.300 \pm .014$	$+1.04$
8	Ne^{18}	$-.104 \pm .006$		-4.02
9	Na^{20}		$-.100 \pm .082$	-2.27
10	Mg^{22}	$-.034 \pm .050$		-5.53
11	Al^{24}		$.045 \pm .012$	-1.92
12	Si^{26}	$-.007 \pm .016$		-5.50
13	P^{28}		$.056 \pm .015$	-1.99
14	S^{30}	$.013 \pm .022$		-4.39
15	Cl^{32}		$-.030 \pm .026$	-1.59
16	Ar^{34}	$.007 \pm .030$		-4.68
17	K^{36}			-1.66
18	Ca^{38}	$.003 \pm .030$		-4.56
19	Sc^{40}		$-.007 \pm .018$	$-.52$
20	Ti^{42}	$-.070 \pm .018$		-3.86

TABLE IV. Comparison to experiment of the predictions of Goldanskii (42) and Kelson & Garvey (44) for the mass differences $M(A, \frac{3}{2}, -\frac{3}{2}) - M(A, \frac{3}{2}, \frac{3}{2})$. Column 2 lists $3b(A, \frac{1}{2})$ (see Equations 9 and 10). Columns 3, 4, and 5 list respectively the values for $b(A-1, 1) + b(A+2, \frac{1}{2})$, $b(A-2, \frac{1}{2}) + b(A, \frac{1}{2}) + b(A+2, \frac{1}{2})$, and $b(A-2, \frac{1}{2}) + b(A+1, 1)$ (see Equations 17 and 18). The last column lists the experimentally determined values of the mass defects for the $T_z = -3/2$ nuclei (see Table I). The use of two values for (1, 16) is discussed in the text.

A	Ref. (42)	Ref. (44)			Exp
7	2.586	1.836	2.155	2.202	1.83
9	3.204	3.913	3.911	3.958	3.951
13	6.663	7.209	6.962	6.753	6.55 ± 40
17	8.228	8.861	8.757	(8.893)	8.599 ± 250
				(8.328)	
21	10.635 ± 24	11.190	10.839 ± 8	10.938	10.99 ± 120
23	12.168	11.701	11.861 ± 12	11.948	11.914 ± 80
25	12.780 ± 24	13.129	13.126 ± 10	13.213	13.216 ± 120
37	18.453	18.632	18.635 ± 20		18.535 ± 50

cur at .436, .000, .736, and 0.200 MeV respectively. Thus directly applying Equation 17 in this case is not correct. The 1^- and 0^- levels are due mostly to a $1p_{1/2}^{-1} 2s_{1/2}$ configuration while the $J^\pi = 2^-$ and 3^- levels arise from the coupling of a $1p_{1/2}$ hole to a $1d_{5/2}$ particle. The Coulomb-energy differences obtained using the $O^{17} - F^{17}$ ground states are appropriate to a $d_{5/2}$ particle, so differences appearing on the left side of Equation 17 should be between the 3^- and 2^- states in $A = 16$. This reduces the deviation encountered in Equation 19 for this case to $+140$ keV. It is also more correct to compare the Coulomb-energy differences of the 0^- and 1^- levels in $A = 16$ to the sum of the $N^{15} - O^{15}$ differences plus the energy differences between the $\frac{1}{2}+$ first excited states of O^{17} and F^{17}. This $A = 17$ pair has a Coulomb-energy difference .371 keV less than that of the $5/2+$ ground states, and employing these levels a deviation of only $-.112$ MeV is had for Equation 19. Thus when the appropriate configurations are identified the larger discrepancies for Equation 17 are removed. The only other significant deviations are noted at $A = 18$ and 42, which represent cases involving a pair of particles outside a closed shell. These Coulomb-energy differences seem to be unusual cases from several points of view (18, 22, 46). Thus it appears from Tables III and IV that the use of Equation 15 yields the Coulomb-energy differences required for the determination of the masses of proton-rich nuclei to an accuracy of better than 200 keV. A rather complete set of predictions has been made for the masses of the proton-rich nuclei with $Z \leq 22$ using the above prescription (44).

General Mass Relationships

Introduction.—We now discuss the various relationships that have been proposed to hold between the masses of neighboring nuclei. Many of these rules which have been uncovered are solely empirical and as such are difficult to discuss (see for example 47–49). One is not, of course, on such firm footing in this section, as the nuclear states involved are no longer related by the well-established approximation of charge independence and one must make assumptions regarding the specifically nuclear interactions and their effects in determining the wavefunctions of the ground states. The most readily understandable mass equations are those that follow from an independent-particle description such as the shell model or a Hartree-Fock calculation. Unfortunately these theories are limited in the range of practical applicability as the configuration space becomes unmanageably large as soon as there are many nucleons beyond a "closed" shell. They therefore have not been used to generate a general expression for nuclear masses though Zeldes et al. (50) have used the notions of a shell model in constructing their nuclear mass equation. It is important to point out that though the range of shell-model mass predictions is limited, they work very well over the regions of applicability. In the $1p$ shell calculations of Cohen & Kurath (51) excellent results were obtained for the masses of these nuclei. Their calculation, however, involves specifying the 15 $1p$ shell two-body matrix elements, and the two single-particle energies. The ground-state wavefunctions and energies are obtained by diagonalization within this space so that a general formula for the masses is impossible. Talmi and his collaborators (52–55) have used a simple expression (55) for the interaction energy within a j^n configuration of neutrons and protons. The expression is

$$\left\langle j^n Tgs \left| \sum_{i>j}^{n} V_{ij} \right| j^n Tgs \right\rangle = \frac{n(n-1)\alpha}{2} + \left[\frac{n}{2}\right]\beta$$
$$+ (T(T+1) - \tfrac{3}{4}n)\gamma \qquad 20.$$

where $[n/2]$ is equal to $n/2$ for even A and $(n-1)/2$ for odd A, and α, β, and γ are related to linear combinations of the nuclear two-body matrix elements within the j shell. The above expression is derived in lowest-seniority approximation for a two-body interaction of the form

$$V_{ik} = x + yg_{ik} + 2z(\mathbf{t}_i \cdot \mathbf{t}_k) \qquad 21.$$

where t_i and t_k are the isospin operators for particles i and k, and g_{ik} is an operator which is nonzero only between $J=0$ pairs. The two-body interaction is characterized by x, y, and z. Equation 20 may not be applied to odd-odd nuclei. To compare this formula to actual nuclear binding energies a term must be added characterizing the interaction of a particle in the j shell, with the other nucleons constituting the core, and the effects of the Coulomb interaction must be removed. Thus an expression of the form

$$BE(j^n T) = nS + \frac{n(n-1)}{2}\alpha + \left[\frac{n}{2}\right]\beta + (T(T+1) - \tfrac{3}{4}n)\gamma \qquad 22.$$

is used to characterize the nuclear binding energy where S is the binding energy of a particle in the j shell due to the field set up by the doubly closed shell of $A-n$ nucleons. Table V presents a fit obtained by Talmi (55) for the binding energies in the $1d_{3/2}$ shell using the above expression and making a one-parameter fit to remove the effects of Coulomb energy. The overall quality of the fit is very good, never deviating from experiment by more than 200 keV. As is evident from the table and was apparent in earlier work by this group (52), the calculated binding energies near the middle of a shell are too small, which reflects the breaking down of the lowest-seniority assumption.

The above expression possesses a rather limited range of validity, as the fitting of this shell-model expression requires the identification of individual j shells and the parameterization of the interaction within them. Further it is usually not the case that configurations of the form $[[\text{closed shell}]^{J=0,T_{cs}}[j^n]^{J_n T_n}]^{J_n,T}$ are sufficient to represent the majority of nuclear ground states.

Mass differences in the supermultiplet model.—Renewed interest in a more

TABLE V. Binding energies in the $1d_{3/2}{}^n$ shell as determined from a fit to Equation 20. The binding energy of S^{32} is subtracted from all numbers.

Nucleus	Binding energy	
	experimental	calculated
$_{16}S_{17}{}^{33}$	8.64	8.68
$_{16}S_{18}{}^{34}$	20.06	20.07
$_{16}S_{19}{}^{35}$	27.05	27.13
$_{16}S_{20}{}^{36}$	36.93	36.89
$_{17}Cl_{16}{}^{33}$	2.29	2.33
$_{17}Cl_{18}{}^{35}$	26.43	26.29
$_{17}Cl_{20}{}^{37}$	45.33	45.17
$_{18}Ar_{17}{}^{35}$	19.67	19.51
$_{18}Ar_{18}{}^{36}$	34.94	34.79
$_{18}Ar_{19}{}^{37}$	43.73	43.90
$_{18}Ar_{20}{}^{38}$	55.57	55.71
$_{19}K_{18}{}^{37}$	36.81	37.03
$_{19}K_{20}{}^{39}$	61.94	61.84
$_{20}Ca_{19}{}^{39}$	54.64	54.54
$_{20}Ca_{20}{}^{40}$	70.28	70.25

global approach through the use of mass relations was brought about by the work of Franzini & Radicati (56). They investigated the ratio

$$R(A, T_3) = \frac{B(A, T_z) - B(A, T_z - 2)}{B(A, T_z - 1) - B(A, T_z - 2)} \qquad 23.$$

where $B(A, T_z)$ refer to the binding energy of a system where the Coulomb effects have been removed They postulated that the nuclear binding energy was given by the following supermultiplet model (14, 30) expression

$$B(A,T_z) = a(A) + b(A)[P^2 + 4P + P'^2 + 2P' + P''^2] \qquad 24.$$

where P, P', and P'' are three numbers specifying the space exchange character of the wavefunction. The smaller the value of the sum in brackets the higher is the space symmetry (more pairs with $T=0$, $S=1$, and $T=1$, $S=0$) which in turn corresponds to lower energy. The three numbers P, P', and P'' are related to the partition quantum numbers Λ_4, Λ_3, Λ_2, and Λ_1 which are the lengths of the rows of the spin-isospin Young's Tableau (see for example 57) associated with the state of interest. They have the obvious character that they are integers satisfying

$$\Lambda_1 + \Lambda_2 + \Lambda_3 + \Lambda_4 = A$$
$$\Lambda_1 \leq \Lambda_2 \leq \Lambda_3 \leq \Lambda_4 \qquad 25.$$

For example, Λ_4 could specify the number of neutrons with spin up, Λ_3 the number of neutrons with spin down, and so forth. The following definitions relate P and Λ

$$2P \equiv \Lambda_4 + \Lambda_3 - (\Lambda_2 + \Lambda_1)$$
$$2P' \equiv \Lambda_4 - \Lambda_3 + \Lambda_2 - \Lambda_1$$
$$2P'' \equiv \Lambda_4 - \Lambda_3 - (\Lambda_2 - \Lambda_1) \qquad 26.$$

The ground state of a nucleus with a given A and T_z is assumed to be in the most spatially symmetric state possible. Thus in every case except odd-odd $T_z = 0$ nuclei, $P = T_z$. If the nucleus has even A, then: $P' = 0$ if the neutron number is even, and $P' = 1$ if the neutron number is odd, and in both cases $P'' = 0$. For odd A: $P' = \frac{1}{2}$ and $P'' = \pm \frac{1}{2}$. An odd-odd nucleus with $T_z = 0$ has exactly the same supermultiplet symmetry as a $T = 1$ even-even nucleus, as it has an $S = 1$, $T = 0$ pair while the $T = 1$ even-even nucleus has a $T = 1$, $S = 0$ pair, so its P numbers are 1, 0, and 0. The expression for the energy (Equation 24) is obtained by assuming only Wigner and Majorana forces. The coefficients $a(A)$ and $b(A)$ depend on the relative strength of the exchange and ordinary forces. Thus the ratios in Equation 23 are now completely determined in this model by the use of Equation 24 and the specification of P, P', and P''. If we specify the value of A and T_z by the integers n and m, the following five cases obtain, excepting $T_z = 0$ with odd neutron number.

$$R(A = 2n + 1, \ T_z = m + \tfrac{1}{2}) = \frac{4(T_z + 1)}{2T_z + 1} \qquad \text{(a)}$$

$$R(A = 4n, \ T_z = 2m) = \frac{2T_z + 2}{T_z + 2} \qquad \text{(b)}$$

$$R(A = 4n, \ T_z = 2m + 1) = \frac{2T_z + 2}{T_z - 1} \qquad \text{(c)} \quad 27.$$

$$R(A = 4n + 2, \ T_z = 2m) = \frac{2T_z + 2}{T_z - 1} \qquad \text{(d)}$$

$$R(A = 4n + 2, \ T_z = 2m + 1) = \frac{2T_z + 2}{T_z + 2} \qquad \text{(e)}$$

In their paper Franzini & Radicati graph the values of $R(A, T_z)$ as function of A for fixed T_z for the various cases presented above. It is rather difficult to judge from the graphs what level of agreement is obtained. The simplest cases to compare to experiment are for $A = 2n + 1$ and as any expression for the binding energy that has quadratic dependence on T_z gives $R \to 2$ as $T_z \to \infty$, the most sensitive and reliable test is for odd A and smallest possible

450 GARVEY

T_z. Table VI presents the values obtained for this ratio with $T_z = 5/2$. The supermultiplet prediction is 2.33 while the average value of the data in Table VI is 2.652 with an average deviation of .220. A model which has the binding energy depending on T_z as $T_z(T_z+1)$ rather than as $T_3(T_3+4)$, the case for the supermultiplet model, gives the value of 2.667.

When the ratio involves even-A nuclei, the effects of pairing come directly into play as the denominator will always have a difference between the

TABLE VI. Values determined for the ratio $R(A, T)$ as defined by Equation 27a for $T_z = 5/2$. The required Coulomb corrections were made using experimentally determined Coulomb-energy differences given in (22). The binding energies were taken from (62).

A	5/2, 1/2	3/2, 1/2	$R(A, 5/2)$	$R(A, 5/2) - \overline{R(A, 5/2)}$
37	15.330	5.104	3.003	.351
39	15.989	6.633	2.410	.242
41	14.508	5.955	2.436	.216
3	12.381	4.169	2.969	.317
5	11.379	4.709	2.416	.236
7	11.469	4.120	2.783	.131
9	11.318	4.908	2.306	.346
51	11.114	4.428	2.509	.143
53	11.053	3.934	2.809	.157
5	12.544	4.758	2.636	.016
7	12.563	5.182	2.424	.228
9	11.211	3.867	2.899	.247
61	9.851	3.419	2.881	.229

$$\overline{R(A, 5/2)} = 2.652$$

binding energies of odd-odd and even-even nuclei. The supermultiplet model has the virtue that there is no free parameter to characterize the strength of this pairing interaction. There exists a single case where the ratio can be tested without Coulomb corrections being applied. That is for the ratio of the excitation energy of the lowest $T = 2$ state in an $A = 4n$, $T_z = 0$ nucleus to the excitation energy of lowest $T = 1$ state in the same nucleus. Table VII shows the values for these ratios. The average value of these ratios is equal to 1.64 with an average deviation of .075; however, there is evidence of a systematic decrease in the value of the ratio with increasing A. The ratios presented in Table VII should be compared to the prediction of Equation 27b with $T_3 = 2$ as the $T = 2$ state in the $T_z = 0$ nucleus is an isobaric analogue to the ground state of the nucleus with $T_z = 2$ and thus the supermultiplet prediction for this ratio is 1.50. However caution should be used when one applies tests to the binding energies of odd-odd nuclei for the simple reason that the J value of the ground state is not specified and it can be anywhere between $|j_p - j_n|$ and $j_p + j_n$ where j_p is the angular momentum of the odd proton and j_n is the angular momentum of the odd neutron. The J weighted value of the energy centroid of all these levels in the odd-odd nucleus is the more appropriate energy to use rather than the position of the lowest state. Usually not all members of this family are known. In the case of $A = 12$ and

TABLE VII. Listing of the experimentally determined values for $R(A, T)$ from Equation 27b with $T = 2$. The excitation energy of the $T = 2$ states is taken from (21). The last column gives the calculated values obtained with Equation 31.

A	$E_x(T=2)$	$E_x(T=1)$	$R(A, 2)$	$R(A, 2) - \overline{R(A, 2)}$	$R(A, 2)_{cal}$
12	27.50	15.11	1.82	.178	1.85
16	22.90	12.96	1.77	.128	1.78
20	16.73	10.27	1.63	.012	1.75
24	15.44	9.52	1.62	.020	1.72
28	15.21	9.32	1.63	.012	1.68
32	12.03	7.01	1.72	.078	1.66
36	10.86	6.61	1.64	.000	1.64
40	11.98	7.66	1.56	.082	1.62
52	8.57	5.78	1.48	.162	1.57
56	9.90	6.33	1.56	.082	1.56

$A = 16$, however, they are known; this reduces the value of the ratio to 1.74 and 1.77 respectively which still does not agree with the model prediction of 1.50. Of course, it may be argued that shell crossings are involved in these cases but for $A = 12$ this argument is not correct within the supermultiplet framework as no spin-orbit force is allowed.

The agreement between the supermultiplet prediction for the value of $R(A, T_z)$ and the experimentally determined values is poor for $T_z = 7/2$ as well as for $T_z = 3$ and 5, $A = 4m$; and $T_z = 4$ and 6, $A = 4m + 2$. In the case of odd A, $T_z = 7/2$ the failure of the model to reproduce the ratio is serious. For even A the cases of failure cited above occur simply because the denominators are smallest in these cases as the binding energy of an even-even nucleus is subtracted from an odd-odd nucleus with more symmetry energy, so the ratio is very sensitive to small effects in the binding energies of these nuclei. In the case of larger T the ratio tends to agree with the prediction of the supermultiplet model. However in the limit of large T, as was mentioned earlier, any model which has quadratic dependence on T yields $R_{T \to \infty}(A, T)$ $\to 2$. Thus it would seem that, contrary to the conclusion drawn by the authors (56), the supermultiplet theory does not work well and only agrees with experimental results insofar as it possesses certain features common to most independent-particle models of the nucleus. An examination of Table VI indicates that the numerator will have to be reduced by $\sim$1.5 MeV or the denominator increased by $\sim$800 keV if agreement is to be had with the supermultiplet theory predictions. These may be taken as the size of the absolute errors involved in this treatment.

Bremond (58) has considered the effects of including a pairing interaction in the supermultiplet scheme. He studies a ratio of mass differences somewhat different from that of Franzini & Radicati but his ratio also is relatively insensitive to the choice of nuclear model and again works best for large T where the value of the ratio is fixed from very general considerations.

Liquid-drop mass relations.—An investigation of the form of the symmetry energy in nuclei was undertaken by Jänecke (59). He again studied ratios of mass differences and used the liquid-drop model of the nucleus (60, 61) as his basic model. In the liquid-drop model the difference in binding energy of two nuclei with the same A and different $T = T_z = (N-Z)/2$, apart from Coulomb-energy differences, is given by

$$\Delta_{T\,T'}(A) = B(A, T) - B(A, T') = (E_{\text{sym}} + E_{\text{pair}})_{T\,A}$$
$$- (E_{\text{sym}} + E_{\text{pair}})_{T'\,A} \qquad 28.$$

where E_{sym} will be given below and $E_{\text{pair}} = \delta(A)$ for even-even nuclei, $E_{\text{pair}} = -\delta(A)$ for odd-odd nuclei, and $E_{\text{pair}} = 0$ for odd-A nuclei. The pairing energy can therefore be written as

$$E_p = \frac{\delta(A)}{2} (1 + (-1)^A)(-1)^{(A/2)+T} \qquad 29.$$

Jänecke assumed that the symmetry energy can be written in general as

$$\frac{V(A)}{A} (T^2 + b(A)T) = E_{\text{sym}} \qquad 30.$$

where $V(A)$ and $b(A)$ are quantities which should not be strongly A dependent. This equation is compatible with several proposed forms of the symmetry energy. The supermultiplet theory gives $T^2 + 4T$ while the jj-coupling shell model discussed earlier gives $T^2 + T$ and so forth. To avoid uncertainties with pairing energies Jänecke studied ratios of the type

$$R_{T'T'}(A) = \frac{B(A, T + 2) - B(A, T)}{B(A, T' + 2) - B(A, T')} \qquad 31.$$

where $B(A, T+2) - B(A, T)$ represents the difference in the binding energy (62) of the indicated nuclei with Coulomb-energy differences removed. The pairing interactions cancel or are not present in these differences. Employing Equations 28 and 30 this ratio is given by

$$R_{T\,T'}(A) = \frac{2T + 2 + b(A)}{2T' + 2 + b(A)} \qquad 32.$$

The value of $R_{T,\,T'}(A)$ (Equation 32) determined from experimental binding energies (62) with Coulomb-difference effects removed agrees best with the prediction of Equation 32 if $b(A) \sim 1$. However, the fluctuations in the values of $R_{T,\,T'}(A)$ as a function of A (see Figure 5 of 59) are so large that this agreement with $b(A) \sim 1$ is only had on the average. It therefore appears that assigning a definite value to $b(A)$ is very difficult till other small effects are properly in hand: for example the pairing energies in the nucleus specified by $(A,\ T)$ could well be different from those in $(A,\ T+2)$ as well as having a difference in their shell structure. As a point of interest the ratio appropriate to the data presented in Table VI from Equations 23, 28, 29, and 30 with $b(A) = 1$ is

$$R(4n,\ T = 2m) = \frac{2T - 1}{T - 1 + \dfrac{\delta(A)A}{V(A)}} \qquad 33.$$

which for $T = 2$ and a reasonable value for $V(A)$, say 100 MeV (63, 64), and $\delta = 34/A^{3/4}$ MeV (65) yields values for $R(A, 2)$ in rather good agreement with experiment. They are presented in the last column of Table VI.

In this same paper (59) Jänecke considered a set of mass relations which were to have a large influence on the work that followed. Using Equations 28, 29, and 30 it is clear that in the expression for $\Delta_{T+2,\,T}(A)$ pairing effects are not present. Hence the following simple mass relation obtains in so far as $\dfrac{V(A)}{A} = \dfrac{V(A)'}{A'}$

$$\Delta_{T+2,T}(A) = 2\Delta_{T+3/2,T+1/2}(A') \qquad 34.$$

The above condition should be approximately satisfied if $A' \sim A$, say $A' = A + 1$. The above relation was pointed out by Jänecke who claimed it to be valid for integer $T > 0$. Of course, in the above treatment no direct account is taken of nuclear shell structure or of the difference in residual interactions within shells and as will be shown this leads to some difficulties.

Figure 2 is taken from (59) and shows the values obtained for $\Delta_{T+2,\,T}(A)$ and $2\Delta_{T+3/2,\,T+1/2}(A)$. The overall agreement with Equation 34 is quite impressive. Having fixed the value of $b(A)$ to equal one (1) from examination of the values of $R_{T,\,T'}(A)$, Jänecke fixed values of $V(A)$ and $\delta(A)$ by equating observed differences in binding energies, with the Coulomb effects removed, to the following equation which follows directly from Equations 28, 29, and 30

$$\Delta_{T\ T'}(A) = \frac{V(A)}{A}\left[T(T+1) - T'(T'+1)\right]$$

$$+ \tfrac{1}{2}(1 + (-1)^A)\left[(-1)^{A/2+T} - (-1)^{A/2+T'}\right]\delta(A)$$

35.

Figure 3 shows the values obtained for $V(A)$ and $\delta(A)$ as presented in (59). The function $V(A)$ shows 15 per cent effects due to shell effects, and a smoothly varying best-fit curve has fluctuations of $\sim$8 per cent. With best values for $V(A)$ and $\delta(A)$ taken from Figure 3, the standard deviation from measured masses (62) is 0.525 MeV (59).

A further feature illustrating the importance of using a difference equation which takes account of shell effects is to be seen by comparing the behaviour of $\Delta_{3/2,1/2}$ (A) in adjacent odd nuclei. These differences are presented in detail in (21) so only an example is cited. In $A = 19$, $\Delta_{3/2,1/2}$ equals

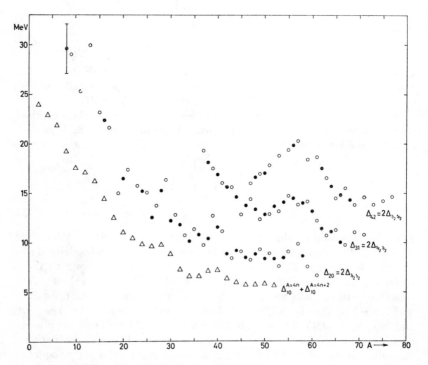

FIG. 2. Plot of values of $\Delta_{T+2,T}(A)$ filled circles and 2 $\Delta_{T+3/2,T+1,2}(A)$ (open circles).

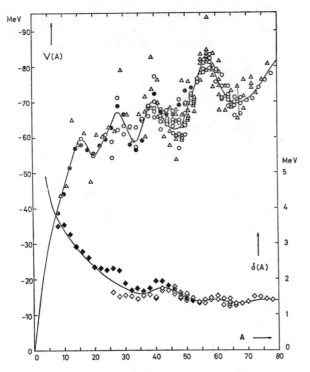

FIG. 3. Symmetry parameters $V(A)$ and pairing energies $\delta(A)$ calculated from the experimental energy differences $\Delta_{T,T'}(A)$. Values of $V(A)$ derived from even- and odd-A nuclei are shown as circles and squares respectively.

7.43 MeV while for $A = 21$ it is 8.86 MeV, thus the predicted value for $\Delta_{2,0}(20)$ varies from 14.86 to 17.72 MeV. Thus though the liquid-drop model provides a useful insight for a proper mass relation, its failure to take proper account of shell and configuration effects results in uncertainties ~ 1 MeV.

Mass relations based on a general independent-particle model.—To overcome the difficulties associated with the neglect of shell effects and the state dependence of the residual nuclear interaction, Garvey & Kelson (66) proposed a set of mass relations. The corresponding equations can be derived very simply by constructing equations among mass differences in which the energy due to all two-body interactions is made to cancel. Consider an equation of the following form

$$\sum_{i=1}^{n} C_i M_i(N_i, Z_i) = 0 \qquad\qquad 36.$$

where $C_i = \pm 1$. The restriction on the set of masses M_i that appear in a given equation is determined by the requirements of cancellation of the number of two-body interactions. Thus cancellation of the neutron-neutron interactions is achieved by requiring

$$\sum_{i=1}^{n} N_i C_i = 0 \qquad\qquad 37a.$$

and similarly the proton-proton interactions cancel if

$$\sum_{i=1}^{n} Z_i C_i = 0 \qquad\qquad 37b.$$

Additionally as the number of neutron-proton interactions in the nucleus represented by $M(N, Z)$ is NZ, cancellation of their effect is obtained by requiring

$$\sum_{i=1}^{n} N_i C_i Z_i = 0 \qquad\qquad 38.$$

For any relation of practical interest n must be an even number. If $n=2$ or 4, only trivial identities result while $n=6$ yields the simplest nontrivial equation. In the case of $n=6$ one has for example

$$M(N_1, Z_1) + m^l(N_2, Z_2) + M(N_3, Z_3) - M(N_3, Z_1)$$
$$\qquad\qquad 39.$$
$$- M(N_1, Z_2) - M(N_2, Z_3) = 0$$

where the Equations 37 are manifestly satisfied. Defining $\Delta Z_i = Z_1 - Z_i$ and $\Delta N_j = N_2 - N_j$, the condition imposed by Equation 38 requires that

$$\Delta Z_2 \Delta N_1 = \Delta Z_3 \Delta N_3 \qquad\qquad 40.$$

to avoid generating identities $\Delta Z_2 \neq \Delta Z_3$ and $\Delta N_1 \neq \Delta N_3$. The simplest values to choose for the products in Equation 40 are ± 1. The choice -1 yields an equation which can be written as

$$M(N + 2, Z - 2) - M(N, Z) + M(N + 1, Z)$$
$$\qquad\qquad 41.$$
$$- M(N + 2, Z - 1) + M(N, Z - 1) - M(N + 1, Z - 2) = 0$$

This equation is diagramed in Figure 4a; it was used as the basis of the results reported in (66). The choice $+1$ for the product gives an equation of the form

$$M(N + 2, Z) - M(N, Z - 2) + M(N + 1, Z - 2)$$
$$\qquad\qquad 42.$$
$$- M(N + 2, Z - 1) + M(N, Z - 1) - M(N + 1, Z) = 0$$

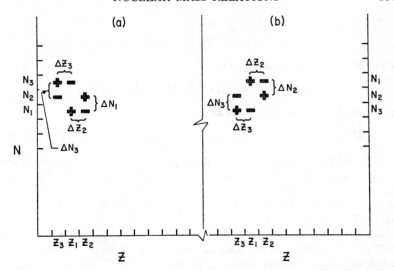

FIG. 4. (a) Display of the form of Equation 41 or 43 in the NZ plane. The sign appropriate to each mass is indicated at its respective location. (b) Display of the form of Equation 42 or 44 in the NZ plane. The sign appropriate to each mass is indicated at its respective location.

and a diagram of this equation in the N, Z plane is shown in Figure 4b. From this figure it is clear that Equations 41 and 42 are related by a 90° rotation in N, Z space. The relationship between these two equations becomes more apparent if they are written in A, T_z representation. Equation 41 becomes

$$M(A, T_z + 2) - M(A, T_z) + M(A + 1, T_z + \tfrac{1}{2}) - M(A + 1, T_z + \tfrac{3}{2})$$
$$+ M(A - \tfrac{1}{2}, T_z + \tfrac{1}{2}) - M(A - 1, T_z + \tfrac{3}{2}) = 0 \quad 43.$$

while Equation 42 is written as

$$M(A + 2, T_z + 1) - M(A - 2, T_z + 1) + M(A - 1, T_z + \tfrac{3}{2})$$
$$- M(A + 1, T_z + \tfrac{3}{2}) + M(A - 1, T_z + \tfrac{1}{2}) \quad 44.$$
$$- M(A + 1, T_z + \tfrac{1}{2}) = 0$$

This cumbersome notation can be made compact by using a modification of a previous definition (see Equation 29) (but note that no Coulomb effects are removed in the definition below).

$$D_{T',T}(A) \equiv M(A, T') - M(A, T) \quad 45.$$

and

$$D_{A',A}(T) \equiv M(A'T) - M(A, T) \qquad 46.$$

Thus we have from Equation 43

$$D_{T+2,T}(A) = D_{T+3/2,T+1/2}(A - 1) + D_{T+3/2,T+1/2}(A + 1) \qquad 47.$$

and Equation 44 now is written as

$$D_{A+2,A-2}(T + 1) = D_{A+1,A-1}(T + 3/2) + D_{A+1,A-1}(T + \tfrac{1}{2}) \qquad 48.$$

Both of these equations can be *directly* applied to known nuclear masses or binding energies as the Coulomb effects cancel out as a result of balancing the proton-proton interactions. Obviously Equation 47 may be used to relate the position of the lowest-lying isobaric analogue states in neighbouring nuclei. The sum of the $T+\tfrac{3}{2}$, $T+\tfrac{1}{2}$ energy splitting in the $A+1$ and $A-1$ systems is equal to the $T+2$, T splitting in the nucleus with A nucleons. This use of Equation 47 has been examined by Jänecke (67) in considerable detail to predict the excitation energy of states $T > T_z + 2$. Further, Equations 47 and 48 hold for integer or noninteger T in contrast to the claim for Equation 34 but differently from Equation 34, Equation 47 does not hold if $T_z = 0$ with $A = 4m + 2$ (i.e. an odd-odd nucleus).

Equation 47 was tested over all known nuclear masses (62) where $N \geq Z$ and if $N = Z$, $N \neq 2n + 1$. This yields 627 test cases, and the average magnitude of the deviation between the right- and left-hand sides of Equation 47 is 205 keV. The sum of the deviations for these 627 cases is .836 MeV which indicates how well the deviations tend to cancel. Figure 5 is a plot of the deviations and the insert shows the distribution of these deviations. When Equation 48 is tested in all possible cases with $N \geq Z - 2$ except $N = Z - 1$, 755 tests can be made and the magnitude of the average deviation encountered is 187 keV. Thus both equations hold very well over the defined ranges with an average deviation $\sim$200 keV, even though the uncertainty in many of the experimentally determined masses used in the tests was larger than this. If the comparisons are restricted to the cases where no mass is uncertain to <250 keV, the magnitude average deviation only decreases slightly and there are, of course, many sound physical reasons for believing that the difference Equations 47 and 48 should not hold exactly.

1. The corresponding single-particle levels and residual interactions all appear in different nuclei. These quantities will clearly change from one nucleus to the next because of changes in nuclear size, the onset of nuclear deformation, and the accompanying changes in the Coulomb energy.

2. The nuclear ground state cannot be described as an independent-particle configuration, particularly because such a specification does not generally yield a state of specific angular momentum.

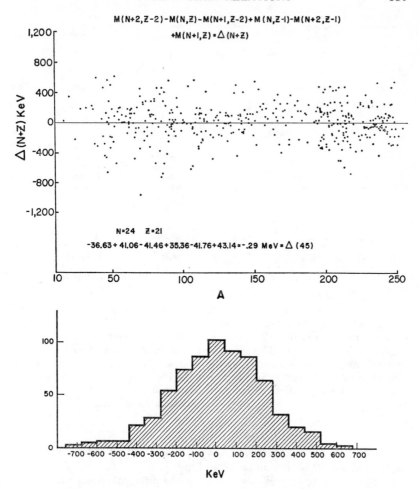

FIG. 5. (a) Plot of the deviations resulting from applying Equation 47 to all measured masses with $N \geq Z$, $Z \geq 6$, and $N \geq 10$. The masses used for this figure are from J. H. E. Mattauch et al., *Nucl. Phys.*, **67**, 1 (1965). An example is also shown. (b) Histogram of the deviations illustrated in (a).

3. The picture presented further assumes a definite parentage relationship between the ground states of the adjacent nuclei. There are some well-known cases of anomalous parentage between adjacent nuclei as for example $Ti^{48} - Ti^{47}$ and $Be^{12} - Be^{11}$. However, states of the proper, normal parentage

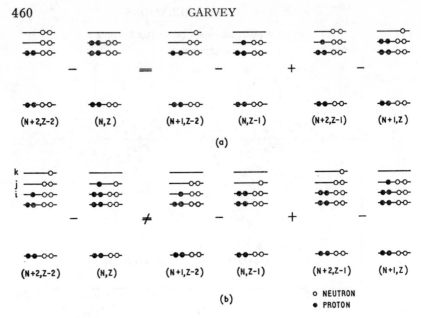

FIG. 6. (a) Diagram of the single-particle energy levels entering into Equation 47 for the case $N=Z=2n$. The single-particle picture illustrated here is not in general correct but is appropriate to the Nilsson model or the supermultiplet model. From the figure it is clear that a set of identities is had for each single-particle level and that all two-body interactions between particles in different levels cancel. (b) Diagram of the single-particle levels entering into Equation 47 with $N=Z=2n+1$. In this case it is evident that the interactions within the level indicated as level j do not cancel. The effect of this noncancellation is observed experimentally (see Table VIII).

lie relatively near the ground states in both cases: within 160 keV in the case of Ti[47] and within 320 keV in the case of Be[11].

The above reasons probably constitute the principal effects entering into the deviations encountered.

Not surprisingly, both of these mass relations have been in the literature, in various guises, for some time. Equation 47 was noted by Feather (68–70) from his study of beta-decay systematics. However he did not emphasize the fundamental nature of its origin but rather stressed its relation to the Weizsäcker liquid-drop model rather than viewing it as a consequence of a general independent-particle model as will be shown. Restricted versions of these equations are also to be found in the work of Brink & Kerman (71) who showed that such equations follow from the assumption of a deformed single-particle field which gives fourfold degenerate orbits. They used the fact that Equations 47 and 48 worked at the beginning of the $2s$-$1d$ shell to illustrate the occurrence of deformation in that region.

In (66) the authors present a simple picture to show why Equation 47 seems to hold so well. Figure 6 shows this picture, the neutrons and protons occupying the lowest possible single-particle states, as is expected for the ground state. Each orbit is taken to be fourfold degenerate because of isospin conjugation and time reversal. Examining a given single-particle orbit one sees that within this orbit all two-body interactions between the particles cancel and that interactions between particles occupying different single-particle orbits also cancel. These interactions are most readily specified by denoting the interaction between two neutrons in levels i and j, say, by $V_{ij}^{n,\,T=1}$, for two protons in levels i and j via $V_{ij}^{p,\,T=1}$ while for a neutron and proton in levels i and j one has $(V_{ij}^{T=1}+V_{ij}^{T=0})/2$.

Thus as is shown in Figure 6a, all these interactions cancel in this pictorial representation of Equation 47. As will be shown later, one need not have such a simple coupling scheme for Equation 47 and 48 to hold exactly, but the picture depicts much of the useful physics. For example, the reason for the exclusion of nuclei with $N=Z=2n+1$, in the case of Equation 45, can readily be seen from Figure 6b. Of all possible cases with $N \geq Z$ this is the only one in which the interaction within a single level does not cancel out. In the level designated by j there is a surfeit of interaction by an amount $\frac{1}{2}(V_{jj}^{T=1}+V_{jj}^{T=0})$. Of course, the relations were set up so that the total number of two-body interactions cancel, thus the interactions between the particles in the levels above and below level j do not cancel either and have an opposite sign $-\frac{1}{2}(V_{ik}^{T+1}+V_{ik}^{T+0})$. Thus subtracting the right-hand side of Equation 47 from the left does not give zero in this case but rather $(V_{jj}^{T=0}+V_{jj}^{T=1}-V_{ik}^{T=1}-V_{ik}^{T=0})/2$. However, interaction energy between particles in the same orbit is so much greater than it is for particles in different orbits that the difference does not cancel. In relations with $N>Z$ involving odd-odd nuclei, the cancellation involves only the difference in the interaction energy between pairs of particles in which both pairs are in *different* single-particle orbits so that the difference cancels to a much better approximation. Table VIII shows the results obtained by applying Equation 47 to all cases with $N=Z$. Where isospin inversion is known to exist (some of the $N=Z$ odd-odd nuclei), the energy of the lowest $T=0$ state is used. The effect mentioned above is clearly evidenced as the relation works quite well for $N=Z=2n$ but a large and consistently negative deviation is had in the odd-odd cases. The recent measurements of Stokes & Young (72) on the mass defects of N^{18} and F^{22} are especially crucial to this table as the previous measurement of F^{22} was off by 1.6 MeV. The deviations in the odd-odd nuclei fall into groups roughly associated with shell structure as would be expected from the arguments presented above. However, the variations in the deviations for $A=46$, 50, and 54 are rather puzzling.

TABLE VIII. A table of the deviations from zero encountered when Equation 47 is employed for $N=Z$. When isotopic inversion occurs (lowest $T=0$ level lies above lowest $T=1$ level) in a self-conjugated odd-odd nucleus, the energy of the lowest $T=0$ state is used rather than the energy of the ground state. Note the systematic deviations from zero encountered when $N=Z$ equals an odd number. The masses used for N^{18} and F^{22} are taken from (61).

					Deviations	
A	$D(-)$(MeV)	$D(+)$(MeV)	Σ(MeV)	D_{20}(MeV)	$e-e$	$o-o$
16	9.77	8.67	18.44	18.43	.01	
18	8.67	4.82	13.49	12.40		$-1.09\pm.03$
20	4.82	5.68	10.50	10.84	.34	
22	5.68	4.38	10.06	$8.01\pm.03$		$2.05\pm.03$
24	4.38	3.83	8.21	7.98	$-.23$	
26	3.83	2.61	6.44	$4.52\pm.3$		$-1.9\ \pm.3$
28	2.61	3.67	6.28	6.47	$+.19$	
30	3.67	1.48	5.15	3.05		$-2.1\ \pm.3$
32	1.48	.248	1.73	1.81	$+.08$	
34	.248	.167	.415	$-.52$		$-.93\pm.3$
36	.167	$-.814$	$-.647$	$-.424$	$+.22$	
38	$-.814$	.565	$-.250$	-1.14		$-.89\pm.3$
40	.565	$-.412$	.153	.191	$+.04$	
42	$-.412$	-2.22	-2.63	-3.49		$-.86$
44	-2.22	-2.06	-4.28	-3.80	$+.48$	
46	-2.06	-2.92	-4.98	-5.51		$-.53$
48	-2.92	-2.56	-5.48	-5.41	.07	
50	-2.56	-3.19	-5.75	-6.82		-1.07
52	-3.19	-3.99	-7.18	-7.08	.10	
54	-3.99	-3.46	-7.45	-7.74		$-.29$
56	-3.45	-3.23	-6.69	-6.69	.00	

In a jj-coupling shell model the uniqueness of the $T=0$ odd-odd nuclei is also evident. The complete expression (55) for the interaction energy amongst n particles in a given j shell with a specific seniority (v) given by

$$E(j^n T\gamma t) = \frac{n(n-1)}{2}\alpha + \left[\frac{n-v}{4}(4j+8-n-v)\right.$$

$$\left. - T(T+1) + t(t+1)\right]\beta + (T(T+1) - \tfrac{3}{4}n)\gamma$$

49.

where the notation was defined below Equation 20 apart from t which is the reduced isospin. In the lowest-seniority approximation the expression simplifies to the form of Equation 20 for even-even nuclei $(v=0)$ and odd-A nuclei $(v=1)$. In the case of odd-odd nuclei the lowest state with $T=T_z$ has either $v=2$ $t=1$ or $v=2$, $t=0$. The parameter β is defined so that the lowest energy occurs for the $t=1$ state; however, in a self-conjugate nucleus $(N=Z)$ the reduced isospin of the unpaired particles in the lowest $T=0$ state must be $t=0$. Thus using Equation 49 (assuming all particles to be in the same j shell), one would expect the deviation Equation 47 for $N+Z$ odd-odd nuclei to be

$$D_{T=2,T=0}(A = 4n + 2) - D_{3/2,1/2}(4n + 1)$$

$$- D_{3/2,1/2}(4n + 3) = \frac{-2(V_0 - V_2)}{2j + 1} \quad 50.$$

With parameters taken from (52) the above expression is equal to -1.65 MeV for the $d_{5/2}$ shell in good agreement with the values shown for $A=22$ and 26 in Table VIII. The uniquely different supermultiplet classification (11) of $N=Z=2n+1$ ground states shows that this model also predicts their anomalous behaviour in the mass relations.

The restriction to $N \geq Z$ for the applicability of Equations 47 and 48 is because the effects of neutron-proton interaction are carried in the form of a symmetry energy. This energy depends on $|N-Z|$ and is therefore discontinuous across the $N=Z$ line. Equation 47 is observed to break down completely, with 4–5 MeV deviations occurring, when it crosses the $N=Z$ line (70). For Equation 48, expressions with $N=Z-1$ must be avoided, else a disruptive odd-odd $T_z=0$ mass is encountered.

As has been pointed out, relations of the form of Equations 47 and 48 are recurrent, and appropriate equations can be summed together to obtain more extended relations. It is easily shown that an equation of the type

$$M(N + \Delta T, Z - \Delta T) - M(N, Z) + \sum_{i=1}^{\Delta T} (M(N - 1 + i, Z - \Delta T + i)$$

$$51.$$

$$- M(N + i, Z - \Delta T - 1 + i)) = 0$$

directly follows from successive application of Equation 47. Suppose that the nucleus with $N+\Delta T$ neutrons and $Z-\Delta T$ protons is extremely neutron rich and its most likely mode of particle decay is via neutron emission. Whether a neutron is emitted or not depends on whether $B(N+\Delta T, Z-\Delta T)$

is greater than $B(N+\Delta T-1, Z-\Delta T)$. $B(N, Z)$ represents the binding energy of the ground state of a nucleus with N neutrons and Z protons. Thus from Equation 51 it follows that

$$B(N + \Delta T, Z - \Delta T) = B(N, Z) + \sum_{i=1}^{\Delta T} [B(N + i, Z - \Delta T - 1 + i) \tag{52.}$$

$$- B(N - 1 + i, Z - \Delta T + i)]$$

and

$$B(N + \Delta T - 1, Z - \Delta T)$$

$$= B(N, Z - 1) + \sum_{i=1}^{\Delta T - 1} [B(N + i, Z - \Delta T - 1 + i) \tag{53.}$$

$$B(N - 1 + i, Z - \Delta T + i)]$$

Upon subtracting Equation 53 from Equation 52

$$B(N + \Delta T, Z - \Delta T) - B(N + \Delta T - 1, Z - \Delta T)$$

$$= B(N, Z) - B(N, Z - 1) + B(N + \Delta T, Z - 1) \tag{54.}$$

$$- B(N + \Delta T - 1, Z)$$

Thus the criterion for neutron emission from a neutron-rich nucleus very far from the valley of stability is given in terms of two binding-energy differences in the valley of stability. Expressing the above results more compactly in terms of separation energies yields

$$S_n(N + \Delta T, Z - \Delta T)$$

$$= S_p(N, Z) + S_n(N + \Delta T, Z - 1) - S_p(N + \Delta T - 1, Z) \tag{55.}$$

where

$$S_n(N, Z) \equiv B(N, Z) - B(N - 1, Z)$$

$$S_p(N, Z) \equiv B(N, Z) - B(N, Z - 1) \tag{56.}$$

Equations identical to Equation 54 can be obtained by a direct use of Equations 39 and 40 with $\Delta N_1 \Delta Z_2 = \Delta N_3 \Delta Z_3 = -(\Delta T - 1)$ with $\Delta N_3 = \Delta Z_2 = -1$. All equations following from setting $-\Delta N_1 \Delta Z = -\Delta N_3 \Delta Z_3 > 1$ can be obtained by successive application of Equation 47. As an example of these more extended mass relations, the neutron separation energy of Ca^{49} is given by

$$S_n(Ca^{49}) = S_n(Cr^{49}) + S_n(V^{52}) - S_p(Cr^{52})$$

the left side is equal to $5.144 \pm .006$ MeV with the right side being equal to $5.017 \pm .015$ MeV, while an instructive example using Equation 48 as a basis yields

$$B(Sb^{130}) - B(Sn^{128}) = B(Sb^{121}) - B(Sn^{120}) + B(Pr^{138}) - B(Pr^{137})$$

which is obeyed to 100 keV. In this case the product in Equation 40 would equal 8.

Extended relationships like Equation 54 have been used to predict the particle stability and the binding energies for the neutron-rich isotopes from He to O (66). There are just two cases of disagreement with experiment in all the work to date (21). He^8 is correctly predicted to be stable but the binding energy is overestimated by 2 MeV while Li^{11}, predicted unstable into $Li^9 + 2n$ by some 2.5 MeV, is actually observed to be stable. The difficulty in He^8 is most certainly due to the large changes, as the number of nucleons changes, in the average nuclear field when so few nucleons are involved. Particularly large changes are known to take place between $A = 7$ and $A = 8$ (51, 73). In Li^{11} the difficulty was occasioned by a wrong estimate of the mass of Li^{10}, which could not be directly obtained because the only mass relation in which it was the sole unknown member was the one involving an odd-odd $T_Z = 0$ nucleus (B^{10}). Therefore the size of the deviation equivalent to the right-hand side of Equation 50 had to be estimated. A value of -2 MeV was used; however for an interaction whose range is short compared to the nuclear size there should be an approximate $1/A$ behaviour in its matrix elements which was neglected so that the -2 MeV estimate was not large enough. Because Li^{10} is unstable and Li^{11} bound one can indirectly infer that a more correct value is -3.5 to -4.1 MeV. Using Equation 55 with $N = 6$, $Z = 5$, and $\Delta T = 2$ yields

$$S_n(Li^{11}) = S_p(B^{11}) + S_n(Be^{12}) - S_p(B^{12})$$

while direct application of Equations 48 and 56 gives

$$S_n(Li^{10}) = S_n(Be^{12}) + S_n(Be^{11}) - S_n(B^{13})$$

Using empirical values of all separation energy except $S_n(Be^{12})$ which is taken from (66) and assigned an uncertainty of 300 keV yields

$$S_n(Li^{11}) = .430 \pm .3 \quad \text{and} \quad S_n(Li^{10}) = -1.1 \pm .3$$

Thus we have Li^{11} bound against decay into Li^{10} and Li^{10} unstable against decay into $Li^9 + n$. The most likely decay mode of Li^{11} is thus to $2n + Li^9$ for which the two-neutron separation energy is $S_{2n}(Li^{11}) = -.6 \pm .6$ so that it does not appear surprising with hindsight that Li^{11} is barely bound. A measurement of the mass of Li^{11} would be useful and exciting and could be done either as a (He^4, Li^{11}) reaction or perhaps more easily via the (Li^7, Li^{11}) reaction. It appears then that the masses of even very light nuclei, far off the valley of stability, can be rather well predicted by using mass relations

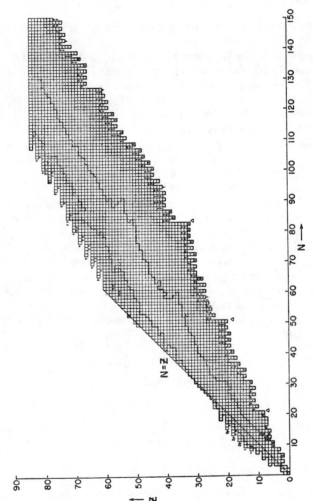

FIG. 7. This figure shows the predicted limits of stability from the mass table obtained using Equation 47. The heavy line indicates the range of the presently known nuclei. The more recently discovered light isotopes are marked with an X. Question marks indicate where the predicted stability is questionable. For completeness the results of (44) are used to predict the limits of stability for the $Z > N$ nuclei up through $Z = 22$. The limit along the $N = Z$ line up to $Z = 50$ is due to limitations in the theory. Above $Z = 50$ the limit shown is determined by proton instability.

essentially based on Equations 47 and 48. This represents a real improvement over what could be achieved with mass equations which are generally unreliable out of the valley of stability and especially so in the light nuclei.

Often, any of several different forms of the extended mass relationships such as Equations 52 or 54 may be used to predict the same unknown mass. If the mass relations were exactly true, then all relations would yield the same value for the unknown mass or binding energy; however, this is clearly not the case and the problem arises as to the relative weight to give to each prediction. This problem has been overcome in an objective fashion in a recent paper (74) wherein two mass tables are generated. One table exactly obeys Equation 47 and all equations derivable from it (Equations 51–53) while the other follows a form dicated by Equation 48; discussion will be limited to the table derived from Equation 47 as it yields results more in accord with experiment (74). It is constructed by least squares fitting, the known nuclear masses (62) to a set exactly obeying Equation 47; its range is $N \geq Z$, and if $N = Z$, $N \neq 2n+1$, with $10 \leq N \leq 154$, $6 \leq Z \leq 100$, and $16 \leq A \leq 253$. Over this range the constructed mass table has an average deviation of .093 MeV from the experimentally determined masses while the standard deviation is $\sigma = .158$ MeV. The retentive reader will note that the average deviation above is appreciably smaller than the .205 MeV average deviation encountered in directly applying Equation 47 over the same set of masses, because in the mass table the deviations are spread over the individual masses so that the expected error should be $\sim .205/\sqrt{6}$ as it is. The table extends to the limits of neutron stability on the neutron-rich side and to the limits of proton stability or the $N = Z$ line, whichever is encountered first.

The limits of particle stability as determined from the table are shown in Figure 7. The nuclei enclosed within the hatchwork should all be stable against single or double nucleon emission. The boundary on the right-hand side is termed the "neutron drip" line. Among neutron-rich nuclei predicted to be stable against particle (neutron) emission are O^{28} and Ca^{70}. A summary of previous work on the limits of particle stability can be found in (21).

An interesting test of mass predictions for neutron-rich heavy nuclei has been made (75) via searching for delayed-neutron emission from neutron-rich fission products. The observation of delayed neutrons ($\tau_n > 1$ min) depends critically on whether or not the beta-decay energy of a parent exceeds the neutron separation energy of its daughter. Thus with a proper identification of the parent beta emitter the presence or absence of ensuing neutron emission serves as a test of predicted mass values. The comparison as reported in (75) between experimental results and the predictions of the table (74) based on Equation 47 shows rather poor agreement. However this is due to the use of incorrect values for the predicted beta-decay energies which should all be increased by .78 MeV. This increase markedly improves the agreement with experiment so that for the 23 cases reported the only case in apparent disagreement with the mass table mentioned above is the non-observation of delayed-neutron emission following the beta decay of $_{33}As^{84}$.

This may not be due to a failure of the mass prediction but simply to the low-energy beta decay that is required to feed the neutron unstable levels not completing sufficiently well with the more energetic decay to the bound levels.

Before concluding it is important to make the following general observations. The validity of Equations 47 and 48 depends only (74) on the separability of the expression governing the nuclear binding energies or masses. If the nuclear masses may be written as

$$M_1(N, Z) = g_1(N) + f_1(Z) + h_1(N + Z) \qquad 57.$$

where $g_1(N), f_1(z)$, and $h_1(N+Z)$ are arbitrary functions of their arguments, then Equation 47 will be exactly obeyed. If the masses are expressible by an equation of the form

$$M_2(N, Z) = g_2(N) + f_2(Z) + h_2(N - Z) \qquad 58.$$

where $g_2(N), f_2(Z)$, and $h_2(N-Z)$ are again arbitrary functions of their arguments, then for Equations 57 and 58 to yield identical results for all masses, i.e. $M_1(N, Z) = M_2(N, Z)$ for all N and Z, the masses must (74) obey the far more restrictive form:

$$M(N, Z) = g(N) + f(Z) + \lambda NZ + \frac{\mu}{2} \left[1 - (-1)^{NZ} \right] \qquad 59.$$

where $g(N)$ and $f(Z)$ are arbitrary functions of N and Z, and λ and μ are constants. Expressions for the nuclear binding energies from various independent-particle models for which the binding energy can be expressed in analytic form can be shown (74) to fall into the class of Equation 59. Thus it would seem that relations not depending on any particular individual-particle model of nucleus but rather stemming from general features will hold to a high order of accuracy (200 keV).

Conversely it also appears that mass relations derived from specific models (56) (58) will hold only insofar as they reflect the more general features of an independent-particle model. Hence mass relations may not be a very useful test to descriminate between various nuclear models.

The real worth of the relations presented above, particularly Equations 47 and 48, is the simplicity and attractiveness of the underlying picture as well as their apparent ability to predict with good accuracy the masses of yet unmeasured nuclei.

ACKNOWLEDGMENTS

The author would like to acknowledge many enlightening discussions over the past few years with Professors Kelson and Talmi on problems in nuclear structure theory. The A. P. Sloan Foundation is thanked for its generous support over the past two years and my appreciation is extended to the Nuclear Physics Laboratory at Oxford for its hospitality this past year.

LITERATURE CITED

1. *Proc. Lysekil Symp. 1966* (Forsling, W., Herrlander, C. J., Hyde, R., Eds., Almqvist & Wiksell, Stockholm, 1966)
2. Poskanzer, A. M., Cosper, S. W., Hyde, E. K., Cerny, J., *Phys. Rev. Letters*, 17, 1271 (1966)
3. Thomas, T. D., Raisbeck, G. M., Boerstling, P., Garvey, G. T., Lynch, R. P., *Phys. Letters*, 27B, 507 (1968)
4. Klapisch, R., Chaumont, J., Philippe, C., Amarel, I., Fergeau, R., Salome, M., Bernas, R., *Nucl. Instr. Methods*, 53, 216 (1967)
5. Burbidge, E. M., Burbidge, G. R., Fowler, W. A., Hoyle, F., *Rev. Mod. Phys.*, 29, 547 (1957)
6. Burbidge G. R., *Ann. Rev. Nucl. Sci.*, 12, 507 (1962)
7. Seegar, P. A., Fowler, W. A., Clayton, D. D., *Ap. J.*, *Suppl. 97*, 11, 121 (1965)
8. Seegar, P. A., in *Proc. 3rd Intern. Conf. Atomic Masses, Winnipeg, 1967*, 583 (Barber, R. C., Ed., Univ. Manitoba Press, 1967)
9. Wing, J., *USAEC Rept. ANL*-6814 (1964) (Unpublished)
10. Wing, J., *Proc. 3rd Intern. Conf. Atomic Masses, Winnipeg, 1967* (Univ. Manitoba Press, 1967)
11. Wing, J., *Nucl. Phys.*, A120, 369 (1968)
12. *Proc. 2nd Conf. Isospin in Nucl. Phys.*, *March 1969* (Anderson, Bloom, True, Cerny, Eds.; to be published)
13. Robson, D. *Ann. Rev. Nucl. Sci.*, 16, 119 (1966)
14. Wigner, E. P., Feenberg, E., *Rept. Progr. Phys.*, 8, 274 (1941)
15. Wigner, E. P., *Proc. Robert A. Welch Found. Conf. Chem. Res. Houston, Texas, 1957*, 1 (1958)
16. Weinberg, S., Treiman, S., *Phys. Rev.*, 116, 465 (1958)
17. Wilkinson, D. H., *Phys. Letters*, 11, 243 (1964)
18. Wilkinson, D. H., *Phys. Rev. Letters*, 12, 348 (1964)
19. Garvey, G. T., Cerny, J., Pehl, R., *Phys. Rev. Letters*, 13, 48 (1964)
20. Garvey, G. T., *Proc. 2nd Conf. Isospin in Nucl. Phys.*, *March 1969* (Anderson, Bloom, True, Cerny; to be published)
21. Cerny, J., *Ann. Rev. Nucl. Sci.*, 18, 27 (1968)
22. Jänecke, J., *Isospin in Nuclear Physics*, Chap. 8 (Wilkinson, D. H., Ed.,
23. Jänecke, J., *Nucl. Phys.*, A128, 632 (1969)
24. Bachall, J. (Private communication)
25. Lauritsen, T., Ajzenberg-Selove, F., *Nucl. Phys.*, 78, 1 (1968)
26. Hecht, K. T., *Nucl. Phys.*, A102, 11 (1967)
27. Hecht, K. T., *Nucl. Phys.*, A114, 280 (1968)
28. Jänecke, J., *Phys. Rev.*, 147, 735 (1966)
29. Jänecke, J., *Nucl. Phys.*, A114, 433 (1968)
30. Wigner, E. P., *Phys. Rev.*, 51, 106 (1937)
31. Feenberg, E., Goertzel, G., *Phys. Rev.*, 70, 597 (1946)
32. Carlson, B. C., Talmi, I., *Phys. Rev.*, 96, 436 (1954)
33. Sherr, R., *Phys. Letters*, 24B, 321 (1967)
34. Wilkinson, D. H., Hay, W. D., *Phys. Letters*, 21, 80 (1966)
35. Nolen, J. A., Jr., Schiffer, J. P., Williams, N., Von Ehrenstein, D., *Phys. Rev. Letters*, 18, 1140 (1967)
36. Inglis, D. R., *Phys. Rev.*, 82, 181 (1951)
37. Hecht, K. T., *Isotopic Spin in Nuclear Physics*, 823 (Fox, J., Robson, D., Eds., Academic Press, New York, 1966)
38. Wilkinson, D. H., *Phys. Rev. Letters*, 13, 571 (1964)
39. Thomas, R. G., *Phys. Rev.*, 88, 1109 (1952)
40. Ehrman, J. R., *Phys. Rev.*, 81, 412 (1951)
41. Lane, A. M., Thomas, R. G., *Rev. Mod. Phys.*, 30, 257 (1958)
42. Goldanskii, V. I., *Nucl. Phys.*, 19, 482 (1960)
43. Goldanskii, V. I., *Ann. Rev. Nucl. Sci.*, 16, 1 (1967)
44. Kelson, I. Garvey, G. T., *Phys. Letters*, 23, 689 (1966)
45. Pehl, R., Cerny, J., *Phys. Letters*, 14, 137 (1965)
46. Bertsch, G. F., *Phys. Rev.*, 174, 1313 (1968)
47. Goldanskii, V. I., *JETP (USSR)*, 38, 1637 (1960); *JETP* (Transl.), 11, 1179 (1960)
48. Yamada, Y., Matumoto, M., *J. Phys. Soc. Japan*, 16, 1497 (1961)
49. Gove, N. B., Yamada, Y., *Nucl. Data*, A4, 237 (1968)
50. Zeldes, N., Grill, A., Simievis, A., *Kbl. Danske Videnskab. Selskab Mat.-Fys. Medd.*, 3, 5 (1968)

North-Holland, Amsterdam, 1969)

51. Cohen, S., Kurath, D., *Nucl. Phys.*, **73**, 1 (1965)
52. Talmi, I., Thieberger, R., *Phys. Rev.*, **103**, 718 (1956)
53. Talmi, I., Unna, I., *Ann. Rev. Nucl. Sci.*, **10**, 353 (1960)
54. Talmi, I., *Rev. Mod. Phys.*, **34**, 704 (1962)
55. Talmi, I., *Lectures in Theoretical Physics 1965*, 39 (Univ. Colorado Press, Boulder, Colo., 1966)
56. Franzini, P., Radicati, L. A., *Phys. Letters*, **6**, 322 (1963)
57. *Group Theory and its Application to Physical Problems* (Hammermesh, M., Ed., Addison-Wesley, Reading, Mass., 1942)
58. Bremond, B., *Nucl. Phys.*, **A113**, 257 (1968)
59. Jänecke, J., *Nucl. Phys.*, **61**, 326 (1965)
60. von Weizsäcker, C. F., *Z. Phys.*, **96**, 431 (1935)
61. Seegar, P. A., Perisho, R. C. *Los Alamos Sci. Lab. Rept. LA3751* (Unpublished)
62. Most studies requiring the use of extensive tabulation of nuclear binding energies use Mattauch, J. H. E., Thiele, W., Wapstra, A. H., *Nucl. Phys.*, **67**, 1 (1965) and Maples, C., Goth, G. W., Cerny, J., *Nucl. Data*, **A4**, 237 (1968)
63. Satchler, G. R., Drisko, R. M., Bassel, R. H., *Phys. Rev.*, **136**, B118 (1964)
64. Hodgson, P. E., *Ann. Rev. Nucl. Sci.*, **17**, 1 (1967)
65. Pryce, M. H. L., *Proc. Phys. Soc.*, **63**, 692 (1950)
66. Garvey, G. T., Kelson, I., *Phys. Rev. Letters*, **16**, 197 (1966)
67. Jänecke, J., *Proc. 3rd Intern. Conf. Atomic Masses, Winnipeg, 1967*, 583 (Barber, R. C., Ed., Univ. Manitoba Press, 1967)
68. Feather, N., *Phil. Mag.*, **43**, 133 (1952)
69. Feather, N., *Phil. Mag.*, **44**, 103 (1953)
70. Feather, N., *Proc. Roy. Soc. Edinburgh*, **57**, Part 2, 104 (1965)
71. Brink, D. M., Kerman, A. K., *Nucl. Phys.*, **12**, 314 (1959)
72. Stokes, R. H., Young, P. G., *Phys. Rev.*, **178**, 1789 (1969)
73. Barker, F. C., *Nucl. Phys.*, **83**, 418 (1966)
74. Garvey, G. T. Gerace, W. J., Jaffe, R. L., Talmi, I., Kelson, I., *Princeton Univ. Rept.*, *937–331* (To be published in *Rev. Mod. Phys.*, 1969)
75. Talbert, W. L., Jr., Tucker, A. B., Day, G. M., *Phys. Rev.*, **177**, 1805 (1969)

COULOMB ENERGIES*

By J. A. Nolen, Jr.

University of Maryland

and J. P. Schiffer

Argonne National Laboratory

CONTENTS

* Research supported by the United States Atomic Energy Commission.

1. INTRODUCTION

The effects of the Coulomb repulsion between protons in a nucleus have been the subject of many studies throughout the past thirty years (1). Given a model of the nucleus, the specifically Coulomb effects can be used to test the model or to extract specific parameters.

One of the first uses of Coulomb-energy differences between mirror-nucleus pairs was to determine nuclear radii. Radii so derived were compared to those determined by other techniques. The simplest model, used originally, was a uniformly charged sphere; this was refined and many microscopic models, mostly based on an independent-particle picture, emerged. These newer models still emphasized either the extraction of charge radii or the parameterization of the rapidly growing body of empirical data on Coulomb-energy differences.

Recently, however, there has been new interest in Coulomb-energy differences. With charge radii now accurately determined by other techniques, Coulomb-energy differences are being used to extract other nuclear-structure information.

We will begin by presenting a review consisting of a history of the interpretation of Coulomb energies, methods used in measuring Coulomb-energy differences, references to recent compilations of empirical Coulomb-energy differences, and references to techniques and empirical information on independently determined nuclear-charge radii. Next a simple model for the calculation of Coulomb-energy differences is presented with a discussion of some possible correction terms. This model is then applied to specific cases. First neutron-density distributions are extracted for nuclei throughout the periodic table, and the "neutron halo" is discussed. Next variations in Coulomb-energy differences are examined through a series of isotopes, to study the dependence of the matter radius on A when Z is constant. Other nuclear-structure effects are also discussed.

2. BACKGROUND

2.1 DEFINITIONS

The total Coulomb energy E_c of a nucleus is the sum of the Coulomb interactions between all the protons in the nucleus. The Coulomb-energy difference Δ between two nuclei in states which are members of the same isospin multiplet is just the difference between their total Coulomb energies. With isospin a good quantum number, the nuclear wavefunctions of mirror pairs, such as $O^{17} - F^{17}$, are identical, and their mass difference is given by:

$$M_{z>} = M_{z<} + \Delta - \Delta_{pn} \qquad 1.$$

where $M_{z>}$ and $M_{z<}$ represent the masses (in MeV) of the members of the isobaric pair with the greater and lesser charges, respectively. The additional Coulomb repulsion Δ increases the mass of the $Z_>$ member of the pair. The proton-neutron atomic mass difference $\Delta_{pn} = 0.7824$ MeV decreases the mass difference because the $Z_>$ member has one more proton and one less neutron than the $Z_<$ member. The $Z_>$ and $Z_<$ members need not be the ground states, nor is Equation 1 restricted to mirror pairs. The only requirement is that the states be members of an isobaric multiplet, i.e., that they have the same T, $T_z(Z_>) = T_z(Z_<) - 1$, and the same structure.

In such a case, the above equation may be considered the definition of the Coulomb-energy difference (Coulomb displacement energy or Coulomb-energy shift) Δ. Strictly speaking, Δ is the Coulomb-energy difference only if the nuclear wavefunctions of the pair are identical and if the nucleon-nucleon force is charge independent and charge symmetric. In mirror pairs there are an equal number of p-n interactions in both nuclei of the pairs and the number of p-p interactions in one equals the number of n-n interactions in the other. Hence, in mirror pairs $(T_z = \pm\frac{1}{2})$ only charge symmetry is required. For $T > \frac{1}{2}$, however, charge independence is necessary for the above equation to be strictly true, if Δ represents only the Coulomb-energy difference. The possibility of using Coulomb energies to test for charge symmetry and charge independence has been frequently suggested. A recent discussion of this problem is given by Henley (2). The use of high-isospin multiplets $(T \geq \frac{3}{2})$ as a test of charge independence was recently reviewed by Cerny (3). The difference in mass between the charged and uncharged pions gives a small violation of charge independence, but preserves charge symmetry. Deviations from charge symmetry are expected to be very small and there seems to be no definite experimental evidence which requires breaking of charge symmetry.

In this paper it will be implicitly assumed that nuclear forces are both charge symmetric and charge independent. Probably the largest uncertainties in present calculations of Coulomb-energy shifts are the effects of isospin mixing and nuclear correlations. In Section 5.1 a model which assumes pure isospin will be presented, while various effects which involve isospin mixing are discussed in Section 5.2. Only when such correction terms are well understood and calculated might one hope to return to the question of charge dependence of nuclear forces.

Throughout this paper the radii of various nuclear distributions and potential wells will be discussed. The root-mean-square (rms) radius of a distribution is denoted by $\langle r^2 \rangle^{1/2}$. The radius of a sphere of uniform density which has the same rms radius is the equivalent radius $R_{eq} = \sqrt{\frac{5}{3}} \langle r^2 \rangle^{1/2}$. Woods-Saxon potential wells and some empirical-density distributions are given in terms of the Fermi shape:

$$f(r) = \frac{1}{1 + e^{((r-R_0)/a)}}$$

where R_0 is the half-way radius and a is the diffuseness of the distribution, both in fermis (10^{-13} cm). For charge distributions the notation used frequently is c and z, instead of R_0 and a, respectively. Radii are also often written in terms of the atomic mass A:

$$R_{eq} = r_{eq} A^{1/3} \quad \text{or} \quad R_0 = r_0 A^{1/3}$$

2.2 MEASUREMENTS AND COMPILATIONS

β^+ decay.—The Coulomb-energy shift Δ between two isobaric nuclear states was defined in Equation 1. The first experimental values of Δ were determined from the β^+ endpoint energy in the positron decay of the $Z_>$ member of $T_z = \pm \frac{1}{2}$ mirror pairs. In this case:

$$\Delta = T_{max} + 2m_e c^2 - \Delta_{pn}$$

where T_{max} is the β^+ endpoint energy. This same method has been used to locate the energies of $T = 1$ states in $T_z = 0$ nuclei from the β^+ decay of its $T_z = -1$ analog, e.g. K^{36} ($T = 1$, $T_z = -1$)→Ar^{36} ($T = 1$, $T_z = 0$) (4).

Charge exchange.—More recently (5) the (p,n) direct charge-exchange reactions have been used extensively to locate the isobaric analog of the target state. The Q value in this reaction is, in fact, equal to the Coulomb-displacement energy: $\Delta = -Q(p,n)$. In heavy nuclei the isobaric analog of the target is a state of relatively simple nuclear configuration (i.e., the same as the ground state of the target nucleus) found at high excitation energy among states of much more complicated structure. The accuracy of the Q-value determination is usually limited to about ± 100 keV by the experimental difficulty of detecting the neutrons with good resolution. One experiment, however, was done at a bombarding energy such that the neutrons which populated the analog state were in the energy range of 3–5 MeV (6) and so Q values accurate to about ± 20 keV could be obtained. In some special cases, e.g. Ca^{42}–Sc^{42}, the (p,n) thresholds have been measured (7) and yielded values accurate to a few keV. The direct (He^3,t) reaction also populates analog states via charge exchange and has been used recently to determine Δ in many nuclei to about ± 8 keV (8).

Analog-state resonances.—With the discovery that isobaric analog states appear as compound-nucleus resonances in proton-elastic scattering and other proton-induced excitation functions, a powerful tool for measuring Coulomb energies became available and many Coulomb-energy shifts have been measured by these techniques. Such results have been compiled, for example, by Long et al. (9), and by Harchol et al. (10). The Coulomb energy is extracted from the center-of-mass resonance energy E_{cm}:

$$\Delta = E_{cm} + B_n$$

where B_n is the separation energy of the neutron from the parent $(Z_<)$.

Forbidden resonances.—Highly accurate Coulomb energies have also been determined in the light nuclei through the use of isobaric-spin forbidden resonances, such as $Si^{28} + p \rightarrow P^{29}$ $(T = \frac{3}{2}) \rightarrow P^{29} + \gamma$ (11), or the "twice-forbidden" scattering such as $Mg^{24} + p \rightarrow Al^{25}$ $(T = \frac{3}{2}) \rightarrow Mg^{24} + p$ (12). The use of these reactions to locate analog states has been reviewed in detail (3).

Direct reactions.—Many analog states have also been located via direct reactions other than charge exchange. For example, Sherr et al. (13) used the (p,d) reaction to measure several Coulomb energies in the $f_{7/2}$ shell. Isospin-coupling rules allow the population of $T = T_z + 1$ states in the residual nuclei with either neutron-pickup $[(p,d)$ or $(He^3,\alpha)]$ or proton-stripping $[(He^3,d)$ or $(\alpha,t)]$ reactions. However, since the yield to the $T_>$ state decreases with increasing neutron excess, this technique seems to be presently limited to light- and medium-mass nuclei $(A \leq 90)$.

Direct two-nucleon transfer reactions such as (p,He^3) and (He^3,p) have also been used to identify analog states. The (He^3,p) reaction, for example, populates the analogs of the states seen in the (t,p) transitions on the same target. A comparison of these reactions on O^{16} was used (14) to identify the $T = 1$ analog of the ground state of O^{18}.

Other reactions, such as (p,t), (He^3,n), and (He^3,He^6), have been used to locate $T = T_z + 2$ states and to help identify members of high-isospin multiplets in light nuclei (3). The isospin-coupling coefficients, useful in giving relative yields, are summarized for various direct reactions in (15), which also presents a review of the use of direct reactions to populate analog states.

Compilations.—Since the original discussions of the Coulomb-energy differences in mirror nuclei, the discovery of isobaric analog states in heavy nuclei has resulted in a tremendously rapid growth in empirical information. This article will emphasize the extraction of nuclear-structure information from Coulomb-energy data (a partial compilation of ground-state data is given in the Appendix). Cerny (3) recently reviewed the data on isospin multiplets in light nuclei. The most complete compilation of Coulomb energies of nuclear ground states as well as isospin multiplets was given by Jänecke (16). Other, more restricted presentations of data are by Harchol et al. (17) and Long et al. (9) who include some data on Coulomb energies of excited states, Sherr (18) for the mass 33–65 region, and Hardy et al. (19) for the $d_{5/2}$ shell.

In a discussion of Coulomb-energy systematics, Jänecke (20) presents plots which bring out various features of the A and T dependence of Coulomb-displacement energies. A pairing term of $120/2T$ keV is subtracted from Δ if $Z_<$ is odd, with nothing subtracted if $Z_<$ is even. The resulting quantity,

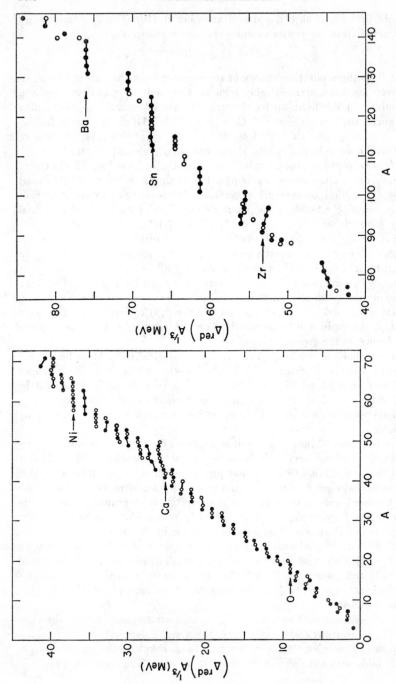

FIG. 1. Reduced Coulomb energies $\times A^{1/3}$ plotted as a function of A. Lines connect points with the same Z. Integral values of T are indicated with open circles.

called the reduced Coulomb energy, is multiplied by $A^{1/3}$ and is plotted as a function of A. Such a plot is given in Figure 1. This shows the gross features of the Coulomb energies, while the fine structure is displayed by subtracting the main A dependence from the reduced Coulomb shifts. Using a uniformly charged sphere, which will be discussed in more detail in the next section, we have

$$\Delta = \frac{6}{5} \frac{Z_< e^2}{r_{eq} A^{1/3}}$$

and with $Z_< = A/2 - T$, and $r_{eq} = 1.20$ we get

$$A^{1/3} \Delta - .722A = - .722(2T)$$

Jänecke (20) suggests plotting this quantity to display fine-structure effects. Such a plot is in Figure 2, with the reduced Coulomb-energy difference in place of Δ to remove pairing oscillations. The $T = \frac{1}{2}$ and $T = 1$ lines show

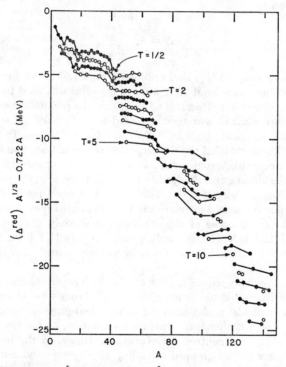

FIG. 2. The quantity $[A^{1/3}\Delta^{(red)} - 0.722\,A]$ plotted as a function of A. Lines are drawn through points of a given T. In the uniformly charged sphere model the lines would be horizontal and equally spaced. Integral values of T are indicated with open circles.

shell-closure effects at $A = 16$ and $A = 40$. Other fine-structure effects, such as deviations of the lines from the horizontal and the unequal spacing of the lines, are also obvious.

3. A HISTORICAL REVIEW (1936–68)

The effects of the Coulomb energy on various aspects of nuclear structure were discussed by Bethe & Bacher (1) over thirty years ago. They wrote down the expression for the total Coulomb energy of a uniformly charged sphere. Using the statistical model, they included the reduction in the Coulomb energy due to antisymmetrization. The calculation was concerned with the form of the Coulomb term in the Weizsäcker (21) semiempirical mass formula.

In 1938 Bethe (22) used Coulomb-energy differences to extract radii for mirror nuclei. In this treatment the effect of antisymmetrization was neglected. The protons in both members of a mirror pair were considered to occupy a uniformly charged sphere of radius R. This led to a Coulomb-energy difference

$$\Delta = \frac{6}{5} \frac{Ze^2}{R_{eq}}$$

For the mirror pair C^{13}-N^{13}, Bethe obtained $R_{eq} = 1.47 \, A^{1/3}$ F for the nuclear radius. He also pointed out that when this radius was used in calculating Coulomb-energy shifts of nearby mirror pairs, the predictions were too low when the last nucleon was tightly bound as in N^{15}-O^{15} ($B_n = 10.8$ MeV) compared to the loosely bound C^{13}-N^{13} ($B_n = 4.9$ MeV) case. To account for this effect, Bethe modified the wavefunction of the last neutron by taking it to be constant inside a radius R_{eq} and attaching an exponential tail, determined by the binding energy, outside R_{eq}. This model actually overcorrected for the binding-energy effect, but it did give a very good qualitative explanation of Coulomb-energy systematics in the light nuclei. For comparison, the currently accepted value of the equivalent uniformly charged sphere for C^{12} is $R_{eq} = 1.35 A^{1/3}$. This is smaller than the $r_{eq} = 1.47$ from the original method and 11 per cent larger than the value obtained with the binding-energy correction.

Most of the many papers written on Coulomb energies between 1940 and 1968 either attempted to extract charge radii through use of an expression based on a statistical model, used the independent-particle model to extract charge radii or other nuclear-structure information, or tried to extract information about the neutron wavefunctions. Hence, in the following, the relevant papers will be grouped according to emphasis and not in chronological order.

3.1 Uniformly Charged Sphere Model

In the expression of Bethe & Bacher (1) for the total Coulomb energy of a nucleus with charge Z, the protons were assumed to occupy a uniformly charged sphere of radius R_{eq}, leading to the direct term

$$E_D = \frac{3}{5} \frac{Z^2 e^2}{R_{eq}}$$

The fact that the ground-state wavefunction must be antisymmetrized gives rise to the exchange term E_{exch}. This was evaluated in the statistical model, using plane waves for the proton wavefunctions:

$$E_{exch} = -0.460 \frac{Z^{4/3} e^2}{R}$$

The net total Coulomb energy then is the sum of these terms. The interpretation of the direct and exchange terms has given rise to much controversy in the literature. The problem was clarified by Carlson & Talmi (23) and was also discussed in detail by Sengupta (24), who pointed out that the above direct term contains the quantum-mechanical direct terms plus a self-energy term, while the exchange term is the quantum-mechanical exchange term minus the self-energy term. The evaluation of the exchange term by Bethe & Bacher was also only strictly applicable to nuclei with even Z where all proton spins would be paired. Sengupta showed that when Z is odd the exchange term will be slightly more negative because the spin of the last proton is not paired. This term is not important in heavier nuclei (where the statistical model is most likely to be valid) but it is relevant to the later discussion of the pairing term in mirror nuclei. The resulting expression for the total Coulomb energy of a nucleus of charge Z occupying a uniformly charged sphere of radius R_{eq} is:

$$E_c = \left[0.60 Z^2 - 0.460 Z^{4/3} - \left\{1 - (-1)^z\right\} 0.15\right] \frac{e^2}{R_{eq}}$$

The Coulomb-energy difference between the ground state of a nucleus of charge $Z_< \equiv Z$ and its isobaric analog of charge $Z_> = Z+1$ is thus given by:

$$\Delta = \left[0.60(2Z + 1) - 0.613 Z^{1/3} - (-1)^Z 0.30\right] \frac{e^2}{R_{eq}} \qquad 2.$$

The first term contains the classical direct contribution of the Coulomb-energy shift and the self-energy of the "last" proton in the analog state. It must be retained because it is subtracted out as part of the second term calculated by Bethe & Bacher. Rewriting the self-energy with the second term the Coulomb-energy shift can be written as the sum of the direct term and a true exchange term Δ_{exch}:

$$\Delta = \Delta_D + \Delta_{exch}$$

where[1]

$$\Delta_{\text{exch}} = - \left[.613Z^{1/3} - .60 + (-1)^Z 0.30\right] \frac{e^2}{R_{\text{eq}}} \qquad 3.$$

For convenience we define a positive quantity, the "exchange factor":

$$\epsilon \equiv - \frac{\Delta_{\text{exch}}}{\Delta_D}$$

so that

$$\Delta = \frac{6}{5} \frac{Ze^2}{R} (1 - \epsilon) \qquad 4.$$

Values of the exchange factor given by this model are listed in column 3 of Table I for analog pairs with $Z_<$ corresponding to closed j shells. Swamy & Green (25) have shown that for $A \geq 20$ Bethe & Bacher's estimate of the total exchange energy E_{exch} is in good agreement with an independent-particle model. However, the change in exchange energy Δ_{exch} between isobaric pairs is model dependent and will be discussed later.

TABLE I

EXCHANGE FACTOR CALCULATED IN VARIOUS MODELS

$$\epsilon(\%) \equiv -\Delta_{\text{exch}}/\Delta_D$$

$Z_<$	A	Statistical model[a]	Oscillator model[b]	Woods-Saxon model[c]	Slater approximation[d]
6	13	11.8	7.1	6.7	
8	17	9.7	5.6	5.4	
14	29	7.0	4.1	3.8	
16	33	6.5	4.3	4.4	
20	41	5.7	3.7	3.7	
28	62	4.6		3.2	
50	120	3.3		2.2	
82	208	2.4		1.8	1.8

[a] Equations 2, 3, and 4, and Reference (24).
[b] Reference (41).
[c] Present.
[d] Reference (67).

Historically, when Coulomb-energy differences were first used to extract the charge radii of mirror nuclei, the exchange effect was neglected (22). For example, Stephens (26) fit the Coulomb-energy differences of the mirror

---

[1] For large neutron excess the pairing term is decreased; replace $(-1)^Z$ by $+1.0$.

nuclei from H^3 to Mg^{26} with the uniformly charged sphere model (neglecting exchange) and found a value for the radius of a uniformly charged sphere $R_{eq} = 1.45A^{1/3}$. This radius of the equivalent uniform sphere was generally accepted (27, 28) and was quoted as late as 1952 by Blatt & Weisskopf (29). This large radius was not initially questioned, probably because it was in reasonable agreement with other measures of the nuclear radius at that time, such as the radii extracted from alpha-decay rates and from neutron cross-section analyses.

However, the measurements of muonic X rays by Fitch & Rainwater (30) and elastic electron scattering by Hofstadter, Fechter & McIntyre (31) led to smaller radii, $R_{eq} \simeq 1.20A^{1/3}$ for heavy nuclei. In reanalyzing the data on mirror nuclei, Cooper & Henley (32) included the exchange term and thereby reduced R_{eq} to$\simeq 1.3A^{1/3}$. By invoking the independent-particle model they indicated that this radius could be decreased still further. This point will be discussed in detail in Section 5.

Using an equation analogous to Equation 2 (without the third term), Peaslee (33) argued that the Coulomb-energy shifts of mirror nuclei must take the form:

$$\Delta = a\left(\frac{Z + \frac{1}{2}}{A^{1/3}}\right) + b \qquad 5.$$

where a is a constant related to R_{eq} and b is the so-called exchange term. Adjusting a and b, Peaslee fit the empirical Coulomb shifts with the average parameters $a = 1.46$ MeV and $b = -1.11$ MeV with an average deviation of 150 keV from $A = 6$ to $A = 39$. This value of a corresponds to a radius $R_{eq} = 1.18A^{1/3}$ and is unreasonably low in view of our present knowledge of charge radii in this region (r_{eq} (O^{16})$\simeq 1.34$ and r_{eq} (Ca^{40})$\simeq 1.30$). Hence in terms of extracting meaningful radii it is not reasonable to treat the exchange term as an overall adjustable constant. In a similar context others (34, 35) claimed the b term to be the Coulomb self-energy of an isolated proton, but as was pointed out by Sengupta (24) it is more correctly that of a proton with its wavefunction spread over the nuclear volume.

However, if the size of the exchange term in Equation 2 is not allowed to vary in fitting the data, then for each known Coulomb-energy shift the radius of the equivalent uniformly charged sphere is determined. Sengupta (24) showed that this equation can yield radii in very good agreement with elastic electron scattering. In Table II we tabulate the results for closed proton shells, extending Sengupta's tabulation to heavier isobaric pairs.

The form used by Peaslee (33), Equation 5, has been used to parameterize (5, 9) in a simple manner the large number of Coulomb-energy shifts measured since the discovery of isobaric anlaog states in heavier nuclei. The parameters extracted by Anderson, Wong & McClure (5) in fitting their (p,n) data from mass 40–165 are very similar to Peaslee's in the light nuclei:

$$a \simeq 1.44 \text{ MeV} \quad \text{and} \quad b \simeq -1.1 \text{ MeV}$$

In extracting radii, however, they used Equation 2 and, therefore, qualitatively agree with the numbers in Table II. Long et al. (9) extended the tabulation by including many new Coulomb energies determined by analog-state resonances. A least squares fit to all data available at that time gave

$$a' = 1.43 \text{ MeV} \quad \text{and} \quad b' = -.85 \text{ MeV}$$

using a modified version of Equation 5:

$$\Delta E_c = a' \frac{Z}{A^{1/3}} + b'$$

This expression is accurate to about ± 100 keV for mass ≥ 50.

TABLE II

CHARGE RADII EXTRACTED FROM COULOMB ENERGIES BY USING
THE STATISTICAL AND OSCILLATOR MODELS
$r_{eq}(F)$

$Z_<$	A	Oscillator[a]	Statistical[b]	Experimental[c]
6	13	1.34	1.30	1.38
8	17	1.26	1.37	1.36
14	29	1.27	1.28	1.30
16	33	1.24	1.27	1.31
20	41	1.18	1.30	1.31
28	62		1.24	1.27
50	120		1.23	1.22
82	208		1.23	1.20

[a] Reference (23), using corrected values of Δ where necessary.
[b] Reference (24).
[c] Extracted from electron scattering or muonic X-ray experiments (for mirror nuclei the charge radius of the core is given).

3.2 INDEPENDENT-PARTICLE MODELS

Independent-particle models have been used in the calculation of Coulomb-energy differences since Bethe (22). As was already mentioned, Bethe considered a mirror pair, such as C^{13}-N^{13}, as having a nucleon outside a core. The $Z_<$ nucleus is converted to the $Z_>$ member by charge exchanging this neutron into a proton. To the extent that the n-n and p-p nuclear forces are equal, the additional mass of the resulting nucleus is due to the Coulomb interaction of this proton with the core protons, less the n-H^1 mass difference. (Corrections to this simple approximation, necessary even if the n-n force equals to p-p force, will be discussed in detail in Section 5.2.) The last nucleon is by definition in the highest shell-model state, and therefore is less bound than an average core nucleon. The radial wavefunction of this particle is, therefore, more extended, and the resulting Coulomb interaction with the core is less than that predicted by a statistical model. Consequently, charge radii extracted on the basis of such an independent-particle model are, in general, less than those from the statistical model (22, 32).

Square-well calculations.—Cooper & Henley (32) used eigenfunctions of an infinite square well to extract charge radii from the experimental mass differences of mirror nuclei. Ironically, this work was stimulated by preliminary muonic X-ray data (30) which indicated $R_{eq} \sim 1.2A^{1/3}$ for nuclei as light as Ti. Hence, at that time, it seemed necessary to go beyond the statistical model. The results of Cooper & Henley, which included the exchange energy, indicated it was possible to lower the radii sufficiently to bring them into agreement with the muonic results. It is now known that electron scattering and muonic X-ray results give $r_{eq} \sim 1.3 - 1.36$ for mass <40 (Table II). Hence this simple independent-particle model actually gives radii which are too small.

Other general features of the experimental Coulomb energies do require the independent-particle model for explanation. Figure 3 is a plot of Δ for the $T = \frac{1}{2}$ mirror pairs. This graph, similar to one given by Kofoed-Hansen

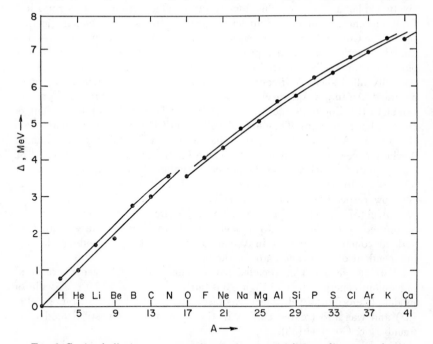

Fig. 3. Coulomb displacement energies of mirror nuclei ($T_z = \pm \frac{1}{2}$) vs. atomic mass. The $Z_<$ member is indicated along the abscissa. The graph shows the pairing effect and shell-closure effects. The lines are smooth curves through the $Z_<$-odd and $Z_<$-even pairs, indicating the magnitude of the pairing effect.

(36), clearly shows two effects: the $Z_<$ odd values fall on a higher curve than the $Z_<$ even values; and there are discontinuities at the shell closures in O^{16} and Ca^{40}.

Jancovici (37) calculated charge radii and Coulomb energies from wavefunctions in a finite square well. The mirror pairs O^{17}-F^{17} and N^{15}-O^{15} were considered in order to study the shell effects. He predicted a 10 per cent discontinuity in Δ at O^{16} in reasonable agreement with experiment (Figure 3), but this agreement is probably misleading. About half of the 10 per cent comes from the neglect of the spin-orbit term in the potential, leading to very different well depths for the two pairs. Also, the results of an oscillator calculation reported in this paper disagree with the corresponding numbers reported by Carlson & Talmi (23). The shell effect at O^{16} is probably due partially to the $p_{1/2}$-$d_{5/2}$ wavefunction difference and partially to the change in sign of the electromagnetic spin-orbit term; the latter, neglected in all papers discussed in this section, will be discussed in Section 5.1.

Oscillator calculations.—A detailed calculation of charge radii and Coulomb energies based on wavefunctions of an infinite oscillator well was carried out by Carlson & Talmi (23). This relatively simple model qualitatively explains many of the fine-structure effects seen in the empirical Coulomb energies of mirror nuclei (Figure 3). In this independent-particle model the Coulomb-energy difference in mirror pairs is the sum of the Coulomb interaction of the last proton (in the $Z_>$ member) with the core of closed proton shells and its interaction with the other protons in the last partially filled shell. A jj-coupling scheme and good proton seniority were assumed. An important result is that the charge radius is inversely proportional to the Coulomb energy: $r_{eq} = \sigma(A)/\Delta$, where $\sigma(A)$ is independent of the oscillator constant. They list the values of $\sigma(A)$ for the mirror pairs from $A = 3$ to 43. A few of the resulting charge radii, with more recent values of Δ where necessary, are included in Table II. In general, the radii based on the oscillator model are $\sim$2–11 per cent smaller than the accepted values. The values for the closed-shell cases O^{17}-F^{17} and Ca^{41}-Sc^{41} are 8 and 11 per cent low, respectively. The closed-shell effect at O^{16}, calculated as specified by Jancovici (37), is only $\sim$2 per cent. Thus, this effect (Figure 3) is not entirely accounted for by the oscillator model alone. It will be shown later that these general conclusions persist in Woods-Saxon calculations, and are largely independent of the exact shape of the potential well.

Pairing effects.—The so-called pairing effect (the odd–even fluctuation) of Coulomb energies seen in Figure 3 is fairly well explained by the oscillator model. The origin of this effect was discussed in detail by Carlson & Talmi (23) and was also treated earlier, in the context of a statistical model, by Feenberg & Goertzel (38).

Feenberg & Goertzel required only that the overall wavefunction of the Z protons in a nucleus be antisymmetric and that the ground state have maximum orbital symmetry. The latter implies that as many proton pairs as possible will have their spins oppositely directed so that their relative wavefunctions will be spatially symmetric. There are $\frac{1}{2}Z$ such pairs if Z is even and $\frac{1}{2}(Z-1)$ pairs if Z is odd. The remaining proton–proton bonds are random or "statistical bonds," three quarters of which will be spin symmetric and one quarter spin antisymmetric. The average Coulomb interaction is larger between protons whose spins are paired off than between protons with spins parallel, since the former can be closer together on the average than the latter. Thus the Coulomb-energy shift for $Z_<$-even mirror pairs is smaller than for $Z_<$-odd pairs. Carlson & Talmi showed that the requirement of maximum orbital symmetry alone does predict pairing, but not enough to explain the empirical Coulomb-energy fluctuations. The additional assumption of lowest possible proton seniority, i.e. the maximum number of relative $L=0$ proton pairs, increases the magnitude of the predicted effect. With jj coupling they obtained slightly less pairing than the data required. An independent discussion of the pairing term was given by Sengupta (24) and is included as the third term in Equation 2. His derivation of the pairing term is based upon the statistical model, and is consistent with the discussion of Feenberg & Goertzel. The magnitude of this term is nearly twice as large in the s-d shell as that predicted by the oscillator model. However, the pairing term arises from fluctuations in the exchange energy, and, as was shown in Table I, the exchange energy is overestimated by the statistical model. Pairing correlations (39) may well account for the remaining discrepancy in the pairing term; this will be discussed with the correction terms in Section 5.2.

The assumption of good proton seniority in the oscillator model was checked by Unna (40) by repeating the calculations of Carlson & Talmi with the requirement of lowest combined p-n seniority and good isospin. The contribution to the Coulomb-energy shift due to the interaction with closed shells is the same in both models. Thus the only differences arise in the intrashell interactions and these are only a small part of the total shift. The largest effects were ≤ 1 per cent in the p shell and ≤ 0.5 per cent in the s-d shell. Some calculations were also made by Sengupta (24) for the p and s-d shells with an oscillator model and LS coupling. A general comparison with the results of Carlson & Talmi (23) shows no advantage of LS over jj coupling; the only significant differences occur in the middle of the s-d shell.

Kofoed-Hansen (36) using simplified oscillator and infinite square-well models came to the same conclusions mentioned above: the independent-particle model gives a qualitative explanation of the shell and pairing effects (Figure 3); but the radii extracted from the model are several per cent too small.

The work of Sood & Green (41) yielded larger radii and, therefore, disagrees less with electron scattering. This represents the first calculation with wavefunctions in a diffuse finite nuclear potential. The main purpose of the paper was to clarify the "proton potential anomaly" (42) problem. They showed that to explain the binding-energy difference of the last proton and neutron in a mirror pair it is not necessary to use different well depths for the two cases. Using identical nuclear wells for the last proton and neutron, the calculated binding energies agreed with experiment to within 500 keV. However, this work was restricted to nuclear potentials that could be solved analytically (43). The resulting accuracy in the extracted charge radii was limited by the "A shifts" that were necessary when the parameters of the potential were varied. Comparing these results with those from the earlier oscillator model (23) and more recent Woods-Saxon bound-state calculations (44) shows discrepancies of $\sim$6 per cent.

More recent shell-model calculations.—In a recent discussion of the systematics of Coulomb-energy shifts, Harchol et al. (17) applied the oscillator model to the mass region $A = 29$ to 64. This extends the treatment of Carlson & Talmi to analog pairs other than mirror nuclei. Lowest proton seniority was assumed (except for Sc^{47}-Ti^{47}, Sc^{49}-Ti^{49}, and V^{51}-Cr^{51} where $v_p = 2$ gave better agreement with the data). They optimized the oscillator constants in each subshell to get a good overall parameterization of the Coulomb energies. In the $d_{3/2}$ and $f_{7/2}$ shells it was necessary to decrease the oscillator constant as neutrons were added to the shell. The Coulomb energies of the S-Cl, Cl-Ar, Ar-K, and K-Ca pairs within the $d_{3/2}$ shell decrease as $A^{-1/3}$ when neutrons are added (18). To fit this trend, the oscillator constant must decrease $\sim$2 per cent per neutron in this shell. In the $f_{7/2}$ shell, on the other hand, the Coulomb energies and the oscillator constant decrease much more slowly with neutron number. Since the $2p$ and $f_{5/2}$ shells fill simultaneously, their oscillator constants are not uniquely determined. The model fits the shell discontinuity at $Z = 20$ and the rapid rate of increase (faster than $ZA^{-1/3}$) of Δ just beyond $Z = 20$.

In a much more detailed calculation, Jänecke (45) has used the equations of Hecht (46, 47) in the analysis of Coulomb energies in the $f_{7/2}$ shell. Such an analysis was also carried out in the $d_{5/2}$ shell (19). In this formalism the total Coulomb energy is written as the sum of isoscalar, isovector, and isotensor parts:

$$E_c(A, T, T_z) = E_c^{(0)}(A, T) - T_z E_c^{(1)}(A, T)$$
$$+ (3T_z^2 - T(T + 1))E_c^{(2)}(A, T)$$

This form is closely related to the isobaric-multiplet mass equation:

$$M(T_z) = a + bT_z + CT_z^2$$

The validity of this equation has been reviewed recently by Cerny (3) and by Jänecke (48). The Coulomb-energy difference between states of isospin T and $T_z = T-1$ and $T_z = T$ is given by:

$$\Delta(A, T, T - 1 \mid T) = E_c^{(1)}(A, T) - 3(2T - 1)E_c^{(2)}(A, T)$$

Hecht (46) has used the formalism of five-dimensional quasispin to derive expressions for the vector and tensor Coulomb energies $E_c^{(1)}$ and $E_c^{(2)}$. These expressions were derived for generalized seniority, $v = 0$, 1, and 2.

Jänecke used these equations to fit the empirical Coulomb energies of the $f_{7/2}$ shell with the assumption of lowest generalized seniority, i.e. $v = 0$ or 1 only. In this approximation the Coulomb energies depend on only three matrix elements: the average Coulomb interaction of an $f_{7/2}$ proton with the core, the $f_{7/2}^2$, $J = 0$ interaction, and an average $f_{7/2}^2$, $J = 0$, 2, 4, 6 matrix element. Electromagnetic spin-orbit and other small effects were allowed for, thereby introducing another parameter. All four parameters were allowed a simple A dependence of the form $A^{1/3\lambda}$, λ being a fifth parameter. By varying the five parameters a least squares fit to the 23 known Coulomb energies was obtained. The fit was very good, the standard deviation being 8.5 keV. The resulting value of $\lambda \sim 0.2$ indicates that the interaction radius grows much more slowly than $A^{1/3}$ as the $f_{7/2}$ shell is filled. The parameters were compared with a harmonic-oscillator calculation. Very good agreement was found if the oscillator constant was decreased by 3 per cent from the beginning of the shell to the end. In this entire treatment, variations within the shell, but not the absolute Coulomb energies, were emphasized. In this calculation, as in all others involving oscillator wavefunctions, the effects of finite binding on the wavefunctions and the Coulomb-energy differences were not included. This raises some questions regarding the absolute significance of the extracted parameters.

The Coulomb-energy equations of Hecht (46, 47) also give the behavior of the paring term for analog pairs with $T > \frac{1}{2}$. The magnitude of the pairing term decreases approximately as $(2T)^{-1}$. This also follows from a qualitative argument comparing $T = T_z = 1$ and $T = T_z = \frac{1}{2}$ parent nuclei. In the latter case a pairing term increases Δ when $Z_<$ is odd, because of the extra $J = 0$ proton pair which is formed in the analog state. In the $T = 1$ case a pairing term also increases Δ if $Z_<$ is odd, but the size of the effect is about one-half the value when $T = \frac{1}{2}$. This is because, of the two neutrons in the excess which can be charge exchanged, only one will go into the magnetic substate occupied by the odd proton to form a $J = 0$ pair. This argument can be extended to $T > 1$ giving an approximate $1/2T$ dependence to the pairing term. This aspect of the pairing term is discussed in detail by Jänecke (20, 45, 49). Such a T dependence gives quantitative agreement between the pairing energies of the $T = \frac{1}{2}$ doublets and the $T = 1$ triplets in the s-d shell (see 45, appendix A). Sherr (18) also shows that there is qualitative agreement with such a term in the $Z_<$-odd isotopic sequences in the $f_{7/2}$ shell. This general behavior also explains the disappearance of the pairing term for $A > 54$ (e.g. 5, 9, 17).

Calculations with Woods-Saxon wavefunctions.—Woods-Saxon wavefunctions were used by Wilkinson & Mafethe (50) and Wilkinson & Hay (51) in analyses of Coulomb energies in $1p$ shell nuclei. They used averaged Coulomb energies in this mass region ($A = 9$–15) to extract charge radii for comparison with values from $(p,2p)$ and (e,e) experiments. Corrections for antisymmetrization and the electromagnetic spin-orbit term were made, but L-S intrashell interactions were neglected. Overall agreement of ~ 10 per cent was found between the radii extracted from Coulomb energies and measured by other techniques.

In an early article on isobaric-spin purity in heavy nuclei (52) Lane & Soper presented calculations of Coulomb-energy shifts performed with Woods-Saxon wavefunctions. These calculations were fairly qualitative, the primary purpose being to show the variation of Coulomb energy for the various single-particle orbits.

Woods-Saxon wavefunctions were used by Jones et al. (53) to calculate Coulomb energies in the Ca isotopes. In Sc^{41} the $2p_{3/2}$ single-particle strength comes 290 keV lower in excitation energy than it does in the parent nucleus, Ca^{41}. Jones et al. showed that this is explained by the difference between the $f_{7/2}$ (ground-state) orbit and the $2p_{3/2}$ orbit as calculated in a Woods-Saxon potential well. Such calculations (54) show that even if the $1f_{7/2}$ and $2p_{3/2}$ neutrons had the same binding energy, the rms radii and Δ can differ by several per cent. Furthermore, Δ for the $2p_{3/2}$ orbit is much more sensitive to binding energy than for the $f_{7/2}$ orbit. The effects of antisymmetrization were ignored in these calculations. This affects the absolute value of the well radius needed to fit the Coulomb energies, but has no significant effect on the discussion of fine structure.

Recently the emphasis in many Coulomb-energy calculations has shifted to a study of the excess nucleons rather than the extraction of charge radii. Using empirical charge distributions in the Coulomb-energy calculations allows the determination of the rms radius of the excess nucleons. This technique has been used to determine (55) the rms radius of the neutrons in Pb^{208}, and in Ca, Ni, and Sn isotopes (44, 55). The same method has recently been applied to other nuclei (57). This essentially model-independent approach to Coulomb-energy calculations is presented in Section 5; numerical results are given in Sections 6 and 7.

4. CHARGE RADII

In the preceding section the past use of Coulomb-energy shifts for the extraction of nuclear charge radii was discussed. This technique had an ultimate accuracy of, at best, several per cent in the rms radius. In recent years other methods of measuring charge radii have become more and more reliable. For example, elastic electron scattering on many nuclei ranging from mass 1 to 209 has been performed with a precision permitting the extraction of rms charge radii to about ± 1 per cent (58, 59). Usually the results are quoted as a two-parameter Fermi distribution. In some cases the experiment justifies the use of a third parameter, usually a "parabolic Fermi shape":

$$f(r) = \left[1 + w \left(\frac{r}{c} \right)^{2} \right]\left[1 + \exp\left(\frac{r - c}{z} \right) \right]^{-1}$$

with the three parameters c, z, and w all in F. In addition to elastic electron scattering, muonic X-ray experiments are yielding rms charge radii with comparable accuracy (58, 59).

It is of some interest to mention the cases in which the independent-particle model has been used to explain electron scattering. The most extensive published work on this subject is that of Elton & Swift (60), who generated charge-density distributions by adding up bound states in a Woods-Saxon potential well. They folded the finite size of the proton charge into the distribution of proton centers of mass to get the nuclear charge distribution. The parameters of the potential were adjusted until agreement with the electron scattering cross section was obtained. Solutions were found for many nuclei, from Li^6 to Ca^{48}, with binding energies, well depths, and well parameters all reasonably consistent with other information. The well depth was allowed to increase for the deeper single-particle states in order to fit the binding energies deduced from $(e,e'p)$ experiments. The rms charge radii extracted in this way differ by no more than 3 per cent from those extracted using the phenomenological Fermi distributions. Perey & Schiffer (61) used Woods-Saxon wavefunctions to explain the so-called "anomalous isotope shift" of the charge radius. They showed, for instance, that the relatively small change in charge radius from Ca^{40} to Ca^{48} is nearly explained by the symmetry term in the potential well even if the overall well radius grows as $A^{1/3}$. However, the fact that the rms radius actually decreases by 1 per cent from Ca^{40} to Ca^{48} could not be explained unless the well grows more slowly than $A^{1/3}$. Similar calculations for the calcium isotopes were done by Gibson & van Oostrum (62). The Woods-Saxon model has also been applied to Pb^{208} by Elton (63) who showed that a single-particle potential well which fit the energies of the particle and hole states near the Fermi surface also produced a charge distribution in agreement with electron scattering.

5. A SIMPLE MODEL FOR CALCULATING COULOMB DISPLACEMENT ENERGIES

In this section a method of calculating the Coulomb-energy shifts, which is largely insensitive to the specific model assumed, is presented. This is possible because certain features, such as the rms radii of the wavefunctions, are much more important than other details. The calculation of some corrections terms requires the assumption of a specific model, but these terms are small, leaving the values of Δ relatively independent of the model. Such a method has the obvious advantage that experimental values of Δ can be converted into nuclear radii and the radii can then be compared to specific models with no confusion in consistency.

In parts 6 and 7 specific results of such calculations are given. These include the so-called "dominant terms": the direct, exchange, and electromagnetic spin-orbit contributions to Δ. Other contributions to the Coulomb-energy shift are more or less arbitrarily referred to as "correction terms." The sizes of these latter terms are discussed in Section 5.2 but have not been included in the results given in parts 6 and 7.

5.1 THE DOMINANT TERMS

The present discussion will be restricted to nuclei with a closed shell of protons. To calculate Coulomb displacement energies of other nuclei, it would be necessary to include the mutual intrashell interaction, discussed as a correction term in Section 5.2 below. The pairing term (23), for example, arises entirely from fluctuations in the intrashell interaction and will not be discussed further here.

The wavefunction of the parent nucleus is taken to consist of a properly antisymmetrized product of eigenfunctions for $Z_<$ protons and N neutrons. The analog of this nucleus if formed by changing one of the $(N-Z_<)$ neutrons in the neutron excess into a proton without changing any of the other nuclear quantum numbers. The initial assumption is one of good isospin. This implies two things: (a) the wavefunctions of the $Z_<$ protons have perfect overlap with the first $Z_<$ neutrons in the core, and (b) the radial wavefunction of the proton in the analog state is identical with that of the corresponding neutron in the parent. It is known that Coulomb effects will cause some violation of both of these assumptions. However, the magnitudes of the violations are fairly easy to estimate and will be discussed as correction terms in Section 5.2. When there is more than one subshell occupied in the neutron excess, the analog state will contain a linear combination of protons from the various shells, weighted by the number of neutrons in each shell. Such a linear conbination formed by operating on the neutron excess of, for example, Ca^{50} with the T-lowering operator can be indicated schematically:

$$T_- | \, [(f_{7/2})^8 (2p_{3/2})^2]_\nu \rangle = \sqrt{\frac{8}{10}} \, | \, [f_{7/2}]_\pi [(f_{7/2})^7 (2p_{3/2})^2]_\nu \rangle$$

$$+ \sqrt{\frac{2}{10}} \, | \, [2p_{3/2}]_\pi [(f_{7/2})^8 (2p_{3/2})]_\nu \rangle$$

6.

In this case Δ for the analog pair Ca^{50}-Sc^{50} is assumed to be the properly weighted average of Δ for the $2p_{3/2}$ and $f_{7/2}$ orbits taken separately.

The Coulomb interaction of atomic electrons in LS coupling is discussed in detail by Condon & Shortley (64). Since we are here concerned with some closed-j-shell nuclei, such as the Ni isotopes, their results are not directly applicable, and we make use of the equations in jj coupling given by de-Shalit & Talmi (65). In their notation the electrostatic Coulomb energy of a proton in the j' shell outside a closed j shell is written:

$$\Delta_{jj'} = \langle j^{2j+1} j' JM \, | \, \sum_i \frac{e^2}{r_{ik}} \, | \, j^{2j+1} j' JM \rangle_a$$

7.

where the sum is over the protons in the closed j shell and k is the proton in the j' shell. The total angular momentum J is equal to j' since the closed shell has no angular momentum. The a indicates the use of antisymmetrized wavefunctions, the e^2/r_{ik} is the Coulomb operator for protons separated by a distance r_{ik}.

By writing the wavefunction $\psi_a(j^{2j+1}j'JM)$ explicitly as a Slater determinant, the energy can be evaluated as a sum of the interactions between particles in the two shells in the various relative angular momentum states J:

$$\Delta_{j'j} = \frac{1}{2j'+1} \sum_J (2J+1)\Delta_{jj'J} \qquad 8.$$

Because the j shell is closed, the summation and the angular integrals can be carried out yielding explicit expressions for the direct and exchange terms Δ_D and Δ_{exch}. In addition to these electrostatic terms, there is the electromagnetic spin-orbit energy. The signs of the proton and neutron magnetic moments are opposite, thereby contributing an energy difference Δ_{so} to the Coulomb energy. The net Coulomb displacement energy is the sum of these three terms, as indicated in Table III.

$$\Delta = \Delta_D + \Delta_{\text{exch}} + \Delta_{so} \qquad 9.$$

The direct term.—The direct term, after integration over the angular variables, can be written in terms of a direct Slater integral (64, p. 177; 65, p. 209):

$$\Delta_D = \sum_j (2j+1)F^0(nl, n'l') \qquad 10.$$

where the sum is over the closed nlj orbits in the core and $n'l'j'$ is the orbit of the proton in the analog state. The Slater integral is:

$$F^0(n_1l_1, n_2l_2) = e^2 \int_0^\infty \int_0^\infty \frac{1}{r_>} u_1{}^2(r_1)u_2{}^2(r_2)(r_1{}^2dr_1)(r_2{}^2dr_2) \qquad 11.$$

where the $u(r)$ are the radial wavefunctions of the protons, and $r_>$ is the larger of r_1, r_2. This can be written in the form:

$$F^0(n_1l_1, n_2l_2) = \int_0^\infty V_{j_1}(r_2)\rho_2(r_2)(4\pi r_2{}^2dr_2) \qquad 12.$$

where $V_c(r_2)$ is the electrostatic Coulomb potential energy (in MeV) of a proton at radius r_2 due to the proton in orbit j_1, and the quantity $\rho_2(r_2)$ is the density of the proton in orbit j_2. The expression for the direct term then reduces to:

$$\Delta_D = \int_0^\infty V_c(r)\rho_{\text{exc}}(r)(4\pi r^2 dr) \qquad 13.$$

TABLE III

Dominant Terms in Coulomb-Energy Shift

$$\Delta = \Delta_D + \Delta_{\text{exch}} + \Delta_{so}$$

DIRECT

$$\Delta_D = \int_0^\infty V_c(r) \rho_{\text{exo}}(r) (4\pi r^2 dr)$$

$V_c(r) \equiv$ experimental Coulomb potential of the core

$\rho_{\text{exo}}(r) \equiv$ density distribution of neutron excess with finite size of nucleon folded in

EXCHANGE

$$\Delta(j')_{\text{exch}} = -\sum_j (2j+1) \sum_k \left\{ \begin{matrix} j & k & j' \\ -\tfrac{1}{2} & 0 & \tfrac{1}{2} \end{matrix} \right\}^2 G^k(nl, n'l')_{(l+l'+k, \text{ even})}$$

where

$$G^k(nl, n'l') \equiv e^2 \int_0^\infty \int_0^\infty \frac{r_<^k}{r_>^{k+1}} u_l(r_1) u_l(r_2) u_{l'}(r_1) u_{l'}(r_2) r_1^2 dr_1 r_2^2 dr_2$$

To be averaged over all excess orbits, j'

$u_l \equiv$ radial wavefunction of nlj orbit

$j \rightarrow$ closed proton shells in core

$j' \rightarrow$ excess orbit

SPIN-ORBIT

$$\Delta_{so} \simeq -(\mu_p - 1/2 - \mu_n) \left(\frac{\hbar^2}{2M_p^2 c^2} \right) \left(\frac{Ze^2}{R_{\text{eq}}^3} \right) (\mathbf{l} \cdot \boldsymbol{\delta})$$

To be averaged over all excess orbits

where $V_c(r)$ represents the total Coulomb potential energy at r due to all the protons in the core.[2] A typical form of $V_c(r)$ and of $\rho_{exc}(r)$ is shown in Figure 4. Writing the analog-state density as ρ_{exc} represents the initial assumption that this is the same as the neutron-excess radial density. The perturbation of ρ_{exc} by the Coulomb repulsion is discussed in the next section.

The exchange term.—The effect of antisymmetrization on the Coulomb energy is to reduce it slightly because of the correlation it requires. This term can be written in terms of exchange Slater integrals (65):

$$\Delta_{\text{exch}} = -\sum_j (2j+1) \sum_k \left\{ \begin{matrix} j & k & j' \\ -1/2 & 0 & 1/2 \end{matrix} \right\}^2 G^k(nl, n'l') \qquad 14.$$

$$(l + l' + k \text{ must be even})$$

where the first sum is over the closed nlj orbits in the core, and $n'l'j'$ is the orbit in the analog state. The second sum involves a 6-j symbol and the ex-

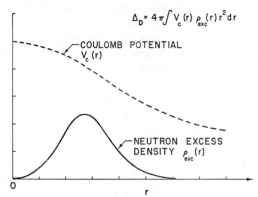

$$\Delta_D = 4\pi \int V_c(r)\, \rho_{\text{exc}}(r)\, r^2 dr$$

COULOMB POTENTIAL
$V_c(r)$

NEUTRON EXCESS
DENSITY $\rho_{\text{exc}}(r)$

FIG. 4. A typical neutron-excess density distribution and the Coulomb potential of the core are shown.

change Slater integral summed over k such that $l+l'+k$ is even and $|j-j'| \leq k \leq |j+j'|$. The Slater integral is:

$$G^k(n_1l_1, n_2l_2) = e^2 \int_0^\infty \int_0^\infty \frac{r_<^k}{r_>^{k+1}} u_1(r_1)u_2(r_2)u_1(r_2)u_2(r_1)(r_1^2 dr_1)(r_2^2 dr_2) \qquad 15.$$

Unlike the direct term, Δ_{exch} depends on the amplitudes of the individual orbits and not just on the overall density distribution. Hence, Δ_{exch} must be calculated with some specific model for the wavefunctions. Fortunately,

[2] In the numerical evaluation of Equation 13 it is necessary to fold the finite size of the nucleon (rms radius = 0.8 F) into the density distribution $\rho_{exc}(r)$. This point is discussed by Elton (66) since it also affects the calculation of charge distributions in the independent-particle model. In our calculations $V_c(r)$ is taken from experiment and therefore already includes the effect of the nucleon size.

however, the total magnitude of this term is relatively small and oscillator wavefunctions give nearly the same exchange factor as Woods-Saxon bound states. Also, the Slater (local density) approximation used by Bethe & Siemens (67) in Pb^{208} agrees with the Woods-Saxon result. These are given in Table I where the exchange factor $\epsilon \equiv -\Delta_{exch}/\Delta_D$ is listed. The statistical model overestimates the exchange factor, especially in the light elements. This occurs because when Equation 15 is evaluated with independent-particle wavefunctions, the last proton has relatively poor overlap with many of the smaller core orbitals. On the other hand, the statistical (Fermi gas) model treats all protons equally and gives a better overlap. In the present work all exchange integrals were evaluated with Woods-Saxon wavefunctions, but it is now clear that oscillator wavefunctions are adequate (57) and using them would reduce computer time substantially.

The spin-orbit term.—In most nuclear-structure calculations the electromagnetic spin-orbit term is neglected, because it is only ~1 per cent of the nuclear spin-orbit splitting. However, it is easy to calculate and should be included in Coulomb-energy calculations, where small effects such as pairing and shell closures are being discussed. This term was used by Inglis (68) to explain part of a discrepancy between the mirror spectra of Li^7-Be^7, and it has been included in other calculations of analog-state energies since then (69, 70).

The spin-orbit interaction for atomic electrons is:

$$W_{so}{}^e = \left\langle \frac{-e\hbar^2}{2m_e{}^2c^2} \frac{1}{r} \frac{dV_c}{r} [1 - 1/2]\mathbf{l}\cdot\mathbf{\sigma} \right\rangle \qquad 16.$$

where the term in brackets shows explicitly how the Larmor precession is reduced a factor of 2 by the relativistic Thomas term (71). This term arises in a relativistic treatment of the acceleration of the electron by the Coulomb force.

For a proton in a shell-model orbit the equation is similar, but the signs of both the Larmor and Thomas terms are changed. The Larmor term is reversed because the proton magnetic moment is parallel to its spin while the Thomas term is reversed because the Coulomb force is repulsive in this case:

$$W_{so}{}^p = \left\langle \frac{e\hbar^2}{2M_p{}^2c^2} \frac{1}{r} \frac{dV_c}{dr} [\mu_p - 1/2]\mathbf{l}\cdot\mathbf{\sigma} \right\rangle \qquad 17.$$

The proton magnetic moment in nuclear nagnetons, μ_p, replaces the magnetic moment (-1 Bohr magneton) of the electron in Equation 16. The corresponding equation for the neutron has no Thomas term since there is no Coulomb force:

$$W_{so}{}^n = \left\langle \frac{e\hbar^2}{2M_p{}^2c^2} \frac{1}{r} \frac{dV_c}{dr} [\mu_n]\mathbf{l}\cdot\mathbf{\sigma} \right\rangle \qquad 18.$$

With the magnetic moments $\mu_p = 2.79$ n.m. and $\mu_p = -1.91$ n.m., the net effect on the Coulomb-energy shift is:

$$\Delta_{so} = (1/2 - \mu_p + \mu_n) \frac{\hbar^2}{2M_p{}^2c^2} (1 \cdot \delta) \int_0^\infty u^2(r) \frac{1}{r} \frac{dV_c}{dr} r^2 dr \qquad 19.$$

where the neutron and proton radial wavefunctions are the same $[u(r)]$, and the electron charge is included in the definition of V_c (in MeV). If the neutron excess contains more than one subshell, Δ_{so} must be averaged over the properly weighted $u_i(r)$. In the present work this term was calculated using the Coulomb potential generated by the empirical charge distribution and Woods-Saxon radial wavefunctions. Typical values of Δ_{so} are given in Table IV for closed-proton-shell nuclei. They are positive when $j < l$ and negative when $j > l$.

TABLE IV

SIZE OF ELECTROMAGNETIC SPIN-ORBIT TERM Δ_{so}

	Orbit	Woods-Saxon[a] (keV)	Approximation[b] (keV)
C^{13}-N^{13}	$1p_{1/2}$	+ 57	+ 53
O^{17}-F^{17}	$1d_{5/2}$	− 58	− 56
Si^{29}-P^{29}	$2s_{1/2}$	0	0
S^{33}-Cl^{33}	$1d_{3/2}$	+114	+ 97
Ca^{41}-Sc^{41}	$1f_{7/2}$	−106	− 98
Ni^{58}-Cu^{58}	$2p_{3/2}$	− 36	− 36
	$1f_{5/2}$	+153	+141
	$2p_{1/2}$	+ 72	+ 70
Pb^{208}-Bi^{208}	$1h_{9/2}$	+199	+205
	$3p_{3/2}$	− 29	− 35

[a] Equation 19.
[b] Equation 21.

A reasonable approximation to Δ_{so} can be made by assuming that $V_c(r)$ is generated by a uniformly charged sphere with an rms radius determined from electron scattering. In this case the electric field $(-dV_c/dr)$ is proportional to r inside the sphere $(r \leq R_{eq})$ and proportional to $1/r^2$ outside. Equation 19 then becomes:

$$\Delta_{so} = (\mu_p - \tfrac{1}{2} - \mu_n) \frac{\hbar^2}{2M_p{}^2c^2} (1 \cdot \delta) \left(\frac{Z_<{}^2}{R_{eq}{}^3} \right)$$
$$\times \left[\int_0^{R_{eq}} u^2(r) r^2 dr + \int_{R_{eq}}^\infty u^2 \left(\frac{R_{eq}{}^3}{r^3} \right) r^2 dr \right] \qquad 20.$$

The radial wavefunction is normalized $\int_0^\infty u^2(r)r^2 dr = 1$, and one can approximate the bracketed term by ~ 1. Since this estimate is slightly high for any wavefunction extending beyond R_{eq}, this approximation would lead us to a slight overestimate of Δ_{so} for a uniform charge distribution. However for a diffuse charge distribution the bracketed term should be ~ 1.2 and we get

$$\Delta_{so} \simeq 1.2(\mu_p - 1/2 - \mu_n) \frac{\hbar^2}{2M_p^2 c^2} \left(-\frac{Z_< e^2}{R_{eq}^3}\right) 1 \cdot \sigma$$

$$\simeq -.15 \frac{Z_<}{R_{eq}^3} (1 \cdot \sigma)$$

21.

where $1 \cdot \sigma = l$ if $j > l$ and $-l - 1$ if $j < l$. Values calculated with this approximation are also given in Table IV.

5.2 Correction Terms

The Coulomb-energy shift calculated by the method suggested in Section 5.1 is clearly subject to several correction terms. Some of these are well-known effects and a reasonable estimate of them can be made. Others are relatively poorly understood and only rough guesses can be given. Such terms are summarized in Table V. They are discussed in the context of our

TABLE V

Correction Terms in the Coulomb-Energy Difference Δ

Effect	Description	Estimate	Comment
Coulomb perturbation	Analog-state rms radius is larger than neutron excess rms	$-(1.0-2.0)\%$	Larger for neutron-excess orbits of low l
Spreading width	Coupling to the $T_<$ states.	~ 0.1	Related to resonance width
Nuclear rearrangement	Core of analog slightly larger than core of parent	$+0.1\%$	Based on nuclear compressibility
Isospin impurity in the core	Possible contribution to analog due to nonorthogonality of proton and neutrons	$+2\%$	Probably extreme overestimate
Intrashell interactions	Depends on coupling scheme in partially filled shell	$\sim \dfrac{100 \text{ keV}}{2T}$	Pairing effect, increases Δ for $Z_<$-odd
Correlations	Arises through the residual interaction	$\sim 10-300$ keV	Preliminary estimates. The net effect is probably positive

basic assumption of good isospin, since most of them imply breakdown of isospin to some degree.

Coulomb perturbation.—It is well known that the radial wavefunction of the proton in the analog state is not identical to that of the corresponding neutron in the parent. If the neutron wavefunction is the eigenfunction of some nuclear potential well, the proton must satisfy the Schroedinger equation in this nuclear potential with a Coulomb term added. The effect of the Coulomb term is to slightly increase the rms radius of the proton wavefunction over that of the neutron. For the pair Ca^{41}-Sc^{41} the $f_{7/2}$ proton in Sc^{41} has 4 per cent larger rms radius than the $f_{7/2}$ neutron in Ca^{41}, calculated in the same nuclear well. The corresponding number for $d_{5/2}$ wavefunctions in O^{17}-F^{17} is 6 per cent.

The effect on Δ, however, is less than one fourth of this percentage. To show this, the Coulomb energy is written as the difference of the particle energies in the mirror states:

$$\Delta = E_p - E_n \qquad 22.$$

where

$$E_n \equiv \langle n | V_N | n \rangle \quad \text{(negative number)}$$

and

$$E_p \equiv \langle p | V_N + V_c | p \rangle$$

i.e. the neutron is in the purely nuclear potential V_N while the proton is in $V_N + V_c$. Thus,

$$\Delta = \langle p | V_c | p \rangle - [\langle n | V_N | n \rangle - \langle p | V_N | p \rangle] \qquad 23.$$

where

$$\langle n | V_N | n \rangle < \langle p | V_N | p \rangle$$

by the variational principle, since $| n \rangle$ is an eigenfunction of V_N. Using the neutron wavefunction to calculate E_c corresponds to assuming

$$\Delta \simeq \langle n | V_c | n \rangle$$

The first term in Equation 23 is smaller than this term because $| p \rangle$ has a larger rms radius than $| n \rangle$. It is shown in Section 6.1 that

$$\frac{\delta(\Delta)}{\Delta} \simeq - \frac{1}{2} \frac{\delta(\langle r^2 \rangle^{1/2}_{\text{exc}})}{\langle r^2 \rangle^{1/2}_{\text{exc}}}$$

The bracketed term in 23 represents a decrease in the nuclear binding which tends to decrease the effect. In fact if the variational minimum in $\langle n | V_N | n \rangle$ is parabolic, then the increase in nuclear energy is one-half the decrease in Coulomb energy and to a good approximation:

$$\Delta(\text{corrected}) \simeq \left(1 - \frac{1}{4} \frac{\delta(\langle r^2 \rangle^{1/2}_{\text{exc}})}{\langle r^2 \rangle^{1/2}_{\text{exc}}}\right) \langle n | V_c | n \rangle \qquad 24.$$

This results in a correction of ~ -1 per cent in the region of oxygen and calcium.[3] For heavier nuclei with large neutron excesses there may be proton-neutron correlations in analog states; these may tend to make the proton wavefunction more similar to the neutrons inside the nucleus, and thereby reduce the Coulomb perturbation effect.

Unbound analog states.—The question of the "Thomas-Ehrman shift" arises in connection with unbound states (72–75). This shift results in R-matrix theory when the internal wavefunctions are matched to the external wavefunctions. It is also known as "boundary condition displacement." The shift is present for all mirror levels, and depends on the difference in the asymptotic behaviors of the proton and neutron. The effect is larger when the asymptotic behaviors are drastically different, such as would be the case

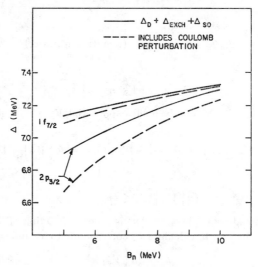

FIG. 5. Calculation of the Coulomb-energy shifts of the $1f_{7/2}$ and $2p_{3/2}$ orbits in Ca[41]. The solid lines include the direct, exchange, and spin-orbit contributions. The dashed lines contain the additional correction for the Coulomb perturbation. For neutron binding <7. MeV the analog state is unbound. The curves extrapolate smoothly into the unbound region.

for a bound neutron and a corresponding unbound proton. The classic example (73, 74) is that the unbound first-excited state of N[13] which is 720 keV lower in excitation energy than its mirror in C[13]. Similarly the $2p_{3/2}$ state in Sc[41] is 290 keV lower than in Ca[41].

The Thomas-Ehrman shift is included in Woods-Saxon calculations in a natural way. No special treatment is needed when the state is unbound. This is indicated in Figure 5 where the quantity plotted (dashed lines) is Δ calculated according to Equation 22. Both Coulomb and nuclear effects were

[3] In practice about 50 keV of the Coulomb perturbation is cancelled because of a kinematic effect arising from the p-n mass difference.

included by solving the Schroedinger equation for both the proton and neutron. The solid lines were calculated as described in Section 5.1 and, therefore, serve to indicate the size of the Coulomb perturbation corrections. When the neutron binding is below $\sim$7 MeV in this example the proton state is unbound; its energy then is taken as the position of the scattering resonance. The Coulomb energy varies smoothly from the bound to unbound region. Thus the "Thomas-Ehrman shift" is accounted for by the difference in the $2p_{3/2}$ and $1f_{7/2}$ wavefunctions, the difference in Δ_{so}, and the Coulomb perturbation corrections mentioned above, whether the state is bound or unbound. Some calculated results for the shift are given in Section 7.1.

Analog states, bound or unbound, do experience a small additional shift not accounted for above. This is a shift arising through the coupling to nearby more complex states. The shift due to this coupling is expected to be only a small fraction of the resonance width (76). The resonance width varies from $\sim$10 keV in the medium-mass region to $\sim$100 keV in heavy nuclei.

Nuclear rearrangement.—This is defined as the change in the total binding energy of the core caused by the extra unit of charge in the analog state. The Coulomb repulsion felt by each proton in the core becomes slightly stronger, resulting in a slight expansion of the core.

The correction to the Coulomb energy due to this effect can be estimated through the use of the semiempirical value of the nuclear compressibility. We first estimated the magnitude of the radius increase in the radius of the core by using Woods-Saxon wavefunctions. A Woods-Saxon well, which binds the $f_{7/2}$ proton in Sc^{41} at the proper energy, was used to calculate the rms radii of all the protons in both Sc^{41} and Ca^{41}, the only difference between the two cases being the extra charge in Sc^{41}. The rms radius of each proton-core orbit in Sc^{41} was $\sim$0.1 per cent larger than in Ca^{41}.

To estimate the change in binding energy we let $E(R)$ be the binding energy as a function of radius with E_0 the variational minimum at R_0. To lowest order $E(R)$ can be expanded:

$$E(R) = E_0 + \frac{1}{2}\left(\frac{\partial^2 E_N}{\partial R^2}\right)\Delta R^2$$

or

$$E(R) = E_0 + \frac{1}{2}KA\left(\frac{\Delta R}{R}\right)^2 \qquad 25.$$

where $K \approx 200$ MeV/nucleon is the nuclear compressibility (77). Using $\Delta R/R = .001$

$$\Delta E = \frac{1}{2}KA\left(\frac{\Delta R}{R}\right)^2 \simeq 10^{-4}A \qquad 26.$$

$$\simeq 4 \text{ keV} \qquad \text{for } Ca^{40}$$

The sign of this term is such that it tends to increase Δ.

Isospin impurity in the core.—The Coulomb perturbation, shifts of unbound states, and core rearrangement all imply violation of our initial assumption of good isospin. However, they do not greatly affect Δ. Further violation of isospin can occur if the proton and neutron orbits in the core do not have perfect overlap. If the core orbits of Ca^{40} are all calculated in the same nuclear well, the average rms radius for protons is 3.5 per cent larger than that for neutrons. The corresponding overlap integrals vary from 99.0 per cent for the $2s$ states to 99.6–99.8 per cent for the other orbits. The neutron states can be expanded in terms of the complete set of orthogonal proton states. When the T_- operator acts on the entire set of neutron orbits in Ca^{41}, there will be contributions to the analog state from that part of each neutron orbit which is not occupied by a proton. However, in every case studied from O^{16} to Pb^{208} the neutrons from the core caused negligible change in Δ.

There is a significant effect due to the nonorthogonality of the neutron excess orbits with proton-core orbits of the same lj. In the medium-mass range this effect is applicable only in special cases where the neutron excess contains $2s$ or $2p$ particles. In the nickel isotopes for example, the $2p$-$1p$ overlap reduces the effective rms radius of the analog state ~ 5 per cent. In Pb where several such orbits are involved the effect is ~ 3 per cent. The effect on the Coulomb energy is roughly one half of this percentage, and in every case calculated *increased* Δ.

This correction term is poorly understood, and the numbers cited here may well be a gross overestimate of its magnitude. We prefer to keep the assumption of good overlap between the core orbits, there being no definite experimental evidence for a large effect of this sort. Hartree-Fock calculations, for example, give a difference betwen the proton and neutron rms radius in Ca^{40} of only ~ 1 per cent (78, 79).

Intrashell interactions.—Whereas the preceding correction terms all implied the breakdown of isospin to some extent, the remaining terms require the use of more detailed models or wavefunctions, but do not necessarily violate isospin. In the independent-particle model the intrashell interaction is required when calculating Δ for a partially filled proton shell. For example, in the Ti^{46}-V^{46} analog pair the difference between the Coulomb matrix element of the $[(\pi f_{7/2})^2(\nu f_{7/2})^4]$, and $[(\pi f_{7/2})^3 (\nu f_{7/2})^3]$, $T=1$ configurations is needed and a coupling scheme must be specified. This type of calculation has been done, for example, by Harchol et al. (17) and by Jänecke (45). For mass ≥ 70 the intrashell interactions are washed out by the large neutron excess, making detailed studies of the intrashell interaction probably only possible in the $f_{7/2}$ shell. In this region the size of the fluctuating part of these terms seems to be ~ 100 keV, decreasing with neutron excess. The average $f_{7/2}$-$f_{7/2}$ Coulomb interaction is ~ 280 keV. The specific calculations discussed in Sections 6 and 7 involve interactions only with closed proton shells, so intrashell interactions do not arise.

Correlations.—The effect of nucleon-nucleon correlations on Δ over and above those required by the Pauli principle has been considered recently by Auerbach et al. (80), and by Bertsch (39). The effect discussed by Auerbach et al. arises when the shell-model residual interaction introduces correlations in the neutron excess of the parent nucleus. The nuclear binding associated with this correlation changes when one of the neutrons is changed into a proton. The change is caused by the Coulomb perturbation of the wavefunction. Their preliminary estimate is that this effect may be on the order of a few tens of keV in the Ca isotopes.

Bertsch (39), on the other hand, has calculated a different type of correlation effect, that of configuration mixing on the intrashell Coulomb interaction. For example, the $(f_{7/2}^2 \ J=0)$ pairing interaction in Ti^{42} is increased by ~ 150 keV when higher configurations $(f_{5/2}^2, p_{3/2}^2, p_{1/2}^2, g_{9/2}^2)$ are mixed in according to better shell-model wavefunctions. If such a calculation were applied to the mirror nuclei it might bring the magnitude of the pairing effect calculated by Carlson & Talmi (23) into better agreement with the empirical pairing term.

More generally, correlation effects would come about from the two-body interaction with its repulsive core and attractive tail. Such effects would behave in a way similar to the exchange term: the repulsive core would tend to increase the magnitude of the exchange term while the attractive part would tend to cancel it. The net correction is probably positive, but the importance of such correlations is not fully known (81).

Core excitation.—There is much controversy over the amount of core excitation in the ground state of closed-proton-shell nuclei, such as the Ca isotopes. Hence a calculation of the effect of core excitation on Δ would be of interest. There exists no accurate estimate of this effect.

If the ground state of Ca^{41}, for example, contains an appreciable three-particle–two-hole component, then Δ would be increased somewhat because the $d_{3/2}$-core Coulomb interaction is ~ 5 per cent larger than the $f_{7/2}$-core interaction. A part of Δ for the three-particle–two-hole component will arise from charge-exchanging a $d_{3/2}$ neutron, and the remainder from an $f_{7/2}$ neutron. For a reasonable amount of core excitation (≤ 20 per cent), a rough estimate increases the Coulomb-energy shift by < 100 keV.[4]

Nonlocality.—It is well known that the self-consistent nuclear potential, which generates the nucleon bound states, is nonlocal. However, the values of Δ tend to be insensitive to this fact. The major contribution (> 95 per cent) to the Coulomb energy is from the direct term, as indicated in Figure 4. The Coulomb potential is determined by experiment and it is primarily the rms radius of the neutron-excess wavefunction that determines the value of Δ. The exchange energy and all other contributions to Δ are small and still not very sensitive to the wavefunctions of the bound states. The type of potential

---

[4] In addition, core excitation can affect the well parameters extracted from electron scattering. Preliminary discussion of this effect has been given (82) but the effect on the radius of the core has not yet been derived.

must be specified, however, if comparison is to be made with other calculations, such as stripping reactions: the density-dependent nonlocal potential used recently by Meldner (83) may be appropriate.

6. NEUTRON-DENSITY DISTRIBUTIONS

The direct Coulomb energy is determined by the integral:

$$\Delta_D = \int_0^\infty V_c(r)\rho_{\text{exc}}(r)4\pi r^2 dr;$$

all other terms contributing to Δ are very much smaller. In the discussion that follows, the numerical results include the exchange and spin-orbit contributions, but the so-called correction terms have been neglected. If the Coulomb potential V_c is taken from electron scattering or muonic X-ray work, then Figure 4 shows that the magnitude of Δ_D is determined by the extent of the radial density distribution of the neutron excess, $\rho_{\text{exc}}(r)$. This distribution is normalized:

$$\int_0^\infty \rho_{\text{exc}}(r)4\pi r^2 dr = 1$$

so that it corresponds to one unit of charge in the analog state. Thus the direct term changes inversely to the radius of ρ_{exc}. In this manner experimental Coulomb energies can be used to determine the radial extent of the neutron excess in any nucleus where the charge radius is known. To eliminate complications from the intrashell shell interactions we limit the calculations to closed-proton-shell nuclei.

6.1 Model Dependence

It is necessary to determine what moments of the charge distribution $\rho_{\text{ch}}(r)$ and the neutron-excess distribution $\rho_{\text{exc}}(r)$ are important in determining Δ_D. The detailed shape of $\rho_{\text{ch}}(r)$ is not determined uniquely by electron scattering or muonic X rays. The rms radius and diffuseness of the $\rho_{\text{ch}}(r)$ are usually the quantities which are well determined, so distributions are normally parameterized in terms of Fermi or parabolic Fermi shapes. Likewise, the detailed shape of ρ_{exc} depends on the wavefunctions contributing to the neutron excess and the well in which they are calculated. In this discussion we will use the uniform model for reference:

$$\Delta_D = \frac{6}{5} \frac{Z_< e^2}{R_{\text{eq}}}$$

where the charge distribution consists of $Z_<$ protons occupying a uniformly charged sphere of radius $R_{\text{eq}} = \sqrt{\frac{5}{3}}\langle r^2 \rangle^{1/2}$. The neutron excess is also a uniform distribution identical with the charge distribution. This can be compared with the case of ρ_{ch} and ρ_{exc} both having Fermi shape with parameters

c and z.[5] The increase of Δ_D with diffuseness is shown in Table VI. Empirical values of z for charge distributions are $\sim$0.5–0.6 F. Roughly half the variation in Table VI comes from z_{ch} and half from z_{exc}, so the dependence on the diffuseness of either distribution near $z = 0.5$ is:

$$\frac{\delta(\Delta_D)}{\Delta_D} \lesssim \frac{1}{20} \frac{\delta(z)}{z}$$

In many nuclei the use of a Fermi shape for the neutron excess is obviously a bad approximation. In Ca^{41}-Sc^{41}, for example, the $f_{7/2}$ neutron wave-

TABLE VI

Coulomb-Energy Shift Calculated for a Fermi Charge Distribution with Diffuseness z

z	0.	.2	.4	.6	.8	1.0
$\langle r^2 \rangle^{1/2}$						
3	1.00[a]	1.00	1.01	1.05	1.13	
4	1.00	1.00	1.005	1.02	1.05	1.10
6	1.00	1.00	1.00	1.01	1.015	1.03

[a] Coulomb-energy shifts given as the ratio to the value for a uniformly charged sphere ($z = 0$) with the same rms radius.

function is a surface-peaked shape, indicated qualitatively in Figure 4. Table VII indicates that the use of a surface-peaked Woods-Saxon wavefunction gives a slightly smaller value of Δ_D than the Fermi shape. Using a charge distribution generated completely from wavefunctions in a Woods-Saxon well does not change Δ_D appreciably.

The direct term in the uniform model varies as $\langle r^2 \rangle^{-1/2}$. This is nearly true for diffuse distributions also; a 2 per cent decrease in either neutron excess or charge radius giving $\sim$1 per cent increase in Δ_D. If the charge radius is decreased by 1 per cent the neutron-excess radius must be increased 1 per cent to keep the Coulomb energy approximately constant. These relationships are summarized in Table VIII.

The neutron core.—Once a radius for the neutron excess is extracted, an additional model dependence enters into the determination of the overall neutron radius. In mirror nuclei this question does not arise because we are

[5] The approximation (66, appendix C):

$$\langle r^2 \rangle = \frac{3}{5} c^2 + \frac{7\pi^2}{5} z^2$$

is useful in relating $\langle r^2 \rangle^{1/2}$, c, z. The accuracy of this approximation is better than 0.01 per cent for $z/c < 20$ per cent.

only interested in the neutron excess. However, in the heavier nuclei this problem is important because there is interest in the overall neutron-proton radius difference ("neutron halo"). In $N=Z$ nuclei, such as Ca^{40}, Woods-Saxon calculations of proton and neutron orbits in the same nuclear well predict the proton radius to be ~ 3 per cent larger than the neutron radius. However, when protons and neutrons occupy the same levels correlations which tend to reduce this difference are probably introduced. Woods-Saxon calculations for Pb^{208} with a symmetry term in the potential also indicate a proton radius ~ 2–4 per cent larger than the radius of the neutron core. And, again, this may be reduced by p-n correlations. Hence in treating the question of the "neutron halo" in heavy nuclei we assume that the neutron core radius is equal to the proton radius.

TABLE VII

COULOMB-ENERGY SHIFTS CALCULATED FOR Ca^{41}-Sc^{41} TO SHOW THE DEPENDENCE
ON THE SHAPES OF THE CHARGE AND NEUTRON-EXCESS DISTRIBUTIONS

Charge distribution $\langle r^2 \rangle^{1/2} = 3.44$ F	Neutron-excess distribution $\langle r^2 \rangle^{1/2} = 3.61$ F	Δ_D (MeV)
Uniform ($R_{eq} = 4.44$)	Uniform ($R_{eq} = 4.66$)	7.58
Fermi $\begin{array}{c} c = 3.39 \\ z = 0.6 \end{array}$	Fermi $\begin{array}{c} c = 3.67 \\ z = 0.6 \end{array}$	7.80
Fermi $\begin{array}{c} c = 3.39 \\ z = 0.6 \end{array}$	$f_{7/2}$ Woods-Saxon	7.71
Woods-Saxon orbits	$f_{7/2}$ Woods-Saxon	7.70

TABLE VIII

SENSITIVITY OF THE COULOMB-ENERGY SHIFTS TO
CHANGES IN THE RMS RADII

$$\frac{\delta(\Delta_D)}{\Delta_D} \simeq -1/2 \frac{\delta(\langle r^2 \rangle_{exc}^{1/2})}{\langle r^2 \rangle_{exc}^{1/2}} \qquad (\text{const } \rho_{ch})$$

$$\frac{\delta(\Delta_D)}{\Delta_D} \simeq -1/2 \frac{\delta(\langle r^2 \rangle_{ch}^{1/2})}{\langle r^2 \rangle_{ch}^{1/2}} \qquad (\text{const } \rho_{exc})$$

$$\frac{\delta(\langle r^2 \rangle_{exc}^{1/2})}{\langle r^2 \rangle_{exc}^{1/2}} \simeq -\frac{\delta(\langle r^2 \rangle_{ch}^{1/2})}{\langle r^2 \rangle_{ch}^{1/2}} \qquad (\text{const } \Delta_D)$$

6.2 Specific Results

The radius of the neutron excess in several closed-proton-shell nuclei has been calculated. With the Coulomb potential calculated from the experimentally determined charge distribution[6] the only adjustable parameter in fitting the Coulomb energy is the radial extent of the neutron-excess wavefunction. In these calculations the neutron-excess wavefunctions were eigenfunctions of a Woods-Saxon potential well whose radius and strength were adjusted to fit the experimental Coulomb displacement energy and the neutron-separation energy. When more than one neutron orbital is filled in the neutron excess, the orbits were weighted according to their occupation numbers. The analog-state density distribution was assumed to be the same as that of the neutron excess ρ_{exc}. The finite size of the nucleon was included in the calculation of Δ_D. The exchange factor ϵ was calculated using Woods-

TABLE IX

NEUTRON-EXCESS DENSITY DISTRIBUTIONS EXTRACTED FROM
COULOMB-ENERGY SHIFTS

Nucleus	Data		Results	
	Δ (MeV)	$\langle r_{ch}^2 \rangle^{1/2}$ (F)	$\langle r_{exc}^2 \rangle^{1/2}$ (F)	$\langle r_{exc}^2 \rangle^{1/2} / \langle r_{ch}^2 \rangle^{1/2}$
C^{13}	3.00	2.50	2.86	1.14
O^{17}	3.54	2.70	3.10	1.15
Si^{29}	5.73	3.10	3.50	1.13
S^{33}	6.35	3.25	3.48	1.06
Ca^{41}	7.28	3.49	3.60	1.03
Ca^{48}	7.18	3.48	3.69	1.06
Ni^{62}	9.38	3.87	4.07	1.05
Sn^{120}	13.70	4.64	4.87	1.05
Pb^{208}	18.87	5.51	5.95	1.08

Saxon wavefunctions for the neutron-excess and proton-core orbits, while the electromagnetic spin-orbit term Δ_{so} was evaluated with the experimental charge-distribution and Woods-Saxon neutron-excess wavefunctions. The answers are then quoted as the rms radius of the neutron excess $\langle r_{exc}^2 \rangle^{1/2}$ which is necessary to fit the empirical Coulomb energy. The particular combination of Woods-Saxon parameters used is of secondary importance.

The values of $\langle r_{exc}^2 \rangle^{1/2}$ extracted with this procedure are given in Table IX. The experimental values of Coulomb energies and charge radii used in the calculation are also listed. The answers are given in fermis and as the ratio of the excess radius to the charge radius.

---

[6] A list of empirical charge parameters used in the present calculations is given in the Appendix.

In the oscillator model the ratio of the rms radius of the last neutron to the radius of the core neutrons depends only on the major shells involved. For example, $\langle r^2 \rangle^{1/2}$ for the $f_{7/2}$ neutron in Ca^{41} is predicted by the oscillator model to be 23 per cent larger than the core. When a local Woods-Saxon potential well is used the prediction is 24 per cent. Such ratios, calculated from Woods-Saxon wavefunctions, are compared with the experimental values in Table X. The experimental ratios are 6 to 20 per cent smaller than the calculated values. More sophisticated calculations have been carried out for some of the nuclei listed in Table X. For example, Hartree-Fock calculations by Tarbutton & Davies (84) for Ca^{48} and Pb^{208} give ratios ~ 2 per cent larger than the Woods-Saxon values in the table.

TABLE X

COMPARISON OF NEUTRON-EXCESS RADII WITH POTENTIAL MODEL

Nucleus	$\langle r_{exc}^2 \rangle^{1/2} / \langle r_{core}^2 \rangle^{1/2}$		
	Experimental[a]	Woods-Saxon[b]	Difference (%)
C^{13}	1.14	1.23	8
O^{17}	1.15	1.37	19
Si^{29}	1.13	1.27	12
S^{33}	1.06	1.12	6
Ca^{41}	1.03	1.24	20
Ca^{48}	1.06	1.23	16
Ni^{62}	1.05	1.17	11
Sn^{120}	1.05	1.18	12
Pb^{208}	1.08	1.14	6

[a] $\langle r_{core}^2 \rangle^{1/2} \equiv$ rms radius of empirical charge distribution.

[b] $\langle r_{core}^2 \rangle^{1/2} \equiv$ rms radius of core neutrons calculated in same potential as excess neutrons.

Another way of stating this result is that the calculated Coulomb energy is too small if a Woods-Saxon well which fits the empirical charge radius is used to calculate the neutron-excess wavefunctions. The results of such calculations for mirror nuclei are given in Table XI. The values of diffuseness for the Woods-Saxon potentials were taken from optical-model work; the exact choice of diffuseness is not important. The parameters of the spin-orbit potential are given in the table. Since the analog-state wavefunction is taken to be identical to the neutron-excess wavefunction, the calculation was only done for the $Z_<$ nucleus. The radius of the proton-core potential well was chosen so that the calculated charge-density distribution (i.e., the proton centers-of-mass distribution with the finite size of the proton folded in) had an rms radius equal to the experimental value. A charge distribution calculated in this way is compared with the parabolic Fermi distribution (85) of Ca^{40} in Figure 6.

The rather large discrepancies (5 to 9 per cent in Table XI) between the calculated and experimental Coulomb energies apparently cannot be explained by any of the correction terms discussed in Section 5.2. Either the

TABLE XI

CoULOMB ENERGIES CALCULATED IN A WOODS-SAXON POTENTIAL WELL
WHICH FITS THE EXPERIMENTAL RMS CHARGE RADIUS

Mirror pair	r_0^a (F)	a (F)	Δ(calc) (MeV)	Δ(exp) (MeV)	Difference (%)
C^{13}-N^{13}	1.30	.5	2.79	3.00	7
O^{17}-F^{17}	1.32	.5	3.23	3.54	9
Si^{29}-P^{29}	1.27	.60	5.53	5.73	4
S^{33}-Cl^{33}	1.29	.60	6.11	6.35	4
Ca^{41}-Sc^{41}	1.29	.65	6.66	7.28	9

[a] $r_0 A^{1/3}$ is the half-way radius and a is the diffuseness of the Woods-Saxon potential well which yields a charge-density distribution with an rms radius equal to the empirical value. A spin-orbit potential of the Thomas type was used with parameters $r_{so} = .95\ r_0$, $a_{so} = a$, and $V_{so} = 6$ MeV. All the proton bound states were calculated with a well depth determined by the binding energy of the last proton in the $Z_<$ nucleus. The neutron-excess wavefunction was calculated with a well depth determined by its binding energy.

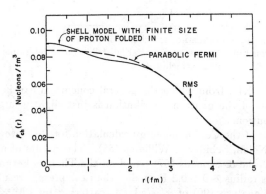

FIG. 6. Comparison of charge-density distributions for Ca^{40}. The parabolic Fermi distribution comes from analysis of 200-MeV elastic electron scattering. The shell-model distribution is the sum of bound states in a Woods-Saxon potential well, adjusted to have the same rms radius as the parabolic Fermi shape. The finite size of the proton is folded into the shell-model distribution.

rms radius of the last neutron is much smaller than expected, as indicated in Table X, or a large correction term has been neglected.

6.3 Agreement with Other Methods of Measuring Neutron Radii

The rms radius of the neutron excess, as determined in the preceding section, can be combined with the radius of the neutron core to give the rms radius of the total neutron distribution of the nucleus. In this way Coulomb energies give useful information on this controversial subject. Historically, Johnson & Teller (86), for example, proposed that neutron distributions in heavy nuclei would be $\sim$1 F larger in radius than the proton distributions. On the other hand, Wilets (87) and Bodmer (88) used a statistical model to argue that the neutron and proton distributions in heavy nuclei should have very nearly the same rms radius. A review of the experimental and theoretical situation in comparison of charge and matter distributions was given by Hill in 1957 (89).

The present results were calculated under the assumption that the distribution of the first Z neutrons is the same as the distribution of the Z protons, i.e., good isospin in the core. Results for some heavier nuclei are

TABLE XII

The Neutron Radius Compared with the Experimental Charge Radius

Nucleus	$\langle r_{\mathrm{ch}}^2 \rangle^{1/2}$ (F)	$\langle r_{\mathrm{n.m.}}^2 \rangle^{1/2}$ (F)	$\langle r_{\mathrm{n.m.}}^2 \rangle^{1/2} - \langle r_{\mathrm{ch}}^2 \rangle^{1/2}$ (F)
Ca^{48}	3.48	3.54	0.06
Ni^{62}	3.87	3.90	0.03
Sn^{120}	4.64	4.71	0.07
Pb^{208}	5.51	5.66	0.15

$\langle r_{\mathrm{ch}}^2 \rangle^{1/2} \equiv$ rms radius of experimental charge distribution.
$\langle r_{\mathrm{n.m.}}^2 \rangle^{1/2} \equiv$ rms radius of "neutron matter" distribution as determined by Coulomb energies (finite size of neutron is included).

given in Table XII, from which the general conclusion can be drawn that the rms radius of the neutron distribution is just slightly larger than the charge distribution.

The results of the Coulomb-energy calculation for Pb^{208} were previously reported by Nolen, Schiffer & Williams (55).[7] The results of a similar calculation by Bethe & Siemens (67) based on the Thomas-Fermi model give a neutron rms radius $\sim$0.1–0.2 F smaller than the present results. A recent optical-model analysis (90) of π^+ and π^- scattering by Pb^{208} (91) gives a neutron rms radius $\sim$0.1 F smaller than the proton radius or $\sim$0.25 F

---

[7] The exchange factor used in (55) was too large by a factor of 2 and the presently accepted Coulomb-energy shift of Pb^{208} is 110 keV less than was used in that work. The corrected neutron radius, given in Table XII, is 0.08 F larger than in (55).

smaller than the present result. Allowing for reasonable uncertainties in the models it appears that all three of these approaches are consistent with approximately equal neutron and proton radii in Pb.

These results can be compared with several recent theoretical calculations of the neutron-proton radius difference. For example, Hartree-Fock calculations (84, 92) predict the neutron rms radius to be $\sim$0.3 F larger than the proton rms radius in Pb^{208}. Using a density-dependent nonlocal single-particle potential, Meldner (83) calculated a difference of 0.07 F.

Greenless, Pyle & Tang (93) have extracted the rms nuclear matter radius of several nuclei, using a reformulated optical model to analyze elastic proton scattering. Comparing the total matter radius with the previously known charge radius, they obtain values for $\langle r_n^2 \rangle^{1/2} - \langle r_p^2 \rangle^{1/2}$ of $\sim$0.6 F for several nuclei from Ni^{58} to Pb^{208}. However, recent revisions (94, 95) of the reformulated optical model are tending to decrease this difference.

A study of the nuclear absorption of K^- mesons from atomic orbits (96) indicates that in heavy nuclei there are about five times more neutrons than protons at the radius of absorption. This result, however, is consistent with equal proton and neutron rms radii. The absorption occurs well outside the nuclear surface (97) and a simple calculation with proton and neutron bound states of a Woods-Saxon potential shows that, even with the rms radii approximately equal, the ratio of neutron density to proton density is $\sim$2 at 8 F and $\sim$25 at 12 F in Pb^{208}, where the nuclear surface is at $\sim$7 F. The results of a Woods-Saxon calculation for Pb^{208} are given in Figure 7 and Table XIII. The densities of the distributions of the protons and neutron core drop off much faster than the neutron excess because of their larger binding enegires. (For the protons it is the binding energy relative to the top of the Coulomb barrier which is relevant.) Recently Wiegand (98) has accurately measured K^- muonic X rays from many elements. A complete analysis of the new data could yield much more quantitative experimental information about the tail of the nuclear-matter distribution.

7. NUCLEAR-STRUCTURE EFFECTS

7.1 BINDING ENERGY AND ORBITAL EFFECTS

In the absence of Coulomb interactions, and if nuclear forces were charge symmetric, the spectra of mirror nuclei would be identical. However, corresponding levels often deviate by $\sim$300 keV or more in mirror pairs. The first two states of three mirror pairs are shown in Figure 8. It is assumed that these states each represent a single particle outside a closed core. (In the case of Ca^{41}-Sc^{41} the centers of gravity of the $2p_{3/2}$ states (99) are given in the figure.) In each case the excited state in the $Z_>$ member comes at lower excitation energy than its parent. The magnitude of this "level shift" is well explained (53, 54, 100) by the different Coulomb energies of the single-particle orbits involved. Results of calculations with wavefunctions in a Woods-Saxon potential are given in Table XIV. The potentials are the same

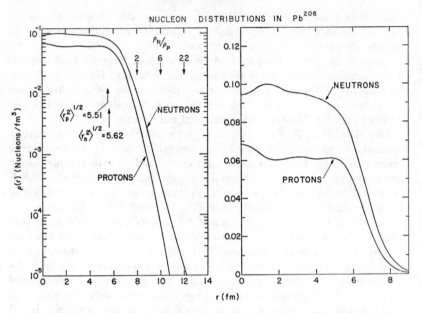

FIG. 7. Proton and neutron densities in Pb[208] calculated in a Woods-Saxon potential well. The rms radii are nearly equal, but the ratio of neutron to proton density in the tail varies from 2 at 8 F to 22 at 12 F, as indicated. Two plots are shown to emphasize the difference at large radii.

as the corresponding ones used in the calculations of the excess radii in Table IX. The net shifts are broken down into four components to show the relative importance of each effect.

The state dependence represents the difference in the two single-particle orbits of different j, calculated for the same binding energy, including $\Delta_D + \Delta_{exch}$. The binding-energy dependence represents the decrease in $\Delta_D + \Delta_{exch}$ when the excited state is recalculated with its experimental binding energy. The spin-orbit term is just the difference in Δ_{so}, and the Coulomb perturbation is the effect of solving the Schroedinger equation for a proton,[8] instead of using the neutron wavefunction, and taking

$$\Delta_D(j) = B_n(j) - B_p(j)$$

What has usually been taken to be the Thomas-Ehrman shift can be seen to have a more complex origin when wavefunctions in a reasonable well are used. As can be seen in Figure 5, Δ is a smooth function of binding energy and does not appear suddenly when the analog state becomes unbound. The magnitude of this shift is more sensitive to the orbital angular momentum than it is to the binding energy; e.g., bound s states can have larger shifts than unbound p states.

[8] The "Coulomb perturbation" as calculated here includes the kinematic effect due to the p-n mass difference.

TABLE XIII

NUCLEON DENSITIES CALCULATED FOR Pb^{208} USING A WOODS-SAXON
POTENTIAL WELL

Radius (F)	Protons	All neutrons	Ratio neutrons/ protons
	(densities[a] in nucleons/$F^3 \times 10^6$)		
5.0	61 200.	87 500.	1.4
5.5	58 200.	81 100.	1.4
6.0	49 000.	69 600.	1.4
6.5	35 300.	52 600.	1.5
7.0	21 300.	34 300.	1.6
7.5	10 700.	19 300.	1.8
8.0	4 590.	9 613.	2.1
8.5	1 709.	4 376.	2.6
9.0	577.	1 891.	3.3
9.5	184.	801.	4.4
10.0	57.4	340.	5.9
10.5	17.9	146.	8.2
11.0	5.7	64.1	11.3
11.5	1.8	28.6	15.7
12.0	0.6	13.0	21.7

Rms radii	protons 5.507 F	*Potential wells*	r_0	a	V
	neutrons 5.622 F	protons	1.265	.7	60.57
		neutrons	1.22	.7	46.97
		spin-orbit	1.10	.7	6.0

NOTE: When the overlap technique is used (cf. Section 3.2) these wavefunctions
fit the Coulomb energy and the rms charge distribution.

[a] The finite size of the nucleon is folded into the density distributions.

TABLE XIV

CALCULATED DIFFERENCE IN COULOMB-ENERGY SHIFTS OF GROUND
STATE AND FIRST-EXCITED STATE IN MIRROR NUCLEI[a]

	C^{13}-N^{13}[b]	O^{17}-F^{17}[c]	Ca^{41}-Sc^{41}[d]
State dependence	230 keV	230 keV	150 keV
Binding-energy dependence	325	90	150
Spin-orbit difference	60	− 60	−70
Coulomb perturbation difference	185	130	75
Net calculated shift	800	390	305
Experimental shift	720	371	290

[a] The calculations were done with a Woods-Saxon well ($r_0 = 1.05$). The small well
radius is required to fit the experimental values of Δ.

[b] Ground state is assumed pure $p_{1/2}$, first-excited state $s_{1/2}$.

[c] Ground state $d_{5/2}$, excited state $s_{1/2}$.

[d] Ground state $f_{7/2}$, excited state $2p_{3/2}$.

It is also possible to compare the analog spectra of heavy nuclei. The isobaric analogs of excited states in many heavy nuclei have been identified as isobaric analog resonances. In heavy nuclei ($T\gg1$) analog spectra are generally much more similar than in mirror nuclei. The low-lying excited states of the parent nucleus are mainly excitations of the last occupied orbitals and thus the proton wavefunctions of the corresponding analog states are a linear combination consisting mainly of terms from the undisturbed orbitals in the neutron excess. Any fluctuations in the Coulomb energy of the levels are reduced to a large degree by the large neutron excess. A systematic study of the Coulomb energies of the low-lying states of the odd-A Ba-La isobaric pairs (101) clearly demonstrates this effect. The Coulomb energies of these states are plotted in Figure 9. The $3s$-$2d$ shells and $2f$-$3p$ shells are being filled in this series of isotopes. While in the $2s$-$1d$ shell region there were variations of $\sim$300 keV or more between the mirror spectra, in the Ba region the largest discrepancy is $\sim$50 keV (0.3 per cent).

7.2 Isotope Shift of Matter Radius

The method of extracting nuclear matter radii from Coulomb energies and empirical charge radii (Section 6) can be applied to series of isotopes where data are available. The results of such an analysis for the isotopes of

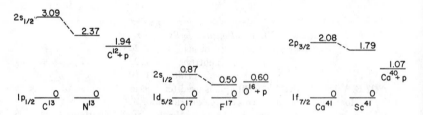

FIG. 8. Level shifts of the first-excited states of three mirror pairs. In N^{13} and Sc^{41} the excited analog states are unbound, while in F^{17} it is bound.

Ca, Ni, and Sn were recently published (56) and are shown in Figure 10. In the Ni and Sn isotopes it was concluded that matter radii increase more rapidly with A than do charge radii but still not quite as fast as $A^{1/3}$. From Ca^{40} to Ca^{48}, however, both the charge radii and Coulomb energies are nearly constant. This results in a nearly constant matter radius in this series up to Ca^{48}.

This qualitative difference between the isotope shift in Ca compared with Ni and Sn is puzzling. If the assumption of good isospin of the core is correct, then the trends shown in Figure 10 may be a difference in core polarization effects (56). In the Ni and Sn isotopes the last neutrons and protons are filling the same major shells. The resulting increased interaction between the excess neutrons and the proton core may cause the matter radius to grow faster than when the excess neutrons are filling the next major shell, as in the Ca isotopes. An implication of these results is indicated in Figure 11. The matter radii of the nuclei with $N = 28$, between Ca[48] and Ni[56] must increase at a rate significantly faster than $A^{1/3}$. If the necessary charge radii were known this prediction could be tested.

7.3 POSSIBLE EFFECTS OF DEFORMATIONS

The effect of deformations on Coulomb energies is of considerable interest. A deformed nucleus would have a larger mean square radius than a spherical one and therefore a smaller value of Δ. This decrease has been estimated (102) at 100–150 keV for the rare-earth region, on the assumption

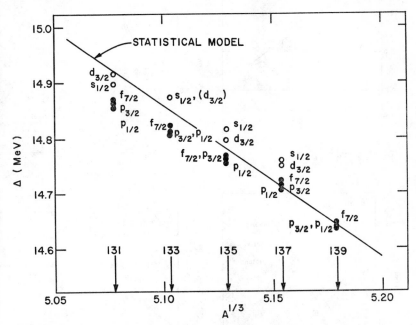

FIG. 9. The Coulomb-energy shifts of states in odd-A Ba-La isobaric pairs.

of a uniform deformation for all orbits. However, if the valence nucleons were to contribute more to the deformation than deeply bound nucleons in closed shells, then this could lead to a larger decrease in the Coulomb energy —perhaps even twice or three times as large a decrease as in a uniform model. It would be desirable to have Coulomb energies for a sequence of iostopes, in which the transition from a spherical to a rotationally deformed shape is well established. Such a set exists in the even-A Sm isotopes: Sm[144] is spherical and Sm[154] exhibits a clear rotational spectrum. Unfortunately, the Coulomb energies of the even Sm isotopes are not yet known. Some trends in this dependence on deformation are shown in Figure 12 where Coulomb energies of $l=3$ states in the vicinity of $N=82$ are plotted. Below $N=82$ the data are for the odd-A Ba isotopes (101). The lightest even-A isotope, Ba[130], exhibits not quite a pure rotational spectrum but approaches one. The nuclei above $N=82$ are odd-A Sm (103) and Nd (104) isotopes and here again the last even isotopes are not quite rotational. Nevertheless the

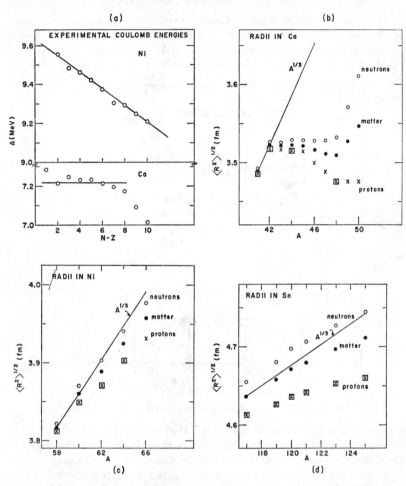

FIG. 10. Coulomb-energy differences, neutron radii, proton radii, and matter radii: (a) the experimental Coulomb-energy differences throughout the Ca and Ni isotopes; (b), (c), and (d) the neutron, proton, and matter radii extracted for Ca, Ni, and Sn isotopes, respectively. The boxed crosses represent measured charge radii, the open crosses extrapolated ones (see references in appendix).

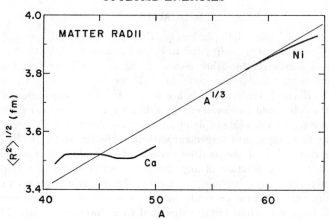

FIG. 11. The matter radii (heavy lines) for Ca and Ni isotopes.
The light line shows an $A^{1/3}$ dependence.

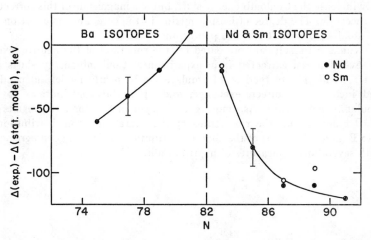

FIG. 12. Coulomb-energy shifts of f states near $N=82$. The deviations of the experimental Coulomb-energy shifts from an $A^{1/3}$ dependence are plotted. The $A^{1/3}$ dependence is calculated in the statistical model (Eq. 2). The error bars shown are typical of the data.

decrease in Δ below the $A^{1/3}$ value on either side of $N=82$ is quite evident; the slight asymmetry may well be caused by a small departure from the $A^{1/3}$ dependence in a sequence of isotopes—as was noted in the Ni and Sn cases. There is need for quantitative calculations for deformed nuclei, such as the Nilsson-model calculations suggested by Kofoed-Hansen (35).

8. SUMMARY

Historically many independent-particle models have been used to calculate Coulomb-energy shifts and to extract charge radii. With charge radii measured accurately by other means it has become clear that the simple Coulomb-energy models are predicting radii too small by $\sim$7 per cent. Alternatively, if the potential models are forced to fit the experimental charge radii, the calculated Coulomb-energy shifts are too low by $\sim$7 per cent.

There are two possible explanations for this effect. The Coulomb-energy shift can be brought into agreement with the experimental value by decreasing the rms radius of the neutron excess $\sim$14 per cent. This seriously disagrees with the prediction of any simple potential model or Hartree-Fock calculation, that the neutron-excess radii be significantly larger than the core.

A different explanation could be the omission of some, previously neglected, large correction to the calculated Coulomb-energy shift. The magnitudes of several possible correction terms were estimated. Most of these were either too small or of the wrong sign. The effect of correlations is probably the least understood of the terms mentioned; correlations will probably increase the Coulomb-energy shift, but the magnitude of this correction is uncertain. The degree of isospin mixing in the core, and its effect on the Coulomb-energy shift are also poorly understood.

Numerical results are presented for the rms radii of the neutron excess of several nuclei extracted from experimental Coulomb-energy shifts and charge radii. The method was formulated to be nearly model independent and includes the direct, exchange, and spin-orbit contributions to the Coulomb-energy shift. Assuming good isospin in the core, the results for Pb^{208} indicate that the rms radius of the overall neutron distribution is 0.15 F larger than that of the charge distribution. This is in good agreement with several other measures of neutron radii.

APPENDIX A. Compilation of Coulomb-energy shifts between ground states and their isobaric analogs.

In Tables A1–A5 all the known experimental Coulomb-energy shifts are listed for ground states, with references. Unless otherwise specified, the necessary binding-energy data are taken from Maples et al.* The experimental errors in the extracted Coulomb-energy shifts are sometimes smaller than those quoted in the original references because more accurate masses were used. In a few cases, where the ground-state analog is unknown, the Coulomb-energy shift of an excited state is given in the tables; the excitation energy of the state is given in the footnote for these nuclei.

* Maples, C., Goth, G. W., Cerny, J., *Nuclear Data*, A2, 429 (1966)

TABLE A1. Experimental Coulomb-energy shifts† between the ground states of $Z_<$ nuclei and their isobaric analogs: $Z_< = 1-19$

$Z_</T_z^*$	1/2	1	3/2	2	5/2
$_1$H	0.7638 (.3)				
$_2$He	1.01 (50)	0.835 (6)°	0.860 (50)°		
$_3$Li	1.644 (2)	1.56 (50)^f	1.559 (20)°		
$_4$Be	1.851 (2)	1.968 (3)^e			
$_5$B	2.763 (1)	2.521 (4)^g	2.453 (4)°		
$_6$C	3.003 (1)	2.938 (1)^h	2.620 (50)°	2.702 (20)°	
$_7$N	3.542 (1)	3.32 (10)^h	3.180 (10)°		
$_8$O	3.542 (1)	3.479 (1)^i	3.524 (20)^p	3.494 (40)°	
$_9$F	4.020 (2)	4.028 (8)^j	3.954 (10)°		
$_{10}$Ne	4.327 (8)	4.282 (3)^k	4.303 (25)°	4.292 (35)°	
$_{11}$Na	4.838 (4)	4.784 (5)^l	4.743 (16)°		
$_{12}$Mg	5.063 (3)^a	5.015 (4)^b	4.984 (5)°		
$_{13}$Al	5.593 (3)	5.467 (6)^l	5.401 (9)°		
$_{14}$Si	5.731 (7)	5.703 (8)^b	5.687 (11)°	5.641 (40)°	
$_{15}$P	6.228 (11)	6.070 (3)^l	6.018 (15)°		
$_{16}$S	6.351 (12)	6.265 (5)^m	[6.285 (30]^x	6.235 (25)^{o,r}	6.161 (40)^t
$_{17}$Cl	6.746 (2)^e	6.683 (4)^n	6.606 (30)°	6.500 (12)^s	6.435 (30)^u
$_{18}$Ar	6.934 (3)^b	6.842 (12)^b		6.66 (70)°	
$_{19}$K	7.305 (7)^{b,d}	7.132 (4)^{l,n}	7.040 (15)^q	7.00 (25)^w	6.980 (25)^v

Where no reference is indicated for the mirror nuclei, the masses are from Mattauch, J. H. E., Thiele, W., Wapstra, A. H., *Nucl. Phys.*, **67**, 1 (1965)

* $A = 2Z_< + 2T_z$.

† Coulomb-energy shift in MeV (error in keV).

^{a} Hardy, J. C., Brunnader, H., Cerny, J., Jänecke, J., *Phys. Rev.* (To be published)

^{b} Endt, P. M., van der Leun, C., *Nucl. Phys.*, A105, 1 (1967)

^{e} Freeman, J. M., Robinson, D. C., Wick, G. L., *Phys. Letters*, 29B, 296 (1969)

^{d} McKenna, C. M., Kemper, K. W., Nelson, J. W., *Bull. Am. Phys. Soc.*, 14, 584 (1969)

^{e} Lauritsen, T., Ajzenberg-Selove, F., *Nucl. Phys.*, 78, 1 (1966)

^{f} The analog of Li^8 (g.s.) is a 300 keV doublet in Be^8. See Marion, J. B., Ludemann, C. A., Roos, P. G., *Phys. Letters*, 22, 172 (1966) and references therein.

^{g} Ajzenberg-Selove, F., Lauritsen, T., *Nucl. Phys.*, A114, 1 (1968)

^{h} Ajzenberg-Selove, F., Lauritsen, T., *Nucl. Phys.*, 11, 1 (1959)

^{i} Blaugrund, A. E., Youngblood, D. H., Morrison, G. C., Segel, R. E., *Phys. Rev.*, 158, 893 (1967)

^{j} MacFarlane, R. D., Siivola, A., *Nucl. Phys.*, 59, 168 (1964); Pearson, J. D., Spear, R. H., *Nucl. Phys.*, 54, 434 (1964)

^{k} Gallmann, A., Frick, G., Warburton, E. K., Alburger, D. E., Hechtl, S., *Phys. Rev.*, 163, 1190 (1967)

^{l} Armini, A. J., Sunier, J. W., Richardson, J. R., *Phys. Rev.*, 165, 1194 (1968)

^{m} Freeman, J. M., Montague, J. H., Murray, G., White, R. E., Burcham, W. E., *Nucl. Phys.*, 65, 113 (1965)

^{n} Berg, R. E., Snelgrove, J. L., Kashy, E., *Phys. Rev.*, 153, 1165 (1967)

^{o} Cerny, J., *Ann. Rev. Nucl. Sci.*, 18 (1969)

^{p} Lennon, C. O., Alderson, P. R., Durell, J. L., Green, L. L., Naqib, I. M., *Phys. Letters*, 28B, 253 (1969)

^{q} Belote, T. A., Dao, F. T., Dorenbusch, W. E., Kuperus, J., Rapaport, J., Smith, S. M., *Nucl. Phys.*, A102, 462 (1967); Lynen, U., Bock, R., Santo, R., Stock, R., *Phys. Letters*, 25B, 9 (1967)

^{r} Broman, L., Fou, C. M., Rosner, B., *Nucl. Phys.*, A112, 195 (1968)

^{s} The Coulomb-energy shift of the $E_x = 0.673$ MeV (5$^-$) excited state in Cl^{38}. Erné, F. C., Veltman, W. A. M., Wintermans, J. A. J. M., *Nucl. Phys.*, 88, 1 (1966)

^{t} Hyder, A. K., Jr., Harris, G. I., *Phys. Letters*, 24B, 273 (1967), using new S^{37} mass from Ajzenberg-Selove, F., *Bull. Am. Phys. Soc.*, 14, 568 (1969)

^{u} Greene, M. W., *Phys. Letters*, 24B, 171 (1967)

^{v} Dorenbusch, W. E., Dao, F. T., Rapaport, J., Belote, T. A., *Phys. Letters*, 26B, 148 (1968); Smith, S. M., Bernstein, A. M., Rickey, M. E., *Nucl. Phys.*, A113, 303 (1968)

^{w} Lynen, U., Oeschler, H., Santo, R., Stock, R., *Nucl. Phys.*, A127, 343 (1969)

^{x} A tentative assignment, assuming the level at 5.67 MeV in Cl^{35} seen via (He^3, d) is the analog of S^{35} (g.s.). Morrison, R. A., *Nucl. Phys.* (To be published, 1969)

$Z<$	$\frac{1}{2}$	1	$\frac{3}{2}$	2	$\frac{5}{2}$	3	$\frac{7}{2}$	4	$\frac{9}{2}$	5	$\frac{11}{2}$
20Ca	7.277(12)[a]	7.214(2)[b]	7.246(15)[c]	7.228(15)[d]	7.236(15)[d]	7.209(15)[d]	7.194(15)[d]	7.175(15)[d]	7.087(15)[e]	7.010(15)[e]	
21Sc	7.632(20)[f]		7.580(15)[g,h,i,j]		7.550(20)[h,j]		7.524(20)[h,j]				
22Ti	7.836(2)[b]	7.818(10)[l]	7.871(10)[l]		7.825(10)[l]	7.806(10)[l]	7.754(30)[m]				
23V			8.109(20)[n]		8.164(20)[o,n]	8.063(10)[o,n]	8.015(30)[n]				
24Cr	8.414(3)[b]	8.413(10)[l]			8.349(10)[l]	8.301(10)[l]					
25Mn	9.033(5)[q]		8.775(20)[h,p]		8.641(20)[l]						
26Fe		8.92(40)[l]	9.000(30)[r]		8.894(20)[l]	8.844(20)[l]					
27Co			9.240(50)[s]				9.050(70)[s]		8.97(80)[t]		
28Ni			9.480(15)[v]		9.151(20)[l]	9.375(15)[v]	9.300(10)[x] / 9.57(40)[h]	9.292(15)[v] / 9.510(15)[α,β]	9.225(15)[v] / 9.49(40)[h]	9.212(15)[v]	
29Cu	9.556(5)[i,q,u]	9.462(15)[w,v]	9.410(15)[v]		9.61(35)[z,h]						
30Zn							9.790(50)[v]	9.760(50)[v]	9.69(20)[δ] / 10.01(50)[z]		
31Ga				9.890(50)[γ]		9.800(50)[γ]				9.620(50)[γ]	9.578(20)[ε]

* $A = 2Z< + 2T_z$.

† Coulomb-energy shift in MeV (error in keV).

a. Mattauch, J. H. E., Thiele, W., Wapstra, A. H., *Nucl. Phys.*, **67**, 1 (1965)

b. Freeman, J. M., Murray, G., Burcham, W. E., *Phys. Letters*, **17**, 317 (1965)

c. Schwartz, J. J., Alford, W. P., Parker, *Phys. Rev.*, **149**, 820 (1966) and (Private communication)

d. Nolen, J. A., Jr., Schiffer, J. P., Williams, N., von Ehrenstein, D., *Phys. Rev. Letters*, **18**, 1140 (1967)

e. Jones, K. W., Schiffer, J. P., Lee, L. L., Jr., Marinov, A., Lerner, J. L., *Phys. Rev.*, **145**, 894 (1966); Grandy, T. B., McDonald, W. J., Dawson, W. K., Neilson, G. C., *Nucl. Phys.*, **A113**, 353 (1968)

f. Aldridge, A. M., Plendl, H. S., Aldridge, J. P. III, *Nucl. Phys.*, **A98**, 323 (1967)

g. Braid, T. H., Meyer-Schützmeister, L., Borlin, D. D., *Isobaric Spin in Nuclear Physics* (Fox, J. D., Robson, D., Eds., Academic Press, New York, 1966)

h. Borlin, D. D. (Thesis, Washington Univ. St. Louis, Mo., unpublished, 1967)

i. Cookson, J. A., Dandy, D., *Nucl. Phys.*, **A97**, 232 (1967)

j. Rosner, B., Pullen, D. J., *Phys. Letters*, **24B**, 454 (1967)

k. Broman, L., Pullen, D. J., Rosner, B., *Bull. Am. Phys. Soc.*, **12**, 682 (1967)

l. Dzubay, T. G., Sherr, R., Dehnhard, D., Becchetti, F. D., Jr., *Bull. Am. Phys. Soc.*, **13**, 1403 (1968); the analog of Fe^{54} and Fe^{58} are 70 keV and 20 keV doublets, respectively.

m. Pullen, D. J., Rosner, B., Hansen, O., *Phys. Rev.*, **177**, 1568 (1969)

n. David, P., Duhm, H. H., Bock, R., Stock, R., *Nucl. Phys.*, **A128**, 47 (1969); the value for V^{48} also requires the new mass of V^{46} given by Hinds, S., Marchant, H., Middleton, R., *Phys. Letters*, **24B**, 34 (1967)

o. Teranishi, E., Furubayashi, B., *Isobaric Spin in Nuclear Physics*, 640 (1966), and ref. n.

p. Trier, A., Gonzalez, L., Rapaport, J., Belote, T. A., Dorenbusch, W. E., *Nucl. Phys.*, **A111**, 241 (1968)

q. Freeman, J. M., Montague, J. H., Murray, G., White, R. E., Burcham, W. E., *Nucl. Phys.*, **65**, 113 (1965)

r. Rosner, B., Holbrow, C. H., Pullen, D. J., *Isobaric Spin in Nuclear Physics*, 595 (1966)

s. Sherr, R., Bayman, B. F., Rost, E., Rickey, M. E., Hoot, C. G., *Phys. Rev.*, **139B**, 1272 (1965)

t. Devins, D. W., Rickey, M. E., Hayakawa, S., Rost, E., *Bull. Am. Phys. Soc.*, **14**, 121 (1969)

u. Harchol, M., Jaffe, A. A., Drory, Ch., Zioni, J., *Phys. Letters*, **20**, 303 (1966)

v. Nolen, J. A., Jr., Schiffer, J. P., Williams, N., Morrison, G. C., *ANL Annual Review* 7481, p. 66 (Unpublished, 1968)

w. Rapaport, J., Young, H. J., *Phys. Letters*, **22**, 466 (1966)

x. Gaarde, C., Wilhjelm, P., Jørgensen, P. B., *Phys. Letters*, **26B**, 143 (1968)

y. Coulomb-energy shift of 0.065-MeV first excited state, ref. x and Lee, L. L., Jr., Marinov, A., Schiffer, J. P., *Phys. Letters*, **8**, 352 (1964)

z. Fou, C. M., Zurmühle, R. W., Joyce, J. M., *Nucl. Phys.*, **A97**, 458 (1967). The Coulomb-energy shift of the analog of the 0.383-MeV excited state in Cu^{66} is given here and in ref. β.

α. Harchol, M., Cochavi, S., Jaffe, A. A., Drory, Ch., *Nucl. Phys.*, **79**, 165 (1966); the lowest level observed was the analog of the 0.85-MeV excited state in Zn^{69}.

β. Couchell, G. P., Balamuth, D. P., Horoshko, R. N., Mitchell, G. E., *Phys. Rev.*, **161**, 1147 (1967)

γ. Dzubay, T. G. (To be published)

δ. Vourvopoulos, G., Fox, J. D., *Phys. Rev.*, **141**, 1180 (1966); the lowest level observed was the analog of the 0.85-MeV excited state in Zn^{69}.

ε. Using resonance energy from ref. β and B^{n} from von Ehrenstein, D., Schiffer, J. P., *Phys. Rev.*, **164**, 1374 (1967).

TABLE A3. Experimental Coulomb-energy shifts† between the ground states of $Z_<$ nuclei and their isobaric analogs: $Z_< = 32-56$, $9/2 \leq T \leq 10$

$Z_<$ \ T_z^*	$\frac{9}{2}$	5	$\frac{11}{2}$	6	$\frac{13}{2}$	7	$\frac{15}{2}$	8	$\frac{17}{2}$	9	$\frac{19}{2}$	10
₃₂Ge	10.51(15)[e]											
₃₃As		10.35(15)[a,d]	10.02(20)[a,b]		9.95(50)[a,c]							
₃₄Se			10.46(15)[e,f]		10.49(12)[e,··]		[e,f]					
₃₅Br												
₃₆Kr							10.860(30)[g]					
₃₇Rb						11.16(90)[f]						
₃₈Sr												
₃₉Y			11.65(25)[h,i,m]	11.45(25)[h,i] 11.610(7)[n-o-p]	11.36(10)[j,k,l]							
₄₀Zr			11.87(25)[g]	11.82(25)[q]	11.72(25)[q]		11.66(25)[q]		11.58(30)[q]			
₄₁Nb				12.01(25)[q,r]								
₄₂Mo		12.36(25)[s]	12.28(25)[s]	12.17(25)[s]	12.15(25)[s]	12.12(25)[s]	12.00(26)[s]		11.94(30)[s]			
₄₃Tc												
₄₄Ru		12.82(25)[t]	12.72(12)[t]		12.62(12)[t]	12.59(12)[t]	12.55(20)[u,t]		12.47(20)[t]			
₄₅Rh												
₄₆Pd			13.15(30)[q]		13.06(30)[q]		12.98(25)[q]		12.91(25)[q]			
₄₇Ag												
₄₈Cd			13.61(30)[w]			13.27(15)[v,b]	13.41(30)[y]	13.20(15)[v,b] 13.39(20)[d]	13.34(35)[y]	13.31(10)[d]	13.28(30)[y,z]	
₄₉In						13.49(60)[x]						
₅₀Sn					13.95(30)[α]		13.89(25)[α]	13.86(20)[n]	13.80(12)[α,o]	13.78(10)[n,α]	13.73(12)[α,o]	13.71(10)[α,n,o]
₅₁Sb												
₅₂Te											14.09(60)[x]	
₅₃I												
₅₄Xe												
₅₅Cs												
₅₆Ba											14.90(20)[β,γ]	

† Coulomb-energy shift in MeV (error in keV).

* $A = 2Z_< + 2T_z$.

a Couchell, G. P., Balamuth, D. P., Horoshko, R. N., Mitchell, G. E., Phys. Rev., 161, 1147 (1967); $E_x(Ge^{77})$ =0.159 MeV.

b Harchol, M., Jaffe, A. A., Zioni, J., Drory, Ch., Nucl. Phys., A90, 473 (1967)

c Harchol, M., Cochavi, S., Jaffe, A. A., Drory, Ch., Nucl. Phys., 79, 165 (1966); $E_x(Ge^{77})$ =0.159 MeV.

d Kernell, R. L., Johnson, C. H., Bull. Am. Phys. Soc., 11, 630 (1966)

e Balamuth, D. P., Couchell, G. P., Mitchell, G. E., Phys. Rev., 170, 995 (1968); in the case of Se^{83} the analog of the ground state was not observed and the observed resonances do not agree well with (d,p) data; for Se^{77} average of E_x =0.956 and E_x =1.134 MeV; $E_x(Se^{79})$ =1.152 MeV.

f Zioni, J., Jaffe, A. A., Drory, Ch., Harchol, M., Nucl. Phys., A109, 401 (1968)

g Hollas, C. L., Hiddleston, H. R., Mistry, V. D., Riley, P. J., Second Conf. Nucl. Isospin, Asilomar, Calif., 1969 and (Private communication)

h Goodman, C. D. (Private communication)

i Ludemann, C. A., Goodman, C. D., Kelly, W. H., Bull. Am. Phys. Soc., 11, 118 (1966)

j Dutt, G. C., McEllistrem, M. T., Obst, A., Bull. Am. Phys. Soc., 12, 697 (1967)

k Cosman, E. R., Joyce, J. M., Shafroth, S. M., Nucl. Phys., A108, 519 (1968)

l Cosman, E. R., Enge, H. A., Sperduto, A., Phys. Rev., 165, 1175 (1968)

m Fou, C. M., Zurmühle, R. W., Joyce, J. M., Phys. Rev., 155, 1248 (1967)

n Johnson, C. H., Kernell, R. L., ORNL-3778, 97 (1965)

o Black, J. L., Islam, M. M., Jones, G. A., Morrison, G. C., Taylor, R. B., Isobaric Spin in Nuclear Physics, 863 (Fox, J. D., Robson, D., Eds., 1966)

p Watson, C., Moore, C. F., Sheline, R. K., Nucl. Phys., 54, 519 (1964)

q Long, D. D., Richard, P., Moore, C. F., Fox, J. D., Phys. Rev., 149, 906 (1966)

r Sheline, R. K., Jernigan, R. T., Ball, J. B., Bhatt, K. H., Kim, Y. E., Vervier, J., Nucl. Phys., 61, 342 (1965)

s Moore, C. F., Richard, P., Watson, C. E., Robson, D., Fox, J. D., Phys. Rev., 141, 1166 (1966); $E_x(Mo^{93})$ =0.994 MeV.

t Friedman, E., Mandelbaum, B., Zioni, J., Jaffe, A. A., Marinov, A., Ginzburg, A., Elitzur, Z. (To be published)

u Fortune, H. T., Kienle, P., Morrison, G. C., Nolen, J. A., Jr.; see also Bull. Am. Phys. Soc., 13, 584 (1968)

v Shugart, C. G., Curry, J. R., Lock, G. A., Moore, P. A., Riley, P. J., Phys. Rev., 178, 1836 (1969); the Coulomb-energy shifts of the 0.081 MeV (2^-) state in Ag^{108} and the ~0.004 MeV (2^-) state in Ag^{110} are given.

w Abranson, E., Plesser, I., Vager, Z., Phys. Letters, 26B, 723 (1968)

x Miller, P. S. (Ph.D. thesis, Princeton Univ., Unpublished, 1968)

y Riley, P. J. (Private communication)

z Hamburger, E. W., Kremenek, J., Cohen, B. L., Moorhead, J. B., Shin, C., Phys. Rev., 162, 1158 (1967)

α Richard, P., Moore, C. F., Becker, J. A., Fox, J. D., Phys. Rev., 145, 971 (1966)

β Williams, N., Morrison, G. C., Nolen, J. A., Jr., Vager, Z., von Ehrenstein, D. (To be published, 1969)

γ von Ehrenstein, D., Nolen, J. A., Jr., Morrison, G. C., Williams, N. (To be published, 1969)

TABLE A4. Experimental Coulomb-energy shifts† between the ground states of $Z_<$ nuclei and their isobaric analogs: $Z_< = 50 - 68$, $\frac{21}{2} \le T \le \frac{31}{2}$

$Z_<$ \\ T_z^*	$\frac{21}{2}$	11	$\frac{23}{2}$	12	$\frac{25}{2}$	13	$\frac{27}{2}$	14	$\frac{29}{2}$	15	$\frac{31}{2}$
50Sn	13.67(15)[a]		13.61(15)[a,b,c]		13.55(15)[a,b,c]						
51Sb		13.86(15)[b]									
52Te	14.09(25)[d]	14.04(20)[b]	14.02(15)[d,b]		13.97(25)[d]	13.91(12)[b,d]					
53I											
54Xe									14.06(40)[e]		
55Cs											
56Ba	14.88(20)[f,g]		14.80(20)[f,g]		14.75(20)[f,g]		14.65(15)[f,h]				
57La					14.8(120)[i]						
58Ce					15.10(25)[h,j,k]						
59Pr			15.56(20)[h,j,k]	15.38(30)[l,h]							
60Nd					15.43(20)[h]		15.33(20)[h]		15.26(20)[h]		15.18(20)[h]
61Pm	15.99(20)[h,m,n]				15.76(25)[m,n]		15.70(25)[m,n]				
62Sm											
63Eu											
64Gd											
65Tb											
66Dy											
67Ho									16.3(150)[o]		16.6(150)[o]
68Er											

† Coulomb-energy shift in MeV (error in keV).

* $A = 2Z_< + 2T_z$.

a Richard, P., Moore, C. F., Becker, J. A., Fox, J. D., Phys. Rev., 145, 971 (1966); $E_x(Sn^{114}) = 0.020$ MeV, $E_x(Sn^{116}) = 0.026$ MeV.

b Harchol, M., Cochavi, S., Jaffe, A. A., Drory, Ch., Nucl. Phys., 79, 165 (1966)

c Nealy, C. L., Sheline, R. K., Phys. Rev., 135, B325 (1964)

d Foster, J. L., Riley, P. J., Moore, C. F., Phys. Rev., 175, 1498 (1968); using the Te$^{116}(d,p)$ Q value of Graue, A., Jastad, E., Lien, J. R., Torvund, P., Nucl. Phys., A103, 209 (1967)

e Moore, P. A., Riley, P. J., Jones, C. M., Mancusi, M. D., Foster, J. L., Jr., Phys. Rev., 180, 1213 (1969); 175, 1516 (1968)

f Williams, N., Morrison, G. C., Nolen, J. A., Jr., Vager, Z., von Ehrenstein, D. (To be published, 1969)

g von Ehrenstein, D., Morrison, G. C., Nolen, J. A., Jr., Morrison, G. C., Williams, N. (To be published, 1969)

h Solf, J., Heusler, A., Wurm, J. P., Max-Planck-Inst. Kernphysik, Heidelberg, Jahresber., 59 (1967)

i Langsford, A., Bowen, P. H., Cox, G. C., Saltmarsh, M. J. M., Nucl. Phys., A113, 433 (1968)

j Wurm, J. P., von Brentano, P., Grosse, E., Seitz, H., Wiedner, C. A., Zaidi, S. A. A., Isobaric Spin in Nuclear Physics, 790 (1966)

k Wiedner, C. A., Heusler, A., Solf, J., Wurm, J. P., Nucl. Phys., A103, 433 (1967)

l Harchol, M., Jaffe, A. A., Zioni, J., Drory, Ch., Nucl. Phys., A90, 473 (1967)

m Jolly, R. K., Moore, C. F., Phys. Rev., 155, 1377 (1967)

n Kenefick, R. A., Sheline, R. K., Phys. Rev., 139, B1479 (1965)

o Anderson, J. D., Wong, C., McClure, J. W., Phys. Rev., 138, B615 (1965)

TABLE A5. Experimental Coulomb-energy shifts† between the ground states of $Z_<$ nuclei and their isobaric analogs: $Z_< = 68-83$, $\dfrac{31}{2} \le T \le \dfrac{45}{2}$

T_z^* Z<	$\frac{31}{2}$	16	$\frac{33}{2}$	17	$\frac{35}{2}$	18	$\frac{37}{2}$	19	$\frac{39}{2}$	20	$\frac{41}{2}$	21
$_{68}$Er					16.69(50)[a]							
$_{69}$Tm												
$_{70}$Yb	17.08(32)[b]		17.01(32)[b]		16.92(32)[b]		16.84(32)[b]					
$_{71}$Lu					17.5(200)[e]							
$_{72}$Hf							17.24(40)[c,d]					
$_{73}$Ta												
$_{74}$W												
$_{75}$Re											17.98(38)[f]	
$_{76}$Os							18.04(32)[f]		17.99(35)[f]			

T_z^* Z∇	$\frac{35}{2}$	18	$\frac{37}{2}$	19	$\frac{39}{2}$	20	$\frac{41}{2}$	21	$\frac{43}{2}$	22	$\frac{45}{2}$	23
$_{77}$Ir												
$_{78}$Pt	18.36(28)[f] 18.6(150)[e]		18.32(29)[f]		18.3(100)[f]							
$_{79}$Au												
$_{80}$Hg												
$_{81}$Tl			19.02(15)[g]		18.95(30)[h]		18.87(15)[g]				18.80(15)[i,j]	
$_{82}$Pb												
$_{83}$Bi												

$_{92}$U^{238} $T_z = 27$, 19.8(120)[e]

† Coulomb-energy shifts in MeV (error in keV).

* $A = 2Z_< + 2T_z$.

[a] Schneid, E. J., Temmer, G. M. (To be published)

[b] Cassagnou, Y., Foissel, P., Gastebois, J., Levi, C., Mittig, W., Papineau, L., *Phys. Letters*, **27B**, 631 (1968); $E_x(\text{Yb}^{172})=1.073$ MeV; $E_x(\text{Yb}^{174})=0.552$ MeV. $E_s(\text{Yb}^{177})=0.379$ MeV.

[c] Allan, D. L., Britt, H. C., Rickey, F. A., Jr., *Phys. Letters*, **27B**, 11 (1968)

[d] Rickey, F. A., Jr., Sheline, R. K., *Phys. Rev.*, **170**, 1157 (1968)

[e] Langsford, A., Bowen, P. H., Cox, G. C., Saltmarsh, M. J. M., *Nucl. Phys.*, **A113**, 433 (1968); Batty, C., Gilmore, R. S., Stafford, G. H., *Nucl. Phys.*, **75**, 599 (1966)

[f] Wiesen, M., Schneid, E. J., Temmer, G. M. (To be published); $E_x(\text{Os}^{89})=0.069$ MeV, $E_x(\text{Os}^{91})=0.074$ MeV.

[g] Lenz, G. H., Temmer, G. M., *Nucl. Phys.*, **A112**, 625 (1968)

[h] Kavaloski, C. D., Lilley, J. S., Richard, P., Stein, N., *Phys. Rev. Letters*, **16**, 807 (1966)

[i] Wharton, W. R., von Brentano, P., Dawson, W. K., Richard, P., *Phys. Rev.*, **176**, 1424 (1968)

[j] Zaidi, S. A. A., Parish, J. L., Kulleck, J. G., Moore, C. F., von Brentano, P., *Phys. Rev.*, **165**, 1312 (1968)

APPENDIX B. Experimental charge-distribution parameters used in the Coulomb-energy calculations.

TABLE A6. Parameters of experimental charge distributions used in the Coulomb-energy calculations

Nucleus	c^a (F)	z^a (F)	w^a (F)	$\langle r^2 \rangle^{1/2b}$ (F)	r_{eq}^c (F)	$r_{1/2}^d$ (F)	Footnotes
C^{12}	2.16	0.50	—	2.50	1.41	(0.94)	e, f, g, h
O^{16}	2.53	0.50	—	2.70	1.38	(1.00)	h, i
Si^{28}	2.78	0.60	—	3.10	1.32	(0.92)	h, j, k
S^{32}	3.05	0.60	—	3.25	1.32	(0.96)	h, j
Ca^{40}	3.68	0.59	−0.10	3.49	1.32	1.04	l
Ca^{42}	3.73	0.59	−0.12	3.52	1.31	1.03	l
Ca^{43}	3.74	0.58	−0.11	3.52	1.30		m
Ca^{44}	3.75	0.57	−0.10	3.51	1.28	1.03	l
Ca^{45}	3.75	0.56	−0.08	3.51	1.27		m
Ca^{46}	3.75	0.55	−0.06	3.50	1.26		m
Ca^{47}	3.75	0.54	−0.04	3.49	1.25		m
Ca^{48}	3.74	0.53	−0.03	3.48	1.24	1.02	l
$Ca^{49,50}$	3.74	0.53	−0.03	3.48			n
Ni^{58}	4.14	0.56	—	3.82	1.27	1.07	o
Ni^{60}	4.20	0.56	—	3.86	1.27	1.07	o, p
Ni^{62}	4.23	0.56	—	3.88	1.27	1.07	p
Ni^{64}	4.24	0.57	—	3.90	1.26	1.06	o
Ni^{66}	4.25	0.57	—	3.92	1.25	1.05	m
Sn^{117}	5.31	0.56	—	4.61	1.23	1.10	
Sn^{119}	5.34	0.56	—	4.63	1.23	1.10	
Sn^{120}	5.35	0.56	—	4.64	1.23	1.10	o, p
Sn^{121}	5.35	0.56	—	4.64	1.22	1.10	
Sn^{123}	5.37	0.56	—	4.65	1.22	1.10	
Sn^{125}	5.38	0.56	—	4.66	1.22	1.09	
Pb^{208}	6.30	0.57	0.40	5.51	1.20	1.13	q

[a] Parameters of the parabolic Fermi shape: $\rho(r) = [1 + w(r/c)^2][1 + \exp{(r - c/z)}]^{-1}$.

[b] The rms radius of the charge distribution.

[c] $R_{eq} = r_{eq}A^{1/3}$ is the radius of the uniformly charged sphere with rms radius equal to that of the experimental charge distribution.

[d] $R_{1/2} = r_{1/2}A^{1/3}$ is the radius at which the charge density is one-half its value at $r = 0$.

[e] In the Coulomb-energy calculations the charge radius of the $Z_<$ member of a $T = \frac{1}{2}$ pair was assumed equal to that of the core.

[f] Meyer-Berkhout, U., Ford, K. W., Green, A. E. S., *Ann. Phys. (N.Y.)*, **8**, 119 (1959)

[g] Elton, L. R. B., Swift, A., *Nucl. Phys.*, **A94**, 52 (1967)

[h] For C^{12}, O^{16}, Si^{28}, and S^{32} the rms radii were fixed at the experimental value, z was chosen arbitrarily; c was thereby determined.

[i] Lacoste, F., Bishop, G. R., *Nucl. Phys.*, **26**, 511 (1961)

[j] Backenstoss, G., in *Proc. Intern. Conf. Electromagnetic Sizes of Nuclei* (Brown, D. J., Sundaresan, M. K., Barton, R. D., Eds., Carleton Univ., Ottawa, Canada, 1967)

[k] For Si^{28} the rms radius was extrapolated as $A^{1/3}$ from Al^{27}.

[l] Frosch, R. F., Hofstadter, R., McCarthy, J. S., Nöldeke, G. K., van Oostrum, K. J., Yearian, M. R., Clark, B. C., Herman, R., Ravenhall, D. G., *Phys. Rev.*, **174**, 1380 (1968).

[m] Values of c, z, and w were extrapolated linearly from nearby isotopes.

[n] Distribution assumed the same as Ca^{48}.

[o] Khvastunov, V. M., Afanasyev, N. G., Afanasyev, V. D., Gulkarov, I. S., Omelaenko, A. S., Savitsky, G. A., Khomich, A. A., Shevchenko, N. G., *Phys. Letters*, **28B**, 119 (1968)

[p] Ehrlich, R. D., *Phys. Rev.*, **173**, 1088 (1968)

[q] Ravenhall, D. G. (Private communication); see also Anderson, H. L., Hargrove, C. K., Hincks, E. P., McAndrew, J. D., McKee, R. J., Barton, R. D., Kessler, D. (To be published)

LITERATURE CITED

1. Bethe, H. A., Bacher, R. F., *Rev. Mod. Phys.*, **8**, 82 (1936)
2. Henley, E. M., *Isobaric Spin in Nuclear Physics* (Fox, J. D., Robson, D., Eds., Academic Press, New York, 1966)
3. Cerny, J., *Ann. Rev. Nucl. Sci.*, **18**, 27 (1968)
4. Berg, R. E., Snelgrove, J. L., Kashy, E., *Phys. Rev.*, **153**, 1165 (1967)
5. Anderson, J. D., Wong, C., McClure, J. W., *Phys. Rev.*, **138**, B615 (1965)
6. Cookson, J. A., Dandy, D., *Nucl. Phys.*, **B97**, 232 (1967)
7. Freeman, J. M., Murray, G., Burcham, W. E., *Phys. Letters*, **17**, 317 (1965)
8. Dzubay, T. G., Sherr, R., Dehnhard, D., Becchetti, F. D., Jr., *Bull. Am. Phys. Soc.*, **13**, 1403 (1968)
9. Long, D. D., Richard, P., Moore, C. F., Fox, J. D., *Phys. Rev.*, **149**, 906 (1966)
10. Harchol, M., Cochavi, S., Jaffe, A. A., Drory, C., *Nucl. Phys.*, **79**, 165 (1966)
11. Youngblood, D. H., Morrison, G. C., Segel, R. E., *Phys. Letters*, **22**, 625 (1966)
12. Teitelman, B., Temmer, G. M., *Phys. Letters*, **26B**, 371 (1968)
13. Sherr, R., Bayman, B. F., Rost, E., Rickey, M. E., Hoot, C. G., *Phys. Rev.*, **139**, B1272 (1965)
14. Jaffe, A. A., Taylor, I. J., Forsyth, P. D., *Proc. Phys. Soc.*, **75**, 940 (1960)
15. Schiffer, J. P., *Isospin in Nuclear Physics*, Chap. 13 (Wilkinson, D. H., Ed., North-Holland, Amsterdam, 1969)
16. Jänecke, J., *Isospin in Nuclear Physics*, Chap. 8
17. Harchol, M., Jaffe, A. A., Miron, J., Unna, I., Zioni, J., *Nucl. Phys.*, **A90**, 459 (1967)
18. Sherr, R., *Phys. Letters*, **24B**, 321 (1967)
19. Hardy, J. C., Brunnader, H., Cerny, J., Jänecke, J. (To be published)
20. Jänecke, J., *Z. Phys.*, **196**, 477 (1966)
21. von Weizsäcker, C. F., *Z. Phys.*, **96**, 431 (1935)
22. Bethe, H. A., *Phys. Rev.*, **54**, 436 (1938)
23. Carlson, B. C., Talmi, I., *Phys. Rev.*, **96**, 436 (1954)
24. Sengupta, S., *Nucl. Phys.*, **21**, 542 (1960)
25. Swamy, N. V. V. J., Green, A. E. S., *Phys. Rev.*, **112**, 1719 (1958)
26. Stephens, W. E., *Phys. Rev.*, **57**, 938 (1940)
27. Wilson, R. R., *Phys. Rev.*, **88**, 350 (1952)
28. Ehrman, J. B., *Phys. Rev.*, **81**, 412 (1951)
29. Blatt, J. M., Weisskopf, V. F., *Theoretical Nuclear Physics* (Wiley, New York, 1952)
30. Fitch, V. L., Rainwater, J., *Phys. Rev.*, **92**, 789 (1953)
31. Hofstadter, R., Fechter, H. R., McIntyre, J. A., *Phys. Rev.*, **91**, 439 (1953)
32. Cooper, L. N., Henley, E. M., *Phys. Rev.*, **92**, 801 (1953)
33. Peaslee, D. C., *Phys. Rev.*, **95**, 717 (1954)
34. Cherry, R. D., *Phys. Rev.*, **115**, 1243 (1959)
35. Kofoed-Hansen, O., *Rev. Mod. Phys.*, **30**, 449 (1958)
36. Kofoed-Hansen, O., *Nucl. Phys.*, **2**, 441 (1956)
37. Jancovici, B. G., *Phys. Rev.*, **95**, 389 (1954)
38. Feenberg, E., Goertzel, G., *Phys. Rev.*, **70**, 597 (1946)
39. Bertsch, G. F., *Phys. Rev.*, **174**, 1313 (1968)
40. Unna, I., *Nucl. Phys.*, **8**, 468 (1958)
41. Sood, P. C., Green, A. E. S., *Nucl. Phys.*, **5**, 274 (1957)
42. Green, A. E. S., *Phys. Rev.*, **102**, 1325 (1956)
43. Green, A. E. S., *Phys. Rev.*, **104**, 1617 (1956)
44. Nolen, J. A., Jr., Schiffer, J. P., *Phys. Letters*, **29B**, 396 (1969)
45. Jänecke, J., *Nucl. Phys.*, **A114**, 433 (1968)
46. Hecht, K. T., *Nucl. Phys.*, **A102**, 11 (1967)
47. Hecht, K. T., *Nucl. Phys.*, **A114**, 280 (1968)
48. Jänecke, J. (To be published)
49. Jänecke, J., *Phys. Rev.*, **147**, 735 (1966)
50. Wilkinson, D. H., Mafethe, M. E., *Nucl. Phys.*, **85**, 97 (1966)
51. Wilkinson, D. H., Hay, W. D., *Phys. Letters*, **21**, 80 (1966)
52. Lane, A. M., Soper, J. M., *Nucl. Phys.*, **37**, 663 (1962)
53. Jones, K. W., Schiffer, J. P., Lee, L. L., Jr., Marinov, A., Lerner, J. L., *Phys. Rev.*, **145**, 894 (1966)
54. Nolen, J. A., Jr., Schiffer, J. P., Williams, N., von Ehrenstein, D., *Phys. Rev. Letters*, **18**, 1140 (1967)
55. Nolen, J. A., Jr., Schiffer, J. P.,

Williams, N., *Phys. Letters*, **27B**, 1 (1968)

56. Schiffer, J. P., Nolen, J. A., Jr., Williams, N., *Phys. Letters*, **29B**, 399 (1969)

57. Friedman, E., Mandelbaum, B., *Nucl. Phys.* (To be published)

58. Collard, H. R., Elton, L. R. B., Hofstadter, R., *Nuclear Radii* (Springer-Verlag, Berlin, 1967)

59. *Proc. Intern. Conf. Electromagnetic Sizes of Nuclei* (Brown, D. J., Sundareson, M. K., Barton, R. D., Eds., Carleton Univ., Ottawa, Canada, 1967)

60. Elton, L. R. B., Swift, A., *Nucl. Phys.*, **A94**, 52 (1967)

61. Perey, F. G., Schiffer, J. P., *Phys. Rev. Letters*, **17**, 324 (1966)

62. Gibson, B. F., van Oostrum, K. J., *Nucl. Phys.*, **A90**, 159 (1967)

63. Elton, L. R. B., *Phys. Letters*, **26B**, 689 (1968)

64. Condon, E. U., Shortley, G. H., *The Theory of Atomic Spectra*, 174–84 (Cambridge Univ. Press, London, 1935, reprinted 1964)

65. de-Shalit, A., Talmi, I., *Nuclear Shell Theory*, 225 (Academic Press, New York, 1963)

66. Elton, L. R. B., *Nuclear Sizes* (Oxford Univ. Press, London, 1961)

67. Bethe, H. A., Siemens, P. J., *Phys. Letters*, **27B**, 549 (1968)

68. Inglis, D. R., *Phys. Rev.*, **82**, 181 (1951); **50**, 783L (1936)

69. Altman, A., MacDonald, W. M., *Nucl. Phys.*, **35**, 593 (1962)

70. Lovitch, L., *Nucl. Phys.*, **62**, 653 (1965)

71. Thomas, L. H., *Nature*, **117**, 514 (1926)

72. Thomas, R. G., *Phys. Rev.*, **81**, 148L (1951)

73. Thomas, R. G., *Phys. Rev.*, **88**, 1109 (1952)

74. Ehrman, J. B., *Phys. Rev.*, **81**, 412 (1951)

75. Lane, A. M., Thomas, R. G., *Rev. Mod. Phys.*, **30**, 257, 329 (1958)

76. MacDonald, W. M., Weidenmüller, H. A. (Private communications, 1969)

77. Falk, D. S., Wilets, L., *Phys. Rev.*, **124**, 1887 (1961)

78. Kerman, A. K., Svenne, J. P., Villars, F. M. H., *Phys. Rev.*, **147**, 710 (1966)

79. Davies, K. T. R., Baranger, M., Tarbutton, R. M., Kuo, T. T. S., *Phys. Rev.*, **177**, 1519 (1969)

80. Auerbach, E., Kahana, S., Scott, C. K., Weneser, J., *Conf. Nuclear Isospin, Asilomar, California, 1969*

81. Kerman, A. K., Talmi, I. (Private communications, 1969)

82. Donnelly, T. W., Walker, G. E., *Phys. Rev. Letters*, **22**, 1121 (1969)

83. Meldner, H., *Phys. Rev.*, **178**, 1815 (1969)

84. Tarbutton, R. M., Davies, K. T. R., *Nucl. Phys.*, **A120**, 1 (1968); (Private communication)

85. Frosch, R. F., Hofstadter, R., McCarthy, J. S., Nöldeke, G. K., van Oostrum, K. J., Yearian, M. R., Clark, B. C., Herman, R., Ravenhall, D. G., *Phys. Rev.*, **174**, 1380 (1968)

86. Johnson, M. H., Teller, E., *Phys. Rev.*, **93**, 357 (1954)

87. Wilets, L., *Phys. Rev.*, **101**, 1805 (1956)

88. Bodmer, A. R., *Nucl. Phys.*, **17**, 388 (1960)

89. Hill, D. L., *Handbuch der Physik*, **39**, 178 (Springer-Verlag, Berlin, 1957)

90. Auerbach, E. H., Qureshi, H. M., Sternheim, M. M., *Phys. Rev. Letters*, **21**, 162 (1968)

91. Abashian, A., Cool, R., Cronin, J. W., *Phys. Rev.*, **104**, 855 (1956)

92. Vautherin, D., Veneroni, M., *Phys. Letters*, **29B**, 203 (1969)

93. Greenlees, G. W., Pyle, G. J., Tang, Y. C., *Phys. Rev.*, **171**, 1115 (1968)

94. Pyle, G. J., Makofske, W., Greenlees, G. W., *Bull. Am. Phys. Soc.*, **14**, 573 (1969)

95. Friedman, E., *Phys. Letters*, **29B**, 213 (1969)

96. Davis, D. H., Lovell, S. P., Csejthey-Barth, M., Sacton, J., Schorochoff, G., O'Reilly, M., *Nucl. Phys.*, **B1**, 434 (1967)

97. Burhop, E. H. S., *Nucl. Phys.*, **B1**, 438 (1967)

98. Wiegand, C. E., *Phys. Rev. Letters*, **22**, 1235 (1969)

99. Bock, R., Duhm, H. H., Stock, R., *Phys. Letters*, **18**, 61 (1965)

100. Tombrello, T. A., *Phys. Letters*, **23**, 134 (1966)

101. Morrison, G. C., Nolen, J. A., Jr., von Ehrenstein, D., Williams, N., *ANL Rept. ANL-7481*, 58 (Unpublished, 1968)

102. Macfarlane, M. H., *Isobaric Spin in Nuclear Physics*, 399 (See Ref. 2)

103. Jolly, R. K., Moore, C. F., *Phys. Rev.*, **155**, 1377 (1967)

104. Solf, J., Heusler, A., Wurm, J. P., *Jahresbericht 1967*, 59 (Max-Planck-Inst. Kernphysik, Heidelberg, 1967)

MUONIC ATOMS AND NUCLEAR STRUCTURE[1]

By C. S. Wu

Columbia University, New York, N. Y.

AND

Lawrence Wilets[2]

University of Washington, Seattle, Washington

CONTENTS

[1] Supported in part by the United States Atomic Energy Commission, the National Science Foundation, and the Higgins Scientific Trust Fund.

[2] Visiting Professor at Columbia University (January 1969) and Princeton University (February–June 1969).

527

INTRODUCTION

The first measurements of muonic X-ray spectra were carried out by Fitch & Rainwater (75) using a NaI scintillation spectrometer in the muon beam of the Nevis synchrocyclotron in 1953. The early phases of its development, the basic features of muonic spectra, and the significant information obtained from these studies were comprehensively reviewed by Rainwater (118), Stearns (126), West (138), and Sens (124). Since the high-resolution Ge(Li) detector was introduced to replace the NaI scintillation spectrometer to investigate the muonic X-ray spectra in 1964, great advances have been made in the accuracy and precision of the X-ray spectra and, consequently, our understanding of the muonic atoms. Short surveys of progress made since the beginning of this second phase of development in muonic atoms can be found in several conference reports from 1965–68 [56, 57, 114, 125, 135, 147–149; references (147) and (114) are the proceedings of the International Conference on Intermediate Energy Physics and Nuclear Sizes, where results on muonic atoms were discussed in detail].

While the γ-detection techniques have taken giant steps forward, the muon beams which produce muonic atoms are still from old types of synchrocyclotrons where low beam intensity has been the main obstacle in limiting the feasibility of many interesting investigations. With the coming new generation of meson factories or factorettes, the muon intensity is expected to be increased by a factor of 10^4 or 10^2. When the anticipated intensive muon beam materializes, many heretofore inhibited experiments on muonic atoms will become possible. Therefore, it is timely to review what we have learned from muonic atoms since the introduction of the Ge(Li) detector into this field, before another exciting era begins.

FORMATION OF MUONIC ATOMS

In this review, there is no space for a detailed discussion of muon production and the mechanism of the formation of muonic atoms. However, a brief account of the main steps which lead to the formation of muonic atoms may help to give a proper perspective of the experimental arrangements and time sequences of the events involved.

Fast pions are produced in a beryllium target bombarded by a high-energy proton beam (a few hundred MeV) in a synchrocyclotron. Negative muons are produced in the decay of negative pions *in flight*. The negative pions and muons are electromagnetically transported from the neighborhood of the cyclotron target through a hole in the shielding walls to the experimental area some distance away. In general, several deflecting magnets and quadrupole magnets are required to bend and to focus the muon beam to the target where the muons are captured to form muonic atoms, as shown in Figure 1. A well-designed muon channel, however, requires several tens, up to more than hundreds, of quadrupole magnets to transport the muon beam efficiently. The muon momenta are generally around 150 MeV/c and cover a spread of 5–10 per cent.

We list below five steps, or energy ranges, in the process of slowing down, capture, radiation, and decay of the muon. Before the muon can be captured into a Bohr orbit near the nucleus, it must be slowed down to near-thermal energies. Fermi & Teller (74) investigated the slowing-down processes in condensed matter theoretically (see below).

High energy to 2 keV.—In the first step while the muon energy slows from several hundred MeV down to 2 keV (during which the velocity of the muon is greater than the velocity of the valence electrons), the slowing-down process can be treated according to the conventional methods ap-

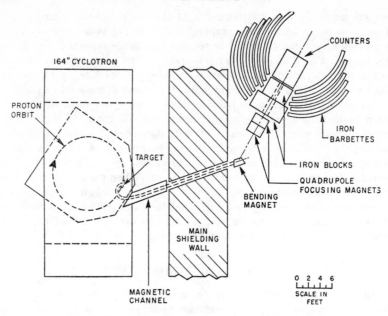

FIG. 1. General layout of an experimental floor plan.

plicable to fast heavy charged particles. They found the time needed to slow down to 2 keV is $\sim 10^{-9}$ to 10^{-10} sec in condensed matter.

From 2 keV to rest.—When the energy of the muon has been reduced below 2 keV and its velocity is therefore less than the velocity of the valence electrons, they considered the muon to be moving with velocity v inside *a degenerate electron gas* which has maximum velocity v_0 where $v \ll v_0$. Their calculation concluded that in condensed substances, a negative muon of 2 keV is slowed down and captured from the orbit closest to the nucleus in $\sim 10^{-13}$ sec, which is very short compared to the muon natural lifetime of 2×10^{-6} sec. In the case of a *metal*, the amount of energy that may be delivered by the low-energy muon to electrons can be arbitrarily small; therefore, the case nearly approaches the ideal conditions of a degenerate gas. In an *insulator* or in *gases*, the loss of energy to electrons must be at least as large as the gap between two Brillouin zones ($\sim$several volts) or the lowest electronic excitation energy of a gas molecule; the energy is transferred in small but discrete amounts and therefore the rate of energy loss to electrons is reduced. The total time needed for energy loss in insulators is a little longer than that in metals but still within the same order of magnitude of 10^{-13} sec. In gases, however, a muon takes 10^{-9} sec to reach its lowest orbit.

Atomic capture.—Once the negative muon loses all its kinetic energy, it is captured by an atom into high orbital momentum states, forming a muonic atom. The distribution of initial states is not well known, although several attempts to calculate it have been made (67, 68, 89, 104). As all muonic states are unoccupied, the muon will cascade down to states of lower energy. The transition is accompanied by the emission of Auger electrons or characteristic electromagnetic radiation or excitation of the nucleus, and takes place within a time $\sim 10^{-13}$ sec.

Electromagnetic cascade.—A muon orbit of the same size as the K electronic orbit $n_e = 1$ will have a principal quantum number

$$n_\mu \simeq \left(\frac{m_\mu}{m_e}\right)^{1/2} \simeq 14$$

The characteristic muonic electric dipole transition $[(n,\ l = n - 1) \rightarrow (n - 1, l - 1)]$ in this region has an energy only a small fraction of the binding energy of the K electron. In cascading down from $n \sim 14$ to lower states, the initial low-energy muonic transitions will interact with outer electrons and yield a strong "Auger effect." However, as n reaches low values, the transition energy increases rapidly according to n^{-3}, so that the interaction with the electron structure is no longer important. For muonic transitions between low n values, radiative transitions dominate. The calculated probability (28) of radiative transitions and Auger transitions in muonic atoms versus the atomic numbers is shown in Figure 2.

Ultimate fate.—Once it reaches the lowest $1s$ level, the muon either decays

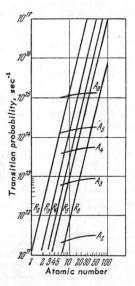

FIG. 2. Calculated radiative (R_n) and Auger (A_n) rates for $(n, l = n - 1) \rightarrow (n - 1, l - 1)$ transitions in muonic atoms (28).

with a half-life of 2.2×10^{-6} sec or is captured by the nucleus in a time of 4×10^{-7} $(82/Z)$ sec. Therefore, only the one-muon atom can be studied.

The binding energies and the linear dimensions of the bound state of muonic atoms can be scaled from the ordinary electronic atoms by the mass factor $m_\mu \simeq 207$ m_e and its inverse. The "K-shell" muon will be ~ 207 times nearer the nucleus than a "K-shell" electron. The close proximity of the K-shell muon in the Coulomb field of a nucleus, together with its exceedingly weak interaction with the nucleus, allows the muon to spend an appreciable fraction of the time $(10^{-7} - 10^{-6})$ sec within the nucleus itself, serving as an ideal probe for the distribution of *nuclear charge* and *nuclear moments*.

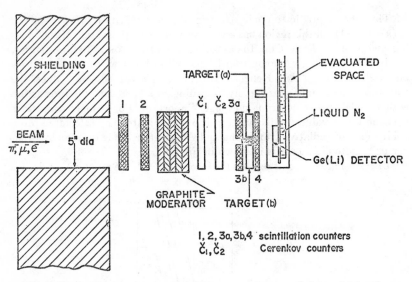

FIG. 3. Typical beam telescope and detector arrangement (schematic) for observing muonic X rays. The configuration shown with 3a and 3b counters is particularly suitable for isotope-shift measurements.

EXPERIMENTAL ARRANGEMENTS

A typical beam-counter telescope system is shown in Figure 3. The "muon stop" signal is given by $[12(3a \text{ or } 3b) \ (\overline{12 \ C_L \text{ or } C_W}) \ 4]$ where $(^-)$ denotes anticoincidence; C_L and C_W are lucite and water Cerenkov counters used to discriminate against electrons which have a greater velocity than the muons, in the momentum selected beam.

Muonic X rays are detected by a Ge(Li) detector. Pulses from the detector are amplified and analyzed by an ADC-type multichannel analyzer interfaced to a computer whose purpose is to serve as a control and central processor for a host of peripheral equipment and to stabilize and calibrate the whole detecting system.

To stabilize the whole system, two pulses from a precision pulser (one for the upper end and one for the lower end of the spectrum) are introduced into the charge-sensitive preamplifier system at the same input at which the pulses from the Ge(Li) detector are introduced. The pulse width is wide enough to cover several channels. If the gain of the amplifier or the discharging current of the ADC ramp fluctuates, the precision pulses will not remain centered in the same channels as initially designated. When the total counts under a peak have reached a prespecified number, the contents of the two adjacent groups of channels, where the pulses are originally introduced, are compared. If the difference of the contents of the two adjacent groups of channels is greater than the statistically significant error, a correction signal is generated by the computer and applied to the servostabilization controls.

The calibration gamma lines are generally from radioactive sources of ^{22}Na, ^{60}Co, ^{24}Na, ^{232}Th, ^{68}Ga, and ^{16}N. To differentiate the calibration pulses from muon X-ray pulses, these cascade gamma lines are identified by the coincidence between the Ge(Li) detector pulses and pulses from a large NaI scintillation counter. The coincidence pulses are used as gating pulses for the ADC as well as identification tags for the calibration pulses. Thus the calibration spectrum is stored simultaneously with the muonic X-ray spectrum but in different parts of the memory.

To observe the small energy differences in the isotope shifts, an additional telescope channel 123b$\overline{4}$ is generally added in parallel to that of 123a$\overline{4}$. This ingenious method was originally introduced by Devons et al. (44). To eliminate any possible scattering of muons from one target ($T1$) to another ($T2$) or vice versa, the plastic scintillation counter 4 is designed to have a ridge which shields $T1$ from $T2$. The coincidence counts 123a$\overline{4}$ and 123b$\overline{4}$ direct the pulses from Ge(Li) detectors to its corresponding submemories. An alternate method in isotope-shift measurements is to measure each spectrum of a series of isotopes of the same element simultaneously with another spectrum of a neighboring element with $Z' = Z - 1$ or $Z' = Z - 2$ and use the lower-Z spectrum as a common reference to measure the isotope shifts.

SIGNIFICANT FINDINGS FROM 1953 TO 1964

From 1953 to 1964, the NaI scintillation spectrometers used as X-ray detectors did not possess enough resolving power to completely resolve even the fine structure of heavy nuclei, let alone the hyperfine structure (Figure 4). Nevertheless, the important conclusions reached from observations in that period were impressive:

Nuclear size.—From the observed $2p \rightarrow 1s$ transition energies, the size of the nucleus was greatly reduced from the previously accepted value of $R = (1.4-1.5)A^{1/3}$ F to $R = 1.2A^{1/3}$ F (75).

Vacuum polarization.—It was shown conclusively by Koslov, Fitch & Rainwater (99) that the vacuum-polarization effect is important and must be included in calculating the binding energies of muonic states. This was

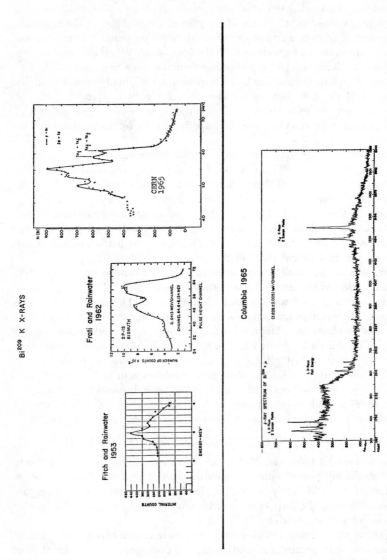

FIG. 4. Chronological progress in the measurement of ^{209}Bi muonic X rays from 1953 to 1965 (148). References in the figure are to Fitch & Rainwater (75), Frati & Rainwater (152), CERN (2), and Columbia (46).

accomplished by a study of the $3d \rightarrow 2p$ transition in phosphorus where finite size effects are unimportant but the longer-range vacuum-polarization potential is significant. This 88.015-keV transition was measured precisely by utilizing the K-absorption edge of Pb.

Muonic mass.—An extension of the above technique led to a more precise determination of the muonic mass, made possible by a determination of the energy dependence of the X-ray absorption coefficient in lead by Bearden (14) and the calculation by Petermann & Yamaguchi (115) of various corrections to the $3d \rightarrow 2p$ transition energies on phosphorus. An average of three experiments (54, 101, 102) yielded

$$m_\mu = (206.76 \pm 0.02)m_e$$

These measurements are the most accurate *direct* determinations of the muon mass at the present time and are in good agreement with the mass ratio of 206.767 obtained from the $(g-2)$ experiments (11, 73) or the f_μ investigation (87).

Isotope shifts and hyperfine structure.—Isotope shifts in muonic atoms began to show promising signs (22, 44), and the presence of hyperfine structure due to the dynamic quadrupole effect was observed but not resolved (1, 7, 62, 119). From the latter studies, it was nevertheless determined that the *intrinsic* quadrupole moment of the first rotational state in an even-even nucleus has the same sign as the transition quadrupole moment, consistent with the rotational model.

CHARACTERISTICS

Salient Features of Muonic Atoms

The main features of the muonic atom are compared with those of electronic atoms in Table I. The finite size effect, which appears in electronic atoms as the isotope shift (classified as hfs), plays a dominant role in muonic atoms, reducing the $1s$ binding energy by a factor of two in heavy atoms. It can also be seen that in the heavy, deformed nuclei, the quadrupole splitting may be of the same order of magnitude as the fine structure of $2p$ or $3d$ levels, which is also comparable to the energies of low-lying nuclear rotational levels. Because of this very large quadrupole interaction energy, a dynamic $E2$ interaction (one form of nuclear polarization) was predicted (91, 142) as early as 1954 and was confirmed experimentally nearly 10 years later. It remains under intensive study. On the other hand, the magnetic hyperfine splittings are not so greatly enhanced since they depend on the magnetic moment of the muon (inversely proportional to its mass). The magnetic hfs is of the order of a few keV, generally two orders of magnitude smaller than the fine structure; however, it is large enough to induce internal conversion in either N or O shell electrons, and this in turn greatly enhances the $M1$ transitions between the magnetic hyperfine states (52, 77). *The*

TABLE I. A COMPARISON OF CHARACTERISTICS OF MUONIC AND ELECTRONIC
ATOMS

Quantity	Expression	Ratio between muonic and electronic atoms	Typical values for muonic atoms
Atomic radius $\langle r_n^{-1}\rangle^{-1}$	$\dfrac{n^2\hbar^2}{Ze^2}\dfrac{1}{m}$	$\dfrac{m_e}{m_\mu}\sim\dfrac{1}{207}$	5 F $(Z\approx 50)$
Energy levels E_n	$-\dfrac{(Z\alpha)^2c^2}{2n^2}m$	$\dfrac{m_\mu}{m_e}\sim 207$	6 MeV $(Z\approx 50, n=1)$
Fine structure $\left\langle\dfrac{(1\cdot\delta)Z}{m^2r^3}\right\rangle$	$\propto Z^4m$	$\dfrac{m_\mu}{m_e}\sim 207$	(550–200) keV in heavy elements
Hyperfine: electric quadrupole $\left\langle\dfrac{e^2Q_N}{r^3}\right\rangle$	$\sim Q_NZ^3m^3$	$\dfrac{m_\mu{}^3}{m_e{}^3}\sim(207)^3\sim 10^7$	(50–500) keV in the highly deformed heavy nuclei
Hyperfine: magnetic dipole $\left\langle\dfrac{\mu_N\mu}{r^3}\right\rangle$	$\sim\mu_NZ^3m^2$	$\dfrac{m_\mu{}^2}{m_e{}^2}\sim(207)^2\sim 4\times 10^4$	$\sim$ a few keV in heavy elements

interdependence of atomic and nuclear structure in the muonic atom is so pronounced as to be readily observable even by present techniques; such studies hold great interest for an understanding of both phenomena.

GENERAL FEATURES OF THE MUONIC X-RAY SPECTRA

The Lyman and Balmer series.—With the high-resolution Ge(Li) detector, the lines in the Lyman and Balmer series of the muonic spectra can be completely resolved. Figure 5 shows the Lyman and Balmer series in Ti observed by Anderson et al. (4, 96). Below each of the experimental spectra in Figure 5 are plotted the spectral intensities calculated by a cascade model (67, 68). The agreements are reasonably good.

In the Lyman series, only transitions of the type $(np\rightarrow 1s)$ contribute. In the Balmer series, the main contribution comes from the transitions $(nd\rightarrow 2p)$. The transitions $(ns\rightarrow 2p)$ have energies too close to $(nd\rightarrow 2p)$ to be resolved. But the contribution from $(ns\rightarrow 2p)$ is only $\sim$1 per cent. However, the transitions of $(np\rightarrow 2s)$ are lower in energy (by 11 keV in Ti) and

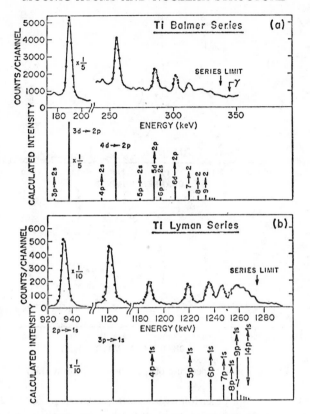

FIG. 5. The Lyman series ($np{\rightarrow}1s$) and Balmer series
($nd{\rightarrow}2p$) of muonic transitions in Ti (96).

have an intensity which is a few per cent of the main peak. The $2s$-$2p$
splitting has now been observed for the first time in muonic atoms (96).

With the excellent resolving power and efficiency of a Si(Li) detector in
the energy region of (5–30) keV, Wetmore et al. (139) obtained the line
spectra in the 8-keV region as shown in Figure 6. Since liquid helium scintil-
lates because of atomic excitations by ionizing particles, the liquid-helium
target was used also as a scintillation counter to detect the passage of the
muon through the target. Muonic K X rays from hydrogen have not been
observed because of the extremely low energy ($\sim$2 keV).

The $2s_{1/2}{\rightarrow}2p_{1/2}$ *transition.*—A particularly interesting feature of the
muonic X-ray spectrum is the $2s_{1/2}{\rightarrow}2p_{1/2}$ transition. The intensities of transitions involving the $2s$ states in the heavy elements are only $\sim$1 per cent of
the intensity of the principal lines. This explains why these transitions have
long eluded detection. Recently, Anderson et al. (5, 6) measured muonic X

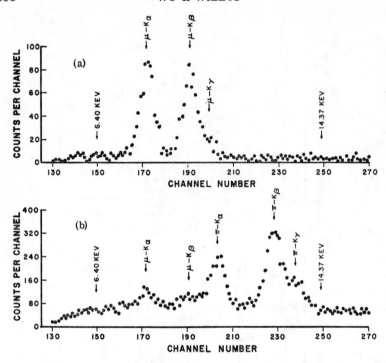

FIG. 6. The line spectrum of the muonic He atom (139).

rays in ^{206}Pb with high precision over a data collection period of 2 months (see Figure 7). They were able to identify the very weak transitions $3p_{3/2}$ $\rightarrow 2s_{1/2}$ and $2s_{1/2} \rightarrow 2p_{1/2}$ by the precisely measured energies (1507.93 ± 80) keV and $(1217.81 \pm .80)$ keV respectively. They fitted the six measured transition energies involving both $1s$, $2s$ (as in $2p_{1/2} \rightarrow 1s_{1/2}$, $2s_{1/2} \rightarrow 2p_{1/2}$, $3p_{3/2} \rightarrow 2s_{1/2}$) and np (as in $np \rightarrow n'd$ and $3d_{3/2} \rightarrow 2p_{1/2}$) with a Fermi charge distribution, determining the parameters c and t. The fit back onto the energy was not very good, the χ^2 per degree of freedom being equal to 3.6. In this fitting, *the nuclear-polarization correction was purposely left out.* Since this correction is largest on the binding energy of the $1s$ muon, they carried out another fitting to obtain the parameters c and t in the Fermi distribution by excluding the transition which involves the $1s_{1/2}$ level $(2p_{1/2} \rightarrow 1s_{1/2})$. Thus the χ^2 was improved to 1.4 per degree of freedom. However, the calculated $2p_{1/2} \rightarrow 1s_{1/2}$ energy difference (5781.54 keV) from these parameters turns out to be (6.8 ± 2.3) keV lower than the experimental value (5788.33 ± 0.48) keV. They attribute this difference to the nuclear-polarization correction on the binding energy of the $1s$ muon. The nuclear-polarization correction on the $1s$ muon in lead has been calculated to be 6 keV (± 30 per cent) by Chen (35) and 5.7 keV (± 50 per cent) by Cole (49, 50). If this interpretation is confirmed, the

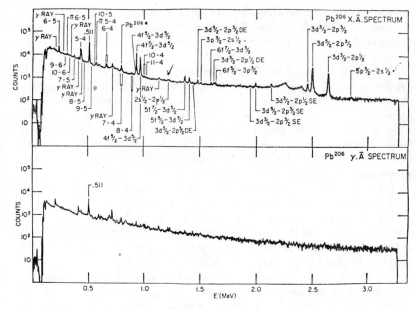

FIG. 7. Muonic spectrum of ^{206}Pb, taken with an anti-Compton device. Note the $2s_{1/2} \rightarrow 2p_{1/2}$ and $3p_{3/2} \rightarrow 2s_{1/2}$ transitions (4).

transitions involving measurements of the $2s$ level offer, for the first time, a direct observation of the nuclear polarization due to the $1s$ muon. However, the effect of nuclear polarization on the binding energy of the $2s$ muon in lead is nearly 1 keV and, therefore, cannot be completely neglected.

CHEMICAL EFFECTS

Relative captures in mixed systems.—It is of interest to study the relative atomic captures of muons in a chemical compound. Fermi & Teller (74) made some crude theoretical estimates based on simple arguments and concluded that the capture probability is proportional to the nuclear charge Z. For instance, in a binary compound A_nB_m, the atomic capture ratio should be nZ_A/mZ_B. Since muons captured by different atoms decay with different decay rates, early experiments (123) in studying this effect were carried out by analyzing the complex time distribution of the μ-decay curve. Recently, the observed relative muonic X ray intensities have been used for this purpose (53, 96, 151) and offer the capability of higher accuracy. It seems that the Z *proportionality* is approximately held in some metallic compounds (alloys) (45) but breaks down completely in *some insulators* (123). The relative atomic capture is not only closer to the stoichiometric ratio n/m but the element with lower Z is occasionally even favored. For instance in Al_2O_3, SiO_2, and P_2O_5, the capture ratio on oxygen to the other elements (according to the Z proportionality) should be 0.92, 1.1, and 1.33 respectively. Instead, ratios of 2.3, 3.5, and 2.7 were observed (123). Probably the arguments used in concluding the Z proportionality are applicable to metals but not to insulators.

Relative X-ray line intensities.—The relative population of upper states can be accounted for approximately by a cascade model. The "missing X-ray paradox" has been solved and long forgotten now. However, noticeable differences due to chemical effects were found in comparing the X-ray intensities in Ti, V, Cr with that in their oxides. For instance, the ratio of the higher transitions $(np{\rightarrow}1s)/(2p{\rightarrow}1s)$ in oxides is very different from that in metals (53, 96, 151). Although there are no adequate interpretations for these observed chemical effects, possible explanations include:

(*a*) The slowing-down process of the very low-energy muon ($E_\mu < 2$ keV) in an insulator is rather different from that in a metal.

(*b*) The Auger transitions, important for high n values, could be affected by the availability of the outer electrons which, in turn, depends on the rate of recapture of the electrons. In other words, small changes in the electron shells could play an important role in the muon cascade.

(*c*) The electrostatic potential seen by a muon in a chemical or molecular environment is not atomic and generally not spherically symmetric. This relaxes the inhibition $\Delta l = \pm 1$ (since l is not a good quantum number) and permits long transitions rather than the usual short-steps ($|\Delta n| = |\Delta l| = 1$) cascade.

It should be interesting to correlate the chemical effects in muonic atoms with those in electronic X rays and in the internal-conversion processes of gamma rays.

NUCLEAR MONOPOLE CHARGE DISTRIBUTION

Muonic atoms have been looked upon as ideal tools to study the distribution of nuclear charge and nuclear moments inside the nucleus. How much information one can obtain from the study of the muonic atoms depends on the accuracy and resolution of the energy measurements. With NaI detectors, even the fine structure of K X rays could not be resolved completely. Therefore, only one parameter of the monopole charge distribution was determined. With Ge(Li) detectors, the fine structure of both the K and L X rays can now be resolved for elements $Z > 25$. In principle, one should be able to determine one parameter of the charge distribution for each transition. However, because of the uncertainties in the measurements and in the theoretical corrections, the possibility of determining even two parameters is still limited to heavy Z isotopes.

Two questions have often been asked of the muonic X rays concerning the nuclear charge distributions: What parameters of the nuclear charge distribution can be determined from the muonic X rays and how model independent are they? How consistent are the results derived from the muonic X rays with that from the electron scattering? We attempt to summarize the answers below.

PARAMETERIZATION OF THE NUCLEAR CHARGE DISTRIBUTION

The energy shift due to the finite size effect is known to be proportional to various moments of the charge distribution. In the low-Z elements, a *perturbation calculation* reveals that the s-state energy shifts are proportional to $\langle r^2 \rangle$, the p-state shifts to $\langle r^4 \rangle$, etc. Since the $2p$ muon in a light element spends very little time inside the nucleus, the energy shift of the $2p$ level is negligible in comparison with experimental uncertainties, thus only $\langle r^2 \rangle$ can be obtained. In reality, even for $Z = 6$, the $2p \rightarrow 1s$ transition does not exactly determine the rms radius. In heavy elements, several higher moments of the charge distribution must be taken into account.

Since a transition yields a single datum (the energy), the question is what characteristic of the charge distribution is being measured. This has been answered very elegantly and practically in a triply remarkable study by Ford & Wills (76). They carried out exact numerical calculations on four models of the charge distribution with various parameters. The functional forms were

"Family II" (*3-parameter*)

$$\rho(r) = \begin{cases} \rho_0[1 + w(r/R)^2]\{1 - \tfrac{1}{2}\exp[-n(R - r)/R]\}, & r < R \\ \rho_0(1 + w)\tfrac{1}{2}\exp[-n(r - R)/R], & r > R \end{cases} \quad 1.$$

Fermi (*3-parameter*)

$$\rho(r) = \begin{cases} \rho_0[1 + w(r/R)^2]\{1 + \exp[(r - R)/a]\}^{-1}, & r < R \\ \rho_0(1 + w)\{1 + \exp[(r - R)/a]\}^{-1}, & r > R \end{cases} \quad 2.$$

Bethe (*2-parameter*)

$$\rho(r) = \begin{cases} \rho_0\{1 - \exp[(r - R)/a]\}^2, & r < R \\ 0, & r > R \end{cases} \quad 3.$$

Bethe-Elton (*4-parameter*)

$$\rho(r) = \begin{cases} \rho_0[1 + w(r/R)^2]\{1 - \tfrac{1}{2}\exp[(r - R)/a]\}^2, & r < R \\ \rho_0(1 + w)\tfrac{1}{4}\exp[-\gamma(r - R)], & r > R \end{cases} \quad 4.$$

They considered reasonable variation in the parameters (keeping the transition energies fixed) as well as discrete variations among the various models. With only small model-dependence, they found the following three remarkable results:

1. A state energy or a transition energy could be characterized by a single (nonintegral) moment of the charge distribution, $\langle r^k \rangle$ (see Figure 8). The radius of the uniform charge distribution yielding the same result (the "equivalent radius") is given by

$$R_k = \left[\frac{3 + k}{3}\langle r^k \rangle\right]^{1/k} \quad 5.$$

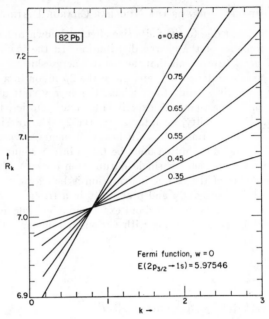

FIG. 8. Graphical determination of the "unique" k value and equivalent radius for the $2p_{3/2} \to 1s_{1/2}$ transition in Pb. This study uses the two-parameter Fermi function (76).

$$k(2p_{1/2} \to 1s_{1/2}) = 2 - Z/68.5$$
$$k(2p_{3/2} \to 1s_{1/2}) = 2 - Z/70.0$$
$$k(3d_{3/2} \to 2p_{1/2}) = 4 - Z/48.0$$
$$k(3d_{3/2} \to 2p_{3/2}) = 4 - Z/65.5 \qquad \qquad 6.$$
$$k(3d_{5/2} \to 2p_{3/2}) = 4 - Z/65.5$$
$$k(4f_{5/2} \to 3d_{3/2}) = 6 - Z/42.2$$
$$k(4f_{7/2} \to 3d_{5/2}) = 6 - Z/68.5$$

2. The exponent k was found to be a nearly linear function of Z for all transitions. For example,

3. The sensitivity of the energy on the effective radius can be expressed through the relationship

$$\delta R_k + C_Z \delta E = 0 \qquad \qquad 7.$$

where again C_Z depends upon Z and the transition involved, but only weakly upon the details of the nuclear charge distribution (Table II).

We proffer the following interpretation of these results: The interaction between the nucleus and the muon may be written in the form

TABLE II. The Ford & Wills Parameter C_Z Defined in Equation 7 (76)

Z	A	C_Z(F/MeV)
13	26.98	196.
20	40.08	44.0
30	65.37	11.9
50	118.7	3.30
82	208.	1.40
92	238.	1.23

$$\int \rho_N(\mathbf{r}) V_\mu(\mathbf{r}) d\mathbf{r} \qquad\qquad 8.$$

where ρ_N is the nuclear charge distribution and V_μ is the potential generated by the muon. (This is an inversion of the usual role of nucleus and muon.) If the muon wavefunction were constant throughout the nucleus, we would have

$$V_\mu = V_\mu(0) + \left(\frac{2\pi e}{3}\psi^2\right)r^2$$

Consider now first-order perturbation theory where we begin with some *reasonable (but not exact)* $\rho_N^{(0)} \Rightarrow \psi^{(0)} \Rightarrow V_\mu^{(0)}$. Then the total (interaction plus kinetic) energy of the system through first order is obtained by evaluating the zeroth-order kinetic energy plus

$$\int \rho_N(\mathbf{r}) V_\mu^{(0)}(\mathbf{r}) d\mathbf{r} \qquad\qquad 9.$$

To this order, V_μ is not reevaluated. Normally, one would express V_μ as a power series [in the nonrelativistic case, only even powers appear]. The Ford-Wills calculation tells us that we can approximate

$$V_\mu^{(0)}(\mathbf{r}) = V_\mu^{(0)}(0) + \text{const } r^k \qquad\qquad 10.$$

and that the form is adequate for all nuclear density models which do not deviate too much from the assumed, reasonable $\rho_N^{(0)}$.

Thus Ford & Wills have given us a very simple method of parameterizing the information contained in each transition energy, $R_k \propto \langle r^k \rangle^{1/k}$, in place of the more usual parameters indicated in the density models of Equations 1–4.

In the case of low- and moderate-Z nuclei we can relate k to the $\langle r^2 \rangle$ and $\langle r^4 \rangle$ moments of the charge distribution (as an example appropriate to s states) as follows. Let

$$V_\mu^{(0)}(r) = V_\mu^{(0)}(0) + a_2 r^2 - a_4 r^4 + \cdots \qquad\qquad 11.$$

The constant term is irrelevant here. We approximate

$$f_1(r) \equiv a_2 r^2 - a_4 r^4 \simeq \text{const } r^k \equiv f_2(r) \qquad 12.$$

We want to determine the constant and k such that

$$\int f_1(r)\rho_N(r)r^2 dr \simeq \int f_2(r)\rho_N(r)r^2 dr \qquad 13.$$

We can write ρ_N as the *faltung*

$$\rho_N(r) = \int \theta(r' - r)g(r')dr' \qquad 14.$$

where θ is the step function and the weighting function

$$g(r') = - d\rho_N(r')/dr' \qquad 15.$$

is peaked around some $r' = c_0$. More precisely, we can define c_0 such that

$$\int (r' - c_0)g(r')dr' = 0 \qquad 16.$$

Then we evaluate $(i = 1, 2)$

$$\int f_i(r)\rho_N(r)r^2 dr$$

$$= \int f_i(r) \left[\theta(c_0 - r) + (r' - c_0)\theta'(c_0 - r) + \frac{1}{2}(r' - c_0)^2 \theta''(c_0 - r) + \cdots \right]$$

$$\cdot g(r')r^2 dr dr' \qquad 17.$$

$$= \int_0^{c_0} f_i(r)r^2 dr \int g(r')dr' + \frac{d}{dr}(f_i(r)r^2)\Big|_{r=c_0}$$

$$\cdot \frac{1}{2} \int (r' - c_0)^2 g(r')dr' + \cdots .$$

The desired relationship is obtained by equating

$$\int_0^{c_0} f_1(r)r^2 dr = \int_0^{c_0} f_2(r)r^2 dr \qquad 18.$$

and

$$f_1'(c_0)c_0^2 + 2f_1(c_0)c_0 = f_2'(c_0)c_0^2 + 2f_2(c_0)c_0 \qquad 19.$$

This yields $(x = a_4 c_0^2/a_2)$

$$(k + 2)(k + 3) = \frac{20 - 30x}{1 - \frac{5}{7}x}$$ 20a.

or

$$k \simeq 2 - \frac{110}{63}x, \quad x \ll 1$$ 20b.

Comparison with Electron Scattering

Muon and electron scattering results have been found to be in quite satisfactory agreement, as can be seen from the following comparisons.

For low-Z elements, high-energy electron scattering measures the form factor

$$F(q) = 1 - \frac{1}{6}q^2\langle r^2\rangle + \frac{1}{120}q^4\langle r^4\rangle \cdots$$ 21.

where q is the momentum transfer. The rms radius can be extracted from $F(q)$ by taking the slope of $F(q)$ vs. q^2 at $q=0$, although several moments are involved at finite q. The muon experiments also measure several moments of the charge distribution, according to Equation 11, and we can relate the a_2 and a_4 to the Ford & Wills parameter k by Equation 20.

It was pointed out by Engfer (70) that a sensitive comparison between the experiments would be at an electron "equivalent momentum transfer" q_{eq} such that the relative weighting of the terms $\langle r^2\rangle$ and $\langle r^4\rangle$ is the same for both. The results in both cases can then be expressed in terms of an equivalent rms radius. Table III shows good agreement between the two types of experiments in light and medium nuclei. Even for heavy nuclei, such as Pb and ^{209}Bi, the agreement is good when the scattering experiments are performed at roughly the equivalent momentum transfer (69, 71).

Calculations of nuclear-polarization effects, called dispersion corrections in electron scattering, have been carried out by several authors, most recently Onley (112) and Rawitscher (120). Although the effects may not be negligible, the parameters of the charge distribution derived from electron scattering to date have not included these corrections.

Summary of Monopole Charge-Distribution Results

$Z < 25$.—The $2p \rightarrow 1s$ translation energies of muonic atoms from $Z = 2$ to 25 have been measured in several laboratories and their results are listed in Table IV. The agreements among them are excellent. The values of the $\langle r^2\rangle^{1/2}$ from muonic atoms and from electron scattering with equivalent momentum transfer q_{eq} are also compared in Table III. No discrepancies could be found. For $Z < 6$ the shift of the $1s$ level is too small to be measured

TABLE III. $\langle r^2 \rangle^{1/2}$ IN F FROM MUONIC ATOMS AND ELECTRON SCATTERING (9, 71, 131)

	From muonic atoms	From electron scatterings
$_2$He	3.1 ±4.7	
$_3^6$Li	3.1 ±3.9	
$_3^7$Li	2.6 ±0.9	
$_4$Be	2.76 ±0.68	2.46 ±0.09
$_5^{10}$B	2.44 ±1.27	2.45 ±0.12
$_5^{11}$B	2.26 ±1.69	2.42 ±0.12
$_6^{12}$C	2.40 ±0.56	2.42 ±0.04
$_7^{14}$N	2.67 ±0.26	2.45 ±0.05
$_8^{16}$O	2.61 ±0.14	2.65 ±0.04
$_8^{18}$O	2.71 ±0.14	2.77 ±0.04
$_9^{19}$F	2.85 ±0.09	
$_{11}^{23}$Na	2.94 ±0.06	
$_{12}$Mg	3.02 ±0.04	2.97 ±0.09
$_{13}$Al	3.025±0.023	2.98 ±0.04
$_{14}$Si	3.086±0.018	3.04 ±0.09
$_{15}$P	3.188±0.018	3.08
$_{16}$S	3.244±0.018	3.18 ±0.09
$_{17}$Cl	3.335±0.018	3.12
$_{19}$K		
$_{20}$Ca		
$_{22}$Ti	3.599±0.009	3.564±0.040

with any precision, so the uncertainties in $\langle r^2 \rangle^{1/2}$ by muonic methods are considerable. However, for $Z > 10$, the muonic method appears to be more accurate in determining the rms radius.

$Z > 25$.—Some of the muonic energies measured in this region are tabulated in Table V.

For muonic atoms of $Z > 25$, the fine-structure splitting can be resolved and the shift of the $2p$ levels can be detected. However, in the beginning of this region, the $2p$ muons spend most of the time outside of the nucleus. Thus, the effect of the finite nuclear size on the $2p$ energy levels is so small that it requires very high precision in the determination of the energies of L X rays to narrow down the limits of the second parameter. If one uses the two-parameter Fermi charge distribution (Equation 2 with $w = 0$, $R = c$, and $a = t/4.39$), then the two parameters to be determined are c and t. The relation between the rms radius to c and t in a Fermi distribution can be approximated by $\langle r^2 \rangle = c^2 + 1.19t^2$. A graphic analysis, known as the c-t iso-energetic plot, was initially introduced by Acker et al. (2) and has been widely adopted in the field. From a measured value of each transition energy, c and t are not separately determined, but a definite relationship between them can be plotted as an isoenergetic curve. These curves will not be ex-

TABLE IV. MUONIC TRANSITION ENERGIES IN keV $(2p \rightarrow 1s)$ FOR $Z < 25$

		Berkeley (17)		William & Mary (139)
${}_{2}$He				8.18 ± 0.04
${}_{3}^{6}$Li		18.1 ± 0.4		18.66 ± 0.07
${}_{3}^{7}$Li		18.1 ± 0.4		18.71 ± 0.06
${}_{4}$Be		33.0 ± 0.2	CERN (9)	33.39 ± 0.05
${}_{5}^{10}$B		51.6 ± 0.3	52.23 ± 0.15	52.18 ± 0.10
${}_{5}^{11}$B	Columbia (46)	51.6 ± 0.3	52.31 ± 0.15	52.23 ± 0.09
${}_{6}$C	75.3 ± 0.5 75.23 ± 0.1 (48)	75.8 ± 0.5	75.25 ± 0.15	75.23 ± 0.08
${}_{7}$N		101.9 ± 0.5	102.29 ± 0.15	
${}_{8}^{16}$O	133.3 ± 0.5	133.4 ± 0.5	133.56 ± 0.15	
${}_{8}^{18}$O			133.57 ± 0.15	
${}_{9}$F	168.07 ± 0.5	168.9 ± 0.5	168.45 ± 0.15	
${}_{11}$Na	250.24 ± 0.5	249.6 ± 0.5	250.21 ± 0.15	Carnegie-Mellon (29)
${}_{12}$Mg	296.88 ± 0.5	296.1 ± 0.5	296.55 ± 0.15	296.40 ± 0.20
${}_{13}$Al	347.21 ± 0.5	Chicago (39)	346.82 ± 0.15	William and Mary (139)
${}_{14}$Si	400.38 ± 0.5	400.2 ± 0.6	400.22 ± 0.15	400.15 ± 0.20
${}_{15}$P	457.06 ± 0.5		456.54 ± 0.20	
${}_{16}$S			516.24 ± 0.25	516.20 ± 0.20
${}_{17}$Cl			578.56 ± 0.30	
${}_{19}$K	712.24 ± 0.5	712.64 ± 0.6		
${}_{20}$Ca	783.8 ± 1.5 (149)	783.56 ± 0.16		783.85 ± 0.15
${}_{22}$Ti	931.57 ± 0.5			

actly parallel and the intersection of two curves $(2p \rightarrow 1s)$ and $(3d \rightarrow 2p)$ should give the unique determination of c and t. In practice, the effectiveness of this analysis is severely limited by the experimental uncertainties, as can be seen from Figure 9 for $Z = 29$ and $Z = 60$, where the uncertainties in both K and L X rays are around ± 1 keV.

$Z > 80$.—In this region, and for spherical nuclei only, both c and t can be determined to higher accuracy. Figure 10 shows the isoenergetic curves for ^{209}Bi, and the interception point gives $c = r_{0}A^{1/3} = 6.63 \pm 0.03$ F, $t = 2.40$

TABLE V. Muonic Transition Energies in keV for $Z > 25$

Nuclide	$(2p \rightarrow 1s)$	$(3d \rightarrow 2p)$	$(2p_{3/2} \rightarrow 1s_{1/2})$	$(3d_{5/2} \rightarrow 2p_{3/2})$	$(2p \rightarrow 1s)$ center of gravity
	Columbia (46)				Chicago (39)
$_{28}$Ni	1422.1 ±0.5 1427.4 ±0.5	309.97			1429.5±0.6
$_{29}$Cu	1506.61±0.5 1512.78±0.5	330.26±0.5 334.8 ±0.5			1510.3±0.6
$_{33}$As	1855.8 ±0.5 1866.9 ±0.5	427.5 ±0.5 436.6 ±0.5			
$_{38}$Sr					2335.6±0.8
$^{90}_{40}$Zr					2528.9±0.8
$_{41}$Nb	2608.8 ±0.8 2630.4 ±0.8	665.4 ±0.5 686.6 ±0.5			
	Columbia (148) (103)				
$^{92}_{42}$Mo	2706.8 ±0.4 2732.3 ±0.4	717.8 ±0.2 696.0 ±0.2	CERN (2)		
$^{115}_{49}$In	3322.67±0.42 3366.27±0.42	943.39±0.48 981.74±0.48			
$^{118}_{50}$Sn	3414.6 ±0.5 3460.4 ±0.5	1022.8 ±0.2 982.6 ±0.2			3413.9±0.5 3459.7±0.5
$_{51}$Sb			3543.3±2.0	1019.6±3.0	
$_{52}$Te			3625.6±2.5	1060.0±3.0	
$_{53}$I			3721.6±2.5	1098.0±3.0	
$^{133}_{55}$Cs	3840.08±0.48 3902.18±0.48	1185.47±0.40 1238.35±0.40	3836.1±3.0 3899.1±3.5	1188.6±3.0 1241.6±3.0	
$_{56}$Ba			3979.8±4.0	1229.2±3.0	
$_{57}$La			4071.2±4.0	1266.8±3.0	
$_{58}$Ce			4160.2±5.0	1314.9±3.0	
$^{141}_{59}$Pr	4184.80±0.34 4262.37±0.34	1358.61±0.45 1424.22±0.45	4184.3±5.0 4258.8±5.5	1356.7±3.0 1422.6±3.0	
$^{144}_{60}$Nd	4254.9 ±0.5 4336.4 ±0.4	1471.8 ±0.2 1402.6 ±0.10			

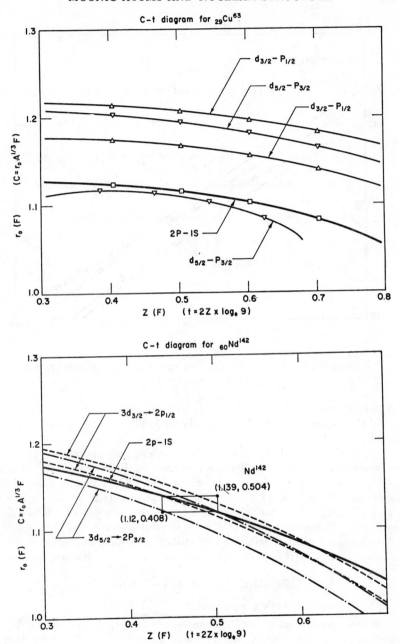

FIG. 9. *c-t* diagrams for $^{63}_{29}$Cu and $^{142}_{60}$Nd (147).

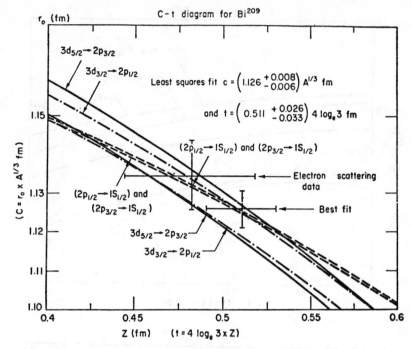

FIG. 10. c-t diagram for ^{209}Bi (9). Electron scattering results (9) are plotted on the same diagram.

± 0.08 F. The values of c and t obtained from the study of electron scattering in ^{209}Bi (110) are given in Table VI and plotted in Figure 10. The agreements are good.

Though the required experimental precision may be eventually achieved, one is still concerned with many other uncertainties. First, there is the absolute energy calibration. There are very few well-calibrated gamma lines in the energy region of $3 < E < 6$ MeV. The best-calibrated line is the ^{16}N line ($E_\gamma = 6.131$ MeV) (34) from ^{16}O$+n \rightarrow ^{16}$N$+p$ with an accuracy of ± 0.5 keV. The energy differences between the photo, the single escape, and the double escape peaks are great aids in linearity calibration within an energy region of $2\ m_e c^2$. Then there remain the uncertainties in the theoretical corrections.

THEORETICAL CORRECTIONS

QUANTUM ELECTRODYNAMIC EFFECTS

Vacuum polarization, or the Uehling effect, is the largest QED effect. To lowest (second) order in αZ, the Coulomb potential is replaced by the Uehling potential, folded into the charge distribution. The vacuum-polarization correction increases the strength of the potential and has a range,

TABLE VI. Parameters of Best Fit with a Fermi-Type Charge Distribution
for ^{209}Bi, Natural Pb, ^{206}Pb, ^{207}Pb, and ^{208}Pb. All
Lengths Are Measured in F

		c	t	$\langle r^2 \rangle^{1/2}$	Ref.
^{209}Bi	μ-atom	6.63 ± 0.03[a]	2.40 ± 0.08[a]	5.513 ± 0.007	12
	e-scattering	6.74 ± 0.08	2.00 ± 0.16	5.49 ± 0.06	71
^{nat}Pb	μ	$6.67 \pm .01$	2.21	$5.493 \pm .007$	2
	e	$6.66 \pm .09$	$2.21 \pm .17$	$5.48 \pm .07$	71
^{206}Pb	μ	$6.639 \pm .010$	2.269 ± 0.030	$5.4894 \pm .0015$	4
^{207}Pb	μ	6.629 ± 0.012	$2.317 \pm .034$	$5.4963 \pm .0016$	4
^{208}Pb	μ	$6.636 \pm .015$	2.320 ± 0.044	$5.5026 \pm .0022$	4
	e	$6.66 \pm .09$	$2.21 \pm .17$	$5.48 \pm .07$	71

[a] Both nuclear polarization and Lamb shift are taken into account in the analysis. Without these corrections, $c = 6.64$ and $t = 2.24$ F, as shown in Figure 10. No dispersion corrections were applied to e^- scattering results.

characteristically, of the electronic Compton wavelength $\lambdabar_e = \hbar/m_e c = 386$ F. The correction is largest for the $1s$ level. The Bohr orbit for the $1s$ muon is $a_\mu = (250/Z)$ F: Except for the very lightest nuclei, we are concerned with distances $r \ll \lambdabar_e$. The vacuum-polarization potential energy between point-charge pairs $+e$ and $-e$ is (cf. 12)

$$V_{vp} = \left(\frac{e^2}{r}\right) \frac{2\alpha}{3\pi} \left[\ln (1.781 r/\lambdabar_e) + \frac{5}{6} + \mathcal{O}\left(\frac{r}{\lambdabar_e}\right) \right] \qquad 22.$$

The correction for the $1s$ state when expressed as a fraction of the $1s$ state energy (both are negative) is a slowly varying, nonmonotonic function of Z, as displayed in Figure 11. For all elements with $Z \geq 18$, the ratio falls between $(6-7) \times 10^{-3}$. Higher-order corrections are down by the relative order $(Z\alpha)^2$ and have been estimated by Wichman & Kroll (143) to yield a correction

$$\left(\frac{\Delta E_{vp}^{(4)}}{E}\right)_{1s} = 2\alpha\{(Z\alpha)^2(0.020940) + (Z\alpha)^4(0.007121)F_0((Z\alpha)^2)\} \qquad 23.$$

where F_0 is a function with numerical value between 1 $(Z\alpha = 0)$ and 4 $(Z\alpha = 1)$. These estimates are based on a point nucleus; the finite size will undoubtedly reduce the value. (In lowest order, finite size reduces the ratio in U to $\frac{1}{4}$ the point-charge value. The reduction might be expected to be even greater here.) For U the Wichman & Kroll estimate yields a ratio of 1.6×10^{-4} or 4 keV. For deformed nuclei in the rare-earth Pearson (153) estimated that there is a small vp correction about $\frac{1}{2}$ per cent in the quadrupole interaction.

Lamb-shift and other corrections.—Barrett et al. (12) have reexamined the

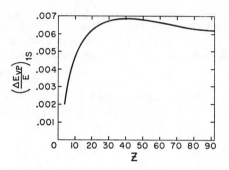

FIG. 11. Vacuum-polarization correction divided by binding energy for
1s muonic level, based on the data of (76).

Lamb-shift corrections to heavy, muonic atoms. The usual expansion in terms of $Z\alpha$ turns out to be valid even for Bi($Z\alpha\simeq0.6$) because the electrostatic potential is nonsingular. Although smaller than the vacuum-polarization correction, the correction is not smaller by $(m_e/m\mu)^2$ as previously assumed. Their calculations also included the effect of muon-pair vacuum polarization and the muon anaomalous magnetic moment. For $_{83}$Bi, Barrett et al. (13) found

level	$\Delta E_{\text{L.s.}}$ (keV)
1s	2.99 ± .16
2s	0.72 ± .15
2$p_{1/2}$	0.35 ± .16
2$p_{3/2}$	0.69 ± .12
2$d_{3/2}$	−0.05 ± .00

ELECTRON SCREENING

For the lower muonic states, electron screening plays a negligible role. In Bi, for example, the effect increases from $+4.6$ eV for the 1s state to $+190$ eV for the 5g state. For $n\geq5$, the effect deserves inclusion. In considering *transitions* between "circular orbits" (n, $l=n-1$), one can consider the nuclear charge to be reduced by the total electronic charge within the mean orbit. (More penetrating oribts are less affected.)

NUCLEAR POLARIZATION

Nuclear polarization describes correlation between the muon and the various nuclear coordinates. In lowest-order approximation, the system is simply written as a product of the (free) nuclear wavefunction and the muon wavefunction calculated from a (nuclear ground-state average) static potential. The perturbation potential (the difference between the full interaction and the static potential) generates distortion in the system wavefunction, which can be described by admixture (or virtual excitation) of higher states.

Polarization is necessarily a second- or higher-order effect, because a first-order distortion in the wavefunction of the system gives rise to a second-order change in the energy. This simple quantum-mechanical principle plays an important role in the interpretation of polarization effects. Since allowing for distortion admits a more general wavefunction, polarization lowers the ground state of the system, although higher levels can be displaced either way.

The earliest discussion of nuclear polarization by electrons was contained in an optical isotope-shift paper by Breit, Arfken & Clendenin (26). Fitch & Rainwater (75) and Cooper & Henley (43) investigated polarization by atomic muons, and several further studies followed (122a); two recent surveys by Chen (35) and Cole (49, 50) have given new and quantitative estimates of the effects.

Except for very special circumstances, polarization effects are of the order of or less than a few keV. The special cases involve either unusually large matrix elements or nearly degenerate levels of the combined system.

Estimates based on simple models.—For orientation, we note here rough estimates based on very simple nuclear models.

(a) *Monopole mode.* The presence of the muon inside the nucleus tends to compress the charge distribution. Two possibilities are given here:

1. *Breathing mode.* We consider the nucleus to be uniformly charged and to undergo uniform contraction. The muon is also considered to have uniform density within the nucleus. The monopole polarization energy is then given by

$$\Delta E_{br}{}^{(0)} \simeq -\frac{9}{50}\frac{e^2}{Kr_0{}^2}\frac{(Zq)^2}{A^{5/2}} \qquad 24.$$

where $KA = R_0{}^2 \partial^2 E_N / \partial R^2$ is the nuclear compressibility and q is the probability that the muon is inside the nucleus. For light nuclei and s-state muons, $q \simeq \frac{4}{3}(ZR_0/n\, a_\mu)^3$, but Equation 24 may be extended to heavy nuclei (where the muon density is not constant) if q is evaluated from available numerical wavefunctions. Thus for the $1s$ state of $^{208}{}_{82}$Pb, if we take $q \sim 0.5$, $K \sim 200$ MeV, and $r_0 = 1.2$ F, we find $\Delta E_{br}{}^{(0)} \simeq 0.3$ keV.

2. *Shape compression.* Here we consider a permanently deformed nucleus which can execute β vibrations. The nuclear deformation energy is described by $\frac{1}{2}C_\beta(\beta - \beta_0)^2$. The muon tends to reduce the deformation to maximize the mutual interaction. The polarization energy is

$$\Delta E_\beta{}^{(0)} \simeq -\frac{1}{2C}\left[\frac{3Ze^2\beta_0 q}{4\pi R_0}\right]^2 \qquad 25.$$

For ^{152}Sm, which is unusually soft to β vibrations, C is estimated from Coulomb excitation (see Equation 72 below) to be 220 MeV ($\beta_0 \approx 0.3$), which yields $\Delta E_\beta^{(0)} \sim 0.22$ keV.

Both of the above estimates yield relatively small values for the 1s displacement. Other modes (see below) give greater contributions.

(b) *Dipole mode.* Polarization through the dipole mode necessarily involves simultaneous atomic excitation because of the change of parity. A crude estimate of coupling to the giant resonance can be effected by invoking the Goldhaber-Teller model which depicts the neutrons and the protons as two rigid spherical units coupled harmonically; and treating the muon-proton coupling as though the muon also experienced a harmonic potential,

$$V(r_\mu) \simeq - \frac{Ze^2}{2R_0} \left(\frac{|\, \mathbf{r}_\mu - \mathbf{R}_p\,|^2}{R_0^2} - 3 \right) \qquad 26.$$

where $\mathbf{R}_p$ is the center of the proton distribution. The latter approximation would be valid for larger-than-real nuclei, where the meson wavefunction is confined completely within the nucleus. The approximation does afford an easy and useful estimate, since the coupled-oscillator problem is exactly soluble. To first order in (m_μ/m_N) the 1s state is lowered by

$$\Delta E^{(1)} \simeq - \frac{3}{4} \frac{\hbar \omega_\mu^3}{\omega_D(\omega_D + \omega_\mu)} \left(\frac{m_\mu}{m_N} \right) \frac{N}{AZ} \qquad 27.$$

For $^{208}_{82}$Pb, we choose the muon oscillator frequency energy to equal the $2p \rightarrow 1s$ transition energy, $\hbar \omega_\mu \simeq 6$ MeV, and the giant dipole frequency to be $\hbar \omega_D \simeq 14$ MeV. This gives $\Delta E^{(1)} \approx -0.5$ keV.

Extensive calculations coupling various collective modes have been performed by Pieper & Greiner (116). Although of considerable interest, they use two approximations which limit their numerical accuracy: the use of a sharp nuclear surface (diffuseness is partially effected by surface vibration); the utilization of the hydrodynamic relationship between the vibrational mass parameter B and the moment of inertia $\mathcal{G} = 3B\beta^2$, whereas these are empirically quite independent parameters.

Detailed calculations.—We consider first spherical nuclei. Deformed nuclei require special consideration, as discussed in the next section.

In perturbation theory,

$$\Delta E = \sum_l \Delta E_l \qquad 28.$$

$$\Delta E_l = \sum_{I\mu} \frac{|\, \langle I_o \mu_o |\, \Delta H_l\, |\, I\mu \rangle\,|^2}{E_{\mu_o} + E_{I_o} - E_\mu - E_I} \qquad 29.$$

where

$$\Delta H_l = - \sum \frac{4\pi e^2}{2l + 1} \frac{r_<^l}{r_>^{l+1}} \mathbf{Y}_l \cdot \mathbf{Y}_l \qquad 30.$$

Here $r_>$ and $r_<$ are the greater and lesser of the muonic and nuclear charge coordinates. To calculate the quantities reliably, we must know both the matrix elements and the energy denominators well.

The matrix element is related to the $B(El)$ associated with the nuclear states, although there is a difference in the radial dependence: While the $B(El)$ value depends on a radial integral of r^l, the matrix element here depends on the integral of $r_<^l/r_>^{l+1}$. This difference has been shown to be not serious (36). That is, the matrix element can be essentially determined by the nuclear $B(El)$ value. In the region where the $B(El)$ values are known experimentally, one can calculate the matrix element rather reliably from any nuclear model which gives the correct $B(El)$ and which has a reasonable radial dependence.

The energy denominator has two parts: E_I of the nuclear state and E_μ of the muonic intermediate state. E_μ can be handled exactly by calculating the correction to the wavefunction

$$\Delta\Psi_I = \sum_\mu \frac{|I\mu\rangle \langle I\mu| \Delta H_l | I_o\mu_o\rangle}{E_{\mu_0} + E_{I_0} - E_\mu - E_I} \qquad 31.$$

from the differential equation

$$(E_{\mu_0} + E_{I_0} - E_I - H_\mu)\Delta\Psi_I = \Delta H_l | I_o\mu_o\rangle \qquad 32.$$

The correction to the energy is then obtained from integration:

$$\Delta E_l = \sum_I \langle I_o\mu_o| \Delta H_l | \Delta\Psi_I\rangle \qquad 33.$$

E_I has to be obtained experimentally. However, it is found (36) that muonic intermediate states in the continuum are the important states, so either E_I is totally negligible compared with E_μ, as in the cases $l \geq 2$; or E_I is comparable to E_μ. Therefore, for $l = 1$ or 0, a little uncertainty in E_I will not affect the overall reliability of the calculation by much.

(a) *Monopole mode.* There is little experimental evidence on the energy and the strength of the monopole states. Chen (40) has used a phenomenological theory of the nuclear surface (144) which gives the correct volume binding energy, surface energy, and surface thickness; he finds a correction of ≈ 1.5 keV for the $1s$ level of Pb. The same value is also obtained from a theoretical nuclear model which gives the nuclear breathing excitation at ≈ 20 MeV, which seems to be a reasonable value.

(b) *Dipole states.* The dipole strength is exhausted by the giant resonance, especially for heavy nuclei. Both E_I and the strength $B(E1)$ of the giant resonant states are known experimentally, so the calculation is reliable. The correction is ≈ 2 keV for $1s$ level of ^{208}Pb (35, 50).

(c) *Nuclear states of $l \geq 2$.* The muonic intermediate states of importance come at very high energies, ≈ 30 MeV or higher, because only at those energies do the radial wavefunctions have significant overlap with the unperturbed muon wavefunctions. Therefore, the nuclear-state energies are un-

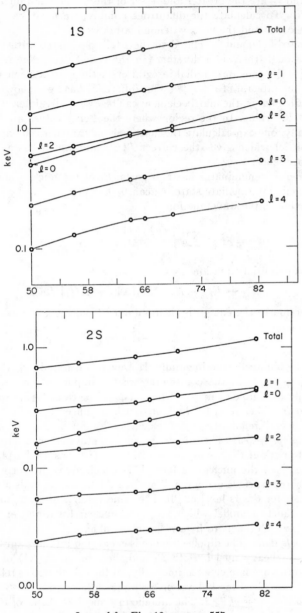

Legend for Fig. 12 on page 557.

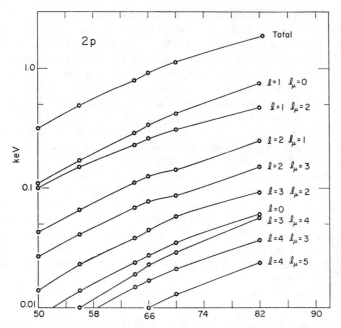

FIG. 12. Nuclear-polarization corrections for "spherical" nuclei (40, 50).

important. One does not have to know the complete spectrum of multiple nuclear excitations. So long as those excitations come at low energy, it does not matter where they are. The closure approximation is ideal for the situation: the main uncertainty in the closure approximation, the choice of the average nuclear energy, is not serious at all, since the main term in the energy denominator, E_μ, has already been handled exactly. The total correction, $l \geq 2$, is calculated rather reliably to be 2.5 keV for the $1s$ level of Pb (35).

(*d*) *Summary.* The corrections to the $1s$ $2p$ and $2s$ levels for various values of l are plotted as a function of Z in Fig. 12 (40, 50). It is noted here that for the $2p$ levels:

1. The calculations are more reliable than the $1s$ levels, since the muon spends more time outside the nucleus, $r_<^l/r_>^{l+1}$ behaves more like r^l, and the monopole contribution drops to minute values, ≈ 0.1 keV.

2. The values in Figure 12 are meaningful only for spherical nuclei. For deformed nuclei with a dynamic $E2$ hyperfine structure, a more careful analysis for the p levels is necessary.

Deformed nuclei.—In deformed nuclei, various ground-state rotational band members are admixed by the presence of the muon. (See pp. 579–86). This introduces extra difficulties since it becomes necessary to know, for example, how the 2^+ states of the ground-state band are coupled to the 1^- giant resonance state, while all the photo-gamma experiments are carried out when the nucleus is initially in the 0^+ state. If one assumes that all states of good angular momentum are projected out from some intrinsic wavefunction, and that there is no mixing of different bands, then the corrections to the various members of the ground-state band are related to one another by simple angular coupling coefficients. For the 1s-state muon, the situation is simple: the nuclear polarization is the same regardless of which member of the band the unperturbed nuclear state is in. For the $2p$ levels, the situation is more complicated, since the dynamic $E2$ effect splits the $2p$ levels into many levels, and each level is shifted a different amount by nuclear polarization. The best way to handle the situation is to use the effective quadrupole matrix element introduced by Chen (41) as follows:

The dynamic $E2$ splittings are obtained from diagonalizing the quadrupole interaction between the lowest rotational band and the spin doublets $2p_{1/2}$, $2p_{3/2}$ of the muon. The quadrupole matrix element is represented schematically by Figure 13. All other nuclear and muon states are very far away in energy and can be handled by perturbation theory. One can introduce the effective quadrupole matrix element by the scheme shown in Figure 14. By diagonalizing H_{eff} instead of H_Q, one includes the effect of nuclear polarization.

In the first of the two second-order graphs, the nuclear state I remains in the ground-state band and the muonic intermediate state μ is allowed to go beyond the spin doublets $2p_{1/2}$, $2p_{3/2}$. This graph takes into account the high muonic states not included in the diagonalization. Such graphs are summed to all orders if, instead of the diagonalization, one solves the coupled-channel equation (108). However, the solution of the coupled-channel equation is time-consuming and the effective matrix element approach is more suitable for actual data analysis.

In the second of the two second-order graphs, the nucleus is scattered into states beyond the ground-state rotational band and then scattered back. The evaluation of the correction to matrix element follows the same line as for the spherical nuclei.

The matrix elements $\langle H_Q{}^2 \rangle$ and $\langle H_l{}^2 \rangle$ are ~ 5–10 per cent of the unperturbed matrix element $\langle H_Q \rangle$. Their inclusion will tend to *increase* the effective H_Q. As is shown in pp. 579–86, this leads to a *smaller* experimental quadrupole moment.

Near-degeneracies.—In addition to the dynamic $E2$ interaction, occasional near-degeneracies may occur which result in line splittings or intensity anomalies.

Nuclear monopole excitations have been considered by Henley & Wilets (83). A 0^+ nuclear excitation (1s muon) might coincide with the muonic $2s$ excitation in the neighborhood of Zn or Kr.

Nuclear dipole excitation has been considered by Sundaresan & Srinivasan (130) who consider degeneracy between the muonic $2p$ states and a 5.9-MeV 1^- nuclear excitation (1s muon) in ^{208}Pb (this is distinct from the higher giant dipole resonance).

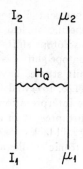

FIG. 13. Diagrammatic representation of the elementary quadrupole interaction.

Although the above degeneracies have not yet been observed, some significant intensity anomalies have been observed and attributed to degeneracy effects. These are discussed in more detail on pp. 596–600.

ISOTOPE SHIFTS

Muonic isotope shifts have now been studied extensively from $Z=8$ to 82. Their main features are in agreement with those found in electronic atoms, but since there are significant differences in the analysis of the data, we give a summary of the electronic isotope shifts with emphasis on the contrast to muonic shifts.

Whereas the energy levels in muonic atoms are displaced by up to 50 per cent from the point-nucleus value, optical electronic displacements ΔE are of the order of one part in 10^4 of the characteristic (transition) energies. Since such accuracy is not attainable in the theoretical evaluation of the spectra, the displacements ΔE are, practically, unobservable. The differences in displacements, δE, among different isotopes are observable even though the magnitude of the shift is down by $\sim A^{-1}$ from ΔE, or a few parts in 10^6 of the transition energy (optical precision is better than $1:10^7$). Even the relative shifts depend upon details of electronic structure which

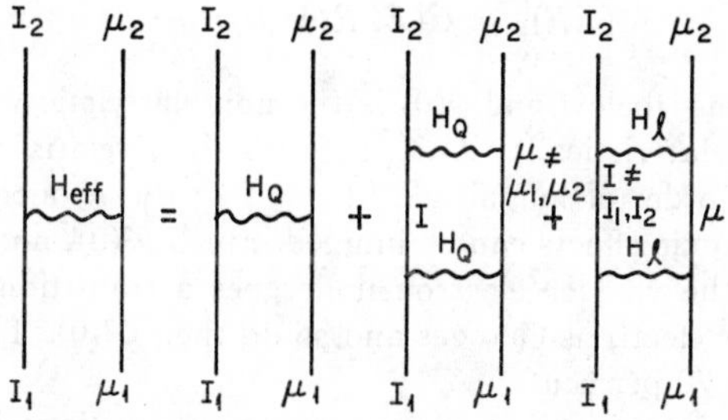

FIG. 14. Diagrammatic representation of the effective quadrupole interaction.

are only known to perhaps 20 per cent. Ratios of shifts are free of most of the uncertainties, except the specific nuclear mass effect (see below). One of the useful roles of the muonic isotope shifts is to provide a normalization and calibration of the optical isotope shifts.

Electronic X-ray isotope-shift studies have been developed into a fine art only recently by Sumbaev (128, 131) and Boehm (21, 24). Because they involve the inner electrons, the interpretation is relatively free of the uncertainties in electronic structure present in optical studies. Although the shifts are relatively larger, they still lack experimental precision.

Both optical and X-ray electronic studies are limited to heavy elements, with $Z \geq 40$.

Optical Isotope Shifts

Major contributions.—The major contributions to optical isotope shifts are classified as follows:

(a) *Normal mass effect.* The electronic mass m_e is replaced by the reduced mass $\mu = m_e M_A/(m_e + M_A)$. All energies are trivially scaled by the reduced mass (for Coulombic interactions). The transition energies of the heavier isotopes are greater than those of the lighter ones.

(b) *Specific mass effect.* The recoil of the nucleus gives rise to a term

$$\frac{1}{2M_A} \sum_{i \neq j} \mathbf{p}_i \cdot \mathbf{p}_j \qquad\qquad 34.$$

where the contribution for $i = j$ has already been identified above as the normal mass effect. This effect exists only for atoms with two or more electrons, and is not present in muonic atoms. The term depends upon two-electron correlations and must be evaluated before nuclear-structure effects can be analyzed. Although there is a factor of A^{-1}, the effect drops off more slowly because of the increase in the possible number of electrons contributing to the sum. Experimental procedures for evaluating this term are given in later sections.

(c) *Volume effect.* Finite nuclear size results in a decrease in binding energy

$$(\Delta E)_v = C(Z, R)\psi^2(0)R_{\text{eff}}^2 \qquad\qquad 35.$$

where R_{eff} is defined below and $\psi(0)$ is the nonrelativistic wavefunction at the origin. Nonrelativistically, $C(Z, R) = 2\pi Z e^2/5$; relativistic corrections increase C by an order of magnitude for high Z; the R dependence of C is weak. Further atomic effects can be lumped into $C\psi^2(0)$, notably electronic shielding: when the valence electron undergoes a transition, the shielding of all of the other electrons changes and so do their $\psi^2(0)$. This reduces the net shift by some 20 per cent.

If the electron wavefunction were constant over nuclear dimensions, the energy shift due to the finite nuclear size would be proportional to $\langle r^2 \rangle$, because the field generated by a uniform charge is of the form $\phi(0) + \phi_2 r^2$. A point nucleus generates a Dirac electron density that is singular at the origin like $\rho_e \propto r^{2(\sigma-1)}$, where $\sigma = (1 - Z^2\alpha^2)^{1/2}$; this, in turn, generates a potential which varies as $\phi(r) = \phi(0) + \phi_2 r^{2\sigma}$. An extended nucleus does not generate a singular electron density. This has led to some confusion in the literature about what moment of the charge distribution is measured by optical electrons. The answer can be obtained as follows: Following Bethe & Negele (20), and in the spirit of Ford & Wills (76), let us consider, in first approximation, that the nucleus can be described by a uniformly charged sphere of radius R. Then the electron density can be expanded in a power series

$$\rho_e(r) = \rho_e(0)\left[1 - \frac{1}{2} Z^2\alpha^2 \left(\frac{r}{R}\right)^2 + \cdots\right] \qquad 36.$$

where we have dropped terms in order $m_e c R / \hbar$ compared with $\frac{3}{4}Z\alpha$. This in turn generates a potential

$$\phi(r) = \phi(0) + \phi_2\left[\left(\frac{r}{R}\right)^2 - \frac{3}{20} Z^2\alpha^2 \left(\frac{r}{R}\right)^4 + \cdots\right] \qquad 37.$$

The constant term $\phi(0)$ is irrelevant. The shift is proportional to various moments of the charge distribution. We can quite accurately express this as a single moment $\langle r^{2\sigma'} \rangle$ by using Equation 20. This gives

$$\sigma' = 1 - .177Z^2\alpha^2 \simeq (1 - .354Z^2\alpha^2)^{1/2} \qquad 38.$$

Optical electrons thus measure $\langle r^{2\sigma'} \rangle$, not $\langle r^2 \rangle$ or $\langle r^{2\sigma} \rangle$, and the effective charge radius is

$$R_{\text{eff}} \equiv \left[\frac{3}{2\sigma' + 3} \langle r^{2\sigma'} \rangle\right]^{1/2\sigma'} \qquad 39.$$

(This is also true of very low-Z muonic atoms, but then $\sigma' \approx 1$ anyway.) Even for Pb, however, σ' is less than unity by only 0.05. In the subsequent discussion, we will take $\sigma' \approx 1$,

    (d) *Deformation.* This effect was first suggested by Brix & Kopfermann (25) and further developed by Wilets, Hill & Ford (141). It is a volume effect, arising from an increase in $\langle r^2 \rangle$ with deformation even though nuclear volume is preserved. The deformation dependence is contained in

$$\langle r^2 \rangle_\beta = \langle r^2 \rangle_o \left(1 + \frac{5}{4\pi} \beta^2\right) \qquad 40.$$

Even though $5/4\pi \, \beta^2$ may be small compared with unity ($\beta \lesssim 0.3$), this is to be multiplied by ΔE. For large changes in β, $(\delta E)_\beta$ can be comparable with $(\delta E)_v$; it can be of either sign and can reverse the normal sense of the isotope shift.

(*e*) *Nuclear polarization.* Polarization effects are similar in muonic and electronic atoms, except that the electronic velocities in the neighborhood of the nucleus (essentially *c*) are at least a factor of two greater than muonic or nucleonic velocities. This suggests treating the nuclear motion adiabatically, that is, solving the electron problem as though the protons were fixed (a candid snapshot), and then averaging over the proton positions with respect to the ground-state wavefunction (equivalent to setting nuclear excitation energies equal to zero and performing closure over nuclear coordinates). This is, in effect, the procedure used by Reiner & Wilets (121) and Church & Weneser (42) to calculate the quadrupole polarization for deformed nuclei. They found the ratio of the polarization shift to be down by more than an order of magnitude relative to the deformation shift. This is probably a fair measure of the maximum electron-polarization effect, and is negligibly small. [The calculations by McKinley (108) for muonic atoms are also relevant here.]

The general features of optical isotope shifts.—The finding can be summarized as follows:

(*a*) The mass effects dominate at low Z, thereby limiting the useful observation of nuclear effects to $Z \geq 40$.

(*b*) On the average, the shifts are smaller than that predicted by the $A^{1/3}$ radius law, which indicates that addition of neutrons expands the radius more slowly than addition of neutrons and protons together in the normal mixture. This has been interpreted as due to finite nuclear compressibility (141) and as a nuclear surface effect (144).

(*c*) There are wide deviations from the average shifts attributed to changes in nuclear deformation, or to shell effects (closed-shell nuclei are anomalously small).

(*d*) Odd-A nuclei appear to be smaller than the mean of the neighboring even–even nuclei (even–odd staggering). This is a kind of pairing effect, implying that an odd neutron does not participate as effectively in polarizing and expanding the (proton) core as does a pair of neutrons.

Extraction of Mass and Field Effects

When a string of several (*e-e*) isotopes is available to both electronic and muonic studies, it is fairly feasible to (149): (*a*) extract the electronic specific mass effects; (*b*) determine the electronic normalization factor $C\psi^2(0)$; and (*c*) test the consistency of the field effect. The first two are atomic-structure effects, and provide a valuable test of atomic calculations as well as calibrating isotope-shift data. A failure of consistency in the field effects might imply significant polarization effects.

The normal mass effect can be removed trivially either before or by the following analysis because the normal and specific mass effects have the same A dependence. Thus in the subsequent discussion we can read either *specific* or *total* mass effect when specific is written.

The specific mass shift is given by

$$(\delta_{12}E)_{s.m.} = C'\delta_{12} A^{-1} \qquad\qquad 41.$$

where $C' = \langle f | 1/2M \Sigma_{i \neq j} \mathbf{p}_i \cdot \mathbf{p}_j | i \rangle$. Similarly, for the field effects

$$(\delta_{12}E)_{field} = C\delta_{12}\langle r^2 \rangle \qquad\qquad 42.$$

In atomic isotope-shift studies, several lines may be available, each characterized by different C' and C; but for a given transition, C' and C are sensibly constant over a string of (e-e) isotopes. For two isotopes, we have an energy shift

$$\delta_{12}E_{opt} = C'\delta_{12}A^{-1} + C\delta_{12}\langle r^2 \rangle \qquad\qquad 43.$$

By taking the ratio of differences, the factor C can be eliminated,

$$\frac{\delta_{12}E_{opt} - C'\delta_{12}A^{-1}}{\delta_{34}E_{opt} - C'\delta_{34}A^{-1}} = \frac{\delta_{12}\langle r^2 \rangle}{\delta_{34}\langle r^2 \rangle} \qquad\qquad 44.$$

leaving only C' and the ratio of the $\delta\langle r^2 \rangle$ undetermined.

From muonic isotope shifts, we can obtain the ratio

$$\frac{\delta_{12}E_\mu}{\delta_{34}E_\mu} = \frac{\delta_{12}\langle r^k \rangle}{\delta_{34}\langle r^k \rangle} \qquad\qquad 45.$$

where k is the Ford & Wills effective-moment parameter, Equation 5.

If we could assume that the nuclear charge distribution scaled uniformly from one isotope to the next, then we would be justified in equating the ratios

$$\frac{\delta_{12}\langle r^k \rangle}{\delta_{34}\langle r^k \rangle} \cdot \frac{\delta_{12}\langle r^2 \rangle}{\delta_{34}\langle r^2 \rangle} \qquad\qquad 46.$$

We need to know how accurate this identification is. We can linearize the dependence of the moments on c and t through the relations

$$\delta\langle r^k \rangle = \alpha_k \delta c + \beta_k \delta t \qquad\qquad 47.$$

where c and t are the half-width and surface thickness of Fermi distribution and

$$\alpha_k = \int \frac{\partial\rho}{\partial c} r^k d\mathbf{r}$$

$$\beta_k = \int \frac{\partial\rho}{\partial t} r^k d\mathbf{r} \qquad\qquad 48.$$

Then the desired ratio is

$$\mathcal{R} = \left(\frac{\delta_{12}\langle r^2\rangle}{\delta_{34}\langle r^2\rangle}\right) \Big/ \left(\frac{\delta_{12}\langle r^k\rangle}{\delta_{34}\langle r^k\rangle}\right)$$

$$= \left(\frac{\alpha_2\delta_{12}c + \beta_2\delta_{12}t}{\alpha_2\delta_{34}c + \beta_2\delta_{34}t}\right) \Big/ \left(\frac{\alpha_k\delta_{12}c + \beta_k\delta_{12}t}{\alpha_k\delta_{34}c + \beta_k\delta_{34}t}\right) \qquad 49.$$

It is interesting that $\mathcal{R}$ *is equal to unity* in the following special cases:

1. if $\delta_{12}t = \delta_{34}t = 0$
2. if $\delta_{12}c = \delta_{34}c = 0$
3. if $\delta_{12}t/\delta_{12}c = \delta_{34}t/\delta_{34}c$

Nevertheless, there are cases when $\mathcal{R}$ can deviate somewhat from unity, as we note below.

[In the case of a trapezoidal charge distribution,

$$\beta_k/\alpha_k \approx \frac{25}{192}\frac{t}{c}(k+5) \quad \text{for } t \ll c \qquad 50.$$

This expression can be corrected to a Fermi distribution by multiplication by 1.3 ($Z = 60$, all k).]

From the study of optical isotope shifts, we know that there are appreciable deviations from the radial $A^{1/3}$ law, and that in many cases these are attributable to changes in deformation. In the present context, a change in deformation is like a change in t. Thus we must expect cases where $\beta_2\delta t$ is comparable to $\alpha_2\delta c$, and of either sign. To obtain some feeling for the sensitivity of $\mathcal{R}$ on δc and δt, let us consider $\alpha\delta c \gg \beta\delta t$. Then we find

$$\mathcal{R} \simeq 1 + \left(\frac{\beta_2}{\alpha_2} - \frac{\beta_k}{\alpha_k}\right)\left(\frac{\delta_{12}t}{\delta_{12}c} - \frac{\delta_{34}t}{\delta_{34}c}\right)$$

$$\approx 1 + \frac{25}{190}\frac{t}{c}(2-k)\left(\frac{\delta_{12}t}{\delta_{34}c} - \frac{\delta_{34}t}{\delta_{34}c}\right) \qquad 51.$$

For the case of $Z = 60$ (Nd) [$k = 1.14$, $c = 6$ F, $t = 2$ F and choosing $\delta_{12}c = \delta_{34}c = .03$ F] we find $\mathcal{R} = 1 + 1.25\ (\delta_{12}t - \delta_{34}t) = 1 + 2.5\ (\delta^2 t)/t$. For this example, the relative fluctuations in t are amplified by a factor of 2.5. This is actually not serious so long as $\delta^2 t$ is small. As an extreme example take $\delta_{34}t = 0$. We then have

$$\mathcal{R} = \frac{1 + (\beta_2/\alpha_2)(\delta_{12}t/\delta_{12}c)}{1 + (\beta_k/\alpha_k)(\delta_{12}t/\delta_{12}c)} \qquad 52.$$

In the limit $(\beta/\alpha)(\delta t/\delta c) \gg 1$, we find $\mathcal{R} \to (\beta_2/\alpha_2)/(\beta_k/\alpha_k) \sim 7/(k+5)$. For Nd, where $k = 1.14$, then $\mathcal{R} \to 1.12$.

If $\delta t/\delta c$ is negative, the isotope shifts can be small, and either the elec-

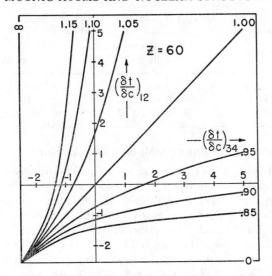

FIG. 15. The ratio $\mathcal{R}$ of Equation 49 is plotted as a function of the two variables $\delta_{12}t/\delta_{12}c$ and $\delta_{34}t/\delta_{34}c$ for a Fermi distribution, $Z = 60$.

tronic or muonic shifts can be zero; very small optical shifts have been observed. In such cases, the deviation of $\mathcal{R}$ from unity could be anomalously large.

In Figure 15, the ratio $\mathcal{R}$ is plotted for $Z \simeq 60$, based on a Fermi distribution. We conclude that $\mathcal{R}$ will usually deviate from unity by less than 10 per cent, *except when the shifts are anomalously small.*

If we tentatively identify the ratios in Equation 46, then we can solve Equation 44 for the specific mass constant C'. If a string of isotopes is available, this can be done for as many *pairs* of *differences* as are available. All should yield the same C'. Any deviation could indicate that the $\delta^2 t$ between isotope pairs is considerable.

A further check on the procedure is available. If several optical lines are available, a C' can be determined for each by the above method. Then if the observed isotope shifts for line a are plotted against the observed shifts for line b, a straight line results, from which a relationship between C_a' and C_b' can be obtained. If we set $\delta A^{-1} = -\delta A/A^2$, this assumes the simple form

$$C_a' - \alpha C_b' + \beta = 0 \qquad\qquad 53.$$

where α is the slope of the line, and β is $A^2/\delta A$ times the intercept of the line with the ordinate. This is known as King's method (97). Where comparisons have been made, the mass constants extracted in this way do satisfy the King relationship.

TABLE VII. CALCIUM ISOTOPE SHIFTS

Isotope pair	Observed muonic K_α shift (keV)	Electronic scattering prediction
40–42	0.69 ± 0.06	0.82 ± 0.20
40–44	0.89 ± 0.05	0.61 ± 0.10
40–48	-0.47 ± 0.12	-0.54 ± 0.08

EXPERIMENTAL MUONIC ISOTOPE SHIFTS

The first isotope shift to receive attention in both electron scattering (86, 113) and muonic studies (22, 44) was Ca in 1964. The ratio of the nuclear charge radii $\langle r_{44}^2 \rangle^{1/2}/\langle r_{40}^2 \rangle^{1/2}$ determined from a combination of these two methods is $1.008 \pm .003$, compared to 1.032 based on the simple $A^{1/3}$ rule. In 1967, Ehrlich et al. (64) measured the isotope shifts among ^{40,42,44,48}Ca with a Ge(Li) detector and obtained 0.69, 0.89, and -0.47 keV as shown in Table VII. Interestingly, the shift between ^{40}Ca and ^{48}Ca is actually inverted as shown in Figure 16 (i.e., $\langle r_{40}^2 \rangle > \langle r_{48}^2 \rangle$). The shell-closure effect has been observed repeatedly whenever added neutrons happened to complete the magic shell of 20, 28, or 50 neutrons. The implication is that the $\langle r^2 \rangle$ of the neutron magic nucleus of mass number A is actually smaller than that of the lighter isotope; this could be due to either an increase in central density or a decrease in the effective surface thickness. The case of $^{136-138}$Ba has been investigated and gives a small and positive shift of $+3.38$ keV (37). The negative shifts of the shell-closure effect observed in the lower shells do not repeat in the shell of $N = 82$. The ratio $\delta E_{exp}/\delta E_{std} = +0.26 \pm .03$ is in excellent agreement with that observed in electron X-ray work $+0.26 \pm 0.06$ (131).

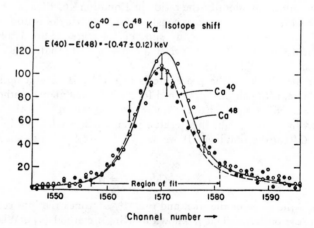

FIG. 16. The muonic K X-ray spectra of ^{40}Ca and ^{48}Ca. Note that the $2p \rightarrow 1s$ transition in ^{48}Ca is higher than that of ^{40}Ca (135).

TABLE VIII. Muonic Isotope Shifts and Electron Scattering Predictions of Same Showing Effect of Shell Closure. A Negative Sign Indicates a Decrease in $\langle r^k \rangle$

	Muonic X rays (keV)	e^- scattering (keV)
$^{40}_{20}Ca_{20}$ $^{44}_{20}Ca_{24}$	$+0.89 \pm 0.05$ (64) $+0.99 \pm .10$ (107)	$+0.61 \pm 0.10$ (113)
$^{40}_{20}Ca_{20}$ $^{48}_{20}Ca_{28}$	-0.47 ± 0.12 (64)	-0.54 ± 0.08 (113
$^{50}_{24}Cr_{26}$ $^{52}_{24}Cr_{28}$	-0.83 ± 0.08 (105)	
$^{52}_{24}Cr_{28}$ $^{53}_{24}Cr_{29}$	$+0.70 \pm 0.08$ (105)	
$^{86}_{38}Sr_{48}$ $^{88}_{38}Sr_{50}$	-1.35 ± 0.12 (65, 66)	
$^{136}_{56}Ba_{80}$ $^{138}_{56}Ba_{82}$	$+3.38 \pm 0.4$ (37, 38)	

A list of measured muonic isotopes shifts with shell-closure effects is shown in Table VIII.

One interesting sequence of isotopes which has been studied is ^{142,143,144,145,146,148,150}Nd. Firstly, the isotopes begin from a nearly spherical shape with neutron magic number ($N = 82$) and proceed to the final member ($N = 90$) which is known to behave as a soft rotor but with the characteristics of permanent deformation. Secondly, the isotope shifts of Nd isotopes have been precisely and extensively studied by the optical method (59, 82, 109, 111), the muonic X-ray method (105–107) and recently, by the electronic X-ray method (23, 24, 128, 129). In this sequence, the shifts due to the normal volume effect are reinforced by the deformation effect; that is, deformation increases with increase of neutron number; the observed isotope shifts are larger than those predicted by the $A^{1/3}$ law. The shifts between two adjacent (even–even) isotopes are generally of the order of tens of keV, and can be easily observed (Figure 17). The line shapes of the odd isotopes are much more complicated than those of the even ones because of the presence of the static quadrupole hyperfine structure. For example, the ground state of ^{143}Nd has spin 7/2 and a positive static quadrupole moment. The interaction of the muon and the nuclear quadrupole moment causes the $2p_{3/2}$ muon state to split into four components (it does not affect the $2p_{1/2}$ state). Below the experimental spectrum of ^{143}Nd in Figure 17, the position of the four hfs

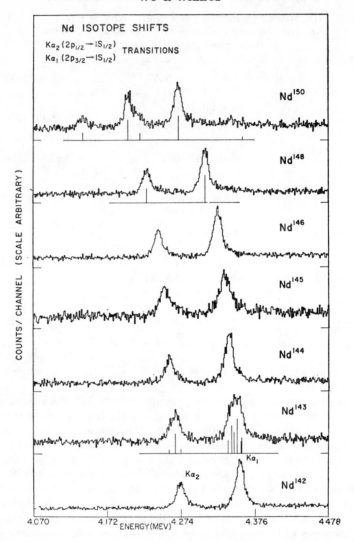

FIG. 17. The muonic K X-ray spectra of Nd isotopes. Note the broadening of the lines in the odd isotopes and the onset of the dynamic hfs in ^{150}Nd (105).

components, as well as the two components due to the isotopic impurities, has been shown schematically. The calculated linewidth agrees roughly with the measured value. The shift between ^{148}Nd and ^{150}Nd is nearly twice as large as that from other (e-e) pairs and the *dynamic* quadrupole hyperfine structure appears suddenly in the deformed ^{150}Nd isotope in contrast to the spectrum of ^{148}Nd.

Isotope shifts in muonic atoms of Nd-isotopes

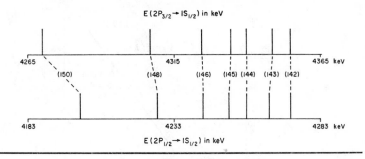

Isotope shifts in optical spectrum of Nd-Isotopes

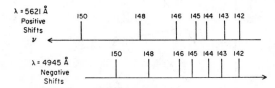

Fig. 18. *Upper plot:* Relative positions of the K_{α_1} and K_{α_2} lines in the Nd isotopes (105). The unusually large isotope shift between 148–150 is due to both the dynamic E2 interaction and the deformation effect (149).

Lower plot: Isotope shifts in the optical spectra of Nd isotopes (82). The large shift between 148–150 is due to the deformation effect.

In Figure 18, the upper portion shows the positions of the observed $K_{\alpha 1}(2p_{3/2}{\rightarrow}1s_{1/2})$ and $K_{\alpha 2}(2p_{1/2}{\rightarrow}1s_{1/2})$ lines from various Nd isotopes. The unusually large difference in the splitting of the $2p_{3/2}$ and $2p_{1/2}$ levels between ^{148}Nd and ^{150}Nd is due mostly to the large difference in the dynamic quadrupole excitation which perturbs the two p levels differently in these two nuclei. However, these perturbations on the various levels of a muonic atom can be calculated using the Wilets (142) or Jacobsohn (91) formulation, with the Chen modification (41). The optical isotope shifts of the lines (5621 Å and 4945 Å) from the Nd isotopes measured by Hansen, Steudel & Walther (82) are reproduced in Figure 18. The general similarity between the muonic and the optical isotope shifts is apparent.

As we have discussed, when a string of several isotopes is available to both electronic and muonic studies, it is fairly feasible to extract the electronic specific mass effects by either equating the ratio Equation 46 or simply plotting δE_{opt} versus δE_{μ}. If a straight line can be obtained, then the intercept on the ordinate gives the specific mass effect or $C'\delta_{12}A^{-1}$. Figure 19 shows two such plots for Nd (optical lines $\lambda = 5262$ Å and $\lambda = 4945$ Å). All the points fall on a smooth straight line for each optical transition line.

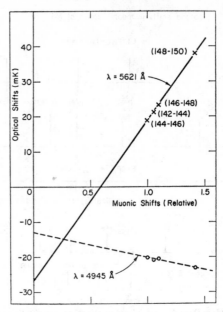

FIG. 19. Muonic isotope shifts (105) plotted against the optical isotope shifts (82) per two neutrons in Nd. The optical data include the total mass effect.

The specific mass effects read from the intercepts give $\Delta E_{\text{s.m.}} \cong -23.55$ mK (mK$\equiv 10^{-3}$cm^{-1}) or $C'_{\text{s.m.}} = -240$ cm^{-1} for $\lambda = 5612$ Å and $\Delta E_{\text{s.m.}} \cong -10.72$ mK or $C'_{\text{s.m.}} = -110$ cm^{-1} for $\lambda = 4945$ Å. Table IX gives the total and the specific mass isotope-shift constants C' and $C'_{\text{s.m.}}$ and $(\Delta E)_{\text{normal mass}}$ in Nd, Mo, and Cr isotope shifts. King's relation for the two lines in optical isotope-shift studies was expressed in Equation 4 of (82) as $C_{4945\,\text{Å}} = C_{5621\,\text{Å}} \times (-0.20) -170$.

The specific mass effect for the line 3283 Å in Sn was obtained by plotting isotope shifts in muonic (65), electronic (23), and optical spectra (127) in traditional Brix-Kopfermann plots and then shifting the optical data vertically as a whole to give a best fit to the muonic or electronic data: C' of (3.3 ± 0.6) mK and (3.0 ± 0.4) mK respectively were obtained.

The general features of the muonic isotope shifts show strong similarities to that exhibited by the optical isotope shifts. On the average, the muonic shifts are also smaller than predicted by the $A^{1/3}$ law and have pronounced deformation-effect and odd–even staggering.

For reference, we have prepared Table X which contains the most recent muonic isotope shifts from $Z = 8$ to 82. For a direct comparison of the isotope-shift results from muonic X rays, electronic X rays, and optical spectra, Table XI has been prepared. It can be seen that the ratios $\delta E_{\text{exp}}/\delta E_{\text{std}}$ between the muonic and electronic results are in good agreement within the

TABLE IX. The Total and Specific Mass Constants
C' and $C'_{\text{s.m.}}$ and $(\Delta E)_{\text{normal mass}}$

		Total mass constant C' in cm^{-1} ($\pm 10\%$)	Specific mass constant $C'_{\text{s.m.}}$ in cm^{-1} ($\pm 10\%$)	$(\Delta E)_{\text{n.m.}}$ in 10^{-3} cm^{-1}
Nd	$\lambda = 5621$ Å	-250	-240	-0.95
	$= 4945$ Å	-121	-110	-1.08
Mo	$\lambda = 5793$ Å	-93.6	-85	-2.0
	$= 6032$ Å	-156.2	-148	-1.9
Cr	$\lambda = 4254$ Å	$+21.48$	$+8.58$	-9.9 (50–52)
		$+22.48$		-9.2 (52–54)
	$= 4274$ Å	$+22.22$	$+9.23$	-9.9 (50–52)
		$+21.22$		-9.2 (52–54)
	$= 4289$ Å	$+22.15$	$+9.16$	-9.9 (50–52)
		$+21.15$		-9.2 (52–54)

experimental uncertainties. This agreement lends support to the identification in Equation 46 of the ratios involving $\delta\langle r^2\rangle$ and $\delta\langle r^k\rangle$.

HYPERFINE STRUCTURE AND THE DYNAMIC
E2 EFFECT

The term hyperfine structure in electronic spectra was bestowed on all interactions involving nuclear structure, such as magnetic dipole, electric quadrupole, and finite size (isotope-shift) effects. In muonic atoms, we have seen that the finite size effect is the most prominent feature of the spectrum. The quadrupole interaction is also greatly enhanced because it contains an r^{-3} dependence. The magnetic dipole also has an r^{-3} dependency but is reduced, relatively, by the appearance of the muon mass in the magnetic moment (Table I). Therefore, the magnetic hfs is generally of only a few keV or less.

Wheeler (140) observed very early that the quadrupole hfs should be comparable to the magnetic fs. Wilets (142) and Jacobsohn (91) suggested that large $E2$ interactions in heavy deformed nuclei could produce an excitation of low-lying rotational states, which are also of comparable energy. This is a dynamic effect which can be depicted as a nuclear tidal wave produced by the orbiting muon. When the dynamic excitation sets in, the static approximation is no longer valid.

TABLE X. Compilation of Measured Isotope Shifts. δE_{std} is the Shift Predicted by the $A^{1/3}$ Law. The Even-Odd Staggering Parameters Are Also Listed

Isotope pair	δE_{field}, keV	$\delta E_{field}/\delta E_{std}$	Ref.
$^{16-18}$O	-0.016 ± 0.02	0.88 ± 0.20	51
$^{28-29}$Si	-0.08 ± 0.04	-0.17 ± 0.09	65, 66
$^{28-30}$Si	0.0 ± 0.04	0.12 ± 0.05	65, 66
$^{39-41}$K	0.35 ± 0.06	0.31 ± 0.04	65, 66
$^{40-42}$Ca	0.69 ± 0.06	0.58 ± 0.04	65, 66
$^{40-44}$Ca	$\{0.89 \pm 0.05$	0.40 ± 0.02	65, 66
	$\{0.94 \pm 0.15$	0.42 ± 0.06	107
$^{40-48}$Ca	-0.47 ± 0.12	-0.02 ± 0.02	65, 66
$^{50-52}$Cr	-0.83 ± 0.08	-0.39 ± 0.04	106
$^{52-54}$Cr	2.17 ± 0.20	1.14 ± 0.01	106
$^{52-53}$Cr	0.70 ± 0.08	0.74 ± 0.08	106
$^{54-56}$Fe	2.97 ± 0.12	0.94 ± 0.05	65, 66
$^{56-57}$Fe	1.25 ± 0.20	0.81 ± 0.12	65, 66
$^{58-60}$Ni	3.14 ± 0.14	0.83 ± 0.04	65, 66
$^{60-61}$Ni	0.97 ± 0.18	0.53 ± 0.09	65, 66
$^{60-62}$Ni	2.00 ± 0.17	0.55 ± 0.04	65, 66
$^{63-65}$Cu	2.31 ± 0.16	0.60 ± 0.04	65, 66
$^{86-87}$Sr	-1.15 ± 0.10	-0.18 ± 0.016	65, 66
$^{86-88}$Sr	-1.35 ± 0.12	-0.192 ± 0.018	65, 66
$^{90-92}$Zr	8.2 ± 0.4	1.14 ± 0.06	65, 66
$^{92-96}$Mo	17.42 ± 0.20	1.30 ± 0.02	106
$^{96-98}$Mo	6.2 ± 0.7	0.83 ± 0.09	33
$^{95-96}$Mo	5.50 ± 0.15	1.64 ± 0.05	106
$^{96-97}$Mo	1.12 ± 0.10	0.33 ± 0.03	106
$^{107-109}$Ag	7.17 ± 0.1	0.78 ± 0.04	32
$^{116-117}$Sn	2.35 ± 0.10	$0.47 \pm .02$	107
$^{116-118}$Sn	$\{6.28 \pm 0.22$	0.630 ± 0.023	65, 66
	$\{6.10 \pm 0.10$	0.61 ± 0.01	107
$^{117-120}$Sn	9.28 ± 0.15	0.62 ± 0.01	65, 66
$^{118-120}$Sn	$\{5.17 \pm 0.10$	0.521 ± 0.014	65, 66
	$\{5.23 \pm 0.08$	0.52 ± 0.01	107
$^{118-119}$Sn	1.84 ± 0.16	0.37 ± 0.03	107
$^{119-120}$Sn	3.42 ± 0.17	0.77 ± 0.03	65, 66
$^{120-122}$Sn	$\{4.95 \pm 0.15$	0.508 ± 0.016	65, 66
	$\{4.44 \pm 0.15$	0.45 ± 0.01	107
$^{122-124}$Sn	4.39 ± 0.10	0.44 ± 0.01	107
$^{142-144}$Nd	16.06 ± 0.20	1.22 ± 0.01	8, 106
$^{144-146}$Nd	15.21 ± 0.20	1.18 ± 0.01	8, 106
$^{146-148}$Nd	16.76 ± 0.20	1.34 ± 0.01	106
$^{148-150}$Nd	21.55 ± 0.25	1.79 ± 0.02	106
$^{142-143}$Nd	6.03 ± 0.30	0.90 ± 0.06	106
$^{144-145}$Nd	5.78 ± 0.50	0.88 ± 0.09	106
$^{162-164}$Dy	10.0 ± 0.50	0.48 ± 0.03	84
$^{168-170}$Er	8.4 ± 0.50	0.38 ± 0.03	84
$^{182-184}$W	9.52 ± 0.40	$0.61 \pm .03$	84
$^{184-186}$W	7.24 ± 0.40	$0.46 \pm .03$	84
$^{206-207}$Pb	3.72 ± 0.32	0.51 ± 0.04	63
$^{206-208}$Pb	9.36 ± 0.30	0.65 ± 0.02	63

TABLE XI. A COMPARISON OF THE RATIO $\delta E_{exp}/\delta E_{std}$ IN ISOTOPE SHIFTS MEASURED BY MUONIC X RAYS, ELECTRONIC X RAYS, AND OPTICAL SPECTRA

Element	Isotope pair	μ X rays	e X rays	Optical (no specific mass correction)
Mo	92–100	$1.06 \pm .01$ (106)	1.02 ± 0.15 (131)	1.21 ± 0.24 (60)
	94–100		1.21 ± 0.31	1.13 ± 0.22 (60)
Sn	116–124	$0.53 \pm .005$ (107)	0.50 ± 0.02 (24)	0.44 (98)
		0.55 (65)		
Ba	136–137	$-0.08 \pm .10$ (37)	-0.01 ± 0.12 (131)	-0.36 (94)
	136–138	$+0.26 \pm 0.03$ (37)	$+0.26 \pm 0.06$ (131)	$+0.15 \pm 0.05$ (94)
Nd	142–144	$1.22 \pm .01$ (106)	1.39 ± 0.07 (24)	
	144–146	$1.18 \pm .01$ (106)	$\{1.29 \pm 0.22$ (131) $\{1.10 \pm 0.15$ (24)	1.30 ± 0.24 (16)
	146–148	$1.34 \pm .01$ (106)	$\{1.34 \pm 0.20$ (131)$\}$ $\{1.30 \pm 0.16$ (24)$\}$	1.56 ± 0.29 (61)
	148–150	$1.79 \pm .02$ (106)	$\{2.27 \pm 0.27$ (131)$\}$ $\{1.91 \pm 0.10$ (24)$\}$	2.15 ± 0.40 (61)
	144–150	$1.38 \pm .01$ (106)	$\{1.65 \pm 0.09$ (131)$\}$ $\{1.66 \pm 0.04$ (24)$\}$	1.67 ± 0.31 (61)
W	182–184	$0.61 \pm .03$ (84)	0.61 ± 0.07 (24)	0.45 ± 0.08 (27)
	184–186	$0.46 \pm .03$ (84)	0.40 ± 0.05 (24)	0.40 ± 0.07 (27)
	182–186	$0.53 \pm .02$ (84)	0.50 ± 0.03 (24)	0.43 ± 0.07 (27)
Pb	206–208	$\{0.65 \pm 0.02$ (70) $\{0.63 \pm 0.01$ (6)	0.51 ± 0.08 (131)$\}$ 0.56 ± 0.08 (36)$\}$	0.64 ± 0.07 (27)

THE STATIC HFS INTERACTION

M1 *interaction.*—The distribution of nuclear magnetization inside of a nucleus is the result of the spin and orbital motions of the nucleons and could, therefore, be quite different from that of charge density. The inclusion of the magnetization distribution in the $M1$ hfs interaction is referred to as the Bohr-Weisskopf effect (38); it leads, in general, to a smaller value of the hfs than that expected for a point nucleus. Furthermore, since various types of nuclear models predict different distributions of the nuclear dipole moment, the experimental measurement of the *B-W* effect should provide a crucial test for various nuclear models.

The *M*1 hfs constant *a* for the $1s_{1/2}$ state of a given particle (e.g., electron or muon) can be expressed as

$$a = (16/3)\pi g_I \mu_a \mu_N \left| \psi_s(0) \right|^2 \qquad \qquad 54.$$

where μ_N and μ_a are the magnetic moments of nucleus and the atomic particle, respectively. For a spin $\frac{1}{2}$ particles $\mu_a = e\hbar/2m_a c$ where m_a is the mass of the atomic particle.

The energy displacement $\Delta W_F{}^{M1}$ due to the magnetic interaction for the state of total angular momentum *F* (**F** = **I** + **J**, **I** and **J** being the total angular momentum for the nucleus and the muon, respectively) is expressed as follows:

$$\Delta W_F{}^{M1} = \frac{1}{2IJ} \left[F(F + I) - I(I + 1) - J(J + 1) \right] A_1 \qquad 55.$$

where A_1 is equal to IJa (*a* is the hfs constant) and is independent of the substate *F*.

The *M*1 hfs can be best studied in the absence of large quadrupole interactions. In particular, it is advantageous to study the hfs of the $2p_{1/2} \to 1s_{1/2}$ transitions ($K\alpha_2$) of a spherical nucleus where the effects of the *E*2 interactions vanish. The ^{209}Bi nucleus is very well suited for the investigation of *M*1 interaction as it can be considered a spherical nucleus. Its spin and moments are given as follows: $I_{g.s.} = 9/2$; $\mu = +4.08$ n.m.; $Q = -0.4$ barn. The A_1 for the $1s_{1/2}$ and the $2p_{1/2}$ states for the point-nucleus, the single-particle, and the configuration-mixing models of ^{209}Bi have been calculated by Le-Bellac (16). Later, Johnson & Sorensen (92) repeated these calculations using a very accurate computer code and introduced some other models. The hfs constant A_1's are tabulated in Table XII. The experimental investigations of the Bi muonic spectrum were carefully carried out in several laboratories (8, 9, 30, 70, 117) and are all in good agreement. The best-determined $2p_{1/2} \to 1s_{1/2}$ line is shown in Figure 20 (30). It can be seen that if A_{1s}/A_{2p} is kept at 1.9, the value of A_{1s} of 2.14 ± 10 keV fits the experimental curve best, which is in better agreement with Johnson & Sorensen's value of 2.09 for configuration mixing.

Very recently, the *M*1 hfs interaction in ^{115}In, ^{133}Cs (103), and ^{141}Pr (103, 136) have also been carefully investigated by analyzing the line shapes of ($2p_{3/2} \to 1s_{1/2}$), ($3d_{3/2} \to 2p_{1/2}$), and of course ($2p_{1/2} \to 1s_{1/2}$). The $(A_1)_{1s}$ and $(A_1)_{2p}$ thus determined are also compared with Johnson's calculations (93) in Table XII. The agreement is satisfactory.

*E*2 *interaction.*—The displacements due to electric quadrupole interaction are given by

$$\triangle W_F{}^{E2} = \frac{\left[6[K(K + 1) - 4/3(I + 1)J(J +)I] \right]}{\left[2I(I + 1)2J(J + 1) \right]} A_2(n, I, J) \qquad 56.$$

TABLE XII. THE MAGNETIC HYPERFINE-SPLITTING CONSTANTS $(A_1)_{1s_{1/2}}$ AND $(A_1)_{2p_{1/2}}$ IN SEVERAL NUCLEI

Nucleus	A_1	Point	Single particle	Arima-Horie	BCS+δ	$G_R=0$	$G_R=Z/A$	Experimental values
$^{93}_{41}$Nb $\left(\begin{array}{c}\mu=6.1671\\I\pi=9/2^+\end{array}\right)$	$1s_{1/2}$	2.32	1.65	1.51	1.51	1.51	1.58	
	$2p_{1/2}$	.436	.368	.335	.335	.335	.350	
$^{115}_{49}$In $\left(\begin{array}{c}5.5351\\9/2^+\end{array}\right)$	$1s_{1/2}$	2.53	1.88	1.56	1.56	1.56	1.84	1.65 $\pm$ .07 (103)
	$2p_{1/2}$	.634	.555	.456	.457	.458	.537	0.55 $\pm$ .13 (103)
$^{127}_{53}$I $\left(\begin{array}{c}2.8091\\5/2^+\end{array}\right)$	$1s_{1/2}$	1.40	1.50	.950	.940	.928	1.11	
	$2p_{1/2}$	.399	.482	.298	.296	.293	.354	
$^{133}_{55}$Cs $\left(\begin{array}{c}2.5789\\7/2^+\end{array}\right)$	$1s_{1/2}$	1.34	.909	1.21	1.15	1.15	1.19	1.11 $\pm$ .18 (103)
	$2p_{1/2}$	.405	.277	.379	.363	.361	.374	.50 $\pm$ .20 (103)
$^{139}_{57}$La $\left(\begin{array}{c}2.7781\\7/2^+\end{array}\right)$	$1s_{1/2}$	1.52	.944	1.26	1.26	1.26	1.26	
	$2p_{1/2}$	.481	.304	.419	.419	.418	.419	
$^{141}_{59}$Pr $\left(\begin{array}{c}4.28\\5/2^+\end{array}\right)$	$1s_{1/2}$	2.41	1.62	1.47	1.47	1.47	1.47	1.53 $\pm$0.15 (103) 1.52 $\pm$0.07 (136)
	$2p_{1/2}$	0.881	.600	.541	.542	.542	.542	0.62 $\pm$ .20 (103)

TABLE XII—(*Continued*)

Nucleus	A_1	Point	Single particle	Arima-Horie	BCS+δ	$G_R=0$	$G_R=Z/A$	Experimental values
$^{151}_{63}$Eu $\left(\dfrac{3.463}{5/2^+}\right)$	$1s_{1/2}$	2.04	1.64	1.13	1.18	1.21	1.41	0.80 ± .27 (30)
	$2p_{1/2}$	.772	.668	.432	.455	.490	.576	
$^{203}_{81}$Tl $\left(\dfrac{1.61169}{1/2^+}\right)$	$1s_{1/2}$	1.19	1.16	.763	.781	.699	.819	.665±0.75 (10)
	$2p_{1/2}$	.635	.617	.405	.414	.370	.433	
$^{205}_{81}$Tl $\left(\dfrac{1.62754}{1/2^+}\right)$	$1s_{1/2}$	1.19	1.16	.771	.788	.720	.819	.580±0.015 (10)
	$2p_{1/2}$	.635	.617	.409	.418	.382	.437	
$^{209}_{83}$Bi $\left(\dfrac{4.09}{9/2-}\right)$	$1s_{1/2}$	3.02	1.32	1.63⎫	By LeBellac (16)			2.0 ±0.2 (93)
	$2p_{1/2}$	1.62	0.71	0.84⎭				
	$1s_{1/2}$	2.97	1.66	2.09⎫	Johnson & Sorensen (92)			1.1 (93)
	$2p_{1/2}$	1.67	.903	1.14⎭				

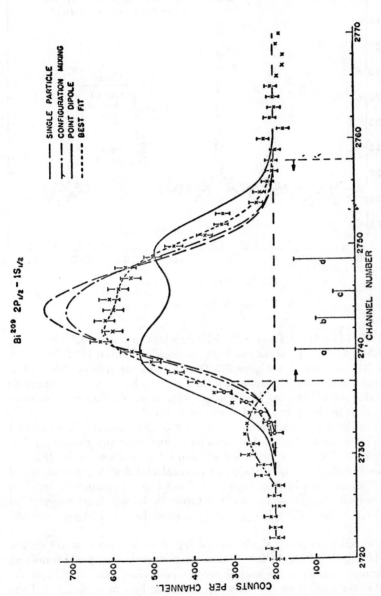

FIG. 20. Muonic transitions $2p_{1/2} \rightarrow 1s_{1/2}$ in ^{209}Bi. The hfs is purely magnetic dipole; predictions of different nuclear models are shown (30).

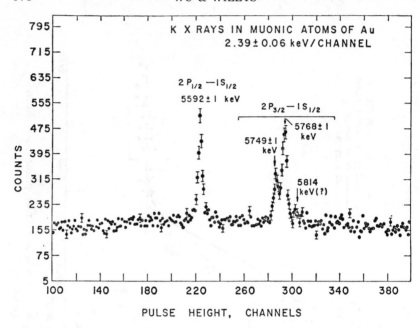

FIG. 21. K X rays in muonic atoms of Au (47).

where $K = F(F+1) - I(I+1) - J(J+1)$. A_2 is hfs constant for $E2$ interaction. The static $E2$ hfs spectrum has been completely resolved in K X rays of ^{197}Au (47) as shown in Figure 21. The ground-state spin of ^{197}Au is 3/2, its measured quadrupole moment is only $+0.596$ barns, and the magnetic moment is $+0.14485$ n.m. The $2p_{3/2}{\rightarrow}1s_{1/2}$ transition should consist of three components in the ratio $5:10:1$ and is so observed.

When the $M1$ hfs interaction is comparable to that of $E2$, then both $E2$ and $M1$ hfs must be taken into account in analyzing the $2p_{3/2}{\rightarrow}1s_{1/2}$ line. Figure 22 illustrates the theoretical $M1$ and $E2$ hfs splitting of the $2p_{3/2}$ and $1s_{1/2}$ levels and for the $2p_{3/2}{\rightarrow}1s_{1/2}$ transition calculated with values of A_1 and A_2 of ^{209}Bi as given in (16). Figure 23 shows a direct comparison between the observed $2p_{3/2}{\rightarrow}1s_{1/2}$ line and the theoretically calculated line of ^{209}Bi with $(A_1)_{1s} = 2.09$ keV, $(A_1)_{2p_{1/2}} = (A_1)_{2p_{3/2}} = 0.66$ keV, and $(A_2)_{2p_{1/2}} = -3.85$ keV.

However, whenever a significant mixing of nuclear-muon states occurs, the relative intensities of the hyperfine components in the $2p{\rightarrow}1s$ transition will be modified. Since the nuclear muon mixing does occur in ^{209}Bi as discussed in the section on Intensity Anomaly, the best fit is obtained if the $F=6$ to $F=5$ component of the $2p_{3/2}{\rightarrow}1s_{1/2}$ transition (designated b in the figure) is reduced in intensity by 30 per cent.

When the quadrupole moment becomes large, then the dynamic $E2$

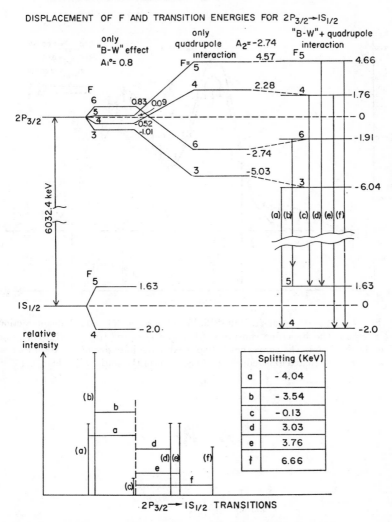

FIG. 22. An illustration of the theoretical $M1$ and $E2$ hfs splitting of the $2p_{3/2}$ and $1s_{1/2}$ levels and for the $2p_{3/2} \rightarrow 1s_{1/2}$ transition calculated with values of A_1 and A_2 as given in (16).

interaction sets in. The pattern of the dynamic hfs spectrum is totally different from that of the static $E2$ interaction. (See Figure 27.)

THE DYNAMIC HFS INTERACTION

Theory.—The largest effects occur for the $2p$ levels. The unperturbed spectrum for ^{182}W is shown in Figure 24. The total angular momentum of

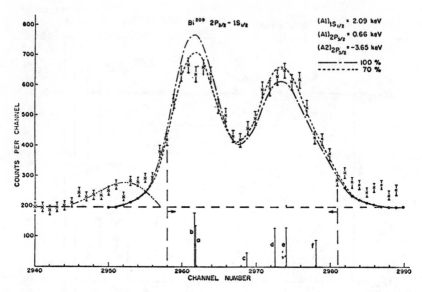

Fig. 23. $2p_{3/2} \rightarrow 1s_{1/2}$ transition in ^{209}Bi. Both the static $M1$ and $E2$ hfs interactions are present. The hyperfine constants $(A_1)_{1s_{1/2}}$, $(A_1)_{2p_{3/2}} = (A_1)_{2p_{1/2}}$ and $(A_2)_{2p_{3/2}}$ are given in the figure. The agreement is improved if the intensity of the $F=6$ to $F=5$ component (designated b) is reduced to 70 per cent of the statistical value (30).

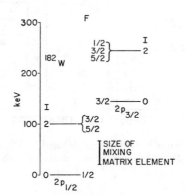

Fig. 24. The unperturbed nuclear-muonic energy levels of ^{182}W. The $I=2$ levels will be split by the "static" quadrupole interaction. Levels of the same F will be admixed and shifted by the "dynamic" $E2$ interaction. The mixing matrix elements are roughly 47 keV.

the system F is composed of the nuclear I and the muonic j. The F degeneracy $(I = 2)$ is removed by the static (diagonal) quadrupole interaction. Those levels of the same F are admixed (and mutually repelled) by the nondiagonal quadrupole interaction. The X-ray transitions are dominated by the muon, to the extent that transitions receive strength essentially only between components of the wavefunction with the same nuclear state. Thus if we assume for the moment that higher n states are pure, only states with $I = 0$ (for even–even nuclei) components will be populated, namely $F = 1/2$ and $3/2$. In fact, dynamic mixing in the $n = 3$ state must also be included to obtain the observed intensities.

Since $2p$ states with $I = 2$ (as well as $I = 0$) components are populated, transitions to the muonic $1s_{1/2}$ state can proceed either to the nuclear ground state $(I = 0)$ or to the first excited state $(I = 2)$. In this way, the muonic cascade can leave the nucleus excited.

For odd-A nuclei, the considerations are similar, but the spectrum can become considerably more complex.

The quadrupole interaction operator is

$$H_Q' = - e^2 \frac{r_>^2}{r_>^3} \rho(\mathbf{r}_N) P_2 (\cos \theta_{N\mu}) \qquad 57.$$

where ρ is the nuclear charge-density operator and $r_<$ and $r_>$ are the lesser and greater of r_μ and r_N.

The required matrix elements are

$$\langle IjF \mid H_Q' \mid I'j'F \rangle = - \frac{e^2}{2}\left(\frac{4\pi}{5}\right)^{1/2} (-1)^{I+j'-F} W(IjI'j'; F2)$$

$$\cdot \langle I \| \bar{Q}_{jj'} \| I' \rangle \langle j \| r^{-3} Y^{(2)} \| j' \rangle \qquad 58.$$

where $\bar{Q}_{jj'}$ is almost, but not quite, the quadrupole moment operator for the nucleus. If the muon wavefunction were excluded from the nucleus, the potential generated by it would have a quadrupole component proportional to r^2. The p wavefunction is proportional to r, for small r, and the density to r^2, giving an r^4 component. This gives for the radial dependence of $r_<^2/r_>^3$

$$\int_0^\infty \mathcal{R}_j(r_\mu) \frac{r_<^2}{r_>^3} \mathcal{R}_{j'}(r_\mu) dr_\mu = a_2 r^2 - a_4 r^4 + \cdots \qquad 59.$$

where

$$a_2 = \int_0^\infty \mathcal{R}_j \mathcal{R}_{j'} r_\mu^{-3} dr_\mu \approx \frac{Z^3}{24 a_\mu^3}$$

$$a_4 = \frac{5}{14}\left[\mathcal{R}_j \mathcal{R}_{j'} r_\mu^{-4} \right]_{r_\mu = 0} \approx \frac{5 Z^5}{14 \cdot 24 a_\mu^5} \cdot \qquad 60.$$

Here $\mathfrak{R}_j$ is the radial function $\propto r_\mu{}^{l+1}$. The final approximations are for $2p$ hydrogen functions $(a_\mu = 256\ F)$. This permits us to write the effective quadrupole operator

$$\tilde{Q} = Q\left(1 - \frac{a_4}{a_2}r^2 + \cdots\right) = Q\left[1 - \frac{5}{14}\left(\frac{Z}{a_\mu}\right)^2 r^2 + \cdots\right] \qquad 61.$$

If we ignore the j and j' dependence in the above, there are two nuclear matrix elements to be considered, $\langle 2\|\tilde{Q}\|0\rangle$ and $\langle 2\|\tilde{Q}\|2\rangle$. In the Bohr-Mottelson model $(K=0)$, they are related to each other and to the intrinsic quadrupole moment $\tilde{Q}_0$ by simple Clebsch-Gordan coefficients. The analysis is usually made in terms of the intrinsic quadrupole moment Q_0 (or in the present context $\tilde{Q}_0$). Although this is model dependent, we can use the correspondence to *define* an *effective* intrinsic moment for each of the transitions involved, $\tilde{Q}_0(0, 2)$ and $\tilde{Q}_0$ $(2, 2)$. These would be equal in the Bohr-Mottelson model.

An important correction to the above development has been given by Chen (41) as described on pp. 557–58.

Experimental results.—A diagram showing splittings of the $2p$ levels and the resultant hfs spectrum of $2p{\to}1s$ transitions of the even–even nucleus ^{182}W is shown in Figure 25. For comparison, the observed hfs spectrum of ^{182}W is displayed at the bottom. With high-resolution Ge(Li) detectors and sufficient counting statistics, the more intense transitions in a dynamic $E2$ hfs are usually completely resolved as shown in Figure 26.

An interesting suggestion of evidence for the dynamic $E2$ was obtained by Belovitskii (15) in 1961 by observing the tracks of the internal-conversion electrons in a nuclear emulsion loaded with ^{238}U. The excited nuclear level (45 keV), which decays mostly by internal conversion, must have been excited by the $2p{\to}1s$ transition. The probability of being excited to the first rotational state in ^{238}U was estimated to be $\sim$50 per cent, very close to the calculated value of 0.56.

Subsequent investigations of $E2$ hfs by Ehrlich et al. (62), Acker et al. (1), Backenstoss et al. (7), and Raboy et al. (119) gave the general outline of this complex spectrum; and several important conclusions were reached from these [pre-Ge(Li)] studies:

1. The intrinsic quadrupole moments of the excited states Q_0 (2, 2) of even–even nuclei were positive ($\langle 2\|Q\|2\rangle$ negative), as predicted from the Bohr-Mottelson model.

2. In ^{238}U, where the $E2$ interaction energy is large in comparison to the low-lying nuclear energies, it was also necessary to take into account mixing in the $3d$ states and to include the $I=4$ nuclear excitation.

Use of Ge(Li) detectors to study $E2$ hfs was initiated in 1965 by Anderson et al. (3) on Th, and Acker et al. (1) on ^{238}U and natural W. The Columbia and Carnegie groups soon joined. The Columbia group (84) studied

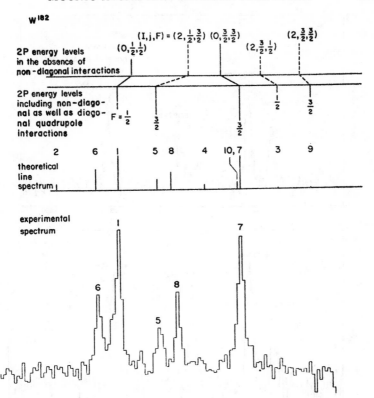

FIG. 25. The effect of the dynamic $E2$ interaction in ^{182}W. The upper lines are the energy levels including diagonal, but neglecting nondiagonal, interactions. Immediately below are the fully perturbed levels. The length of the lines gives the theoretical populations of the states. (The broken lines are unpopulated, the length is just proportional to $2F+1$.)

The theoretical and experimental K X-ray spectra are compared below (148).

separated even–even deformed nuclei in the rare-earth region extensively, and the Carnegie group (132) concentrated on the odd nuclei and on ^{238}U and Th. The hfs of ^{238}U and ^{232}Th has also been studied in detail by the Chicago group (107a). Figure 26 shows the hfs spectra of the three even isotopes of ^{182,184,186}W. Five intense lines (lines 1, 5, 6, 7, 8) are clearly exhibited. The isotopic variations among the three spectra are interesting. The spacings between 5–6 and 7–8 should *approximately* represent the first rotational energies, which are 100, 111, and 123 keV in ^{182}W, ^{184}W, and ^{186}W respectively. (See however, the next section on the isomer effect.)

Figure 27 shows the K X-ray hfs spectra (30) of the odd nuclei ^{151}Eu

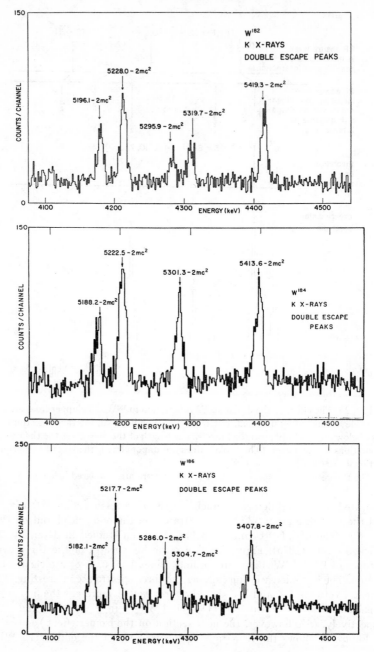

FIG. 26. K X-ray spectra of three W isotopes exhibiting the dynamic $E2$ effect (84).

and ^{153}Eu which were measured with a Ge(Li) detector with a resolution of 5 keV. The striking contrast in complexity between these two spectra illustrates the onset of permanent deformation in $^{153}_{63}$Eu$_{90}$ and thus the mixing of states. The same phenomenon was observed between $^{148}_{60}$Nd$_{88}$ and $^{150}_{60}$Nd$_{90}$ (Figure 17).

These more precise studies permit a determination of the two parameters $\tilde{Q}_0(0, 2)$ and $\tilde{Q}_0(2, 2)$. Although each of these depends upon the j and j' of the

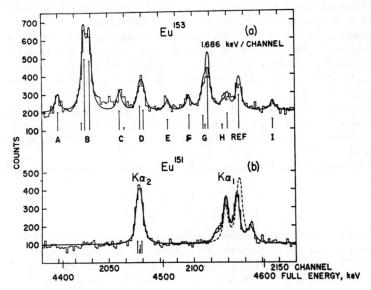

FIG. 27. K X-ray spectra of the odd-A nuclides ^{151}Eu and ^{153}Eu (30).

muon, such refinements can barely be extracted from the data. The following further conclusions can be drawn:

3. The ratio

$$\frac{\tilde{Q}_0(2, 2)}{\tilde{Q}_0(0, 2)} \qquad\qquad 62.$$

does not deviate from unity within experimental limits (of several per cent) in strongly deformed nuclei. This is in agreement with the Bohr-Mottelson strong coupling ($K=0$) prediction. Deviations could arise from a variety of causes in less strongly deformed, or vibrational, nuclei.

4. If a value is assumed for the ratio (say 1.00), then the mean $\tilde{Q}_0$ is determined more precisely than the ratio of the $\tilde{Q}_0$. Attempts have been made to extract from $\tilde{Q}_0$ an intrinsic deformed charge distribution utilizing the full K and L X-ray energies. Such a description presupposes the Bohr-Mottelson model. Several density distributions were introduced by Acker et al. (2), and the Columbia group (84) introduced the rather flexible form

$$\rho(r) = \rho_0 \left[1 + \exp \frac{r - c(1 + \beta Y_{20})}{a(1 + \beta' Y_{20})} \right]^{-1} \qquad 63.$$

where β' Y_{20} gives the angular variation in the surface thickness. If the $2p$ and $3d$ energy splitting are calculated by diagonalizing the quadrupole interaction between the lowest rotational band and the $2p$ and $3d$ fine-structure doublets, it is in general necessary to introduce large negative values of the β' parameters (skin thickness greater at the nuclear "equator") to obtain a quadrupole moment in agreement with the $Q_0(0.2)$ values obtained from Coulomb excitation. When the quadrupole interaction is renormalized to include polarization effects as described by Chen (40) (see p. 558), good agreement with Coulomb excitation results is obtained with a uniform skin thickness ($\beta' \equiv 0$). See Table XIII (85) where the hyperfine results of deformed nuclei are summarized.

The polarization effect is also important in determining the mean surface thickness (t or a). If no correction to the $2p$ levels is applied, the resultant t is unrealistically small. A correction of $\sim$2 keV (see pp. 557–58) returns t to a value of 2–2.5 F.

5. The data are not sufficiently precise to determine intrinsic Y_{40} moments.

THE ISOMER EFFECT

HISTORICAL BACKGROUND

In the dynamic $E2$ interaction between a muon and a strongly deformed nucleus, it was shown that the first rotational levels can be excited. The probability that the nucleus will be left in the $I = 2^+$ state by the dynamic $E2$ excitation depends on the $E2$ interaction energy, $2p$ fine structure, and the rotational energy. For example, it is 0.30 for muonic atoms of ^{152}Sm and 0.56 for ^{238}U. The subsequent deexcitation of the nuclear state occurs in $\sim$10^{-9} sec or less, and in the presence of the $1s$ muon, since the muon capture lifetime from the $1s$ state is $\sim$10^{-6} sec in the rare-earth region.

Devons (55) suggested in 1963 that the presence of the spherically symmetric field of the $1s$ muon in the vicinity of the nucleus may affect the deformation of the nucleus, and consequently its rotational energy, in some measurable way. At this suggestion, Wilets & Chinn (146) calculated the energy shift of the rotational γ line in a deformed nucleus due to the polarization effect of the $1s$ muon and obtained shifts of possibly up to (but not greater than) 1 keV. A similar, independent calculation was carried out by Hüffner (88) .Although this shift occurs also in the muonic K X-ray spectrum, it is more difficult to measure a shift of $\sim$1 keV in the 5-MeV region than in the $\sim$100-keV nuclear γ-ray region.

EXPERIMENTAL INVESTIGATIONS

For reliable measurement of such a small energy shift of the nuclear gamma ray with and without the influence of a muon, the Columbia group

TABLE XIII. Summary of $E2$ Hyperfine Results on
Even–Even Deformed Nuclei

| | Parameters of the charge distribution of deformed nuclei | | | | | | | | | |
| | without nuclear polarization and Lamb-shift corrections | | | | | with nuclear polarization (41) and Lamb-shift (13) corrections, $\beta'=0$ | | | | |
	c, F	t, F	β	β'	Q_0, b	c, F	t, F	β	Q_0, b	(Q_0)C.E., b
(a) Columbia (84)						**(a′) (85)**				
150Nd	6.083	1.68±.02	0.274	0	5.27	5.87	2.34	0.278	5.15	5.17±.12
152Sm	6.093	1.77±.02	0.302	−0.22	5.85	5.90	2.36	0.296	5.78	5.85±.15
162Dy	6.26	1.59±.02	0.337	−0.28	7.38	6.01	2.40	0.338	7.36	7.12±.12
164Dy	6.34	1.30±.03	0.329	−0.12	7.53	6.11	2.19	0.334	7.42	7.50±.20
168Er	6.34	1.51±.02	0.354	−0.92	7.77	6.17	2.18	0.333	7.77	7.66±.15
170Er	6.42	1.26±.03	0.341	−0.87	7.80	6.27	1.94	0.326	7.75	7.45±.13
182W	6.47	1.85±.02	0.272	−0.51	6.57	6.41	2.12	0.248	6.56	6.58±.06
184W	6.49	1.84±.03	0.269	−0.78	6.19	6.42	2.17	0.237	6.27	6.21±.06
186W	6.55	1.75±.02	0.243	−0.54	6.01	6.46	2.10	0.222	5.90	5.93±.05
(b) CERN (58)										
159Tb	6.2	1.5±0.4	0.30						7.6±0.4	
165Ho	6.27	1.5±0.4	0.30						7.9±0.5	
181Ta	6.51	1.5±0.4	0.25						7.5±0.4	
232Th	7.10	1.49±0.14	0.23						9.8±0.3	
233U	7.11	1.50±0.5	0.24						10.3±0.3	
235U	7.14	1.44±0.17	0.241						10.6±0.2	
238U	7.15	1.46±0.12	0.253						11.25±0.15	
239Pu	7.18	1.35±0.3	0.26						12.0±0.3	
(c) Carnegie-Mellon-Argonne-Binghamton (30)										
151Eu	6.27	1.3±0.4							2.75	
153Eu	6.06	1.78±0.4							7.01±0.10	
158Gd	6.33	0.714							7.15	
160Gd	6.35	0.57							7.38	
175Lu	6.24	2.07±0.3							8.02±0.12	
232Th	7.07	1.75							11.3	
238U	7.10	1.75							11.3	

Remarks: Argument of the Fermi charge-distribution function:

In (a)   (a′)

$$\frac{r-c[1+\beta Y_{20}(\theta)]}{a[1+\beta' Y_{20}(\theta)]}; \qquad \frac{r-c[1+\beta Y_{20}(\theta)]}{a}.$$

In (b) and (c)

$$\frac{r[1+\beta Y_{20}(\theta)]-c}{a}.$$

(18, 19) compared directly and simultaneously the energy of the nuclear gamma ray which is in coincidence with a stopped muon with that of the unshifted gamma ray following $\beta^{\pm}$ decay from a radioactive $Z\pm1$ nucleus in the same detecting system and under exactly the same conditions. Refer here to Figure 3. Pulses corresponding to the two γ rays were stored separately in designated parts of the computer according to the tagging signals: The nuclear gamma rays from the muonic atoms were gated by the muon stopping signals (123$\overline{4}$); those from a radioactive source were gated by the coincidence of the cascade nuclear gamma rays $(4^+\xrightarrow{\gamma_2}2^+\xrightarrow{\gamma_1}0^+)$ with the aid of a large NaI detector. Where no radioactive source for direct comparison was available, the deexcitation line in the muonic atom was measured with respect to several well-known calibration lines. The energy of the unshifted line, produced by Coulomb excitation, was subsequently determined by using the same set of calibration lines. Any possible shifts were repeatedly checked for by comparing the atomic K X rays of Pb from a Pb target in the muonic spectrum with the K X rays of Pb from ^{207}Bi (K capture) in the calibration spectrum. No shifts were found within the experimental uncertainty (± 20 eV). The first gamma-ray energy shift observed was in ^{152}Sm (122 keV) (18). The corresponding radioactive source was ^{152}Eu which decays by β^- to ^{152*}Sm

$$^{152}\text{Eu} \rightarrow {}^{152*}\text{Sm} + \beta^- + \nu$$
$$\xrightarrow{\quad\quad} {}^{152}\text{Sm} + \gamma(122 \text{ keV})$$

An energy shift of $(+0.56 \pm 0.06)$ keV (19) was observed. The $+$ sign implies that the γ-ray energy from the muonic atom ^{152}Sm is higher than that from the radioactive source ^{152}Eu.

At that time, the unusually large increase in nuclear radius from the ground state to the first rotational level in ^{152}Sm was observed by Grodzins et al. (150) using the Mössbauer isomer effect. From the large difference in the radii, Telegdi (133) pointed out that the isomer effect alone could account for the energy shift observed in the muonic Sm, without invoking the polarization effect. The polarization effect suggested by Devons was nearly zero because of the similarity of the rotational-band members. In fact, it follows from elementary quantum-mechanical considerations (18, 134) that the observed shift is necessarily a first-order perturbation effect and should be attributed to a change in radius of the *unperturbed* nucleus in the $I = 2^+$ excited state. This has been conventionally called an isomeric shift, analogous to the isotope shift.

Following Columbia's first work on ^{152}Sm, CERN (95) reported the observation of the gamma-ray energy shift in ^{181}Ta (136 keV). Since then, the energy shifts of gamma rays in ^{150}Nd, ^{154}Gd, ^{166}Er, ^{182}W, ^{184}W, ^{186}W, ^{188}Os, ^{192}Os have been measured (10, 19); the typical nuclear gamma shifts in muonic ^{152}Sm, ^{184}W, and Os isotopes are shown in Figures 28 and 29. The measured gamma-ray energy shifts are summarized in Table XIV; the agreement between Columbia's and CERN's results is excellent. It is to be especially noted that several of these measured shifts have negative signs. Negative shifts imply that the effective radius of the excited state is smaller than that of ground state.

THEORETICAL CONSIDERATIONS

Comparison of Mössbauer and muonic shifts.—Isomeric shifts are quite analogous to isotope shifts. Although a few such shifts have been observed

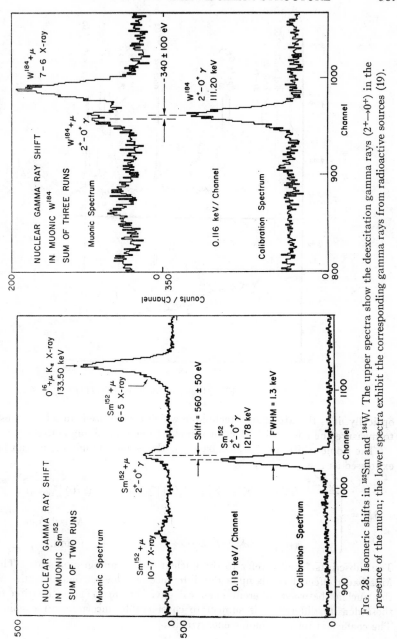

FIG. 28. Isomeric shifts in ^{152}Sm and ^{184}W. The upper spectra show the deexcitation gamma rays ($2^+ \rightarrow 0^+$) in the presence of the muon; the lower spectra exhibit the corresponding gamma rays from radioactive sources (19).

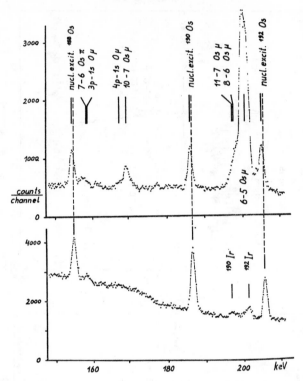

FIG. 29. Same as Fig. 24, except for ^{188,190,192}Os (13).

optically, it is the Mössbauer technique which can be used (in select cases) to measure the effect with electrons in short-lived, low-lying states.

The isomer shifts in the Mössbauer methods and muonic atoms can be expressed as

$$\delta E = \begin{cases} \dfrac{2\pi Ze^2}{5}\,(\langle r^2\rangle_* - \langle r^2\rangle_0)\sum_e\,(\,|\,\psi_e(0)\,|^2{}_{\text{absorber}} - \,|\,\psi_e(0)\,|^2{}_{\text{source}}) & \\[4pt] & \text{Mössbauer} \qquad 64. \\[8pt] (\langle r^k\rangle_* - \langle r^k\rangle_0)\delta E/\delta\langle r^k\rangle & \text{muon.} \end{cases}$$

The subscripts * and 0 refer to the nuclear excited and ground states. The Mössbauer effect depends upon the difference in the electronic densities at the nucleus between different chemical environments. The evaluation of these density differences is subject to considerable uncertainty for solids. The coefficient for the muonic energy shift,

$$\delta E/\delta\langle r^k\rangle = -\,\frac{R_k}{k\langle r^k\rangle}\,C_z \qquad\qquad 65.$$

VII, THE BRACKETED VALUES INDICATE RANGE OF VARIATION DUE TO ASSUMING INTERNAL CONVERSION IN EITHER N OR O SHELLS

I Isotope	II $E(2^+ \to 0^+)$ (keV)	III Comparison source	IV Exp. energy shift δE_{obs}, eV	V Center of gravity shift (77) $\delta E_{c.g.}$, eV	VI δE_{isomer} eV	VII $\dfrac{\delta \langle r^2 \rangle}{\langle r^2 \rangle} \times 10^4$ Muonic	Mössbauer	Marshalek (106)
$^{150}_{60}$Nd	130.17	Coulomb excitation (78)	$+570 \pm 120$ (19)	-270	$+840 \pm 120$	$+5.8 \pm 0.8$	No measurement	
$^{152}_{62}$Sm	121.78	^{152}Eu(β^+, E.C.)	$+560 \pm 60$ (19) $+500 \pm 40$ (12)	-360 -240	$+920 \pm 70$ $+800 \pm 70$	$+5.9 \pm 0.4$ $+5.1 \pm 0.4$	$+3.7$ (79) (80)	$+19.7$
$^{154}_{64}$Gd	123.07	^{154}Eu(β^-)	$+670 \pm 150$ (19)	-310	$+980 \pm 150$	$+5.9 \pm 0.8$	$+7.5 \pm 2.3$ (122)	$+8.8$
$^{166}_{66}$Er	80.56	^{166}Ho(β^-)	-350 ± 150 (19)	-320	-30 ± 150	-0.16 ± 0.8	No measurement	$+0.51$
$^{182}_{74}$W	100.10	^{182}Ta(β^-)	-320 ± 100 (19) -290 ± 90 (12)	-290 -240	-30 ± 100 -80 ± 100	-0.13 ± 0.5 -0.33 ± 0.5	$+0.16 \pm 0.05$ (122)	$+2.45$
$^{184}_{74}$W	111.20	^{184}Re (E.C.)	-340 ± 100 (19) -350 ± 50 (12)	-314 -250	-25 ± 100 -90 ± 100	-0.11 ± 0.5 -0.40 ± 0.5	No measurement	$+2.88$
$^{186}_{74}$W	122.57	^{186}Re (E.C.)	-350 ± 100 (19) -400 ± 40 (12)	-340 -250	-10 ± 100 -100 ± 100	-0.04 ± 0.5 -0.50 ± 0.5	No measurement	$+5.5$
$^{188}_{76}$Os	155.0	^{188}Re(β^-)	-400 ± 40 (12)	-210	-190	-0.8		
$^{190}_{76}$Os	186.7	^{190}Ir (E.C.)	-470 ± 40 (12)	-110	-360	-1.4		
$^{192}_{76}$Os	205.79	^{192}Ir (E.C.)	-610 ± 50 (12)	-96	-514	-2.0		

can be determined quite precisely. As in the case of isotope shifts, however, the Mössbauer effect measures $\langle r^2 \rangle$ and the muon measures $\langle r^k \rangle$.

Nuclear polarization.—As in the case of isotope shifts, nuclear polarization (35, 77) enters only through the *change* in the polarization energy between the two states. The effects are probably even smaller for isomer shifts than for isotope shifts for the following reasons. For the cases in question, the excited state is a member of the same rotational band. In first approximation, the polarization shift is the same for both levels. A slight difference between the 0^+ and 2^+ state occurs because of Coriolis band mixing in the 2^+ state. Interesting effects lurk here, but both theory and experiment must be improved to explore them. It has been assumed in analyses to date that the change in nuclear polarization is zero.

Shift of center of gravity of the magnetic hfs.—In the even–even deformed nuclei, the magnetic dipole interaction between the $1s$ muon and the excited nucleus splits the excited nuclear state $I = 2^+$ into a hyperfine doublet with $F = 5/2^+$ and $3/2^+$. The $F = 5/2^+$ members of the doublet are not populated statistically [i.e., proportional to $(2F+1)$]. Their feeding depends on the mechanism by which these rotational states are excited and varies from nucleus to nucleus. The calculated ratios of feedings to the two hyperfine states $(F = 5/2 : F = 3/2)$ are more than the statistic ratio $3:2$, which implies a shift of center of gravity towards higher energy.

A further effect was pointed out by Gal, Grodzins & Hüfner (77) and by Daniel (52): The energy separation between the $F = 5/2^+$ and $F = 3/2^+$ states is ~ 600 eV; therefore, it is possible for the $F = 5/2$ state to decay, mainly through an $M1$ transition, to the $F = 3/2$ state, via a strong Auger emission from the N or O electron shells. [A similar effect of a strong inter-doublet transition has been shown by Winston & Telegdi (145) to be important for μ^- capture.] This intense $M1$ interdoublet transition lowers the observed γ-ray energy, since most of the nuclear deexcitation takes place now from the $F = 3/2$ substates (see Figure 30).

The ratio of the $M1$ transition rate to the $E2$ nuclear deexcitation rate from the $F = 5/2$ level is given by Gal et al. (77) in the form of

$$R_\Gamma(M1:E2) = \frac{\Gamma_\gamma(M1)}{\Gamma_\gamma(E2)} \frac{1 + \alpha(M1)}{1 + \alpha(E2)} \qquad 66.$$

For those deformed nuclides whose muonic γ-ray shifts have been studied, the ratio $\Gamma_\gamma(M1)/\Gamma_\gamma(M2)$ can be calculated with good accuracy and gives values ranging from 3 to 13 times 10^{-4}. However, the internal-conversion coefficients in the N and O electron shells can be evaluated only with large uncertainty and they vary from 2 to 25 10^{+4}. The corrected isomer effects are also listed in Table XIV. The muonic isomer shifts in the Os isotopes measured by the CERN group are negative even with the largest calculated

Magnetic Hyperfine Interaction in 2^+ State

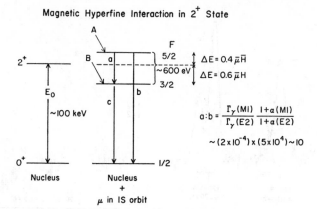

FIG. 30. Schematic representation of the hyperfine interaction of the $1s_{1/2}$ muon with an even-even nucleus in the 2^+ state (80).

correction possible (infinitely fast $M1$ transition). Unfortunately, there are no measured Mössbauer isomer effects in Os with which the corrected muonic isomer effects can be compared.

Magnetic hfs in the nuclear γ ray.—A well-resolved magnetic hyperfine splitting of the nuclear gamma ray due to the presence of the $1s$ muon has been observed in muonic ^{203}Ti and ^{205}Ti isotopes by the Darmstadt group at CERN (13). The ^{203}Ti and ^{205}Ti both have ground-state spin $1/2^+$ and the first-excited state spin $3/2^+$. Therefore, the F values for the excited states are 2 and 1 for the ground states 1 and 0. Because of the near-degeneracy of the muonic $2p_{1/2} - 2p_{3/2}$ splitting and of the excitation energy of the first $3/2^+$ state of thallium, the nucleus can be excited during the cascade of the muon. Both the nuclear ground state and the $3/2^+$ excited state are split by the magnetic interaction with the $1s$ muon, and four nuclear γ lines should appear. However, because of the fast $M1$ transition between the magnetic substrates $F=2 \rightarrow F=1$ by internal conversion in the N and O electron shells, the half-life of the $F=2$ state is estimated to be $\sim 2 \times 10^{-11}$ sec; therefore only the nuclear transitions from the $F=1$ component of the excited state to the two hyperfine states of the nuclear ground state are observable (Figure 31). The observed ΔE^{hfs} in ^{203}Ti and ^{205}Ti are (2.66 ± 0.30) keV and (2.32 ± 0.06) keV respectively, in fair agreement with the magnetic hfs splitting of 2.87 keV calculated by Engfer & Scheck (72). Their treatment of the configuration mixing follows closely the Arima method. A detailed discussion on the isomer effect in thallium can also be found in this reference (72).

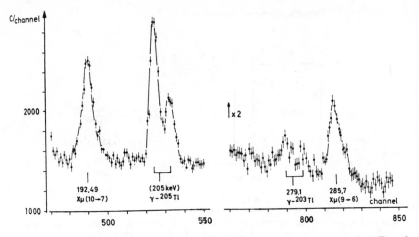

FIG. 31. Spectra of the hyperfine components of the nuclear γ ray in muonic ^{203}Tl and ^{205}Tl. The $X\mu(n_1 \rightarrow n_t)$ denote muonic X-ray lines in Tl (10).

RESULTS AND INTERPRETATION

The fractional changes of $\delta\langle r^2 \rangle / \langle r^2 \rangle$ (δt assumed to be small) are plotted against mass number A as shown in Figure 32. It is very striking to see that all three nuclei $^{150}_{60}$Nd, $^{152}_{62}$Sm, and $^{154}_{64}$Gd, known as soft rotators, have about equally large stretching. The isomer effects for the rest of the nuclei, known as rigid rotors, drop down an order of magnitude from that of the soft rotors. In a hydrodynamic model, where the energy of the system is given by (18, 146)

$$T_{rot} + V(\beta) = I(I+1)A(\beta) + \tfrac{1}{2}C(\beta - \beta_0)^2$$
$$= I(I+1)[A(\beta_0) + A'(\beta_0)(\beta - \beta_0) + \cdots] \qquad 67.$$
$$+ \tfrac{1}{2}C(\beta - \beta_0)^2$$

where

$$A'(\beta_0) = \frac{\partial A(\beta)}{\partial \beta}\bigg|_{\beta_0}$$

The change in the equilibrium β between the state $I=0$ and a state I is

$$\delta\beta = -I(I+1)A'/C \qquad 68.$$

From Equation 40, we can relate this to $\delta\langle r^2 \rangle$:

$$\frac{\delta\langle r^2 \rangle}{\langle r^2 \rangle} = \frac{5}{2\pi}\beta_0\delta\beta = -\frac{5}{2\pi}\frac{\beta_0 A'}{C}I(I+1) \qquad 69.$$

The critical quantities here are $A'/A = -\mathcal{G}'/\mathcal{G} \equiv -\mathcal{G}^{-1}\partial\mathcal{G}/\partial\beta$ and C; β_0 is fairly well known.

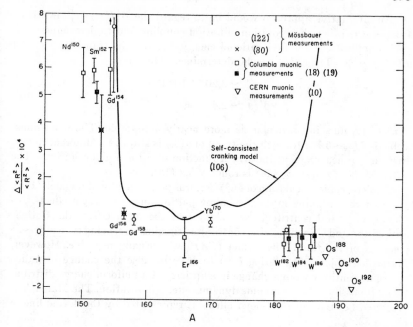

FIG. 32. $\delta\langle r^2\rangle/\langle r^2\rangle$ plotted against mass number A for several even-even nuclei in the rare-earth region. The two values of $\delta\langle r^2\rangle/\langle r^2\rangle$ shown for each of the W isotopes and ^{152}Sm isotopes represent two values of δE corresponding to N_I or O_I electron shell conversion as discussed in (77). The Mössbauer isomer shifts are taken from (80, 122).

If one attributes the nuclear rotational correction term $B[I(I+1)]^2$ to rotation-vibration coupling, then

$$B = -A'^2/2C \qquad\qquad 70.$$

Then, using the experimental values of B and $\delta\langle r^2\rangle/\langle r^2\rangle$, C and $\mathcal{I}'/\mathcal{I}$ can be determined.

For ^{152}Sm, the Columbia group (18, 19) found

$$\delta\langle r^2\rangle/\langle r^2\rangle = (5.9 \pm .4) \times 10^{-4}$$

which, along with

$$B = -0.20 \text{ keV}$$
$$\beta_0 = 0.3 \qquad\qquad 71.$$

give then

$$C = 2.45 \times 10^6 \text{ keV}$$
$$\mathcal{I}'/\mathcal{I} = 46$$

An alternative approach would be to regard the $B[I(I+1)]^2$ as arising from effects other than rotation-vibration coupling and to determine C experimentally from the β-vibrational energy $\hbar\omega_\beta$ and the transition element $B(E2) \propto \omega_\beta/C$. Then $\mathscr{I}'/\mathscr{I}$ can be determined. This gives

$$C \simeq 2.20 \times 10^5 \text{ keV}$$
$$\mathscr{I}'/\mathscr{I} \simeq 2.4$$

72.

The $\mathscr{I}'/\mathscr{I}$ value, in particular, is more nearly acceptable. The irrotational value is $2/\beta_0 \sim 6.7$. Although the trend of $\mathscr{I}(\beta_0)$ is known for different nuclei, there is no measurement for $\mathscr{I}$ as a function of β for a given nucleus. The value $\partial\mathscr{I}(\beta_0)/\partial\beta_0$ near $\beta_0 \sim 0.3$ is roughly 3.2 (18).

A microscopic calculation of $\delta\langle r^2\rangle/\langle r^2\rangle$ was performed by Marshalek (106). He used the cranking model, including pairing and centrifugal effects. Although he found centrifugal stretching was the main effect, the Coriolis antipairing and Coriolis mixing contributed significantly, being typically $\sim$25 per cent of the total. The former (CAP) was usually negative. However, Marshalek objects to relating $\delta\langle r^2\rangle$ to $\delta\langle\beta^2\rangle$ because the change in radius comes not entirely from a change in ellipticity of a uniform charge distribution but possibly also from some dynamic effects of rotation. The Marshalek-predicted behavior of the isomer effects is represented by the dotted line in Figure 32.

The isomer effects observed by muonic atoms and Mössbauer technique are plotted in Figure 32 for comparison. It is gratifying to see that the agreement between them is good within the limit of experimental precision and under the *ad hoc* assumption of $\delta t = 0$ and $\delta E_{\text{nucl. pol.}} = 0$. However, the agreement between the experimental and theoretical values is very poor, particularly at the upper end of the deformed region. In fact, the isomer effects of Os isotopes all give negative values while the predicted values are large and positive. The reason for this discrepancy is not clear.

ANOMALOUS INTENSITIES

It has been observed that in certain muonic spectra (e.g., in [209]Bi, [205]Tl, and [127]I), the intensity ratio

$$\frac{I(2p_{3/2} \rightarrow 1s_{1/2})}{I(2p_{1/2} \rightarrow 1s_{1/2})}$$

73.

is considerably smaller than the theoretically expected value of 1.9. The anomaly comes from at least *two different sources:* One may be due to effects in the cascade processes and the other may have its origin in a resonance process involving the nuclear levels.

In the cascade calculation, if the initial distribution of the muonic states is known, then the transition rates can be calculated using the general expression for an electromagnetic transition of multipolarity L from an initial state of energy E and total angular momentum F to a lower $(2F'+1)$-

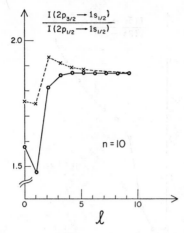

FIG. 33. The intensity ratio $I(2p_{3/2}\rightarrow 1s_{1/2})/I(2p_{1/2}\rightarrow 1s_{1/2})$ as a function of the angular momentum l of the captured muon (89).

degenerate state of energy E'. If the initial states are assumed to be statistically populated, then the intensity ratios of the K doublets and L triplets are given approximately as

$$(2p_{3/2} \rightarrow 1s_{1/2}):(2p_{1/2} \rightarrow 1s_{1/2}) = 2:1 \qquad 74.$$

$$(3d_{5/2} \rightarrow 2p_{3/2}):(3d_{3/2} \rightarrow 2p_{3/2}):(3d_{3/2} \rightarrow 2p_{1/2}) = 9:1:5 \qquad 75.$$

$$(4f_{7/2} \rightarrow 3d_{5/2}):(4f_{5/2} \rightarrow 3d_{5/2}):(4f_{5/2} \rightarrow 3d_{3/2}) = 20:1:14 \qquad 76.$$

If the dependence of the transition rate on $(\Delta E)^3$ and the effect of the finite nucleus on the $E1$ matrix element are taken into account, then the ratio is slightly modified but not more than a few per cent.

The main difficulty in a cascade calculation is the large uncertainty in the initial distribution. To see what types of initial distributions could give the observed anomalously low intensity ratio, J. Hüfner (89) evaluated the intensity ratio of the K doublets as a function of the quantum numbers n and l of that state which is assumed to be populated by the free muon ($n \leq 14, l \leq n-1$). The fine-structure doublet of each Bohr orbit was assumed to be populated proportional to $(2j+1)$. The calculated intensity ratio (for $n=10$) is shown in Figure 33: For $l \geq 2$, the ratio is almost constant at 1.9. To obtain a ratio as low as 1.5, as experimentally observed in Bi, the muon must be captured initially into states with $l=0$ or 1. However, Hüfner also showed that if the muon is captured into $l=0$ or $l=1$, one should find the $2s \rightarrow 2p$ transition stronger than the $3d \rightarrow 2p$ transition, as shown in Figure 34.

Ever since the observation of the intensity anomaly in ^{209}Bi in early muonic measurements, attempts have been made to attribute this anomaly to nuclear excitation, but with no experimental support. However, the

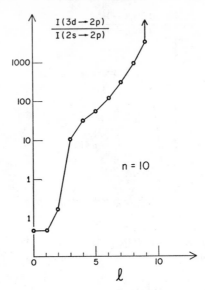

FIG. 34. The intensity ratio $I(3d \rightarrow 2p)/I(2s \rightarrow 2p)$ as a function of the angular momentum l of the captured muon (89).

recent discovery of certain excited states of the nucleus ^{209}Bi has lent strong support to a proposal made by Hüfner (90) that the anomaly does have its origin in nuclear excitation. Inelastic scattering, Bi(p,p')Bi* (81), has revealed a septuplet of nuclear states centered around 2.6 MeV (Figure 35).

The most probable interpretation for these states is given in terms of model of core excitation. The $h_{9/2}$ proton weakly couples to the first-excited 3^- state in ^{208}Pb (at 2.615 MeV) giving rise to states with spins and parities between $3/2^+$ and $15/2^+$. Hüfner (90) proposed that the states at 2.739 MeV and at 2.560 MeV seem to be of particular interest for the muonic Bi:

^{209}Bi.—The states

$$| 1 \rangle = | 3d_{5/2}, 9/2^- \rangle \qquad\qquad 77a.$$

and

$$| 2 \rangle = | 2p_{1/2}, 15/2^+ \rangle \qquad\qquad 77b.$$

are eigenstates of $H_N + H_\mu$ in the notation $|\mu, N\rangle$. However, these two states are nearly degenerate and can be mixed by electric octupole interaction between the two states of Bi:

$$\Delta E_{12} = E_1 - E_2 = \left[E_\mu(3d_{5/2}) - E_\mu(2p_{1/2}) \right]$$
$$- \left[E_N(9/2^-) - E_N(15/2^+) \right] \qquad\qquad 78.$$
$$= (2.744 \pm 4) - (2.739 \pm 3) = 5 \pm 5 \text{ (keV)}$$

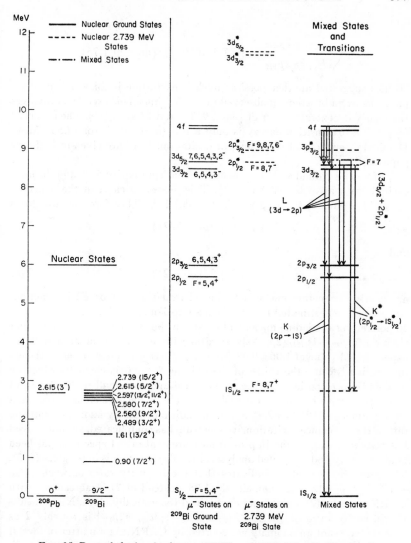

FIG. 35. Part of the level scheme of ^{208}Pb and ^{209}Bi (89, 90). The levels are not drawn to scale.

The strength of this mixing can be estimated from the known lifetime of the first 3^- state in ^{208}Pb. This resonant excitation increases the intensity of the $(2p_{1/2} \rightarrow 1s_{1/2})$ line and reduces the intensity of $(3d_{5/2} \rightarrow 2p_{3/2})$ transitions, in approximate agreement with the observed ratios:

$$\frac{2p_{3/2} - 1s_{1/2}}{2p_{1/2} - 1s_{1/2}} = 1.39 \pm 0.10 \quad \text{(normal 1.92)}$$

and

$$\frac{3d_{5/2} - 2p_{3/2}}{3d_{3/2} = 2p_{1/2}} = 1.55 \pm 0.10 \quad \text{(normal 1.75)}$$

Hüfner suggested another possible nuclear excitation in Bi which may explain the extra broadening observed in the K_{α_1} line. This excitation involves the nearly degenerate states of $|3d_{5/2}, 9/2^-\rangle$ and $|2p_{3/2}, 9/2^+\rangle$. The observed $(3d_{5/2} \rightarrow 2p_{3/2})$ transition energy is only a few (6 ± 6) keV from 2.560 MeV. Detailed discussions of these nuclear excitations in Bi are given in Hüfner's paper (90).

^{127}I.—The ground state of this nucleus has spin-parity $5/2^+$ and the first-excited state at 57.6 keV has $7/2^+$. This energy is close to the expected $2p_{3/2} - 2p_{1/2}$ doublet splitting which is $\sim$55.2 keV. Therefore the two states

$$|1\rangle = |2p_{3/2}, 5/2^+\rangle \qquad \qquad 79a.$$

and

$$|2\rangle = |2p_{1/2}, 7/2^+\rangle \qquad \qquad 79b.$$

are nearly degenerate and can be mixed by either $M1$ or $E2$ interaction. Acker et al. (2) estimated that for an assumption of maximum mixing, some 10 per cent of the iodine nuclei will be left in the excited 57.6-keV state and that the $(2p_{3/2} \rightarrow 1s_{1/2})/(2p_{1/2} \rightarrow 1s_{1/2})$ ratio will be $\sim$1.3, compared with an experimental value of 1.06 ± 0.08. However, these estimates are nevertheless in question because the values of the $M1$ and $E2$ matrix elements which he used (137) are too large in the former case and too small in the latter in comparison with recently measured values (100). Because of the large magnetic moment of ^{127}I ($\mu = 2.81$ n.m.), in addition to its dynamic hfs splitting due to the resonance excitation, its spectrum has complicated, yet unresolved line structure. A new and improved measurement of ^{127}I spectrum has been made recently and a detailed analysis of its hfs spectrum is in progress (103). The preliminary analysis indicates that the intensity ratio between $(2p_{3/2} \rightarrow 1s_{1/2}):(2p_{1/2} \rightarrow 1s_{1/2})$ is anomalously 1.08 (instead of 2) in agreement with CERN's results (2) and it can be explained theoretically (103). Nevertheless the intensity ratio of $[(3d_{5/2} + 3d_{5/2}) \rightarrow 2p_{3/2}]:[3d_{3/2} \rightarrow 2p_{1/2}]$ is roughly 2 as expected and not an anomaly as reported by CERN. As a matter of fact, if the mixing is in the $2p$ states only, this ratio should not be affected at all.

^{205}Tl.—The evidence of the connection between the anomalous intensity ratio and the nuclear excitation has been found in ^{205}Tl and is discussed in the section on the isomer effect.

SUMMARY AND LOOKING FORWARD

We have reviewed some highlights of recent progress in the study of muonic atoms. The present precision and accuracy in energy determination give only one parameter of the charge distribution in light and medium nuclei and two parameters for very heavy nuclei. The pertinent question of what characteristic of the charge distribution is being measured has been elegantly answered by Ford & Wills by a single moment $\langle r^k \rangle^{1/k}$ where k is almost model independent.

The theoretical investigation of the nuclear-polarization correction has been greatly extended to several multipole modes and to more realistic nuclear models. The Lamb shifts have been recalculated and turn out to be no longer insignificant. The profound effects of these corrections on the interpretation of muonic results are rather striking as discussed in the section on Dynamic $E2$ Effect. The possible observation of the nuclear polarization on the $1s$ level has been evidenced in the study of the transitions involving the $2s$ level.

Isotope shifts in muonic atoms have been applied to extract the specific mass effects and the field effects of the optical isotope shifts. The general agreements between muonic, electronic, and optical isotope shifts are satisfactory.

The muonic magnetic hfs spectra in ^{209}Bi, ^{115}In, ^{113}Cs, and ^{141}Pr have been studied and compared favorably with the theoretical calculations of the B-W effect with configuration-mixing models.

The dynamic $E2$ hfs spectra in many deformed nuclei have been studied in great detail. A more exact nuclear-polarization calculation is necessary; otherwise the extracted parameters c and t in Fermi distribution become rather unreasonable.

The riddle of the intensity anomaly in ^{209}Bi, which had puzzled people for a long time, is now finally solved by Hüfner's suggestion of resonance excitation. The intensity anomaly in ^{205}Tl and ^{127}I can also be explained by the same treatment.

The observation of the very small energy shifts of the rotational gamma rays in muonic atoms of deformed nuclei offers a new method to determine the isomer effects. In the interpretation of this effect, a strong $M1$ transition between the magnetic doublets of the excited state must be taken into account. The magnetic hfs splitting on the nuclear gamma rays has been observed in ^{203}Tl and ^{205}Tl.

Being limited by space, we regret that many other interesting subjects closely related to muonic atoms were not discussed in this review. Among these are: muonic hydrogen, muonic molecules, depolarization of muons, isotone shifts, and nuclear excitation (including fission) by radiationless muonic transitions or muon capture.

With the coming of modern meson facilities, our understanding of the nuclear structure through the investigation of the muonic atoms will undoubtedly be elevated to a much higher level. Energy determination and resolution will not be limited by statistical precision. In particular, curved-crystal spectrometers may then be used. *In order to utilize these more precise experiments, it will be necessary to have available theoretical calculations of comparable accuracy.* It ought then be possible to extract nuclear-polarization effects as well as further parameters of the charge distribution. In the case of isotope shifts, a much smaller target of separated isotopes will be possible, thus the investigation can be extended to many more isotopes.

For the dynamic $E2$ excitation, with the present muonic intensity, only the first rotational level can be studied extensively. With an intense muon beam, not only the γ rays from first, but also from the second and third excited rotational states may be studied for their isomer effects and their magnetic hfs splittings, etc. Since the B-W effect of the $M1$ hfs relates closely to the nuclear model, this potentiality should greatly enrich our knowledge on nuclear structure.

We have mentioned that the only muonic X ray of light nuclei which has not been observed is that of hydrogen. The hfs splitting of the muonic hydrogen would be a most interesting subject to investigate. The lowest hfs level of $F=0$ can be formed by stopping μ^- in H_2 gas at very low pressure. If polarized infrared light at the wavelength of 6.79 μ is applied, a transition can be induced from $(F, M_F) = (0, 0)$ to $(1,1)$. This transition could be detected through the change in angular distribution of the decay electrons as used in muonium hfs experiment.

ACKNOWLEDGMENTS

The authors are grateful to many colleagues for continual discussions of the material presented. Special thanks are due to Dr. M. Y. Chen for his contributions to the section on nuclear polarization and to the Columbia Muonic Group (D. Hitlin, E. R. Macagno, W. Y. Lee, B. Budick, S. Bernow, J. W. Kast, S. C. Cheng, R. C. Barrett, S. Devons, and J. Rainwater) for communicating to us their results before publication.

LITERATURE CITED

1. Acker, H. L., Marshall, J., Backenstoss, G., Quitmann, D., *Nucl. Phys.*, **62**, 477 (1965)
2. Acker, H. L., Backenstoss, G., Daum, C., Sens, J. C., DeWit, S. A., *Nucl. Phys.*, **87**, 1 (1966); *Phys. Letters*, **14**, 317 (1965)
3. Anderson, H. L., *Proc. Williamsburg Conf. Intermediate Energy Phys.*, *Feb. 1966;* Anderson, H. L., McKee, R. J., Hargrove, C. K., Hincks, E. P., *Phys. Rev. Letters*, **16**, 434 (1966)
4. Anderson, H. L., *Proc. Intern. Conf. Electromagnetic Sizes of Nuclei, Ottawa, Canada, May 1967*
5. Anderson, H. L., Hargrove, C. K., Hincks, E. P., McAndrew, J. D., McKee, R. J., Kessler, D., *Phys. Rev. Letters*, **22**, 221 (1969)
6. Anderson, H. L., Hargrove, C. K., Hincks, E. P., McAndrew, J. D., McKee, R. J., Barton, R. D., Kessler, D. (Preprint, April 1969)
7. Backenstoss, G., Goebel, K., Stadler, B., Hegel, U., Quitmann, D., *Nucl. Phys.*, **62**, 449 (1965)
8. Bardin, T. T., Barrett, R. C., Cohen, R. C., Devons, S., Hitlin, D., Nissim-Sabat, C., Rainwater, J., Runge, K., Wu, C. S., *Phys. Rev. Letters*, **16**, 718 (1966)
9. Backenstoss, G., Charalambus, S., Daniel, H., Koch, H., Poelz, G., Schmitt, H., Tauscher, L., *Phys. Letters*, **25B**, 547 (1967)
10. Baader, R., Backe, H., Engfer, R., Kankeleit, E., Schröder, U., Walter, H., Wien, K., *Phys. Letters*, **27B**, 428 (1968)
11. Bailey, J., Bartl, W., von Bochmann, G., Brown, R. C. A., Farley, F. J., Joestlein, H., Picasso, L., Williams, R. W., *CERN preprint* (1968)
12. Barrett, R. C., Brodsky, S. J., Erickson, G. W., Goldhaber, M. H., *Phys. Rev.*, **166**, 1589 (1968)
13. Barrett, R. C., *Phys. Letters*, **28B**, 93 (1968)
14. Bearden, A. J., *Phys. Rev. Letters*, **4**, 240 (1960)
15. Belovitskii, G. E., *Soviet Phys. JETP*, **14**, 50 (1962)
16. LeBellac, M., *Nucl. Phys.*, **40**, 645 (1963)
17. Jenkins, D. A., Kunselman, R., Simmons, M. K., Yamazaki, T., *Phys. Rev. Letters*, **17**, 1 (1966)
18. Bernow, S., Devons, S., Duerdoth, I., Hitlin, D., Kast, J. W., Macagno, E. R., Rainwater, J., Runge, K., Wu, C. S., *Phys. Rev. Letters*, **18**, 787 (1967)
19. Bernow, S., Devons, S., Duerdoth, I., Hitlin, D., Kast, J. W., Lee, W. Y., Macagno, E. R., Rainwater, J., Wu, C. S., *Phys. Rev.*, *Letters*, **21**, 457 (1968)
20. Bethe, H. A., Negele, J. W., *Nucl. Phys.*, **A117**, 575 (1968)
21. Bhattacherjee, S. K., Boehm, F., Lee, P., *Phys. Rev. Letters*, **20**, 1295 (1968) (Nd isotope shifts.)
22. Bjorkland, J. A., Raboy, S., Trail, C. C., Ehrlich, R. D., Powers, R. J., *Phys. Rev.*, **136**, B341 (1964)
23. Boehm, F., Chesler, R. B., Brockmeier, R. T., *Phys. Rev. Letters*, **18**, 953 (1967)
24. Boehm, F., Chesler, R. B., *Phys. Rev.*, **166**, 1206 (1968)
25. Brix, P., Kopfermann, H., *Nach. Akad. Wiss. Göttingen, Math. Phys. Kl.*, **31** (1967)
26. Breit, G., Arfken, G. B., Clendenin, W. W., *Phys. Rev.*, **78**, 390 (1950)
27. Brix, P., Kopfermann, H., *Rev. Mod. Phys.*, **30**, 517 (1958)
28. Burbidge, G. R., de Borde, A. H., *Phys. Rev.*, **89**, 189 (1953)
29. Carnegie-Mellon 1967: Suzuki, A., *Phys. Rev. Letters*, **19**, 1005 (1967)
30. Carrigan, R. A., Jr., Gupta, P. D., Sutton, R. B., Suzuki, M. N., Thompson, A. C., Cole, R. E., Prestwich, W. V., Gaigalas, A. K., Raboy, S., *Bull. Am. Phys. Soc.*, **13**, 65 (1968)
31. Carrigan, R. A., Jr., Gupta, P. D., Sutton, R. B., Suzuki, M. N., Thompson, A. C., Cole, R. E., Prestwich, W. V., Gaigalas, A. K., Raboy, S., *Phys. Rev. Letters*, **20**, 874 (1968)
32. Carrigan, R. A., Jr., Gupta, P. D., Sutton, R. B., Suzuki, M. N., Thompson, A. C., Cole, R. E., Prestwich, W. V., Gaigalas, A. K., Raboy, S., *Phys. Letters*, **27B**, 622 (1968)
33. Chasman, C., Ristinen, R. A., Cohen, R. C., Devons, S., Nissim-Sabat, C., *Phys. Rev. Letters*, **14**, 181 (1965)
34. Chasman, C., Jones, K. W., Ristinen, R. A., Alburger, D. E., *BNL 11082*

35. Chen, M. Y. (Ph.D. thesis, Princeton Univ., *PUC-937-291*); (Private communication)
36. Chesler, R. B., Boehm, F., *Phys. Rev.*, **166**, 1206 (1968)
37. Cheng, S. C., Macagno, E. R., Hitlin, D., Bernow, S., Lee, W. Y., Kast, J. W., Rushton, A., Wu, C. S. (Preprint, 1969)
38. Bohr, A., Weisskopf, V. F., *Phys. Rev.*, **77**, 94 (1950)
39. Chicago 1968: Ehrlich, R. D. (Thesis, Univ. Chicago, 1968). Also Ehrlich, R. D., Fryberger, D., Densen, D. A., Nissim-Sabat, C., Powers, R. J., Telegdi, V. L., *Phys. Rev. Letters*, **18**, 959 (1967); [Erratum, **19**, 344 (1967)]
40. Chen, M. Y. (Preprint and private communication)
41. Chen, M. Y. (Preprint and private communication)
42. Church, E. L., Weneser, J., *Bull. Am. Phys. Soc.* **11**, 338 (1966); (To be published)
43. Cooper, L. N., Henley, E. M., *Phys. Rev.*, **75**, 1315 (1953)
44. Cohen, R. C., Devons, S., Kanaris, A. D., Nissim-Sabat, C., *Phys. Letters*, **11**, 70 (1964)
45. Cohen, R. C., Devons, S., Kanaris, A. D., Nissim-Sabat, C., *Bull. Am. Phys. Soc.*, **9**, 652 (1964)
46. Columbia 1965: *A progress report on high resolution studies of muonic X-rays. Jan 1, 1966. GEN-72; ONR-666* (72)
47. Cole, R. E., Guso, R., Raboy, S., Carrigan, R. A., Jr., Gaigalas, A., Sutton, R. B., *Phys. Letters*, **19**, 18 (1965)
48. *Progr. Rept., Pegram Lab., NYO-Gen-72-132*
49. Cole, R. K., Jr., *Phys. Letters*, **25B**, 178 (1967)
50. Cole, R. K., Jr., *Phys. Rev.* (To be published)
51. Daniel, H., Poelz, G., Schmitt, H., Backenstoss, G., Koch, H., Charalambus, S., *Z. Physik*, **205**, 472 (1967)
52. Daniel, H., *Naturwissenschaften*, **55**, 339 (1968)
53. Daniel, H., Koch, H., Poelz, G. Schmitt, H., Tauscher, L., Backenstoss, G., Charalambus, S., *Phys. Letters*, **26B**, 281 (1968)
54. Devons, S., Gidal, G., Lederman, L. M., Shapiro, G., *Phys. Rev. Letters*, **5**, 330 (1960)
55. Devons, S., at *1st Intern. Conf. High Energy Phys. and Nucl. Struct.,* CERN, *1963* (Unpublished)
56. Devons, S., *Proc. Topical Conf. Use of Elementary Particles in Nucl. Struct. Res., Brussels, 14–16 Sept. 1965*
57. Devons, S., Duerdoth, I., *Advan. Nucl. Phys.*, **2**, 295 (1968)
58. DeWitt, S. A., Backenstoss, G., Daum, C., Sens, J. C., Acker, H. L., *Nucl. Phys.*, **87**, 657 (1967)
59. Dontsov, Yu. P., Morozov, V. A., Striganov, A. P., *Opt. Spectry.*, **8**, 391 (1960)
60. Dontsov, Yu. P., *Opt. Spectry.*, **8**, 446 (1960) [*Opt. Spektry.*, **8**, 236 (1960)]
61. Dontsov, Yu. P., Morozov, V. A., Striganov, A. P., *Opt. Spektry.*, **8**, 741 (1960)
62. Ehrlich, R. D., Powers, R. J., Telegdi, V. L., Bjorkland, J. A., Raboy, S., Trail, C. C., *Phys. Rev. Letters*, **13**, 550 (1964)
63. Ehrlich, R. D., et al., *Phys. Letters*, **23**, 469 (1966)
64. Ehrlich, R. D., Fryberger, D., Jensen, D. A., Nissim-Sabat, C., Powers, R. J., Telegdi, V. L., Hargrove, C. K., *Phys. Rev. Letters*, **18**, 959 (1967); **19**, 334 (E) (1967)
65. Ehrlich, R. D., *Phys. Rev.*, **173**, 1088 (1968). (Si to Sn)
66. Ehrlich, R. D. (Thesis, Univ. Chicago, 1968)
67. Eisenberg, Y., Kessler, D., *Nuovo Cimento*, **6**, 1195 (1961); *Phys. Rev.*, **123**, 1472 (1961)
68. Eisenberg, Y., Kessler, D., *Phys. Rev.*, **130**, 2349 (1963)
69. Elton, L. R. B. (See Ref. 114), 267
70. Engfer, R., *Proc. Intern. Sch. Phys., Enrico Fermi Inst., June 1966*
71. Engfer, R., Theissen, H., Van Niftrik, G. J. C. (See Ref. 114), 184
72. Engfer, R., Scheck, F., *Z. Physik*, **216**, 274 (1968)
73. Farley, F. J. M., Bailey, J., Brown, R. C. A., Giesch, M., Joestlein, H., van der Meer, S., Picasso, E., Tannenbaum, M., *Nuovo Cimento*, **15A**, 281 (1966)
74. Fermi, E., Teller, E., *Phys. Rev.*, **72**, 399 (1947)
75. Fitch, V. L., Rainwater, J., *Phys. Rev.*, **92**, 789 (1953)
76. Ford, K. W., Wills, J. G., *LASL-Preprint, LA-DC-11393* (1968) and Wills, J. G. (Private communication)
77. Gal, A., Grodzins, L., Hüfner, J., *Phys. Rev. Letters*, **21**, 453 (1968)
78. Greenbaum, E., Hsu, F., Chow, Z. Y., Howes, R., *Columbia Rept. The*

Coulomb excitation on ^{150}Nd was carried out on the 8-MeV He^{++} beam of the Pegram Van de Graaf accelerator.

79. Grodzins, L., et al., *Phys. Rev. Letters*, **18**, 791 (1967), J. Hüfner has measured the 5 f shielding effect in solids by observing the isotope shifts of ^{151}Eu and ^{153}Eu in CaF$_2$ and found an increase of the electron density by a factor of 2.09 compared with the free atom. Hence $\delta\langle r^2 \rangle / \langle r^2 \rangle = +4.8 \times 10^{-4}$.

80. Grodzins, L., *Proc. Dubna Symp. Nucl. Struct., July 1968, IAEA, Vienna*

81. Hafele, J. C., Woods, R., *Phys. Letters*, **23**, 579 (1966)

82. Hansen, J. E., Steudel, A., Walther, H., *Z. Physik*, **203**, 296 (1967). (Most extensive and precise.)

83. Henley, E. M., Wilets, L., *Phys. Rev. Letters*, **20**, 1389 (1968)

84. Hitlin, D., Bernow, S., Devons, S., Duerdoth, I., Kast, J. W., Macagno, E. R., Rainwater, J., Runge, K., Wu, C. S., Barrett, R. C. (See Ref. 114); Hitlin, D. (Thesis, Columbia Univ., 1968)

85. Hitlin, D., Bernow, S., Kast, J. W., Macagno, E. R., Devons, S., Rainwater, J., Wu, C. S., Barrett, R. C. (Preprint and private communication)

86. Hofstadter, R., Nöldeke, G. K., Van Oostrum, K. J., Suelzle, L. R., Yearian, M. R., Clark, B. C., Herman, R., Ravenhall, D. G., *Phys. Rev. Letters*, **15**, 758 (1965)

87. Hutchinson, D. P., Menes, J., Shapiro, G., Patlach, A. M., Penman, S., *Phys. Rev. Letters*, **7**, 129 (1961)

88. Hüfner, J., *Nucl. Phys.*, **60**, 427 (1964)

89. Hüfner, J. (See Ref. 147); 87

90. Hüfner, J., *Phys. Letters*, **25B**, 189 (1967)

91. Jacobsohn, B. A., *Phys. Rev.*, **96**, 1637 (1954)

92. Johnson, J., Sorensen, R. A., *Phys. Letters*, **26B**, 700 (1968)

93. Johnson, J. A., *Bull. Am. Phys. Soc.*, **14**, 538 (1969)

94. Kaliteevsky, N. I., Fradkin, E. E., Chaika, M. P., *Izv. Akad. Nauk SSSR, Ser. Fiz.*, **25**, 1178 (1961)

95. Kankeleit, E., Backe, H., Backenstoss, G., Daniel, H., Engfer, R., Poelz, G., Schmitt, H., Tauscher, L., Wien, K., *Proc. Intern. Conf. hfs Interaction by Nucl. Radiation, Asilomar, Calif., Aug. 1967*

96. Kessler, D., Anderson, H. L., Dixit, M. S., Evans, H. J., McKee, R. J.,

Hargrove, C. K., Barton, R. D., Hincks, E. P., McAndrew, J. D., *Phys. Rev. Letters*, **18**, 1179 (1967)

97. King, W. H., *Proc. Roy. Soc. A*, **280**, 430 (1964)

98. King, W. H., Kuhn, H. G., Stacey, D. N., *Proc. Roy. Soc.*, **296**, 24 (1967)

99. Koslov, S., Fitch, V. L., Rainwater, J., *Phys. Rev.*, **95**, 291 (1954)

100. Kownacki, J., Ludziejewski, J., Moszynski, M., *Nucl. Phys.*, **A107**, 476 (1968)

101. Lathrop, J., Lundy, R., Telegdi, V. L., Winston, R., Yovanovich, D. D., *Nuovo Cimento*, **17**, 109 (1960)

102. Lathrop, J., Lundy, R., Telegdi, V. L., Winston, R., Yovanovich, D. D., Breaden, A. J., *Nuovo Cimento*, **17**, 114 (1960)

103. Lee, W. Y., Budick, B., Bernow, S., Chen, M. Y., Cheng, S. C., Hitlin, D., Kast, J. W., Macagno, E. R., Rushton, A., Wu, C. S., *Phys. Rev. Letters* (In press) (I) ^{133}Cs, ^{115}In, ^{141}Pr; (II) ^{127}I

104. Mann, R. A., Rose, M. E., *Phys. Rev.*, **121**, 293 (1961)

105. Macagno, E. R., Bernow, S., Devons, S., Duerdoth, I., Hitlin, D., Kast, J. W., Rainwater, J., Runge, K., Wu, C. S. (See Ref. 114), 71; Macagno, E. (Thesis, Columbia Univ., 1968)

106. Marshalek, E. R., *Phys. Rev. Letters*, **20**, 214 (1968)

107. Macagno, E. R., Bernow, S., Cheng, S. C., Hitlin, D., Kast, J. W., Rushton, A., Lee, W. Y., Devons, S., Rainwater, J., Wu, C. S. (Preprint and private communication)

107a. McKee, R. J., *Univ. Chicago Publ. EFINS 66-39* (Thesis)

108. McKinley, J. M., *Bull. Am. Phys. Soc.*, 678 (1968); (Preprint, 1969)

109. Murakawa, K., *Phys. Rev.*, **96**, 1543 (1954)

110. Van Niftrik, G. J. C., Engfer, R., *Phys. Letters*, **22**, 490 (1966)

111. Nöldeke, G., Steudel, A., *Z. Physik*, **137**, 632 (1954)

112. Onley, D. S., *Nucl. Phys.*, **A118**, 436 (1968)

113. Van Oostrum, K. J., Hofstadter, R., Noldeke, G. K., Yearian, M. R., Clark, B. C., Herman, R., Ravenhall, D. G., *Phys. Rev. Letters*, **16**, 528 (1966)

114. *Proc. Intern. Conf. Electromagnetic Sizes of Nuclei, Ottawa, May 22–24, 1967*

115. Petermann, A., Yamaguchi, Y., *Phys. Rev. Letters*, 2, 359 (1959)
116. Pieper, W., Greiner, W., *Phys. Letters*, 24B, 377 (1967)
117. Powers, R. J., *Phys. Rev.*, 169, 1 (1968)
118. Rainwater, J., *Ann. Rev. Nucl. Sci.*, 1, 1 (1957)
119. Raboy, S., Trail, C. C., Bjorkland, J. A., Ehrlich, R. D., Powers, R. J., Telegdi, V. L., *Nucl. Phys.*, 73, 353 (1965)
120. Rawitscher, G. H., *Phys. Rev.*, 151, 846 (1966)
121. Reiner, A. S., Wilets, L., *Nucl. Phys.*, 36, 452 (1962)
122. Rehm, K. E., Henning, W., Kienle, P., *Phys. Rev. Letters*, 22, 790 (1969); Also Kienle, P., et al., *Proc. Intern. Conf. Nucl. Struct.*, 195 (Tokyo, 1967); Kienle, P., *Hyperfine Structure and Nuclear Radiations* (Mathias, E., Shirley, D. A., Eds., North-Holland, Amsterdam, 1968)
122a. Scheck, F., *Z. Phys.*, 172, 239 (1963)
123. Sens, J. C., Swanson, R. A., Telegdi, V. L., Yovanovich, D. D., *Nuovo Cimento*, 7, 4 (1958)
124. Sens, J. C., Invited Talk at *1st Intern. Conf. High Energy Phys. and Nucl. Struct.*, *CERN*, 1963
125. Sens, J. C., *Proc. 2nd Intern. Conf. High Energy Phys. and Nucl. Struct.*, *Rehovoth, 1967* (North-Holland)
126. Stearns, M. B., *Progr. Nucl. Phys.*, 6, 108 (1957)
127. Stacey, D. N., *Proc. Roy. Soc. A*, 280, 459 (1964)
128. Sumbaev, O. I., Mezentsev, A. F., *Soviet Phys. JETP*, 22, 323 (1966)
129. Sumbaev, O. I., Petrovich, E. V., Zykov, V. S., Rylinkov, A. S., Grushko, A. I., Transl.: *Soviet J. Nucl. Phys.*, 5, 387 (1967)
130. Sundaresan, M. K., Srinivasan, V. S., *Phys. Rev. Letters*, 21, 1509 (1968)
131. Sumbaev, O. I., *Proc. Dubna Symp. Nucl. Struct.*, *July 1968*, *IAEA*, *Vienna*
132. Sutton, R. B., *Ottawa Conf. Electromagnetic Sizes of Nuclei, 1967*
133. Telegdi, V. L. (Private communication)
134. Telegdi, V. L., Wilets, L. (Unpublished)
135. Telegdi, V. L., *Proc. Intern. At. Phys. Conf.*
136. Thompson, A. C., Adler, D. T., Gupta, P. D., Sutton, R. B., Raboy, S., *Bull. Am. Phys. Soc.*, 14, 538 (1969)
137. Wapstra, A. H., Nijgh, G. J., van Lieshout, R., *Nuclear Spectroscopy Tables* (North-Holland, Amsterdam, 1959)
138. West, D., *Rept. Progr. Phys.*, 21, 271 (1958)
139. Wetmore, R. J., Buckle, D. C., Kane, J. R., Siegel, R. T., *Phys. Rev. Letters*, 19, 1003 (1967)
140. Wheeler, J. A., *Phys. Rev.*, 92, 812 (1953)
141. Wilets, L., Hill, D. L., Ford, K. W., *Phys. Rev.*, 91, 1488 (1953)
142. Wilets, L., *Kgl. Danske Mat. Fys. Medd.*, 29, 3 (1954)
143. Wichman, E. H., Kroll, N. M., *Phys. Rev.*, 96, 232 (1954); 101, 843 (1956)
144. Wilets, L., *Rev. Mod. Phys.*, 30, 542 (1958)
145. Winston, R., Telegdi, V. L., *Phys. Rev. Letters*, 1, 104 (1961)
146. Wilets, L., Chinn, D. (Unpublished)
147. *Proc. Williamsburg Conf. Intermediate Energy Phys.*, *Feb. 10–12, 1966* (Two volumes, printed by College of William and Mary)
148. Wu, C. S., *Proc. Intern. Nucl. Phys. Conf.*, *Gatlinburg, Tenn., Sept. 12–17, 1966* (Academic Press)
149. Wu, C. S. *Proc. Intern. Symp. Phys. One or Two Electron Atoms* (Arnold Sommerfelt Centennial Memorial Meeting), *Sept. 1968* (North-Holland); *Proc. Dubna Symp. Nucl. Struct.*, *July 1968* (IAEA, Vienna)
150. Yeboah-Amankwah, D., Grodzins, L., Frankel, R., *Phys. Rev. Letters*, 18, 791 (1967)
151. Zinov, V. G., Konin, A. D., Mulshin, A. I., Polyakava, R. V., *Yad. Fiz.*, 5, 591 (1967) [Engl. transl.]: *J. Nucl. Phys.*, 5, 420 (1967)
152. Frati, W., Rainwater, J., *Phys. Rev.*, 128, 2360 (1962)
153. Pearson, J. M., *Nucl. Phys.*, 45, 401 (1963)
154. Gershtein, S. S., Petrukhin, V. I., Ponomarev, L. I., Prokoshkin, Yu. D., *Ob'ed. Inst. Yad. Issled.*, *Dubna, P4-3860* (1968)

SOME RELATED ARTICLES APPEARING
IN OTHER *ANNUAL REVIEWS*

From the *Annual Review of Astronomy and Astrophysics*, Volume 7 (1969)
    Meyer, Peter: Cosmic Rays in the Galaxy, 1–38
    Neupert, W. M: X Rays from the Sun, 121–48
    Wallerstein, George, and Conti, Peter S: Lithium and Beryllium in Stars, 99–120
From the *Annual Review of Entomology*, Volume 14 (1969)
    Proverbs, M. D: Induced Sterilization and Control of Insects, 81–102
From the *Annual Review of Genetics*, Volume 3 (1969)
    Drake, John W: Mutagenic Mechanisms, 247–68
    Witkin, Evelyn M: Ultraviolet-Induced Mutation and DNA Repair, 525–52
From the *Annual Review of Medicine*, Volume 20 (1969)
    Gottschalk, Alexander: Technetium 99M in Clinical Nuclear Medicine, 131–40
From the *Annual Review of Microbiology*, Volume 23 (1969)
    Witkin, Evelyn M: Ultraviolet-Induced Mutation and DNA Repair, 487–514
From the *Annual Review of Physical Chemistry*, Volume 20 (1969)
    Arnold, James R., and Suess, Hans E: Cosmochemistry, 293–314
    Shirley, D. A: Topics in Mössbauer Spectroscopy, 25–44
From *The Excitement and Fascination of Science*
    Libby, W. F: Thirty Years of Atomic Chemistry, 241–54

AUTHOR INDEX

CUMULATIVE INDEXES

VOLUMES 10-19

INDEX OF CONTRIBUTING AUTHORS

INDEX OF CHAPTER TITLES

VOLUMES 10-19

PARTICLE PHYSICS

Spin and Parity Determination of